May 2–3, 2013
Paris, France

Association for Computing Machinery

Advancing Computing as a Science & Profession

GLSVLSI'13

Proceedings of the ACM International Conference of the

Great Lakes Symposium on VLSI

Sponsored by:

IEEE CEDA, IEEE, SIGDA, and ACM ECRC

Association for
Computing Machinery

Advancing Computing as a Science & Profession

The Association for Computing Machinery
2 Penn Plaza, Suite 701
New York, New York 10121-0701

ISBN: 978-1-4503-1902-7 (Digital)
ISBN: 978-1-4503-2279-9 (Print)

Additional copies may be ordered prepaid from:

ACM Order Department
PO Box 30777
New York, NY 10087-0777, USA

Phone: 1-800-342-6626 (USA and Canada)
+1-212-626-0500 (Global)
Fax: +1-212-944-1318
E-mail: acmhelp@acm.org
Hours of Operation: 8:30 am – 4:30 pm ET

Printed in the USA

Foreword

Welcome to the 23rd edition of the *Great Lakes Symposium on VLSI (GLSVLSI) 2013* held at Palais des Congrès de Paris, in Paris (France). GLSVLSI is a premier venue for the dissemination of manuscripts of the highest quality in all areas related to VLSI, devices and system-level design. The venue of this year's GLSVLSI is Paris, which does not even hold any worldly famous lake, it shelters several astonishing jewels like the lakes at Bois de Vincennes and Bois du Boulogne. Moreover, we truly believe that Paris is a *great location* for a symposium on VLSI (another way to build the acronym of GLSVLSI) for 2013 and you will enjoy the beautiful city as well as the program over the two days of this year's GLSVLSI activity.

This year, GLSVLSI is co-located with the 1st ACM European Computing Research Congress (ACM ECRC) and the SIGCHI conference. The goal of the ACM European Computing Research Congress (ACM ECRC) is to co-locate multiple (ACM) European research conferences, workshops, and/or research meetings to create a significant gathering of European computing researchers. Moreover, SIGCHI is the premier international society for professionals, academics and students who are interested in human-technology & human-computer interaction (HCI). The co-located event brings a great opportunity for GLSVLSI's attendees to interact with other first-class researchers from around the world, making this year's symposium a great event.

As for the technical meeting, GLSVLSI 2013 was a resounding success: 238 papers were submitted, including authors from 36 different countries, of which 51 papers were accepted for oral presentation at the symposium (a 21% acceptance rate). Including poster papers, a total of 76 papers, with authors from 27 different countries, will be included in the symposium and published in the conference proceedings. Of these papers in the program, 34% are from North American, 30% from Asia, 29% from Europe, 5% from Africa/Middle East, and 2% from Central/South America. The final technical program consists of 30 full presentations and 21 short presentations in 11 oral sessions and 24 posters in two poster sessions. GLSVLSI 2013 starts on Thursday, May 2nd, with an exciting line up of speakers on a broad range of issues related to test, low power design, CAD, VLSI circuits, emerging technologies, and a session on VLSI for Biology. The technical program continues on Friday, May 3rd, presents the latest industrial and academic research in VLSI design and CAD with additional sessions on test, emerging technologies, and a session dedicated to new VLSI designs for functional units. During these two days, we will have the opportunity to attend two keynote speeches organized by the ACM ECRC, and a GLSVLSI keynote talk discussing the secrets of electronics for a greener planet. The technical program of GLSVLSI 2013 includes two parallel tracks to allow longer presentations and discussions during the two days of the conference. Overall, there are 14 regular sessions in the technical program and a session where the "best paper award" candidates will have the opportunity to present their outstanding work.

Moreover, the social program of GLSVLSI 2013 includes two special events this year. The first event consists on a reception cocktail on Thursday evening. Then, on Friday, May 3rd, a Cruise Gala Dinner will be the second social event, which will be held on The Captain Fracasse boat on the beautiful Seine river. The boat will depart at 20:30 from the Bir-Hakeim bridge, which is located just 3 minutes away from the Eiffel Tower. During the cruise, you will discover the most beautiful landmarks of Paris such as the Eiffel Tower, the Pont de l'Alma and the Musée d'Orsay. The boat will go to the Bercy area and come back to the starting point. During the trip, you will enjoy an elegant dinner of 2 hours in duration.

Finally, we would like to thank all the members of the Technical Program Committee and the secondary reviewers, who have done a great job in evaluating the submissions and determining the final high quality technical program of GLSVLSI 2013; the Track Chairs have done an outstanding job in ensuring that each submitted paper was consistently reviewed and helping to prepare the final program. In addition, the other members of the Organizing Committees (Katzalin Olcoz, Ayse K. Coskun and Theocharis Theocharides) should be also commended for providing excellent service with respect to publication, registration, and organization of the event and web presence to make this new edition of GLSVLSI a great success.

We trust that you will find GLSVLSI2013 rewarding in meeting your highest expectation of quality and technical advancement as reflected in these Proceedings and the technical presentations.

José L. Ayala *General Chair*	**Alex K. Jones & Patrick Madden** *Program Chairs*

Table of Contents

Session 15: CAD for 3D

Session 16: Design and Modeling

Poster Session 1

Poster Session 2

GLSVLSI 2013 Conference Organization

General Chair: Jose L. Ayala *(Complutense University of Madrid, Spain)*

Program Chairs: Alex Jones *(University of Pittsburgh, USA)*
Patrick Madden *(Binghamton University, USA)*

Proceedings Chair: Ayse K. Coskun *(Boston University, USA)*

Local Arrangements Chair: Katzalin Olcoz *(Complutense University of Madrid, Spain)*

Web Chair: Theocharis Theocharides *(University of Cyprus, Cyprus)*

Steering Committee: David Atienza *(EPFL, Switzerland)*
Iris Bahar *(Brown University, USA)*
Sanjukta Bhanja *(University of South Florida, USA)*
Erik Brunvand *(University of Utah, USA)*
Joseph Cavallaro *(Rice University, USA)*
David Garrett *(Broadcom, USA)*
Yehea Ismail *(America Univ. Cairo, Egypt)*
Fabrizio Lombardi *(Northeastern University, USA)*
Enrico Macii *(Politecnico di Torino, Italy)*
Vijaykrishnan Narayanan *(Penn State University, USA)*
Gang Qu *(University of Maryland, USA)*
Mircea R. Stan *(University of Virginia, USA)*
Ken Stevens *(University of Utah, USA)*
Yuan Xie *(Penn State University, USA)*
Zhiyuan Yan *(Lehigh University, USA)*
Tong Zhang *(Rensselaer, USA)*
Hai Zhou *(Northwestern University, USA)*

Program Committee: ***VLSI Design Track***
Track Chair: Martin Margala *(University of Massachusetts, Lowell)*
Track Chair: James Stine *(Oklahoma State University, USA)*
Andreas Burg *(ETH Zurich, Switzerland)*
Yangdong Deng *(Inst. of Microelectronics, Tsinghua University, China)*
Manish Goal *(Texas Instruments, Inc.)*
Antonio Jimenez *(IMSE-CNM, Spain)*
Rabi Mahapatra *(Texas A&M University, USA)*
Saraju Mohanty *(University of North Texas, USA)*
Vassilis Paliouras *(University of Patras, Greece)*
Sudeep Pasricha *(Colorado State University, USA)*
Zhijie Shi *(University of Connecticut, USA)*
Aida Todri-Sanial *(National Center for Scientific Research, France)*
Miroslav Velev *(Aries Design Automation)*

Program Committee (continued):

Testing Track

Track Chair: Jennifer Dworak *(Southern Methodist University, USA)*
Track Chair: Erik Larsson *(Lund University, Sweden)*
Chih-Tsun Huang *(National Tsing Hua University, China)*
Nicola Nicolici *(McMaster University, Canada)*
Sule Ozev *(Arizona State University, USA)*
Haralampos Stratigopoulos *(TIMA Laboratory, France)*
Spyros Tragoudas *(Southern Illinois University, USA)*
Xiaoqing Wen *(Kyushu Institute of Technology, Japan)*

Circuits Track

Track Chair: Theocharis Theocharides *(University of Cyprus, Cyprus)*
Track Chair: Lei Wang *(University of Connecticut, USA)*
Jinghong Chen *(Southern Methodist University, USA)*
Bozena Kaminska *(SFU, Canada)*
Selcuk Kose *(University of South Florida, USA)*
Chrysostomos Nicopoulos *(University of Cyprus, Cyprus)*
Emre Salman *(Stony Brook University, USA)*
Diego Vázquez-García De La Vega *(Inst. de Microelec. de Sevilla, Spain)*

Emerging Technologies Track

Track Chair: Hai (Helen) Li *(University of Pittsburgh, USA)*
Track Chair: Dmitri Strukov *(UCSB, USA)*
Shamik Das *(The MITRE Corporation, USA)*
Haykel Ben Jamaa *(CEA - LETI, France)*
Dhireesha Kudithipudi *(Rochester Institute of Technology, USA)*
Sébastien Le Beux *(Lyon Institute of Nanotechnology, France)*
Kartik Mohanram *(University of Pittsburgh, USA)*
Guangyu Sun *(Peking University, China)*
Mehdi Tahoori *(Karlsruhe Institute of Technology, Germany)*
Chris Winstead *(Utah State University, USA)*
Wei Zhang *(Nanyang Technological University, Singapore)*

Low Power Track

Track Chair: Alberto Macii *(Politecnico di Torino, Italy)*
Track Chair: Vijaykrishnan Narayanan *(Penn. State University, USA)*
Antonio Acosta *(IMSE-CNM/University of Seville, Spain)*
Alberto Bocca *(Politecnico di Torino - DAUIN, Italy)*
Andrea Calimera *(Politecnico di Torino, Italy)*
Koushik Chakraborty *(Utah State University, USA)*
Minsu Choi *(Missouri University of Science & Technology, USA)*

Program Committee (continued):

Ayse Coskun *(Boston University, USA)*
Domenik Helms *(OFFIS Research Institute, Germany)*
Kyung Ki Kim *(Daegu University, South Korea)*
Marisa Lopez-Vallejo *(Universidad Politecnica Madrid, Spain)*
Chao Lu *(Purdue University, USA)*
Alberto Nannarelli *(Technical University of Denmark, Denmark)*
Marvin Onabajo *(Northeastern University, USA)*
Joaquin Recas *(Universidad Complutense de Madrid, Spain)*
Baris Taskin *(Drexel University, USA)*
Woo Kim *(Advanced Micro Devices, USA)*

CAD Track

Track Chair: Jens Lienig *(TU Dresden, Germany)*
Track Chair: Gabriela Nicolescu *(Ecole Polytech. de Montreal, Canada)*
Nadine Azemard *(LIRMM / CNRS, France)*
Laleh Behjat *(University of Calgary, Canada)*
Giovanni Beltrame *(Ecole Polytechnique de Montreal, Canada)*
Youcef Bouchebaba *(ST Microelectronics)*
Francesco Bruschi *(Politecnico di Milano, Italy)*
Maciej Ciesielski *(University of Massachusetts, USA)*
Philippe Coussy *(Université de Bretagne Sud / Lab-STICC, France)*
Rolf Drechsler *(University of Bremen, Germany)*
Fabrizio Ferrandi *(Politecnico di Milano, Italy)*
Niraj Jha *(Princeton University, USA)*
Brett Meyer *(McGill University, Canada)*
Massimo Poncino *(Politecnico di Torino, Italy)*
Matteo Sonza Reorda *(Politecnico di Torino, Italy)*
Ashoka Sathanur *(IMEC/Holst, Netherlands)*
Biplab K Sikdar *(Bengal Engineering and Science University, India)*
Jelena Trajkovic *(Concordia University, Canada)*
Evangeline Young *(The Chinese University of Hong Kong, China)*

GLSVLSI 2013 Sponsors & Supporters

Sponsor:

Technical Supporters:

A Novel Statistical and Circuit-Based Technique for Counterfeit Detection in Existing ICs

Rashmi Moudgil[1], Dinesh Ganta[1], Leyla Nazhandali[1], Michael Hsiao[1], Chao Wang[1], Simin Hall[2]
[1]Department of Electrical and Computer Engineering, Virginia Tech
[2]Department of Mechanical Engineering, Virginia Tech
(rmoudgil, diganta, leyla, hsiao, chaowang, thall57)@vt.edu

ABSTRACT
Previously used ICs, which are resold as new, result in undue lost revenue, cause lower performance, reduced life span, and even catastrophic failure of platforms and systems. Non-invasive and inexpensive techniques are needed to establish the authenticity of such ICs that do not have special in-built structures for counterfeit detection. Although delay of circuit increases with its age, it cannot directly reveal the age of the chip, as it is also greatly influenced by process variation. In this work, we show that the relationship between two or more paths within the chip is a great indicator of its age. Using the proposed statistical and circuit-level technique, we observe over 97% correct detection of an aged IC from a new IC.

Categories and Subject Descriptors
B.8.1 [**Performance and Reliability**]: Reliability, Testing, and Fault-Tolerance

Keywords
Counterfeit, Aging, Process Variation

1. INTRODUCTION
Globalization has provided us with a vast choice of hardware suppliers at various levels of the design and manufacturing flow. Designs and fabricated chips can now come from practically anywhere in the world. However, this has not come without costs. The trustworthiness of the received device is no longer a guarantee. The reliability of a system is directly dependent on the reliability of the components that it is built of. One unreliable IC in a system possibly can result in a catastrophic system failure. An area where this is particularly critical is in defense and medical related systems where the reliability of a system is of utmost importance. Recently, many cases have come to light, which show ICs being scavenged from electronic waste, repackaged and sold as genuine new ICs [1]. The focus of this paper is this specific form of IC counterfeiting.

A report on the counterfeit ICs by IHS iSuppli estimates a $169 billion in potential annual risk to the global electronics business [2-5]. These studies suggest that IC counterfeiting is a serious and growing problem. Some of the direct impacts due to IC counterfeiting are enumerated here: Firstly, consumers do not get what they pay for. The devices get slower and are more prone to failure.

GLSVLSI'13, May 2–3, 2013, Paris, France.

Secondly, semiconductor manufacturers incur a significant loss due to lost sale, which eventually hurts everyone related to this industry including consumers. Thirdly, system manufacturers suffer bad reputation when the components fail. In addition, they endure loss for providing warranty. And lastly, in case of critical applications such as avionics, defense systems, and medical devices, untimely IC failures can result in catastrophic events. As a result, it is extremely important to be able to distinguish new authentic ICs from counterfeits and/or used parts.

For the future designs we might have the luxury of inserting specialized circuitry to the chip in order to help establish the age or usage of the chip (e.g., ROM-based fuses can be implemented to act as IC seals or special protected registers can be employed to produce the manufacturing date of the chip). However, determining whether an existing chip is a counterfeit is much harder. For such cases, statistical and testing techniques are needed. Figure 1 illustrates these two categories of counterfeit detection methods.

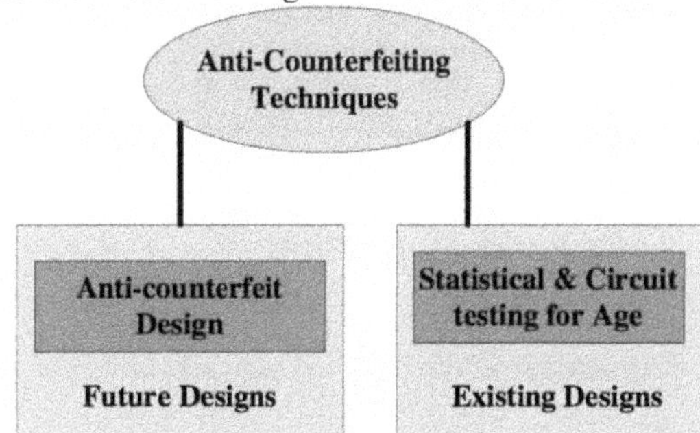

Figure 1: Anti-Counterfeit Techniques

Although finding anti-counterfeit methods for future designs is not a trivial problem, it poses a less challenge compared to detecting counterfeits amongst existing chips, where one has to depend on just the original circuit itself to establish the age of the device. This paper focuses on solutions for existing ICs.

The major contributions of this paper are as follows:

1. Based on detailed simulation results, we show that although the delay of a circuit is greatly affected by aging, it cannot be used as an age indicator by itself since delay is also significantly affected by process variation.
2. We show that the relationship between two or more paths in the circuit can be used as an age marker and we propose a statistical method based on this fact, which can correctly predict the age of an IC.
3. We provide an extensive analysis of our proposed method and show its accuracy by implementing on variety of sample circuits, including few cores and ISCAS benchmarks.

The rest of this paper is organized as follows: In Section 2, the related work is discussed. Section 3 presents the background and motivation of this work. Section 4 introduces the anti-counterfeiting techniques proposed in this work and Section 5 highlights the simulation framework. Results and conclusion are presented in Section 6 and Section 7, respectively.

2. RELATED WORK

In regard to counterfeit prevention and detection, traditional techniques such as printing serial numbers or barcodes to prevent counterfeiting in integrated circuits are not effective as they can be easily cloned or faked. IC Metering [6-8] provides techniques that allow post fabrication control on the ICs. Although metering techniques can be leveraged to prevent counterfeiting, most such techniques require modifying the design significantly, for example, its state machine [6]. In [9], the authors present a technique to prevent piracy of ICs, which renders the ICs inoperative upon fabrication, and it requires a device unique key to take the IC to the functional mode. Recently, some anti-counterfeiting methodologies were presented which use Physical Unclonable Functions (PUFs) [10]. Another odometer based technique to estimate age of ICs has been proposed in [11]. Although, most of these methods are designed to monitor reliability of ICs, they can be deployed for anti-counterfeiting purposes as they provide information on the age of the ICs. Again, an important limitation with all techniques mentioned above is that they are only applicable to designs for the future. Many applications including a wide range of critical applications heavily depend on designs that were created years ago due to the fact that the test cost and time can be prohibitive of they keep deploying newer designs. As a result, it is necessary to distinguish between unused and used (counterfeit) for ICs already manufactured with no anti-counterfeiting capabilities. There are few companies, which deploy inspection techniques using X-ray and Infrared microscopy for this category of ICs, but special high magnification machinery is needed for it [21]. Instead, our paper aims to investigate cost-effective anti-counterfeiting techniques for this purpose.

3. Background and Motivation

This section provides a brief overview of transistor aging, and process variation. Their collective effect on circuit behavior establishes the motivation for our proposed technique.

3.1 *Aging*

In simple terms, aging can be defined as slow but eventually permanent variations that generally deteriorate circuit performance over time. It results from continuous degradation of transistor characteristics, which causes slower operation of circuits, irregular-timing characteristics, increase in power consumption and sometimes even functional failures [13-15]. As shown in Figure 2, the major physical mechanisms behind aging in ICs are the Bias Temperature Instability (BTI), Hot Carrier Injection (HCI) and Time Dependent Dielectric Breakdown (TDDB). They arise both within the gate di-electric and at the boundary of silicon and gate oxide. These effects are particularly prominent at lower technologies due to increasing electric fields in the devices, and decreasing thickness of oxide. Negative Bias Temperature Instability (NBTI) is prominent in PMOS transistors, and can shift (degrade) the PMOS threshold voltage by more than 50mV over ten years. This translates to more than 20% degradation in circuit speed [18-19]. Similarly, Positive Bias Temperature Instability (PBTI) effect occurs in NMOS. In both BTI cases, the amount of charges in the gate dielectric changes with the gate bias, because of charge trapping and de-trapping. HCI effect is more dependent on the switching activity of the transistors, as it is related to carrier injection in presence of high electric fields. TDDB is a consequence of aging, which results in shorting of gate dielectric leading to gate failure. Detailed studies on the factors that contribute to aging of transistors are presented in [12-16]. One of the important facts in regard to aging of transistors is that the amount of activity of the transistors has a direct effect on the aging process.

3.2 *Process Variation*

Process variation (PV) is the random and permanent deviation from the designed, nominal value of a circuit structure, caused by random effects during manufacturing [12]. PV can be separated into two categories: The first category covers variations in process parameters, such as impurity concentration densities, oxide thicknesses, and diffusion depths. These result from non-uniform conditions during the deposition and/or the diffusion of the dopants. The second category covers variations in the dimensions of the devices. These result from limited resolution of the photolithographic process, which in turn causes width and length variations in transistors.

It is important to note, that variation in manufacturing, can introduce mismatch between two transistors sitting within a single die, and on a higher level of abstraction, from one wafer to other, and between manufacturing lots. This spatial classification of process variation is widely grouped as intra-die (within the die), and inter-die (across dies). In general, inter-die variation can have more severe effect on the design compared to intra-die variation.

3.3 Intertwined Effect of PV and Aging

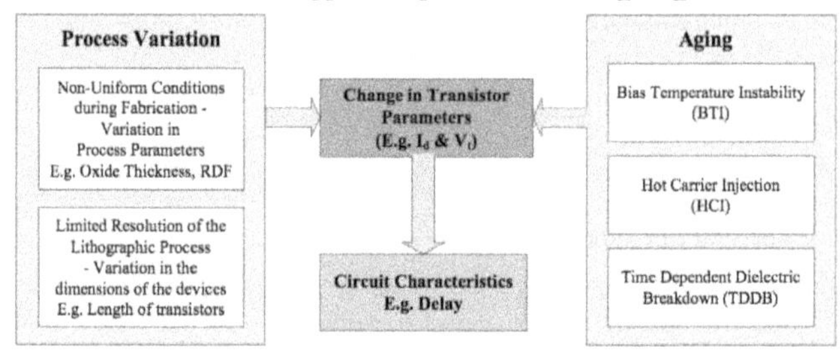

Figure2. Impact of PV and Aging on an IC

Figure 2 summarizes how both Process Variation (PV) and Aging impact common electrical parameters such as threshold voltage, saturation current and eventually circuit characteristics like delay or power, of an IC. This means the effects of PV and aging are intertwined and very hard to isolate from each other. This makes the counterfeit detection a very challenging problem.

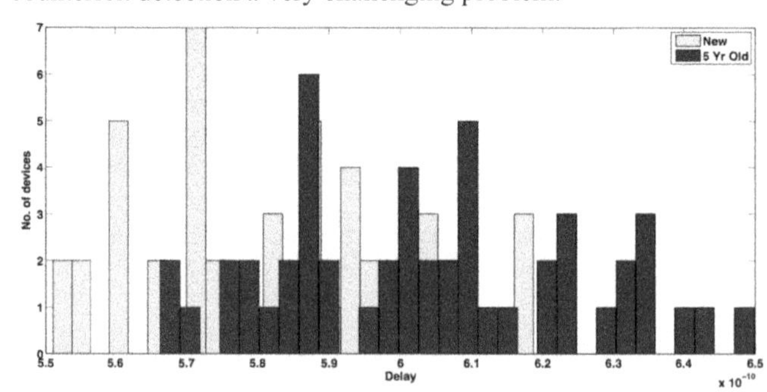

Figure3. Delay Distribution of a path in a new v/s old IC

Figure 3 shows this intertwined effect from simulation point of view. A simple 50-stage inverter chain is simulated both as a new circuit (light-blue histogram) and as a 5-year old circuit (dark-blue histogram). In both cases, the circuit is also subjected to process variation using Monte Carlo simulation. More details of our experimental setup are provided in Section 5. Although the aged circuits are typically slower than new circuits, there is a significant overlap in the range of delay values for the two groups. In fact, more than 70% of both the new and old circuits have delay between 0.565ns to 0.615ns. Suppose we pick a delay value of 0.6ns, it would be impossible to tell if it corresponds to an aged IC – which was originally faster and now has increased delay due to aging or to a new one, which is just slower due to process.

The above observation demonstrates that delay of a single path from the chip is not effective as a direct indicator of the age of an IC. To make matters more complicated, there is recent submicron data showing that aging itself can be affected by process variation [20]. Specifically, there is significant dependence of NBTI and PBTI on device geometry, e.g., channel length, which is variable as virtue of process also. These effects have begun to be modeled by commercial tools [20], but reverse engineering the cumulative effect to isolate age becomes a tedious problem. In this paper, we propose a method that overcomes this challenge by combining the characteristics of more than one path and isolating the effect of aging.

4. ANTI-COUNTERFEIT TECHNIQUES

This section discusses the method we propose for detecting old counterfeit ICs, which lack special anti-counterfeiting. We begin with the goal of just being able to distinguish an old device from a new one and later extend the methodology to accurate predict the age of a chip.

As we saw in Section 3, although delay of a single path is significantly affected by aging, it is a poor indicator of the age of the chip as it is also greatly affected by process variation. Therefore, instead of using a single path, we propose the use of a set of paths from within a chip, such that each path individually has a characteristic aging behavior. The relationship between the delays of the chosen paths is used as an indicator of the age of the chip. The premise of our method is that the pattern that governs the relationship between two delay paths for new ICs are distinguishably different from the pattern governing that of old ICs. This means, if we have a trusted set of new chips and we are able to study them before and after (accelerated) aging, we can characterize these relationship patterns at different ages. Now, when an untrusted chip is under test, the relationship of the same paths is evaluated and it is determined whether it follows the pattern belonging to the new chips or older chips. This is illustrated in Figure 4.

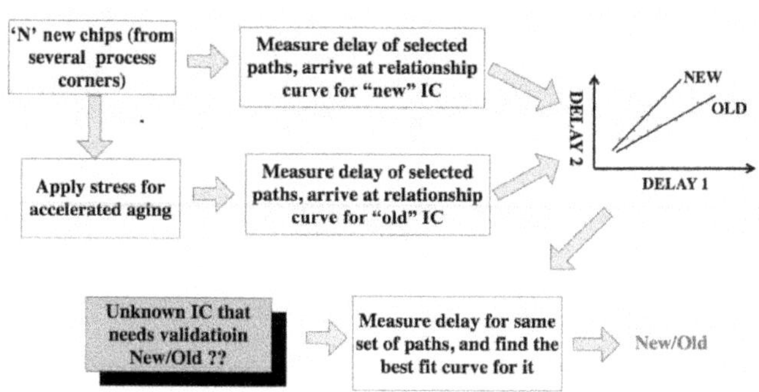

Figure4. Detection method for New/Old in existing designs

The details of our technique for a 2-path method are described as follows, where the relationship between the two paths is used for determining the age.

1. Two different paths are selected from the chip.
2. The delays of the two paths are measured for a variety of trusted new chips. Ideally the trusted chip pool is big enough to present a realistic mix of chips affected by process variation. Curve fitting is used to identify a pattern relating the two paths to each other, similar to the "AGE0" line shown in Figure 5.
3. The trusted chips undergo accelerated aging by being subjected to more than nominal operating voltages and high temperatures.
4. The delays of the two paths are again measured for all the trusted chips. Once more curve fitting is used for relating the paths to each other, similar to the "AGE5" line in Figure 5.
5. When an untrusted chip is under test, the delays of the same two paths are measured. Based on mathematical methods (e.g. shortest perpendicular distance), it is determined whether the "new ICs" curve is a better fit for the untrusted IC or the "older ICs".

This method can be extended to a 3-path or 4-path method, which uses more than two paths, and the relationship between their delays is fitted to a surface plot instead.

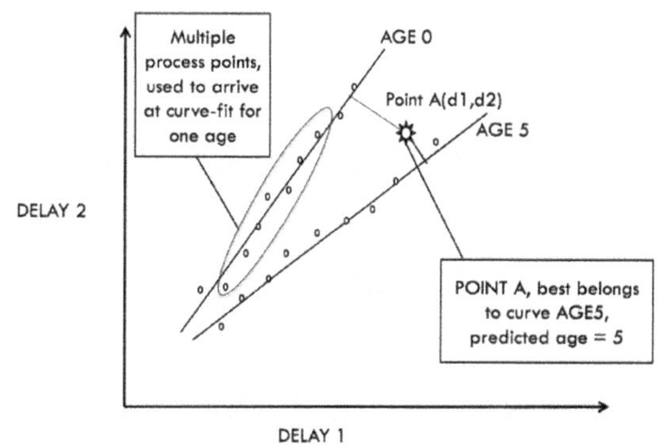

Figure5. Two-Delay method for detection of Age

The important aspect is to decide on the kind of paths to choose. Our method is successful if the paths chosen resemble each other in nature of their delay, e.g., paths with fairly similar capacitive loading, and similar length, but have a distinct aging profile. The latter is achieved by using paths of dissimilar activity – i.e. paths undergoing different number of transitions. The path with higher activity is more affected by age, which means the relationship of the two paths changes significantly over time. In a real chip, some portion of the chip logic is not as active during its lifetime, as other portions. For example, the test paths around the memory test logic will be quite less active or totally inactive, during the lifetime, relative to a functional clock tree path. Such paths, which significantly differ in their activity, are good candidates for this scheme. We provide a detailed sensitivity analysis in Section 6, which details more on the best candidate paths for this method.

4.1 *Just-in-Time Voltage Reduction*

We will show in the Section 6 that the above-proposed technique can achieve only up to 87% accuracy for detecting counterfeit ICs, and almost 1 out of every 3 new chips get falsely detected as old. In order to improve our method, we propose an augmentation technique that we call Just-in-Time Voltage Reduction. In this approach, we reduce the operating voltage of the chip at the time of testing. The rationale is that the effect of aging can be more easily detected at lower voltages. This is explained in more detail in the rest of this section.

The first order equations for gate delay (inverter with symmetrical nmos and pmos), shown in equations (1) and (2), provide the basis for modeling delay in both strong-inversion and sub-threshold regions [22]. The dependence of the delay on the difference of VDD and V_T changes from a linear relation to exponential one as we move into sub-threshold region.

$$t_d = \frac{K C_G V_{DD}}{(V_{DD}-V_T)^{\alpha}} \quad (1)$$

$$t_{d,sub} = \frac{K C_G V_{DD}}{I_o \exp(\frac{V_{DD}-V_T}{n V_{th}})} \quad (2)$$

On the other hand, the equations of aging models [17] show how HCI and BTI result in increase to the threshold voltage of the transistor over time. So, it is expected that operating the circuit at lower voltages, especially near or lower than threshold voltage will result in more drop in circuit speed after aging compared to a circuit that is run at nominal voltage.

Simulation results support this hypothesis. Figure 6 shows the plot of the delay of a path, as it is aged from zero to five years. While all circuits are operated at nominal voltage during their lifetime, each is subjected to a different voltage, i.e., 0.5, 0.8 and 1.2V at the time of delay measurement. As expected, the delays are higher at lower voltages. In addition, we observe that at 0.5V, the effect of aging is more distinguishable, as seen by the clear increase in delay after aging. We intend to use this observation to augment our proposed method for improved detection of counterfeit ICs. Section 6 shows the improvement in results using this.

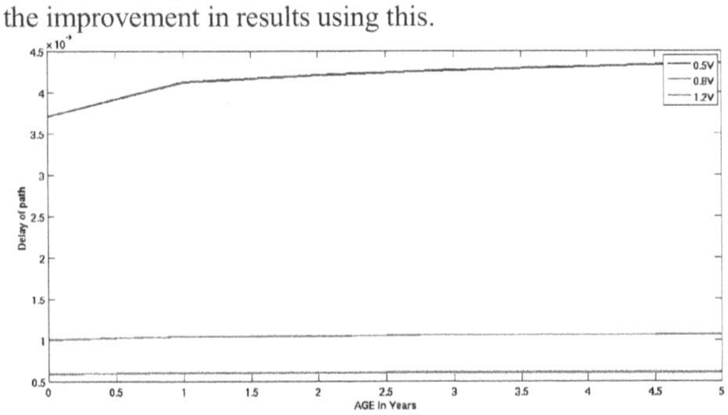

Figure6. Delay v/s Age for diff supply voltage used during post-stress phase (testing phase only)

5. EXPERIMENTAL SETUP

All the circuits are simulated by HSPICE tool using publicly available 90nm Predictive Technology Model (PTM). The effect of aging is simulated through MOS Reliability Analysis (MOSRA) model provided by HSPICE. MOSRA accurately models the HCI and BTI aging mechanisms and analyzes their impact on circuit performance using actual circuit operation and stimulus. It operates in two phases of simulation: pre-stress and post-stress. Different operating voltages can be used for each phase. In our analysis, we use the nominal voltage for the pre-stress phase, which represents the typical usage of the circuit. We use different voltages for the post-stress phase (testing stage).

The effect of process variation is modeled using Monte Carlo (MC) simulation. Specifically, we vary the threshold voltage of transistors (V_{th0}) following Gaussian distributions with means obtained from nominal values in the technology file and 3σ variation (30mV). Using MC, 100 random process points each are evaluated for all age groups. The characterization step, to arrive at reference curves, uses 70 points each for every age group. For the testing phase, the remaining 30 points each (in all 150 unknown chips) are used for validation of prediction scheme.

Test Circuits: A variety of test circuits are used in this paper. The results of Section 6 are based on two sets of circuits. The first set is a variety of inverter chains that differ from each other in their typical delay and activity. Typical delay is the characteristic delay of the path, which has not been affected by either process variation or aging. We modify the depth of the chain and the load capacitance at gate outputs in order to arrive at different typical delays. The input stimulus is controlled to bring desired variation in the activity for different paths. The second set of test circuits for Section 6 consists of typical circuits such as dividers, discrete-cosine-transform circuits and few ISCAS benchmarks.

6. Results, Analysis and Improvements

This section is organized in three parts. In the first part, we use the set of inverter-based circuits for establishing the results of our base method, and the augmented reduced voltage method. In the second part, using the same set of circuits, we show a detailed sensitivity analysis of the method of detection to the type of paths selected. As a remedy to this sensitivity, we introduce 3-path approach, and show the improved results. In the third part, we share our results on some standard circuits. Test results are categorized into two sets: Plain New/Old detection and Exact Prediction of age of the design under test. For exact age analysis, we consider 4 possible ages: 0 (new), 2, 5 and 10 years.

6.1 Basic 2-path method

Table 1 shows the results achieved for the 2-path method as described in Section 4. It also provides some basic information about the two paths involved in this experiment.

Candidate Paths	Transition Activity Ratio	**100:1**
	Typical Delay Difference	**None**
	Post-Stress Voltage	**1.2V**
Prediction of New/Old	Overall Correct Prediction	**87%**
	False Negatives	**6 out of 90**
	False Positives	**10 out of 30**
Exact Prediction of Age	Overall Correct Prediction	**58%**

Table1. Prediction results for Basic 2-Delay method

The result shows that if the two paths have an equal typical delay (in the absence of PV effect) and one of them is 100 times more active than the other, we are able to predict with 87% accuracy, whether a chip is old or new. This method is seen to have a high false positive rate, which means we will be discarding more than 30% authentic chips as potential counterfeits. It is also shown that the exact age prediction is correct only 58% of times.

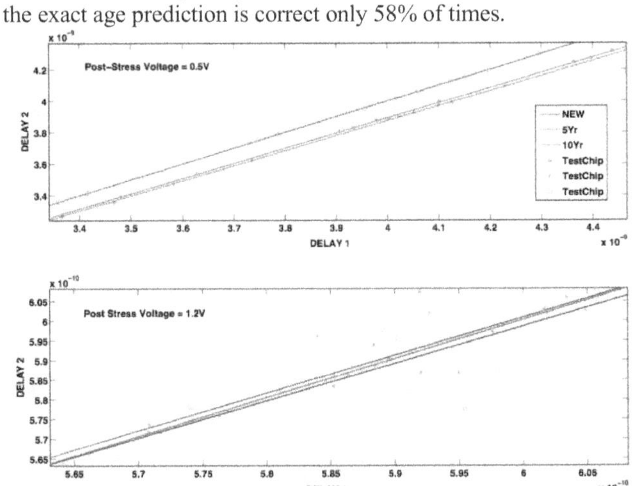

Figure7. Curve-fits across different ages, for different post-stress voltages

Now for the same experimental setup, instead of making the post-stress measurements at 1.2V, we use an operating voltage of 0.5V. As illustrated in Figure 7, the relationship curves obtained at 0.5V are more distinct from one age group to other, and therefore, result in higher correct prediction rate.

With "Just-in-time voltage reduction" method, both false positives and false negatives are reduced to zero, and hence, we are able to achieve 100% accurate detection of old counterfeits from new ICs. Also, the exact age prediction improves to as high as 99%. Nevertheless, this method is very sensitive to the characteristics of the two paths under test. Details of this sensitivity along with a proposed remedy are provided in the next part of this section.

6.2 Sensitivity analysis and Improvements

Broadly, we have analyzed the sensitivity of our detection method, to variation in two characteristics of the paths under test: 1) activity ratio between the selected paths and 2) difference in the typical delays of selected path. Just-in-time voltage reduction is employed for the results of this part.

6.2.1 Sensitivity Analysis

Difference in the activity of the paths is the key reason behind the working of our method. We use the fact that the activity directly influences the aging profile of a path or the degree of aging in a path. One goal is to find out the amount by which they should be different, in order to yield high predictability. In addition, we would like to relax the constraint that the chosen paths should be of same typical delay. For a comprehensive view of the sensitivity to the two factors together, the entire variation matrix is plotted, to show the impact on both test aspects.

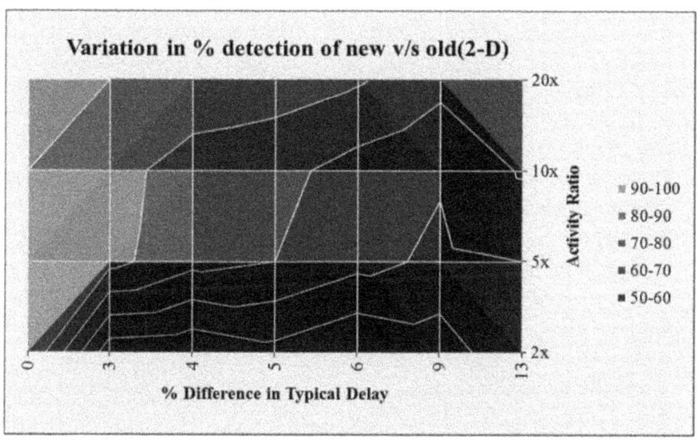

Figure8. Sensitivity of 2-D prediction method (new/old)

For the variation matrix in Figure 8, the typical delay is varied between 0 to 13% and the activity ratio is varied from 2x to 20x. On the activity axis, it is clear that for a fixed set of paths, the prediction result improves as we move from a lower activity difference to a higher one. As for variation to difference in typical delay, it is seen that any deviation from similar typical delay causes deterioration of the detection rate from 100%(as stated in Section 6.1), to even about 50-60%. Similar sensitivity trend is seen for the second test category of exact age prediction, as shown in Figure 9.

In conclusion, for the 2-path method, if we choose paths, which have similar typical delay, an activity difference of 10 times or more is able to produce good prediction results. However, in order to relax the constraint for path selection in terms of typical delay, it is seen that extending the method to include more delay paths can help mitigate the variation, as discussed in following section.

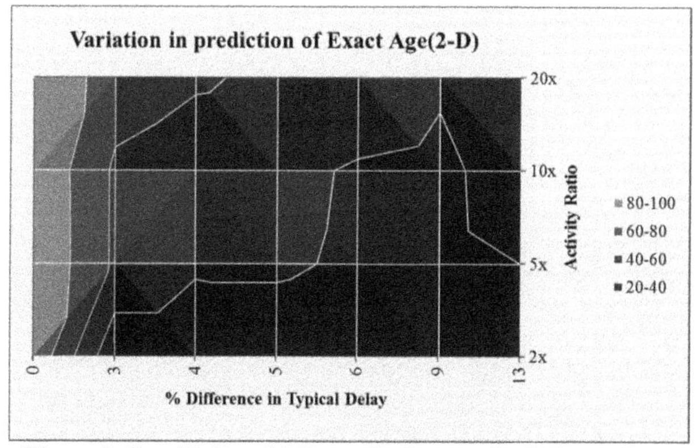

Figure9. Sensitivity of 2-D prediction method (Exact Age)

6.2.2 Further Improvement: 3-path Approach

In the 3-path method, an additional path is chosen. Since the search space for how these three paths should be related to each other in terms of their typical delay and activity is very large, we impose some restrictions by emulating a scenario to a typical design. In this scenario, we assume that we are able to evaluate two critical paths with similar (but not the same) typical delay from a circuit with some moderate to high activity and then, a third path from a test circuit with much lower activity is chosen. Similar to our 2-path method, the relationship curve is defined between the three delay variables using surface fit techniques. And for the unknown test point, minimum residual is used to identify the best-fit surface it belongs to.

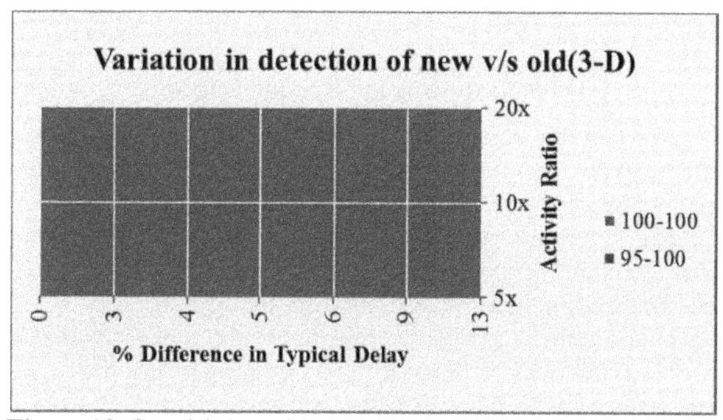

Figure10. Sensitivity with 3-D prediction method (new/old)

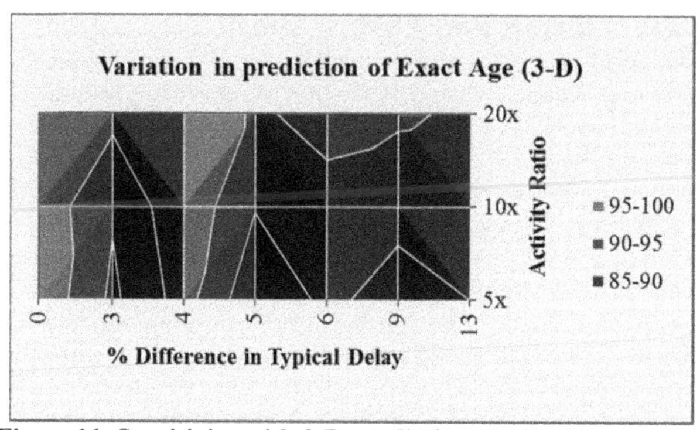

Figure11. Sensitivity with 3-D prediction method (Exact Age)

The improvement in the results for both test types is shown in Figures 10-11. These can be compared to Figures 8-9, as a third path is added to the same couplets that were used to evaluate those results, such that, the following is satisfied in the chosen triplet:

1) Minimum activity ratio between the high activity paths to low activity path is 5x. The activity ratio between the higher activity paths should be minimal.
2) In terms of typical delay, all paths are within 15% difference in typical delay to each other.

In Figures 10-11, for any data cell of the variation matrix, 10 possible triplets are evaluated. On an average, large improvements are observed in the prediction results for all triplets, except for a few outliers. Hence, each data cell value records an average of the %prediction over the chosen triplets. The sensitivity of the plain detection is almost reduced to nil, and percent detection is regained to 100%, and the exact age prediction is also correct over 90% of times, if the activity ratio between high activity and low activity path is at least 10x (minimum activity ratio).

6.3 Real circuits

The 3-D method is deployed on a few standard circuits like multiplier, divider, discrete-cosine-transform (DCT) module, and combinational circuits from ISCAS benchmark. Top critical paths were extracted from the circuits. As expected, the degree of variation in typical delays among the critical paths is not more than 10%. Some sample results for each flavor of circuit, is shown in Table2. The typical delay difference between the 3 chosen paths is also captured, and minimum activity ratio is 10x.

Circuit	% Diff in typical delay	Detection of Old/New	Exact Age Prediction
DCT	6-7%	100%	97.5%
Divider	<1%	100%	98.3%
C6882 (ISCAS)	15%	97.5%	91%

Table 2: Results for standard circuits

This idea can be easily extended to real designs. Finding the candidate paths, as per the requirement of the detection scheme should not be very difficult. Critical paths from a functionally active block can be chosen. And the third path can be chosen from a test path of the chip. If the critical test paths are not as long, we can re-choose the high activity paths such that the typical delay is in the range of the critical test path. Eventually, the aim is to not have more than 10% typical delay difference between the paths. If the typical delays of the chosen paths are almost equal, then even 2 paths of different activity are sufficient for successful detection.

7. Conclusion and Future Work

In this paper, we presented a technique for successful detection of counterfeits among existing ICs which do not have in-built age-detection circuits. For indicating the age of the design, we used a simple parameter of circuit delay, and filtered out the impact of process variation by using the relationship between the delays of paths, instead of using the delay of a single path. Our results that were based on industry-enabled simulation framework show a detection rate of over 97% for identifying an old IC from a new IC. As no circuit study is complete without real chip validation, as future work, we plan a chip tapeout. A variety of paths with varied typical delay will be fabricated. All the chips will be extensively characterized in terms of delay while running at different operating voltages. Later, the chips will undergo accelerated aging using heat and high operating voltage. The chips will be consequently characterized several times and the results will be aggregated. The proposed methodology will then be employed to detect aged versus new devices and prediction rates will be compared to those from simulation results.

8. REFERENCES

[1] http://armed-services.senate.gov/statemnt/2011/11%20November/Sharpe%20Slides%2011-08-11.pdf

[2] http://spectrum.ieee.org/riskfactor/computing/hardware/the-financial-risks-of-counterfeit-semiconductors

[3] http://www.isuppli.com/Semiconductor-Value-Chain/News/pages/Top-5-Most-Counterfeited-Parts-Represent-a-$169-Billion-Potential-Challenge-for-Global-Semiconductor-Market.aspx

[4] http://www.ihs.com/info/sc/a/combating-counterfeits/index.aspx?tid=t4.

[5] http://www.reuters.com/article/2010/10/26/us-china-counterfeit-defence-idUSTRE69P3GH20101026.

[6] F. Koushanfar and G. Qu, "Hardware metering," in *Design Automation Conference (DAC),* pp. 490–493, 2001 2001.

[7] F. Koushanfar, "Integrated circuits metering for piracy protection and digital rights management: An overview," Invited Paper, 2011.

[8] Y. M. Alkabani and F. Koushanfar, "Active hardware metering for intellectual property protection and security," in *Proceedings of 16th USENIX Security Symposium on USENIX Security Symposium,* SS'07, (Berkeley, CA, USA), pp. 20:1–20:16, USENIX Association, 2007.

[9] W. Griffin, A. Raghunathan, and K. Roy, "Clip: Circuit level ic protection through direct injection of process variations," *Very Large Scale Integration (VLSI) Systems, IEEE Transactions on,* vol. 20, pp. 791–803, may 2012.

[10] S. Devadas, E. Suh, S. Paral, R. Sowell, T. Ziola, and V. Khandelwal, "Design and implementation of puf-based "unclonable" rfid ics for anti-counterfeiting and security applications," in *RFID, 2008 IEEE International Conference on,* pp. 58–64, april 2008.

[11] T.-H. Kim, R. Persaud, and C. Kim, "Silicon odometer: An on-chip reliability monitor for measuring frequency degradation of digital circuits," *Solid-State Circuits, IEEE Journal of,* vol. 43, pp. 874–880, April 2008.

[12] P. Gupta and A. B. Kahng, "Manufacturing-aware physical design," in *ICCAD '03: Proceedings of the 2003 IEEE/ACM international conference on Computer-aided design.* Washington, DC, USA: IEEE Computer Society, 2003, p. 681.

[13] A. H. Baba and S. Mitra, "Testing for transistor aging," VLSI Test Symposium, IEEE, vol. 0, pp. 215–220, 2009.

[14] X. Chen, Y. Wang, Y. Cao, Y. Ma, and H. Yang, "Variation-aware supply voltage assignment for minimizing circuit degradation and leakage," in ISLPED '09: Proceedings of the 14th ACM/IEEE international symposium on Low power electronics and design. New York, NY, USA: ACM, 2009, pp. 39–44.

[15] T.Douseki, M.Harada, and T.Tsuchiya,"Ultra-low-voltagemtcmos/simoxtechnologyhardenedtotemperaturevariation," Solid-State Electronics, vol. 41, pp. 519–525, 1997.

[16] D. Lorenz, G. Georgakos, and U. Schlichtmann, "Aging analysis of circuit timing considering nbti and hci," in On-Line Testing Symposium, 2009. IOLTS 2009. 15th IEEE International, June 2009

[17] M. Alam and S. Mahapatra, "A comprehensive model of pmos nbti degradation," Microelectronics Reliability, vol. 45, no. 1, pp. 71 – 81, 2005.

[18] S. Borkar, "Electronics beyond nano-scale CMOS," DAC, 2006.

[19] D. K. Schroder, and J. A. Babcock, "Negative bias temperature instability: Road to cross in deep submicron silicon semiconductor manufacturing," Journal of Applied Physics, pp.1-18, July 2003.

[20] Bogdan Tudor, Joddy Wang, Zhaoping Chen, Robin Tan, Weidong Liu and Frank Lee , Synopsys Inc., "An Accurate and Scalable MOSFET Aging Model for Circuit Simulation", 12th Int'l Symposium on Quality Electronic Design, 2011

[21] http://www.chipscalereview.com/white_papers/DetectingCounterfeitComponents.pdf

[22] "Sub-Threshold Design for Ultra Low-Power Systems", By Alice Wang, Benton H. Calhoun, Anantha P. Chandrakasan, Pg 75

Unexcitability Analysis of SEUs Affecting the Routing Structure of SRAM-based FPGAs

Cinzia Bernardeschi
Department of Information Engineering
University of Pisa, Italy
cinzia.bernardeschi@ing.unipi.it

Luca Cassano
Department of Information Engineering
University of Pisa, Italy
luca.cassano@ing.unipi.it

Andrea Domenici
Department of Information Engineering
University of Pisa, Italy
andrea.domenici@ing.unipi.it

Luca Sterpone
Dipartimento di Automatica ed Informatica
Politecnico di Torino, Italy
luca.sterpone@polito.it

ABSTRACT

Testing SEUs in the configuration memory of SRAM-based FPGAs is very costly due to their large configuration memory, therefore it is necessary to optimize the generation of test patterns. In particular, in order to reduce the effort required of automatic test pattern generators, it is useful to identify early the unexcitable faults, i.e., those faults that cannot be excited by any combination of input signals. In this paper, the unexcitability of SEUs affecting the configuration bits controlling the routing resources of SRAM-based FPGAs is considered. Since this part of the configuration memory contains the largest number of configuration bits, its testing is particularly onerous. Faults in the routing resources are modeled considering the actual electrical behavior of the affected interconnections, thus the resulting fault model is more accurate than the classical open/short model usually considered. This paper introduces a methodology to prove the unexcitability of these faults. The methodology has been implemented in a tool based on a formal specification language (SAL) and a model checker (SAL-SMC). Results from the application of the tool to some circuits from the ITC'99 benchmark are reported.

Categories and Subject Descriptors

B.8.1 [**Hardware**]: Performance and Reliability—*Reliability, Testing, and Fault-Tolerance*; B.7.2 [**Hardware**]: Design Aids—*Verification*

Keywords

SRAM-based FPGA, Model Checking, SAL, Single Event Upset, Untestability Analysis

1. INTRODUCTION

SRAM-based FPGAs are increasingly being employed in safety-critical applications such as avionic or space ones, where a harsh radiation environment causes a high incidence of *Single Event Upsets (SEUs)* that may corrupt the functionality of the device [1, 6]. The industrial use of electronic devices in safety-critical systems is regulated by application-related safety standards that impose strict safety requirements on the system. In particular, safety standards, such as [8], [5], and [7], require in-service testing activities for safety-related systems.

Automatic test pattern generation (ATPG) for integrated circuits is a hard task, since in modern VLSI systems the total number of faults that need to be detected may be very large. In FPGA-based systems, faults in the configuration memory must be tested, in addition to those in user resources. A number of these faults may be demonstrated to be untestable, thus reducing the effort required of ATPG tools. Moreover, demonstrating the untestability of faults in a VLSI design provides a way to assess the degree of testability of the system.

A number of works addressing various aspects of the analysis of untestability of faults in digital systems can be found in the literature. In [10] two algorithms (FILL and FUNI) to demonstrate untestability of stuck-at faults are presented. FILL identifies large subsets of illegal states in synchronous sequential circuits, and FUNI finds untestable faults that require illegal states previously found by FILL to be detected. In [16] a preprocessing method for accelerating SAT-based ATPGs by eliminating untestable faults is presented. The method takes into account the stuck-at fault model and it addresses only *easy-to-classify* untestable faults. In [13] and [11] a new subclass of untestable faults, called *register enable stuck-on* is defined and a method for generating property specification language (PSL) assertions for proving the untestability of this class of faults is presented. Stuck-at faults on the clock-enable signals of registers at the register transfer level (RTL) are addressed. The same authors propose in [12] a hierarchical untestability identification method. The method addresses untestable faults in functional units, such as adders and multiplexers, at the RTL level. Untestability analysis of SEUs in the configuration bits controlling logic components of SRAM-based FPGAs was proposed in [3], assuming the fault model proposed in [14], much more accurate than the stuck-at fault model.

As shown in [17], a more accurate fault model than the open/short (usually assumed for interconnection faults) has to be considered for SEUs in the configuration memory controlling routing resources of FPGA-based systems. This detailed fault model, together with the large number of configuration bits, makes the issue of fault untestability more relevant.

The present work introduces a methodology to prove the unexcitability of SEUs in the configuration memory of SRAM-based FPGAs controlling routing resources. The methodology relies on the *Electrical Effects Static Analyzer* (E^2STAR) tool [4] for the identification of the effects of the considered SEUs, and on the SAL environment [2] for the modeling of netlists and for the proof of the unexcitability of faults. The main contribution of this paper consists in presenting the first automatic tool for the analysis of the unexcitability of SEUs in the routing structure, at the post place-and-route netlist level, and with an accurate model [17] of the effects of SEUs.

The remainder of this paper is organized as follows: in Section 2 the considered fault model is presented; Section 3 introduces some background information on the E^2STAR tool and the SAL environment; Section 4 describes the proposed approach; Section 5 reports results from the application of approach to some circuits from the ITC'99 benchmark; Section 6 concludes the paper.

2. EFFECTS OF SEUS IN THE ROUTING STRUCTURE

An FPGA is an array of programmable logic blocks, interconnected through a programmable routing architecture and communicating with the output through programmable I/O pads [9]. Logic blocks may be simple combinational or sequential functions, called *soft* logic blocks, e.g., lookup tables, multiplexers and flip-flops, or more complex structures, called *hard* logic blocks, e.g., memories, adders, and microcontrollers. The routing architecture in an FPGA consists of wires of various length and programmable switches that form the desired connections among logic blocks and I/O pads. Finally, the I/O architecture is composed of I/O pads disposed along the perimeter of the device, each one implementing one or more communication standards.

Programming an FPGA device consists in downloading a programming code, called a *bitstream*, in its configuration memory. The bitstream determines the hardware structure of the system to be implemented in the FPGA, and thus the functionality performed by the system. In particular the bitstream determines the functionalities performed by programmable logic blocks, the internal connections among logic blocks and the external connections among logic blocks and I/O pads.

Each path of a circuit implemented in an FPGA device is realized by the connection of logic modules through *Switch-boxes* (programmable routing components that can be configured through *Programmable Interconnect Points (PIPs)*). An SEU occurring in a configuration bit controlling a PIP may alter or interrupt the propagation of one or more signals to the logic block at the downstream of the affected location.

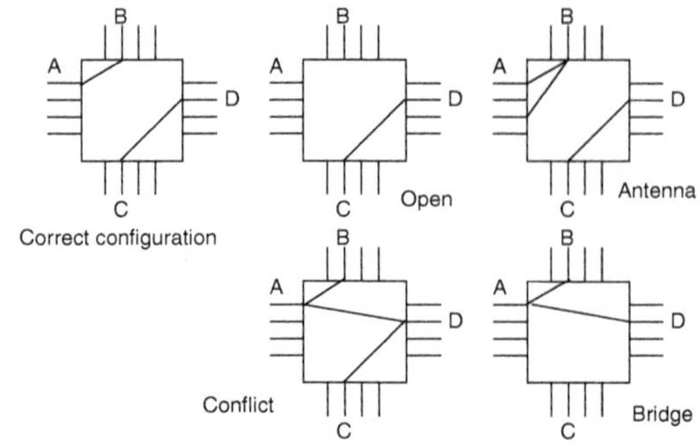

Figure 1: Effects of SEUs in the routing components.

To show the model of SEUs [17], let us consider the switch block shown in Fig. 1, where two PIPs connect wire A to B and wire C to D, respectively. Depending on the position and the electrical properties of the affected PIP, an SEU in the routing structure can cause the following topological modifications (also shown in Fig. 1): (i) *Open*, where the PIP is no longer programmed and thus the corresponding connection ($A \to B$) is deleted; (ii) *antenna*, where a new connection (unused $\to B$) is added between an unused input node and a used output node; (iii) *conflict*, where a new connection ($A \to D$) is added between a used input node and a used output node; and (iv) *bridge*, where an existing connection is deleted ($C \to D$) and a new one ($A \to D$) is added between a used input node and the output node of the deleted connection. Effects of SEUs in PIPs involving unused connections are not considered as they do not cause a faulty behaviour.

3. BACKGROUND

In this section we provide background information about E^2STAR [4] and the SAL environment [2].

3.1 The E²STAR Tool

The purpose of the E^2STAR tool is to provide an accurate analysis of the effects of SEUs in the FPGA configuration memory. E^2STAR not only reports information about the routing topology modifications induced by an SEU, but it also determines the generated electrical effect and the corresponding propagation points in the routing and logic resources used by the circuit mapped on the FPGA. To characterize the logical effects induced by routing faults, fault injection experiments were carried out on a physical prototype of an FPGA application, and the corresponding outputs were observed. Then, logical faults were simulated in a behavioral model of the same application, and by comparing the outputs of the prototype with those of the simulation it was possible to map routing faults to logical effects [17].

For each given configuration memory bit the E^2STAR tool is able to find the routing segments through which an SEU in the configuration bit is propagated and the logic effects

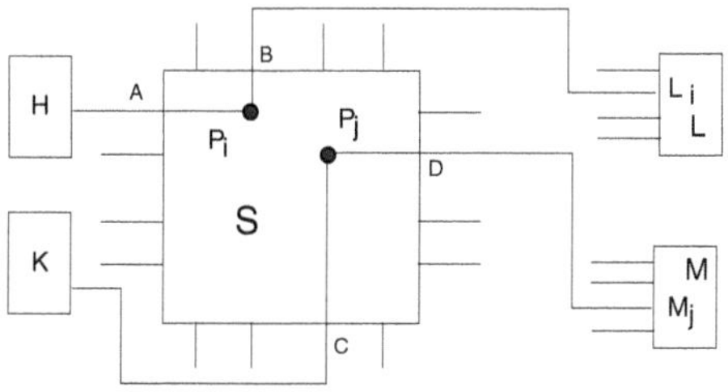

Figure 2: An example switch box.

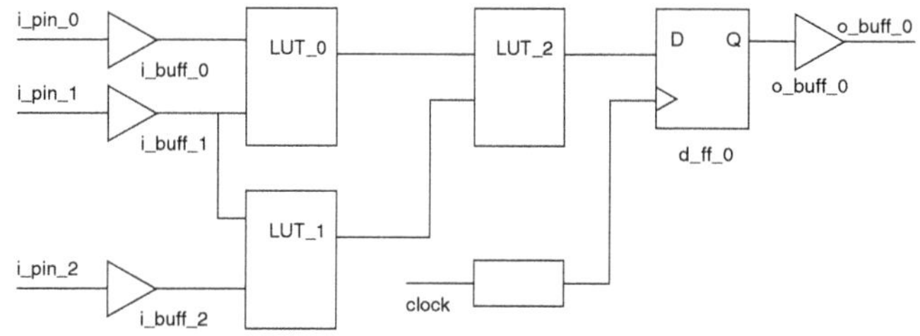

Figure 3: An example netlist.

that the SEU will have in the various propagation routing segments.

With reference to Fig. 2, if S is a switch box, H, K, L, and M are components connected through S, L_i and M_j are the input pins of L and M connected to S through the PIPs P_i and P_j, respectively, the possible logical effects of an SEU in the PIP configuration bits are the following:

- *Stuck-at on P_i*: The logic signal on L_i is set at a fixed value.
- *Bridge between P_i and P_j*: The logic values on L_i and M_j are exchanged.
- *Wired-AND (Wired-OR) between P_i and P_j*: The logic signals on L_i and M_j are set to the AND (OR) of the outputs of P_i and P_j.
- *Wired-MIX between P_i and P_j*: if the outputs of P_i and P_j are different, the logic signals on L_i and M_j are set to 0 and 1, respectively; they are left unaltered otherwise.

It may be observed that a given SEU in the configuration bit associated with a PIP can propagate to different routing segments, and that the same SEU can have different effects on the routing segments through which it propagates.

3.2 The SAL Environment

The Symbolic Analysis Laboratory (SAL) is a framework for combining different tools for the specification and verification of concurrent systems [2].

The SAL language is a strongly-typed description language. The basic concept in the SAL specification language is the `Module`. A SAL module is a self-contained specification of a transition system. A module consists of a `State` and an `Initialization` on the state and a list of `Transitions` on the state. The state is defined by four disjoint sets of `Input`, `Output`, `Global`, and `Local` variables. The input and global variables are *observed*, i.e., they can only be read. The output and local variables are *controlled*, i.e., they can be both read and written. Each SAL variable has two values, the *current* value (denoted, e.g., by `x`) and the *next* value (denoted, e.g., by `x'`), valid in the current and the next state (respectively) of the module.

The initialization is used to specify an initial value for all or some of the controlled state variables of the module. Like initializations, a `Definition` is a simple assignment of a value to a controlled variable. Transitions are assignments of a value to a next-state variable. A guarded command is composed of a guard, i.e., a Boolean condition defined on state variables, and one or more transitions. The transitions can be performed only if the guard is satisfied.

The SAL Symbolic Model Checker (SAL-SMC) uses LTL (Linear Temporal Logic) as an assertion language [15]. LTL formulas state properties about each linear path induced by a transition system. Typical LTL operators are: $\mathbf{G}(p)$ states that p is always true; $\mathbf{F}(p)$ states that p will eventually be true; $\mathbf{U}(p,q)$ states that p is true until a state is reached where q is true; $\mathbf{X}(p)$ states that p is true in the next state.

For a formal definition of LTL see [15]. Typical properties expressed with LTL formulas are *safety*, in the form $\mathbf{G}(\neg\chi)$, stating that the undesired condition χ is never satisfied, and *liveness*, in the form $\mathbf{G}(\mathbf{F}(\psi))$ stating that the desired condition ψ will eventually be satisfied. An expression of the form $M \vdash F$ means that a SAL specification M is a *model* of the LTL formula F, i.e., property F holds for the system specified by M.

4. THE PROPOSED METHODOLOGY

The SAL language has been used to model FPGA-based systems starting from a description of the circuit at the netlist level before the place-and-route phase. Each netlist is described by a SAL `MODULE`. Inputs and outputs of the system are modeled by SAL `Input` and `Output` variables. Each component in the netlist is modeled as a SAL `Local` variable that represents the output of the component itself.

The behavior of each component except flip-flops is described by `Definitions`. The behavior of an input buffer can be described as an assignment between a local variable, modeling the buffer, and an input variable, modeling the associated input pin. Similarly, the behavior of an output buffer can be described as an assignment between two local variables, one modeling the buffer and the other modeling the connected component. The behavior of LUTs is described by the corresponding logic functions. To show how we modeled the behavior of components we refer to the simple example of netlist shown in Fig. 3, where LUT_0 implements the OR function, while LUT_1 and LUT_2 implement the AND function. Examples of SAL models for input and output buffers are shown below:

```
i_buff_0 = i_pin_0;
o_buff_0 = d_ff_0;
```

The lookup tables of the circuit can be modeled as follows:

```
LUT_0 = i_buff_0 OR i_buff_1;
LUT_1 = i_buff_0 AND i_buff_1;
LUT_2 = LUT_0 AND LUT_1;
```

Multiplexers can be described by a conditional clause, as follows (`s` represents the select signal for the multiplexer):

```
y IN IF(s = FALSE) THEN {x1} ELSE {x2} ENDIF;
```

Flip-flops are described by `Transitions`. D-flip-flops can be described as a simple assignment between the next value of the flip-flop and the current value of its input:

```
d_ff_0' = LUT_2;
```

Other types of flip-flops (FDC, FDP, FDCE and FDPE) can be described by a conditional clause.

4.1 Unexcitability Theorems for SEUs in the Routing Structure

The analysis carried out by the E^2STAR tool makes it possible to model the effects of SEUs in the routing structure as observed at the netlist (functional) level, instead of modeling them explicitly, i.e., at the level of the routing structure itself.

Proving that a routing fault is unexcitable means proving that the signal propagated by a given faulty routing line is always the same as the signal that the line would have propagated if no fault had occurred. More precisely, with reference to Fig. 2, the conditions for unexcitability of an SEU s in the configuration bits controlling the PIPs depend on its logical effects, as follows:

- *Stuck-at* 0 *(*1*) in P_i*: s is unexcitable if the output of H is always 0 (1).
- *Bridge between P_i and P_j*: s is unexcitable if the output of H always equals the output of K.
- *Wired-AND between P_i and P_j*: s is unexcitable if the output of H always equals $H \wedge K$ and the output of K always equals $H \wedge K$.
- *Wired-OR between P_i and P_j*: s is unexcitable if the output of H always equals $H \vee K$ and the output of K always equals $H \vee K$.
- *Wired-MIX between P_i and P_j*: s is unexcitable if the output of H always equals the output of K or the output of H is 1 and the output of K is 0.

Further, let $\hat{H}$ and $\hat{K}$ be the SAL variables modeling H and K respectively. The unexcitability theorems associated with the six possible logical effects of an SEU in the configuration bits controlling P_i and P_j are shown in Table 1, where C is the SAL model of the circuit (see the discussion about logical effects in Section 3.1). Note that unexcitability theorems are safety theorems of the form $\mathbf{G}(\neg(\chi))$, where χ is the fault excitation condition.

The table shows that in order to prove a given fault as unexcitable, we try to prove that the components feeding the putatively affected components through a switchbox will never produce the combinations of values that excite the fault.

It may be observed that a given SEU in the configuration bit associated with a PIP can propagate to different routing segments, and that the same SEU can have different logical effects on the routing segments through which it propagates. In such a case the unexcitability theorem associated with the SEU will be composed of the disjunction of a number of unexcitability lemmas, each associated with one of the propagation segments of the SEU itself, since an SEU with multiple propagation segments is excited if and only if at least one of the faulty propagation segments is excited.

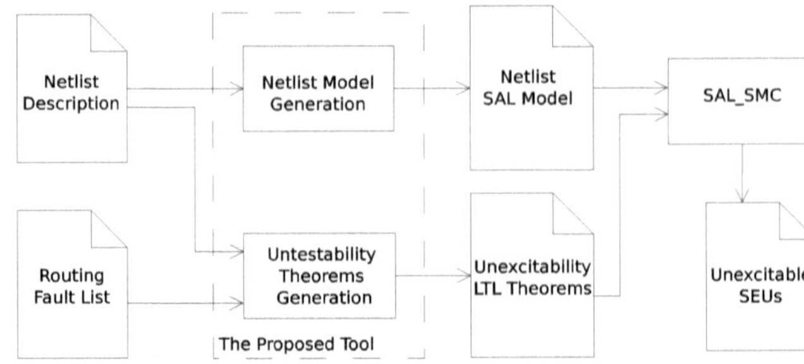

Figure 4: The execution flow of the proposed tool.

As an example, let us consider a PIP P. Let Y_1, Y_2, Y_3 and Y_4 be four destination components connected to P. Let us suppose that an SEU s in P propagates as a stuck-at-0 towards Y_1, as a stuck-at-1 towards Y_2, and as a wired-mix between Y_3 and Y_4. Let X_1, X_2, X_3, and X_4 be the netlist components generating the signals directed to Y_1, Y_2, Y_3, and Y_4, respectively, and let $\hat{X}_1$ $\hat{X}_2$, $\hat{X}_3$, and $\hat{X}_4$ be the corresponding SAL variables. The unexcitability theorem associated with s is:

$$C \vdash \mathbf{G}(\neg(\hat{X}_1 \vee (\neg\hat{X}_2) \vee ((\hat{X}_3 \neq \hat{X}_4) \wedge (\neg\hat{X}_3 \vee \hat{X}_4))))$$

The theorem above is expressed in SAL as follows:

```
F: THEOREM
  circuit |- G(NOT((X1=TRUE) OR (X2=FALSE) OR
                  (X3/=X4 AND (X3=FALSE OR X4=TRUE))));
```

4.2 The Execution Flow

The overall execution flow of the proposed tool is shown in Fig. 4. The netlist description file contains an intermediate description of the netlist, generated by a parser designed for the EDIF language. The routing fault list file contains the list of the routing faults and the associated logical effects and propagation routing segments for each effect. This file is produced by the E^2STAR tool starting from the description of the netlist after the place and route phase. In this way we are able to interact with commercial CAD tools used in the standard design process of FPGA-based systems.

The first step performed by the proposed tool is the creation of the SAL specification of the netlist under analysis starting from the netlist description file. Then, for each possible SEU in the routing fault list an unexcitability theorem is specified. At the end of these two steps a SAL specification of the netlist and the list of the LTL unexcitability theorems associated with all the SEUs in the configuration bits are created.

Finally the SAL-SMC model checker is invoked to prove each unexcitability theorem, one at a time. If a theorem is proved, the corresponding fault is demonstrated to be unexcitable and it is logged. These last steps are repeated until all theorems have been processed by SAL-SMC. At the end of this analysis a report file containing the number and the list of the unexcitable faults is created.

5. EXPERIMENTAL RESULTS

The tool has been applied to some circuits from the ITC'99 benchmark. These designs provide various test cases with

Table 1: **Unexcitability theorems for routing faults.**

s-a-0 on P_i	$C \vdash \mathbf{G}(\neg(\hat{H}))$
s-a-1 on P_i	$C \vdash \mathbf{G}(\neg(\neg\hat{H}))$
bridge between P_i and P_j	$C \vdash \mathbf{G}(\neg(\hat{H} \neq \hat{K}))$
Wired-AND between P_i and P_j	$C \vdash \mathbf{G}(\neg((\hat{H} \neq (\hat{H} \wedge \hat{K})) \vee (\hat{K} \neq (\hat{H} \wedge \hat{K}))))$
Wired-OR between P_i and P_j	$C \vdash \mathbf{G}(\neg((\hat{H} \neq (\hat{H} \vee \hat{K})) \vee (\hat{K} \neq (\hat{H} \vee \hat{K}))))$
Wired-MIX between P_i and P_j	$C \vdash \mathbf{G}(\neg((\hat{H} \neq \hat{K}) \wedge (\neg\hat{H} \vee \hat{K})))$

Table 2: **Characteristics of the benchmarks.**

Circuit	LUTs	FFs	MUXs	IBuffs	OBuffs
b01	5	5	0	3	2
b02	4	4	0	2	1
b03	22	30	0	5	4
b06	8	8	0	3	6
b07	135	53	15	2	8
b08	23	21	0	10	4
b09	35	29	0	2	1
b10	40	17	1	12	6
b11	77	32	9	8	6
b13	106	59	11	11	10

Table 3: **Effects of SEUs in the routing structure.**

Circuit	Faults	s-a-0	s-a-1	W-AND	W-Mix	Bridge
b01	547	708	2,944	5	7	0
b02	304	118	339	5	7	102
b03	5,910	8,105	21,661	1,423	1,431	2,320
b06	566	372	790	0	18	305
b07	10,431	18,331	47,762	4,085	3,911	4,739
b08	2,689	3,074	8,061	464	496	1,217
b09	3,872	6,569	15,948	567	512	1,908
b10	3,942	4,603	10,727	482	692	1,498
b11	10,104	14,059	35,749	3,537	3,536	4,480
b13	7,203	10,390	27,720	1,143	1,387	3,602

Table 4: **Results from the application of the tool**

Circuit	RF	RUn	T(min)	RUn%	LUn%
b01	547	0	2.45	0	0
b02	304	0	0.80	0	0.11
b03	5,910	170	219.36	2.87	48.42
b06	566	0	1.33	0	7.69
b07	10,431	385	1,863.53	3.69	31.33
b08	2,689	0	37.10	0	4.16
b09	3,872	0	93.29	0	27.45
b10	3,942	0	30.70	0	32.27
b11	10,104	13	1,347.54	0.12	21.56
b13	7,203	321	251.91	4.45	39.72

circuits characterized by single clock signal, no tri-state buses nor internal memories. We synthesized the VHDL code of the circuits using the Xilinx ISE CAD tool and mapping the benchmark circuits on the Xilinx Virtex-II SRAM-based FPGAs. The characteristics of the circuits in terms of number of Look-Up Tables (LUTs), Flip-Flop (FFs), MUXs, Input and Output buffers are shown in Table 2.

Table 3 shows the results from the application of the E^2STAR tool to the considered circuits. The table shows the number of identified critical configuration memory bits (Faults) and the number of propagation points classified by logical effects (s-a-0, s-a-1, W-AND, W-Mix, Bridge). From Table 3 it can be noticed that the number of propagation points per SEU is much higher than the actual number of SEUs. In particular the number of fault propagation points is on average 5.3 times larger than the number of faults. Moreover, as it is also discussed in [17], the largest number of effects of SEUs is stuck-at 0 and stuck-at 1.

Results of the analysis of the considered circuits are shown in Table 4. The table shows the total number of SEUs affecting the routing structure (RF), the number of unexcitable SEUs affecting the routing structure (RUn), and the time (in minutes) needed to carry out the analysis. Moreover, the second part of Table 4 shows the percentage of unexcitable SEUs affecting the routing structure (RUn%) and compares it with the percentage of unexcitable SEUs affecting logic resources (LUn%) calculated using the tool presented in [3]. The computer used for the experiments was equipped with an Intel Core i5 (QuadCore) 2.67 GHz, 256 KB L1 Cache, 1 MB L2 Cache, 8MB L3 Cache, 4 GB RAM.

The experiments show that, in spite of the very large number of propagation points (that make SEUs much easier to excite), four out of ten circuits have unexcitable SEUs in configuration bits controlling routing resources. The average unexcitability is 1.1%, with a peak of about 4.4% for b13. If we look at the second part of Table 4, we can see that SEUs affecting routing resources are much easier to excite than faults affecting logic resources. This may be explained, if we take two points into account: (i) the excitation of an SEU in a configuration bit controlling an LUT depends on the values of all the inputs of the LUT while, as discussed in Section 4.1, the excitation of an SEU in a configuration bit controlling a PIP depends on the value of one or two signals; and (ii) as we previously discussed, each SEU in the routing structure has a very large number of propagation points. Moreover, it can be argued that stuck-at effects seem to be much easier to excite than wired-AND, wired-MIX and Bridge effects. In fact, among the considered circuits, the ones showing unexcitable SEUs are those circuits with the highest number of wired-AND, wired-MIX and Bridge effects.

The time required for the analysis ranges from a few seconds up to a few minutes for very small circuits. The analysis time for medium size circuits increases to about a few hours while the required time for larger circuits is about one day. These times are reasonable, taking three points into account: (i) this analysis should be performed just once during the design of the system; (ii) by demonstrating the unexcitability of faults, the results of this analysis can reduce the time required for the generation of test patterns; and (iii) the proposed analysis addresses a very accurate fault model for SEUs, thus making the number of faults to be analysed very large.

Finally, from the point of view of fault tolerance identifying a subset of SEUs to which the system is not sensitive allows an early assessment of the robustness of the system to SEUs in the configuration memory.

6. CONCLUSIONS AND FUTURE WORK

In this paper a tool has been proposed for the unexcitability analysis of SEUs affecting the configuration memory controlling routing resource of SRAM-based FPGAs. The tool implements a fault model that is more accurate than the open/short model usually assumed for the analysis of interconnections in digital circuits. The application of the tool to some circuits from the ITC'99 benchmark has shown that, in spite of the large number of propagation points, also SEUs in routing components can be demonstrated to be unexcitable. Such an analysis may have a twofold benefit for system designers: on the one hand it can reduce the effort required of ATPG tools; on the other hand it can express a measure of sensitivity to SEUs of the system.

Future work will address the analyis of unpropagability, with the aim of detecting those faults that cannot be tested because they are always masked before reaching the output of the system. Further, the counterexamples produced by SAL-SMC for excitable faults can be used to generate test patterns. Substituting the SAL model checker with an on-the-fly model checking tool, can be considered, in order to improve the scalability of the tool.

7. ACKNOWLEDGMENTS

The authors thank the anonymous reviewers for their useful comments and suggestions.

8. REFERENCES

[1] R. Baumann. Radiation-induced Soft Errors in Advanced Semiconductor Technologies. *IEEE Transactions on Device and Materials Reliability*, 5(3):305 – 316, September 2005.

[2] S. Bensalem, V. Ganesh, Y. Lakhnech, C. Munoz, S. Owre, H. Rueß, J. Rushby, V. Rusu, H. Saidi, N. Shankar, E. Singerman, and A. Tiwari. An Overview of SAL. In *Proceedings of the Fifth NASA Langley Formal Methods Workshop (LFM 2000)*, pages 187–196, 2000.

[3] C. Bernardeschi, L. Cassano, and A. Domenici. SEU-X: a SEu Un-uXecitbility prover for SRAM-FPGAs. In *Proceedings of the 18th IEEE International On-Line Testing Symposium (IOLTS2012)*, June 2012.

[4] C. Bernardeschi, L. Cassano, A. Domenici, and L. Sterpone. Accurate Simulation of SEUs in the Configuration Memory of SRAM-based FPGAs. In *Proceedings of the IEEE International Symposium on Defect and Fault Tolerance in VLSI and Nanotechnology Systems (DFT 2012)*, 2012.

[5] European Committee for Electrotechnical Standardization (CENELEC). EN 50129: Railway applications - Communications, signaling and processing systems - Safety related electronic systems for signaling, February 2003.

[6] P. Graham, M. Caffrey, J. Zimmerman, D. E. Johnson, P. Sundararajan, and C. Patterson. Consequences and Categories of SRAM FPGA Configuration SEUs. In *Proceedings of the 6th Military and Aerospace Applications of Programmable Logic Devices (MAPLD'03)*, September 2003.

[7] International Atomic Energy Agency (IAEA). NS-G-1.3: Instrumentation and Control Systems Important to Safety in Nuclear Power Plants, 2002.

[8] International Organization for Standardization (ISO). 26262-5: Road vehicles - Functional safety - Part 5. Product development: hardware level, December 2009.

[9] I. Kuon, R. Tessier, and J. Rose. FPGA Architecture: Survey and Challenges. *Foundations and Trends Electronic Design Automation*, 2(2):135–253, Feb. 2008.

[10] D. Long, M. Iyer, and M. Abramovici. FILL and FUNI: Algorithms to Identify Illegal States and Sequential Untestable Faults. *ACM Transaction on Design Automation of Electronic Systems*, 5(3):632–657, 2000.

[11] J. Raik, H. Fujiwara, R. Ubar, and A. Krivenko. Untestable Fault Identification in Sequential Circuits Using Model-Checking. In *Proceedings of the 17th Asian Test Symposium (ATS'08)*, pages 21–26, 2008.

[12] J. Raik, A. Rannaste, M. Jenihhin, T. Viilukas, R. Ubar, and H. Fujiwara. Constraint-Based Hierarchical Untestability Identification for Synchronous Sequential Circuits. In *Proceedings of the 16th European Test Symposium (ETS'11)*, pages 147–152, 2011.

[13] J. Raik, R. Ubar, A. Krivenko, and M. Kruus. Hierarchical Identification of Untestable Faults in Sequential Circuits. In *Proceedings of the 10th Euromicro Conference on Digital System Design Architectures, Methods and Tools (DSD'07)*, 2007.

[14] M. Rebaudengo, M. Sonza Reorda, and M. Violante. A new functional fault model for FPGA application-oriented testing. In *Proceedings of the 17th IEEE International Symposium on Defect and Fault Tolerance in VLSI Systems (DFT 2002)*, pages 372 – 380, 2002.

[15] K. Rozier. Linear Temporal Logic Symbolic Model Checking. *Computer Science Review*, 5(2):163 – 203, 2011.

[16] D. Tille and R. Drechsler. A Fast Untestability Proof for SAT-based ATPG. In *Proceedings of the 12th International Symposium on Design and Diagnostics of Electronic Circuits & Systems (DDECS'09)*, pages 38–43, 2009.

[17] M. Violante, N. Battezzati, and L. Sterpone. *Reconfigurable Field Programmable Gate Arrays for Mission-Critical Applications.* Springer Science & Business Media, 2011.

A Novel Intermittent Fault Markov Model for Deep Sub-Micron Processors

Babak Saghaie
School of Computer Science and Engineering
The University of New South Wales
Sydney, Australia
babaks@cse.unsw.edu.au

Roshan Ragel
Department of Computer Engineering
The University of Peradeniya
Peradeniya, Sri Lanka
roshanr@pdn.ac.lk

Sri Parameswaran
School of Computer Science and Engineering
The University of New South Wales
Sydney, Australia
sridevan@cse.unsw.edu.au

Aleks Ignjatovic
School of Computer Science and Engineering
The University of New South Wales
Sydney, Australia
ignjat@cse.unsw.edu.au

ABSTRACT

Intermittent faults (IF) in chips are becoming commonplace with the current technology trend and the process scaling. In this paper, we first modify the well known birth-death Markov model so that *availability* can be calculated. We then show that the standard birth-death Markov model does not capture IF correctly, and create a novel Markov model for intermittent faults that is derived from the specific nature of such faults. The proposed model, for the first time, differentiates risky and normal components and therefore does not waste processing time for unnecessary testing procedures. Consequently, the availability of processors with the proposed model increases significantly compared to the traditional model (from 0.90 to 0.99 with a typical parameter set). In addition, the proposed model facilitates parameter space exploration. Positive effects were observed with varying parameters such as error rate, recovery time and test program length. It was concluded that choice of right testing parameters are vital for gaining optimal system availability and the new model supports achieving the same.

Categories and Subject Descriptors

B.8.1 [**Performance and Reliability**]: Reliability, Testing, and Fault-Tolerance

General Terms

Theory, Reliability

Keywords

Markov Model; Intermittent Fault; Process Variation; Availability; Online Testing

1. INTRODUCTION

The main driving force behind advances in computer hardware is process scaling. As a result of scaling, computation power has increased by more than five orders of magnitude in the past decades and this trend is continuing [3]. Also, power consumption per transistor has been dramatically reduced, making it possible to compute intensive tasks on devices that run on battery. As a result of scaling, and low power nodes, computer hardware has to be designed not only for performance, but for low energy consumption and reliability as well. As we scale down to the deep sub-micron feature sizes, interconnects and transistors are miniaturized and are more prone to physical defects. Therefore, circuits become more susceptible to intermittent faults. This paper models the intermittent fault for dependable computers.

Intermittent fault is caused by marginal and unstable hardware. In deep sub-micron circuits, a small manufacturing residue that was tolerated by previous fabrications, will violate the design boundaries. The impending process variation (PV) in deep sub-micron, will increase the occurrence of IF in future circuits [3]. Static variation happens during fabrication (i.e. sub-wavelength lithography, random dopant fluctuations). Dynamic variation depends on timing and type of application being run on processor and will result in temperature and voltage variations across chip die.

Table 1: Notations

$1/\lambda$	Mean time to failure
$1/\mu_r$	Mean time to recovery
$1/\lambda_t$	Time interval between test
$1/\mu_t$	Test program length in time

(a) Traditional Model

$1/\mu_r$	Mean time for recovery
$1/\lambda_{t1}$	Time interval between test for normal components
$1/\lambda_{t2}$	Time interval between test for risky components
$1/\lambda_1$	Mean time to failure for normal component
$1/\lambda_2$	Mean time to failure for risky component
$1/\mu_t$	Test program length in time
$1/\mu_n$	Mean time to move from risky to normal state

(b) New Model

Because chip manufacturers want to achieve the highest frequencies possible in each technology, the voltage is rarely scaled down. Furthermore, the gap between supply and threshold voltage is shrinking, making it harder to retain the noise margin. Subsequently, the voltage does not scale down ideally resulting in power consumption density and processor temperature increase [13]. IF rises from a complex interaction of static variation (fabrication time) and dynamic (run time) variation. Component wear out will slow down the switching speed [17], as a consequence, IF will compromise the design safety margins. When combined with the mentioned factors, aging of transistors and wirings will first manifest as intermittent fault. Gradually, the wear out will turn into a permanent defect.

GLSVLSI'13, May 2–3, 2013, Paris, France.

Therefore, with a proper intermittent fault detection model, it is also possible to detect permanent faults at early stages.

Another type of fault is the transient fault (TF). It is caused by temporary environmental factors like cosmic rays, electromagnetic interference, power supply and interconnect noise. The main cause of transient fault are particles, if undetected, TF will cause soft-error. When cosmic rays enter the atmosphere, they produce a high energy particle that can alter bits in computer circuits. After a time interval, the circuit will come back to its normal operation.

Similar to the transient fault, the intermittent fault will appear and disappear in a period of time. In the past, these two were combined and modeled as one fault. Constantinescu studied the IF in deep sub-micron technologies [5]. The study observed that as transient and intermittent faults have different underlying causes, they produce different patterns of errors. First, the transient fault occurrence is fairly distributed among the chip as they are equally likely to receive colliding particles and environmental noise while the intermittent fault is focused around areas with marginal and problematic circuitry. Second, IF has a pattern of bursts and once activated, this type of fault has a higher occurrence frequency than TF.

In [5], Constantinescu states that there is no available tool for accurately simulating the IF occurrence in hardware and for that reason Constantinescu gathered field dependability data to study IF. Failure log of 310.7 server years of data from 256 servers were collected. 6.2% of servers experienced intermittent faults [4]. While a small portion of devices log more than thousands of IF induced errors, rest of devices rarely if ever experienced any error. It was concluded that a small minority of components were responsible for majority of fault onsets. The author predicts that challenges imposed by scaling and increased complexity will elevate the probability of IF and shorten the time interval between activations. He reminds the reader of the necessity for solutions that are fast enough to handle this high frequency of error recurrence. Furthermore, a recent large-scale study on hardware failure revealed that IF occurrence rate in processor and dynamic RAM increases by two orders of magnitude after first failure [14].

The rest of the paper is organized as follows: Section II decribes Markov modeling of intermittent faults as well as depenadbility measures. Related work is presented in section III. In Section IV the four state model is presented. Section V introduces our novel Markov model for IF. The state space for reliability parameters is explored in Section VI and the paper is concluded in Section VII.

2. BACKGROUND

2.1 Markov Model for Intermittent Fault

Markov model is a special case of stochastic processes where transition from present state has an exponential distribution and is only dependent on current state and not on previous states or when it arrived at the present state [2]. In other words, past is implicitly embedded in current state. This memorylessness property makes complex problems tractable. In literature, a two-state continuous-time Markov model is used as standard for IF modeling [10, 12, 18]. The famous birth-death process is shown in Figure 1. When the component is in operational state, the chance that a failure will occur has the rate of λ, and the component will move to the failed state as shown in Figure 1.

Similarly once in the failed state, component will return to operational state with the rate of μ, and the mean time of remaining in failed state will be $\frac{1}{\mu}$. Once the model enters the failed state, if there is no repair, it is impossible to leave, and such a state is called *absorbing*.

The actual failure rate is dependent on circuit conditions such as physical design, load, variations, wear out, etc. and it is not possible to accurately predict the failure rate beforehand [5]. Hence, the expected mean time to failure is considered with a pessimistic view to be in the range of tens to hundreds of milliseconds [6, 10]. Obviously, in healthy components, intermittent faults do not happen at such a high rate. As a single rate is assigned to all components; if

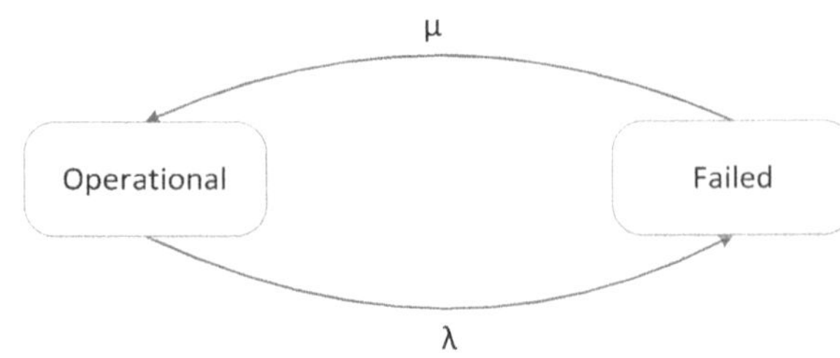

Figure 1: Traditional birth-death model

this pessimistic rate is lowered, it will compromise the error detection in problematic components. For birth-death process, according to steady state convergence theorem [2], the long-term probabilities denoted as $\Pi_0 + \Pi_1$are:

$$\lambda * \Pi_0 = \mu * \Pi_1 \qquad \Pi_0 + \Pi_1 = 1$$

yields:

$$\Pi_0 = \frac{\mu}{\lambda + \mu} \tag{1}$$

$$\Pi_1 = \frac{\lambda}{\lambda + \mu} \tag{2}$$

2.2 Dependability Measures

In dependability study, we are interested in knowing the system up time. Reliability R(t) is the probability that system operates free of fault in time interval $[0, t]$ given the condition that system was healthy at time zero. Availability A(t), is the probability that the system is operational and in good state at the instant of time t [8]. Both R(t) and A(t) are a function of failure rate. Availability also depends on the time that it takes to detect and recover from fault. Consequently, availability is a more meaningful dependability measure for systems with repair [8, 9]. Because the failure and repair rates are in the order of milliseconds, after a short time (in the order of few seconds), the model will have thousands of iterations and dependability measures will virtually reach steady state. Hence, we are primarily interested in steady state availability A_{SS}. Formulas 1 and 2 represent steady state probabilities of operational and failed states respectively. However, they do not separate the portion of time processor was executing ordinary program (available time) from the time it was executing the test code (unavailable time). To the best of our knowledge, no clear availability analysis has been given based on previous IF Markov models.

3. RELATED WORK

Hsu and Hsu [7] combined permanent fault and the two state IF model to form a new three state model. Since permanent fault represents an absorbing state in the Markov model in [7], the steady state probability of two other states will become zero. Regardless of quality of component, even a minute failure rate will eventually transit the model to permanent fault and it will remain there for good. Therefore, for analysis of long term system dependability, Markov model must be free of absorbing state(s).

A dependable system needs to operate according to specification in spite of expected failures. Online periodic Built-In-Self-Test (BIST) can be utilized to detect the fault [13]. CASP is a concurrent periodic self test implemented on multi-core OpenSPARK T1 [20]. The Test Controller selects processors in a round-robin fashion, then stalls the execution of normal instructions in under test core and saves the status. Next a pre-generated compact test pattern which is stored in an on-chip memory is extracted and the result is compared with stored values. A disparity indicates an error. Paschalis and Gizopoulos implemented a software-based self-test (SBST) [16]. Its compact test program with a small number of memory references contribute to a relatively short test execution time. The online periodic testing proposed in [16] is suitable for low-cost embedded processors. After fault detection, a proper recovery technique is used to restore the last correct state. Typical fault recovery techniques can be classified either into a roll-forward technique (usually adapting redundant computation [1]) or a roll-backward technique (usually utilizing time redundancy [11]).

In previous work, researchers tried to derive the effect of frequency and timing of intermittent fault testing on performance and reliability. Nagakawa and Yasui [12] calculated the mean number of tests to detect a fault and mean time to detection, and afterwards, they determined the optimal time between tests that minimizes the cost to detect a fault. Kranitis et al. [10] formulated optimal testing frequency to minimize the cumulative amount of processing time wasted in recovery and detection. As the resulting formulas for the optimal testing time in [10] do not have a closed form format, the corresponding equations are solved numerically for each set of values.

Heron et al. [6] implemented online periodic testing in real-time embedded systems to examine performance impact of different test scheduling policies in multiprocessing environment. It was shown that if processor utilization stays below 85% the overhead remains stable, because usually there is an idle processor to which the task can be migrated. After that utilization threshold, with the increase of processor load, overhead rises sharply. The authors in [6] concluded that compared to a strict round-robin scheduling, a policy that is aware of system idle times has a lower number of preemption count and provides a good compromise between performance impact and dependability. For the reason that processor workload affects dynamic factors like voltage and temperature fluctuations, under stress marginal components have a higher chance of IF onset. Therefore, focus of testing idle units while improving performance, will give a false sense of dependability.

Depending on fault site and application code being run at the time of activation, the fault could be masked by architecture or propagated to output and cause an error. Pan et al. [15] introduced a new metric called Intermittent vulnerability Factor (IVF) which is defined as the probability that an intermittent fault would result in a visible error (failure). They computed IVF of register file and reorder buffer, with the latter having substantially higher vulnerability factor than the former [15].

Although earlier research proposed test policies for IF, they did not distinguish the minority of components that are repeatedly causing it. Besides, the effects of parameters like error rate, recovery rate and test program length on availability were not evaluated.

3.1 Contribution

The contribution of our research is as follows: we modify traditional model of intermittent fault to determine system available times. Then, we introduce our novel model. Rather than round-robin or a scheduling policy that utilizes idle time, our proposed model is designed to better address the specific nature of IF (ie a burst of errors recurring at the same site). Dependability formulas for models are calculated and presented. Next, we explore the parameter space (error and test rates, recovery and test program length) to contrast the traditional model with the new one and observe the effect of varying system specifications and dependability factors on availability.

4. TRADITIONAL MODEL WITH TEST AND FAULTY STATES

Here we propose the first stage of our model for intermittent fault by adding two states to the birth-death process model, as shown in Figure 2. Let us assume that the component has a failure rate of λ. At the beginning, the component is in the normal operational state. As listed in Table 1, a periodic testing is scheduled with mean time between testing of $\frac{1}{\lambda_t}$. Similarly, the online periodic test program lasts for an average time of $\frac{1}{\mu_t}$. In fault detected state, appropriate

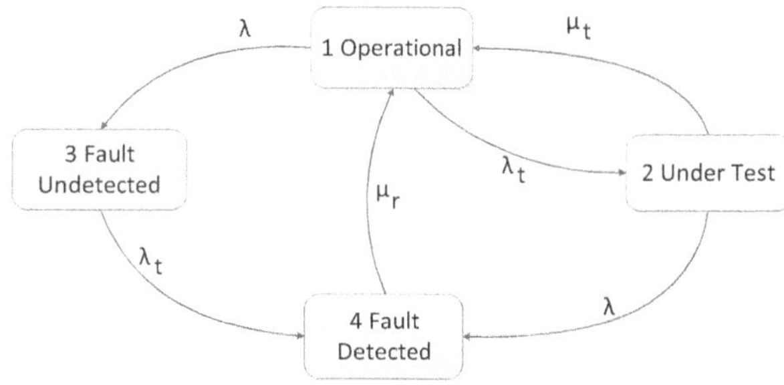

Figure 2: Traditional model with test state

fault recovery mechanisms such as check-pointing and roll-back is implemented. The device will go back to operational state with mean recovery time of $\frac{1}{\mu_r}$. The online periodic testing program, similar to [10], has a perfect detection ability. At the activation of IF, it is assumed that if the device is under test, the fault is detected straightaway and the model moves to fault detected state. Likewise, if the device is in operational state and the fault becomes active, the device will go to fault undetected state and it will remain there until the periodic testing program is run to detect the fault. As we will illustrate in section VI, if the testing frequency is too high, the device will be in under test state more than needed. On the other hand, if the testing frequency is below the desired rate, the fault will remain undetected for a longer period of time. The aim is to find the optimal interval between running the test program. We can explore the parameter space and calculate availability depending on error rate, test frequency, test program length and recovery time. The four state model facilitates this approach as all of the required factors are integrated in one mathematical model and a separate set of complex analytical formulations is not necessary. For the model in Figure 2, all four states are in one recurrent class. We write balance and normalization equations for steady state probability [2, 19].

The balance equations are:

$$\text{state one(operational): } (\lambda_t + \lambda) * \Pi_1 = \mu_r * \Pi_4 + \mu_t * \Pi_2$$

$$\text{state two(under test): } (\mu_t + \lambda) * \Pi_2 = \lambda_t * \Pi_1$$

$$\text{state three(fault undetected): } \lambda_t * \Pi_3 = \lambda * \Pi_1$$

$$\text{state four(fault detected): } \mu_r * \Pi_4 = \lambda_t * \Pi_3 + \lambda * \Pi_2$$

$$\text{The normalization equation is: } \Pi_1 + \Pi_2 + \Pi_3 + \Pi_4 = 1$$

We solve these equations to get the steady state availability denoted as A_{ss}. When the model is in state one, the processor performs its normal operations while the remaining three states are considered down time.

$$A_{ss} = \Pi_1 = \frac{1}{\frac{\lambda_t}{\lambda+\mu_t} + \frac{\lambda}{\lambda_t} + \frac{\lambda\,(\lambda+\lambda_t+\mu_t)}{\mu_r\,(\lambda+\mu_t)} + 1} \quad (3)$$

we differentiate availability with respect to λ_t

$$\frac{\partial(Availability)}{\partial(\lambda_t)} = 0 \quad (4)$$

Formula 4 is solved to get the root which is

$$\lambda_t = \left(\sqrt{\frac{\mu_r\,\lambda^2 + \mu_r\,\mu_t\,\lambda}{\lambda+\mu_r}} \right) \quad (5)$$

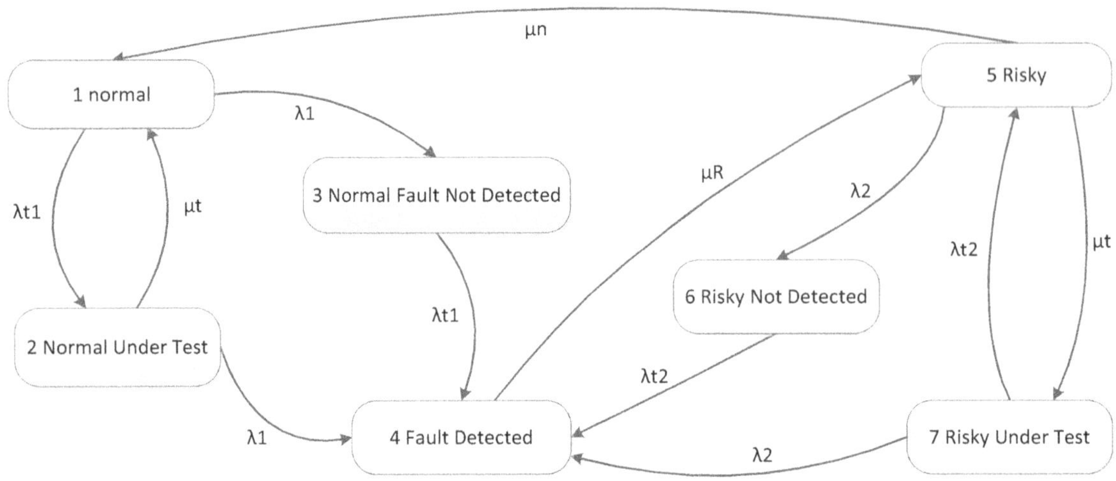

Figure 3: Seven state model

when μ_r, μ_t and λ are available, the optimal duration between testing $\frac{1}{\lambda_t}$ can be easily calculated by Formula 5.

5. NEW MODEL WITH RISKY STATE

As mentioned before, in the traditional model a flat failure rate is assigned to all components, regardless of their health and condition. Designers either have to pessimistically increase the rate that impacts performance, or decrease it which has a negative effect on fault coverage. In reality the probability of this type of fault is not evenly distributed.

Previous research cited in Section I indicate that intermittent fault has its specific spatial and temporal patterns. Based on these findings, we propose a novel Markov model that differentiates between healthy components with ones that had a recent history of failures. The new proposed model is depicted in Figure 3. The model consists of seven states. The operational state is divided into two states: normal (number 1) and risky (number 5). Unlike the last model, when the device recovers from failure it does not go back to the normal state. Instead, it will be labeled risky. A risky element has a higher error rate, as a result the mean time between testing ($\frac{1}{\lambda_{t2}}$) is scheduled to be shorter compared to a normal component. If for a consecutive number of tests $\lceil \frac{\lambda_{t2}}{\mu_n} \rceil$, the identified risky device shows no symptom of a fault, it will be brought back to the normal state. All of seven states are in one recurrent class. Hence, we write balance equations for the proposed model depicted in Figure 3:

state one(normal operational):

$(\lambda_t + \lambda_1) * \Pi_1 = \mu_t * \Pi_2 + \mu_n * \Pi_5$

state two(Normal Under Test)

$(\mu_t + \lambda_1) * \Pi_2 = \lambda_{t1} * \Pi_1$

state three(Normal Fault Not Detected):

$\lambda_{t1} * \Pi_3 = \lambda_1 * \Pi_1$

state four(Fault Detected):

$\mu_R * \pi_4 = \lambda_1 * \Pi_2 + \lambda_{t1} * \Pi_3 + \lambda_2 * \Pi_7 + \lambda_{t2} * \Pi_6$

state five(Risky):

$(\mu_n + \lambda_2 + \lambda_{t2}) * \Pi_5 = \mu_R * \Pi_4 + \mu_t * \Pi_7$

state six(Risky Not Detected):

$\lambda_{t2} * \Pi_6 = \lambda_2 * \Pi_5$

state seven(Risky Under Test):

$(\lambda_2 + \mu_t) * \Pi_7 = \lambda_{t2} * \Pi_5$

The normalization equation is:

$$\sum_{i=1}^{7} \Pi_i = 1$$

We solve these equations to calculate the steady state availability of Π_1 to Π_7. In states one and five, the system is up and running tasks, While in other states, it's either under test or faulty. Thus, the steady state availability is calculated by summation of normal and risky states:

$$A_{ss} = \pi_1 + \pi_5 = \frac{\lambda_{t1}\lambda_{t2}\mu_n\mu_r(\lambda_2 + \mu_t)(\lambda_1 + \lambda_{t1} + \mu_t)}{\Omega} \quad (6)$$

where $\Omega =$

$$\begin{aligned}
&\lambda_1^2(\lambda_2^2\lambda_{t1}(\lambda_{t2} + \mu_r) + \lambda_{t2}(\mu_n\mu_r\mu_t + \lambda_{t1}(\lambda_{t2}\mu_r + (\mu_n + \mu_r)\mu_t))+\\
&\lambda_2(\lambda_{t2}\mu_n\mu_r + \lambda_{t1}(\lambda_{t2}^2 + \mu_r\mu_t + \lambda_{t2}(\mu_n + \mu_r + \mu_t))))+\\
&\lambda_{t1}\lambda_{t2}\mu_n\mu_r(\lambda_2 + \mu_t)(\lambda_{t1} + \mu_t)+\\
&\lambda_1(\lambda_{t1} + \mu_t)(\lambda_2^2\lambda_{t1}(\lambda_{t2} + \mu_r)+\\
&\lambda_{t2}(\mu_n\mu_r\mu_t + \lambda_{t1}(\lambda_{t2}\mu_r + (\mu_n + \mu_r)\mu_t))+\\
&\lambda_2(\lambda_{t2}\mu_n\mu_r + \lambda_{t1}(\lambda_{t2}^2 + \mu_r\mu_t + \lambda_{t2}(\mu_n + \mu_r + \mu_t))))
\end{aligned} \quad (7)$$

6. CASE STUDY PARAMETERS

We explore the parameter space to examine different situations and compare models. Parameters in Table 2 were chosen based on previous studies performed in [12, 6, 16]. The entire system is modeled within one mathematical structure which makes design space exploration easier. Depending on system characteristics that define performance and reliability, parameters vary from one system to another, however the underlying structure is the same. Models were implemented with the help of Matlab.

6.1 Test Rate

We start with test rate. Table 2 lists the set of parameters. To compare seven and four state models, a system is given in set one. For the traditional model, λ_t ranges from 0.01 to 0.3. The availability is calculated by Formula 3 and the result is shown in Figure 4. The optimal test rate is calculated from Formula 5. For given parameters in set one, the optimal test rate is: $\lambda_t = \frac{\sqrt{331}\sqrt{10010}}{10010} = 0.1818$.

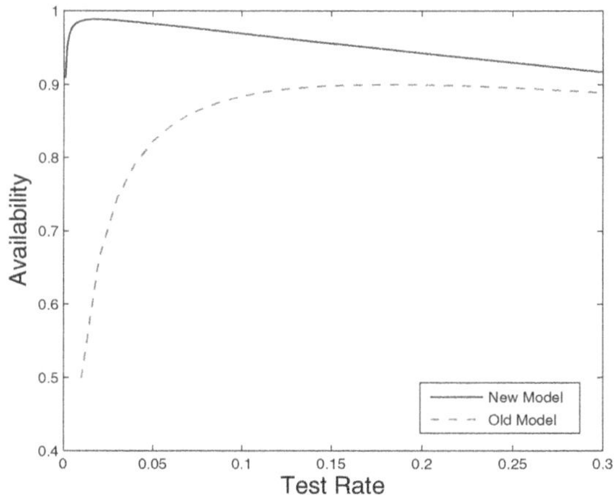

Figure 4: Contrast of old and new models

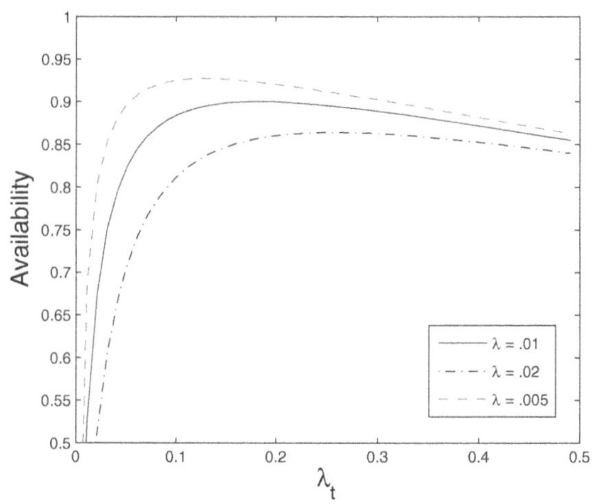

Figure 5: Availability of four state model with different error rates

The availability for the optimal test rate is 0.90. Subsequently, for traditional model with the right choice of mean time between testing, the A_{SS} will be 0.90.

The test is repeated for the new model. Formula 6 is differentiated to calculate the optimal test rate. λ_{t2} has a range from 0.01 to 0.3. The availability response is added to Figure 4 as a function of test rate. Here the healthy components are separated from the ones having a history of problems. As a result, the pessimistic error rate which is two orders of magnitude higher than normal is confined to risky states. For set one in table 2, the optimal time between test for the new model is 0.0182 and availability is 0.9887. In the traditional model, the test routine is called roughly each $\frac{1}{0.18185} = 5.5$ms, while for the new model, the optimal time between testing is $\frac{1}{0.0182} = 59$ms. For this reason, it takes longer to detect the first error and to move to the fault detected state. State three represents the probability that the considered normal component becomes faulty but yet to be detected. Matlab calculates the steady state probability of this state for parameters in set one to be $\Pi_3 = 0.005$. The saved time well outweighs the time consumed in state three and as a result, the overall A_{SS} is improved. In Figure 4, availability of four and seven state models are compared with the latter reaching the higher peak at lower test rate. For the seven state model, the period of running normal programs without interruption

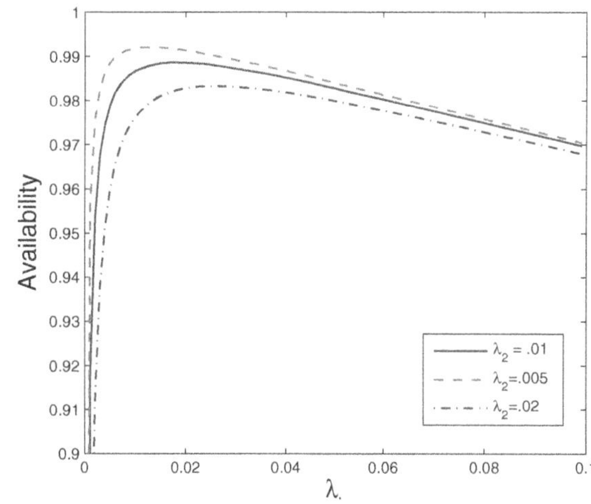

Figure 6: Availability of seven state model with different error rates

is more than ten times longer. This lower test frequency will result in better performance. The processing time of healthy components is not wasted for unnecessary interruptions by testing pessimistically. Consequently, the system up time has improved from 0.9 to about 0.99.

6.2 Error Rate

Different error rates are examined in sets one, two and three. The results are depicted in Figures 5 and 6. It is evident from figures that with the increase of error rate, while availability response retains its overall concave shape, the peak will be lower and shifts to the right. The reason is that higher probability of failure translates into more frequent checking and recovery from failure. For all three sets, in comparison to the traditional model, the new model reaches significantly higher availability with lower number of tests needed.

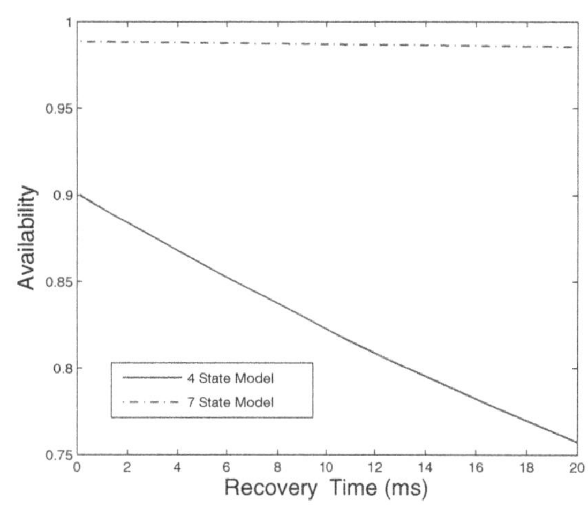

Figure 7: Effect of recovery time

6.3 Recovery Rate

Independent of the recovery mechanism you choose in your fault tolerant system, the recovery rate (or the mean time to recovery) of the mechanism would have an influence on availability. In this subsection, we compare the effect of the varying recovery rate on availability of four and seven state models. The mean time to recovery in set four varies from 0.02 to 20 milliseconds. For each

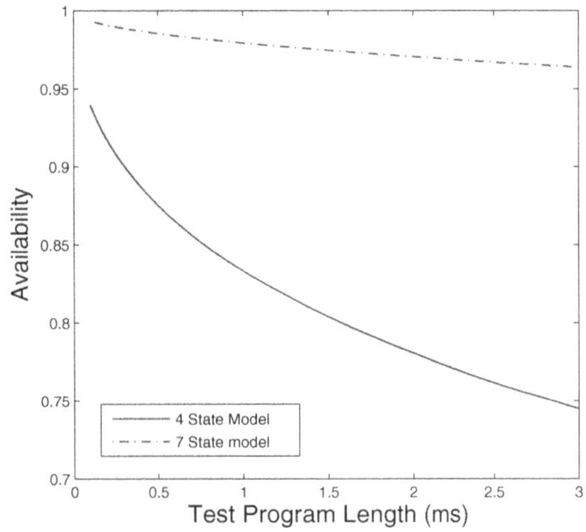

Figure 8: Effect of test program length

Table 2: Set of Parameters

Set One	$\lambda = 0.01, \mu_r = 10, \mu_t = 3.33, \lambda_t$ =0.01 to 0.5
Set two	$\lambda = 0.02, \mu_r = 10, \mu_t = 3.33, \lambda_t$ =0.01 to 0.5
Set three	$\lambda = 0.005, \mu_r = 10, \mu_t = 3.33, \lambda_t$ =0.01 to 0.5
Set four	$\lambda = 0.01, \mu_t = 3.33, \mu_r$ =0.05 to 50
Set five	$\lambda = 0.01, \mu_r = 10, \mu_t$ =0.33 to 10

(a)Parameters for Traditional Model

Set One	$\lambda_1 = 0.0001, \lambda_2 = 0.01, \mu_r = 10, \mu_t = 3.33, \mu_n = 0.02, \lambda_t$ =0.001 to 0.2
Set two	$\lambda_1 = 0.0002, \lambda_2 = 0.02, \mu_r = 10, \mu_t = 3.33, \mu_n = 0.02, \lambda_t$ =0.001 to 0.2
Set three	$\lambda_1 = 0.00005, \lambda_2 = 0.005, \mu_r = 10, \mu_t = 3.33, \mu_n = 0.02, \lambda_t$ =0.001 to 0.2
Set four	$\lambda_1 = 0.0001, \lambda_2 = 0.01, \mu_t = 3.33, \mu_n = 0.02, \mu_R$ =0.05 to 50
Set five	$\lambda_1 = 0.0001, \lambda_2 = 0.01 \mu_n = 0.02, \mu_r = 10, \mu_t$ =0.33 to 10

(b)Parameters for New Model

recovery time, the optimum test rate and corresponding availability is calculated. While for $\mu_r = 50$, A_{SS} is calculated to be 0.9887, for a $\mu_r = 1$, the availability becomes 0.9885. Normal recovery time for a modern processor is in the range of 0.1 to 10 ms. As we can see in Figure 7, with the increase of recovery time, availability decreases linearly. However, for the seven state model, this change is only in the range of thousandths and is negligible for variation of time in typical recovery techniques.

6.4 Test Program Length

Similar to the recovery rate, the effect of test program length is examined. For set five, the test length varies between 0.1 to 3 ms. Again the corresponding test rate and A_{SS} are calculated. As illustrated in figure 8, in the new model, the availability decreases inversely proportional to the increase of test time. Subsequently, for an online periodic test routine that lasts for 3 ms, the system uptime will drop to near 0.96. Compared to previous subsection, for μ_t this change is more significant. That is because system regularly checks for error whereas it only occasionally requires to recover from fault. Hence, the effect of recovery process on availability is not significant for the given values. For the contrast of the two models, the traditional one runs test programs more frequently, leading to more dramatic drop of the availability.

7. CONCLUSION

In this paper, we introduced a novel Markov model for intermittent faults. First, two new states were added to birth-death process to assist calculating dependability measures. Next, the model was expanded to seven states, separating normal healthy components from components that are more prone to IF. The availability of new and traditional models were compared. We concluded that because the new model does not waste time for unnecessary testing, it achieves significantly better availability. Moreover, choice of right test interval timing is crucial for system performance. We also explored design space with variation of different parameters (such as $\lambda, \mu_r, \mu_t, etc...$) and observed their effect on dependability measures. However, we have to bear in mind that models cannot encompass all the factors leading to faults. Besides, the included ones usually are not fully understood and/or are simplified to make the problem tractable. In future, more field data and a detailed simulation of IF will facilitate the effort to plan a better mathematical abstract model.

8. REFERENCES

[1] A. Avizienis, J.-C. Laprie, B. Randell, and C. Landwehr. Basic Concepts and Taxonomy of Dependable and Secure Computing. *IEEE Trans. Dependable Secur. Comput.*, 1(1):11–33, Jan. 2004.

[2] D. P. Bertsekas and J. N. Tsitsiklis. *Introduction to probability*. Athena Scientific, 2002.

[3] S. Borkar. Designing reliable systems from unreliable components: the challenges of transistor variability and degradation. *Micro, IEEE*, 25(6):10–16, 2005.

[4] Constantinescu. Impact of Intermittent Faults on Nanocomputing Devices. In *WDSN*, 2007.

[5] C. Constantinescu. Impact of deep submicron technology on dependability of VLSI circuits. In *Dependable Systems and Networks, 2002. DSN 2002. Proceedings. International Conference on*, pages 205–209, 2002.

[6] O. Heron, J. Guilhemsang, N. Ventroux, and A. Giulieri. Analysis of on-line self-testing policies for real-time embedded multiprocessors in DSM technologies. In *On-Line Testing Symposium (IOLTS), 2010 IEEE 16th International*, pages 49–55, 2010.

[7] Y.-T. HSU and C.-F. HSU. Novel model of intermittent faults for reliability and safety measures in long-life computer systems. *International Journal of Electronics*, 71(6):917–937, 1991.

[8] B. W. Johnson. *Design and analysis of fault-tolerant digital systems*. Addison-Wesley Pub. Co., 1989.

[9] I. Koren and C. M. Krishna. *Fault-Tolerant Systems*. Elsevier/Morgan Kaufmann, 2007.

[10] N. Kranitis, A. Merentitis, N. Laoutaris, G. Theodorou, A. Paschalis, D. Gizopoulos, and C. Halatsis. Optimal Periodic Testing of Intermittent Faults In Embedded Pipelined Processor Applications. In *Design, Automation and Test in Europe, 2006. DATE '06. Proceedings*, volume 1, pages 1–6, 2006.

[11] T. Li, R. Ragel, and S. Parameswaran. Reli: Hardware/software Checkpoint and Recovery scheme for embedded processors. In *Design, Automation Test in Europe Conference Exhibition (DATE), 2012*, pages 875–880, Mar. 2012.

[12] T. Nakagawa and K. Yasui. Optimal testing-policies for intermittent faults. *Reliability, IEEE Transactions on*, 38(5):577–580, 1989.

[13] M. Nicolaidis and Y. Zorian. On-Line Testing for VLSI\—A Compendium of Approaches. *J. Electron. Test.*, 12(1-2):7–20, 1998.

[14] E. B. Nightingale, J. R. Douceur, and V. Orgovan. Cycles, cells and platters: an empirical analysisof hardware failures on a million consumer PCs, 2011.

[15] S. Pan, Y. Hu, and X. Li. IVF: Characterizing the Vulnerability of Microprocessor Structures to Intermittent Faults. *Very Large Scale Integration (VLSI) Systems, IEEE Transactions on*, PP(99):1–14, 2011.

[16] A. Paschalis and D. Gizopoulos. Effective software-based self-test strategies for on-line periodic testing of embedded processors. *Computer-Aided Design of Integrated Circuits and Systems, IEEE Transactions on*, 24(1):88–99, 2005.

[17] J. C. Smolens, B. T. Gold, J. C. Hoe, B. Falsafi, and K. Mai. Detecting emerging wearout faults. In *In Proceedings of the IEEE Workshop on Silicon Errors in Logic - System Effects*, 2007.

[18] S. Y. H. Su, I. Koren, and Y. K. Malaiya. A Continuous-Parameter Markov Model and Detection Procedures for Intermittent Faults. *Computers, IEEE Transactions on*, C-27(6):567–570, 1978.

[19] K. S. Trivedi. *Probability and Statistics with Reliability, Queuing, and Computer Science Applications*. Wiley, 2002.

[20] L. Yanjing, S. Makar, and S. Mitra. CASP: Concurrent Autonomous Chip Self-Test Using Stored Test Patterns. In *Design, Automation and Test in Europe, 2008. DATE '08*, pages 885–890, 2008.

High-Endurance Hybrid Cache Design in CMP Architecture with Cache Partitioning and Access-Aware Policy

Shun-Ming Syu, Yu-Hui Shao, and Ing-Chao Lin
Dept. of Computer Science & Information Engineering
National Cheng Kung University, Tainan, Taiwan 701
Email: {p76004520, p76011373, iclin}@mail.ncku.edu.tw

ABSTRACT

In recent years, NVM (non-volatile memory) technologies, such as STT-RAM (spin transfer torque RAM) and PRAM (phase change RAM), have drawn a lot of attention due to their low leakage and high density. However, both NVMs suffer from high write latency and limited endurance problems. To overcome these problems, the SRAM/NVM hybrid cache architecture has been proposed, and the write pressure on NVM can be mitigated with appropriate write management policy. Moreover, many wear leveling techniques have been proposed to extend the lifetime of NVM in the hybrid cache. In this paper, we proposed a hybrid cache design that includes SRAM cache, STT-RAM cache, and STT-RAM/SRAM hybrid cache banks for CMP (chip multi-processors) architecture. We also propose a partition-level wear leveling scheme and access-aware policies to mitigate unbalanced wear-out of STT-RAM lines within a partition and among different cache partitions. Experimental results show that, our proposed scheme and policies can achieve an average of 89 times improvement in cache lifetime and are able to save 58% power consumption compared to SRAM cache.

Categories and Subject Descriptors

B.3.2: [**Memory Structures**]: Cache memories

General Terms:

Algorithms, Management, Design

Keywords

STT-RAM; hybrid cache; wear leveling; cache partitioning

1. INTRODUCTION

Improving processors' performance by increasing frequency has become more difficult due to the overheat problem. To overcome this problem, computer architecture shifts from single-core chip to multi-core chip design which is also referred to as chip multi-processors (CMP). By dividing the workload to different cores, higher performance can be achieved even if the frequency of each core is not increased significantly.

In a CMP architecture, the last-level cache (LLC) is typically shared by multiple processors and is composed of SRAM. Recently, NVM (non-volatile memory) technologies, such as PRAM (phase change RAM) and STT-RAM (spin transfer torque RAM), have drawn a lot of attention due to their low leakage and high density and are good candidate for LLC. However, both NVMs suffer from high write latency, high write power, and limited write endurance problems. TABLE 1 compares SRAM, STT-RAM, and PRAM. Compared to PRAM, STT-RAM has faster access time, lower power consumption, and higher write endurance. Thus, STT-RAM is more suitable for on-chip last-level cache design. The simulated result in [3] shows that the maximum lifetime of PRAM-based cache is only about a few months while STT-RAM-based cache can last for more than tens of years. Therefore, we focus on the LLC with the STT-RAM as NVM in this paper.

TABLE 1. Comparison between SRAM and NVMs[3]

	SRAM	STT-RAM	PRAM
Density	1X	4X	16X
Read time	Very fast	Fast	Slow
Write time	Very fast	Slow	Very slow
Read power	Low	Low	Medium
Write power	Low	High	High
Leakage power	High	Low	Low
Endurance	10^{16}	4×10^{12}	10^{9}

To mitigate the high write latency and energy of STT-RAM, hybrid cache architecture that consists of SRAM and STT-RAM is proposed [11][12], as shown in Figure 1. Most of the last level cache is STT-RAM, and only some of them are SRAM. The STT-RAM banks that are closest to a core is the local bank of the core, and the core has minimum access latency to the banks. With the help of small SRAM lines and an appropriate write management, the write pressure on STT-RAM lines can be reduced and the lifetime can be improved. In addition, large power consumption can be reduced when compared to SRAM-only cache.

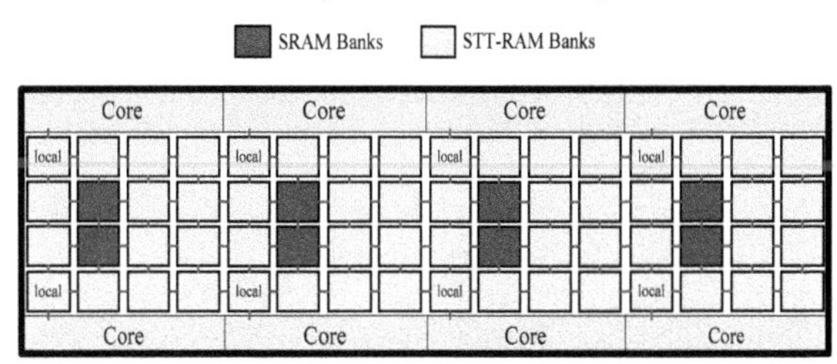

Figure 1. Hybrid cache for CMP architecture

However, the interference among competing processes leads to increased shared resource contention and conflict misses in LLC, reducing the performance. One solution to mitigate this problem is to partition the shared cache into several private partitions for each core, and the cache contention is reduced. Furthermore, to improve cache utilization, dynamic cache partitioning schemes have been proposed to manage the resource through allocating specific portions of cache to each core according to their workload profiling information[5][9][10].

This work was partially supported by National Science Council of Taiwan under Grant No.100-2221-E-006-177

Current related cache partitioning works focus on increasing CMP performance, and the private partition size for a core is determined by cache miss rate and instructions per cycle (IPC). However, when using STT-RAM as last-level cache, cache partitioning should also consider write endurance problem of STT-RAM. Otherwise, unbalanced write traffic to each partition will reduce the lifetime of STT-RAM cache.

To demonstrate the unbalanced write situation, we simulate a 4-core CMP machine using a modified M5 simulator [1]. The benchmarks are selected from PARSEC 2.1 [2], and the details of system configuration are listed in TABLE 2. The L2 cache is dynamically partitioned according to the workloads, but the write traffic to each bank is not considered. Simulation results show the write accesses to each bank are non-uniformly distributed over the sixteen banks in all benchmarks and four most unbalanced cases are illustrated in Figure 2. As a result, the cache banks that suffer from large writes accesses will wear out faster than other banks.

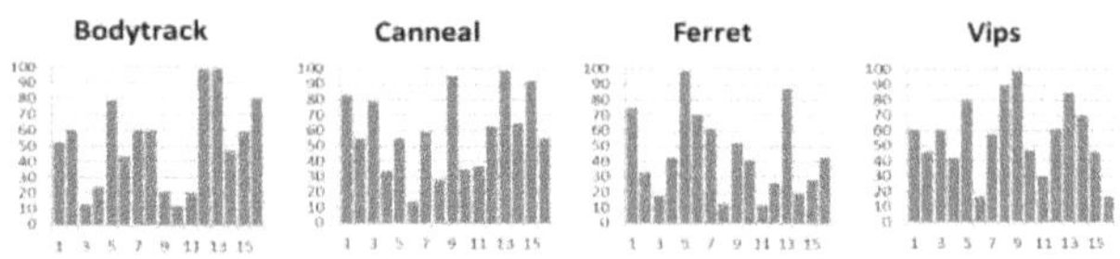

Figure 2. Non-uniformity of writes among the sixteen banks

TABLE 2. System configuration used in simulation

Processor	4 cores, 2 GHz
L1 I/D cache	32/64 KB per core, 4-way, 64B block size
Shared L2 cache	16 MB(1MB × 16 banks), 16-way, 64B block size

Therefore, it is critical to consider the write pressure to each bank during cache partitioning. However, the current cache partitioning techniques are not suitable for hybrid cache since the unbalanced writes traffic among different banks is not considered. Besides, local banks suffer from more serious lifetime problem since they normally receive more write pressure than other cache banks. To solve these problems, we propose a novel hybrid cache architecture that contains three types of cache banks: SRAM bank, STT-RAM bank, and STT-RAM/SRAM hybrid bank. In the meantime, two access-aware policies and partition-level wear leveling scheme are proposed. To the best of our knowledge, this paper is the first work to investigate the dynamic cache partitioning and unbalanced write pressure among different partitions for a hybrid cache in CMP architecture. The contributions of this paper can be summarized as follows:

- **A novel hybrid cache architecture for chip multi-processors:** In order to reduce write pressure on the local bank of a core, we propose a new hybrid bank architecture that combines both STT-RAM and SRAM lines. With the help of SRAM lines in the local bank, the write pressure can be significantly reduced.
- **Access-aware policies to manage hybrid cache:** In order to reduce the write traffic to STT-RAM regions and uniform the write distribution over STT-RAM lines, we propose a STT-RAM write management policy. In addition, to improve the write utilization of SRAM, a SRAM read management policy that moves less-written data from SRAM to STT-RAM is proposed.
- **Partition-level wear leveling scheme to improve lifetime:** In order to uniform the write pressure among different cache partitions, we propose a partition-level wear leveling scheme. The scheme considers the different workload write-pressure from each core to change the capacity of SRAM and STT-RAM regions in each partition. Therefore, the write pressure among cache partitions is balanced.
- Experimental results show that, the proposed hybrid cache with access-aware policies and partition-level wear leveling scheme can improve the lifetime of hybrid cache by 89 times on average. In addition, 58% total power consumption can be reduced when compared to SRAM cache.

The rest of this paper is organized as follows. Section 2 presents our proposed hybrid cache architecture, access-aware policies, and partition-level wear leveling scheme. Section 3 details the experimental results, and Section 4 introduces the related work for hybrid cache. Section 5 concludes the paper.

2. METHODOLOGY

This section first presents our proposed hybrid cache architecture, and then describes the details of proposed access-aware policies and partition-level wear leveling scheme.

2.1 Hybrid Cache Architecture

As mentioned in Section 1, a hybrid cache contains both SRAM and STT-RAM banks. The write access to STT-RAM banks can be redirected to SRAM banks to reduce the write pressure. However, to reduce power consumption, the capacity of total SRAM banks is normally much smaller than that of total STT-RAM banks in hybrid cache. Hence, when a STT-RAM would like to redirect its write traffic to a SRAM bank, there is high chance that the SRAM bank is already occupied by the write traffic from other STT-RAM banks and the redirection will fail. This issue is more serious on the local bank since the bank has less latency and normally receives significant amount of write traffic from its core. If the write traffic on the local bank cannot be redirected, the lifetime of the local bank will be reduced. Therefore, to improve the lifetime of the local bank, we propose to integrate STT-RAM with SRAM to construct a local hybrid bank. Only the write traffic to the local hybrid bank can be redirected to the SRAM region in the local hybrid bank. This can reduce the write pressure of the local bank and the unbalanced wear-out between local and non-local STT-RAM banks can be improved.

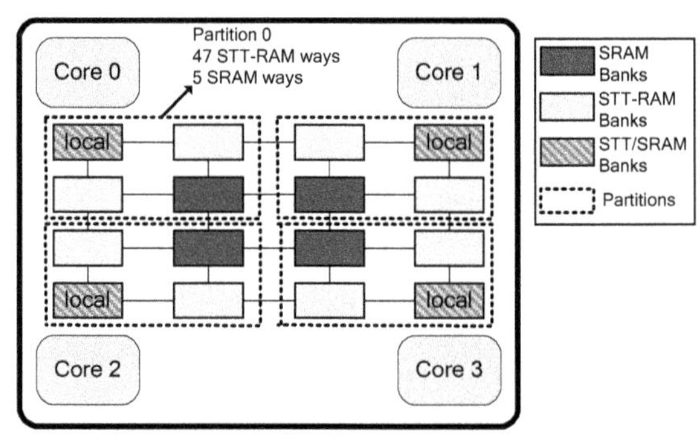

Figure 3. Proposed CMP hybrid cache architecture

Figure 3 shows our proposed architecture which includes 4-core CMP system and a shared hybrid L2 cache. The private L1 cache in each core is not shown. The hybrid cache consists of four SRAM cache banks, eight STT-RAM cache banks, and four STT-RAM/SRAM hybrid cache banks, which are connected through interconnection network. The capacity density of STT-RAM cache is about four times higher than SRAM cache [11]. To achieve similar area for each bank, the cache capacity of each STT-RAM and SRAM banks used in our work are 2MB (16-way) and 512KB (4-way), respectively. Also, the proposed hybrid bank contains 1920KB STT-RAM and 128KB SRAM, which provides a total of 2MB cache (15-way + 1-way). Another way to see the shared hybrid L2 cache is as a 26MB and 208-way equivalent cache that consists of STT-RAM and SRAM ways.

2.2 Access-aware Policies

To improve the endurance of the STT-RAM lines and improve the write utilization of SRAM lines, two access-aware policies are proposed to distribute the write traffic in each cache partition: STT-RAM write management policy and SRAM read management policy.

2.2.1 STT-RAM write management policy

The main ideas of our proposed STT-RAM write management policy are (1) to reduce the write traffic to STT-RAM (2) to uniform the write distribution over STT-RAM lines in a cache partition. Algorithm 1 shows the details of STT-RAM write management policy. When a write request has cache hit of STT-RAM line and there is an invalid SRAM line, the write request will be redirected to the invalid SRAM line and the STT-RAM line is invalidated (lines 2 to 4). Note that the SRAM lines in hybrid bank can only receive the write request from the STT-RAM line of the same hybrid bank. Next, if there is no invalid SRAM line, then the data will be written into STT-RAM lines. To avoid large amount of write requests on the same STT-RAM line, the write request is redirected to another invalid STT-RAM line instead of the original cache line (lines 5 to 7). If there are no other invalid STT-RAM lines, the write request will be performed on the original cache line (lines 8 to 9). Lines 11 to 16 describe the situation when a write miss occurs in STT-RAM lines. When a cache miss in STT-RAM lines occurs, and there is an invalid SRAM line, the write request is redirected to SRAM lines; otherwise, a STT-RAM line is selected as a victim based on LRU policy and must be written back to the lower level memory for a write-back cache. Then, the write operation is performed on the selected victim.

Algorithm 1: STT-RAM write management policy

1. **if** a write request hits on a STT-RAM line, *line* {
2.   **if** there are invalid SRAM lines
3.   Redirect the write request to invalid SRAM lines
4.   Invalidate *line*
5.   **else if** there are another invalid STT-RAM lines
6.   Redirect the write request to other STT-RAM lines
7.   Invalidate *line*
8.   **else**
9.   Perform the write operation on *line*
10. }
11. **else** { //cache miss in STT-RAM lines
12.   **if** SRAM has invalid lines
13.   Redirect the write request to invalid SRAM lines
14.   **else**
15.   Select LRU STT-RAM line to be the victim
16.   Perform the write operation on the victim
17. }

2.2.2 SRAM read management policy

In order to further increase the write utilization of SRAM, the less-written data that occupies the SRAM lines should be removed. When less-written data is removed from SRAM, more space can be left for write operation. Based on LRU policy, we define a *Less Recently Written Set* (LRWS) which has n SRAM lines. The set contains the least recently written (LRW_1), the second least recently written (LRW_2), the third least recently written until the nth least recently written (LRW_n) cache lines in SRAM. All the cache lines in Less Recently Written Set are considered as candidates that should be moved to STT-RAM. In our work, n is set to 10% of SRAM capacity in each partition.

Algorithm 2: SRAM read management policy

1. When a read request hits on a SRAM line, *line*
2.   **if** *line* is in LRWS **and** all SRAM lines are in use
3.   **if** there are invalid STT-RAM lines
4.   Copy the data from *line* to invalid STT-RAM line
5.   Perform read operation on invalid STT-RAM line
6.   Invalidate *line*
7.   **else**
8.   Select LRU STT-RAM line to be the victim
9.   Copy *line* to the victim
10.   Perform read operations on the victim
11.   Invalidate *line*
12.   **else** //*line* is not in LRWS
13.   Perform the read operation on *line*

Algorithm 2 shows the details of SRAM read management policy. Lines 1 to 11 describe the situation that a read request hit on the SRAM line in LRWS and all other SRAM lines are in use. If there is an invalid STT-RAM line, the read request will be redirected to the invalid STT-RAM line and the SRAM line is invalidated (lines 3 to 6). Otherwise, the LRU STT-RAM line is selected to be the victim and the read request is redirected to the victim. Then, the SRAM line is invalidated. (lines 8 to 11)

2.3 Partition-Level Wear Leveling Scheme

As mentioned in Section 1, partitioning hybrid cache should consider unbalanced write traffic to each partition for the endurance problem of STT-RAM. Figure 4 depicts our framework for dynamic cache partitioning. Each core has its own private L1 instruction cache and L1 data cache. The hybrid L2 cache accesses from each core are monitored by the corresponding monitor. Then, with the collecting information from each monitor, proposed partitioning algorithm decides how to allocate shared cache to each partition.

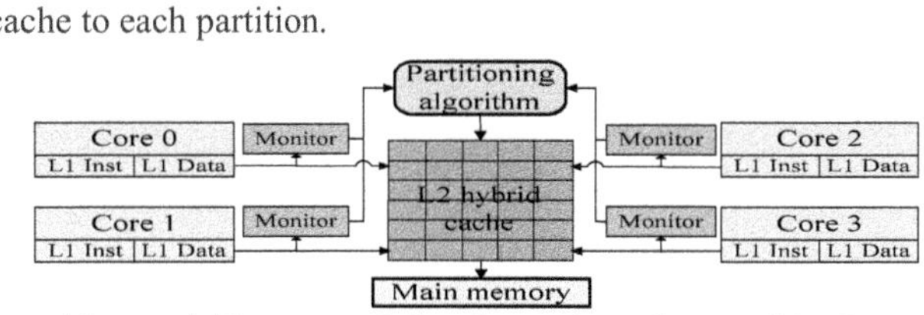

Figure 4. Framework for dynamic cache partitioning

The proposed algorithm partitions a fixed number of ways in each set for scheduled processes, and all ways in L2 cache will be assigned to a corresponding core. A partition example is shown in Figure 3. The private partition of core 0 consists of a hybrid bank, a SRAM bank, and two STT-RAM banks. In other words, the partition can be regarded as a 52-way hybrid cache that consists of forty-seven STT-RAM ways and five SRAM ways. The partitioning algorithm includes two dynamic decisions while partitioning a hybrid cache: (1) which partition needs to increase or decrease cache ways? (2) which way should be reallocated, SRAM way or STT-RAM way? To determine how to better partition a hybrid cache, we monitor the write counts on SRAM and STT-RAM for each core over time. In addition, the number of SRAM and STT-RAM ways that owned by each core is maintained. With the information of write counts and way number for each partition, two variables are defined as follows:

$$WPSW = \frac{write\ counts\ in\ SRAM}{number\ of\ SRAM\ ways}$$

$$WPNW = \frac{write\ counts\ in\ STT\text{-}RAM}{number\ of\ STT\text{-}RAM\ ways}$$

The variables *WPSW* and *WPNW* are the writes per SRAM way and the writes per NVM way, respectively. The large *WPSW* value of a partition means the write utilization for SRAM region in the partition is high. Similarly, if a partition has large *WPNW* value, then the write utilization for STT-RAM region in the partition will be high, leading to a fast wear-out than other partitions with small *WPNW* value. By detecting the write utilization for SRAM and STT-RAM in each cache partition, appropriate cache ways are allocated to the partition whose STT-RAM region suffering from large write pressure. Therefore, the unbalanced write pressure among cache partitions is improved. Pseudo-codes for our partitioning algorithm are shown in Algorithm 3.

At each time interval period, the values of *WPSW* and *WPNW* for each partition are computed. Then, two arrays W_{SRAM} and $W_{STT\text{-}RAM}$ are used to store their value and sort them by non-increasing order. (lines 1 to 4) Next, we find the partition$_i$ that has most writes per STT-RAM way (line 6). To reduce the *WPNW* value of partition$_i$, the allocated cache ways for partition$_i$ need to increase. Since more SRAM ways can mitigate more write pressure on STT-RAM ways with our access-aware policies. Therefore, compared with the STT-RAM way, allocating a SRAM way to partition$_i$ can reduce more *WPNW* value and is a better choice. Hence, a SRAM way of partition$_j$ that has minimum *WPSW* value will be allocated to partition$_i$ if they are not the same core (lines 7 to 10). Otherwise, a SRAM way of another core that has the second minimum *WPSW* value can be allocated to partition$_i$. However, if partition$_i$ has the minimum *WPSW* value, which means it has low write utilization for SRAM, increasing SRAM ways for partition$_i$ cannot mitigate the write pressure on STT-RAM ways effectively. Therefore, a STT-RAM way of partition$_k$ that has minimum *WPNW* value is allocated to partition$_i$. (lines 11 to 14)

Algorithm 3: cache partitioning algorithm

```
1.  W_SRAM = {WPSW_0, WPSW_1, ..., WPSW_n}
2.  W_STT-RAM = {WPNW_0, WPNW_1, ..., WPNW_n}
3.  Sort array W_SRAM by non-increasing order
4.  Sort array W_STT-RAM by non-increasing order
5.
6.  Assume WPNW_i is head of array W_STT-RAM
7.  Assume WPSW_j is tail of array W_SRAM
8.  if (i != j)
9.      SRAM_Way[partition_j] -= 1
10.     SRAM_Way[partition_i] += 1
11. else
12.     Assume WPNW_k is tail of array W_STT-RAM
13.     STTRAM_Way[partition_k] -= 1
14.     STTRAM_Way[partition_i] += 1
```

3. EXPERIMENTAL SETUP AND RESULT

We use M5 Simulator to simulate a 2GHz processor with four cores. The classic memory model in M5 is modified to implement the hybrid cache with non-uniform access time, access-aware policies, and the partition-level wear leveling scheme described in Section 2. TABLE 3 details the main simulation parameters applied in our experiments. We use PARSEC 2.1 benchmark suite [2] as the parallel workload to evaluate the proposed scheme. In order to obtain a reasonable evaluation, all benchmarks are fast forwarded to the beginning of region of interest, and then the next 100M instructions are considered as warm up period. The latency and energy values of SRAM cache used in this work are obtained from CACTI [13], and for STT-RAM cache we use modified NVSim [4] and the model reported in [11] to obtain the parameters in 65nm technology, which are given in TABLE 4.

TABLE 3. System configuration used in simulation

Processor	4 cores, 2 GHz
L1 I/D Cache	32/64 KB per core (private), 4-way, 2cycles latency, 64B block size, write back
L2 Hybrid Cache	8 STT-RAM banks, 4 SRAM banks, 4 STT-RAM/SRAM hybrid banks, write back
SRAM bank	512KB, 4-way set associative, 64B block size
STT-RAM bank	2MB, 16-way set associative., 64B block size
STT-RAM/SRAM hybrid bank	2MB(1920KB+128KB), 16-way(15+1) set associative, 64B block size
L2 Cache Access Lat.	SRAM: 6 cycles STT-RAM read/write: 6/23 cycles
Coherence Mechanism	MESI directory protocol
Network	Ring network, one router per bank, 3 cycles router and 1 cycle link latency
Repartitioning Period	100M cycles

3.1 Wear Leveling Result

To evaluate the wear leveling result on hybrid cache, we implement a baseline cache partitioning scheme named *Equal-partitions*, which is equivalent to assigning hybrid cache partitions of equal SRAM and STT-RAM sizes to each core as shown in Figure 3. Each core has two STT-RAM banks, a SRAM bank, and a hybrid bank. We first show the simulated results with and without our proposed access-aware policies in Section *3.1.1*, and then compare results between partition-level wear leveling scheme and *Equal-partitions* in Section *3.1.2*.

TABLE 4. Timing and latency parameters of SRAM, STT-RAM, and STT/SRAM cache banks

Bank type	Tech.	Op.	Lat.(ns)	Energy(nJ)	Leak.(W)
SRAM bank (512KB)	SRAM	R/W	1.74	0.609	0.3691
STT-RAM Bank (2MB)	STT-RAM	R	2.99	0.598	0.0686
		W	11.41	4.375	
STT/SRAM Hybrid bank (1920/128KB)	SRAM	R/W	1.57	0.219	0.0764
	STT-RAM	R	1.71	0.569	
		W	11.32	4.215	

3.1.1 Access-aware policies

With *Equal-partitions* scheme, for each benchmark we compute the write counts per STT-RAM set (*WPNS*) which represents the write traffic to be reduced. Figure 5 shows that the write traffic to STT-RAM of proposed access-aware policies is lower than that of the baseline scheme without wear leveling scheme. On average, proposed access-aware policies can effectively redirect write traffic from STT-RAM to SRAM and significantly reduce the *WPNS* value about 43% compared to the baseline scheme. Then, we compute the average deviation of write counts on STT-RAM sets, which represents the non-uniform degree of writes distribution. Figure 6 shows about 84% reduction of average deviation of writes on STT-RAM sets, meaning the non-uniformly distribution of writes over STT-RAM cells is highly improved. From Figure 5 andFigure **6**, it can be seen that our hybrid cache management policies can uniform the write distribution over STT-RAM cells in all workloads and reduce large write traffic to STT-RAM in most cases. However, there are only 18% and 12% reduction of write traffic for *canneal* and *vips*, respectively. It means more capacity of SRAM is needed for them to reduce the write pressure over STT-RAM sets. Therefore, to allocate more SRAM to the core that running the application like *canneal* and *vips*, a dynamic cache partitioning is needed.

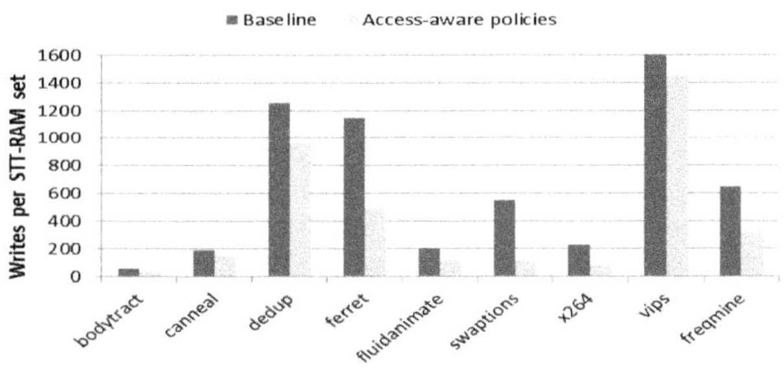

Figure 5. The comparison of write traffic to STT-RAM region with and without proposed access-aware policy

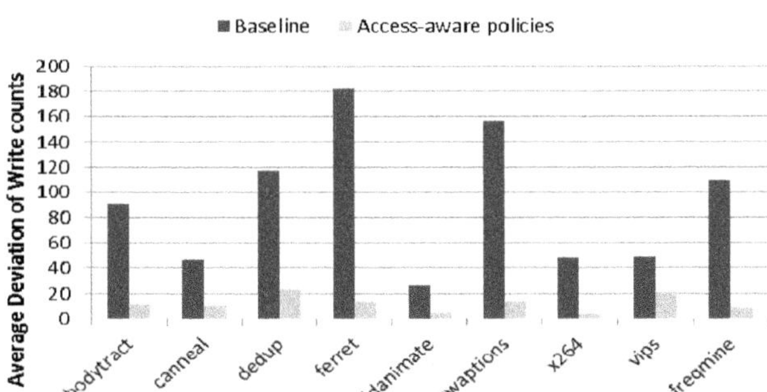

Figure 6. The comparison of average deviation of write counts on STT-RAM sets with and without access-aware policy

3.1.2 Partition-level wear leveling scheme

Figure 7 depicts the percent of the L2 cache write requests of each core. It can be seen that write requests are not balanced, which is a strong indication of the need for partitioning the cache considering write traffic for endurance concern. Figure 8 shows the percent of write traffic to each cache bank with *Equal-partitions* and access-aware policies. The write distribution among banks is non-uniform due to the write traffic to each partition is unbalanced. As mentioned before, to increase the efficiency of proposed access-aware policies, it is necessary to dynamically change the size of partitions according to the write traffic to each partition. As shown in **Error! Reference source not found.**Figure 9, compared with *Equal-partitions*, our partition-level wear leveling scheme can mitigate the unbalance of writes among cache banks, especially for *bodytract*, *ferret*, and *x264* benchmarks that have significant unbalanced writes among each core. Note that both cache partitioning schemes use our proposed access-aware policies and only the writes in STT-RAM region are counted. The write counts of four SRAM banks are not shown in the bank-level comparison.

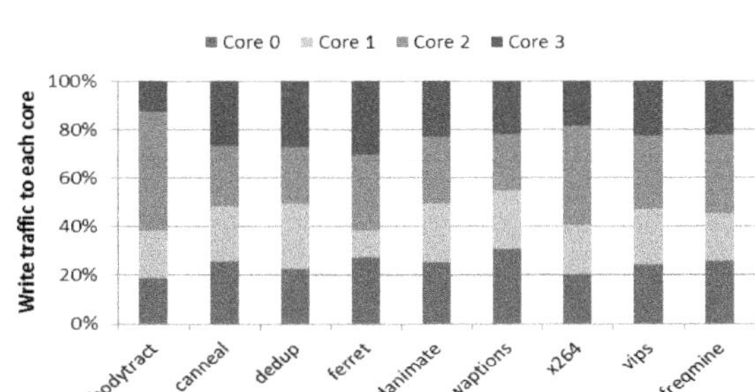

Figure 7. The unbalanced write distribution among four cores in CMPs

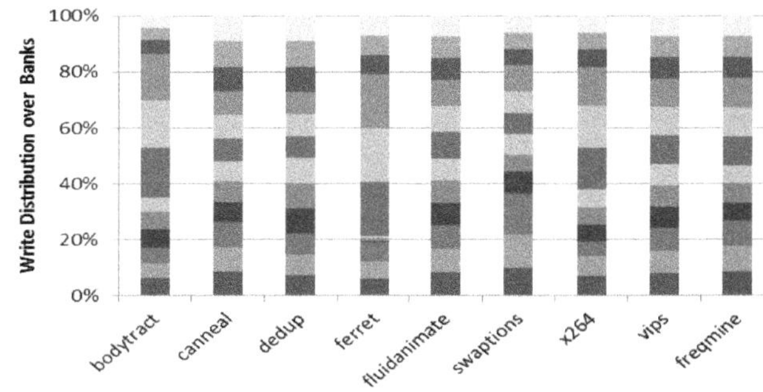

Figure 8. Write distribution among banks for Equal-partitions

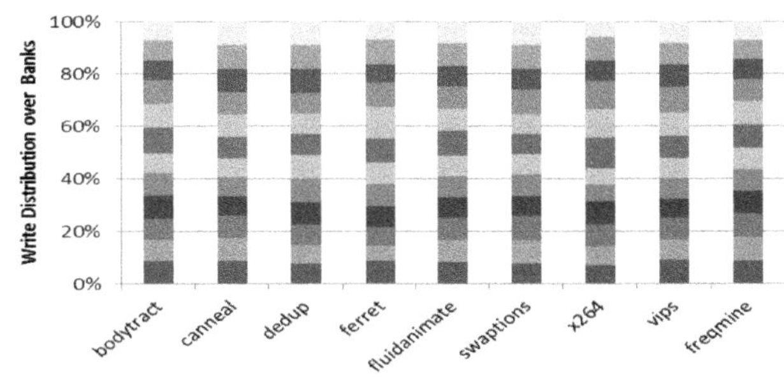

Figure 9. Write distribution among banks for partition-level wear leveling scheme

3.2 Lifetime Evaluation

In this work, we assume the endurance limit of a STT-RAM cell is 4×10^{12}. The STT-RAM cell cannot be used if its write counts exceed the endurance limit during the simulation. To evaluate the lifetime of STT-RAM region, we repeat the behavior of the write distribution in each workload, and the system fails if the total number of defective lines is greater than 16. Figure 10 shows the estimated lifetime of hybrid cache with proposed partition-level wear leveling, access-aware policies and equal partitions over the baseline scheme without wear leveling. For all workloads, the normalized lifetime when policies and equal partition are used improves by more than 80% compared to the baseline scheme. Moreover, combining our proposed policies and partition-level wear leveling scheme can achieve further 15% increase in lifetime over the equal partition scheme on average. On average, with our access-aware policies and partition-level wear leveling scheme, the lifetime of proposed hybrid cache can be improved by 89 times.

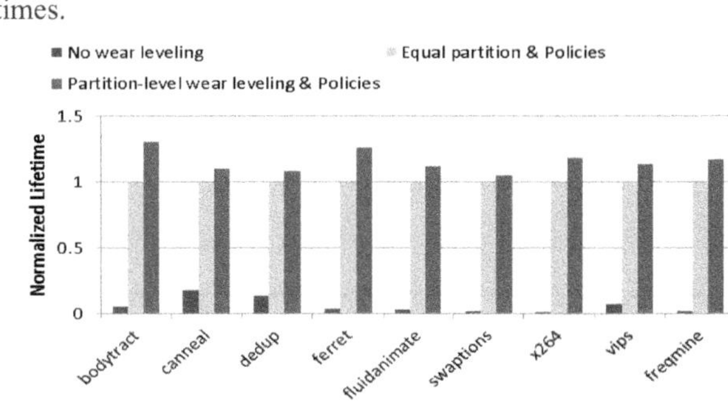

Figure 10. The comparison of lifetime between baseline and proposed wear leveling scheme

3.3 Energy Consumption Evaluation

Figure 11 shows the comparison of power consumption for SRAM L2 cache and hybrid L2 cache. The SRAM L2 cache consists of sixteen SRAM banks and provides total 8MB capacity, which has similar area as the hybrid cache with 26MB. Due to the low leakage power feature of STT-RAM, the hybrid cache consumes less power than SRAM cache, and saves 58% power consumption on average.

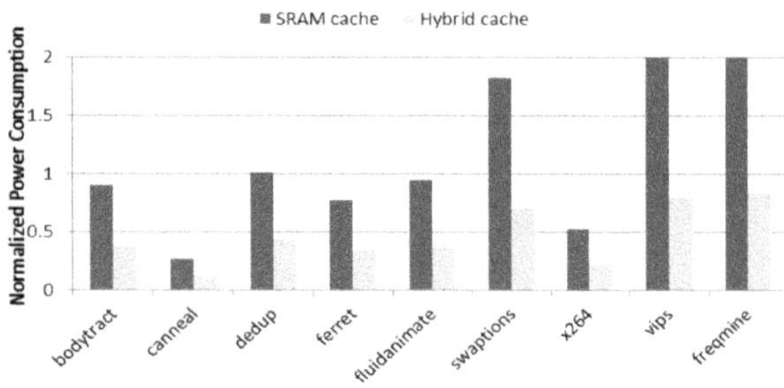

Figure 11. Power consumption for SRAM and hybrid cache

3.4 Performance Evaluation

Since the capacity of hybrid cache is about three times larger than that of SRAM cache, the access miss rate is reduced as shown in Figure 12. The average reduction of miss rate across all the workloads is about 26%. However, the long write latency of STT-RAM can decrease the performance even though the access miss rate is reduced. Figure 13 shows the performance averagely increased about 8.5% in five workloads while decreased about 5% in rest four workloads.

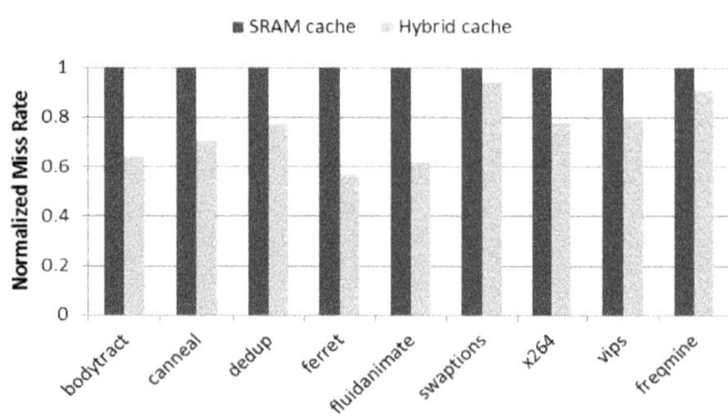

Figure 12. Access miss rate for SRAM and hybrid cache

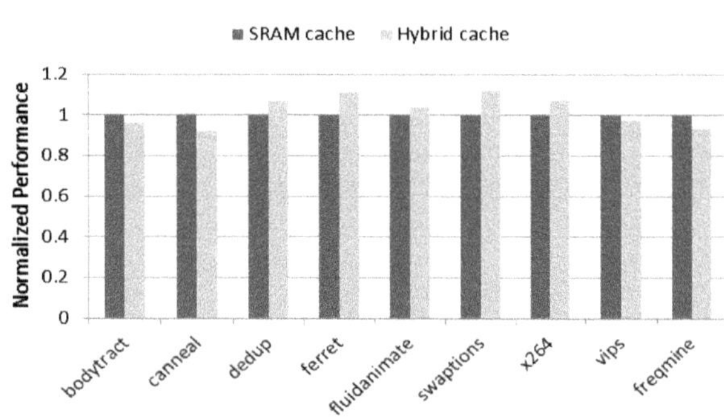

Figure 13. Normalized performance for SRAM and hybrid cache

4. RELATED WORK

The concept of integrating NVMs with SRAM to costruct hybrid cache was previously proposed in [8][11][12]. Sun et al. [11] proposed read-preemptive write buffer and SRAM/STT-RAM hybrid L2 cache to improve the performance reduction problem due to the long write latency of NVM. Wu et al. [12] proposed a read-write aware hybrid cache architecture design, and the cache was partitioned into read and write regions, which comprised of STT-RAM and SRAM respectively, to improve cache performance. Li et al. [8] proposed energy-aware read and write mechanisms for the hybrid cache to improve performance for workloads with different write patterns. However, these works did not consider the endurance issues of NVMs due to the unbalanced write distribution. Jadidi et al. [7] pointed out that the non-uniformity of writes in STT-RAM cache lines results in fast wear-out of some lines than others, and a remapping method was proposed to balance the write accesses among each STT-RAM cache sets and lines to extend cache lifetime.

5. CONCLUSION

In this paper, we proposed low energy consumption and high capacity hybrid cache design consists of SRAM and STT-RAM for CMP architecture. For lifetime extension and performance improvement, two access-aware policies were proposed to increase the write utilization of SRAM and uniform the write distribution over STT-RAM cells. In addition, considered with the unbalanced write-pressure over each partition, we proposed a partition-level wear leveling scheme to further improve the lifetime of hybrid cache. The experimental result showed that the hybrid cache can save 58% power compared to SRAM cache. In the meantime, the cache lifetime was improved by 89 times on average compared to baseline configuration.

6. REFERENCES

[1] N. Binkert, R. Dreslinski, L. Hsu, K. Lim, A. Saidi, and S. Reinhardt, "The M5 Simulator: Modeling Networked Systems," MICRO, vol 26, pp. 52–60, 2006.

[2] C. Bienia, S. Kumar, J. Singh, and K. Li, "The Parsec Benchmark Suite: Characterization and Architectural Implications," in Proc. of PACT, pp. 72–81, 2008.

[3] Y.-T. Chen, J. Cong, H. Huang, B. Liu, M. Potkonjak, and G. Reinman, "Dynamically Reconfigurable Hybrid Cache: An Energy-Efficient Last-Level Cache Design," in Proc. of DATE, pp. 45–50, 2012.

[4] X. Dong, C. Xu, Y. Xie, and N. Jouppi, "NVSim: A Circuit-Level Performance, Energy, and Area Model for Emerging Nonvolatile Memory," TCAD, vol 31, pp. 994–1007, 2012.

[5] J. Huh, C. Kim, H. Shafi, L. Zhang, D. Burger, and S. Keckler, "A NUCA Substrate for Flexible CMP Cache Sharing," TPDS, vol. 18, pp. 1028–1040, 2007.

[6] J.-W. Hsieh, and Y.-H. Kuan, "Double Circular Caching Scheme for DRAM/PRAM Hybrid Cache," in Proc. of RTCSA, pp. 469–472, 2012.

[7] A. Jadidi, M. Arjomand, and S.-A. Hamid, "High-endurance and Performance-Efficient Design of Hybrid Cache Architectures through Adaptive Line Replacement," in Proc. of ISLPED, pp. 79–84, 2011.

[8] J. Li, C. Xue, and Y. Xu, "STT-RAM based Energy-Efficiency Hybrid Cache for CMPs," in Proc. of VLSI-SoC, pp. 31–36, 2011.

[9] K. Nesbit, M. Moreto, F. Cazorla, A. Ramirez, M. Valero, and J. Smith, "Multicore Resource Management," in Proc. of MICRO, pp. 6–16, 2008.

[10] M. Qureshi, and Y. Patt, "Utility-Based Cache Partitioning: A Low-Overhead. High-Performance, Runtime Mechanism to Partition Shared Caches," in Proc. of MICRO, pp. 423–432, 2006.

[11] G. Sun, X. Dong, Y. Xie, J. Li, and Y. Chen, "A Novel Acrhitecture of the 3D Stacked Mram L2 Cache for CMPs," in Proc. of HPCA, pp. 239–249, 2009.

[12] X. Wu, J. Li, E. Speight, and Y. Xie, "Power and Performance of Read-Write Aware Hybrid Cache with Non-Volatile Memories," in Proc. of DATE, pp. 737–742, 2009.

[13] CACTI: An integrated cache and memory access time, cycle time, area, leakage, and dynamic power model. Version 5.3 available at http://www.hpl.hp.com/research/cacti/

Harvesting-Aware Energy Management for Multicore Platforms with Hybrid Energy Storage

Yi Xiang and Sudeep Pasricha
Department of Electrical and Computer Engineering,
Colorado State University, Fort Collins, CO, USA
E-mail: {yix, sudeep}@colostate.edu

ABSTRACT

In this paper, we propose a novel framework for energy and workload management in multi-core embedded systems with solar energy harvesting and a periodic hard real-time task set as the workload. Compared to prior work, our energy management framework possesses several advantages, including (i) a battery-supercapacitor hybrid energy storage module for more efficient system energy management, (ii) a semi-dynamic scheduling heuristic that continuously adapts to run-time harvested power variations without losing the consistency of the periodic task set, and (iii) a coarse-grained core shutdown heuristic for additional energy savings. Experimental studies show that our framework results in a reduction in task miss rate by up to 61% and task miss penalty by up to 65% compared to the best known prior work.

Categories and Subject Descriptors

C.3 [**Computer Systems Organization**]: Special-Purpose and Application-Based Systems -- *Real-time and embedded systems*. B.8.2 [**Performance and Reliability**]: Performance Analysis and Design Aids

General Terms

Algorithms, Design, Performance

Keywords

Energy Harvesting, Supercapacitor, Dynamic Power Management, Dynamic Voltage and Frequency Scaling, Task Scheduling

1. INTRODUCTION

Power and energy constraints have led to significant changes in the design of contemporary computing systems. In the last decade, the concept of thread-level parallelism (TLP) to improve performance within a power budget has seen widespread adoption across various computing platforms, ranging from high-end servers to desktops, as well as embedded devices. Recent years have also witnessed a significant increase in the use of multi-core processors in low-power embedded devices [1]. With advances in parallel programming and power management techniques, embedded devices with multi-core processors and TLP support are outperforming single-core platforms in terms of both performance and energy efficiency [2]. But as core counts rise to cope with increasing application complexity, techniques for workload distribution and energy management are the key to achieving energy savings in emerging multi-core embedded systems.

For some applications, we need energy autonomous devices that utilize ambient energy to perform computations without relying entirely on an external power supply or frequent battery charges. As the most widely available energy source, solar energy harvesting has attracted a lot of attention and is rapidly gaining momentum. [3][4] To fully exploit the capability of energy harvesting systems, a considerable amount of work has explored task scheduling, primarily for embedded systems with real-time task sets. An early work [8] proposed the lazy scheduling algorithm (LSA) that executed tasks as late as possible, reducing deadline miss rates when compared to the EDF algorithm. However, the approach in [8] does not consider DVFS and always executes tasks at full speed. Because a processor's dynamic power is generally a convex function of frequency, operating the processor at a lower frequency often results in higher energy efficiency. Liu et al. [9] proposed the EA-DVFS technique that takes processor DVFS into consideration. EA-DVFS utilized task slack to slow down execution speed, thus achieving more energy savings than LSA especially when total task utilization is low. Later the same authors proposed a more intelligent technique called HA-DVFS [10], which improved energy efficiency mainly by distributing multiple arriving tasks as evenly as possible over time and executing them with more uniform frequency. However, these works focus on uni-processor systems and have not considered execution on multi-core platforms. Recently, a utilization-based technique (UTB) was proposed in [11] to better address periodic task scheduling in energy-harvesting system. UTB takes advantage of predictability provided by periodic task information for better task distribution. Moreover, UTB proposed a simple extension to support multi-core platforms by allocating a subset of tasks to each core and executing the single-core UTB algorithm separately on each core.

All of these prior efforts on harvesting aware task scheduling assume an ideal battery as the energy storage medium limited merely by it capacity, ignoring other factors such as rate capacity, recovery effect, and lifetime in terms of recharge cycles [6]. When applied to real-world platforms, overlooking these factors can result in suboptimal or even false scheduling that diminishes system efficiency, stability and lifespan. For example, rate capacity effect leads to decreasing battery capacity when discharging current increases [12]. Supercapacitors present an interesting alternative to batteries for energy storage. A substantial amount of research on supercapacitors has demonstrated their benefits over electrochemical batteries, including orders of magnitude higher recharge cycles, much less charge overhead, and significantly higher efficiency with high current discharge. However, high capacity supercapacitors are not practical for small-package low-power embedded systems due to their significantly lower energy density and higher leakage overhead than an electrochemical battery, even with the state-of-art supercapacitor technology [14]. Recent work by Ongaro et al. [5] and Mirhoseini et al. [12] has shown that a battery-supercapacitor hybrid system can overcome the limitations of both types of energy storage mediums.

In this paper, *we propose a novel framework based on a semi-dynamic algorithm (HY-SDA) for energy and workload management in multi-core embedded systems with solar energy harvesting and a hybrid battery-supercapacitor energy storage system*. HY-SDA aims to minimize deadline miss rate or penalty of periodic tasks in the presence of variant and insufficient energy harvesting. Compared to prior work, HY-SDA reacts to run-time energy shortages and fluctuations proactively to find significantly greater scope for energy savings, especially in multi-core platforms. At the system level, HY-SDA is triggered at specified time epochs to adjust inter-core task allocation and set a per-core execution strategy based on the energy budget provided by the hybrid energy storage system. Our experimental studies show that HY-SDA outperforms the best known prior work (UTB [11]), achieving superior task drop rate reduction and energy efficiency.

2. PROBLEM FORMULATION

2.1 System Model

Our focus is on the problem of effective workload and energy management of real-time multi-core embedded systems with periodic tasks, powered by solar energy, as shown in Figure 1.

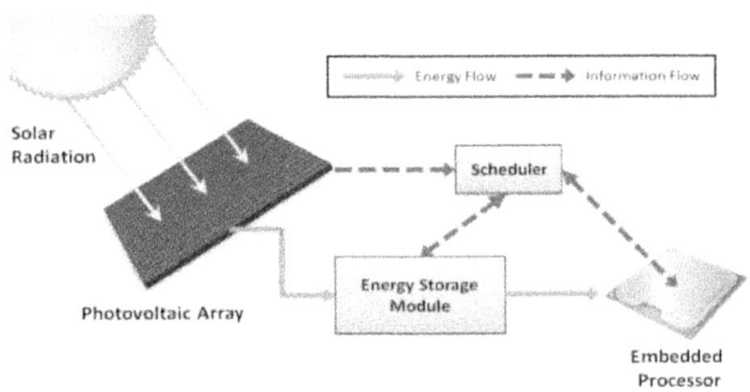

Figure 1. Real-time embedded processing with energy harvesting

2.1.1 Energy Harvesting and Energy Storage Module

A photovoltaic (PV) array is used as a power source for our embedded system, converting ambient solar energy into electric power. Naturally, the amount of harvested power varies over time due to changing environmental conditions, like angle of sunlight incidence, cloud density, temperature, humidity, etc. To cope with the unstable nature of the solar energy source, rechargeable batteries and/or supercapacitors can be used to buffer solar energy collected by photovoltaic cells. In our study, the converted solar power at time t is denoted as $P_H(t)$. The energy E_H charged into energy storage system between time instances t_1 and t_2 is given by

$$E_H(t_1 \sim t_2) = \eta_{\text{harv}} \int_{t1}^{t2} P_H(t)\,dt \qquad (1)$$

where η_{harv} is a coefficient between 0 and 1 to represent charging efficiency of the energy storage system. The capacity of the energy storage device is limited. Clearly harvested energy will be wasted if the energy storage device is already fully charged. We assume that task execution must be halted when the remaining energy in the system is less than 10 percent, thus reserving enough storage to maintain system status and ensure graceful shutdown. In this paper, we focus on the scenario where nominal power of the PV array exceeds peak power consumption of the embedded processors. Thus it is necessary to have a storage system that can preserve a considerable amount of energy for later system execution when solar energy is insufficient. A hybrid battery-supercapacitor storage system meets these goals, and is described in Section 3.2.1.

2.1.2 Periodic Real-Time Task Model

We assume a task set of N independent periodic real-time tasks ψ: $\{\tau_1, \ldots, \tau_N\}$, in which each periodic task τ_i has a characteristic triplet (C_i, D_i, T_i), $i \in \{1, \ldots, N\}$. C_i is the maximum number of CPU clock cycles needed to finish a job instance of task τ_i, referred to as worst-case execution cycles (WCEC). The relative deadline of the task, D_i, is the time interval between a job's arrival time and its deadline. A job instance is missed if it is not finished before its deadline. T_i is the period of the task. At the beginning of each period, a new job instance of that task will be dispatched to the system. Like most recent works on periodic task scheduling (e.g., [11]) we assume that D_i equals T_i, with all jobs finishing before the arrival of the next job instance of the same task. In addition, we define an attribute X_i, which is the miss penalty associated with each task. Each time that a task's job misses the deadline, the job will be aborted and the penalty applied to the system. Thus, we can refine the triplet for task τ_i as (C_i, T_i, X_i). The relative importance of a task can be characterized by a *penalty density*, defined as the ratio of the task miss penalty and WCEC (X_i/C_i) [15].

2.1.3 DVFS-Enabled Processor Model

We consider an embedded system with homogeneous multi-core processors. The processors possess DVFS capability at the core level and have support for task preemption. Each core has M discrete voltage and frequency levels: φ: $\{L_0, \ldots, L_M\}$. Each level is characterized by L_j: (v_j, p_j, f_j), $j \in \{1, \ldots, M\}$, which represents voltage, average power, and frequency respectively. We consider power-frequency levels of the Xscale processor as shown in Table 1. Here level 0 represents the idle power of the processor when no task is executed while the system stays in active state. Typically, the dynamic power-frequency function is convex. Thus, a processor running at lower frequency can execute the same number of cycles with lower energy consumption. However, this is not always the case when static power is considered. To find an energy optimal frequency, we represent energy efficiency of a v-*f* level L_i by δ_i = cycles executed/energy consumed = f_i/p_i. From Table 1 we can conclude that level 2 is the most energy efficient because executing at this level consumes the least energy for a given number of cycles. We call the most energy efficient level as the *critical level* and thus $f_{crt} = f_2$. Although it is desirable to execute tasks at this critical frequency level, due to unique task timing constraints executing tasks at the critical level may end up being insufficient to finish all task instances by their deadlines.

We also define utilization of a periodic task (U) with respect to the full speed provided by the processor; i.e., a task's utilization is its execution time under highest frequency divided by its period,

$$U_i = \frac{C_i/f_{max}}{T_i} \qquad (2)$$

The utilization for an entire task set is simply the accumulation of all tasks' utilization. In preemptive real-time system, a task set is schedulable by the EDF algorithm under a frequency level j if it meets the following condition:

$$U_{total} \leq \frac{f_j}{f_{max}} \qquad (3)$$

When total utilization is known, the most energy efficient frequency can be deduced from the equation, assuming $f_j \geq f_{crt}$. In this paper, we focus on a fully subscribed system with U_{total} = 100% to maximally utilize the available system resources. Our proposed framework aims to finish as many tasks as possible in such a system under any given energy harvesting profile.

Table 1. XScale Processor [13] Power and Frequency Levels

Level	0	1	2	3	4	5
Voltage(V)	-	0.75	1.0	1.3	1.6	1.8
Power(mW)	40	80	170	400	900	1600
Frequency(MHz)	idle	150	400	600	800	1000
Energy Efficiency	0	1.875	2.353	1.5	0.889	0.625

2.1.4 Energy Manager Module

A scheduler module is an important component of the system for information gathering and execution control. The scheduler gathers information by monitoring the energy storage medium and multi-core processor state (Figure 1). The gathered data, together with profiled periodic task set information, informs a management algorithm in our scheduler that coordinates operation of the multi-core platform. Each core is eventually assigned a strategy by the scheduler to guide intra-core task execution.

2.2 Scheduling Objective

Our primary objective is to *reduce total task miss rate or penalty* under variant harvested energy conditions at run-time. Our technique should react to changing harvested energy dynamics to complete as much (critical) work as possible, thus maximizing overall system utility and cost effectiveness.

3. PROPOSED FRAMEWORK

3.1 Motivation

Most prior work deals with dynamic solar energy variation by halting, dropping, or speeding up execution of a current task, changing instantly from an initial schedule deduced offline. For energy harvesting aware periodic task set scheduling, the most recent work, UTB [11], also follows this route. Although UTB deduces an optimal initial schedule offline assuming sufficient energy, it does not cope well with run-time energy variations, and there is scope for significant improvement as discussed below:

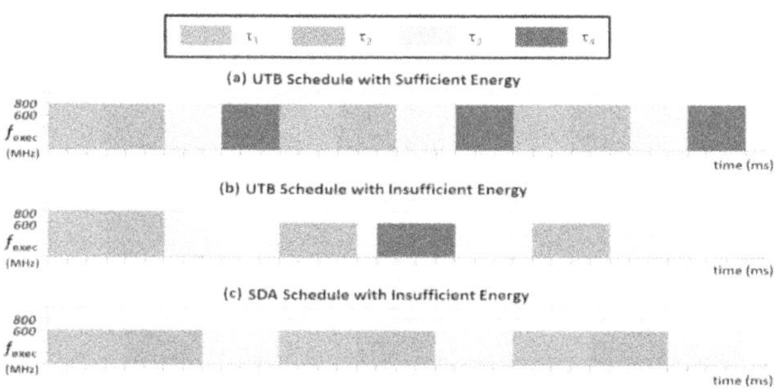

Figure 2. Motivation for proposed semi-dynamic approach

1) The task drop mechanism in UTB reacts to run-time energy shortages *passively*, only when the current task lacks sufficient energy to finish in time. In the motivational example shown in Figure 2, we assume a task set with four periodic tasks ($\tau_1 \sim \tau_4$), where each task has WCEC of 2.4 million CPU cycles and task period of 12ms. According to Table 1, Eq. (2) and Ineq. (3), UTB initially sets execution frequency to 800MHz so that all tasks can finish with the best efficiency if energy is sufficient, as shown in Figure 2(a). However, the real issue lies in the run-time energy management with an insufficient energy budget. Let us assume remaining energy in the energy storage is 7200μJ and harvested power in the next 36ms (3 periods) is 200mW, i.e., 200μJ of incoming energy per microsecond. After finishing three jobs, the energy storage is depleted, and UTB has to drop jobs due to insufficient energy, as shown in Figure 2(b). Only 6 out of 12 job instances are finished with UTB, resulting in a 50% miss rate. With the same energy budget, our proposed semi-dynamic algorithm (SDA) copes with energy shortage by *proactively* dropping tasks. It drops one task, τ_4, based on the energy budget which helps to execute the remaining tasks steadily at a lower frequency of 600 MHz. According to Table 1, executing at 600MHz has power consumption of 400mW, which is dramatically lower than 900mW at 800MHz due to the nonlinear relation between frequency and power consumption. As can be seen in Figure 2(c), all accepted job instances for $\tau_1 \sim \tau_3$ are finished and the overall miss rate is 25%, which is significantly lower than 50% achieved by UTB.

2) UTB encourages dropping tasks with longer execution time, because finishing them requires more energy than others. This biased dropping may be undesirable for real-world applications, as tasks with longer execution time may represent complex applications of high importance. Furthermore, the fundamental problem is the absence of flexibility in UTB to schedule tasks with controlled priority. In spite of the fact that SDA drops all job instances of τ_4 in Figure 2(c), its semi-dynamic framework provide flexibility to deal with this issue; e.g., by dynamically increasing miss penalty of τ_4 later to give it more priority in the next time epoch when rescheduling becomes possible again.

3) On multi-core platforms, UTB partitions tasks into separate sets and then executes each set on a core using a single-core scheduling algorithm. However, as all cores are dependent on the same energy source, such isolated run-time adjustment is not amenable to learning upcoming energy requirements of other cores, leading to suboptimal or even faulty schedules. In addition, static task partitioning in UTB wastes the flexibility provided by a multi-core platform. In contrast, SDA triggers task rescheduling to exploit multi-core flexibility.

In summary, there are many limitations with the best known prior work on harvesting-aware task scheduling. Our HY-SDA approach (described in the next section) addresses these limitations.

3.2 Semi-Dynamic Algorithm

In this section, we describe our novel energy and workload management framework (HY-SDA) based on a semi-dynamic algorithm, for systems with hybrid battery-supercapacitor energy storage, solar energy harvesting, and periodic task sets.

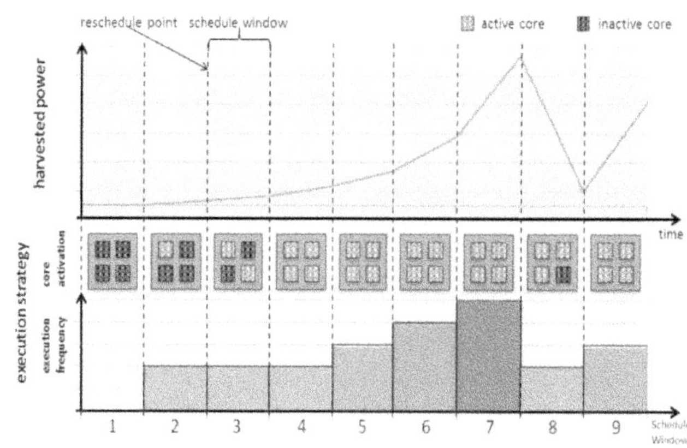

Figure 3. Illustration of semi-dynamic framework

One of the underlying ideas behind HY-SDA is to exploit time-segmentation during energy management, as illustrated in Figure 3. At each specified time interval (epoch), there is a reschedule point, where the execution strategy can be adjusted based on the energy budget available in the hybrid energy storage system. A time frame between two reschedule points is called a *schedule window*, within which a strategy specified at the beginning is in effect until the next reschedule point. Thus reschedule points provide dynamic adaptivity needed by the energy harvesting aware system to adjust the task execution strategy, while the schedule window enables stable execution so that periodic task information can be utilized for better energy savings, as in Figure 2(c). From schedule window 1 to 4 in Figure 3, it can be seen that under low energy conditions, HY-SDA maintains execution at optimal low (critical) frequency and increases the number of active cores to finish more tasks as harvested energy increases. Cores only execute at higher frequency when the energy harvested is abundant. Thus HY-SDA has better energy efficiency as execution frequency within a schedule window remains stable.

At each reschedule point, our technique is composed of three stages: *(i) energy budgeting*, to take advantage of our proposed hybrid storage system, *(ii) active core count selection*, which selects the number of processing cores to activate, and *(iii) penalty-aware task rejection*, to filter out subset of tasks that are less important and cannot be supported by the energy budget. These three stages are organized in an order that successor stages make use of efforts made by previous stages, rather than diminishing them, and are described in the following sections (3.2.1-3.2.3).

3.2.1 Hybrid Storage System and Energy Budgeting

In this section, we introduce our hybrid energy storage system and an energy budgeting heuristic to make use of its properties.

Battery-Supercapacitor Hybrid Energy Storage: Inspired by Ongaro et al.'s work [5], we propose a hybrid energy storage system with one Li-Ion battery and two separate supercapacitors connected by a dc bus, as shown in Figure 4. During each schedule window, one capacitor is used to collect energy extracted from the PV array, while the other one is used as a power source for system operation or battery charging. At each reschedule point, the two supercapacitors switch their roles. Supercapacitors charge the battery only when their saved energy exceeds peak requirements of processors running at full speed. The PV array, battery and supercapacitors are coupled with bidirectional dc-dc converters to serve the purpose of voltage conversions between components with maximum power point tracking (MPPT) [7] and voltage level compatibility. This hybrid battery and dual-supercapacitor design has many advantages:

- It requires small capacity for the two supercapacitors, as each of them is only used to keep energy harvested during one single schedule window and is discharged in the next schedule window;
- The supercapacitor with energy buffered during the last schedule window acts as a known stable energy source for the system, which filters out short term solar energy variation. Thus, no energy prediction scheme is needed in our design, avoiding complexity and inaccuracy introduced by non-ideal prediction mechanisms. Moreover, the stable energy source makes it possible to charge the battery with a steady constant current for more effective charging [6];
- The supercapacitors can support embedded processors directly, taking advantage of a much lower charging/discharging overhead compared to an electrochemical battery;
- The battery offers high capacity to preserve energy especially in scenarios with excessive harvested energy.

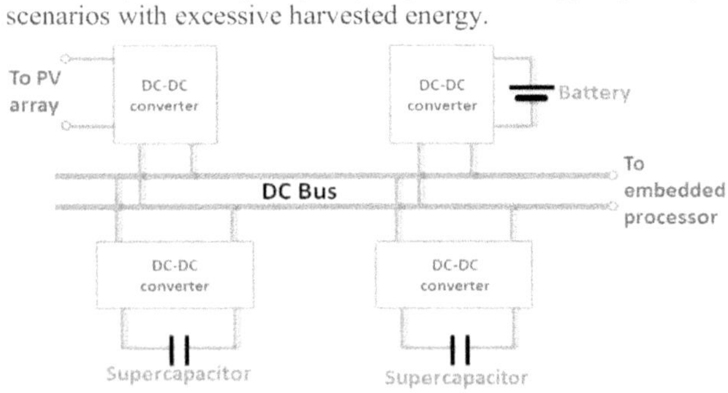

Figure 4. Proposed hybrid energy storage module

Energy Budgeting: Our energy budgeting heuristic selects among energy sources (supercapacitors and battery), sets the amount of energy to charge the battery for (E_{chrg}), and, assigns energy budget for system execution in the upcoming schedule window (E_{window}), as shown in Algorithm 1. The heuristic is based on storage levels of the battery (LV_B) and supercapacitor (LV_C). We assume that LV_B is provided directly by the energy storage, while the storage level of the supercapacitor can be classified into three levels of LV_C, (lines 1-3). As we want to avoid battery charging/discharging overhead, there are only 2 scenarios where the battery is selected as a power source: one is when energy harvested in the supercapacitor is below a critical level ($LV_C = 1$); the other is when battery storage level is high ($LV_B = 3$) such that battery overflow becomes a possibility (line 4). On the other hand, the battery is charged only when energy in the supercapacitor exceeds peak requirements of the processors (lines 12-14).

Algorithm 1 Energy Budgeting Heuristic

Input: (i) harvested energy in charged capacitor, E_C; (ii) battery energy storage level, LV_B; (iii) energy budget to execute one core at critical frequency, E_{crt}; (iv) energy budget to execute one core at maximum frequency, E_{max}; (v) number of cores in embedded processor, NUM_CORE
Output: (i) assigned energy budget for next schedule window, E_{window}; (ii) decision to activate battery as power source, B_{on}; (iii) energy to be charged into battery during next window with constant current, E_{chrg}

```
1.  if E_C < E_crt then     LV_C = 1
2.  else if E_C > E_max × NUM_CORE then     LV_C = 3
3.  else   LV_C = 2
4.  if LV_B > LV_C then
5.      B_on = true
6.      if LV_B = 2 then     E_window = E_crt × NUM_CORE
7.      if LV_B = 3 then     E_window = E_max × NUM_CORE
8.  else
9.      B_on = false
10.     if LV_C = 1 then     E_window = 0
11.     if LV_C = 2 then     E_window = E_C
12.     if LV_C = 3 then
13.         E_window = E_max × NUM_CORE
14.         E_chrg = E_C - E_window
```

3.2.2 Active Core Count Selection

The main reason for having an active core count selection heuristic is that running a processor below its critical frequency actually decreases energy efficiency, as can be seen in Table 1. This situation can occur when the energy budget is so low that only a small subset of tasks can be accepted, i.e., after evenly distributing these tasks to all cores, utilization on each core is smaller than maximum utilization supported by the critical frequency. With our active core count selection heuristic, we can shut down some cores at each reschedule point based on the estimated energy budget. The power dissipated by inactive cores is negligible and the remaining cores can then receive enough workload to run at critical frequency. Also the associated power state switching overhead is minimal as we only trigger core shutdown at reschedule points. Also our heuristic should compare resulting efficiencies before making shutdown decisions.

The pseudo code of the active core count selection heuristic is given in Algorithm 2. Initially, the scheduler gets the energy budget for the upcoming schedule window from the hybrid energy storage system. Then, the core shutdown procedure is triggered when the energy budget is unable to support all active cores to execute at the critical frequency (line 2). Subsequently (lines 3-10) if one less active core results in a better efficiency, then the scheduler shuts down one core. If the energy budget for the current schedule window is extremely low, eventually all cores in the system will be shut down to save harvested energy for future execution. Recursively, these steps set the number of cores to keep active. Finally, the objective task-set utilization for penalty-aware task rejection is obtained by summing up supported utilization of each core (line 11). As a result of this selection, the number of cores activated is tightly related to the energy budget available.

Algorithm 2 Active Core Count Selection Heuristic

Input: (i) energy budget for coming schedule window, E_{window}; (ii) energy budget to execute one core at critical frequency, E_{crt}; (iii) dual-speed method energy efficiency profile for task utilizations from 0 to 1, $\delta(U)$; (iv) number of cores in embedded processor, NUM_CORE

Output: (i) number of cores to active in next schedule window, num_active_core; (ii) objective utilization for next window, U_{obj}

1. num_active_core ← NUM_CORE
2. **while** $E_{per\_core} < E_{crt}$ **and** num_active_core > 0 **do**
3. $E_{num\_core} \leftarrow E_{window}$ / num_active_core
4. $E_{num\_core-1} \leftarrow E_{window}$ / (num_active_core-1)
5.   calculate f_{num_core-1} and f_{num_core}, maximum frequencies supported by E_{num_core-1} and E_{num_core}
6.   based on Inequality (3), calculate U_{num_core-1} and U_{num_core}, maximum utilization supported by f_{num_core-1} and f_{num_core}
7.   look up profile for δ (U_{num_core}) and δ (U_{num_core-1})
8.   **if** $\delta(U_{num\_core}) < \delta(U_{num\_core-1})$ **then**
9.   num_active_core ← num_active_core – 1
10.   update E_{per_core} , U_{per_core}
11. $U_{obj} \leftarrow U_{per\_core} \times$ num_active_core

3.2.3 Penalty-Aware Task Rejection and Assignment

To add task priority control in HY-SDA, we distinguish a task's importance by assigning a miss penalty to each task [15]. In this step, our scheme rejects tasks with lower penalty density (see Section 2.1.2) first, rather than simply drop tasks with longer execution time to allocate the limited energy budget to more important tasks for miss penalty reduction. In particular, for the case when all tasks are assigned an identical miss penalty, this scheme reduces miss penalty equivalent to miss rate. We describe our task rejection heuristic below in Algorithm 3.

Algorithm 3 Penalty Aware Task Rejection and Assignment Heuristic

Input: objective utilization from algorithm 2, U_{obj}

Output: optimal execution frequency for each core, f_{opt}(core_id)

1. sort task set T in non-decreasing order of tasks' penalty densities
2. $T_{accepted} \leftarrow T$
3. **for** n = 1:N **do**
4.   **if** $U_{accepted} > U_{obj}$ **then**
5.   reject n^{th} task
6.   **else**
7.   done with task rejection, **break**
8. sort accepted task set $T_{accepted}$ in non-increasing order of task utilization
9. **for** n = 1:$N_{accepted}$ **do**
10.   assign n^{th} task to active core with the lowest utilization
11. get assigned task utilization for each active core, $U_{(core\_id)}$
12. based on Inequality (3), calculate f_{opt} (core_id)
13. execute assigned tasks on each core with dual-speed heuristic

In lines 1-7, we sort all tasks in non-decreasing order of tasks' penalty densities so that we can then reject tasks one by one until the remaining tasks' total utilization is lower than objective utilization given by Algorithm 2. The remaining tasks form the accepted task set and are assigned to all active cores using a simple but effective heuristic in lines 8-10. This heuristic not only enables priority control among tasks, but also evenly distributes workload to each core for execution under a stable frequency for better efficiency. After all accepted tasks are assigned, we get actual utilization and optimal frequency for each core. A dual-speed heuristic is implemented to approximate each core's designated optimal frequency by switching between two discrete frequency levels available in our XScale model [17].

4. EXPERIMENTAL STUDIES

4.1 Experiment Setup

We developed a simulator in C++ to evaluate the effectiveness of our proposed semi-dynamic energy and workload management algorithm with hybrid energy storage (HY-SDA). We use the approach given in Mirhoseini et al.'s work to model rate capacity effect of batteries [12]. We compared our scheme to the state of the art Utilization-Based Algorithm (UTB) [11], which we modeled in our environment. In addition, we implemented three variants of HY-SDA, namely (i) BA-SDA: SDA for battery-only system with doubled battery capacity; (ii) CA-SDA: SDA for supercapacitor-only system with doubled supercapacitor capacity, and (iii) MISS-SDA: a modified version of HY-SDA to focus on miss rate reduction. UTB, BA-SDA and CA-SDA rely on *moving average* algorithm for energy harvesting prediction [10] as they do not have a supercapacitor to buffer harvested energy in each schedule window. The processor power model is shown in Table 1. The energy harvesting profile is obtained from historical weather data from Golden, Colorado, USA, provided by the Measurement and Instrumentation Data Center (MIDC) of National Renewable Energy Laboratory (NREL) [16]. Our system only executes during daytime over a span of 750 minutes, from 6:00 AM to 6:30 PM, when solar energy is available. We randomly generate 50 task sets with full utilization. In each task set, tasks are assigned with miss penalty ranging from 1 to 100 with uniform distribution. Finally, we set schedule window size (for all SDA based algorithms) to be five minutes compared to the average task execution time of a few hundred milliseconds – not too short to cause frequent changes in the execution policy and not too long to lead to high capacity requirements from the supercapacitors in the system.

4.2 Experiment Results

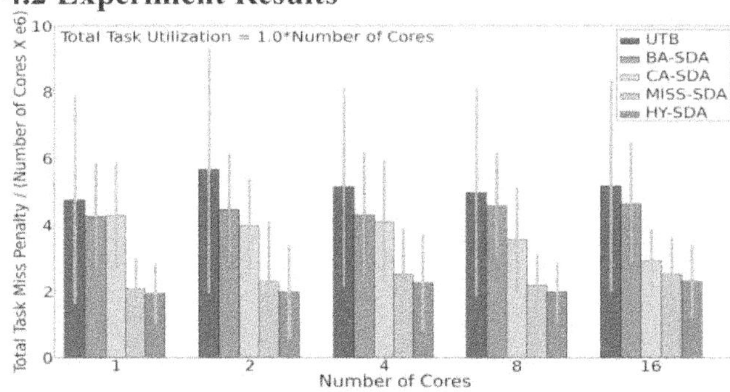

Figure 5. Overall miss penalty comparison

We first compared average overall miss penalty for the various techniques, with increasing multi-core platform complexity (1 to 16 cores). Capacities of batteries and supercapacitors, and nominal harvested energy scale linearly with number of cores in the processors. The results for this experiment are shown in Figure 5. UTB, BA-SDA and CA-SDA can be seen to have a much higher miss penalty. Respectively, the performances of these techniques mainly suffer from unstable execution frequencies, lower charging/ discharging efficiency of the battery, and limited capacity of the supercapacitor. Moreover, UTB has much higher variation in results, due to the limitation that its run-time task dropping scheme does not take task miss penalty into consideration. Also note that CA-SDA has an advantage over BA-SDA with increasing number of cores in system. This is due to the rate capacity effect of the battery and the fact that more cores means higher current demand. On the other hand, our HY-SDA scheme leads to the lowest task miss penalty, while MISS-SDA results in slightly higher miss penalty than HY-SDA because it instead focuses on miss rate reduction. Figure 6 shows a comparison of the miss rate for the various techniques. The miss rate of MISS-SDA is lower and has less variation compared to that of HY-SDA, which is to be expected as MISS-SDA aims to reduce miss rate. Note that although HY-SDA is aimed at miss penalty reduction, it still achieves much lower miss rate than UTB, BA-SDA and CA-SDA.

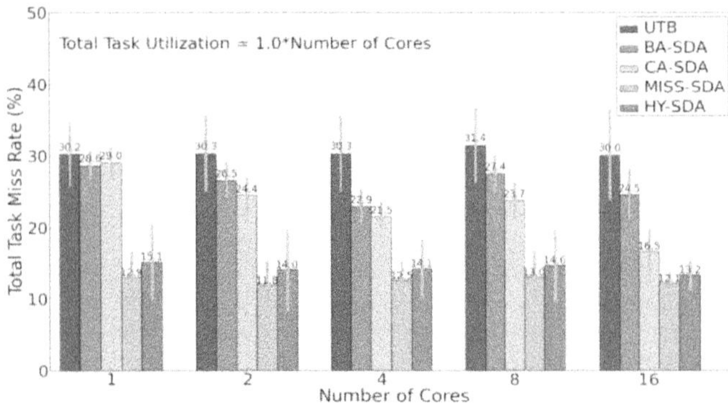

Figure 6. Overall miss rate comparison

We also compared HY-SDA with UTB for a scenario where harvested power fluctuates over time. The result of this experiment on a 16 core system is shown in Figure 7. First of all, we can see that the HY-SDA actually results in a higher miss rate than UTB at the beginning, because it waits until the supercapacitor is charged to the critical level. Subsequently, higher miss rate reduction for HY-SDA is achieved when harvesting power is low or changes dramatically, reflecting the advantage that HY-SDA has over UTB to cope with stringent energy budgets and filter out solar harvesting variations. Moreover, HY-SDA results in a more significant miss rate reduction after 12 PM. The reason for this is that HY-SDA's high energy efficiency leads to more energy savings in the battery, which enables more tasks to be executed and meet their deadlines.

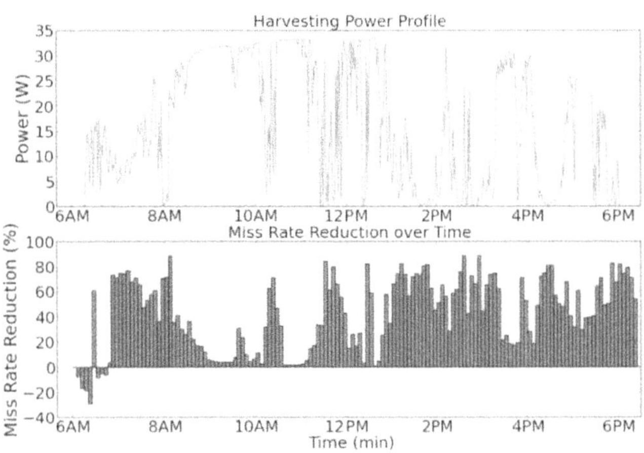

Figure 7. Miss rate reduction of HY-SDA compared to UTB

To highlight the advantage of the dual-supercapacitor design in our proposed hybrid energy storage system, we define a new metric, *budget violation rate*, which is the percentage of unfinished jobs for the accepted tasks. Results for a 16-core configuration are shown in Figure 8. HY-SDA and MISS-SDA, which make use of the hybrid energy storage system, have much lower budget violation rates, because for most of the time their energy budgeting is based on the known amount of energy buffered in supercapacitor. In contrast, other schemes' energy budgeting can be misled by inaccuracy in their solar energy harvesting prediction, which explains their higher energy budget violation rates.

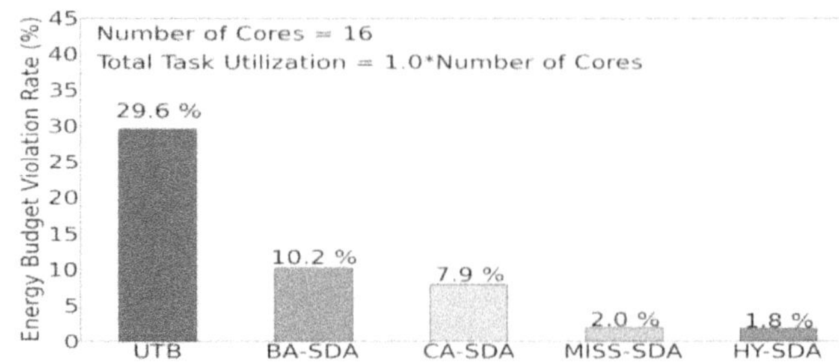

Figure 8. Budget violation rate comparison

5. CONCLUSION

In this paper, we proposed a new framework for energy and workload management (HY-SDA) based on a semi-dynamic algorithm, for real-time multiprocessor embedded systems with solar energy harvesting and a hybrid battery-supercapacitor energy storage system. Compared to the best known previous work, our approach is very promising, reducing miss rate by up to 61 % and miss penalty by up to 65% for high intensity workloads.

ACKNOWLEDGEMENTS

This research is sponsored in part by grants from NSF (CCF-1252500) and SRC.

REFERENCES

[1] Nvdia Tegra 3 processor, http://www.nvidia.com/object/tegra-3-processor.html.

[2] The benefits of multiple CPU cores in mobile devices, http://www.nvidia.com/content/PDF/tegra_white_papers/Benefits-of-Multi-core-CPUs-in-Mobile-Devices_Ver1.2.pdf.

[3] V. Raghunathan et al., "Design considerations for solar energy harvesting wireless embedded systems," in IPSN, 2005, pp. 457-462.

[4] X. Jiang, J. Polastre, and D. Culler, "Perpetual environmentally powered sensor networks," IPSN, 2005, pp. 463-468.

[5] F. Ongaro, S. Saggini, and P. Mattavelli, "Li-Ion battery-supercapacitor hybrid storage system for a Long Lifetime, Photovoltaic-Based Wireless Sensor Network," IEEE Trans. Power Electron., vol. 27, issue 9, pp. 3944-3952, Sept. 2012.

[6] B. Carter, J. Matsumoto, A. Prater, and D. Smith, "Lithium ion battery performance and charge control," in IECEC, 1996, vol. 1, pp. 363-368.

[7] N. Femia et al., "Distributed maximum power point tracking of photovoltaic arrays: novel approach and system analysis," IEEE Trans. Indust. Electron., vol. 55, no. 7, pp. 2610-2621, Jul. 2008.

[8] C. Moser, D. Brunelli, L. Thiele, and L. Benini, "Lazy scheduling for energy-harvesting sensor nodes," in DIPES, 2006, pp. 125-134.

[9] S. Liu, Q. Qiu, and Q. Wu, "Energy aware dynamic voltage and frequency selection for real-time systems with energy harvesting," in DATE, 2008, pp. 236-241.

[10] S. Liu, J. Lu, Q. Wu, and Q. Qiu, "Harvesting-aware power management for real-time systems with renewable energy," IEEE Trans. VLSI Syst., vol. 20, no. 8, pp. 1473-1486, Aug. 2012.

[11] J. Lu, Q. Qiu, "Scheduling and mapping of periodic tasks on multi-core embedded systems with energy harvesting," in IGCC, 2011.

[12] A. Mirhoseini, F. Koushanfar, "HypoEnergy: hybrid supercapacitor-battery power-supply optimization for energy efficiency," DATE '11.

[13] Intel XScale, http://download.intel.com/design/intelxscale/27347302.pdf

[14] Z. Xu et al., "Electrochemical supercapacitor electrodes from Sponge-like graphene nanoarchitectures with ultrahigh power density", J. Phys. Chem. Lett., pp. 2928-2933, Oct, 2012 (3).

[15] J. Chen, T. Kuo, C. Yang, and K. King, "Energy-efficient real-time task scheduling with task rejection", in DATE, 2007, pp. 1-6.

[16] NREL Measurement and Instrumentation Data Center (MIDC), http://www.nrel.gov/midc/.

[17] D. Rajan, R. Zuck, and, C. Poellabauer, "Workload-aware dual-speed dynamic voltage scaling", in RTSCA, 2006, pp. 251-256

Early Stage Power Management for 3D FPGAs Considering Hierarchical Routing Resources

Krishna Chaitanya Nunna
Department of Advanced Informatics
Kyushu University
3-8-33-309 Momochi, Sawara-ku,
Fukuoka 814-0001, Japan
krishna@soc.ait.kyushu-u.ac.jp

Farhad Mehdipour
E-JUST Center
Kyushu University
3-8-33-305 Momochi, Sawara-ku,
Fukuoka 814-0001, Japan
farhad@ejust.kyushu-u.ac.jp

Kazuaki Murakami
Department of Advanced Informatics
Kyushu University
744 Motooka, Nishi-ku
Fukuoka 819-0395, Japan
murakami@ait.kyushu-u.ac.jp

ABSTRACT

Designing for low power consumption demands power-efficient devices and good design practices to leverage the architectural features without compromising performance. Power estimation at an early stage of electronic design automation (EDA) flow is essential in order to handle the design issues much earlier. In this paper, we are proposing a methodology for evaluating the power in three-dimensional field-programmable gate arrays (3D FPGAs) at an early stage of the design cycle namely at the partitioning step and making it a power-aware stage. As a part of the work, we also estimate the routing resources needed for the power evaluation. Our estimated power values are compared against the values obtained from a 3D place and route tool, TPR along with the added power calculations, which is demonstrating acceptable accuracy. Our results show that there is a scope for achieving desired distribution of power among the layers well before the placement with reasonable deviation in estimation and proves that our methodology is providing opportunity for power management at earlier stages of the design flow.

Categories and Subject Descriptors

B.7.1 [**Integrated Circuits**]: Types and Design Styles — *Gate arrays*; J.6 [**Computer Applications**]: Computer-Aided Engineering — *Computer-aided design (CAD);* I.6.5 [**Simulation and Modeling**]: Model Development.

Keywords

3D FPGA, power management, power-aware partitioning, early estimation, hierarchical routing resources.

1. INTRODUCTION

Three-dimensional field-programmable gate array (3D FPGA) is one of the promising innovations which can provide benefits like increasing transistor density, reduced form factor, heterogeneous architectures and improvement in delay by significantly reducing the wire lengths of integrated circuits [1]. Compared to application-specific integrated circuit (ASIC), power consumption in FPGA is most critical due to its large number of resources. Each of the available resources (logic, routing, clock etc.) will have significant contribution to the overall power consumption. In spite of advantages over two-dimensional (2D), thermal effects are expected to be significantly higher in 3D chips due to higher power density and this can cause greater degradation in device performance and chip reliability [2]. The main focus of commercial vendors and also the majority of published research on FPGA architecture and electronic design automation (EDA) are improving area efficiency and performance. Though area and speed have been the main research focus to date, power is likely to be a key consideration in the design of future FPGAs especially in 3D platform.

Fast growing industry requirements result in rapid increase of circuit complexity. This is putting challenges to EDA tool developers for reducing the design time that can help make design decisions early in the process. Concentration on power at the early stage of design cycle will reduce the burden on later stages which are already having bigger problem size. Especially for 3D platform, as the design sizes increase rapidly, better handling of the critical parameters like power and temperature at the partitioning step will provide enough opportunities to the EDA developers to have more options in solving interdependent issues related to the later stages. Through our work, we are proposing a methodology for evaluating the power before placement and a power-aware partitioning algorithm for 3D FPGA considering hierarchical routing resources. As a part of the work, we study the variation in routing and logic resources between partitioning routine iterations and their effect on total capacitance and switching activity for each layer in order to evaluate the dynamic power consumption. Our methodology can be helpful in distributing the power based on the demand such as considering the location of heat sink for efficient thermal management. And also it is flexible for interfacing different types of partitioning routines available in the research domain.

The past and recent work related to this area is encouraging. In [3] authors introduce a TSV-aware partitioning algorithm for efficient utilization of the inter-layer connectivity. Ababei *et al.* [4] introduced 3D place and route tool for FPGAs in which the partitioning is done explicitly. The work proposed by [5] introduces a multi-level multi-layer partitioning for 3D integrated circuits (ICs). For both [4] and [5], the objective for partitioning is minimizing the number of cuts between the partitions targeting for better performance. A 3D IC partitioning routine is proposed by [6] using integer linear programs (ILPs) for minimizing the usage of vertical interconnects subject to the footprint and power consumption constraint. Work presented in [7] is performing the partitioning implicitly within placement. But as the circuit size is growing, explicit partitioning can provide opportunities for handling the critical issues at early stage of the design cycle. The

GLSVLSI'13, May 2–3, 2013, Paris, France.

works presented above are trying to show the benefits of 3D integration in terms of reduced wire-length. But none of them dealt directly with the critical issues related to the power consumption at the early stage of design cycle. To the best of our knowledge, our work on early estimation of power consumption namely at the partitioning stage for 3D FPGAs is first of its kind. In the following, section 2 covers background on the preliminary experiments for power consumption in FPGA, 3D FPGA architecture specification and hierarchical routing resources in FPGA. Our methodology for power evaluation and routing estimation is explained in section 3. Section 4 covers the power-aware partitioning. Experiment results are discussed in Section 5 and the paper concludes along with the future work in section 6.

2. BACKGROUND

Power in FPGA comprises of dynamic and static components in which dynamic power depends on switching activities of the logic and static power is due to the leakage pertaining to the transistor structure. Earlier work in [8] and [9] shows that the dynamic power consumption in FPGA due to routing resources is much higher than others because of their large share in chip area. To get the knowledge on power distribution among FPGA resources, we have conducted preliminary experiments and results were shown in our work in [10]. We have observed that the power consumed by routing resources is much higher than the other resources and it is growing with the increase in circuit size. On the other hand, power due to logic is low and almost unchanged. From these observations it is evident that the role of routing fabric in FPGA power is momentous.

The target architecture for our work is similar to Xilinx Virtex-II device [11], which is a commercial 2D FPGA, for each layer and such identical device layers are aligned vertically using through-silicon-vias (TSVs) as shown in Fig. 1. Each layer consists of programmable resources like logic block (LB) array, connection box (CB), switch box (SB) and routing channels in X and Y directions. Each TSV is having a channel width set by the architecture definition. The hierarchical routing resources in FPGA provide a greater flexibility for routing the input circuit. In commercial FPGAs, for example, a Xilinx Virtex II device has fully buffered programmable interconnections, with a number of resources counted between any two adjacent SBs at which each vertical and horizontal routing channel interacts. The widely used academic tool VPR has similar resources with a combination of single-length (SL), double-length (DL), hex-length (HL) segments and long length (LL) wires. An example for the hierarchical routing segments in a 2D FPGA is shown in Fig. 2. Here a segment is defined as a metal wire that can span over a specific number of LBs and terminate at SBs. SL segments intersect at each SB to form a grid of interconnections. An example connection between a source LB and a sink LB routed by SL

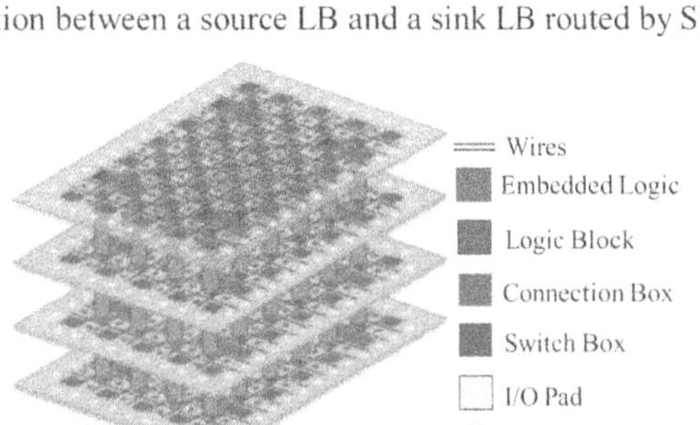

Figure 1. Schematic of 3D FPGA.

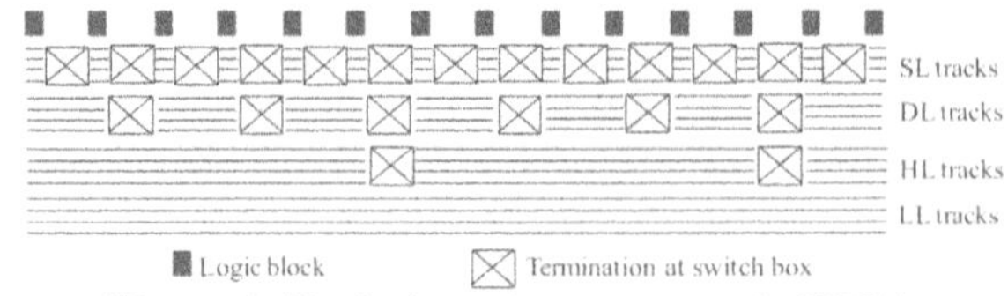

Figure 2. Typical routing segments in FPGA.

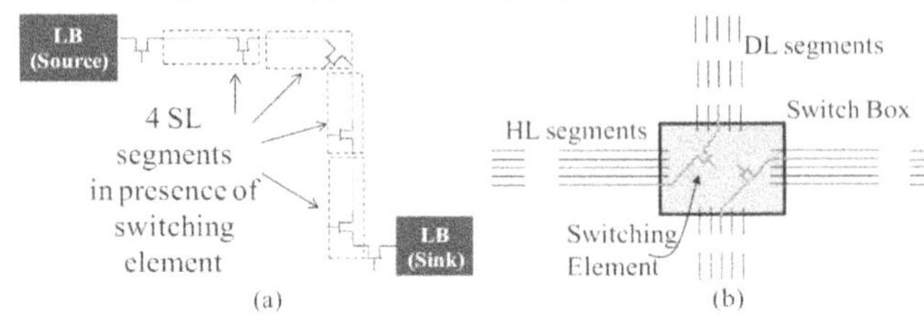

Figure 3. Example connection using (a) only SL segments (b) combination of HL and DL segments.

segments is shown in Fig. 3(a). DL segments are twice as long as SL ones and run past two LBs before entering a SB. Similarly HL segments run past six LBs. And example connection using DL and HL segments is shown in Fig. 3(b). LL wires bypass all SBs and form a grid of connections that runs vertically and horizontally the entire length and width of FPGA array.

3. POWER EVALUATION AND ROUTING ESTIMATION

Due to remarkable complexity of recent designs, it seems necessary to estimate performance metrics in early design phases. The availability of accurate early estimates can reduce the required number of iterations through the entire design flow, which allows design trade-offs to be evaluated at an abstract level resulting in reduction in the design time and cost. And there is an opportunity of refining the outcome of the early stage through incremental procedures. As a part of the power-aware partitioning (shown in Fig. 4(a)) we are proposing a power evaluation methodology (zoom-out) as well. As the configuration of partitions changes during the partitioning process, the amount of logic and routing resources also changes. Within the process, the variation in dynamic power consumption is measured accordingly. As the routing power dominates the other components, our work is mainly based on estimating the dynamic power consumed by the interconnections. However the power consumed by LBs is added because of variations in the number of nodes in each partition between the iterations of the partitioning routine. The dynamic power depends on the switching activity (α), switching capacitance (C), frequency of operation (f) and supply voltage (V) and are related as:

$$P_{dyn} = \alpha * f * C * V^2 \qquad (1)$$

Here f and V are constants for a given architecture. On the other hand, α and C are the critical parameters in which we need to update α for each layer and measure C time to time. For α, probabilistic estimation is typically faster than simulation based approach since it involves a one-time calculation for each net in a circuit and it doesn't change for a single net during the partitioning routine. Hence in our work, α is measured using a probabilistic approach similar to ACE-2.0 [12]. For power estimations, we need capacitance C of the net as well as the associated switching elements (e.g. pass transistors). Capacitance of each net depends on its length while the exact length of each net is known only after the routing phase. At the partitioning stage, since no geometrical information is available, there should be

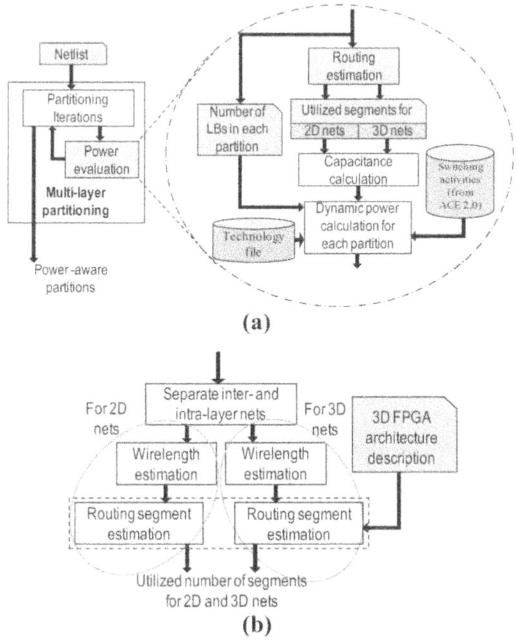

Figure 4. (a) Power evaluation methodology inside partitioning (b) Methodology for early estimation of hierarchical routing resources [10].

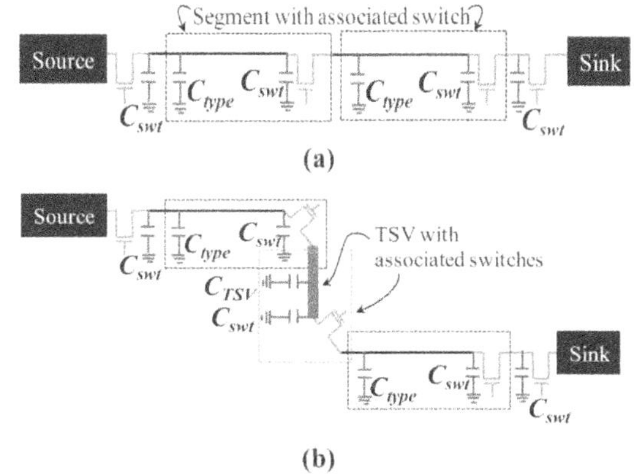

Figure 5(a). Capacitances between source and sink LBs on FPGA plane. (b) Capacitances between source and sink LBs located on two FPGA planes.

proper estimation criteria for capacitance. Already existing routing estimation techniques in FPGA research community provide only the probability of each net having a specific length. For more accurate power calculations, it is observed that the wirelength probability is not enough due to existing various segment types available in FPGA and various possibilities for routing a net using those hierarchical routing segments. Along with the routing segments, their associated switches also contribute to the total power consumption and hence it is necessary to know the used segments as well as used switches. In order to have such kind of routing estimation, we have already proposed a methodology for early estimation of hierarchical routing resources in 3D FPGA [10] as shown in Fig. 4(b). As each segment on FPGA plane terminates at SBs as shown in Fig. 3(a), it is straightforward to find the number of switches associated with each segment. After performing the multi-layer recursive bi-partitioning routine [13] on the input netlist, our routing estimation methodology [10] separates the obtained nets into 2D and 3D nets. As shown in Fig. 4(b), we get the probability of each net to have a specific length (wirelength distribution) for both 2D and 3D nets separately using Stroobandt's model [14]. Depending on these values, segment estimation model calculates the number of segments used by the given circuit. Here the actual geometry of the nets is predicted depending on the possible combinations of routing the nets in different directions (X, Y and Z) separately for 2D and 3D nets. For more details on the implementation of this methodology, it is recommended to the reader to see our work in [10].

After obtaining the number of segments used by the given circuit, the distributed switching capacitance $C_{D,2D}$ for 2D nets is calculated as:

$$C_{D,2D} = \sum_{type} \frac{N_{type}}{\sum_{type} N_{type}} \left(C_{type} + C_{swt}\right) + 2C_{swt} \quad (2)$$

In Eq. 2, N_{type} is the number of segments of specific type used by the circuit, C_{type} is the capacitance of each type of segment and C_{swt} is the associated switch capacitance. Both C_{type} and C_{swt} are typically obtained based on the technology node. The term $N_{type} / \sum_{type} N_{type}$ is the probability of using a specific type of segment out of the total segment types. The term $2C_{swt}$ is the capacitance associated with the switches at the source's output and sink's input terminals as shown in Fig. 5(a). Since the typical capacitance of a TSV in 3D devices is several times of capacitance of a wire on horizontal plane, $C_{D,3D}$ for 3D nets need to be obtained separately by considering capacitance of TSV, C_{TSV} (Fig. 5(b)) as given below:

$$C_{D,3D} = \sum_{type} \frac{N_{type}}{\sum_{type} N_{type}} \left(C_{type} + C_{TSV} + 2C_{swt}\right) + 2C_{swt} \quad (3)$$

Since we are having the number of TSVs in the form of cutsize from the partitioning routine, an extra switch C_{swt} is introduced separately for TSV (can be seen in Fig. 5(b)). Now these distributed capacitances for 2D and 3D nets and switching activities are supplied to the dynamic power calculation in Eq. 1. The power associated with the LBs is also calculated by tracking the nodes in each partition.

4. POWER-AWARE PARTITIONING

Our power-aware partitioning algorithm is explained in Fig. 6. The routine starts with a random selection and moving the node between the initial partitions such that the number of nodes in each partition is less than the available nodes (LBs). At every movement, the partition number for each node is updated. After the current movement, we will apply our power evaluation methodology given in Fig. 4(a) (zoom-out) for dynamic power calculations.

While moving the nodes, the switching activity associated with each net is captured instantaneously. An example movement of a node between partitions is shown in Fig. 7. As shown, when node 4 is moving from partition P_1 to P_2, even though the cutsize remains same, contributing switching activities and capacitances for each partition will be affected due to the alterations in the nets' configurations. As shown in Fig. 7(a), for example, before the movement of node 4 from P_1 to P_2, net h is contributing to the power of P_1 due to its source node 4. As shown in Fig. 7(b), after the movement, net h is contributing to the power of P_2 due its source node 4 position in P_2. At the same time, even though nets c and d crossed the partitions, they contribute to the power of P_1 only as the source nodes of these nets are located in P_1. Similar way is followed for assigning α and C to a particular partition. After each node movement, the dynamic power for each partition is calculated. As we are targeting for desired distribution of power

Algorithm: Power-aware partitioning
Terms: $N_{nodes,partition}$: Number of nodes in each partition after movement, $P_{dyn,T}$: total dynamic power, P_{net}: net power, P_{logic}: logic power, P_{diff}: percentage power difference, ΔR_D: difference between desired distribution of layers, P_{TOL}: power tolerance, $N_{cut,T}$: total cutsize, $N_{cut,MIN}$: minimum cutsize, G: maximum allowed cut percent, N_{TSV}: available number of TSVs.
Objective: Partitioning the circuit in to N parts based on power
Inputs: Netlist
Outputs: power-aware partitions
Constraints: $N_{cut,T} < N_{TSV}$; $N_{cut,T} < N_{cut,MIN} * (1+G)$;
Function power aware partitioning (Nodes, Hyperedges)
While (1) {
Select a random node;
If($N_{nodes,partition}$< given array size for each FPGA layer)
Move the node between partitions;
Update the partition number for the node;
Separate the nets into 2D and 3D for each partition (Fig. 5(b));
Estimate the routing segments for all partitions (Fig. 5(b));
Calculate C_D for with each routing segment (Eq. 2 and Eq. 3);
for each partition {
for each net i { Capture α; Calculate P_{net}; }
Calculate P_{logic} for each node;
$P_{dyn} = \sum_{nets} P_{net} + \sum_{LBs} P_{logic}$;
}
Calculate P_{diff} between partitions and P_{DD};
if(P_{DD} > previous P_{DD}) and ($N_{cut,T}$ > ($N_{cut,MIN}$ *(1+G)) and ($N_{cut,T} > N_{TSV}$) {
Unmake the move;
Update the partition number for the node;
}
if((ΔR_D - P_{TOL})<P_{diff}<(ΔR_D+P_{TOL})) then terminate the procedure;
}

Figure 6. Power-aware partitioning algorithm.

among layers, the decision on fixing the nodes after movement depends on the desired power deviation P_{DD}. The final objective is to minimize P_{DD} subject to certain constraints. It is given as,

$$P_{DD} = \sqrt{\sum_{i=1}^{N} \frac{(R_i - R_{Di})^2}{N}} \quad (4)$$

In Eq. 4, R_i represents the ratio of power for each layer P_i to the total power among N layers. R_{Di} represents the desired distribution set by the user for each layer. This P_{DD} is compared with the P_{DD} of the previous movement. The current movement will be canceled under the conditions: P_{DD} is greater than previous P_{DD}; cutsize is greater than $N_{cut,MIN}*(1+G)$, where minimum cutsize $N_{cut,MIN}$ is obtained by traditional partitioning routine with objective function as minimizing the cutsize, and G is the additional cut percent allowed which is determined through empirical analysis; total cutsize is greater than the available

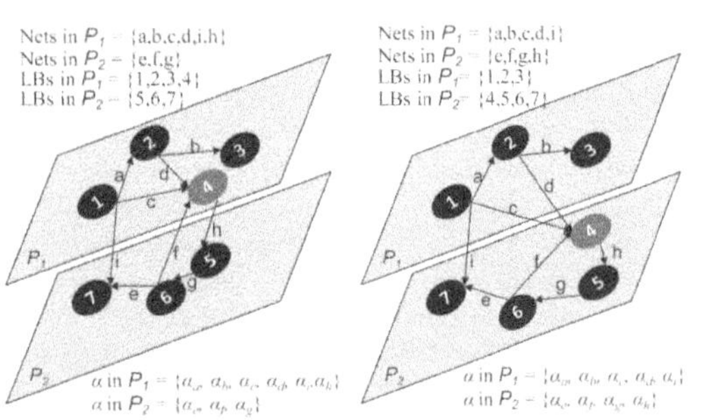

(a) **(b)**

Figure 7. Assigning switching activities during the movement of nodes.

number of TSVs, N_{TSV} in the given architecture. The routine will stop once the condition (ΔR_D - P_{TOL})<P_{diff}<(ΔR_D+P_{TOL}) satisfies. Here ΔR_D is the difference between the desired distributions of layers, P_{TOL} is the tolerance of power that can be expected for a given circuit and P_{diff} is the percentage power difference among layers. This parameter can be varied according to minimum power difference that we can achieve for each and individual circuit.

5. EXPERIMENT RESULTS

Our proposed power-aware partitioning algorithm, power evaluation methodology and routing estimation methodology are implemented using C++ on Ubuntu Linux machine. We have considered a 4-layer 3D FPGA with an array size of 50 x 50 LBs and a routing channel width of 65 wiring tracks for each layer such that all the selected 14 MCNC benchmark circuits [15] will be successfully placed and routed. The share of each type of segment in the routing channel is similar to Xilinx Virtex II device. In order to observe the variation of power during the partitioning we have conducted experiments on sensitivity analysis prior to the actual work. We ran all the circuits on traditional partitioning routine with the objective of minimizing the cutsize and used our power evaluation methodology. As shown in Fig. 8, over a number of iterations, the variation of total power due to the movement of nodes and nets is significantly higher for a particular partition. And we have also observed that this variation is increasing with respect to the increase in the circuit size.

In the next phase of experiments, we have compared our power values estimated at the partitioning stage against the values obtained by running TPR [4] on 14 MCNC benchmark circuits for the validation purpose. TPR is a 3D place and route tool targeted for academic research on 3D FPGAs. It is a 3D extension to the highly successful VPR [15]. In TPR, explicit partitioning is done by hMeTiS [16]. We appended our power calculation technique to TPR and then performed placement and routing and obtained the actual length of the nets followed by our power calculations. Power values and percentage error for the individual layers of all 14 circuits are shown in Table 1. As shown, we have achieved absolute error of 12.25%, 13.47%, 14.41% and 13.43% on average for layer 1, layer 2, layer 3 and layer 4 respectively. Here we showed the absolute errors for individual layers separately instead of total array because we are targeting for power distribution among layers. It should be noted that the power is under-estimated for some circuits and over-estimated for others. This comes from the mean error difference obtained from the routing segments estimation methodology. From our observations, we believe that these under- and over-estimations can be mitigated by characterizing the placement and routing algorithms at the early stage of the design cycle which would be a part of our future work.

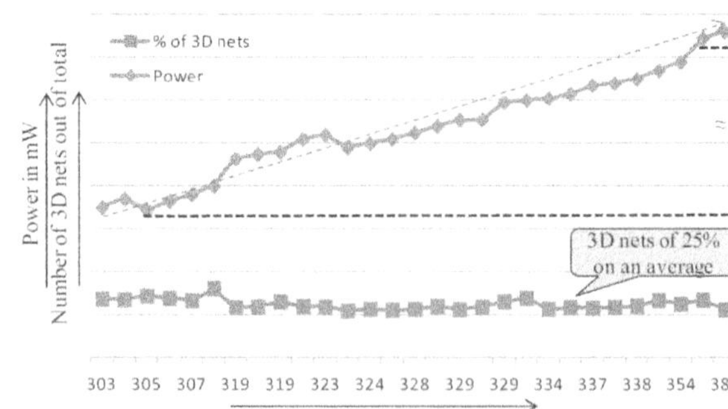

Figure 8. Variation of power and number of 3D nets over the partitioning iteraitons.

Table 1. Percentage estimation error validated against TPR

Circuit	LBs	Nets including multi-terminal nets	Power at TPR output (mW)				Power from our estimations obtained at partitioning stage (mW)				Percentage Error			
			p1	p2	p3	p4	p1	p2	p3	p4	E1	E2	E3	E4
alu4	1522	5408	14.2	248.2	10.8	259.4	12.8	278.6	9.1	289.4	9.86	-12.25	15.74	-11.57
apex2	1878	6692	12.3	129.3	4.5	98.6	10.1	109.2	5.1	82.9	17.94	15.54	-14.38	15.88
bigkey	1707	6537	64.3	25.1	13.5	14.2	72.5	22.1	15.1	12.5	-12.76	11.65	-12.17	11.96
clma	8383	30495	135.3	1028.3	121.7	35.2	154.6	922.4	101.3	41.2	-14.26	10.30	16.76	-17.05
des	1591	6110	58.1	19.6	20.4	191.2	52.8	16.3	22.5	211.2	9.20	16.83	-10.33	-10.45
diffeq	1497	5673	6.3	9.9	7.2	4.4	7.0	8.1	5.8	5.0	-10.46	17.69	19.61	-15.37
elliptic	3604	13756	15.8	29.5	12.8	71.4	14.0	25.8	14.9	79.5	11.15	12.53	-16.51	-11.26
frisc	3556	13658	61.0	9.5	35.3	14.1	69.7	8.2	30.7	12.1	-14.27	13.74	12.85	14.35
pdc	4575	17193	248.1	12.2	289.6	13.2	226.3	14.2	251.8	11.3	8.79	-16.39	13.05	14.34
s298	1931	6959	62.5	132.2	429.2	74.6	52.3	148.3	379.7	64.3	16.22	-12.15	11.53	13.89
s38417	6406	22807	234.1	198.7	22.1	26.2	263.1	172.4	25.8	29.1	-12.39	13.26	-16.82	-11.19
s38584	6447	22100	26.2	23.7	31.1	12.9	29.6	26.3	35.2	14.9	-12.78	-11.28	-13.20	-15.88
seq	1750	6193	98.7	104.1	46.0	50.0	88.6	93.2	38.5	56.2	10.22	10.45	16.45	-12.44
spla	3690	13808	259.0	12.9	8.1	132.7	230.1	14.7	9.1	149.2	11.15	-14.44	-12.32	-12.40
Absolute Average											**12.25**	**13.47**	**14.41**	**13.43**

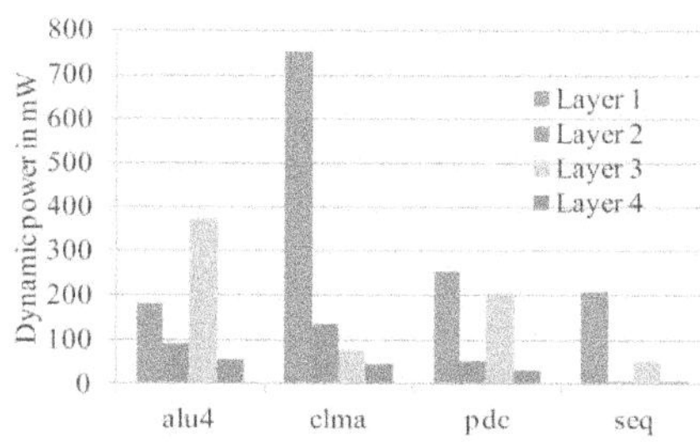

Figure 9. Non-uniform power distribution obtained from TPR with conventional partitioning.

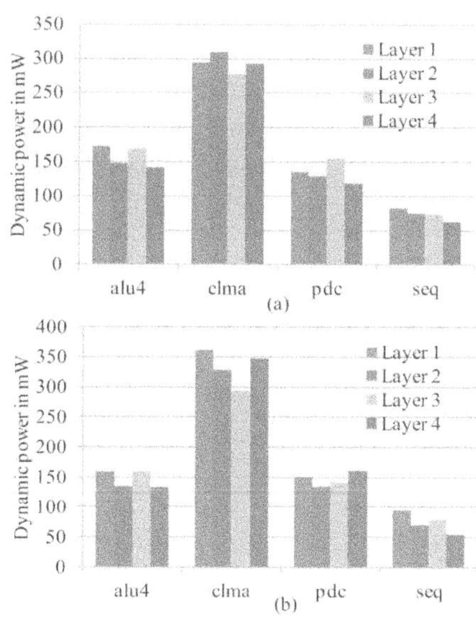

Figure 10. Uniform power distribution obtained from (a) our proposed power-aware partitioning and (b) TPR running with power-aware partitions as input.

To investigate the performance of our power-aware partitioning algorithm, at first we ran five MCNC benchmark circuits, having different circuit sizes, using TPR with conventional partitioning routine which is targeting cutsize minimization objective. As shown in Fig. 9, non-uniform distribution of power is obtained with a variation of power from 20% (for example, between layer 2 and layer 3 for *clma*) to 90% (for example, between layer 1 and layer 2 for *clma*). These are the values obtained after successful routing. It is observable that conventional partitioning which is targeting cutsize minimization objective leads to a non-uniform power distribution among the layers. This kind of variation may results in heat problems on 3D platform. Note that in *clma*, the big difference for layer 1 compared to the others is coming from the high fan-out pins incurring large number of nets.

After that, we ran the same circuits on our proposed power-aware partitioning routine. The goal of this experiment is to achieve uniform/desired power distribution among four layers in considered 3D FPGA at the partitioning stage. This can be targeted to avoid high peak temperatures for thermal management at the early stage of design flow. As shown in Fig. 10(a) we have achieved around 12% power difference between layers for the proposed power-aware partitioning algorithm targeting uniform power distribution. Our results are compared against TPR shown in Fig. 10(b). These values are obtained by feeding the power-aware partitions, generated using our proposed power-aware partitioning algorithm, to TPR and running the placement and routing steps. As shown in the figure, percentage variation in power between layers is around 16% for TPR case where more accurate power evaluation is performed on the placed and routed circuit. This rather small difference shows that our power-aware partitioning is able to distribute the power among the layers such that the later stages can have less burden on resolving the issues related to power/thermal management.

Our power-aware partitioning routine is able to distribute the power among the layers according to the user requirement (e.g. when a non-uniform power density distribution is demanded depending on the location of the heat sink for the sake of thermal management). This distribution can be changed by setting the parameter R_{Di} in Eq. 4 as per the user requirement. Fig. 11(a) shows the desired power distribution with R_{Di} as 20%. It is showing that around 16% to 22% difference between the any layers is obtained from our power-aware partitioning routine. For example *pdc*, layer 1 is having the highest power consumption and layer 2 is having power consumption around 18% less than the power consumed by layer 1. Similarly power consumption in layer 3 is around 18% less than the power consumed by layer 2 and similarly for layer 4 also. This is an appropriate power distribution when the heat sinks are closer to the first layer. As shown in Fig. 11(b), our values are at comparable level with TPR results and our power-aware partitioning is able to distribute the power according

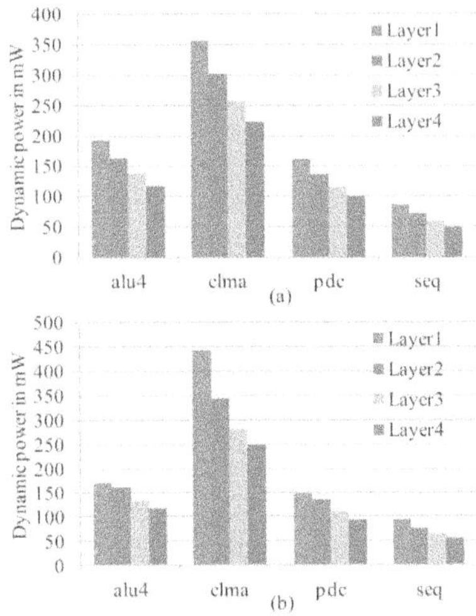

Figure 11. Desired power distribution obtained from (a) our proposed power-aware partitioning and (b) TPR running with power-aware partitions as input.

to the user requirement. Our results showing that the power distribution can be achieved at the early stage of the design cycle in order to consider the power/thermal issues well before the placement stage. Our methodology can be easily interfaced with TPR tool in which only performance based partitioning; placement and routing on the given design are done. We achieved the uniform/desired power distribution at a cost of cutsize increment. As shown in Fig. 6, one of the constraints given to the power-aware routine is $N_{cut,T} < (N_{cut,MIN}\ (1+G))$. That means, the total cutsize after each iteration must be less than $1+G$ times of the minimum cutsize that we can obtain from the traditional partitioning algorithm with the objective of cutsize minimization. The best suitable value for G is selected empirically such that the overall cutsize should be comparable to $N_{cut,MIN}$.

6. CONCLUSION

We proposed a power evaluation methodology by considering the hierarchical routing resources at an early stage of the design cycle namely at the partitioning and a power-aware partitioning algorithm for 3D FPGAs. Our results show that the estimated power values at partitioning stage have acceptable deviation compared with the values obtained after doing actual placement and routing of the circuit using TPR. In addition, it is observed that the power can be distributed according to the demand produced by the user for the purpose of power/thermal management well before placement stage. We achieved uniform/desired power distribution among the layers at a cost of increased cutsize. We believe that the power-aware methodology at the early stage of design cycle can help the designers to handle multiple objectives more efficiently especially in 3D platform. The immediate future work would be extending the proposed power-aware technique to the early thermal-aware partitioning methodology for 3D FPGAs depending on the dynamic power as well as the static power consumption.

7. ACKNOWLEDGMENTS

The authors wish to thank Nobuaki Miyakawa (Honda Research Institute, Japan), Kai KJ, Ohashi Masahiro (Panasonic, Japan), Koji Inoue, Masayoshi Yoshimura and Takaki Hanada (Kyushu University, Japan), for their valuable comments on the work.

8. REFERENCES

[1] J. Alexander, J. P. Cohoon, J. L. Colflesh, J. Karro, E. L. Peters, and G. Robins, "Placement and Routing for Three-Dimensional FPGAs," in *Proc. 4th Canadian Workshop Field-Programmable Devices*, pp. 11–18, 1996.

[2] S. Im and K. Banerjee, "Full Chip Thermal Analysis of Planar (2-D) and Vertically Integrated (3-D) High Performance ICs", in *IEDM Tech. Dig.*, pp. 727–730, 2000.

[3] Siozios K and Soudris D, "A Tabu-based Partitioning and Layer Assignment Algorithm for 3D FPGAs", in *IEEE Embedded System Letters*, Vol. 3, pp. 97-100, Sept. 2011.

[4] Cristinel Ababei, Hushrav Mogal and Kia Bazargan, "Three-Dimensional Place and Route for FPGAs", in *IEEE Trans. on Computer-Aided Design of Integrated Circuits and Systems (TCAD)*, Vol. 25, Issue 6, pp. 1132-1140, June 2006.

[5] Yu Cheng Hu, Yin Lin Chung and Mely Chen Chi, "A Multilevel Multilayer Partitioning Algorithm for 3D ICs", in *Proc. 11th International Symposium on Quality Electronic Design (ISQED)*, pp. 483-487, March 2010.

[6] Jiang I.H.-R, "Generic Integer Linear Programming Formulation for 3D IC Partitioning", in *Proc. International SoC Conference*, pp. 321-324, Sept. 2009.

[7] Jason Cong, Guojie Luo, Jie Wei and Yan Zhang, "Temperature-Aware 3D IC Placement Via Tranformation", in *Proc. Asia and South Pacific Design Automation Conference*, pp. 780-785, Jan 2007.

[8] L. Shang, A. Kaviani, and K. Bathala, "Dynamic Power Consumption in the Virtex-II FPGA Family", in *Proc. ACM/SIGDA International Symposium on Field Programmable Gate Arrays*, pp. 157-164, 2002.

[9] T. Tuan and B. Lai, "Leakage Power Analysis of a 90nm FPGA", in *Proc. IEEE Custom Integrated Circuits Conference*, pp. 57-60, Sep. 2003.

[10] Krishna C Nunna, F. Mehdipour, M. Yoshimura and K. Murakami, "Methodology for Early Estimation of Hierarchical Routing Resources in 3D FPGAs", in *Proc. 20th IFIP/IEEE International Conference on Very Large Scale Integration*, pp. 213-218, Oct. 2012.

[11] Xilinx Virtex-II Platform FPGAs: Complete Data Sheet, http://www.tl.mit.edu/Courses/6.111/labkit/datasheets/virtex2datasheet.pdf.

[12] Lamoureux J and Wilton S.J.E, "Activity Estimation for Field-Programmable Gate Arrays" in *Proc. International Conference on Field Programmable Logic and Applications*, pp. 1-8, 2006.

[13] Trifunovic. A and Knottenbelt. J, "A Parallel Algorithm for Multilevel K-way Hypergraph Partitioning", in *Third International Symposium on Algorithms, Models and Tools for Parallel Computing on Heterogeneous Networks*, pp. 114 – 121, 2004.

[14] D. Stroobandt, *A priori wire length estimates for digital design*, Kluwer, first ed., 2001.

[15] VPR-Source code and benchmarks http://www.eecg.toronto.edu/~vaughn/vpr/vpr.html.

[16] G. Karypis, R. Aggarwal, V. Kumar and S. Shekhar, "Multi-level Hypergraph Partitioning: Applications in VLSI design," in *Proc. ACM/IEEE Design Automation Conf. (DAC)*, pp. 526–529, 1997.

A Low Power 6T-SRAM Using Negative Bit-Line for Variability Tolerance beyond 22nm node

Pablo Royer and Marisa López-Vallejo
Departamento de Ingeniería Electrónica, Universidad Politécnica de Madrid
Ciudad Universitaria s/n, 28040 Madrid, Spain
{proyer, marisa}@die.upm.es

ABSTRACT

The expected large variations of electrical characteristics of sub-22 nm devices represents a limitation on future electronic circuits. This is particularly relevant on RAM memories that have to ensure both read stability and write ability of all the cells. In this paper we present a 6T-SRAM designed with 14nm FinFETs that makes use of the negative bit-line write assist technique allowing a reduction of the supply voltage without a degradation on neither speed nor stability. A new metric is introduced to quantify and control the drawbacks related to negative bit-line voltage. Additionally a new approach to predict the tails of the read current distribution under variability has been presented. Experimental results show that power consumption is reduced by 25% due to the decrease on the supply voltage.

Categories and Subject Descriptors

B.8.2 [**Performance and Reliability**]: Performance Analysis and Design Aids

Keywords

SRAM; Mismatch; Reliability; Low Power; Near Threshold Operation

1. INTRODUCTION

With the continuous reduction on transistor sizes, the variability of their electrical characteristics is expected to keep growing. Predictive models allow us to foresee the behaviour of the transistors beyond the 22nm node. These models are especially helpful for those circuits that are more sensitive to variations, like SRAMS. Since SRAMs occupy large areas, local mismatches appear more frequently what significantly reduces their reliability [4]. Every cell in a memory array has to ensure read stability, that is the ability of the cell to keep its content unchanged during a read access. Meanwhile the cell must also ensure write ability, that is, its content must be able to be flipped during a write access.

In addition to stability, process variations also present a bound to the supply voltage reduction of memories [11], that is being reduced at a much slower rate than the combinational logic supply voltage. On top of that, the low activity factors that memories present makes that the optimal energy consumption point is reached for high supply voltages [4]. As a consequence it is important to find SRAM designs that are able to work at lower voltages with small standby power so that they scale well for low activity factors.

Designs that enhance the cell stability usually come at the cost of a loss in write ability figures. The conventional design process can be done by first ensuring the stability and introducing later write-ability enhancements. Word-line boosting and negative bit-line write-assist have been proposed [3] to enhance write ability especially at lower supply voltages while they significantly reduce the impact of process variations. Negative bit-line write assist seems to be a promising candidate since it does not necessarily entail a degradation of the stability neither an increase in the main supply voltage which is incompatible with low power systems [6].

Different schemes have been proposed to obtain the negative voltage. [10] uses a capacitive coupling-based transient negative voltage while [6] generates a negative voltage power source. While [6] focuses on the periphery of the memory in order to implement the negative bit-line scheme using a conventional cell array, this work scope is in the design of the memory array at cell level, exploring the delay-energy tradeoff that can be achieved by changing the cell design, and paying more attention to the effects of variability on the yield of the memory for technologies beyond the 22nm node. For that purpose we have designed and compared two 256-Kbit memories working at 1GHz with a 14nm FinFET predictive technology, one of them making use of negative bit-line write assist. The main contributions of this work are the following:

- A new metric, *Hold Trip Point*, has been introduced to help quantifying and keeping under control the drawbacks of using negative voltages in the bit-lines. Additionally, the required simulations are faster.
- A new approach to predict the tails of the read current distribution under variability has been also presented.
- A deep variability-aware design space exploration for the FinFET transistors used in the 6T-SRAM cell has been carried out following a systematic methodology.
- A significant power reduction (25%) is obtained for the proposed SRAM. Furthermore, we have analyzed the extended range of main supply voltage that the in-

creased stability may provide and the speed and power consumption results it brings.

The rest of the paper is organized as follows: section 2 introduces the metrics used in this work to evaluate our SRAM performance and stability and explains how negative bit-line helps improving the later. Section 3.1 explains how we model the variability in order to measure the spread in the metrics of the memory. The flow to choose the possible memory cells that are going to be compared is shown in section 3.2. The results are detailed in section 4 while section 5 summarizes the paper.

2. METRICS USED IN THIS WORK

2.1 Stability Metrics

2.1.1 Read Stability

The read stability is the ability of the cell to hold the stored bit. An accidental flip of the contents may occur because of an asymmetry in the cell due to transistor mismatch. A read access is the weakest point as far as stability is concerned since it increases the disturbance in the node storing a zero.

The usual metric to measure read stability is the *static noise margin*, which tells the maximum static noise that can be injected in the internal nodes of the cell with its content not flipping. It is measured by finding the biggest square between the two transfer curves of the inverters of the memory cell, as explained by [13]. Both hold and read static noise margin exist, hereafter we will refer to the read static noise margin as it is the most restrictive.

2.1.2 Write Ability

On the other hand, the memory cell has to be able to change its contents when being written. The usual metric that is used here is the *write trip point* [2], that is the value to which the bit-line needs to be discharged in order to flip the contents of the cell.

2.1.3 Hold trip point

For regular SRAM cells, the write trip point and static noise margin metrics introduced before are the industry standard to estimate the stability of the cells. However, when using the negative bit-line write enhancement the drawbacks it entails have to be kept under control.

Therefore we introduce a new metric that we call *hold trip point* that tells us how far we can discharge the bit-line —including negative values— without changing the contents of non-accessed cells.

The static noise margin is computed following [13] doing a sweep of one of the internal nodes of the cell and this is only for one particular voltage in the bit-line. Several simulations are required to have an overview of how this metric changes as the bit-lines are discharged to negative values. Besides, the hold trip point metric only requires one simulation since the sweep is done directly on the voltage of the bit-line, allowing a faster computation.

2.2 Write Ability enhancements

While read stability is a hard constraint —when the static noise margin has a negative value it means that the cell is unable to keep the data leading to the failure of the memory— a cell with a negative write trip point can still be written if we bring the bit-line to a negative voltage as proposed by [3], lower than the value of the write trip point.

This presents a drawback: the word-lines of the cells that are not accessed are kept to zero while there is a negative voltage in the bit-line, this can lead to the pass gate driving some current that could flip the cell content, specially if the voltage difference approaches the threshold voltage of the transistor. This is measured by the hold trip point.

2.3 Performance Metrics

The performance of the memory is determined by its delay and energy. While energy metrics consider mean values regardless of changes from one cell to another, delay metrics are set by the slowest cell determined with a 5-sigma margin.

2.3.1 Energy Consumption

We separate the energy consumption of the memory in two parts, the stand-by or leakage energy, that is the energy consumed by the array while no read or write access occurs, and active energy, corresponding to charging or discharging the capacitances during a read or write access.

The stand-by energy corresponds to the leakage current of all the cells while none of them is being accessed:

$$E_{stand-by} = I_{stand-by} \cdot Vdd \cdot N_{cols} \cdot N_{rows} \cdot T_{clock} \quad (1)$$

where $I_{stand-by}$ is the current leaked by one cell and N_{rows} and N_{cols} is the size of the whole memory array. T_{clock} is the period of the clock at which the circuit operates. This current is leaked by the full cell and comes from the power source, and does not have to be confused with the leakage current that usualy only takes into account the current leaked by the pass gates from the bit-lines.

The read energy corresponds to the discharge of the bit-lines, even if there are sense amplifiers and during a read operation they only need to be discharged ΔV_{sa}, this value is obtained for the cell with the lowest read current, while for power consumption the mean read current has to be considered. In addition depending on the sense amplifier we use it can lead to completely discharge the bit-line, then we will consider this worst case leading to the following formula:

$$E_{read} = N_{cols} \cdot N_{rows} \cdot C_{bl,cell} \cdot Vdd^2 \quad (2)$$

The write energy is mainly the energy required to discharge the bit-line capacitance to the required voltage:

$$E_{write} = N_{cols} \cdot N_{rows} \cdot C_{bl,cell} \cdot |Vdd - Vss|^2 \quad (3)$$

where N_{cols} is the number of columns of the array being written and N_{rows} is the number of rows of the memory array. $C_{bl,cell}$ is the contribution of one cell to the capacitance of the bit-line. Vss is the power source used to discharge the bit-line that has to be lower than the required write trip point; it can be zero if the cell works as a standard memory or negative if it requires negative bit-line write assist. The formula comes as the addition of the energies required to discharge the bit-line capacitance from Vdd to Vss with the Vss power source and the one required to charge it again from Vss to Vdd with the Vdd power source. In addition to this energy the cells that are not written are activated and discharge their bit-lines just as if they were being read, this value has to be added. Its calculation is the same as the one for the read energy but taking only the number of columns that are not being written.

Only the energies due to the charge and discharge of the bit-lines are considered. Given that during both read and write accesses a large number of bit-lines are switched and only the capacitance of one word-line is switched, the contribution of the word-line to the total energy is negligible.

2.3.2 Delay

We consider the memory delay will be limited by the read access since it corresponds to a single cell discharging a huge capacitance, while the discharge of that capacitance is not a problem during the write access since it is done by peripheral circuitry. Nevertheless this assumption will be verified once the final memory cells are selected.

The read access time corresponds to the time required by the slowest cell to discharge ΔV_{sa} of the voltage of the bit-line:

$$T_{read} = \frac{N_{rows} \cdot C_{bl,cell} \cdot \Delta V_{sa}}{I_{read}} \quad (4)$$

where I_{read} is the read current of the slowest cell, that is the current at which the pull-down discharges the bit-line through the pass-gate, it is measured at the beginning of the discharge. This approximation assumes the discharge of a capacitor by a constant current, it can be applied as long as the discharge ΔV_{sa} is small with respect to the supply voltage.

The delay estimated by the time it takes to discharge the bit-line represents only the contribution of the cell array and it is used as an estimation of the other factors affecting the delay. While the delay introduced by the periphery —time to decode the address, time to charge the word-line as well as the delay introduced by the sense amplifiers— is also an important part of the total delay of the memory, this delay presents a smaller spread and can be tuned independently. Our target is to keep the array delay below 50% of the aimed clock cycle.

3. METHODOLOGY

3.1 Variability Modelling

While a nominal memory cell may meet time, power and stability constraints, some will fail after considering intra-die variability affecting the individual transistors of the cell. Our target is to obtain stable cells in both write-ability and read stability metrics up to 5 standard deviations (sigmas), which ensures more than a 90% yield for the 256-Kbit memory array we are using as a benchmark. Other common figure for the yield is 95% but a relaxed target of 90% seems to be reasonable for a predictive technology. We assume that the stability metrics present a normal distribution[11].

Although the variability of the transistors is a consequence of physical parameters such as roughness of the edges of the gate or of the active zone, we are only interested on how they translate to the electrical response of the transistor. Two main parameters are affected: the threshold voltage of the gate and the drain-source current. Intra-die variability is thus simulated by injecting threshold voltage and drain-source current variations in all transistors [8]. In the test circuit this is achieved with a voltage generator in the gate and a current source between the drain and the source, dependent on the current of the nominal device between those two nodes of the cell, as shown in figure 1. This injector-based method allows us to abstract the physical parameters

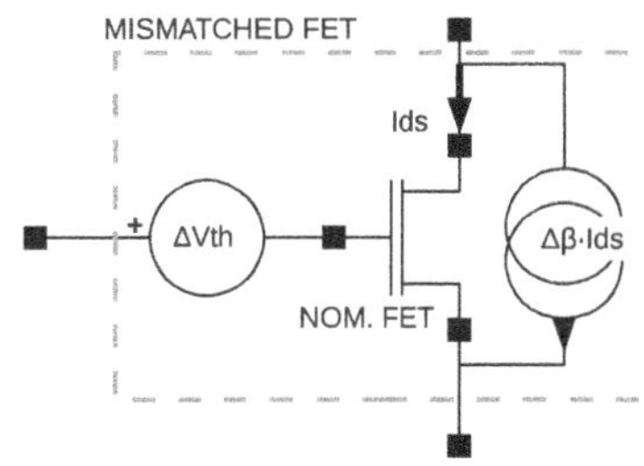

Figure 1: Voltage and Current generators used to simulate the variability of a transistor.

variations that are behind the variability observed at an electrical level.

For the value of threshold voltage variation we have assumed a variability coefficient $A_{vt} = 1mV\mu m$ which seems to be a converging limit of future technologies [14]. The voltage value is computed depending on the transistor size (Pelgrom's Rule [12]) while the active current mismatch is computed according to the correlation presented in [9].

We have explored 6T SRAM cells using the predictive technology models for low stand-by power 14nm FinFET provided by Arizona State University [1]. This technology presents a fin height of $23nm$ with a fin width of $10nm$, that throws fin effective width of $56nm$ for a length of $18nm$. The transistor area required to infer the mismatch if computed taking as the effective width of a 1fin-transistor the sum of two fin heights and one fin width[5], as a result the area of one fin is:

$$\sqrt{A_{fin}} = \sqrt{L_{fin} \cdot (2H_{fin} + W_{fin})} = 31.75nm \quad (5)$$

3.2 Design Space Exploration

We have explored different cells whose main difference is the number of fins of the pass-gate, pull-up and pull-down transistors. Among all the cells we have retained the configuration with 2 pull-up, 3 pass-gate and 3 pull-down fins. Even if cells with other fin configurations presented better nominal performance, the retained cell presents less variability thanks to a larger area of the transistors. Smaller layouts with less fins in the pass-gate pull-down path still complied with read and write metrics but the spread in the read delay was such that the 1GHz target was not met for low supply voltages. For smaller cells with relaxed speed constraints the power consumption conclusions seemed to go in the same direction as the selected layout.

Despite a fully-depleted channel the V_{th} of the transistors can still be tuned for FinFET technology using the gate work function [7]. The impact of V_{th} tuning was simulated by adding a constant voltage source at the gate. This is done in addition and independently to the dV_{th} injected to model the variability of the transistors.

The cell topology (number of fins) and the threshold voltage of the transistors are thus the design parameters explored, every combination has been simulated for a voltage supply ranging from $100mV$ to $900mV$ with a *fast SPICE* simulator including 150 points Monte-Carlo mismatch simulations to obtain a fast estimation of the spread of the metrics we use. The resulting cells were first checked to pass the stability test with a 5 sigmas confidence. That is, if for a particular metric $\mu_{metric} - 5\sigma_{metric} < th_{metric}$ the

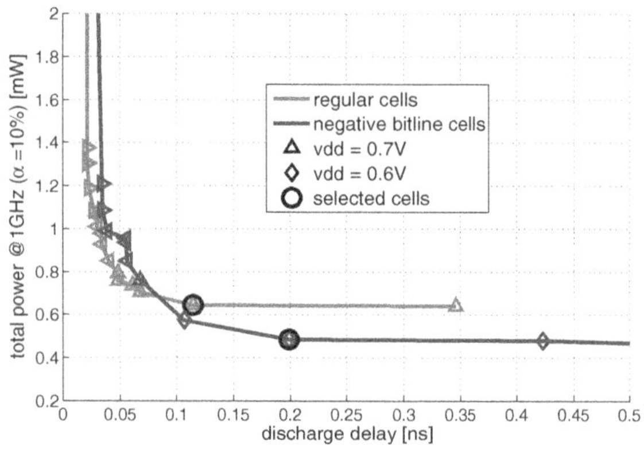

Figure 2: Candidate cells presenting an optimal delay-energy tradeoff. The shape of the markers represents the main supply voltage of the cells. The two cells chosen for comparison are circled.

cell does not comply with that metric. The th_{metric} is equal to zero in the case of the static noise margin, for the write trip point it can be zero for a normal working cell, or negative equal to the value to which the bit-line is discharged, if we consider negative bit-line write enhancement. They were filtered according to stability as belonging to two categories:

1. Cells that complied with both static noise margin and write trip point (zero threshold) were able to work as regular memory cells.
2. Cells that complied with static noise margin but not with write trip point were allowed to use a negative bit-line voltage if this voltage was still above the hold trip point, which means the negative bit-line voltage would not disturb the contents of the non-accessed cells. Those cells were able to work using negative bit-line write enhancement.

From the two categories only the cells presenting an optimal delay-energy balance were kept (shown in figure 2).

The results show that negative bit-line does not bring any improvement for supply voltages at which there are already some cells able to work, since the power consumption is increased by the larger bit-line swing during write operation as shown in equation 3. On the other hand some cells using negative bit-line techniques are able to work at supply voltages lower than any standard-working cell could achieve. This is shown in figure 2 as the curve for regular memory cells ends with Vdd values of $0.7V$ while the negative bit-line cells can go down to $0.6V$.

Some of the negative bit-line cells compensate the increase in power consumption due to the second power source by a more important power saving due to the decrease of the main supply voltage. Lowering the main supply voltage reduces the read energy while its impact on the write energy depends on whether the voltage difference between Vdd and Vss is increased or not.

Table 1 shows the two retained cells, a regular one, and another using negative bit-line write enhancement. The table summarizes the supply voltage at which the cell operates (Vdd), the negative voltage required in the bit-line if so (Vss), and the threshold voltage tuning of transistors in the

Table 1: Selected cells.

cell	regular cell	negative bit-line cell
Vdd	700 mV	600 mV
Vss	0	-50 mV
ΔVt_{pu}	0	0
ΔVt_{pg}	0	0
ΔVt_{pd}	100 mV	0

cell (ΔVt_{trt}). Detailed results are presented in the results section.

3.3 Read current tail estimation

The cells discharge the bit lines with a read current also affected by variability. Thus some of the cells may not meet the memory speed constraints. As a consequence it is important to model the read current distribution.

For read currents close to the mean the read current seems to follow a normal distribution but as soon as we look for values at 5 sigmas the tails of a normal distribution would throw negative current values which are impossible. This problem increases as we use low voltage supplies and high threshold voltage transistors. The reason why this happens is because at low read currents the pass gate transistor switches to the sub-threshold region so the tails of the distribution do no longer follow a normal distribution but a log-normal distribution instead.

The tail of the distribution will be in the logarithmic region while simulation data are in the linear region or in the transition and therefore cannot be used to find out the parameters of the log-normal distribution. This is solved by using the response curve of the transistors that discharge the bit-line to translate the read current to an equivalent gate voltage that does follow a normal distribution for a larger range of standard deviations, since the current and threshold voltage injectors can be translated to a gate voltage shift in the sub-threshold region. Then we find the gate voltage 5 sigmas away for that equivalent distribution, and finally take the read current corresponding to this gate voltage. The process is shown in figure 3 comparing the read currents we obtain using read current normal distribution and equivalent gate voltage distribution. The results we obtain using this method are almost independent from the V_{ds} value we use to obtain the transfer curve: halving V_{ds} leads to a change in the inferred read current below 1%. For our results we have used a V_{ds} equal to the main supply voltage and simulated the transfer curve of the pass-gate and pull-down transistors in series.

4. RESULTS

The two retained cells have gone through a more accurate 5000-point Monte-Carlo mismatch simulation. We use the results of this simulation to obtain the stability and performance metrics that we present in this section. We have assumed a 512x512-bit square memory array, all the 512 rows would be accessed during a read operation —512 bit-lines are discharged even if not all the bits are actually read— while 128 bits are written during a write operation and the remaining 384 bits are read as their cells are activated even if the results are not used. The sense amplifier design is assumed to be such that the required bit-line swing is $\Delta V_{sa} = 100mV$.

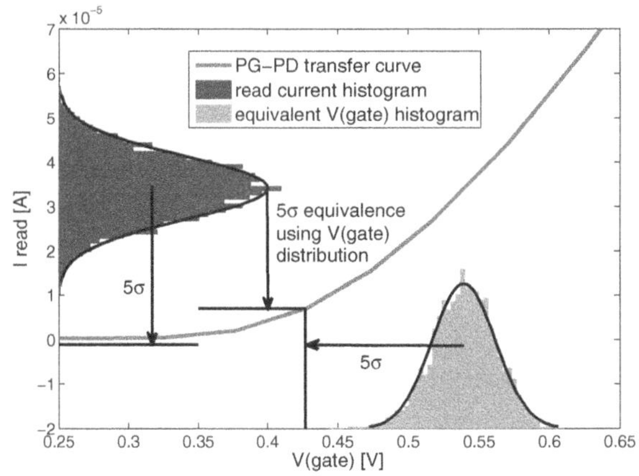

Figure 3: Transfer curve (Ids vs Vg) of the pass gate transistor used to find the equivalent Vg distribution from the read current distribution.

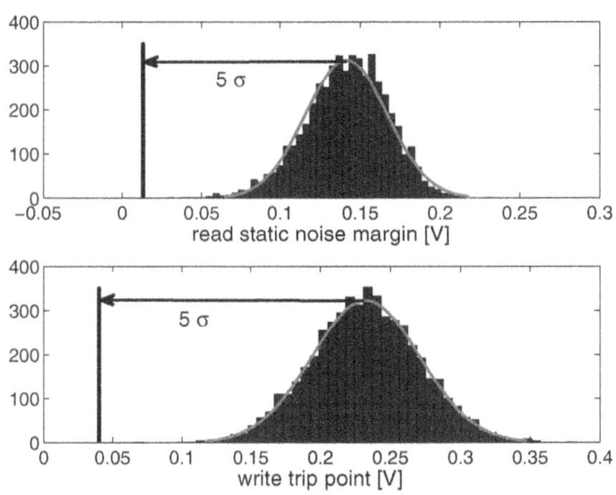

Figure 4: Read and Write Stability metrics for the conventional cell.

4.1 Stability Results

The regular cell presents positive metrics with 5 sigma margins for both read and write margins as shown in figure 4 which ensures that the cell is stable during read and write and that our array can be built with a sufficient yield.

The stability while the cell is only retaining the data is ensured by the hold static noise margin whose plot is omitted since it is less restrictive than the read static noise margin.

The cell using negative bit-line write-assist also presents a read static noise margin 5 sigmas above zero as shown in figure 5, but the write trip point constraint is no longer met for normal operation. We can see in figure 6 that in order to ensure the write-ability of the cell with a 5 sigma margin, a negative voltage is required in the bit-line. Using this negative voltage in the bit-line while keeping a zero voltage in the word-line of the non-accessed cells does not disturb their contents since the negative voltage is far above the hold trip point with a 5 sigma margin. The space between the 5 sigma limits of the two metrics represents the range of voltages that are safe to use in the bit-line, in our case we take a Vss value of $-50mV$.

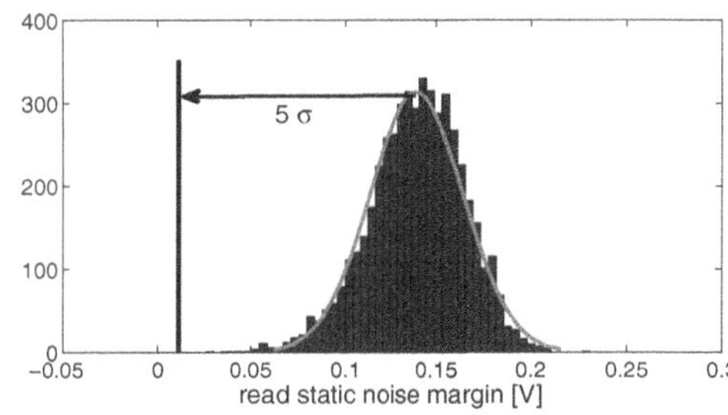

Figure 5: Read Static Noise Margin of the negative bit-line cell.

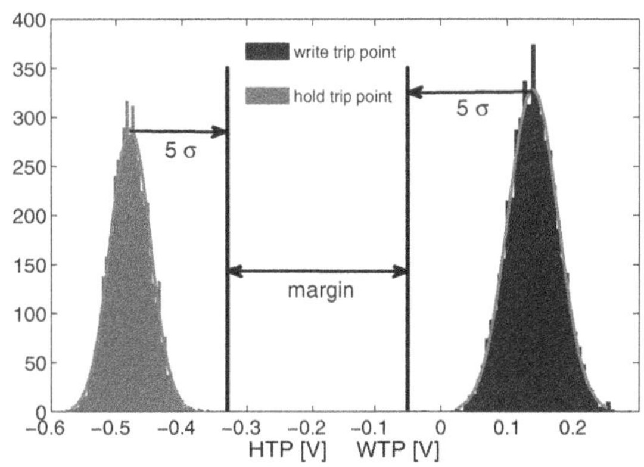

Figure 6: Write and Hold trip points for the negative bit-line cell.

4.2 Performance results

The speed performance is determined by the read current at which the cells are able to discharge the bit-line during a read access.

Using the method presented in section 3.3 we obtain a read current at 5 sigmas margin of $6.9\mu A$ for the regular cell instead of the over-pessimistic value of $-1\mu A$ we would have obtained assuming a pure normal distribution of the read current or the over-optimistic value of $11\mu A$ for a pure log-normal distribution. For the cell using negative bit-line we obtain a read current of $6.8\mu A$ instead of $2\mu A$ or $10\mu A$ respectively. This results on a delay of $0.36ns$ for the regular cell and $0.37ns$ for the negative bit-line cell. This meets the requirement of the discharge delay being less than 50% of the aimed delay of the memory. Both cells would be able to make the memory work at the target frequency of 1GHz.

Even if the capacitances of the bit-lines are the same in the two cells the read and write energies are reduced for the cell with negative bit-line write improvement since the main supply voltage is reduced as well as the difference between Vdd and Vss. For the write energy we obtain a reduction from $6.40pJ$ to $4.91pJ$ and a reduction from $6.40pJ$ to $4.70pJ$ for the read energy.

Two factors affect the standby power in a different direction, while the supply voltage lowering tends to decrease the standby power, the fact that transistors with lower threshold voltage are required for negative bit-line cell increases the leakage current when the cell is not accessed. In the end the

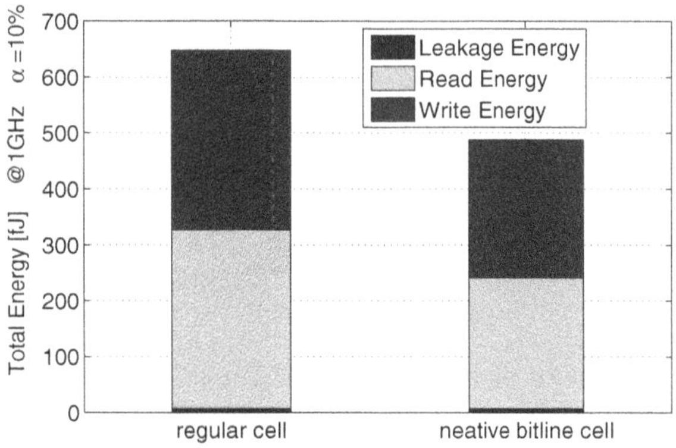

Figure 7: Energy comparison of the two memories.

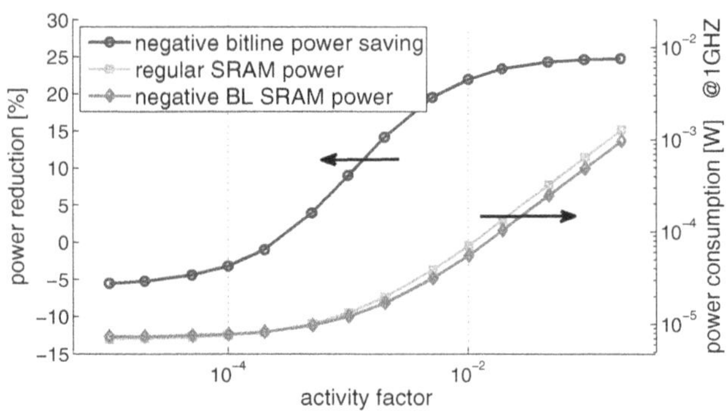

Figure 8: Power consumption of the memories (right) and negative bit-line improvement over regular memory (left) depending on the activity factor.

leakage power is $6.9\mu W$ and $7.3\mu W$ for the the full memory array using regular and negative bit-line cells respectively.

Considering a 10% activity factor, shared equally by write and read accesses, at a frequency of 1GHz we obtain a power consumption of $647\mu W$ for the regular memory array and of $488\mu W$ for the memory using negative bit-line which represents the 25% improvement shown in figure 7.

As figure 8 shows, despite having a lower voltage supply, the negative bit-line cell keeps presenting better energy consumption figures for activity factors below 10^{-3}.

5. CONCLUSION

The negative bit-line write enhancement has shown to be able to increase the stability of the 6T-SRAM memory cells using predictive FinFET technology beyond $22nm$. This allows negative bit-line cells to work at lower main supply voltages than regular memory cells.

We have introduced a new metric *hold trip point* that tells us if a cell is able to work with negative bit-line values and how low we can discharge it keeping the other cells stable. We have also implemented a new method to estimate the read current distribution.

The delay-energy design space has been explored showing how negative bit-line biasing presents benefits in the low power region while it is unable to bring improvements in the high speed region, corresponding to higher voltages, at which there are already regular cells able to work.

The two retained cells have shown that negative bit-line cells are able to maintain the speed performance while the main supply voltage is reduced. This results on lower read and write energies. The reduction on the power consumption is effective even for activity factors as low as $2.5 \cdot 10^{-4}$.

6. ACKNOWLEDGEMENTS

This work was funded by the CICYT project TOLERA TEC2012-31292 of the Spanish Ministry of Economy and Competitiveness.

The authors would like to thank Stefan Cosemans and Paul Zuber from IMEC Belgium for their valuable feedback.

7. REFERENCES

[1] `http://ptm.asu.edu/`.

[2] K. Agarwal and S. Nassif. Statistical analysis of SRAM cell stability. In *DAC 2006*.

[3] V. Chandra, C. Pietrzyk, and R. Aitken. On the efficacy of write-assist techniques in low voltage nanoscale SRAMs. In *DATE 2010*.

[4] G. Chen, D. Sylvester, D. Blaauw, and T. Mudge. Yield-driven near-threshold SRAM design. *Very Large Scale Integration (VLSI) Systems, IEEE Transactions on*, 2010.

[5] H. Dadgour, K. Endo, V. De, and K. Banerjee. Grain-orientation induced work function variation in nanoscale metal-gate transistors ;part II: Implications for process, device, and circuit design. *Electron Devices, IEEE Transactions on*, 2010.

[6] A. Goel, R. Sharma, and A. Gupta. Process variations aware area efficient negative bit-line voltage scheme for improving write ability of SRAM in nanometer technologies. *Circuits, Devices Systems, IET*, 2012.

[7] M. Jurczak, N. Collaert, A. Veloso, T. Hoffmann, and S. Biesemans. Review of finfet technology. In *SOI Conference, 2009 IEEE International*.

[8] P. Kinget. Device mismatch and tradeoffs in the design of analog circuits. *Solid-State Circuits, IEEE Journal of*, 2005.

[9] M. Miranda, P. Zuber, P. Dobrovolny, and P. Roussel. Variability aware modeling for yield enhancement of sram and logic. In *DATE'2011*.

[10] S. Mukhopadhyay, R. Rao, J.-J. Kim, and C.-T. Chuang. SRAM write-ability improvement with transient negative bit-line voltage. *Very Large Scale Integration (VLSI) Systems, IEEE Transactions on*, 2011.

[11] H. Park and S. e. a. Song. Accurate projection of vccmin by modeling dual slope in finfet based SRAM, and impact of long term reliability on end of life vccmin. In *Reliability Physics Symposium (IRPS), 2010 IEEE International*.

[12] M. Pelgrom, A. Duinmaijer, and A. Welbers. Matching properties of mos transistors. *Solid-State Circuits, IEEE Journal of*, oct 1989.

[13] E. Seevinck, F. List, and J. Lohstroh. Static-noise margin analysis of MOS SRAM cells. *Solid-State Circuits, IEEE Journal of*, oct 1987.

[14] O. Weber and O. e. a. Faynot. High immunity to threshold voltage variability in undoped ultra-thin FDSOI MOSFETs and its physical understanding. In *IEDM 2008*.

Virtual Register Renaming: Energy Efficient Substrate for Continual Flow Pipelines

Komal Jothi and Haitham Akkary
Department of Electrical and Computer Engineering
American University of Beirut
Beirut, Lebanon
+961 1 350 000
{kmj04, ha95}@aub.edu.lb

ABSTRACT

Continual Flow Pipelines (CFP) allow a processor core to process instruction windows of hundreds of in-flight instructions while keeping its cycle critical scheduler and register file small. CFP defers the execution of instructions that depend on cache misses, moving these instructions to a single ported SRAM buffer outside the pipeline, and continues execution of instructions independent of the cache miss until the miss data is fetched into the cache. This way, the CFP processor core avoids pipeline execution stalls due to cache misses while keeping multi-ported power hungry structures in the pipeline small. However, CFP cores have to wake up and replay the deferred miss dependent instructions through the execution pipeline, again, when the miss data arrives, causing these instructions to increase the circuit activity of the execution pipeline and consequently core energy consumption. In this paper, we present and evaluate virtual register renaming as a substrate for CFP cores that significantly shortens the replay loop and reduces the circuit activity of deferred miss dependent instructions, thus increasing the energy efficiency of Continual Flow Pipelines architectures.

Categories and Subject Descriptors

C.1.1 [**Processor Architectures**]: Single Data Stream Architectures – *pipeline processors.*

General Terms

Performance, Design.

Keywords

Superscalar Processors, Instruction Level Parallelism, Continual Flow Pipelines, Latency Tolerant Processors, Virtual Register Renaming.

1. INTRODUCTION

Power constraints on future multicore processors pose a difficult challenge to processor architects: how to provide energy efficient single-thread performance for applications that are hard to parallelize, while placing on a single die as many cores as possible for high throughput performance.

Improving single-thread performance by increasing the capacity of instruction buffers (reorder buffers, reservation stations, and load and store queues) and the size of on-chip caches come at the expense of area, power, and complexity. Larger associative caches and multi-ported, timing critical instruction buffers have become increasingly difficult to design while maintaining high clock rates.

Continual Flow Pipeline architectures [10][11][13][19] reduce the impact of data cache misses on performance without having to increase instruction buffers and physical register files sizes. These architectures have common characteristics, but vary in implementation details. They all break away from conventional reorder buffer mechanisms for managing speculative out-of-order execution and handling exceptions and branch mispredictions [18]. Instead of using a reorder buffer for tracking, buffering and sequentially committing one by one the current set of in-flight instructions, also called the instruction window, these architectures use bulk commit of execution results, and register state checkpoints to recover from branch mispredictions and exceptions [1][12][15]. Instead of residing completely in the hardware instruction buffers, the instruction window becomes virtual and composed physically of a partial, discontiguous subset of the conventional and sequential instruction window. As such, instructions that do not encounter cache misses, called miss-independent instructions, enter the instruction window, execute and leave quickly, freeing all their pipeline resources. On the other hand, instructions that depend on data cache misses, called miss-dependent instructions, temporarily leave the reorder buffer and reservation stations and enter a waiting buffer outside the execution pipeline, where they wait as long as necessary for their input operands without blocking execution. Afterwards, when the miss data cache block is fetched, the miss-dependent instructions are replayed from the waiting buffer, i.e. renamed and re-inserted again into the reservation stations to be scheduled for execution.

Previous Continual Flow Pipeline architectures that target tolerating cache misses at all levels of the cache hierarchy require two waves of miss dependent instructions processing through the pipeline. In the first wave, all miss dependent instructions are poisoned and then scheduled for pseudo execution, writeback, and finally pseudo-retirement in the reorder buffer from which they move to the waiting buffer. In the second execution wave, miss dependent instructions are replayed from the waiting buffer when the miss data is loaded into the L1 data cache. During this replay wave, the miss dependent instructions are fetched from the waiting buffer, allocated instruction buffer entries, renamed using a separate dedicated register renaming map table, and then re-enter the reservation stations where they are scheduled for execution. Finally, after they execute, they write back their results to the reorder buffer which retires them in program order into the retirement register file (RRF).

Since load misses to the L1 data cache are frequent, a significant percentage of instructions in typical programs are dependent on load misses and are therefore moved from the CFP core pipeline into the waiting buffer and then later replayed. This causes a significant increase in energy required to process miss dependent instructions and subsequently reduces the previously claimed power advantage of CFP small instruction buffers [19].

1.1 Paper Contribution

This work presents and evaluates a Continual Flow Pipeline core architecture with virtual register renaming (VRR-CFP) using the Simplescalar simulation tools (www.simplescalar.com) and benchmarks from Spec 2000 and Spec 2006. Compared to CFP without virtual register renaming, our simulation results show significant reduction in circuit activity when processing instructions that depend on loads that miss the L1 data cache, including those that miss all the way to DRAM.

The simulated VRR-CFP architecture is based on a simultaneous continual flow pipeline (Sim CFP) [13]. Like simultaneous CFP, it uses a set of reservation stations (RS) sized appropriately to handle the common case of instructions that hit the L1 data cache, augmented with a single-ported SRAM waiting buffer for instructions that miss the L1 data cache. However, unlike simultaneous CFP, which pseudo executes and pseudo retires dependent instructions in the reorder buffer before moving them to the waiting buffer, VRR-CFP uses an order list of instructions in the reservation stations to move miss-dependent instructions from the reservation stations to the waiting buffer whenever their allocated buffer entries are needed to unblock the execution pipeline. VRR-CFP improves performance and reduces circuit activity over previous CFP architectures due to three advantages:

- VRR-CFP keeps miss dependent instructions in the reservation stations as long as possible after the miss before they are evicted to the waiting buffer. This reduces the number of miss dependent instructions that are replayed in case of medium latency load misses, which are those loads that miss the L1 data cache but hit the L2 data cache.
- VRR-CFP removes the reorder buffer and the register renaming stages from the replay loop.
- On previous CFP architectures, all load miss dependent instructions have to be moved into the waiting buffer and then replayed. This is necessary since when the miss load pseudo retires and moves into the waiting buffer, it releases its renamed physical destination register. This breaks the dependence links between the miss load and its dependents, requiring the full dependent instructions sequence to be replayed and renamed again to re-establish the dependence links. VRR-CFP renames registers using virtual register IDs that are persistent throughout the life time of the dependent instructions. Moving the miss load or any dependent instructions into the waiting buffer does not break the dependence links between instructions, thus allowing partial replay of the load and its dependents. This reduces significantly the number of replayed instructions, execution time and the energy consumed by dependent instructions.

The rest of this paper is organized as follows. Section 2 presents an overview of VRR-CFP architecture. Section 3 outlines the simulation methodology and simulated machine configurations. Results are presented in section 4. Section 5 discusses related work. Finally, the paper concludes in Section 6.

2. CFP WITH VIRTUAL REGISTER RENAMING

In his section, we describe the core architecture of CFP with virtual registers renaming (VRR-CFP), its key mechanisms, and its differences from previous continual flow pipeline architectures.

2.1 VRR-CFP Architecture Overview

Fig. 1 shows a block diagram of the VRR-CFP core. VRR-CFP core microarchitecture uses Tomasulo's algorithm [20] and reservation stations to execute instructions out-of-order.

Like all other superscalar architectures, VRR-CFP uses a reorder buffer to commit instructions and update architecture register and memory state in program order. However, VRR-CFP does not use the reorder buffer for register renaming. Instead, it performs register renaming using virtual register IDs generated by a special counter. These virtual register IDs are not mapped to any fixed storage locations in the core, and therefore can be large in number and allocated to instructions throughout their life time, including miss dependent instructions evicted to the waiting buffers. Since the virtual register IDs are plentiful and are not associated with actual physical storage, VRR-CFP does not run the risk of execution pipeline buffer-full stalls resulting from miss dependent instructions holding on to their renamed registers for a long time while waiting for the long latency load miss.

The virtual register ID counter is finite in size and cannot be allowed to overflow for correctness reason. VRR-CFP architecture opportunistically resets the virtual register ID counter whenever it can, e.g. when the pipeline is flushed to recover from a mispredicted branch. Otherwise, a pipeline stall and drain is forced to reset the counter when it overflows. With 10-bit virtual ID counter, our simulations show that the impact on performance of forced pipeline stalls to reset the counter is negligible.

Virtual register renaming gives VRR-CFP a significant advantage over previous CFP architectures. Past CFP architectures require all miss dependent instructions to be replayed and renamed again to re-establish dependence links, which is necessary for the reservation stations to re-dispatch the miss dependent instructions in correct data flow order. In contrast, since the virtual register renaming IDs are permanent from the time the miss dependent instructions are renamed until they execute and commit, VRR-CFP can do partial replay of dependent instructions. What this means is that if the load miss data is fetched from memory after the load is moved to the waiting buffer but before its dependents have been moved, VRR-CFP simply replays the load, saving significant delay that would be encountered if all the miss dependent instructions that are still in the reservation stations had to be replayed through the waiting buffer to be renamed again.

2.1.1 Miss Independent Execution

Like previous CFP cores, The VRR-CFP core is capable of executing cache miss dependent and independent instructions concurrently in the pipeline, supported by two retirement register file contexts (RRF), one for retiring miss independent instructions and the other for retiring miss dependent instructions.

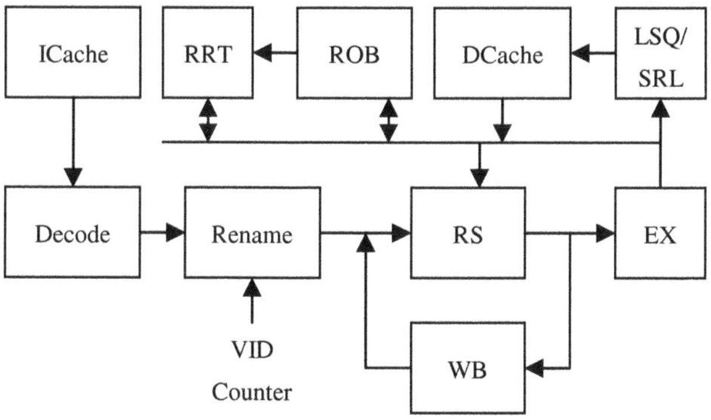

Fig. 1. VRR-CFP Block Diagram

In VRR-CFP, execution initially starts using a retirement register file context (RRF) that we call the independent retirement register file. When an L1 data cache load miss occurs, a poison bit is set in the destination reorder buffer entry of the load. Load-dependent instructions in the reservation stations (RS) capture the poison bit from the common write back data bus. They are then woken up, as if the load completed, and are scheduled by the reservation stations control logic for pseudo-execution. Pseudo-execution of poisoned instructions does not actually use any execution units. However, pseudo-execution consumes RS dispatch ports and writeback bus cycles to propagate poison bits through instruction dependences and to identify all instructions in the reservation stations that depend on the load miss data. After pseudo-execution, miss dependent instructions stay in their reservation stations until they are waken up for real execution when the load miss data arrives, or until they are moved into the waiting buffer in case their resources are needed to unblock the execution pipeline and execute miss independent instructions.

Poison bits propagate through instruction dependences, and identify all instructions that depend on the load miss and their descendants including memory operations. Wherever register data resides or flows in the core, poison bits extend and qualify the data. For this purpose, VRR-CFP has new poison bits added to the reservation stations, writeback buses, bypass network, reorder buffer entries (ROB), retirement register file (RRF) and store queue entries (SQ).

2.1.2 Replay Loop and Miss Dependent Execution

Fig. 1 shows the reduced replay loop in VRR-CFP consisting of two stages: the reservation stations (RS) and the waiting buffer (WB). The waiting buffer basically acts as a second level storage for the reservation stations. With virtual register renaming, entries can be freely evicted from the RS to the WB and then loaded back again to the RS to be scheduled for execution at a later time.

In VRR-CFP, miss dependent instructions are evicted from the reservation stations to the waiting buffer only when their buffer resources are needed to unblock the execution pipeline. Therefore, when a miss is processed and its data fetched to the L1 data cache, the miss dependent instructions may still be in the reservation stations. In this case, when the data is written back, the miss data is captured by all the reservation stations that have instructions that depend on the miss. The captured writeback data sets the source operand "ready" state bits and clears the poison bits of the dependent reservation station entries, making the instructions in these entries ready for scheduling and dispatch to the execution units.

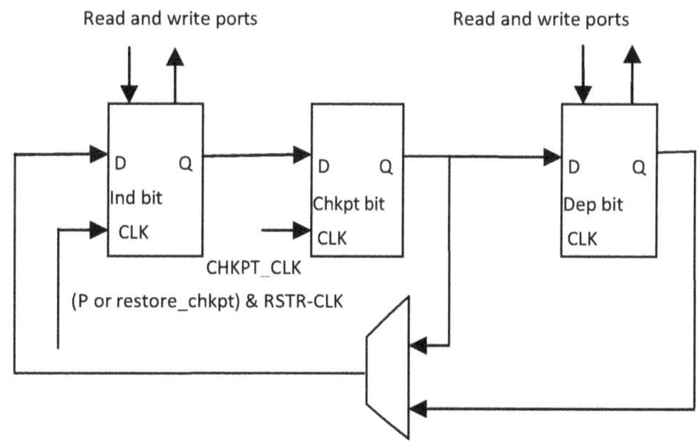

Fig. 2. Register File Cell with Checkpoint and two Contexts Bits

Evicting miss dependent instructions to the waiting buffer on resource need basis significantly reduces the number of replayed instructions, especially in the case of medium latency load misses, which are those loads that miss the L1 data cache but hit the on-chip L2 cache.

In case of a load miss that goes all the way to DRAM, it is often the case that the long miss latency causes the instruction buffers to fill up and some or all of the in-flight miss dependent instructions to evict to the waiting buffer. When the load miss is serviced, the miss load and its dependents are re-inserted from the waiting buffer back to the reservation stations, from which they are scheduled for execution.

Notice that even though replayed miss dependent instructions do not need to be renamed again, they sometimes, as in Tomasulo's algorithm, need to read source operands that have already been computed and retired from the ROB to the register file (RRF). State bits that track whether the last instructions to write logical registers have been retired are stored in a special storage structure. These state bits are checked during replay to determine if the operands are ready in the RRF when instructions are moved from the WB to the RS.

2.1.3 VRR-CFP Reservation Stations

VRR-CFP uses a centralized array of conventional data-capture reservation stations [16]. Each reservation station entry is extended with a poison bit per source operand and L1-DCache-miss bit. The L1-DCache-miss bit is set to 1 if the entry contains a load instruction that has missed the L1 data cache. We say an instruction is poisoned if one of its source poison bits or the L1-DCache-bit is set to 1. A source operand of an instruction is poisoned if and only if it is the destination of another poisoned instruction. In other words, the poison bits propagate the dependences from L1 data cache misses to later instructions in the program to identify instructions that may encounter long data cache miss delays. These instructions are candidates to move to the waiting buffer to avoid pipeline stalls that could occur if any of the reservation stations, reorder buffer load queue or store queue arrays becomes full.

The reservation stations array is augmented with a free list and an order list. VRR-CFP uses the free list to track the occupancy of the reservation stations. The order list tracks the allocation order of the reservation stations, which matches the program order of the instructions, since reservation stations are allocated in the rename pipeline stage. The RS order list could be implemented as part of the reorder buffer by adding the reservation station ID of each instruction to its allocated reorder buffer entry. It also could be implemented as a special array separate from the ROB.

Four conditions are checked to determine if an instruction should be moved to the waiting buffer: 1) the instruction is at the head of the RS order list, 2) the instruction is poisoned, 3) one of the RS, reorder buffer, load queue or store queue arrays is full, and 4) every source operand of the instruction is either poisoned or ready. The last condition ensures that the miss dependent instructions carry their non-poisoned input values with them when they are replayed, since there is no guarantee that these values would not be overwritten in the register file by replay time.

In addition, each reservation station has a state bit that indicates if the instruction in the entry has been replayed once before, i.e. it has been moved earlier in time to the waiting buffer and then back to the RS array. Each reservation station also contains a load miss identifier, in case it has a load instruction that misses the data cache. An implementation could use for this purpose the ID of the L1 data cache fill buffer used to handle the load miss.

2.1.4 Waiting Buffer

The waiting buffer is a wide single ported SRAM array managed as a circular buffer using head and tail pointers. Miss dependent RS entries at the head of the RS array moves to the tail of the waiting buffer when any of the instruction buffers fills up due to data cache misses. When a data cache miss is completed, VRR-CFP replays the miss dependent entries by loading them back from the head of the waiting buffer to the tail of the RS. Ideally, the width of the two buses connecting the RS and the waiting buffer would match the pipeline width. Narrower interconnect can also be used, trading some performance for reduced hardware complexity.

A key to the efficiency of VRR-CFP large window implementation is the fact that the waiting buffer has no CAM ports, connections to write back buses for capturing data operands or conventional ready/schedule logic. All these functions are handled in the RS array after the data cache miss completes and the miss dependent instructions are replayed. Therefore, the waiting buffer array can be designed using non-tagged SRAM and made significantly larger than the RS array at much lower area and power cost than if the RS array is enlarged to completely hold the large instruction window.

In order to wake up miss dependents from the waiting buffer and replay them, the L1 data cache fill buffer handling a load miss has to receive and save the waiting buffer ID of its load miss. When the miss is completed, VRR-CFP replays the load miss and its dependents in program order, as described earlier, from the head of the waiting buffer back into the RS allocate/write stage of the execution pipeline. These replayed instructions do not need to be renamed again. Their virtual register renames are still valid, thus can be used by the RS to schedule these instructions and to grab their results from the writeback bus into the reservation stations of any dependent instructions, including any instructions that have not been replayed but still waiting in the RS.

2.1.5 Register File and Results Integration

VRR-CFP uses the reorder buffer to recover from miss independent branch mispredictions and to handle exceptions incurred by miss independent instructions. On the other hand, it uses checkpoints to handle branch mispredictions or exceptions encountered by miss dependent instructions.

Like prior CFP proposals, VRR-CFP has a specialized register file for checkpointing register state at the load miss, for later use to handle miss dependent branch mispredictions and exceptions. The register file also has special logic for integrating the results of independent and dependent instructions and to restore precise register state after all miss dependent instructions execute.

Fig. 2 shows VRR-CFP retirement register file cell with checkpoint flash copy support. VRR-CFP uses a flash copy of the RRF for creating checkpoints. In one cycle every independent RRF state bit (leftmost latch) is shifted into a checkpoint latch within the register cell (center latch). The register file can be restored from the checkpoint in one cycle by asserting RSTR_CLK.

In addition to the checkpoint bit and the independent RRF context bit, VRR-CFP register file cell contains one context bit for the dependent RRF state (rightmost latch). The independent and dependent instruction results integration is a special restore cycle. At the end of dependent execution and when all instructions in the WB have replayed and retired, the RRF has all the live-out registers, some computed by the independent instructions and some computed by the dependent instructions, as determined by the poison bits in the RRF. To integrate these results back into one context, a restore cycle is performed from the dependent context into the independent context. However, not all registers are copied. Fig. 2 shows that only poisoned registers are copied by using the poison bits to enable the clock of the copy operation. A 2-to-1 multiplexer in the cell restores either the checkpoint bit or the dependent bit during a RSTR_CLK cycle.

2.1.6 Handling Multiple Load Misses

The waiting buffer could contain at any time multiple load misses and their dependents, stored in program order. Miss load wakeup and replay from the waiting buffer is done in program order, even though miss loads wakeups could arrive to the WB out of order. If a load miss occurs after it is moved to the waiting buffer and then back to the RS, the load stalls until the data is fetched into the L1 data cache. In other words, dependent instructions do not enter the waiting buffer more than once. As a consequence, pointer chasing code that encounter dependent misses do not fully benefit from the VRR-CFP scheduling window hierarchy. However, we have chosen this approach to keep hardware simple.

2.1.7 Load and Store Execution in VRR-CFP

To maintain proper memory ordering of loads and stores from the independent and dependent instructions execution, VRR-CFP, like previous CFP proposals, uses load and store queues (LSQ), a Store Redo Log (SRL) [7] and a store-set memory dependence predictor [3]. All stores, dependent and independent, are allocated entries (and IDs) in the SRL in program order at the rename stage of the pipeline. Every load, dependent or independent, carries the SRL ID of the last prior store. The SRL IDs assigned to loads and stores are unique and determine the order of memory instructions.

VRR-CFP performs load-store memory dependence prediction in the front end of the pipeline with a store-set predictor [3] to determine the store on which a load may depend. The predictor uses SRL IDs in the prediction and writes store poison bits in the SRL, thus allowing propagation of poison bits from stores to predicted-dependent loads.

In order to support concurrent, speculative execution of dependent and independent loads and stores, VRR-CFP L1 data cache has two new states: Speculative Independent (Spec_Ind) and Speculative Dependent (Spec_Dep). A block that is not in one of these two states is considered to be committed and would be in one of the states defined by the cache coherence protocol, e.g. Shared, Exclusive, or Modified in a MESI coherence protocol.

A detailed description of VRR-CFP Store Redo Log and the speculative L1 data cache is outside the scope of this work but has been described in great detail in previous CFP work [7][13].

3. EVALUATION METHODOLOGY

We built our VRR-CFP architecture model on the Simplescalar ARM ISA simulation infrastructure (www.simplescalar.com) and used all 14 "C" benchmarks from SPEC 2000 and Spec 2006 that we succeeded in compiling using the Simplescalar cross compiler tool. After skipping the initialization code, we simulated 200 million instructions from each benchmark, consisting of four different samples carefully selected from representative execution phases. The selected execution phases display wide variation in the cache miss rate as well as the branch misprediction rate, both of which significantly impact CFP execution behavior. The results section below shows the average performance of these selected samples for each benchmark.

Table I shows the simulated machine configurations. We used two-level cache hierarchy with sizes and latency representative of current multicore processors, scaled down proportionally to the number of cores, since we simulate a single VRR-CFP core.

Table I. Simulated Machine Configurations

Pipeline and buffers	**Baseline core**: 16 stage, 4-wide, 128-entry ROB, 64-entry RS, 64-entry LQ, 48-entry SQ **VRR-CFP/Sim-CFP core**: 16 stage, 4-wide, 128-entry ROB, 64-entry RS, 64-entry LQ, 48-entry SQ, 256 entry WB, 256 entry SRL
L1 DCache	16KB , 4-way, 3 cycles 64-byte line
L1 ICache	16KB , 4-way, 3 cycles 64-byte line
L2 cache	Unified, 1M/core, 16-way, 35 cycles, 64-byte line
DRAM latency	300 cycles L2 to data return
Branch predictor	Combined bimodal and gshare, 4K Meta, 4K bimodal, 4K gshare, 4K BTB, 16 entry Return Address Stack

4. SIMULATION RESULTS

We present in this section two key results. We first compare VRR-CFP instruction/cycle rate (IPC) to simultaneous CFP (Sim-CFP). Second, we compare the activity event counts of various key circuit components of the two architectures. We use counts of various events to compare the two architectures as an indirect measure of energy consumption since we do not have currently a complete full core power model, leaving direct energy comparison for future work.

4.1 VRR-CFP versus Sim-CFP Performance

Fig. 3 compares IPC of VRR-CFP to simultaneous CFP when applied to L1 data cache misses using our Spec 2000 and Spec 2006 benchmarks. VRR-CFP, with its virtual register renaming and its short replay loop, outperforms simultaneous CFP by an average speedup of 5.1%. The maximum improvement occurs on gcc (24.8% speedup), which displays high rate of L1 cache misses as well as high branch misprediction rate. The reduced replay loop of VRR-CFP is very favorable to benchmarks like gcc, since reducing the amount of replay also significantly reduces the number of costly miss-dependent branch mispredictions. The variation in the improvement between benchmarks is mainly due to the variation in the cache miss rates and branch misprediction rates of our simulation samples.

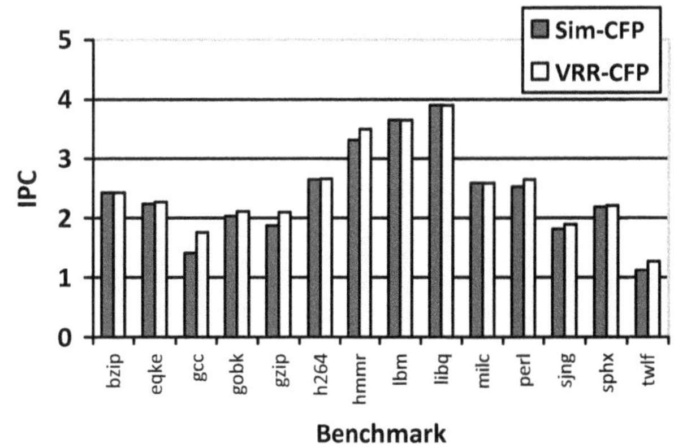

Fig. 3. Comparing Simultaneous CFP and VRR-CFP IPC

4.2 VRR-CFP and Sim-CFP Circuit Activity

Table II shows various circuit activity statistics for VRR-CFP and Sim-CFP models. Columns 2 and 3 show the percentage of total retired instructions that enter the waiting buffer and then replay. Columns 4 and 5 show the percentage of all retired instructions that are executed twice due to roll back to checkpoints for recovery from miss dependent mispredicted branches that are replayed.

Table II. Checkpoint Rollback and Replay Statistics

	VRR-CFP Percent Replay	Sim-CFP Percent Replay	VRR-CFP Percent Rollback	Sim-CFP Percent Rollback
Bzip2	0.0%	0.1%	0.0%	0.0%
Equake	0.9%	2.1%	0.6%	0.6%
Gcc	4.2%	19.3%	2.0%	7.1%
Gobmk	0.7%	3.9%	0.3%	3.0%
Gzip	1.2%	5.1%	0.8%	5.1%
H264	0.5%	2.2%	0.1%	0.3%
Hmmr	0.4%	4.5%	0.0%	0.1%
Lbm	0.0%	0.1%	0.0%	0.0%
Libq	0.0%	0.0%	0.0%	0.0%
Milc	0.0%	0.0%	0.0%	0.0%
Perl	0.9%	6.7%	0.0%	0.4%
Sjeng	0.9%	4.5%	0.4%	2.1%
Sphnx	0.3%	1.4%	0.0%	0.1%
Twolf	6.6%	13.3%	5.0%	8.2%

Columns 4 and 5 show significant reduction in the percentage of instructions that are replayed on VRR-CFP compared to Sim-CFP. For example, replayed instructions in gcc benchmark drop from 19.3% to 4.2%. Similarly, execution waste due to rollback to checkpoints due to recovery from replayed miss dependent mispredicted branches drop significantly as well. On gcc for example, execution rollback drops from 7.1% to 2.0% of total retired instructions. Notice that miss dependent mispredicted branches that are not replayed but are waken up from the RS do not release their reorder buffer entries and therefore can be corrected using the less costly reorder buffer branch recovery.

5. RELATED WORK

Proposals for latency tolerant out-of-order cores include the Waiting Instruction Buffer [14], Virtual ROB [6], Cherry [15], Checkpoint Processing and Recovery [1], Continual Flow Pipelines [13][19] and Out of Order Commit Processors [5]. None of these however deal with L1 data cache misses or execute miss-dependent and independent instructions simultaneously.

iCFP [10] tolerates cache misses at all levels in the cache hierarchy, but uses an in-order pipeline, which is less suitable for the performance needs of conventional single-thread applications.

BOLT [11] utilizes additional map tables in Simultaneous Multithreading architecture to re-rename L2 miss dependent slice, combined with a program order slice and a unified physical register file that supports aggressive register reclamation. BOLT reuse of SMT hardware is for improving energy-efficiency, but does not extend the use of SMT to simultaneous execution of the dependent and independent slices to improve performance. Neither does BOLT use virtual register renaming or a streamlined replay loop to improve energy efficiency.

Sun Microsystems Rock is a single die multicore processor for high throughput computing. Rock uses Simultaneous Speculative Threading [2] to defer dependent instructions into a buffer, and executes the deferred instructions from the checkpoint after the miss data returns. The deferred instructions execution uses a simultaneous hardware thread and merges the results into the scout thread future file. Rock uses an in-order pipeline, while VRR-CFP core is out-of-order and thus provides better performance than Rock on single-thread applications.

Gonzalez et al. [8][9] proposed using virtual registers to shorten the lifetime of physical registers. Kilo instruction processors [4] also used virtual renaming and ephemeral registers to do late allocation of physical registers. In contrast to virtual-physical registers and ephemeral registers, VRR-CFP uses Virtual Register Renaming (VRR) [17] and does not require any physical registers to be late allocated for instruction results.

6. CONCLUSION

This paper presents a Continual Flow Pipeline architecture that uses virtual register renaming to significantly improve performance (5.1% average) and reduce replay loop circuit activity (up to 15.1%) and checkpoint rollback execution (up to 5.1%) when compared to previous CFP designs. It achieves this by reducing the replay latency associated with the processing of miss dependent instructions after the miss data arrives into the L1 data cache and wakes up the miss load. The architecture achieves this reduction in the miss dependent instructions processing latency by 1) keeping these instructions as long as possible in the reservation stations and moving them to the waiting buffer only when instruction buffer resources are needed and 2) using virtual register renaming which allows partial replay of miss dependent instructions, requiring only the subset of these instructions that have already moved to the waiting buffer to be replayed.

7. ACKNOWLEDGMENTS

This work has been supported by a grant from Intel Corporation.

8. REFERENCES

[1] H. Akkary, R. Rajwar, and S. Srinivasan, "Checkpoint processing and recovery: towards scalable large instruction window processors," in Proceedings of MICRO-36, Dec 2003.

[2] S. Chaudhry, R. Cypher, M. Ekman, M. Karlsson, A. Landin, S. Yip, H. Zeffer, and M. Tremblay, "Simultaneous speculative threading: a novel pipeline architecture implemented in Sun's Rock processor," in Proceedings of ISCA-36, June 2009.

[3] G. Z. Chrysos and J. Emer, "Memory dependence prediction using store sets," in Proceedings of ISCA-25, June 1998.

[4] A. Cristal, O. J. Santana, M. Valero and J. F. Martinez, "Toward kilo-instruction processors," in ACM Transactions on Architecture and Code Optimization, vol. 1, issue 4, pp 389-417, Dec 2004.

[5] A. Cristal, D. Ortega, J. Llosa, and M. Valero, "Out-of-order commit processors," in Proceedings of HPCA-10, Feb 2004.

[6] A. Cristal, M. Valero, J. Llosa, and A. Gonzalez, "Large virtual ROBs by processor checkpointing," Tech. Report, UPC-DAC-2002-39, Department of Computer Science, Barcelona, Spain, July 2002.

[7] A. Gandhi, H. Akkary, R. Rajwar, S. T. Srinivasan and K. Lai, "Scalable load and store processing in latency tolerant processors," in Proceedings of ISCA-32, June 2005.

[8] A. Gonzalez, J. Gonzalez and M. Valero, "Virtual-physical registers," in Proceedings of HPCA-4, Feb 1998.

[9] A. Gonzalez, M. Valero, J. Gonzalez, and T. Monreal, "Virtual registers," in Proceedings of HPCA-3, Feb 1997.

[10] A. Hilton, S. Nagarakatte, and A. Roth, "Tolerating all-level cache misses in in-order processors," in Proceedings of HPCA-15, 2009.

[11] A. Hilton and A. Roth, "BOLT: energy-efficient out-of-order latency-tolerant execution," in Proceedings of HPCA-16, Feb 2010.

[12] W. W. Hwu and Y. N. Patt, "Checkpoint repair for out-of-order execution machines," in Proceedings of ISCA-14, June 1987.

[13] K. Jothi, H. Akkary, and M. Sharafeddine, "Simultaneous continual flow pipeline architecture," in Proceedings of ICCD-29, Oct 2011.

[14] R. Lebeck, J. Koppanalil, T. Li, J. Patwardhan, and E. Rotenberg, "A large, fast instruction window for tolerating cache misses," in Proceedings of ISCA-29, May 2002.

[15] J. F. Martinez, J. Renau, M. C. Huang, M. Prvulovic, and J. Torrellas, "Cherry: checkpointed early resource recycling in out-of-order microprocessors," in Proceedings of MICRO-35, Nov 2002.

[16] D. B. Papworth, "Tuning the Pentium Pro microarchitecture," in IEEE MICRO, vol. 16, no. 2, April 1996.

[17] M. Sharafeddine, H. Akkary and D. Carmean, "Virtual register renaming," in Proceedings of the 26th International Conference on Computing Systems, Feb 2013.

[18] J. E. Smith and G. S. Sohi, "The microarchitecture of superscalar processors," in Proceedings of the IEEE, vol. 83, no. 12, Dec 1995.

[19] S. T. Srinivasan, R. Rajwar, H. Akkary, A. Gandhi, and M. Upton, "Continual flow pipelines," in Proceedings of ASPLOS-11, Oct 2004.

[20] R. M. Tomasulo, "An efficient algorithm for exploiting multiple arithmetic Units,'' IBM Journal of Research and Development, vol. 11, pp. 25-33, Jan 1967.

Skew-Bounded Low Swing Clock Tree Optimization

Can Sitik
Electrical and Computer Engineering
Drexel University
Philadelphia, PA, 19104 USA
E-mail: as3577@drexel.edu

Baris Taskin
Electrical and Computer Engineering
Drexel University
Philadelphia, PA, 19104 USA
E-mail: taskin@coe.drexel.edu

ABSTRACT

This paper introduces a methodology that optimizes the performance of a low swing clock tree under a skew bound. Low-swing clock trees are preferred for a reduction in the clock switching power, with an expected trade-off in clock slew and skew. In this paper, a heuristic optimization process is introduced that keeps the clock skew under the same skew budget of the originating full-swing clock tree. In this low swing clock optimization, the low power consumption property is preserved. The effect of slew on the logic timing, which is naturally degraded due to low-swing operation, is analyzed within timing slack of some paths in order to highlight the effectiveness of the low swing clock trees in lowering power consumption with limited impact on timing constraints. The experiments performed with the 4 largest ISCAS'89 benchmark circuits operating at 500 MHz, 90 nm technology and 4 different V_{dd} levels show that the optimized low swing clock tree can achieve an average of upto 11.0% reduction in the power consumption with no more than a skew degradation of 0.5% of the clock period (i.e. within the practical skew budget).

Categories and Subject Descriptors

B.7.1 [**Hardware**]: INTEGRATED CIRCUITS—*Types and Design Styles*

General Terms

Design

Keywords

VLSI, clocking, low-power, low-swing

1. INTRODUCTION

High performance designs, such as high end microprocessors, target tight performance constraints with minimum overhead whereas low-power ASIC designs target tight power constraints with maximum performance [1–4]. Clock distribution networks constitute an important part of the IC design due to their direct effect on the performance and the power consumption. Operating the clock network with low swing is one of the techniques that is explored in order to reduce the power consumption attributed to the clock network (of a microprocessor or an ASIC) [5–9].

Low-swing operation can be adopted at varying levels of a clock tree with different implications. Low swing operation on the entire clock tree without any modifications at the sink level would lead to negative impacts on clock slew, skew and logic path timing. Thus, a previous approach considers to use level shifters at the last level of the clock tree, assuming the overhead of level shifters, in order to keep the signal integrity for logic path timing at the sink level [6]. Another approach considers to use special kind of D flip-flops or buffers designed to work with low swing clock signals, proposing to use custom cells for lower power [7, 9]. All alternative approaches of low swing clock trees yield promising results. However low-swing applicability remains limited in practice due to a number of factors (based on adopted approach) including (*i*) degradation in the skew performance, (*ii*) degradation in expected power reduction, (*iii*) degradation in data timing due to slew degradation, (*iv*) necessitating level shifters of varying sizes, (*v*) necessitating low-swing FF designs. The approach in this work keeps the method highly practical for short-term applicability and integration with standard industrial-strength automation tools and design libraries. To this end, the optimization method in this work is proposed in the conversion of a full-swing clock tree built by an industrial-strength clock tree synthesis (CTS) routine to low swing clocking. The method does not utilize low-swing FFs, or depend on the existence of explicitly sized level shifters, in order to remain compliant with standard cell libraries, addressing factor (*iv*) and factor (*v*) above. The proposed optimization method also eliminates the degradation in skew performance [factor (*i*)], preserves the expected power reduction of low-swing operation [factor (*ii*)] and satisfies the data path timing requirements by controlling clock slew [factor (*iii*)]. More importantly, the optimization routine is built into a novel automation tool within the industrial-tool flow, that is the first automated routine to synthesize low swing clock network with practical slew and skew budgets.

The proposed methodology is not a straight-forward procedure due to the following challenges: 1) The effect of each upsizing/downsizing must be characterized as standard cell library models are available only at certain levels (1.2V, 0.8V etc.), 2) The slew at the clock pins of register sinks must be analyzed, because slew definition must be modified to include the effect of low swing driver buffers, 3) The change in skew and slew must be considered at all process corners for variation-awareness. The proposed methodology starts with a synthesized full-swing tree, follows a library characterization stage in order to analyze the effect of upsizing and downsizing of buffers in the given standard cell library, and finally minimizes its skew overhead by in-place upsizing/downsizing of clock buffers depending on the V_{dd} level selected and the target skew budget. To that end, this methodology can target any point on the power vs. skew trade off curve depending on the application, assuming presence of another power supply for the low swing voltage level. The methodology provides the desirable output of a low swing clock network with power savings obtained through lower V_{dd}, with (almost) the same performance of a full-swing network with a low run time, implemented as a new low-swing-CTS tool.

The rest of the paper is organized as follows. The preliminaries of the low swing clock networks are introduced in Section 2. The methodology of this work is explained in detail in Section 3. The results of the performed experiments on benchmark circuits are presented in Section 4. The paper is finalized with concluding remarks in Section 5.

2. PRELIMINARIES

In this section, the preliminaries of the low-swing clock networks are briefed. The change in the insertion delay and the power consumption depending on the V_{dd} level of the clock tree is presented in Section 2.1. The effect of lower V_{dd} and higher slew of the clock signal on the register-to-register logic path timing path is analyzed in Section 2.2.

2.1 Insertion Delay and Power in Low Swing Clock Trees

The delay and the power consumption of a clock tree depend on the supply voltage and the load capacitance including the capacitance at the sinks. Simple first order and higher order models are known that estimate the gate delay and the power consumption using the voltage and capacitance information [10, 11]. A dedicated model is needed for low-swing operation, particularly for the topology target in this paper: A low swing clock tree driving (e.g. register) sinks operating with full swing data input/output but with low swing clock input. To that end, SPICE-accurate simulations are used in this step in order to profile the delay and the power characteristics of low swing clock trees. A sample circuit, s35932 of ISCAS'89 benchmarks, is used to observe the change in the insertion delay at each sink and the total power consumption of the design with the change in the supply voltage on the clock tree at SPICE accuracy.

The profile of insertion delay at each sink vs. supply voltage is shown in Figure 1. As expected, the insertion delay at each sink increases with decreasing supply voltage due to the RC effect of the physical buffers and wires on the clock tree. The insertion delays form a monotonic behaviour at every individual sink. However, it is observed that each sink scales differently with the scaled supply voltage, which increases the difference between the maximum and the minimum insertion delay, which is also shown in Figure 2. Thus, the clock skew increases with decreasing voltage driving the (buffers of the) clock tree. This key observation leads to the fact that it is essential to monitor the skew when a low swing clock tree is used as the clock skew may not satisfy the target skew bound of the design at the low swing clock operation.

The power consumption, on the other hand, is expected to decrease with decreasing supply voltage. However, this decrease in power consumption is only until a point due to the worsening of the power profiles of the D flip-flops of the register sinks, shown in Figure 3. Although the power consumption of the clock tree scales quadratically with the supply voltage, driving D flip-flops with low swing buffer drivers increases the power consumption at the driven register sinks whose datapath is at full swing. The reason of this increase is the effect of slower switching of internal transistors of the

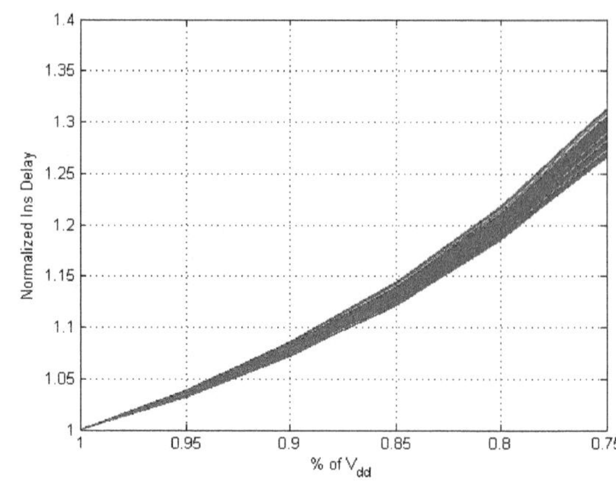

Figure 1: Normalized insertion delay of all sinks in the design, interpolated using 5 different low swing values

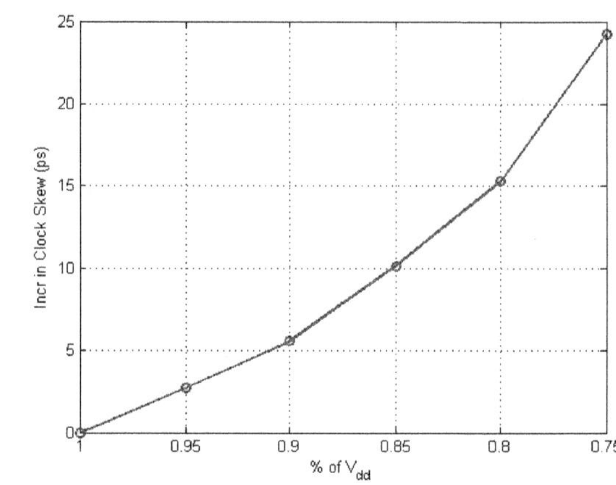

Figure 2: The increase in skew, interpolated using 5 different low swing values

D flip-flops (higher short circuit power). Thus, this increase outweighs the savings obtained at the clock tree after the 80% point. This observation implies that scaling clock tree voltage swing is not advantageous after a certain point.

2.2 Register-to-register Timing in Low Swing Clock Trees

In the standard IC Design flow, clock slew is constrained to a practical value for timing and power performance. In terms of timing performance, for instance, slew directly impacts the clock-to-output delay of the sink sequential-element (e.g. register). With this (worst-case) known impact on clock-to-output delay, a slew bound permits more accurate analysis of register-to-register timing early in the design cycle.

At low-swing clocking, the slew definition from traditional full-swing operation needs to be revised in order to define its equivalent low-swing operation form. The equivalence is defined as the identical effect on the register-to-register timing of the logic paths at full swing. Then, sufficient bounds are set for low swing voltage levels with the minimum degradation of timing slack. When a low swing clock tree drives register sinks operating at full *data* swing, the definition of slew as the time elapsed between the 10% and the

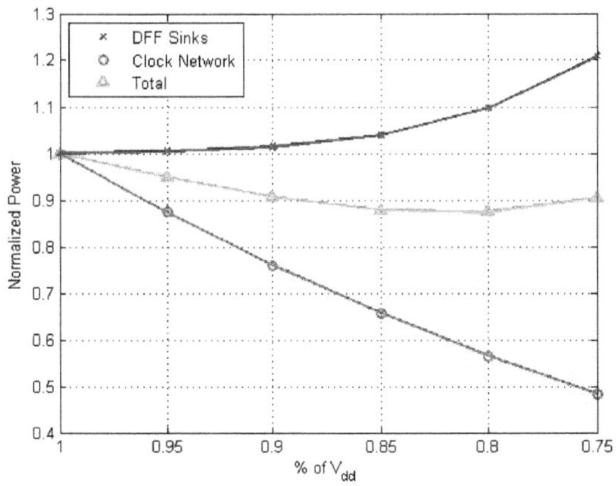

Figure 3: Normalized power for DFF sink registers, clock tree and the total power consumption, interpolated with 5 different low swing values

90% point of the *clock* signal is meaningless, because the voltage swing will likely not reach at 90% of full swing. To that end, a low swing slew must be defined considering its effect on the clock-to-output delay, in order to constraint the optimization stage with a new meaningful slew. In this work, low swing slew is defined as the time elapsed when a signal switches from 10% point to 50% point, as the voltage swing can not be scaled down to 50% of its nominal value, which is the switching point of the DFF sinks operating at full scale. It is argued in this work that this low-swing slew should be bounded so as to provide the equivalence to the original full-swing slew bound that provides similar (or constrained) clock-to-output delays at the sinks.

While the proposed definition for the equivalent low swing clock slew is sufficient to present the proposed low-swing clock optimization methodology, a complete empirical verification of the methodology necessitates standard cell library support for timing, which does not readily exist. The necessitated support is the timing data for standard cells at varying voltage levels of low swing (e.g. 0.85V, 0.9V, etc.) which typically do not exist in libraries: Instead gates are characterized only at supported voltage nodes, e.g. 0.8 V for low voltage and 1.2 V for high voltage. The static timing analysis tools are not applicable to measure the timing slack (i.e. impact of clock slew) due to the lack of these circuit models at low swing in the standard cell libraries. To fill this gap in empirical verification, 20 random paths, which are created using the available types of DFFs in the selected target library and logic cells, are analyzed at 4 different voltage levels (95%, 90%, 85%, 80% levels of the nominal V_{dd}). This analysis is performed in order to observe the relation between the low swing slew and the local timing (voltage levels beyond 80% point are excluded because they are shown to be disadvantageous in Section 2.1). Naturally, these 20 random paths do not theoretically guarantee timing but are empirically demonstrative of the effectiveness of the low-swing slew bounding methodology. It is also important to note that, this change in the clock signal only changes the clock-to-output delay of the D flip-flop, and the delay on the rest of the local path (that contributes to slack) is affected by less than 1 ps. Therefore, it is sufficient to analyze the effect of low swing slew on the clock-to-output delay of different types of DFFs in the library (there are four different DFFs in the selected SAED Library [12] of *Synopsys*, therefore 20 random paths are sufficient), independent of the logic path in the design. The

Table 1: Low swing slew values, and the effect of a low swing clock supply on local timing. The decreased slack, induced by increased clock-to-output delay, is traded off for the power savings of low swing clocks

	Low Swing Slew (ps)	Avg. Decrease in Slack (ps)	Max Decrease in Slack (ps)
V_{dd}	79.65	-	-
$0.95 \times V_{dd}$	83.83	1.26	1.78
$0.90 \times V_{dd}$	88.51	2.94	3.39
$0.85 \times V_{dd}$	93.74	5.08	5.58
$0.80 \times V_{dd}$	99.54	7.84	8.76

additional cell characterization study of standard cells, in order to develop timing models at varying voltage levels of low swing, is not presented here or used in experiments, aiming not to divert the discussion from low swing clock tree design. Without lack of generality, these studies can be performed or cell libraries with such information can be used instead, in a practical application of proposed novelties in low swing clock design.

For demonstration purposes, 200 ps is selected as the slew, which is a typical value (10% of the clock period at 500 MHz operation). The local path (logic and the DFFs) is supplied by full swing and the clock signal is applied with 4 different voltage levels of low swing, with 200 ps slew. As the low swing slew is defined as the transition between the 10% and 50% of the full swing (not the applied low swing) voltage, it is expected to have larger low swing values, and their effect on the local timing must be investigated in order to have a clear understanding of the parameters traded off against each other. The low swing slew values, the average increase and the worst case increase in the clock-to-output delay of 20 random paths at 4 voltage levels with 200 ps slew are shown in Table 1.

The measured low swing slew values, shown in Table 1 are set as the new constraints on the design, as the effect on the logical timing is limited (as low as 8.76 ps, 0.439% of the clock period, at the worst case). It is shown in Figure 3 that the power savings of the design are proportional to the scaled voltage level until 80% point, therefore it is shown here that the savings in the power consumption are successfully traded off with an insignicifant degradation in the logic timing (slack).

3. METHODOLOGY

In Section 2, it is shown that significant power improvements can be obtained through low swing clocking with a degradation in clock skew (Section 2.1) and an insignicant degradation in logic timing due to low swing slew (Section 2.2). Therefore, an optimization that minimizes the skew overhead of low swing clocks without degrading the power savings and the slew can obtain clock trees that perform (almost) the same as full swing clocks with significant power savings. To address that, a new methodology that optimizes low swing clock trees generated from a full-swing clock tree without any placement or clock topology changes, is proposed in this paper. The methodology only involves in-place buffer sizing, aiming not to alter the logic placement or full-swing clock tree topology for minimal disturbance of legacy designs. The proposed methodology has 3 mains steps:

1. Synthesizing an initial clock tree at full swing,

2. Characterizing the buffers of the library,

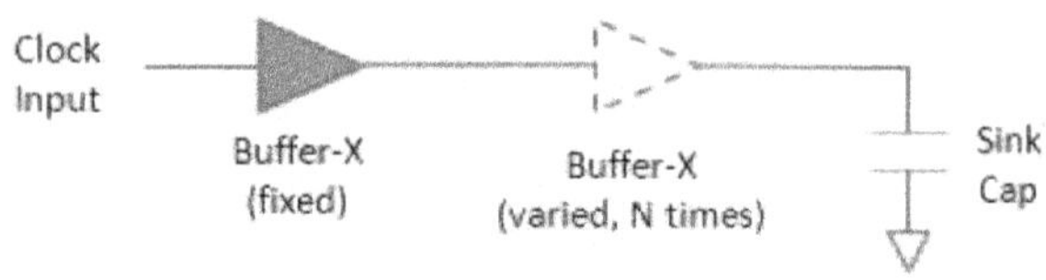

Figure 4: 2-level model for buffer characterization

3. Applying low swing optimization.

The initial CTS can be performed with any of the industrial tools at the full voltage swing. This CTS creates a clock tree that is optimized for a number of metrics such as routing congestion, tree depth, etc., therefore it is a good starting point for the proposed optimized methodology. In the second step, a buffer characterization is performed in order to observe the effect of in-place buffer sizing on the delay for the buffers available in the library, which is explained in Section 3.1. In the final step, the low swing clock tree optimization is performed at the given V_{dd} value on the clock tree that is initially synthesized for full swing, in order to decrease the skew under the skew budget without degrading the power and the slew, which is presented in Section 3.2.

3.1 Buffer Characterization

In standard cell libraries, the granularity of buffer sizes vary significantly. Convex methods for sizing of buffers are not always feasible due to highly discrete characteristic of the problem. To that end, the characterization of buffers is a necessary step, which provides the data set for a look-up table based optimization proposed in the following step (Section 3.2). Note that these look-up tables need to be generated once for each standard cell library, and the utilization of look-up tables in optimization has advantages in the run-time profiles.

In the buffer characterization step, a discrete set of candidate voltage levels for low swing operation and the discrete set of buffers in a target standard cell library are considered. The characterization analyzes the effect of discretely upsizing and downsizing of buffers on the insertion delay at each selected low swing voltage node. The challenge in this process is in maintaining the accuracy of timing of buffers with respect to those in a typical clock tree, as the sizing of one buffer affects the delay of all other buffers in the clock tree. To address this challenge, the number of sizing allowed for each buffer is defined as N, a user-specified number to be selected depending on the granularity of the buffer library so as to keep the inaccuracy of buffer characterization reasonable. The inaccuracy in this step can result in more number of iterations in the next step (Section 3.2), however this is not observed in the experiments, therefore the buffer characterization step proposed here is sufficient in practice. In order to achieve a good look-up table, a simple 2-level tree is created with the same type of buffer as the driver in the first level and a typical capacitance value that creates a slew rate of the slew constraint of the design to model a sink, shown in Figure 4. Having a 2-level tree structure enables to observe both the effect of loading to the previous level and the effect of driving a load on the insertion delay. As the number of sizings allowed is N, each buffer is simulated $2 \times N$ times, N for upsizing and N for downsizing. If there is r number of buffers in the library, this step has a complexity of $O(N \times r)$. The parameter N depends on the size of the buffer library r, therefore the complexity scales quadratically with the size of the buffer library. However, this step takes on the order of seconds, as only a 2-level model is simulated for each buffer for a small finite number of buffers.

Algorithm 1 Low Swing Clock Tree Optimization

Input: Buffer Characterization, Initial Clock Tree, Voltage Level
  Selected Buffers Queues, $Q_{max} = Q_{min} = \emptyset$
  Obtain the initial skew, $skew_{real}$
  while $skew_{real} < skew_{const} + skew_{margin}$ **do**
    $skew_{curr} = skew_{real}$
    while $skew_{curr} < skew_{const}$ **do**
      Find the driving buffer of the registers with min insertion delay B_{min}
      Find the driving buffer of the registers with max insertion delay B_{max}
      Select the buffer, which when removed, improves the skew the most, set empty set for the other
      $Q_{max} = Q_{max} \cup B_{max}$
      $Q_{min} = Q_{min} \cup B_{min}$
      Update $skew_{curr}$ with the remaining set
    end while
    while $Q_{max}! = \emptyset$ **do**
      Pop one buffer
      Analyze its neighboring buffers
      if *neighboring buffers have the same delay profile* **then**
        Size their driving buffer
      else
        Size only the corresponding buffer
      end if
    end while
    while $Q_{min}! = \emptyset$ **do**
      Pop one buffer
      if *the neighboring buffers have the same delay profile* **then**
        Size their driving buffer
      else
        Size only the corresponding buffer
      end if
    end while
    Run a SPICE simulation to calculate $skew_{real}$
  end while

3.2 Low Swing Clock Tree Optimization

Low swing clock tree optimization defines the major novelty of this paper. The objective of optimization is to preserve the power and the slew properties of low swing clock tree with a better skew in order to meet the skew budgets as closely as that of the full swing clock tree with the same topology. In this step, a low swing clock tree is optimized for skew under a skew budget via in-place buffer sizing without degrading the slew and the power consumption, with the guidance of the characteristics of the buffers in the library (Section 3.1). The proposed algorithm is outlined in Algorithm 1.

In this algorithm, the clusters of registers driven by the same clock buffer are identified and the clusters whose insertion delay is outside a window (defined by the skew budget) are marked. In order to achieve this goal, the clusters are selected one at a time from both the maximum and the minimum insertion delay paths (the algebraic different of which is the clock skew), and their driving (i.e. parent) buffer is sized up or down in order to improve the skew. In order not to disturb the rest of clock tree, the buffers that have the same driving buffer, as the (i.e. corresponding) buffer that is being analyzed, are also identified and their insertion delays are investigated. If their insertion delay have the same characteristic with the corresponding buffer, their driving buffer is sized, considering those buffers may also be in our region of interest, so as to limit the number of buffers that are modified. If not, only the corresponding buffer is sized. Once the iterations of sizing neces-

Table 2: **Power, Worst Skew (WS) and Low Swing Slew (LSS) Comparison over a Full Swing Clock, and an Unoptimized Low Swing Clock Synthesized by *IC Compiler*, at** $0.90 \times V_{dd}$

Circuits	Full Swing CT Synthesized by *ICC*			Low Swing CT before Optimization			Proposed Work		
	Power (mW)	WS (ps)	LSS (ps)	Power (mW)	WS (ps)	LSS (ps)	Power (mW)	WS (ps)	LSS (ps)
s38584	30.56	60.55	63.42	28.50	72.40	75.66	28.58	61.69	75.54
s38417	34.50	62.43	59.05	31.84	73.22	70.73	31.81	61.40	70.90
s35932	28.75	83.87	62.55	26.09	101.96	74.75	26.02	82.58	74.66
b1	58.93	78.65	63.35	53.89	90.51	75.40	54.00	78.29	75.53
	Avg. Improvement Compared to Full Swing			8.1%	-13.61 ps	-12.04 ps	8.0%	0.39 ps	-12.07 ps

Table 3: **Power, Worst Skew (WS) and Low Swing Slew (LSS) Comparison over a Full Swing Clock, and an Unoptimized Low Swing Clock Synthesized by *IC Compiler*, at** $0.85 \times V_{dd}$

Circuits	Full Swing CT Synthesized by *ICC*			Low Swing CT before Optimization			Proposed Work		
	Power (mW)	WS (ps)	LSS (ps)	Power (mW)	WS (ps)	LSS (ps)	Power (mW)	WS (ps)	LSS (ps)
s38584	30.56	60.55	63.42	27.87	81.02	84.20	28.03	61.83	82.79
s38417	34.50	62.43	59.05	30.95	80.61	78.87	30.92	66.54	78.84
s35932	28.75	83.87	62.55	25.25	113.94	83.23	25.23	88.85	80.76
b1	58.93	78.65	63.35	52.25	98.22	84.16	52.34	83.21	84.01
	Avg. Improvement Compared to Full Swing			10.7%	-22.54 ps	-20.52 ps	10.6%	-3.73 ps	-19.51 ps

sary buffers are completed, a SPICE simulation is run for SPICE-accurate timing verification. The clock skew can be decreased more than an allowed margin ($skew_{const} + skew_{margin}$) due to the change in the insertion delay of the registers outside of the region of interest affected by the sizing of other buffers. If clock skew degradation is not within the allowed margin, another iteration is performed. It is observed that the number of iterations is 1 in most of the cases, and no more than 3 iterations are necessary in the experiments performed for this work. The complexity of the proposed algorithm depends on the number of buffers that reside in the region of interest (defined by the neighboring buffers analyzed in optimization). In the worst case, it is $O(n)$ where n is the number of buffers at the last level of the clock tree. However, it is highly unlikely to size all the buffers at the last level (in practice, it is observed to be at most 6 buffers) therefore the complexity is approximated to be constant.

4. EXPERIMENTAL RESULTS

The proposed methodology is implemented with TCL in order to inter-operate with the existing industrial tools (e.g. Synopsys in this case) and tested on 3 largest benchmark circuits of ISCAS'89 benchmark circuits (s35932, s38417, s38584) and 1 more benchmark circuit that is created (and called b1) by combining s35932 and s38584 in order to observe the scalability. Note that ISPD'10 clock contest benchmark circuits cannot be used in this analysis as ISPD'10 clock contest benchmarks do not have data to complete the analysis in the Synopsys flow. Nonetheless, note that the largest ISCAS benchmark circuits selected here have similar number of register sinks as that in the largest ISPD'10 clock contest circuits. The RTL level designs are synthesized using *Design Compiler* of *Synopsys*, the physical placement and the synthesis of the initial full swing clock tree is performed with *IC Compiler* of *Synopsys*. The power and the skew analysis is performed using *CustomSim XA* simulator of *Synopsys* at the SPICE accuracy with 90 nm technology, operated at 500 MHz. In order to show the effectiveness of the proposed methodology, the skew, the slew and the power numbers of the full swing clock tree synthesized by *IC Compiler*, the low swing clock tree before and after the optimization are compared. Also, the comparisons are performed at 4 different low swing levels at 95%, 90%, 85% and 80% of the full swing voltage to target different levels of trade-off on the power vs. timing curve. The number of allowed buffer sizing N is set to 2, considering the number of available buffer sizes is 5 in the selected *SAED 90nm EDK Library* of *Synopsys*. The allowed degradation margin $skew_{margin}$ is set to 10 ps, 0.5% of the clock period, in order to preserve the overall skew lower than 100 ps. Note that, 100 ps is 5% of the clock period at 500 MHz, which is a typical value. In order to account for the PVT variations, the optimizations are performed at the worst case corner of the process, and the final circuit is verified at 3 different corners of process, which are:

1. Best Corner (BC): V=V_{dd}+10%, T=-40°C, fast transistors
2. Nominal Corner (NC): V=V_{dd}, T=25°C, typical transistors
3. Worst Corner (WC): V=V_{dd}-10%, T=125°C, slow transistors

The experimental results and the comparisons to the full swing clock tree and an unoptimized low swing clock tree at voltage levels of $0.90 \times V_{dd}$, $0.85 \times V_{dd}$ and $0.80 \times V_{dd}$ are shown in Table 2, Table 3 and Table 4, respectively. The comparisons at $0.95 \times V_{dd}$ are excluded because the degradation in the skew is less than the allowable margin. The experimental results show that the proposed methodology is applicable at all available voltage levels (whenever applicable to satisfy the skew and the slew constraints) in order to obtain skew values as low as the corresponding full swing clock tree while preserving the substantial decrease in the power consumption. The decrease in power consumption is achieved through the degradation in the slew, however the low swing slew constraints defined in Section 2.2 are satisfied, which guarantees an insignificant decrease in the timing slack (at 80% of supply voltage, 11.0% power savings obtained with a slack decrease of $\approx$ 8 ps, as shown in Table 1). Furthermore, it is also important to note that, the proposed methodology can obtain different power savings (8.0% to 11.0%) and different slew degradation (12.07 ps to 30.82 ps) at different low swing voltage levels with (almost) the same skew value (at most 3.73 ps increase) within the budget. Thus, the proposed methodology can target a wide range of applications by changing the applied low swing voltage level in order to budget the available timing slack for power savings.

Table 4: **Power, Worst Skew (WS) and Low Swing Slew (LSS) Comparison over a Full Swing Clock, and an Unoptimized Low Swing Clock Synthesized by *IC Compiler*, at** $0.80 \times V_{dd}$

Circuits	Full Swing CT Synthesized by *ICC*			Low Swing CT before Optimization			Proposed Work		
	Power (mW)	WS (ps)	LSS (ps)	Power (mW)	WS (ps)	LSS (ps)	Power (mW)	WS (ps)	LSS (ps)
s38584	30.56	60.55	63.42	27.72	92.40	94.85	27.95	66.40	93.63
s38417	34.50	62.43	59.05	30.70	88.34	89.47	30.65	55.71	89.33
s35932	28.75	83.87	62.55	25.13	127.87	93.96	25.17	91.54	93.73
b1	58.93	78.65	63.35	51.88	108.94	95.15	51.95	77.52	94.94
	Avg. Improvement Compared to Full Swing			11.2%	-33.48 ps	-31.27 ps	11.0%	-1.42 ps	-30.82 ps

Table 5: Number of Iterations and Number of Buffers Modified for each Benchmark Circuit at All Voltage Levels

	No of Sinks	No of Clock Buffers	No of Iterations			No of Buffers Modified		
			$0.90 \times V_{dd}$	$0.85 \times V_{dd}$	$0.80 \times V_{dd}$	$0.90 \times V_{dd}$	$0.85 \times V_{dd}$	$0.80 \times V_{dd}$
s38584	1238	134	1	1	1	1	3	4
s38417	1463	150	1	1	3	1	3	2
s35932	1728	170	1	2	2	2	3	6
b1	2976	315	1	2	2	3	4	6

The run time of the proposed algorithm is also analyzed. The number of iterations (each takes about $\approx$ 5 mins) for each benchmark circuit at each voltage level is shown in Table 5 with the circuit size information. Even with the largest benchmark circuit with 2976 sinks, the proposed methodology converges in 2 iterations at most. The number of iterations is 1 in most of the cases, and the max number of iterations observed is 3, which occurs only once for s38417. Note that, the allowed skew degradation margin $skew_{margin}$ can be increased for better run time with less performance, or can be decreased for better performance by degrading the run time, which also shows the practicality of the algorithm for automation purposes. This phenomenon actually can be seen when the voltage level is at its 80%, shown in Figure 4, the skew is less compared to the skew at 85%, shown in Figure 3, but the number of iterations is higher at 80%. Moreover, as the delay characteristics of the neighboring buffers are considered, the clusters in the region of interest are optimized with a single sizing at the (n-1)st level and the number of buffers that are modified (sized) are limited to at most 6 buffers among the 100+ buffers of the clock trees.

5. CONCLUSION

A new methodology for optimization of low swing clock trees is proposed. Low swing clock trees are known to be effective for power savings but with a penalty in performance. In this work, a fast heuristic is proposed in order to optimize the clock skew while preserving the power savings of the low swing clock tree. Moreover, the degradation in the slew is also analyzed and constrained in order to minimize the degradation in the timing slack. Proposed methodology can achieve a skew as low as full swing clock trees with limited number of in-place buffer sizing, therefore it is highly practical. The allowed skew degradation margin can be used in order to trade-off the skew and the run time, also the applied low swing voltage level can be used to trade-off the power savings and the timing slack. Thus, the proposed methodology has a wide range of applications depending on the timing and power constraints.

6. REFERENCES

[1] G. Shamanna, N. Kurd, J. Douglas, and M. Morrise, "Scalable, sub-1w, sub-10ps clock skew, global clock distribution architecture for Intel Core i7/i5/i3 microprocessors," in *Proceedings of the IEEE Symposium on VLSI Circuits (VLSIC)*, June 2010, pp. 83–84.

[2] N. Kurd, J. Barkarullah, R. Dizon, T. Fletcher, and P. Madland, "A multigigahertz clocking scheme for the Pentium 4 microprocessor," *IEEE Journal of Solid-State Circuits (JSSC)*, vol. 36, no. 11, pp. 1647–1653, Nov. 2001.

[3] D.-J. Lee, M.-C. Kim, and I. Markov, "Low-power clock trees for CPUs," in *Proceedings of the IEEE/ACM International Conference on Computer-Aided Design (ICCAD)*, Nov. 2010, pp. 444–451.

[4] J. Lu, X. Mao, and B. Taskin, "Integrated clock mesh synthesis with incremental register placement," *IEEE Transactions on Computer-Aided Design (TCAD)*, vol. 31, no. 2, pp. 217–227, Feb. 2012.

[5] Q. Zhu and M. Zhang, "Low-voltage swing clock distribution schemes," in *IEEE International Symposium on Circuits and Systems (ISCAS)*, May 2001, pp. 418–421.

[6] J. Pangjun and S. Sapatnekar, "Low-power clock distribution using multiple voltages and reduced swings," *IEEE Transactions on Very Large Scale Integration (TVLSI) Systems*, vol. 10, no. 3, pp. 309–318, June 2002.

[7] F. Haj Ali Asgari and M. Sachdev, "A low-power reduced swing global clocking methodology," *IEEE Transactions on Very Large Scale Integration (TVLSI) Systems*, vol. 12, no. 5, pp. 538–545, May 2004.

[8] D. Markovic, J. Tschanz, and V. De, "Feasibility study of low-swing clocking," in *the International Conference on Microelectronics*, vol. 2, May 2004, pp. 547–550.

[9] C. Kim and S.-M. Kang, "A low-swing clock double-edge triggered flip-flop," *IEEE Journal of Solid-State Circuits*, vol. 37, no. 5, pp. 648–652, May 2002.

[10] S. Raja, F. Varadi, M. Becer, and J. Geada, "Transistor level gate modeling for accurate and fast timing, noise, and power analysis," in *Proceedings of the ACM/IEEE Design Automation Conference (DAC)*, June 2008, pp. 456–461.

[11] P. Li and E. Acar, "A waveform independent gate model for accurate timing analysis," in *IEEE International Conference on Computer Design (ICCD)*, Oct. 2005, pp. 363–365.

[12] *Synopsys 90nm Generic Library*, Synopsys Inc., 2009.

Coordinating Prefetching and STT-RAM based Last-level Cache Management for Multicore Systems*

Mengjie Mao, Hai (Helen) Li, Alex K. Jones, Yiran Chen
Department of Electrical and Computer Engineering, University of Pittsburgh
{mem231, hal66, akjones, yic52}@pitt.edu

ABSTRACT

Data prefetching is a common mechanism to mitigate the bottleneck of off-chip memory bandwidth in modern computing systems. Unfortunately, the side effects of prefetching are an additional burden on off-chip communication and increased cache write operations. With the proposal of *spin-transfer torque random access memory* (STT-RAM) based *last-level caches* (LLCs) for their high density and low power consumption, the increase of write pressure to the cache from prefetching coupled with the characteristically long write access compared with traditional SRAM caches exacerbates the performance cost of prefetching schemes. In this work, we propose two orthogonal techniques to reduce the negative performance impact induced by aggressive prefetching on multicore systems employing STT-RAM based LLC. First, *basic priority assignment* prioritizes the different types of access requests of LLC by their criticality and responds to them based on priority. Second, *priority boosting* differentiates requests by application and prioritizes the relatively few requests from applications with non-intensive accesses to the LLC, which usually creates the most severe performance degradation in multi-core systems. Combining these two prioritization policies can alleviate the negative effect induced by aggressive prefetching. Our results show that these techniques can achieve an 8.3% average application speedup compared to a baseline, prefetch only design without prioritization.

Categories and Subject Descriptors

C.1.0 [**Processor Architectures**]: General

Keywords

STT-RAM; Prefetching; Cache management

1. INTRODUCTION

In modern computer architecture, data prefetching is commonly used in order to alleviate the impact of long memory accesses on system performance [1, 7]. Based on the prediction of future data usage, the corresponding memory blocks are preloaded into caches. Correct predictions that bring in the required data in advance of their request can significantly accelerate the system performance. Because an access to main memory can require several hundred clock cycles, existing prefetcher designs typically focus on prefetching data from main memory to the *last-level cache* (LLC). However, in *chip-multiprocessors* (CMPs), aggressive prefetch configurations coupled with limited prefetching accuracy can induce more access conflicts among various shared memory resources [5] and thus, limit the potential performance of the system.

Recently, emerging *spin-transfer torque random access memory* (STT-RAM) has been proposed for LLC in CMPs to relieve increasing power trends [6, 16]. Benefiting from its non-volatility, STT-RAM has nearly zero leakage energy, which is the major portion of LLC energy consumption. Moreover, its simple cell structure dramatically enhances cache density and same area replacement capacity compared with SRAM. However, programming an STT-RAM cell requires higher dynamic energy and longer access latency than SRAM [8, 14]. To relax the impact of the prolonged access latency of writes in STT-RAM LLCs, several strategies have been proposed including a hybrid cache architecture composed of both STT-RAM and SRAM, which requires a complex replacement policy and hence requires significant hardware redesign and complexity [6]. In this work, we focus on homogeneous STT-RAM based LLC designs.

CMP systems with prefetching place extra stress on STT-RAM-based LLCs. Aggressive prefetching can generate a considerable amount of prefetch requests. In order to reduce the implementation complexity, prefetch requests are usually treated as common fill requests fetched from main memory and written into LLC [11]. Thus, more write operations will be served. Due to the high write latency, this will increase access pressure in the form of increased bank conflicts and consequently, degrade performance, potentially reducing or eliminating the performance gains of prefetching.

In this paper, we propose two simple and implementable techniques to reduce the negative impact on system performance of CMPs with STT-RAM based LLC induced by aggressive prefetching. First, *basic priority assignment* (BPA) is a static scheme, which prioritizes different types of LLC requests (read, fill, prefetch fill, and write back) based on their criticality. Critical requests are assigned a high priority, allowing them to preempt the memory ports and get served earlier. Second, we propose *priority boosting* (PB) based on the observation that when multiple programs are running together, the LLC requests sent by *LLC non-intensive* (LLCNI) applications can be starved by the frequent accesses initiated by *LLC intensive* (LLCI) applications, consequently degrading the performance of LLCNI applications. PB solves this issue by prioritizing the requests of LLCNI

*This work is supported in part by NSF awards CNS-1116171 and EFRI-1038139.

GLSVLSI'13, May 2–3, 2013, Paris, France.

applications to be served promptly. These two techniques are orthogonal and hence can be integrated together seamlessly: BPA addresses the priority among different types of LLC requests, and PB differentiates the priority among different applications. In the paper, we report the best priority configuration obtained through exhaustive simulations. We also analyze system fairness in detail and demonstrate the impact on each individual benchmark in our workload. Our simulation results show that combining BPA and PB can improve CMP system performance up to 28% over the baseline CMP design without any prioritization. The average performance improvement across the entire workload is 8.3%.

The remainder of this paper is organized as follows: Section 2 briefly introduces STT-RAM fundamentals and the related work on prioritization in STT-RAM caches and prefetchers. Section 3 presents the detailed methodologies of BPA and PB. Section 4 describes the experimental setup and the simulation results are presented and analyzed in Section 5. Finally, we conclude the paper in Section 6.

2. BACKGROUND & RELATED WORK

2.1 STT-RAM Fundamentals

Recently, STT-RAM has gained significant attention for its high density, better scalability, relatively fast read access, and extremely low leakage power [8]. Figure 1 illustrates the structure of a *one-transistor-one-MTJ* (1T1J) STT-RAM cell, in which a *magnetic tunnel junction* (MTJ) is used for binary data storage. A MTJ consists of two ferromagnetic layers, namely, the *reference layer* and *free layer* with a tunnel barrier layer (MgO) sandwiched between. When the two magnetization directions of the two ferromagnetic layers are parallel or anti-parallel, the MTJ exhibits a low or high resistance state, representing logic '0' or '1', respectively. Though the magnetization of the reference layer is fixed, the free layer can be flipped by applying a switching current with the corresponding polarization, which in general involves a long write period and high write energy consumption.

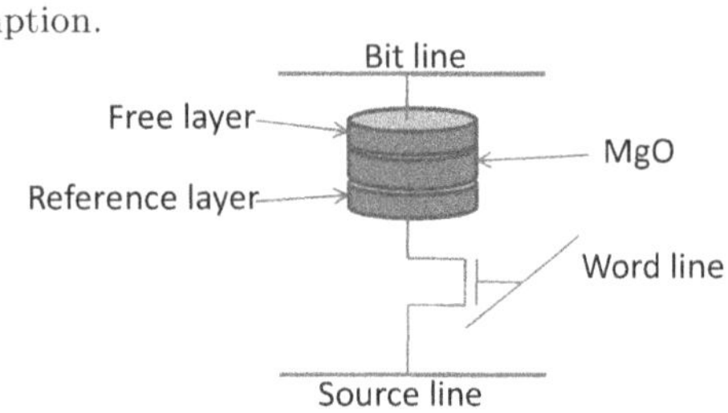

Figure 1: The schematic of one STT-RAM cell.

2.2 Prioritization in STT-RAM Cache

The request prioritization in STT-RAM based LLC designs was first proposed by Sun *et al.* to alleviate the performance degradation induced by slow STT-RAM write operations [6]. When read and write back requests occur at a bank at the same time, the read always takes priority and is served promptly. To maximize the performance benefit, *retirement accomplishment degree* (RAD) is used to determine at what stage an existing write back request can be preempted by a read request. In this work, we extend the prioritization assignment to all other types of LLC requests after considering prefetching in CMP systems including RADs for different types of requests based on their criticality and the tradeoff between performance and energy consumption.

2.3 Previous Work on Prefetch Controlling

Since data prefetchers with low accuracy can unnecessarily occupy memory resources and even potentially degrade system performance, filter techniques are commonly used [13, 17]. For example, Zhuang and Lee [17] proposed a hardware-based cache pollution filter for prefetchers. In this mechanism, a history table consisting of saturation counters is used to record the accuracy of each prefetch request. A prefetch request with low accuracy will be prevented from issuing. Additionally, by collecting information such as temporal accuracy, prefetch timeliness, cache pollution, and memory bandwidth utilization, it is possible to dynamically tune a prefetcher's aggressiveness [11, 3]. Dynamic prefetcher aggressiveness tuning mainly focuses on shared main memory resource contention in the presence of multiple prefetchers.

3. METHODOLOGY

To address the write pressure from aggressive prefetchers in the context of STT-RAM-based LLC design, we propose two priority techniques, namely, BPA and PB. BPA is a static scheme that differentiates the LLC access priorities of different cache requests. In contrast, PB primarily focuses on distinguishing the different cache access behaviors of various applications. Since an application changes behaviors across its execution phases, PB can dynamically adjust the priority assignments, accordingly. The two techniques are orthogonal to each other, and hence, can be seamlessly integrated together.

3.1 Basic Priority Assignment (BPA)

The shared LLC is a key resource in CMP systems that supports several types of accesses and operations including: (1) *read* requests to LLC due to misses in upper level caches; (2) *write back* requests to LLC due to dirty evictions in upper level caches; (3) *fill* operations from main memory due to LLC misses; and (4) *prefetching fill* requests from upper level caches that must also be served by LLC. During system execution, these operations result in a considerable number of access requests at the LLC, often introducing severe bank conflicts. The long write latency of STT-RAM LLC further aggravates this situation.

In Figure 2, we show the distributions of different LLC requests involving in writing STT-RAM cells in a four-core CMP system before and after applying prefetching. The evaluation was based on ten workloads, each of which consists of four different applications. The details of system configuration and workload makeup for evaluation are contained in Tables 1 and 2 in Section 4. With the presence of prefetching, the total write operation related requests increase by 35%. The extra write operations are introduced by the wasted prefetches that are never being used and by the prefetching-induced cache pollution. The increased STT-RAM write number makes bank conflicts more severe, limiting the potential of system performance improvement achieved through prefetching techniques.

Figure 3 (a) details the average latency of each type of LLC access request in a baseline system with prefetching, and Figure 3 (b) shows the increased stall cycles over a non-prefetching system of three types of requests incurred due to bank conflicts. Interestingly, the read and fill requests that are critical to application performance have to wait longer on average to obtain bank accesses compared to write back operations. Therefore, it can be inferred that to alleviate

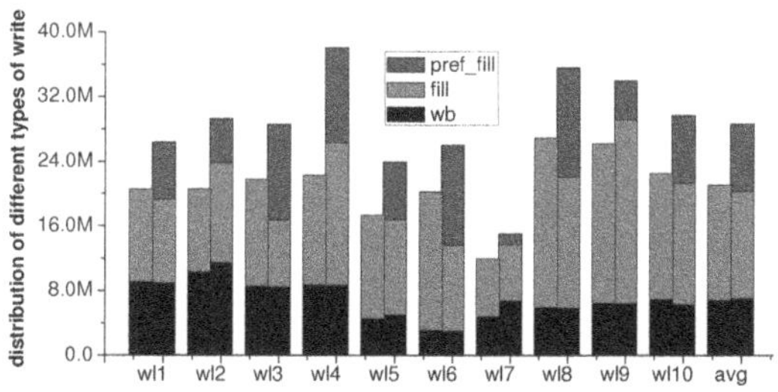

Figure 2: The distribution of different types of LLC write requests before (left bars) and after (right bars) applying prefetching.

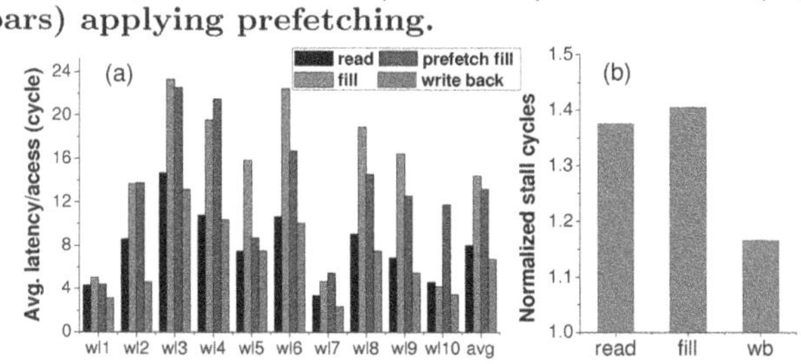

Figure 3: (a) The average latency in cycles of different types of LLC accesses; (b) the increased average latency in cycles of read, fill and write back requests.

the bank conflicts in a CMP system requires prioritization of different requests to the LLC by their criticality.

A read request sent from upper level cache is the most critical to be serviced quickly considering that the following data-dependent execution must stall until the data arrives. A fill request is slightly less critical than a read as it is triggered by a LLC miss and hence has relative long delay[1]. Compared with read and fill requests, a prefetch fill request is less critical since the prefetched data is less likely to be used immediately. Write backs are the least critical because the corresponding data is evicted from caches and not expected to be needed again soon. Based on the above observation, we propose a *basic priority assignment* (BPA) technique which prioritizes LLC requests based on their criticality. When multiple requests compete for an access port, the one with the highest priority will take the bank first.

Unfortunately, BPA alone may not be sufficient. A low-priority write back request once issued can still block future requests with higher priority due to the long write period of STT-RAM. For instance, a read could be blocked by a write back request if the read arrives at the same access bank servicing the write back request only a few cycles later. In such a situation, we allow the high-priority request to conditionally preempt the bank occupied by the low-priority requests. Similar to [6], we use RAD to determine at what stage of completion a low-priority request can be preempted. For example, a RAD of 25% of write back requests means that read/fill/prefetch fill requests can preempt the write back if the elapsed time the write back has occupied the bank is still less than a quarter of the total clock cycles required by an STT-RAM write operation. If the write back operation has completed more than 25%, it will continue to occupy the bank until the operation is finished.

Similar with [6], we augment each entry of the write buffer and *miss state handling register* (MSHR) with a 6-bit counter to record the percent completed of each write back request.

[1] In our simulations, for cache consistency, a fill request cannot forward its data to the upper level cache until the corresponding data write to the LLC is successfully completed.

We also add a 64-byte entry for each MSHR to buffer a preempted prefetch fill request. In a 20-entry write buffer with 128 MSHRs this requires an additional 8.1 KB of storage, which is less than 0.1% capacity of the LLC data array.

3.2 Priority Boosting (PB)

In multi-program environments, a *memory non-intensive* application can experience a severe slowdown or short-starvation while others are unfairly prioritized by the memory scheduler [5]. In traditional SRAM-based multiprocessors, short-starvation is only associated with main memory access. However, coupling aggressive prefetching techniques together with STT-RAM caches, which exhibit long write latencies, makes LLC another source of short-starvation in LLCNI applications in the following scenarios:

- Consider a LLCNI application is running with other LLCI applications, which frequently access LLC and/or generates prefetch requests. Those rare requests initiated by the LLCNI application could be flooded and blocked at the LLC bank by the numerous requests from the LLCI applications. Though our proposed BPA technique can potentially accelerate certain types of requests, it does not differentiate between their sources. Therefore, when several requests pending at an access bank have the same priority, the LLC controller will issue them *first-come-first-served.* On average, the requests from LLCNI applications have to wait longer compared with those from LLCI applications.
- Because a LLCNI application does not access LLC frequently, its corresponding cache blocks in LLC could move up to *least recently used* (LRU) position and get evicted more easily. Thus, a LLNCI read request has a higher probability to incur a cache miss and require a fill request compared with a LLCI program read. Hence, when reads are prioritized over fills as in BPA, the performance penalty against LLCNI will be further exacerbated as the LLCI read requests will starve the LLCNI fill requests.

In this work, we propose the PB technique to accelerate the requests sent by LLCNI applications and hence alleviate the short-starvation situation: once an application is identified as LLCNI compared with the other programs running together, we will assign the highest priority to its read and fill requests over all the other requests regardless of their types and resources. The PB technique assumes that the amount of LLC requests from a LLCNI application is much less than those from LLCI applications. Accelerating these small amount of LLC requests has little impact on the performance of LLCI applications but could significantly speedup the LLCNI requests. Consequently, the performance of the entire system improves.

Considering that the data access behavior of an application changes across its execution phases [12], a dynamic PB scheme is applied to accommodate such changes. Dynamic PB divides the execution of a multi-program workload into contiguous intervals. Within an interval i, the LLC access number to a single application $CurrentCount_i$ is counted and recorded. At the end of an interval, the application with the lowest $CurrentCount_i$ is selected. If its $CurrentCount_i$ is also lower than a pre-set threshold, the application is taken as having a low LLC access degree (i.e., it is LLCNI) and its read and fill requests will obtain the highest priority in the next interval $i+1$.

In order to reflect the temporal trend of an application's execution and give a higher weight to the most recent inter-

Table 1: System Configuration

Execution	4GHz, OOO 4 issues (up to 2 mem and 2 FP), 20 stages pipeline, gshare branch predictor, 1024-entry BTB, global history length 14, 20-cycle branch misprediction penalty, 256-entry ROB
Private caches	32KB L1IC, 4-way assoc, 4 banks, 4 ports, 1-cycle latency, 64B line 32 KB L1DC, 4-way assoc, 4 banks, 4 ports, 2-cycle latency, write back, 64B line 256 KB L2, 8-way assoc, 4 banks, 1 r/w port, 8-cycle latency, write back, 64B line
Prefetcher	Stream prefetcher with 64 streams, max prefetch degree 4, max prefetch distance 64
Shared L3 cache	8 MB STT-RAM shared L3, 16-way assoc, 8 banks, 1 r/w port, 20-cycle read, 48-cycle write, 64B line, write back, 0.34 nJ/read, 1.72 nJ/write
Main memory	Fixed 500-cycle latency

vals, we use Equation (1) [11] to calculate the access count of an application at the end of interval i, which will be propagated to the next interval. Two 10-bit counters per-core are required to record the LLC access degree of the applications [11]. There is a calculation latency to reclassify applications and start applying PB in a new interval. This delay (<100 cycles) is negligible compared to the length of the interval (typically 0.5 million cycles).

$$AccessCount_i = 0.5 \cdot AccessCount_{i-1} + 0.5 \cdot CurrentCount_i \quad (1)$$

4. SIMULATION SETUP

In our evaluation we used MacSim [18] to simulate a CMP system with four IBM POWER4 cores [7] with an integrated stream prefetcher. The stream prefetcher can trace multiple access streams between the LLC and main memory simultaneously and boost system performance significantly for applications with regular memory access patterns [4]. We obtained the STT-RAM cell parameters from measured data [8] and extracted the cache latency and energy parameters at the 65nm technology node from NVsim [15]. To better understand the interference between prefetching and LLC data management, we set a fixed delay for requests traveling between LLC and main memory. Table 1 summarizes the system configuration in detail.

We selected two and 16 benchmarks from SPEC CPU 2000 and 2006, respectively, to form ten workloads used in our evaluations. The pin-based [2] trace extraction tool combined with the MacSim tool set was modified to collect the execution trace of each benchmark. For each benchmark, we skipped the initialization phase ranging from 5 billion to 100 billion instructions and then execute up to 400 million instructions. Each multi-programmed workload consists of four different benchmarks, which include at least two benchmarks whose LLC *accesses per kilo-instructions* (APKI) are greater than eight, at least one benchmark with LLC APKI less than eigth, and at least one benchmark that issues more than one *prefetch requests per kilo-instructions* (PPKI). Table 2 summarizes the characteristics of each benchmark and which workload to which it belongs. Here, the data used for a benchmark characterization was collected by running it alone on the 4-core CMP system with stream prefetching.

5. SIMULATION RESULTS

5.1 Evaluation Metrics

To measure the overall performance of multiple programs running on a CMP system, we used the *harmonic mean of speedups* (HS) [10] defined as:

$$HS = \frac{N}{\sum_{i=0}^{N-1} \frac{IPC_i^{alone}}{IPC_i^{together}}}, \quad (2)$$

where IPC_i^{alone} is the *instruction per cycle* (IPC) of benchmark i when it runs exclusively on the CMP system. Accordingly, $IPC_i^{together}$ represents its IPC measured when running with other benchmarks, simultaneously. We also evaluated the *unfairness* of a CMP system [9] defined by

$$Unfairness = \frac{MAX\{IS_0, IS_1, \ldots IS_{N-1}\}}{MIN\{IS_0, IS_1, \ldots IS_{N-1}\}}. \quad (3)$$

Here, IS_i is the *individual slowdown* of benchmark i when running with other programs, which can be quantified as

$$IndividualSlowdown_i = \frac{IPC_i^{together}}{IPC_i^{alone}}. \quad (4)$$

Table 2: Benchmark Characterization

benchmark	APKI	PPKI	Workload
art	34.98	0.25	1, 2, 7
ammp	8.10	0.32	7, 10
bzip2	6.46	0.01	6, 10
gcc	0.56	0.34	3, 7
mcf	86.44	12.01	1, 9, 10
milc	21.03	.71	1, 8
zeusmp	14.78	2.15	2, 6, 7
gromacs	1.32	0.15	4, 9
cactusADM	5.24	0.73	3, 8
leslie3d	22.39	3.82	6, 8, 9
dealII	3.36	0.09	1
hmmer	3.03	0.35	5
GemsFDTD	27.22	1.65	3, 4, 5
h264ref	1.27	0.16	2
lbm	28.69	8.94	2, 3, 4
omnetpp	22.39	5.58	8, 9, 10
astar	2.67	0.02	5
xalancbmk	24.77	10.49	4, 5, 6

5.2 Enhancement by BPA

We evaluated the effectiveness of three BPA priority assignments depicted in Figure 4. For example, *prior 2* is a simple assignment which grants the highest priority to read and fill requests, the second highest priority to prefetch fill. A new request at higher level can preempt the access bank if the request at lower level completes 60% or less. From *prior 2* to *prior 4*, the aggressiveness of read requests increases.

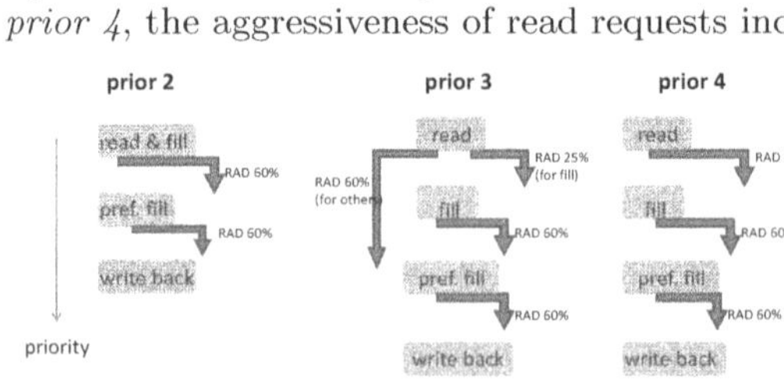

Figure 4: Three priority assignments.

Figure 5 shows the system performance of the different BPA assignments compared with the baseline design without request prioritization in the LLC. All three priority configurations achieve improved system performance. *Prior 2* performs the best among the three configurations: it obtains on average ~ 6.5% speedup over the baseline.

The key observations from Figure 5 are: (1) Giving read requests some form of higher priority provides a benefit to allow stalled execution to proceed more quickly, improving system performance even when LLC bank conflicts are intensive. (2) From the system perspective, a fill request turns out to be as critical as a read. Thus, increasing the priority of fill requests, as in *prior 2*, further improves performance. (3) Some workloads (*e.g.*, wl3, wl4, wl6, and wl8) are sensitive to fill requests. In such a situation, preempting fills with reads can actually hinder performance benefits. The system can operate more efficiently by assigning the same priority to fill and read requests. (4) Extremely aggressive priority to read requests, as in *prior 4*, helps minimally in overall system performance. As the system is also sensitive to buffer resource usage, additional preemption by read requests may result in overflows in the write buffer and MSHR.

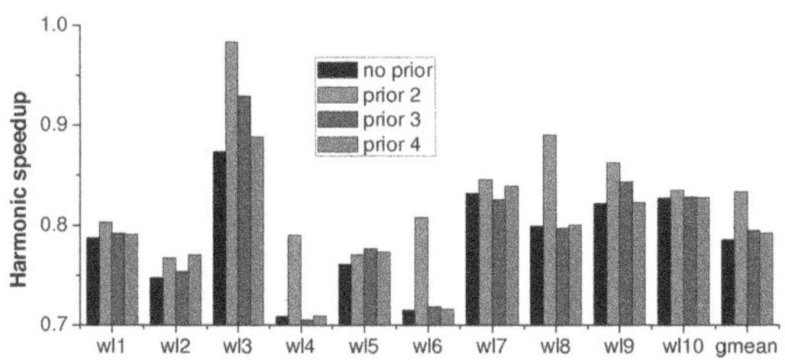

Figure 5: Harmonic mean of speedup (HS) comparison under different BPA assignments.

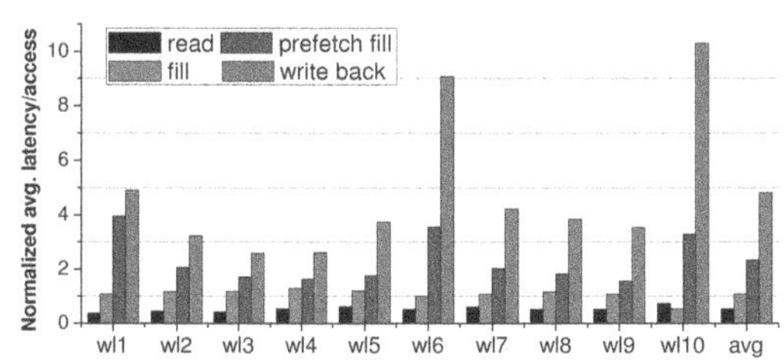

Figure 6: Average latency of different types of LLC requests when applying *prior 2*.

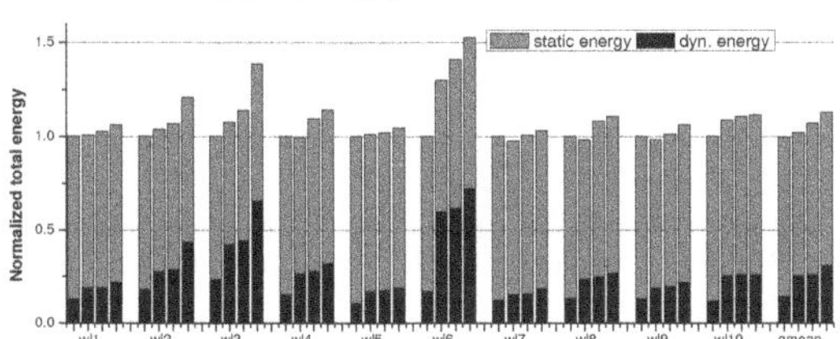

Figure 7: Comparison of LLC energy under different BPA assignments. From left to right in each subgroup: baseline, *prior 2*, *prior 3*, *prior 4*

To get more insight of performance improvement, we demonstrate the average latency of different types of LLC requests of *prior 2* in Figure 6. All the results are normalized to the corresponding numbers measured in the baseline design without prioritization. The results indicate that a large portion of the contention overhead has been shifted from reads to prefetch fill or write back requests while the latency of fill requests remains relatively unchanged.

Figure 7 compares the total energy consumption of LLC of the three priority assignments normalized to the baseline design without prioritization. In general, the retries of write operations induced by preemption unavoidably increases LLC dynamic energy. *Prior 2*, which does not stall fill requests, consumes less dynamic energy compared to the other two BPA assignments. In contrast, the dynamic energy in *prior 4* climbs dramatically because almost all the write operations experience at least one retry. Among all the workloads, *workload 6* is an outlier, suffering a dramatic penalty in dynamic energy from prioritization due to an abnormally high number of read requests and corresponding preemptions. Even with a STT-RAM LLC, static energy still dominates the total energy due to the peripheral circuitry. Thus, improved execution time from prioritization reduces static energy to counteract the dynamic energy increase. Our simulation results show that *prior 2* gracefully balances the system performance and energy consumption. Thus, we select it as the optimized BPA configuration.

5.3 Effectiveness of PB

To evaluate PB, we determined an interval for calculating the LLC access degree of applications by examining the total eviction numbers. Based on prior work [11], 8192 evictions from LLC were used for an interval. We empirically chose 819 (10%) as the threshold to define the access degree.

Figure 8 shows the normalized system performance when applying only PB, only BPA, and BPA+PB. The configuration with only PB achieves an average 4.4% improvement in performance. The gain mainly comes from boosting the speed of LLCNI applications. PB coupled with BPA speeds up by about 8.3% against the baseline design without prioritization. The additional improvement compared with the PB only scheme is from the accelerations of read and fill requests in LLCI applications, which efficiently reduce the waiting time on the critical path of execution.

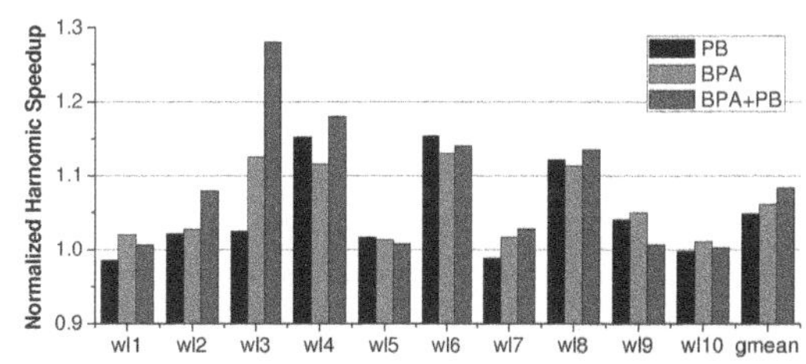

Figure 8: HS for PB-only, BPA-only and BPA+PA, normalized to a baseline w/o prioritization.

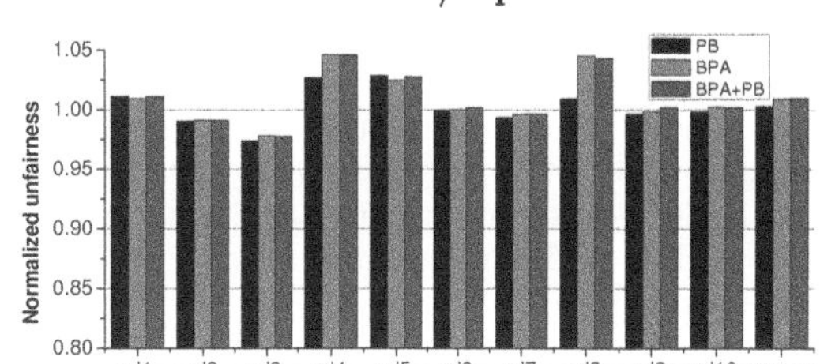

Figure 9: Unfairness comparison.

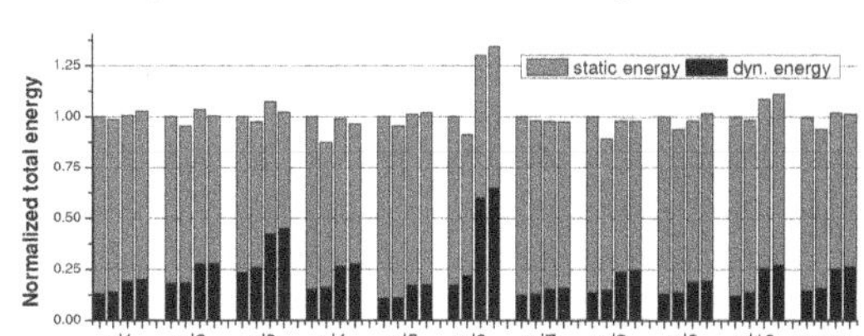

Figure 10: Distribution of total LLC energy consumption for, from left to right in each subgroup: baseline, PB, BPA, and BPA+PB.

PB is somewhat biased to LLCNI applications at the expense of the LLCI applications, which potentially increases

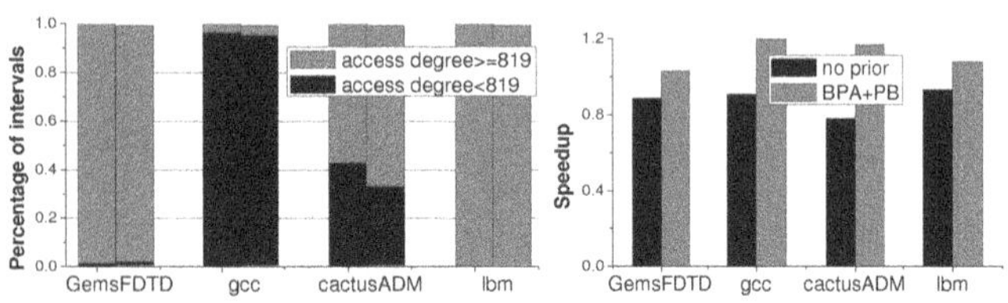

(a) (b)

Figure 11: (a) Distribution of LLC access degree. Left: baseline. Right: BPA+PB; (b) Individual speedup of each application in wl 3.

the unfairness of the overall system: in some extreme situations, the individual slowdown of a LLCNI application is already most heavily impacted due to the multiprogram execution prior to applying PB and the unfairness after applying PB could further increase as denoted in Equation 3. Figure 9 shows the unfairness of each workload with different configurations normalized to that of baseline design. Fortunately, in most workloads, before boosting, the individual slowdown of LLCNI applications is not most heavily impacted by the multiprogram environment. Thus, the unfairness under all configurations increases only nominally compared with the system without PB and BPA.

Figure 10 shows the normalized total energy distribution. PB slightly increases the dynamic energy due to prioritizing the requests of the LLCNI application resulting in more fills; thanks to the performance improvement, the corresponding static energy is reduced such that the total energy consumption of LLC is lower than the baseline. For PB+BPA, the dynamic energy geometric mean increases by 81% and the static energy is reduced by 13%. Due to the tradeoff, the overall energy increase is under 2%.

In Figure 11, we provide a case study of *workload 3* before and after applying BPA+PB to the LLC management. As shown in Figure 11(a), `gcc` exhibits a low access degree in almost all the intervals during execution while `GemsFDTD` and `lbm` generate a high volume of requests across most intervals for both configurations. `cactusADM` displays $\sim 40\%$ and 35% of low access degree in systems without and with prioritization, respectively. Our proposed dynamic PB is able to apply boosting to `cactusADM` as needed, without blindly treating it as a LLCNI or LLCI application during the entire execution. Figure 11(b) shows the individual speedup for both configurations. All the benchmarks speed up benefiting from BPA while `gcc` and `cactusADM` also enjoy the merit provided by PB. Moreover, the unfairness of this workload decreases from 1.19 to 1.16.

5.4 Prefetch Filtering

It is possible to combine PB+BPA with a prefetch filter [17] to reduce extraneous prefetch requests. This can further alleviate LLC access conflicts to provide additional performance improvements while decreasing LLC energy consumption as shown in Figure 12. When compared with PB+BPA alone the number of write requests is reduced by 10% yielding an additional 3.2% improvement in performance and 4% reduction in total energy. This experiment indicates that prioritization can cooperate with other prefetch control mechanisms delivering high performance and energy efficiency to multicore systems with prefetching.

6. CONCLUSION

In CMP systems with aggressive prefetching, STT-RAM based LLCs suffer from increased latency due to higher write

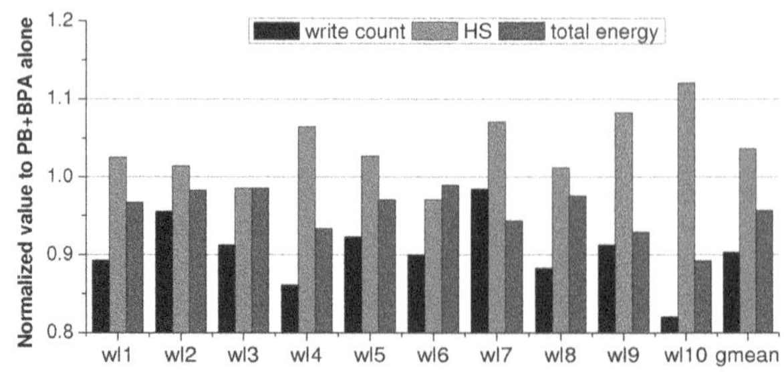

Figure 12: Impact of adding prefetch filter to PB+BPA. From left to right number of write requests, HS, and total LLC energy consumption.

pressure. To solve this problem, we propose two prioritization policies to reduce the waiting time of critical requests: 1) basic priority assignment (BPA) grants different priorities to different types of LLC requests, and 2) priority boosting (PB) dynamically identifies LLC non-intensive applications and accelerates their requests. Combining these two techniques together, the negative impact induced by bank conflicts can be significantly alleviated. Our simulation results show that the system performance can be improved by 8.3% on average and a peak performance of 28% compared to the baseline design without any prioritization.

7. REFERENCES

[1] J. Doweck. Inside intel core microarchitecture and smart memory access. *Intel White Paper*, 2006.

[2] C.-K. Luk et al. Pin: Building customized program analysis tools with dynamic instrumentation. In *PLDI*, pages 190–200, June 2005.

[3] E. Ebrahimi et al. Coordinated control of multiple prefetchers in multi-core systems. In *MICRO*, pages 316–326, 2009.

[4] E. Ebrahimi et al. Techniques for bandwidth-efficient prefetching of linked data structures in hybrid prefetching systems. In *HPCA*, pages 7–17, 2009.

[5] E. Ebrahimi et al. Prefetch-aware shared resource management for multi-core systems. In *ISCA*, pages 141–152, 2011.

[6] G. Sun et al. A novel architecture of the 3d stacked mram l2 cache for cmps. In *HPCA*, pages 239–249, 2009.

[7] J. Tendler et al. Power4 system microarchitecture. *IBM Technical White Paper*, Oct. 2001.

[8] M. Hosomi et al. A novel non-volatile memory with spin torque transfer magnetization switching: Spin-ram. In *IEDM*, pages 459–462, 2005.

[9] O. Mutlu et al. Stall-time fair memory access scheduling for chip multiprocessors. In *MICRO*, pages 146–160, 2007.

[10] S. Eyerman et al. System-level performance metrics for multiprogram workloads. *IEEE Micro*, 28(3):42–53, 2008.

[11] S. Srinath et al. Feedback directed prefetching: Improving the performance and bandwidth-efficiency of hardware prefetchers. In *HPCA*, pages 63–74, 2007.

[12] T. Sherwood et al. Basic block distribution analysis to find periodic behavior and simulation points in applications. In *PACT*, pages 3–14, 2001.

[13] W.-F. Lin et al. Filtering superfluous prefetches using density vectors. In *ICCD*, 2001.

[14] W. Zhao et al. Macro-model of spin-transfer torque based magnetic tunnel junction device for hybrid magnetic-cmos design. In *IBMSW*, pages 40–43, 2006.

[15] X. Dong et al. Nvsim: A circuit-level performance, energy, and area model for emerging nonvolatile memory. *IEEE TCAD of ICS*, 31(7):994–1007, 2012.

[16] X. Wu et al. Hybrid cache architecture with disparate memory technologies. In *ISCA*, pages 34–45, 2009.

[17] X. Zhuang et al. A hardware-based cache pollution filtering mechanism for aggressive prefetchers. In *ICPP*, 2003.

[18] MacSim. http://code.google.com/p/macsim/.

Variability-Aware Design of Energy-Delay Optimal Linear Pipelines Operating in the Near-Threshold Regime and Above

Qing Xie, Yanzhi Wang, and Massoud Pedram
University of Southern California
Department of Electrical Engineering
Los Angeles, California, United States, 90089
{xqing, yanzhiwa, pedram}@usc.edu

ABSTRACT

Soft-edge flip-flop based pipelines can improve the performance and energy efficiency of circuits operating in the super-threshold (supply voltage) regime by allowing opportunistic time borrowing. The application of this technique to near-threshold regime of operation, however, faces a significant challenge due to large circuit parameter variations that result from manufacturing process imperfections and substrate temperature changes. This paper thus addresses the issue of variability-aware design of energy-delay optimal linear pipelines that are aimed at operating in both the near-threshold and super-threshold regimes. Precisely, this goal is achieved by deriving optimal delay line configuration included in the soft-edge flip-flop design for the near-threshold and the super-threshold operations. The key is to ensure that the same transistor sizes result in effective operation of the delay line (and hence appropriate setting of the transparency window size) in both regimes of operation under process induced variations. Experimental results demonstrate the efficacy of the proposed solution.

Categories and Subject Descriptors

B.6.1 **[Hardware]:** Logic Design – *sequential circuits*

Keywords

Energy-delay optimal pipeline design, soft edge flip-flop, time borrowing, near-threshold, ultra-low voltage, parameter variability

1. INTRODUCTION

With the increase in demand for battery-powered devices and wireless equipments, the need for energy-efficient design gains growing attention. The ultra-low voltage operation, in particular, *near-threshold* (NT) operation, is very effective in minimizing energy consumption by reducing supply voltage to near the threshold voltage V_{th} with a sacrifice of the performance of the circuits [1][2]. It is especially beneficial for those applications with relaxed performance requirements, such as medical monitoring and building health monitoring. Previous work on NT operation proved the existence of the *minimum energy* (*operation*) *point* (MEP), which is the optimal supply voltage level to minimize energy consumption, and derived the MEP voltage analytically for some integrated circuits [3][4].

Although offering promising energy reduction, one of the major concerns of NT operation is that digital circuits operating in the NT regime are very sensitive to the process variation. For example, $(3\sigma/\mu)$ delay variation of a combinational logic block at 0.5 V increases by a factor of 2.5 compared to that at 1 V using 90nm technology [5].

Pipelined data path in a modern processor is a major contributor to the total energy consumption of the processor. Many techniques such as pipeline gating [6], clock gating [7], and voltage scaling [8] are proposed to reduce the energy consumption in traditional *super-threshold* (ST) regimes. In the NT regime, the pipelined circuits should be designed not only energy-efficient, but also variability-tolerant. The authors in [11] developed a statistical methodology to enhance the yield of the pipelined circuits, considering the inter-die and intra-die variation. Recently, a two-phase latch-based design strategy is proposed for the synchronous pipelined circuits in the NT operation regime to derive the optimal ratio between the total width of sequential logic and that of combinational logic [10]. However, the latch-based design has many limitations including hold time violation issues, design difficulties using standard EDA tools, and requirement of an extra clock network, which is power and area inefficient.

An SEFF is the same as a D-flip-flop except that a *delay line* (DL) is added to postpone the clock edges of the master latch to create a *transparency window*, during which both master and slave latches are turned on. Using the transparency window, the SEFFs can pass the slack time across the pipeline stages. However, the DLs mentioned above consume extra amount of energy. Authors in [9][12] showed that the larger *transparency window size* (also known as *softness*) is, the more energy overhead it brings. Applying soft-edge flip-flops (SEFFs) is a useful technique to combat the process variation and improve the operating frequency. The authors of [12] proposed to utilize soft-edge flip-flops in sequential circuits to improve the yield in the presence of process variation. In [9], the authors proposed an SEFF-based pipeline design methodology jointly considering the voltage scaling and time borrowing, and demonstrated noticeable reduction in the *energy-delay product* (EDP) in the ST regime.

In this work, we present an SEFF-based pipeline design methodology for both ST and NT operation regimes targeting at minimizing the EDP. The stage delay increases exponentially with the supply voltage of pipelined circuits in the NT regime. However, we show that the SEFF-based pipeline designed and optimized in the ST regime [9][12] cannot provide enough softness to achieve the minimum EDP in the NT regime due to the delay variation in delay lines embedded within the SEFFs. To design a single delay line that can minimize the EDP in both the ST and NT regimes, we present a novel delay line structure. Precisely, we add a PMOS header on top of the traditional delay line (beneath the supply voltage rail). The header PMOS results in a slight supply voltage drop on the delay line, which is negligible in the ST regime but has a significant impact on the transparency window size in the NT regime. In this way, the proposed pipelined circuits can provide appropriate softness in both ST and NT regimes and subsequently achieve the optimal EDP.

This research is sponsored in part by grants from the Defense Advanced Research Projects Agency and the National Science Foundation.

GLSVLSI'13, May 2–3, 2013, Paris, France.

We formulate the EDP minimization problem for the SEFF-based pipelined circuits as a mathematical programming problem, where the clock period, sizes of the PMOS header, and configurations of the delay lines are the optimization variables. We construct a Pareto-optimal energy/power consumption vs. flip-flop softness tradeoff curve with respect to PMOS header size and the delay line configuration. We take into account the delay variation of both combinational and sequential logics, and derive the joint distribution of clock-to-q delay, setup time, stage delay, and flip-flop softness. The timing constraints of pipelined circuits are enforced by using the $3\sigma/\mu$ delay. Experimental results show that the reduction in the EDP is 14% and 16.5% in the ST and NT operation regimes, respectively.

2. SEFF-BASED PIPELINE

2.1 Soft-Edge Flip-Flops

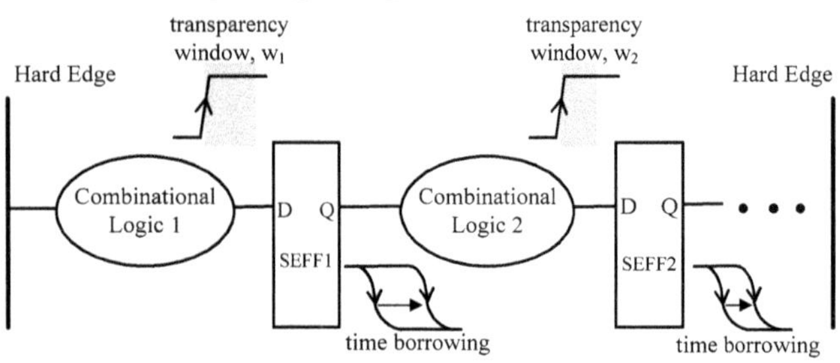

Figure 1. A linear pipeline with soft-edge flip-flops.

Figure 1 depicts a general synchronous SEFF-based linear pipelined circuit. Considering the data consistency between the SEFF-based pipeline and the input and output environments, we impose hard boundary conditions using traditional hard-edge flip-flops at the beginning and end of the pipelined circuits. Between the two hard edges, the pipelined circuit has multiple combinational logic stages whose delay is affected by process, voltage, and temperature (PVT) variations. We build the stage register using the SEFFs. The key idea is to postpone the clock signal for the master latch to create a transparency window, which enables time borrowing across the pipeline stages.

The SEFF-based pipeline has the advantages of higher operating frequency and variability tolerance due to the following two reasons. First, the operating frequency of a traditional hard-edge pipeline is always limited by the delay of the critical stage. While in SEFF-based pipelines, the slack time from the non-critical stages can be effectively passed to the critical stage so that the overall performance can be improved [2]. Second, we utilize the fact that the local random process variations are alleviated for deeper combinational logics since the variations cancel out when the logic depth increases [5][12]. Thus, the soft edges between the stages reduce the sensitivity of the circuits on process variations.

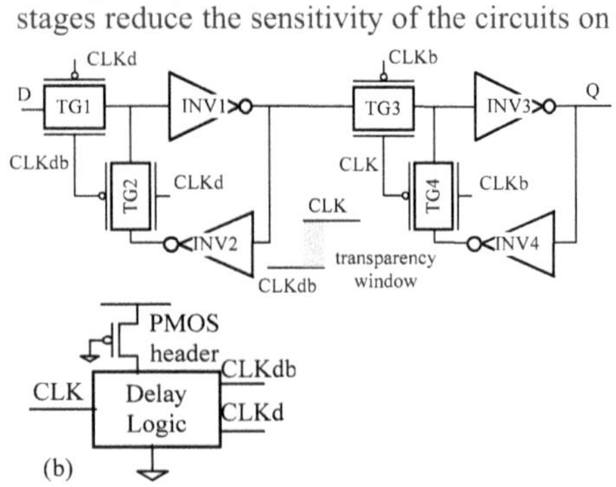

Figure 2. Design of the positive-edge triggered soft-edge master slave flip-flops (a) and the proposed delay line (b).

The SEFFs come with the price of extra amount of power consumption. As shown in Figure 2, the DLs are typically a series of properly sized inverters. We tune the softness, which is the delay of DL, through sizing or adding/removing the inverters in DL. In general, the power consumption of the SEFF satisfies a positive relationship versus the softness in both ST and NT regimes, as shown in Figure 3. We fit the power consumption of an SEFF versus softness using a linear function.

SEFF also affects the timing conditions the pipeline stages. Intuitively, all three *critical times*: the setup time, hold time and clock-to-q delay, are postponed by the transparency window size, denoted by w. Since the data can be captured at the end of the transparency window, the setup time decreases by w. The hold time increases by w since the data needs to be stabilized during the transparency window. In the worst case, the clock-to-q delay also increases by w since the input data may come at the end of the transparency window. Thus, we have the modified timing conditions for the i-th stage of the SEFF-based pipeline as follows

$$\begin{gathered} t_{cq,i-1} + w_{i-1} + D_{max,i} + t_{se,i} - w_i \leq T_{clk}, \\ t_{cq,i-1} + D_{min,i} \geq t_{h,i} + w_i \end{gathered} \tag{1}$$

where $D_{max,i}$ and $D_{min,i}$ are the worst-case and best-case delay of the i-th combinational logic, respectively. T_{clk} is the clock period of the pipeline.

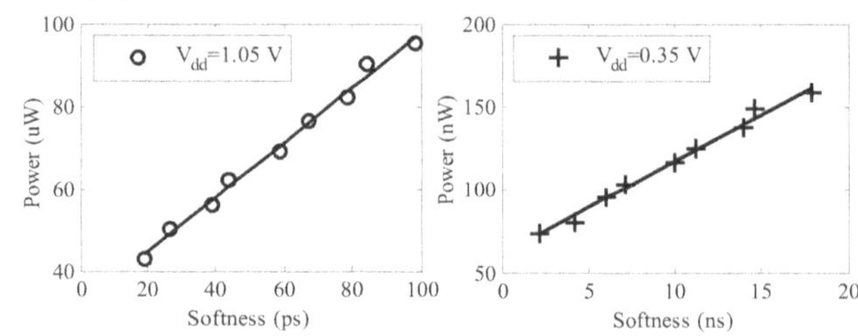

Figure 3. Power consumption of an SEFF versus softness.

2.2 Energy-Delay Product

Energy per throughput is typically used as the metric of the pipelined circuits. The throughput is defined as the number of observed output data divided by the observation duration, which is the product of number of clock cycles and the clock period,

$$thruput = \frac{\#\, of\ output\ data}{\#\, of\ clock\ cycles \times T_{clk}} \propto \frac{1}{T_{clk}}. \tag{2}$$

For a pipeline in steady-state, throughput defined in (2) is proportional to the inverse of the clock period. To account for energy consumption, we define the *energy per throughput* as,

$$\frac{Energy}{thruput} \propto E_c \cdot T_{clk}, \tag{3}$$

where E_c is the total energy consumption per clock cycle of the entire pipeline. We use the *energy-delay product*, $E_c \cdot T_{clk}$, as the cost function in this work.

Although the SEFFs consume extra energy, especially for large transparency windows, they can be utilized to improve the energy efficiency if properly designed. The reason is that the leakage energy per clock cycle decreases as the clock period is reduced. This is extremely helpful for pipelined circuits operating in the NT operation regime due to the fact that the leakage energy consumption plays a more important role in the NT operation regime [1][2], as shown in Figure 4.

Figure 4 depicts the leakage energy consumption per cycle (here a cycle is defined as the worst-case delay of that combinational circuit at that voltage level) for some selected ISCAS'85 benchmarks at different supply voltage levels, under the 32/28nm technology [13]. The circuit delay increases exponentially and the leakage power decreases linearly when the supply voltage drops [1][2] thus the leakage energy consumption increases. In contrast, the dynamic energy consumption drops continuously. In the ST operation regime ($V_{dd} > 0.6$ V), the leakage energy as a percentage in total energy is low compared with the dynamic energy, while in the NT operation regime ($V_{dd} < 0.6$ V), it plays an important role.

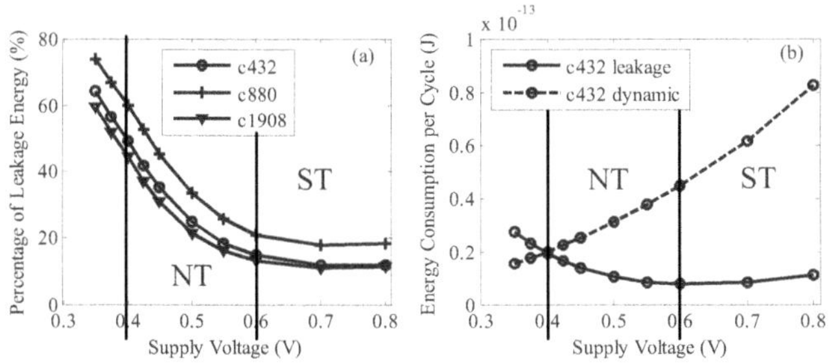

Figure 4. Leakage energy consumption as a percentage of total energy consumption (a) and the comparison of leakage and dynamic energy consumption (b) for some ISCAS'85 benchmarks operating at different supply voltages.

3. NEAR-THRESHOLD REGIME

The authors in [9] addressed the transparency window assignment problem to minimize the EDP for SEFF-based pipelined circuits in the traditional ST regime, considering the voltage scaling and delay variability. In this work, we focus on the pipelined circuits that are designed to operate in both NT and ST regimes. Thus the issues that we need to address include: 1) how to jointly determine the transparency window sizes and the clock period considering the delay variations and the timing yield; and 2) how to obtain the optimal DL design that minimizes the EDP in both the NT and ST regimes, accounting for the fact that the circuit delay follows different relations versus the supply voltage (i.e., α-power law in the ST regime and exponential relation in the NT regime).

3.1 Timing Variability

The process variation, such as Random Dopant Fluctuation (RDF), results in variation of some important parameters such as the threshold voltage v_{th}. Both of the inter-die (global) and intra-die (local) threshold voltage variations cause the delay variation in logic circuits. The inter-die variation shifts the delays of all logic circuits in the same direction, i.e., either all increase or all decrease. In contrast, the delay variations caused by the intra-die variation are completely shifted to random directions. These two types of variations are handled in different ways. The inter-die variation can be effectively mitigated by body biasing [14]. Thus, in this work, we account for the intra-die variation only and consider the delays follow the Gaussian distribution $N(\mu,\sigma)$.

3.1.1 Timing variability in combinational circuits

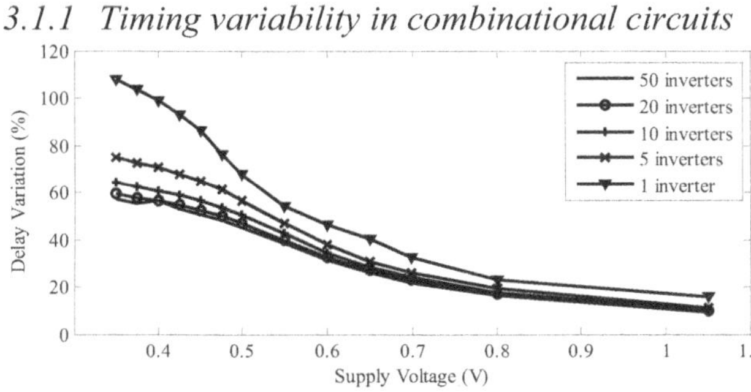

Figure 5. 3σ delay variations of an inverter chain versus the supply voltage obtained using Monte Carlo simulation.

It is known that the on-current of a CMOS gate, which determines the circuit delay, is very sensitive to the variation of threshold voltage in the NT operation regime. Precisely, the on-current of a circuit in the NT regime is exponentially proportional to the threshold voltage [1][2]. Thus, the variation in delay becomes much more significant in the NT regime [5]. We perform 5000 Monte Carlo simulations using 32/28nm technology and assume 10% intra-die v_{th} variation. Figure 5 shows the $3\sigma/\mu$ delay variation of several FO4 inverter chains, where μ and σ are the mean and standard variation of the delay distribution. The delay variation increases by 5X at $v_{dd} = 0.35$ V, compared to that at $v_{dd} = 1.05$ V. The results also show that the delay variation reduces as the length of inverter chain increases. This is due to the effect that the random v_{th} variations have more chance to cancel out with each other for a long inverter chain.

3.1.2 Timing variability in flip-flops

Similar to the combinational circuits, the flip-flops also have delay variations in their components, which results in the variations of the critical times [15][16]. For example, according to [15], the setup time is the summation of the delays of gates TG1, INV1, and INV2. Figure 6 shows the distributions of the setup time and clock-to-q delay under same v_{th} variation as above.

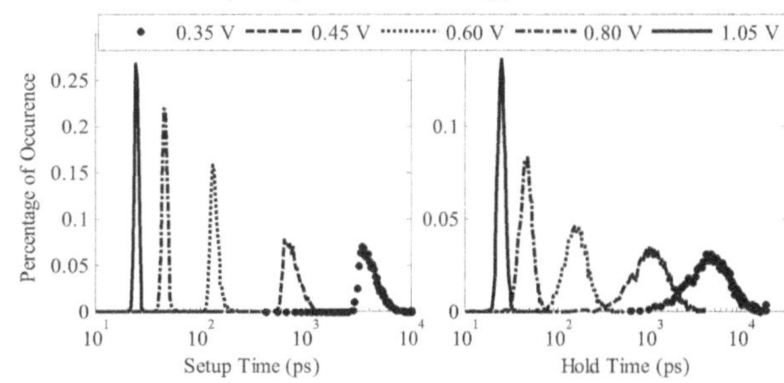

Figure 6. Distributions of the setup time (left) and clock-to-q (right) time under threshold voltage variation.

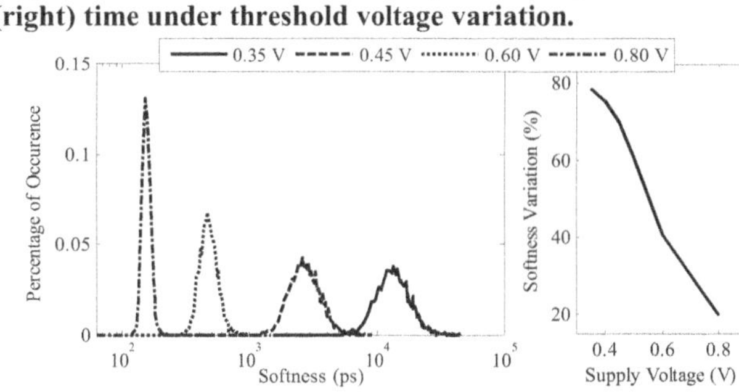

Figure 7. Distribution (left) and $3\sigma/\mu$ variation (right) of the transparency window under the threshold voltage variation. This DL provides 84 ps transparency window at 1.05V.

In addition, the SEFFs incur another type of the delay variation on the DLs, that is, the transparency windows are also affected by the delay variation. Since the DL is an independent part in the SEFF and only postpones the clock signal of the mater latch, the variation in DLs does not affect the critical times. Figure 7 shows the distribution and $3\sigma/\mu$ variation of the transparency window size under the threshold voltage variation.

3.2 Yield of the Pipeline

The timing yield of the pipelined circuits is defined as the probability that all the stages in the designed pipeline meet a certain delay target. Under the process variation, the delay is a random variable and can be modeled using a normal distribution $\boldsymbol{N(\mu,\sigma)}$ with mean value of $\boldsymbol{\mu}$ and standard variation of $\boldsymbol{\sigma}$. In (1), we define two new random variables for setup time constraint and hold time constraint as follow,

$$D_{setup,i} = t_{cq,i-1} + w_{i-1} + D_{max,i} + t_{se,i} - w_i \quad (4)$$

$$D_{hold,i} = t_{h,i} + w_i - t_{cq,i-1} - D_{min,i}$$

To enforce the timing yield, we rewrite (1) as follows and set the timing conditions using the $\pm 3\sigma$ delays of $D_{setup,i}$ and $D_{hold,i}$,

$$\begin{aligned} D_{setup,i}(3\sigma) &\le T_{clk} \\ D_{hold,i}(-3\sigma) &\le 0 \end{aligned} \tag{5}$$

In this work, we apply frequency scaling to find the minimum T_{clk} so that condition (5) is met.

Note that all random variables in $D_{setup,i}$ are independent of each other due to the intra-die variation. First, the distribution of the combinational delay $D_{max,i}$ is independent of SEFF. Second, the critical times in SEFF are also independent of the transparency window, because the DL is a separate part in SEFF and is not involved in determining the critical times. Finally, $t_{se,i}$ and $t_{cq,i-1}$ comes from different SEFFs, and thereby they are independent. Similarly, all random variables in $D_{hold,i}$ are also independent of each other. Since the sum of several independent normal random variables is still a normal random variable, we calculate the mean value and standard variation of $D_{setup,i}$ and $D_{hold,i}$ as follows:

$$\begin{aligned} \mu_{total} &= \sum_{i=1}^{N} \mu_i \\ \sigma_{total} &= \left(\sum_{i=1}^{N} \sigma_i^2\right)^{1/2} \end{aligned} \tag{6}$$

3.3 PMOS Header in Delay Line

As we mentioned above, the low supply voltage in NT regime significantly slows down the circuits, including both the combinational logics in the pipeline stages and the DLs that determine the transparency window sizes. In addition, the delay variation of the circuit depends on the logic depth. However, the critical paths in the combination circuits are typically much longer than the DLs. Thus, a fundamental issue is that the DLs, which are traditionally designed and optimized in the ST regime to provide the optimal transparency windows, may not work optimally in the NT regime. In this work, we propose a novel structure of the DL and a design methodology to optimize the EDP of the pipeline in both NT and ST regimes, in order to ensure that the SEFF-based pipeline works optimally in these two regimes.

Figure 7 shows that the $3\sigma/\mu$ variation of the transparency window size increases significantly in the NT regime due to its simple structure, i.e., a few inverters. To minimize the EDP, we have to design a relatively long DL, which can provide enough softness to tolerate the delay variation. For example, to provide 10ns softness at 0.35V, we need to over-design the DL to have the mean softness of 50ns if the 3σ delay variation is 80%. Thus, the SEFF-based pipelines that are designed and optimized in the ST regime typically cannot provide enough softness in the NT regime since the delay variation increases significantly.

We propose to add a PMOS header on top of the traditional DL, as shown in Figure 2 (b). The PMOS header delays the DL in two ways. First, the on-current decreases so that the DL takes more time to make the transition. Second, the existence of the PMOS header results in a voltage drop at the source terminal of DL, which is equivalent to reducing the supply voltage. This effect is small in the ST regime since the typical supply voltage is much larger than the voltage drop. However, in the NT regime, this voltage drop is no longer negligible compared to the low supply voltage. In contrast, it plays an important role in extending the softness since the delay is exponentially related to the supply voltage in the NT regime. Therefore, the proposed DL structure provide us a large transparency window size in the NT regime while only slightly affects that in the ST regime.

Figure 8 depicts the normalized transparency window size of two modified DLs with different PMOS header widths. We normalize the window size to the window size of the original header-less DLs. Figure 8 shows that with the proposed PMOS header we can increase the transparency window size by more than 3X in the NT regime while only slightly affect that in the ST regime. Another advantage of the proposed DL structure is the DL power savings due to the stacking effect. We determine the width of the PMOS header using the following design methodology.

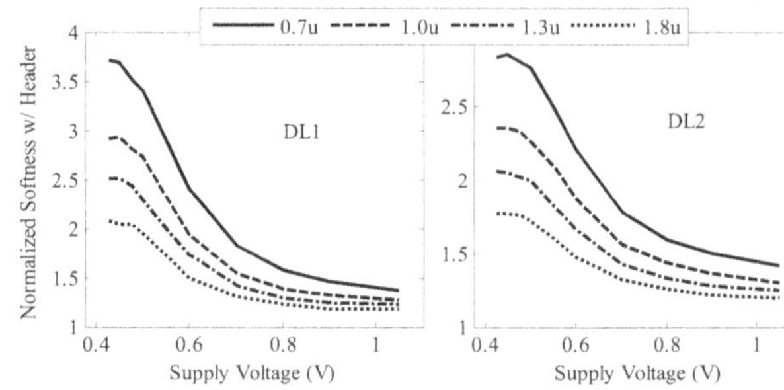

Figure 8. Normalized softness versus supply voltage for DL1 (left, 20ps at 1.05V) and DL2 (right, 84ps at 1.05V).

4. OPTIMIZATION OF ENERGY-DELAY PRODUCT IN SEFF-BASED PIPELINE

4.1 Formulation of the Optimization Problem

We focus on minimizing the EDP in the SEFF-based pipelined circuits. The energy consumption per clock cycle, denoted by E_c, contains two components: the dynamic energy consumption E^{dyn} and the leakage energy consumption E^{leak}. The component E^{dyn} is determined by the total capacitance being charged and discharged during each transition, the switching factor and the supply voltage, and thus it is independent of the clock period. In contrast, E^{leak} is linearly proportional to the clock period. For each of them, we separate the energy consumed by the combinational logics and flip-flops. Therefore, E_c is given by,

$$\begin{aligned} E_c &= E^{dyn} + E^{leak} = E_{cb}^{dyn} + E_{FF}^{dyn} + E_{cb}^{leak} + E_{FF}^{leak} \\ &= \sum_{i=1}^{N} (E_{cb,i}^{dyn}(V_{dd}) + P_{cb,i}^{leak}(V_{dd}) \cdot T_{clk}) \\ &+ \sum_{j=1}^{N-1} \Big(E_{dff,j}^{dyn}(V_{dd}) + E_{DL,j}^{dyn}(V_{dd}, w_j) + P_{dff,j}^{leak}(V_{dd}) \\ &\quad \cdot T_{clk} + P_{DL,j}^{leak}(V_{dd}, w_j) \cdot T_{clk}\Big) \end{aligned} \tag{7}$$

The subscript *cb* in (7) stands for combinational logics. We separate the energy consumed by SEFFs into two parts: *dff* stands for a traditional hard-edge D-flip-flop (DFF), and *DL* stands for delay line. We formulate the EDP optimization problem in SEFF-based pipelined circuits as follows.

- Given: Characterized distribution of w_j, t_{se}, t_{cq}, t_h, $D_{max,i}$, $D_{min,i}$; energy consumption $E_{cb,i}^{dyn}$, $E_{dff,j}^{dyn}$, $E_{DL,j}^{dyn}$; power consumption $P_{cb,i}^{leak}$, $P_{dff,j}^{leak}$, $P_{DL,j}^{leak}$ for $i \in [1, \dots, N], j \in [1, \dots, N-1]$, at each specific supply voltage V_{dd}.
- Find: w_j^0, s_j, and T_{clk}, for $j \in [1, \dots, N-1]$.
- Minimize: $EDP = E_c \cdot T_{clk}$.
- Subject to: Constraints (4), (5), and (6).

Note that in (7), w_j stands for the transparency window size of the j-th SEFF. It depends on the window size w_j^0 of the original header-less j-th DL, and PMOS header width s_j. w_j^0's and s_j's are the actual design variables for the DL. Thus, we find the appropriate w_j^0 and s_j values as our final results.

4.2 Pareto Optimal Curve of Energy/Power-Softness Trade-off

To solve the EDP optimization problem, we first derive the relation between w_j and $E_{DL,j}^{dyn}(V_{dd}, w_j)$, $P_{DL,j}^{leak}(V_{dd}, w_j)$, respectively. The former term depends on w_j since the PMOS header introduces extra capacitance and reduces the voltage swing of the DLs, while the latter one does due to the stacking effect. To characterize these relations, we sweep different original DL window sizes w_j^0's and PMOS header sizes s_j's, as shown in Figure 9. Since the energy overhead is positive related to the transparency window size, it would be more desirable if we can achieve larger window size with smaller energy overhead. In Figure 9, the circle markers represent the original header-less DL windows and the cross markers represent the proposed DLs using different PMOS headers. We find that by adopting PMOS headers, we reduce the energy/power overhead and simultaneously extend the window size. We achieve better energy-softness trade-off by applying relatively small PMOS header. However, the PMOS header cannot be too small otherwise it increases the skew between the delayed CLKd and CLKdb signals. We limit the minimum width of PMOS to be 0.7um so that the skew between CLKd and CLKdb signals is acceptable, according to our simulation results.

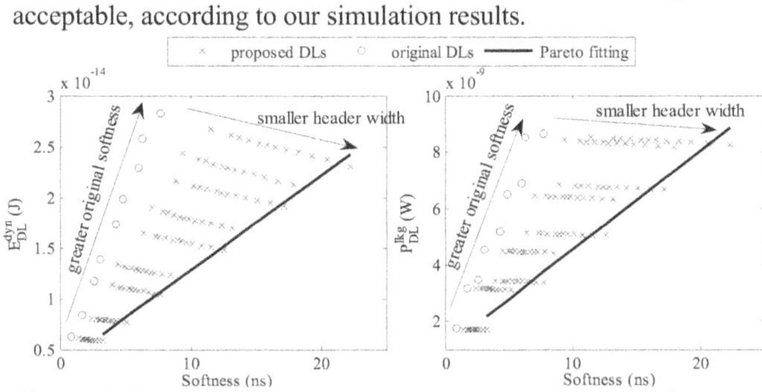

Figure 9. Dynamic energy and leakage power overhead versus softness for different original header-less DLs and header width at $V_{dd} = 0.4\ V$.

Figure 9 shows many energy/power-softness trade-off options at a specific V_{dd}, however, we are only interested in those with large transparency window size and small energy/power consumption. Therefore, we create Pareto-optimal curves for both (dynamic energy consumption vs. softness) and (leakage power vs. softness), as shown in Figure 9. We fit the two Pareto-optimal curves using linear relations so that, we have,

$$\begin{aligned} Pareto(E_{cb,i}^{dyn}(w_j)) &= a_E \cdot w_j + b_E \\ Pareto(P_{DL,j}^{leak}(w_j)) &= a_P \cdot w_j + b_P \end{aligned} \tag{8}$$

where a_E, b_E, a_P, b_P are fitting parameters. The Pareto-optimal curves provide the optimal energy/power-softness trade-off points that we can achieve using the proposed DLs. We use (8) to substitute $E_{DL,j}^{dyn}$ and $P_{DL,j}^{leak}$ in (7).

4.3 Solution Method

The cost function in (7) includes the product of T_{clk} and w_j, and thereby is not convex. To solve the optimization problem, we perform a ternary search on T_{clk}. During each search, we solve a linear programming (LP) problem for a fixed T_{clk} to obtain the optimal w_j and the corresponding cost function value, based on which we narrow down the search range of T_{clk}. The optimal clock period is determined where we find the minimal value of $E_c \cdot T_{clk}$. Note that although the Pareto-optimal curves in (8) provide optimal trade-off points, the DL configurations are discrete so that not all points along the Pareto curves are feasible. In general, we pick the feasible point (each feasible point gives a combination of w_j^0 and s_j) that has the closest but smaller softness value compared with the optimal w_j, since a larger value of w_j could potentially cause hold time violation, which is more difficult to handle. However, smaller values of w_j could also potentially lead to setup time violations. Thus we check the setup time inequality (5) and slightly extend the clock period to solve the setup time violation, if there is. The design flowchart is given in Figure 10.

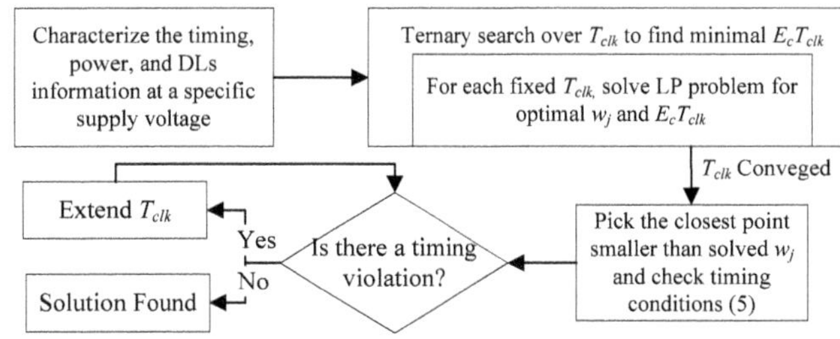

Figure 10. Proposed design flowchart of SEFF-based pipeline.

Figure 10 show the proposed design flowchart of the SEFF-based pipeline at a specific supply voltage. To design a pipelined circuits working in both the ST and NT operation regime, we perform the design flow for two desired supply voltage in these regimes, e.g. 0.4V and 0.8V, and configure the DLs and PMOS header as,

$$\begin{aligned} w_j &= \eta \cdot w_{j,NT} + (1-\eta) \cdot w_{j,ST} \\ s_j &= \eta \cdot s_{j,NT} + (1-\eta) \cdot s_{j,ST} \end{aligned} \tag{9}$$

where the η is the ratio of the time that the pipeline is being operated in the NT operation regime.

5. EXPERIMENTAL RESULTS

We apply the proposed design flow to some example pipelined circuits, while the stages are synthesized using ISCAS'85 benchmarks. We adopt the Synopsys 32/28nm technology and explore the supply voltage ranging from 0.35V to 1.05V. Based on these combinational benchmarks, we create four example pipelines, as shown in Table 2. We compare the proposed design method with the method proposed in [9], which designs and optimizes the SEFF-based pipeline only in the ST regime. We use hard-edge DFFs-based pipelines as the baseline, in which the clock period is simply determined by the slowest stage, and normalize the optimized EDP to it.

Table 1. Distribution of maximum and minimum delay (ps) of selection benchmarks at nonimal voltage 1.05V.

Benchmarks	$\mu(D_{\max})$	$\sigma(D_{\max})$	$\mu(D_{\min})$	$\sigma(D_{\min})$
c432	803	19.3	119	3.2
c499	614	9.5	259	2.4
c880	759	20.5	144	3.6
c1355	1047	31.9	387	10.8
c1908	994	31.6	281	8.6

Table 2. Four example pipelined circuits.

Pipeline	Configuration
TB1	c1908, SEFF, c880
TB2	c432, SEFF, c1908, SEFF, c499
TB3	c1908, SEFF, c432, SEFF, c1355, SEFF, c880
TB4	c432, SEFF, c880, SEFF, c1908, SEFF, c499

Table 3 shows the results of the EDPs obtained by using different design methods. The percentage reductions of the EDP range from 6.0% to 18.4%. Compared to the hard-edge DFFs-based pipelines, the SEFF-based pipelines achieve better EDPs using the frequency scaling. Precisely, the SEFF-based pipeline improves the operating frequency via time borrowing when there are slacks in some of the pipeline stages. Furthermore, since the leakage energy consumption is the product of the leakage power and the clock period, the leakage energy consumption is also reduced as the clock period decreases.

The leakage energy reduction mentioned above becomes more noticeable in the NT regime and below where the leakage energy consumption is dominant. As seen in Figure 4, the leakage energy is only a small portion of the total energy consumption in the ST regime, while it dominates the total energy consumption in the NT regime. Figure 11 compares the delay reduction for TB1 achieved by the SEFF-based pipelines. Notice that the normalized delay in Figure 11 is lower than the corresponding normalized EDP in Table 3 at high supply voltages, which indicates that the total energy consumption of SEFF-based pipeline is slightly higher than hard-edge DFFs-based pipeline due to the energy overhead of the SEFFs. However, at low supply voltage levels, the normalized EDP of SEFF-based pipeline is lower because both energy and delay are reduced.

Our design method consistently outperforms the design method proposed in [9]. In the ST regime, our method outperforms the method in [9] by achieving better energy-delay trade-off curves, as shown in Figure 9. In other words, we provide the same softness with lower energy consumption by using the proposed DL design. In the NT regime, where the DL in [9] cannot provide enough softness, our method achieves better EDPs by providing more softness with acceptable energy consumption. Thus, we achieve lower EDP in the NT regime.

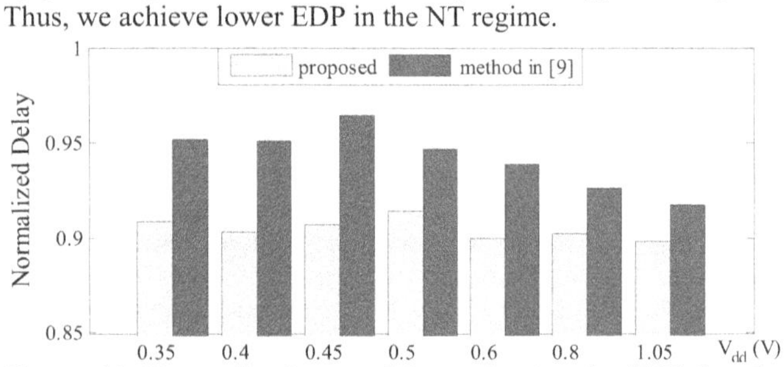

Figure 11. Normalized stage delay reduction for TB1 for the proposed method versus the baseline method of [9].

6. CONCLUSION

Previous work on soft-edge flip-flop (SEFF)-based pipeline mainly focused on the super-threshold operation regime. In this work, we proposed a design methodology for SEFF-based pipeline in both the super- and near-threshold (NT) operation regimes. We considered the high process variation in NT regime and maintained the timing yield of the pipeline by setting the timing constraints using the 3σ delay. We modified the structure of the delay lines by adding a PMOS header to achieve better energy-softness trade-off, which is demonstrated to be very effective in NT regime due to the high process variation. We applied the proposed method to some example pipelines constructed using ISCAS'85 benchmarks and demonstrated significant energy-delay savings in both the ST and NT regimes.

REFERENCES

[1] Dreslinski, R. G., et al. 2010. Near-threshold computing: Reclaiming moore's law through energy efficient integrated circuits. *Proceedings of the IEEE*, 98(2).

[2] Markovic, D., Wang, C. C., Alarcon, L. P., Liu, T. T., and Rabaey, and J. M. 2010. Ultralow-power design in near-threshold region. *Proceedings of the IEEE*, 98(2).

[3] Calhoun, B. H., Wang, A., and Chandrakasan, A. 2005. Modeling and sizing for minimum energy operation in subthreshold circuits. In *Journal of Solid State Circuits*, 40(9).

[4] Zhai, B., Hanson, S., Blaauw, D., and Sylvester, D. 2005. Analysis and mitigation of variability in subthreshold design. In *Proc. of Int'l Symp. on Low Power Electronics and Design*.

[5] Seo, S., et al. 2012. Process variation in near-threshold wide SIMD architectures. In *Proc. of Design Automation Conf.*

[6] Manne, S., Klauser, A., and Grunwald, D. 1998. Pipeline gating: speculation control for energy reduction. In *ACM SIGARCH Computer Architecture News*.

[7] Jacobson, H., et al. 2005. Stretching the limits of clock-gating efficiency in server-class processors. *In Proc. of High-Performance Computer Architecture*.

[8] Partovi, H., Burd, R., Salim, U, Weber, F., DiGregorio, L., and Draper, D. 1996. Flow-through latch and edge-triggered flip-flop hybrid elements. *In Proc. of Solid-State Circuits Conf.*

[9] Ghasemazar, M. and Pedram, M. 2008. Minimizing the energy cost of throughput in a linear pipeline by opportunistic time borrowing. In *Proc. of Int'l Conf. on Computer Aided Design*.

[10] Seok, M., Jeon, D., Chakrabarti, C., Blaauw, D., and Sylvester, D. 2011. Pipeline strategy for improving optimal energy efficiency in ultra-low voltage design. In *Proc. of Design Automation Conf.*

[11] Datta, A., Bhunia, S., Mukhopadhyay, S., Banerjee, N., and Roy, K. 2005. Statistical modeling of pipeline delay and design of pipeline under process variation to enhance yield in sub-100nm technologies. In *Proc. of Design and Test in Europe*.

[12] Joshi, V., Blaauw, D., and Sylvester, D. 2007. Soft-edge flip-flops for improved timing yield: design and optimization. In *Proc. of Int'l Conf. on Computer Aided Design*.

[13] Synopsys 32/28 nm Generic Library: https://sso.synopsys.com/idp/Authn/UserPassword.

[14] Hanson, S., et al. 2007. Performance and variability optimization strategies in a sub-200mV, 3.5 pJ/inst, 11nW subthreshold processor. In *Proc. of Int'l Symp. on VLSI Circuits*.

[15] Fisher, S., Dagan, R., Blonder, S., and Fish, A. 2011. An improved model for delay/energy estimation in near-threshold flip-flops. In *Proc. of Int'l Symp. on Circuits and Systems*.

[16] Lotze, N., Ortmanns, M., and Manoli, Y. 2008. Variability of flip-flop timing at sub-threshold voltages. In *Proc. of Int'l Symp. on Low Power Electronics and Design*.

[17] Pu, Y., et al. 2010. Misleading energy and performance claims in sub/near threshold digital systems. In *Proc. of Int'l Conf. on Computer Aided Design*.

Table 3. The energy-delay products of the proposed design method for four example pipelines.

V_{dd} (V)	Normalized Energy-Delay Product (%)											
	TB1			TB2			TB3			TB4		
	Hard-edge	Method in [9]	Proposed	Hard-edge	Method in [9]	Proposed	Hard-edge	Method in [9]	Proposed	Hard-edge	Method in [9]	Proposed
1.05	100	92.70	89.93	100	96.45	90.98	100	95.83	92.89	100	95.53	89.94
0.8	100	93.19	90.01	100	96.53	91.03	100	95.27	93.11	100	95.91	90.18
0.6	100	93.21	88.47	100	96.70	88.79	100	96.71	91.64	100	96.21	87.25
0.5	100	93.82	89.89	100	97.22	88.05	100	97.57	92.03	100	96.83	86.02
0.45	100	95.52	88.30	100	98.48	87.18	100	99.49	94.00	100	98.25	84.67
0.4	100	93.12	86.58	100	96.39	86.55	100	98.74	92.82	100	95.55	83.54
0.35	100	92.56	86.02	100	95.79	85.29	100	98.15	92.60	100	94.65	81.65

An Energy-Efficient Truly All-Digital Temperature Sensor for SoC Applications

Tzu-Yuan Kuo, Keng-Jui Chang, Jen-Hsiang Lee, Zong-Wu He, Jinn-Shyan Wang
Dept. of EE and SoC/AIM-HI Centers, National Chung-Cheng University, Taiwan

ABSTRACT

This paper presents a truly all-digital one-point-calibrated temperature sensor for SoC applications. Several design techniques are developed to achieve small area, low power, low energy per sample, and high accuracy. First, in contrast to dual DLL's plus charge pump used in the traditional design, single DLL is used in both calibration and measurement operations to avoid locking mismatch and to increase process-variation tolerance, resulting in advantages in terms of area, power, energy, and accuracy. Second, a normalization-after-averaging design is developed to reduce computation cycles, also leading to low power and low energy. Finally, to meet different applications, clock chopping is used not only for adjustable conversion rate but also for low power when a high conversion rate is not needed. Compared to the state-of-the-art design, the proposed 90nm temperature sensor has 65% less normalized area and 26% higher accuracy with 79% less energy per sample. As the sampling rate is adjusted 1/50 lower, 90% of the power consumption is saved.

Categories and Subject Descriptors

VLSI circuit, low power

Keywords

Temperature sensor, calibration, energy-efficient, delay locked loop, SoC

1. INTRODUCTION

System-on-a-chip (SoC) designs often integrate many intellectual properties (IPs) such as the RISC controller and multimedia accelerator to increase the design efficiency. With the advanced nanometer CMOS technology, the complexity and the operating frequency of the SoC nowadays rise to a new high record, and thereby the issues of reducing power consumption and making power management become the most important design concerns. Among those techniques for power or thermal management [1]-[4], the temperature sensor (TS) plays an important role to generate one of the needed in-time information about the run-time conditions. The POWER7 processor in [4] even integrates 40 distributed TS's to construct the thermal map and to report the locations of hotspots induced by high-workload circuitry. With the thermal information, the processor, for example, could cool down the hotspots by using techniques such as power gating or adaptive workload control.

Conventionally, high-precision analog temperature sensors (ATS) [5][6] were designed by measuring temperature-dependent current of the substrate p-n-p bipolar transistor. In the nanometer era, it usually needs to take a long time to design an ATS that should take serious consideration on the challenges caused by the ever increasing process variations. Other drawbacks of the ATS include large area and high power consumption. In recent years, time-domain digital temperature sensors (DTS) [7]-[9] are developed for SoC applications that don't need high-precision temperature information. For these applications, DTS's have been shown to own benefits of small area, low power, and low design effort.

The basic operation procedures of DTS's are firstly comparing the delay time of a temperature-dependent inverter chain to that of a temperature-independent reference delay line and secondly quantizing and converting the comparison result into a digital code by a time-to-digital converter (TDC). Traditionally, the time-domain DTS's should perform two-point calibration to have their accuracy approaching to that of the voltage-domain ATS's. For reducing calibration time and cost, a dual-DLL based DTS [8] was recently proposed to adopt one-point calibration that can still removes the effects of process variations. However, this design has some shortcomings. First, the two always-active DLLs under an over-high conversation rate (5000 samples/s) not only result in high circuit complexity and area cost but also induce high power consumption at the mili-watt level. Second, the process-induced locking mismatch between the two DLLs will degrade its accuracy. Third, this design is not a truly all-digital circuit because it requires a power-hungry charge-bump circuit in the reference DLL. The mixed-signal design should take larger design effort as compared to all-digital circuitry.

In this paper, a one-point-calibrated truly all-digital 90nm temperature sensor is presented. Several design techniques are developed to achieve small area, low power, low energy per sample, and high accuracy. First, in contrast to dual DLLs plus charge pump used in the traditional design, single DLL is used in both calibration and measurement operations to avoid locking mismatch and to increase process-variation tolerance, resulting in advantages in terms of area, power, energy, and accuracy. Second, a normalization-after-averaging design is developed to reduce computation cycles, also leading to low power and low energy. Finally, to meet different applications, clock chopping is used not only for adjustable conversion rate but also for low power when a high conversion rate is not needed.

The rest of this paper is organized as follows. Section 2 describes the basic design concept of the proposed single-DLL-based temperature sensor. The detailed implementation of the proposed energy-efficient DTS is discussed in Section 3. The silicon implementation results of the chip are shown in Section 4. Finally, conclusions are drawn in Section 5.

2. THE PROPOSED TEMPERATURE SENSOR

The conceptual architecture diagram of the proposed single-DLL based truly all-digital temperature sensor is shown in Figure 1(a). It consists of a DLL (in the shaded region) and other logic blocks for doing calibration (to be described later). A reference clock *clk* with a fixed cycle time t_{cyc} is sent to the DTS, and then propagates along the delay chain in the DLL. The multiplexer taps all the outputs of the delay chain. The counter determines which path in the multiplexer will conduct so that the propagated *clk* in the delay chain can pass through to reach the phase detector (PD). The PD then compares the phase of the external *clk* and that of the delayed clock, and the output of the PD will guide the counter to find the right control word for the delay chain until the two clock signals get locked, as shown in Figure 1(b). Once the DLL enters the locking state, the output of the counter, i.e. the control code for the delay chain, will be kept in one of the registers, Reg_C or Reg_M for further calculation.

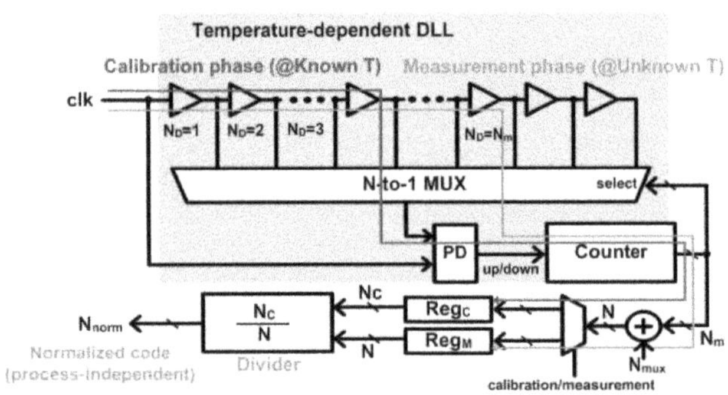

(a)

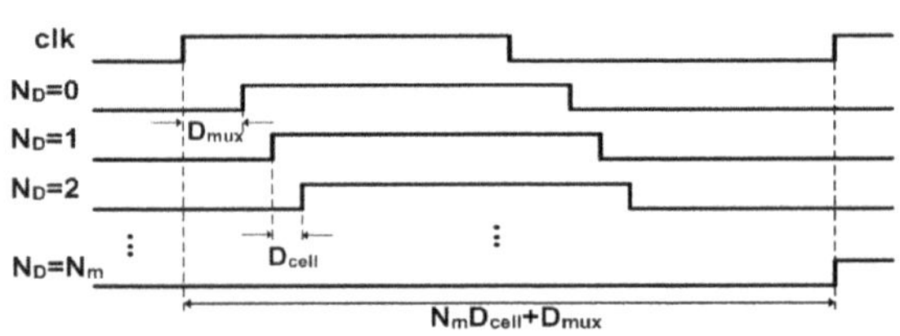

(b)

Figure 1. (a) The conceptual architecture diagram and (b) the operation waveforms of the proposed DTS.

Undoubtedly, the delay of the inverter chain varies with both temperature and process. In [8], the delay, D, of the inverter chain is simplified and expressed as

$$D(T, P) = T^{-\alpha}G(P) \quad (1)$$

where G(P) varies only with process variations and T stands for the temperature with α a given constant for a CMOS process. In the locking state, the clock period, i.e. t_{cyc}, can be expressed as

$$t_{cyc} = N_m \times D_{cell}(T, P) + D_{mux}(T, P). \quad (2)$$

N_m is the number of the delay cells that *clk* passes through in the locking state, $D_{cell}(T, P)$ is the delay of a delay cell, and $D_{mux}(T, P)$ is the delay of the multiplexer. $D_{cell}(T, P)$ and $D_{mux}(T, P)$ can be expressed as $D_{cell}(T, P) = T^{-\alpha}G_{cell}(P)$ and $D_{mux}(T, P) = T^{-\alpha}G_{mux}(P)$, respectively, where $G_{cell}(P)$ and $G_{mux}(P)$ denote process variations of the delay cell and the multiplexer, respectively. The multiplexer can be viewed as a delay cell to simplify the analysis. Therefore, $D_{mux}(T, P)$ can be rewritten as $D_{mux}(T, P) = N_{mux} \times T^{-\alpha}G_{cell}(P)$, where N_{mux} is the ratio of $G_{mux}(P)$ to $G_{cell}(P)$. N_{mux} is a fixed number and can be obtained by simulation for a given technology. By substituting $D_{cell}(T, P)$ and $D_{mux}(T, P)$, eq. (2) becomes

$$t_{cyc} = (N_m + N_{mux}) \times D_{cell}(T, P) = (N_m + N_{mux}) \times T^{-\alpha}G_{cell}(P). \quad (3)$$

Therefore, we have

$$N \equiv (N_m + N_{mux}) = t_{cyc} / [T^{-\alpha}G_{cell}(P)], \quad (4)$$

where N is the effective total number of the delay cells that *clk* passes through in the locking state.

The proposed DTS is operated in two phases, calibration and measurement. In the calibration phase, the temperature T_c is fixed. The DLL will store N_c (= $t_{cyc} / T_c^{-\alpha}G_{cell}(P)$) in register Reg_C. In the measurement phase, N (= $t_{cyc} / T^{-\alpha}G_{cell}(P)$), indicating the unknown temperature T, will be stored in register Reg_M. Finally, by dividing N_c by N with a series divider, we have

$$Q_{norm} = N_c / N = (T/T_c)^{-\alpha} \quad (5)$$

The effects of process variations are now removed by one-point calibration, and thus Q_{norm} varies only with the temperature.

The proposed DTS is implemented in 90nm CMOS process. Figure 2(a) shows simulation results of N in different process corners. As the temperature rises, N_m is decreased because D_{cell} is increased. We note that process variations makes N vary across different corners remarkably. Figure 2(b) shows simulation results of Q_{norm} with the calibration temperature specified at 50°C. It is found that Q_{norm} is almost the same across different process corner.

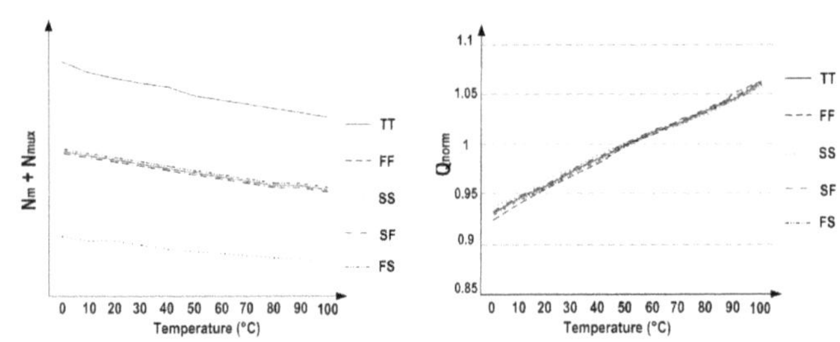

(a) (b)

Figure 2. (a) N and (b) Q_{norm} at different process corners.

3. DESIGN TECHNIQUES FOR LOW POWER

Besides not using mixed-signal circuits, the proposed truly all-digital TS only reuse the same DLL in both phases. Therefore, locking mismatch in calibration and measurement operations can be avoided, resulting in increased process-variation tolerance and getting advantages of reducing area and power while increasing accuracy. Now, we will introduce other design techniques for low power in this section. The architecture diagram is depicted in Figure 3 again with more details.

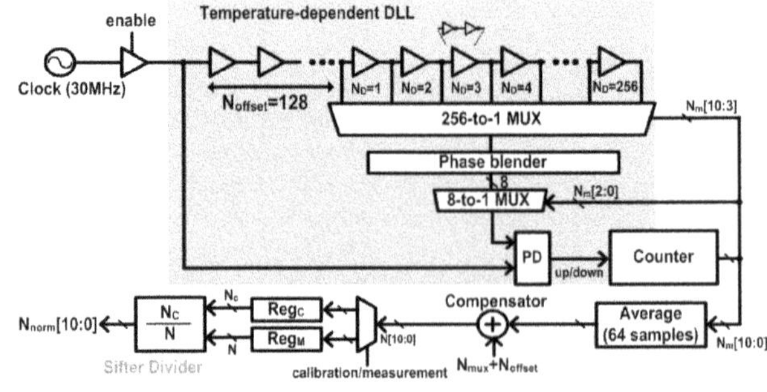

Figure 3. The implementation of the proposed DTS.

3.1 The Delay Locked Loop

A 30 MHz reference clock is used in this work. A delay cell is composed by two inverters, and the minimum required number of the delay cells in the DLL for this clock is 384 when considering the worst-corner operation. However, simulation results as shown in Figure 4(a) indicate that N is between 128 and 384 across different process corners and over a temperature range of 0°C and 100℃. A straight forward implementation of the delay chain includes 384 delay cells with outputs of all delay cells tapped by the multiplexer, as shown in Figure 4(b). The left most 128 delay cells and the associated multiplexing circuitry do nothing but make the conversion time longer and waste more power. The delay chain of this work is designed as shown in Figure 4(c), where only the delay cells after the left most 128 bits are tapped by the multiplexer, resulting in reduced circuit complexity, reduced conversion time, and reduced power consumption. Besides, the bit number of the counter is also reduced from 9 bits to 8 bits. Note that with this design, $N_{mux}+N_{offset}$ should be added back to obtain a right value of N, as shown in Figure 4(c).

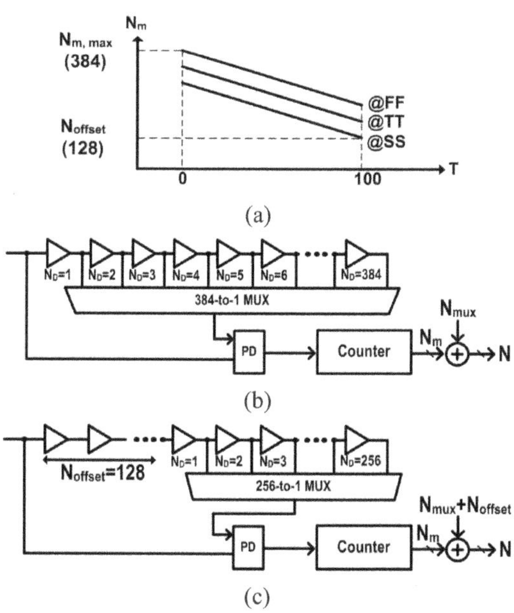

Figure 4. (a) Simulation results of N_m. (b) Straight forward implementation, and (c) real implementation of the DLL.

Following the 256-to-1 multiplexer (MUX), a phase blender is used to increase the delay resolution. Please refer to [10] for the detailed design of the phase blender. In this work, the phase blender has 8 outputs, and is followed by an 8-to-1 MUX. The counter needs another 3 bits for controlling the added MUX.

3.2 Normalization after Averaging

For SoC applications, several noise sources induce high jitter to the reference clock, and thereby degrading the accuracy of temperature sensing. The simplified architecture of the dual-DLL based DTS in [9] is illustrated in Figure 5(a), where an operation of averaging 64 normalization results (Figure 5(b)) is used to reduce the effect of variation caused by jitter. Considering the long computation time of normalization for every sample, our design only performs averaging on Nm's and then adds on $N_{mux}+N_{offset}$ for getting the final normalization result (Figure 5(c)). Therefore, less computation leads to low power, low energy per sample, and short conversion time. The maximum conversion rate of the proposed DTS is as high as 14.6 KHz with a 30 MHz clock.

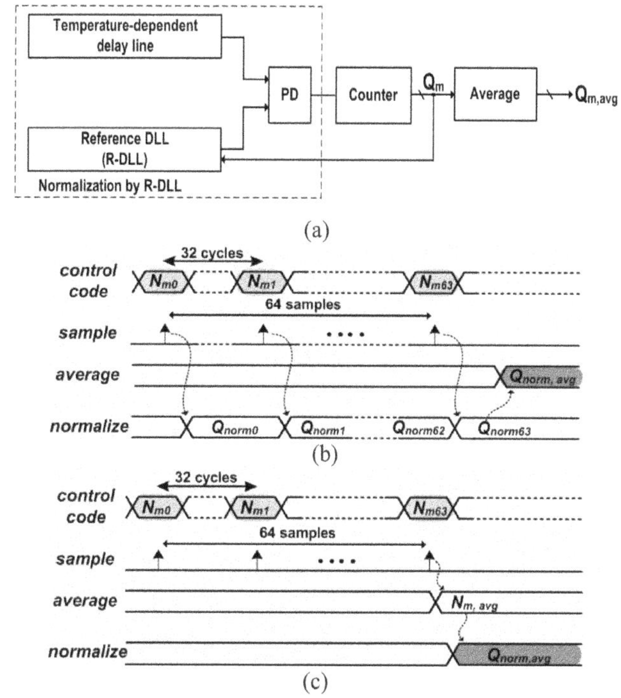

Figure 5. (a) Simplified architecture of the dual-DLL DTS in [8] and (b) operations of normalization and averaging. (c) The operation of normalization after averaging of the proposed DTS.

3.3 Clock Chopping

The DTS in [8] has a sampling rate of 5000 samples/s to enable a fast temperature tracking. However, the high sampling rate may be not helpful for thermal control but cause high dynamic power consumption in the always-active dual DLL's. In this work, as shown in Figure 3, a tri-stage is inserted in the incoming path to enable clock chopping decided by the system. Therefore, adjustable conversion rate (as illustrated in Figure 6,) can be implemented for different applications, and low power can be achieved when a high conversion rate is not needed.

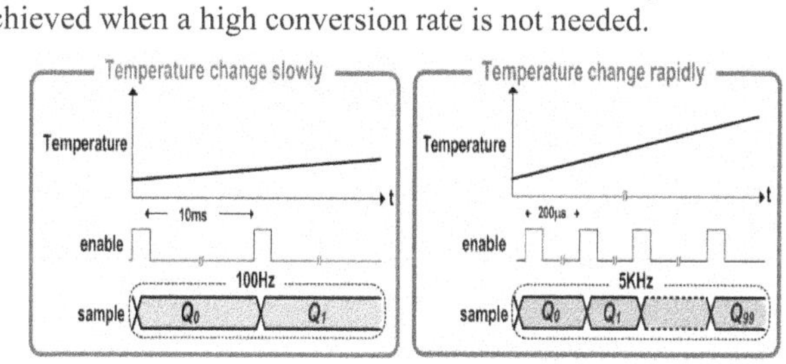

Figure 6. Feasibility of adjustable conversion rate.

4. TEST CHIP AND MEASUREMENT RESULTS

The chip was fabricated in a standard 90nm 1.0V digital CMOS process. The die micrograph is show in Figure 7, and the core area of the DTS is $0.027mm^2$. The layout of the DLL is fully customized and the other circuits are realized by cell-based tools. Figure 8 show the one-point-calibrated measurement results for a temperature range of 0°C ~ 100°C. With 11-bit output from the divider, the temperature resolution is 0.145°C. Figure 9 shows the temperature error ranges from -1.64°C to +1.69°C. The conversion rate can be scaled from 14.6k S/s to 100 S/s, and power consumption ranges from 502μW to 115.1μW. The leakage power with this process is as high as 112.8μW.

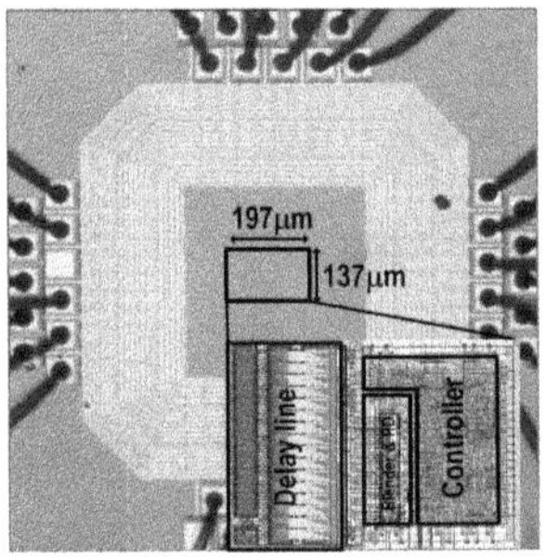

Figure 7. Die micrograph of the proposed DTS.

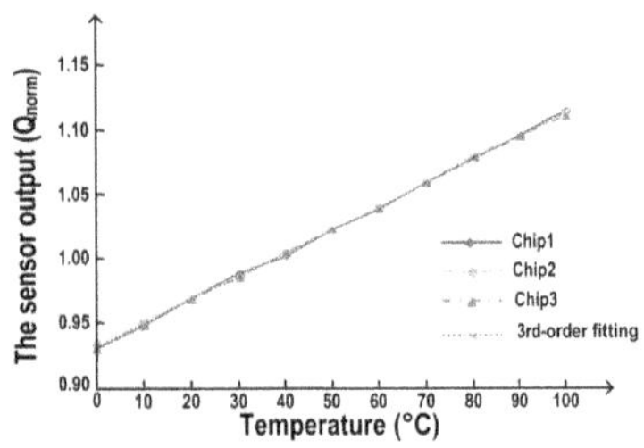

Figure 8. Measurement results of three test chips.

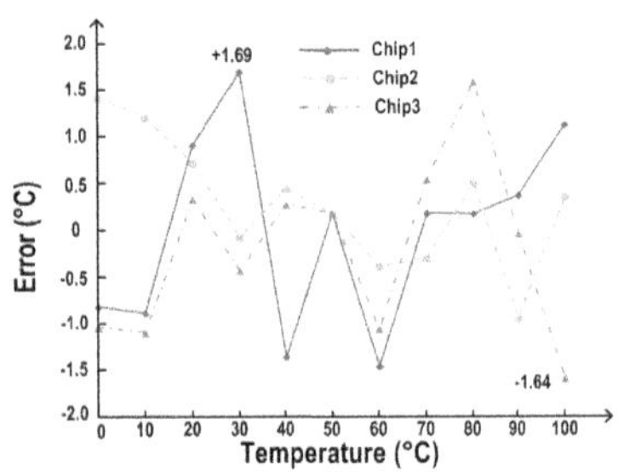

Figure 9. Measurement error.

Table I lists comparison results of different DTS's. Although the DTS in [9] claims to achieve auto-calibration, it suffers from large error even over a smaller temperature range. Compared to the DTS in [8], the proposed design with 5k samples/s has 65% less normalized area and 26% smaller error with 79% less energy per sample. When the sampling rate is adjusted 1/50 lower, 90% of the power consumption is saved.

5. CONCLUSIONS

The design and implementation of an energy-efficient truly all-digital 90nm temperature sensor is presented. The main features of the design include single-DLL based one-point calibration, power-efficient DLL design, normalization after averaging operation, and adjustable conversion rate. The core area is only 0.27 mm^2, and power consumption is as small as 115.1μW at a sampling rate of 100S/s. The measured temperature inaccuracy ranges from -1.64 °C to +1.69 °C, over 0 °C to 100 °C range. Compared to the state-of-the-art design, the proposed DTS has 65% less normalized area and 26% higher accuracy with 79% less energy per sample. When the sampling rate is adjusted 1/50 lower, 90% of the power consumption is saved.

6. ACKNOWLEDGMENTS

This work was supported in part by the National Science Council and in part by the Science Park Administration under research grants of NSC 98-2221-E-194-043-MY3 and 101A23, respectively. The authors also thank the SoC/AIM-HI Center of Chung Cheng University, Ministry of Education, and CIC, Taiwan, for their support on research work and chip fabrication.

Table I. Comparison results of temperature sensors.

Sensor	[8]	[9]	This work		
Process / V_{DD}	130nm@1.2V	65nm@1V	**90nm@1.2V**		
Area(mm^2)	0.16 (1)	0.010 (0.25)	**0.027** **(0.35)**		
calibration	One-point	Auto	**One-point**		
Range (°C)	0 ~ 100	0 ~ 60	**0 ~ 100**		
Resolution (°C)	0.66	0.139	**0.145**		
Error (°C)	-1.8 ~ +2.3	-5.1 ~ +3.4	**-1.64 ~ + 1.69**		
Conversion rate (samples/s)	5k	10k	**14.6k**	**5k**	**100**
Power (μW)	1200 (1)	150 (0.125)	**502** **(0.42)**	**249** **(0.21)**	**115** **(0.1)**
Normalized Power #	575 (1)	414 (0.72)	**502** **(0.87)**	**249** **(0.43)**	**115** **(0.2)**
Energy (μJ/sample)	0.24 (1)	0.015 (0.06)	**0.03** **(0.125)**	**0.05** **(0.21)**	**1.51** **(6.29)**

#Normalized Power = Power × (90nm / process)^2 × (1.2V / V_{DD})^2

7. REFERENCES

[1] R. Mukherjee and S. O. Memik, "Systematic temperature sensor allocation and placement for microprocessors," *in Proc. ACM/IEEE Design Autom. Conf.*, Jul. 2006, pp. 542–547.

[2] C. Poirier, et al., "Power and temperature control on a 90nm Itanium®-family processor," *in Proc. ISSCC Dig. Tech. Papers*, Feb. 2005, pp. 304-305.

[3] D. E. Duarte, et al., "Temperature sensor design in a high volume manufacturing 65nm CMOS digital process," *CICC*, Sept. 2007, pp. 221–224.

[4] M. Ware, et al., "Architecture for power management: The IBM Power7 approach," *in Proc. IEEE Int. Symp. HPCA*, Jan. 2010, pp. 1–11.

[5] G. Wang and G. C. M. Meijer, "Temperature characteristics of bipolar transistors fabricated in CMOS technology," *Sensors and Actuators A*, vol. 87, pp. 81–89, Sept. 2000.

[6] M. A. P. Pertijs, K. A. A. Makinwa, and J. H. Huijsing, "A CMOS smart temperature sensor with a 3σ inaccuracy of ±0.1°C from -55 °C to 125 °C," *IEEE J. Solid-State Circuits*, vol. 40, pp. 2805–2815, Dec. 2005.

[7] P. Chen, et al., "A time-to-digital-converter-based CMOS smart temperature sensor," *IEEE J. Solid-State Circuits*, vol. 40, no. 8, pp. 1642-1648, Aug. 2005.

[8] K. Woo, S. Meninger, T. Xanthopoulos, E. Crain, D. Ha, and D. Ham, "Dual-DLL-based CMOS all-digital temperature sensor for microprocessor thermal monitoring," *in Proc. ISSCC Dig. Tech. Papers*, Feb. 2009, pp. 68–69a.

[9] C.-C. Chung and C.-R. Yang, "An autocalibrated all-digital temperature sensor for on-chip thermal monitoring," *IEEE Trans. Circuits Syst. II, Exp. Briefs,* vol. 58, no2, pp.105-107, Feb. 2011.

[10] K. Sung and L.-S. Kim, "A high-resolution synchronous mirror delay using successive approximation register," *IEEE J. Solid-State Circuits*, vol. 39, no. 11, pp. 1997-2004, Nov. 2004.

Efficient Characterization of TSV-to-Transistor Noise Coupling in 3D ICs

Hailang Wang
Department of ECE
Stony Brook University
Stony Brook, NY 11794
haiwang@ece.sunysb.edu

Mohammad H. Asgari
Department of EE
Columbia University
New York, NY 11794
mha2135@columbia.edu

Emre Salman
Department of ECE
Stony Brook University
Stony Brook, NY 11794
emre@ece.sunysb.edu

ABSTRACT

A methodology is proposed to characterize TSV induced noise coupling in three-dimensional (3D) integrated circuits. Different substrate biasing schemes (such as a single substrate contact versus regularly placed substrate contacts) and TSV fabrication methods (such as via-first and via-last) are considered. A compact π model is proposed to efficiently estimate the coupling noise at a victim transistor. Each admittance within the compact model is approximated with a closed-form expression consisting of logarithmic functions. The methodology is validated using a 3D transmission line matrix (3D-TLM) method, demonstrating, on average, 4.8% error. The compact model and the closed-form expressions are utilized to better understand TSV induced noise as a function of multiple parameters such as TSV type and placement of substrate contacts.

Categories and Subject Descriptors

B.7.m [**Integrated Circuits**]: [Miscellaneous]

Keywords

Compact model; TSV; Noise; 3D IC

1. INTRODUCTION

Three-dimensional (3D) integrated circuits (ICs) address some of the critical issues encountered in planar technologies such as the adverse effects of global interconnects [1]. Furthermore, heterogeneous integration of diverse circuits and materials is facilitated. Through silicon via (TSV) based 3D technology is among the most promising vertical integration technologies due to relatively high density vertical interconnects [2]. International Technology Roadmap for Semiconductors identifies three phases for the application of 3D technology: (1) memory stacks, (2) processor-memory stacks, and (3) heterogeneous 3D integration with sensing and communication blocks.

An important challenge in each of these applications is to ensure system wide signal integrity. In addition to traditional noise coupling and propagation mechanisms such as crosstalk, power supply noise, and substrate coupling, 3D circuits also suffer from TSV induced noise coupling [3]. Specifically, during a signal transition within a TSV, noise couples into the substrate due to both dielectric and depletion capacitances. The coupling noise propagates throughout the substrate and affects the reliability of nearby transistors. This issue is exacerbated for TSVs that carry signals with high switching activity factors and fast transitions such as clock signals.

Existing works have investigated the effect of TSV geometry and substrate type on noise propagation [4]. The efficacy of traditional techniques (such as guard rings) in reducing TSV induced noise has also been investigated [5]. These studies, however, do not consider different substrate biasing schemes and distinct TSV fabrication methods. Furthermore, computationally expensive approaches (such as 3D electromagnetic extraction) are utilized, prohibiting the use of these approaches for fast evaluation of different physical structures during the early stages of TSV floorplanning.

The primary contributions of this paper are: (1) a compact model and a closed-form expression for each admittance within this model are developed to efficiently estimate TSV induced noise at a victim transistor, (2) these models (validated using a 3D transmission line matrix method) are sufficiently flexible to consider different substrate biasing schemes and both via-first and via-last TSV characteristics, and (3) design guidelines are provided to reduce TSV induced noise based on the results obtained from these models.

The rest of the paper is organized as follows. A highly distributed electrical model (used as a reference) to analyze noise injection and propagation is described in Section 2. A compact π model is proposed in Section 3 for efficient estimation of TSV induced noise at a victim transistor. To consider different substrate biasing schemes and TSV types, each admittance within the compact model is expressed in Section 4 as a function of multiple physical parameters. The results are discussed and design guidelines are provided in Section 5. Finally, the paper is concluded in Section 6.

2. DISTRIBUTED MODEL FOR TSV INDUCED NOISE COUPLING

The TSV and substrate models are provided, respectively, in Sections 2.1 and 2.2. These models are based on discretizing the physical structure into unit cells and modeling each unit cell with lumped parasitic impedances. Also referred to as 3D transmission line matrix method (3D-TLM) [6], the

GLSVLSI'13, May 2–3, 2013, Paris, France.

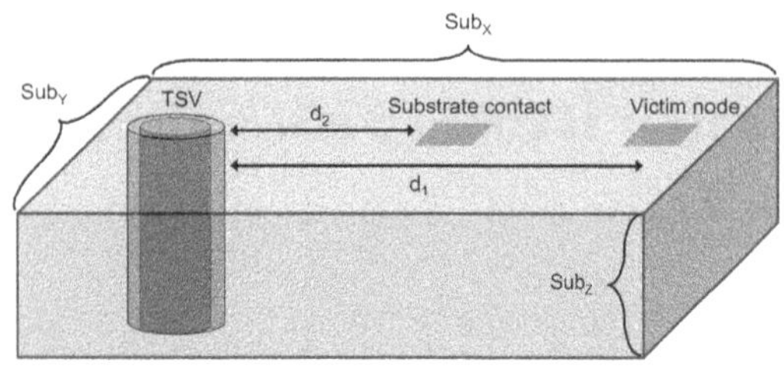

Figure 1: Physical structure discretized based on 3D TLM method to analyze TSV induced noise coupling.

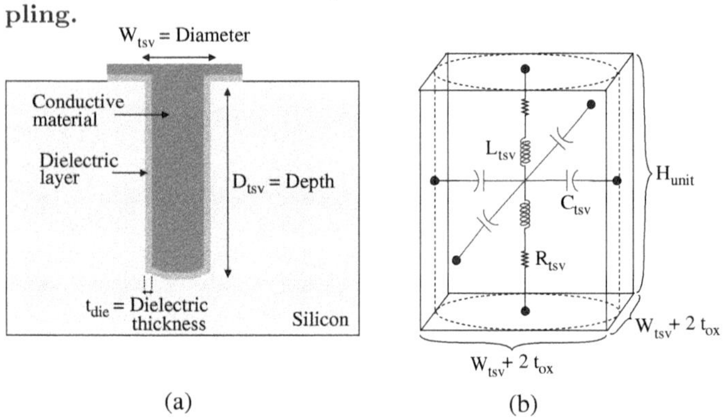

Figure 2: TSV representations: (a) cross-section of a TSV consisting of a conductive material and dielectric layer, (b) electrical model of a unit cell TSV used for discretization.

accuracy of this technique has been previously verified by comparing the results with 3D field solvers [5] and experimental data [7]. This distributed model is used as a reference in this paper to validate the proposed compact model (see Section 3) and closed-form expressions (see Section 4).

2.1 TSV Model

A typical TSV is represented as a cylinder with a diameter and depth, as illustrated in Fig. 2(a). The two primary components of a TSV are (1) conductive filling material such as polysilicon (for via-first TSVs), tungsten (for via-middle TSVs), and copper (for both via-middle and via-last TSVs), (2) a dielectric layer that surrounds the conductive part to prevent the filling material from diffusing into the silicon [8]. In this study, noise coupling analysis is achieved for both via-first and via-last TSVs.

A TSV unit cell consisting of a parasitic resistance R_{tsv}^{unit}, parasitic inductance L_{tsv}^{unit}, and capacitance to substrate C_{tsv}^{unit} is illustrated in Fig. 2(b) [9]. The x and y dimensions of the unit cell are both equal to $W + 2t_{ox}$ as determined by the TSV diameter W and thickness of the oxide layer t_{ox}. The z dimension is equal to H_{unit} as determined by the TSV height and the required resolution in the transmission line matrix method.

Considering the skin effect, the unit TSV resistance R_{tsv}^{unit} is determined by [10]

$$R_{tsv}^{unit} = \frac{1}{2}\sqrt{(R_{AC}^{tsv,unit})^2 + (R_{DC}^{tsv,unit})^2}, \quad (1)$$

where the DC resistance $R_{DC}^{tsv,unit}$ and AC resistance $R_{AC}^{tsv,unit}$ are, respectively,

$$R_{DC}^{tsv,unit} = \frac{1}{2}\frac{\rho_f H_{unit}}{\pi (W/2)^2}, \quad (2)$$

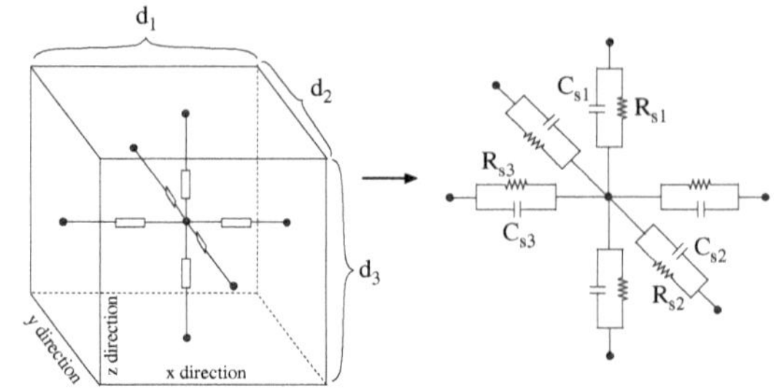

Figure 3: Distributed model of a substrate network where each unit cell is represented by six resistances and capacitances.

$$R_{AC}^{tsv} = \frac{\rho_f H_{unit}}{4\pi (W/2)\delta_{tsv}}. \quad (3)$$

ρ_f is the conductivity of the filling material and the skin depth δ_{tsv} is

$$\delta_{tsv} = \frac{1}{\sqrt{\pi f \mu_f \rho_f}}, \quad (4)$$

where f is the frequency and μ_f is the permeability of the filling material. The unit TSV inductance L_{tsv}^{unit} is

$$L_{tsv}^{unit} = \frac{1}{2}\frac{\mu_o}{4\pi}[2H_{unit}\ln(\frac{2H_{unit} + \sqrt{(W/2)^2 + (2H_{unit})^2}}{W/2}) + (W/2 - \sqrt{(W/2)^2 + (2H_{unit})^2})], \quad (5)$$

where μ_o is vacuum permeability. The unit TSV capacitance C_{tsv}^{unit} is determined from the cylindrical capacitor formula as [11]

$$C_{tsv}^{unit} = \frac{1}{4}\frac{2\pi\epsilon_{ox} H_{unit}}{\ln(\frac{W/2+t_{ox}}{W/2})}, \quad (6)$$

where ϵ_{ox} is the oxide permittivity.

2.2 Substrate Model

A lightly doped bulk type substrate is assumed. Note that an epi type substrate with a heavily doped bulk beneath the lightly doped silicon layer typically produces greater noise coupling and is therefore less applicable to 3D heterogeneous integration where circuits with distinct electrical characteristics coexist.

A similar discretization technique is applied to model the lightly doped substrate. A unit substrate cell consisting of six parallel RC admittances is illustrated in Fig. 3. Referring to this figure, the three substrate resistances R_{s1}, R_{s2}, and R_{s3} are, respectively,

$$R_{s1} = \frac{1}{2}\frac{\rho_s d_3}{d_1 d_2}, \quad (7)$$

$$R_{s2} = \frac{1}{2}\frac{\rho_s d_2}{d_1 d_3}, \quad (8)$$

$$R_{s3} = \frac{1}{2}\frac{\rho_s d_1}{d_2 d_3}, \quad (9)$$

where ρ_s is the substrate resistivity. Similarly, the three substrate capacitances C_{s1}, C_{s2}, and C_{s3} are, respectively,

$$C_{s1} = 2\frac{\epsilon_s d_1 d_2}{d_3}, \quad (10)$$

$$C_{s2} = 2\frac{\epsilon_s d_3 d_1}{d_2}, \quad (11)$$

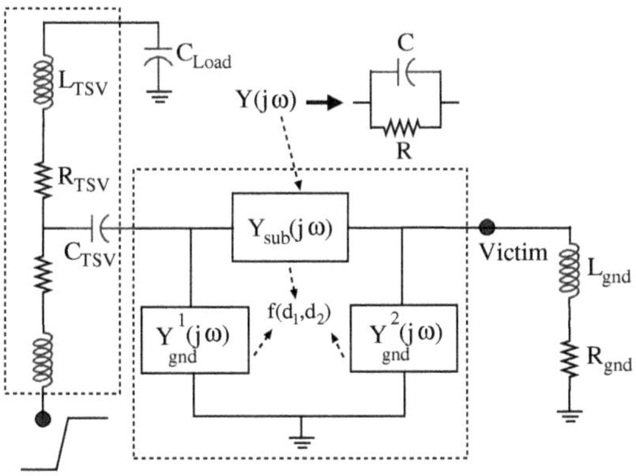

Figure 4: Compact π model to efficiently estimate the noise at the victim node in the presence of a TSV and substrate contacts.

$$C_{s3} = 2\frac{\epsilon_s d_2 d_3}{d_1}, \tag{12}$$

where ϵ_s is the dielectric permittivity of silicon.

The TSV and the substrate unit cells are combined to produce a highly distributed mesh based on 3D TLM. Substrate contacts are also considered in the model to properly bias the substrate. Despite the reasonable accuracy achieved by this model [5, 7], the computational complexity is significantly high, particularly when the dimensions of the unit cells are small. This issue is exacerbated as the distance between the TSV and victim transistor increases. Furthermore, the number and location of the substrate contacts play an important role in characterizing the TSV safe zone. Reanalysis of the distributed structure when these characteristics change is computationally prohibitive. A compact model is proposed to alleviate these limitations, as described in the following section.

3. COMPACT Π MODEL FOR EFFICIENT TSV NOISE COUPLING ANALYSIS

A two-port, linear time-invariant network can be generally characterized with four admittances: $Y_{11}(j\omega)$, $Y_{12}(j\omega)$, $Y_{21}(j\omega)$, and $Y_{22}(j\omega)$. Utilizing this characteristic, the proposed compact model consists of a single cell TSV and an equivalent two-port π network to model noise propagation, as depicted in Fig. 4. Each electrical element within the π network consists of a parallel RC circuit, producing an admittance $(1/R) + j\omega C$. These admittances are obtained from the distributed model (based on 3D TLM) described in the previous section. Specifically, the four $Y(j\omega)$ parameters of the distributed mesh is obtained through an AC analysis. The resistances and capacitances of the π network are determined such that the four $Y(j\omega)$ parameters of the two networks are equal. According to this procedure, the $Y(j\omega)$ parameters of the π network $Y_{sub}(j\omega)$, $Y^1_{gnd}(j\omega)$, and $Y^2_{gnd}(j\omega)$ are determined as follows:

- $Y_{sub}(j\omega) = (1/R_{sub}) + j\omega C_{sub} = Y_{21}(j\omega)$: represents the equivalent substrate admittance between the TSV and victim transistor.
- $Y^1_{gnd}(j\omega) = (1/R^1_{gnd}) + j\omega C^1_{gnd} = Y_{11}(j\omega) - Y_{21}(j\omega)$: represents the equivalent substrate admittance between the TSV and ground node.

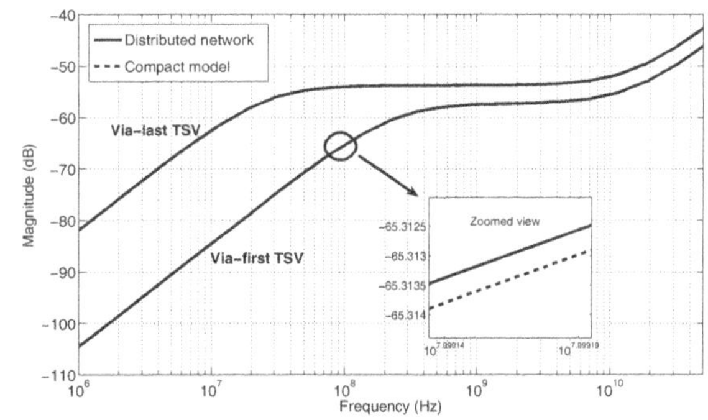

Figure 5: Comparison of the proposed compact π model with high complexity distributed mesh for both via-first and via-last TSVs. The solid line represents noise at the victim node obtained from distributed mesh whereas the dashed line represents noise at the victim node obtained from the compact π model.

- $Y^2_{gnd}(j\omega) = (1/R^2_{gnd}) + j\omega C^2_{gnd} = Y_{22}(j\omega) - Y_{21}(j\omega)$: represents the equivalent substrate admittance between the victim node and ground node.

Note that Y_{11}, Y_{21}, and Y_{12} are obtained by simulating the distributed mesh. Also note that a single RC value is chosen for each admittance since the variation of the resistance and capacitance with frequency is negligible in the frequency range of interest. Specifically, the maximum change is less than 0.1% up to 100 GHz.

The accuracy of the compact model is demonstrated by comparing the transfer function of the compact model with the transfer function of the distributed mesh with significantly higher complexity. Assuming a lightly doped substrate (with 10 Ωcm resistivity and 103.4×10^{-12} F/m absolute permittivity), the two transfer functions are compared in Fig. 5 for both via-first and via-last TSVs. In the distributed model, the dimensions L_{sub}, W_{sub}, and H_{sub} of the unit substrate cell are, respectively, 1 μm, 1 μm, and 1.25 μm. The distance between the TSV and victim node is 10 μm and a single substrate contact is placed between the two ports. The overall length of the substrate is 100 μm. The height of the substrate (determined by the TSV height) is 10 μm for a via-first TSV and 50 μm for a via-last TSV. Alternatively, the width of the substrate (partly determined by the TSV diameter) is 5 μm for a via-first TSV and 12 μm for a via-last TSV. Note that via-last TSVs have greater dimensions as compared to via-first TSVs, significantly affecting the noise at the victim node, as further discussed in Section 5.

Comparison of Complexity: For a via-first TSV, the x and y dimensions of the unit TSV cell are both equal to 4.4 μm ($W + 2t_{ox}$ where $W = 4$ and $t_{ox} = 0.2$), whereas the z dimension is 1 μm. Alternatively, for a via-last TSV, the x and y dimensions of the unit TSV cell are equal to 10.4 μm ($W + 2t_{ox}$ where $W = 10$ and $t_{ox} = 0.2$) and the z dimension is 5 μm. In the distributed model with a via-first TSV, these dimensions produce 48,080 number of circuit elements (resistance, capacitance, and inductance). For a via-last TSV, this number increases to 576,080 due to greater y and z dimensions of the substrate. Alternatively, the compact pi model contains only 11 number of elements for both via-first and via-last TSVs.

Comparison of Accuracy: As illustrated in Fig. 5, noise

Table 1: Fitting coefficients for the function F that approximates the admittances within the compact model (see Fig. 4) for each case. The function F is given by (13).

Cases	Admittances	Fitting coefficients A	B	C	D	E	Average error (%)
Case 1	$R_{sub} = 1000/F$ (kΩ)	27.09	0	0	-0.98	-5.11	6.6
	$C_{sub} = F$ (aF)	326.7	0.41	0.48	-17.26	-67.55	2.8
	$R^1_{gnd} = 1000/F$ (kΩ)	28.31	0	0	-8.14	1.98	1.9
	$C^1_{gnd} = F$ (aF)	301.3	-0.28	0.66	-96.94	30.09	1.6
	$R^2_{gnd} = F$ (kΩ)	45.91	2.30	3.08	-218.3	154.9	9.6
	$C^2_{gnd} = 1000/F$ (aF)	8.05	0.28	0.29	-20.39	12.49	10.4
Case 2	$R_{sub} = 1000/F$ (kΩ)	29.18	0.28	0.38	1.34	-11.78	8.3
	$C_{sub} = F$ (aF)	317.1	2.99	4.24	14.2	-127.6	10.8
	$R^1_{gnd} = 1000/F$ (kΩ)	69.24	-0.046	1.97	-35.05	4.73	0.7
	$C^1_{gnd} = F$ (aF)	758.5	-0.49	21.13	-380.8	51.33	0.7
	$R^2_{gnd} = 1000/F$ (kΩ)	9.50	-0.19	-0.99	-1.95	9.31	1.6
	$C^2_{gnd} = F$ (aF)	106.4	-2.05	-10.99	-20.83	100.7	1.6
Case 3	$R_{sub} = 1000/F$ (kΩ)	27.35	-0.082	0.036	-2.11	-1.12	2.4
	$C_{sub} = F$ (aF)	296.6	-0.89	0.83	-20.78	-13.12	1.9
	$R^1_{gnd} = 1000/F$ (kΩ)	36.16	0.028	0.39	-10.16	1.18	2.4
	$C^1_{gnd} = F$ (aF)	235.6	-1.18	-2.16	-61.86	51.91	6.9
	$R^2_{gnd} = 1000/F$ (kΩ)	11.57	0.11	-0.19	4.43	-5.86	11.6
	$C^2_{gnd} = F$ (aF)	129.1	1.18	-2.22	49.1	-65.23	10.9
Case 4	$R_{sub} = 1000/F$ (kΩ)	24.41	0.13	0.42	1.69	-8.36	8.5
	$C_{sub} = F$ (aF)	265.3	1.40	4.67	17.88	-91.2	7.2
	$R^1_{gnd} = 1000/F$ (kΩ)	117.6	-0.03	0.40	-35.66	3.64	0.3
	$C^1_{gnd} = F$ (aF)	998	0.03	-68.78	37.31	32.11	2.9
	$R^2_{gnd} = 1000/F$ (kΩ)	14.56	-0.01	-0.73	-3.49	5.93	2.4
	$C^2_{gnd} = F$ (aF)	162	-0.13	-8.17	-37.74	64.26	2.5

coupling due to TSVs is accurately estimated by the compact model with negligible error within the frequency range of interest. Note that the noise magnitude at the victim node is higher for via-last TSVs due to higher TSV capacitance and greater substrate dimensions. This difference is more than 20 dB at low frequencies and decreases to approximately 4 dB in the gigahertz range.

The proposed compact model supports the analysis of TSV induced coupling noise. To be able to consider the effect of various design parameters on coupling noise, each *RC* element within the compact model is expressed as a function of two physical distances, as described in the following section.

4. TSV SAFE ZONE CHARACTERIZATION

To determine TSV safe zone, the dependence of TSV induced noise on design parameters such as distance between TSV and victim node, and the number and location of substrate contacts should be characterized. Two substrate biasing schemes are considered. In the first scheme, as depicted in Fig. 6(a), a single substrate contact is placed between the TSV and victim node. The physical distance between the TSV and victim node is d_1 and the distance between the TSV and substrate contact is d_2. In the second scheme, as depicted in Fig. 6(b), substrate contacts are regularly placed between the TSV and victim node. In this case, d_2 refers to the distance between each substrate contact. The second scenario is considered since substrate contacts can be regularly placed in an automated manner based on latch-up constraints of the technology. These two scenarios are separately investigated for both via-first and via-last TSVs, producing four different cases, as summarized below:

- Case 1: via-first TSV with a single substrate contact between TSV and victim node
- Case 2: via-first TSV with regularly placed substrate contacts between TSV and victim node

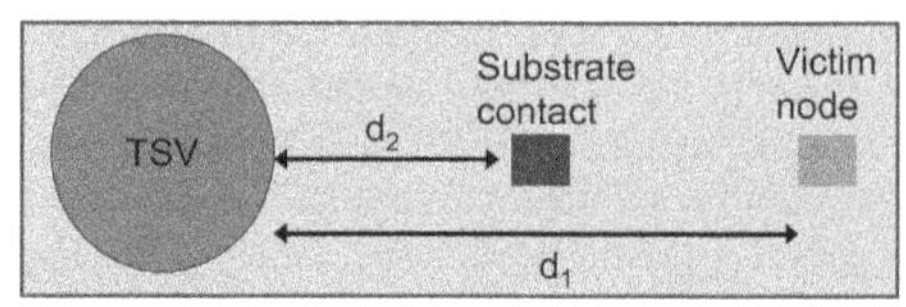

(a)

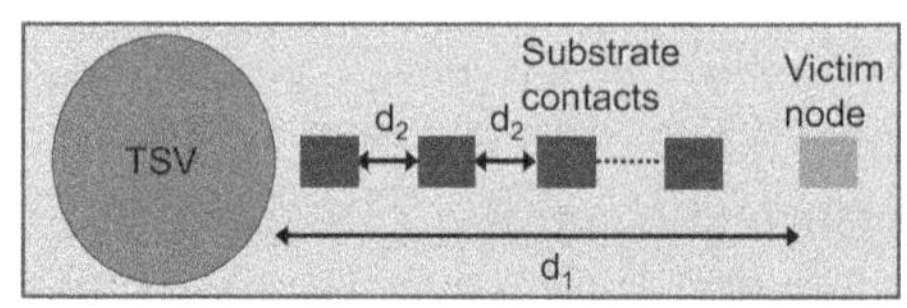

(b)

Figure 6: The two substrate biasing schemes used to characterize noise coupling: (a) single substrate contact between a TSV and victim node, (b) regular placement of the substrate contacts between a TSV and victim node.

- Case 3: via-last TSV with a single substrate contact between TSV and victim node
- Case 4: via-last TSV with regularly placed substrate contacts between TSV and victim node

For each case, the $Y(j\omega)$ parameters of the π network are characterized as a function of d_1 and d_2. To evaluate these dependencies, AC analyses of the distributed mesh (based on 3D TLM) described in Section 2 are performed at different d_1 and d_2. The data obtained in this step are used to generate a 3D surface for each resistance and capacitance within $Y_{sub}(j\omega)$, $Y^1_{gnd}(j\omega)$, and $Y^2_{gnd}(j\omega)$. This surface is approximated with a logarithmic function using a 3D least square regression analysis. The logarithmic function F used to approximate the admittances of the π network as a function of the physical distances d_1 and d_2 is

$$F(d_1, d_2) = A + Bd_1 + Cd_2 + D \ln d_2 + E \ln d_1, \quad (13)$$

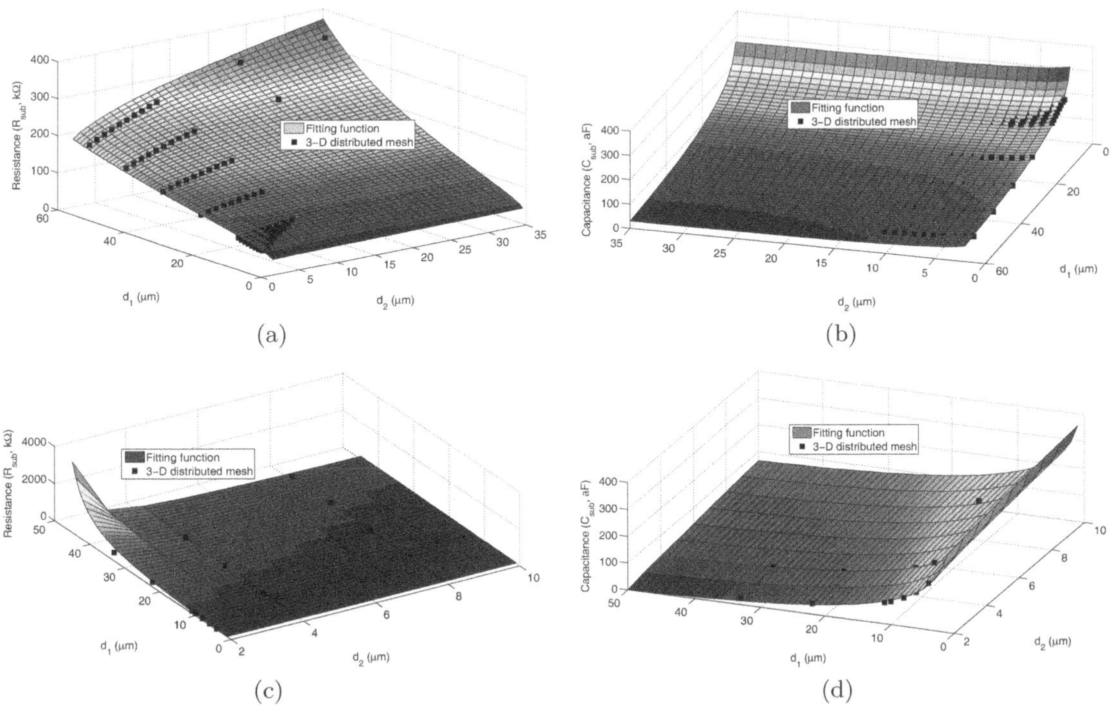

Figure 7: Comparison of the data obtained from the analysis of the distributed mesh with the function F that approximates these data: (a) resistance R_{sub} of the $Y_{sub}(j\omega)$ with a via-first TSV and a single substrate contact (case 1), (b) capacitance C_{sub} of the $Y_{sub}(j\omega)$ with a via-first TSV and a single substrate contact, (c) resistance R_{sub} of the $Y_{sub}(j\omega)$ with a via-last TSV and regularly placed substrate contacts, and (d) capacitance C_{sub} of the $Y_{sub}(j\omega)$ with a via-last TSV and regularly placed substrate contacts.

where A, B, C, D, and E are fitting coefficients. Note that both the resistance (in kilo Ωs) and capacitance (in atto Farads) of each $Y(j\omega)$ within the π network are represented by the function F. Also note that the distances d_1 and d_2 are in μm. Since the π network has three admittances each consisting of a parallel RC circuit, six logarithmic functions are developed for each case, producing a total of 24 functions. The fitting coefficients for each function are listed in Table 1.

As an example, the resistance R_{sub} and capacitance C_{sub} of the $Y_{sub}(j\omega)$ are plotted, respectively, in Figs. 7(a) and 7(b) for a via-first TSV with a single substrate contact (case 1). The same parameters are plotted for a via-last TSV with regularly placed substrate contacts (case 4) in, respectively, Figs. 7(c) and 7(d). The dotted points represent the data obtained from the analysis of the distributed mesh and the surface represents the function F that approximates these data. The procedure is similar for other cases and the RC elements of the remaining admittances ($Y^1_{gnd}(j\omega)$ and $Y^2_{gnd}(j\omega)$) within the compact model. Note that in all cases, d_1 is greater than d_2 since substrate contacts are placed between the TSV and victim node.

The sufficient accuracy of the fitting method is demonstrated by quantifying the average percent error (as compared to the distributed mesh based on 3D TLM) for each resistance and capacitance within the π network. These values are listed in the last column of Table 1. For case 1 and case 3 (a single substrate contact exists between the TSV and victim node), d_1 (distance between TSV and victim node) varies from 4 μm to 55 μm and d_2 (distance between TSV and substrate contact) varies from 2 μm to 33 μm. Alternatively, for case 2 and case 4 (multiple substrate contacts are regularly placed between the TSV and victim node), d_1 varies from 4 μm to 44 μm and d_2 (distance between two substrate contacts) varies from 2 μm to 8 μm. Note that the maximum average error is slightly over 10% for certain resistances and capacitances. This error, however, does not significantly affect the electrical characteristics (and noise estimation at the victim node) since the maximum error occurs at the extreme cases when the resistance is sufficiently large and capacitance is sufficiently small. Also note that the average error over four cases is 4.8%.

The proposed model and the function F can be used to efficiently characterize TSV-to-transistor noise coupling for (1) different design parameters such as the distance between TSV and the victim transistor, number and location of substrate contacts and (2) different TSV fabrication technologies such as via-first and via-last.

5. DESIGN GUIDELINES

Peak-to-peak noise at the victim transistor due to TSV activity is analyzed using (13) and the compact model illustrated in Fig. 4. This noise is depicted in Figs. 8(a) and 8(b) as a function of d_2 when d_1 is constant at 30 μm.

According to Fig. 8(a), where a single substrate contact exists between the TSV and victim node, switching noise is reduced as the substrate contact is placed closer to the victim node as opposed to the TSV. This characteristic is due to TSV height and distributed TSV capacitance to substrate. Thus, a single substrate contact closer to the TSV is not sufficiently effective. Note that based on Fig. 8(a), this characteristic is stronger in via-last TSVs since the height of a via-last TSV is five times greater than a via-first TSV. In traditional 2D circuits, it is typically a physical design deci-

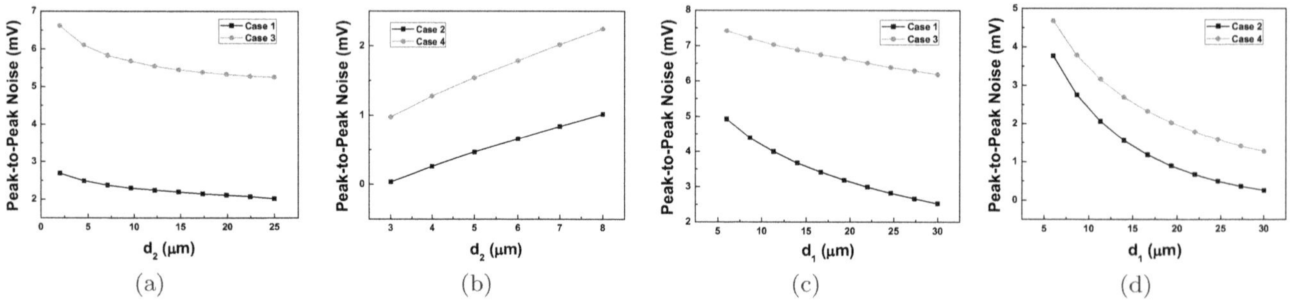

Figure 8: TSV induced switching noise at the victim node: (a) as a function of d_2 at constant d_1 for case 1 and case 3, (b) as a function of d_2 at constant d_1 for case 2 and case 4, (c) as a function of d_1 at constant d_2 for case 1 and case 3, and (d) as a function of d_1 at constant d_2 for case 2 and case 4.

sion to place the substrate contacts (or guard rings) around an aggressor noise source or around a sensitive victim block. In 3D circuits where TSVs are primary source of switching noise, placing the substrate contacts closer to the victim block is more advantageous, as demonstrated in Fig. 8(a).

According to Fig. 8(b), where multiple, regularly placed substrate contacts exist between the TSV and victim node, switching noise is significantly less as compared to Fig. 8(a) and is further reduced as d_2 decreases, *i.e.*, number of substrate contacts increases. Also note that in both figures, switching noise due to via-last TSVs is significantly greater than via-first TSVs since the diameter is larger and height is longer.

Peak-to-peak switching noise at the victim transistor is shown in Figs. 8(c) and 8(d) as a function of d_1 when d_2 is constant at 4 μm. As illustrated in Fig. 8(c), when only a single substrate contact exists, placing the victim transistor farther from the switching TSV is an effective method for via-first TSVs. Alternatively, for via-last TSVs, the noise exhibits low sensitivity to the distance between TSV and victim transistor. This phenomenon is due to longer height (therefore smaller substrate resistances) and larger diameter (therefore larger capacitances) of via-last TSVs.

According to Fig. 8(d), when multiple substrate contacts are regularly placed, increasing the physical distance between the switching TSV and victim transistor is helpful for both via-first and via-last TSVs. In this case, the effective impedance between the TSV and substrate contact becomes significantly lower since the number of substrate contacts increases as d_1 is increased.

6. CONCLUSIONS

TSV-to-transistor noise coupling has been evaluated and quantified in 3D ICs. A compact π model has been proposed to estimate noise at the victim transistor as a function of different substrate biasing schemes (single substrate contact versus multiple, regularly placed substrate contacts) and TSV fabrication methods (via-first versus via-last). A closed-form expression has been developed to approximate each admittance within the π model with a logarithmic function. Both the compact model and the closed-form expression have been validated using a 3D transmission line matrix method with an average error of 4.8%. These expressions and the model have been utilized to better understand the effect of different physical parameters on noise for both via-first and via-last TSVs.

7. REFERENCES

[1] E. Salman and E. G. Friedman, *High Performance Integrated Circuit Design*. McGraw-Hill, 2012.

[2] V. F. Pavlidis and E. G. Friedman, *Three-Dimensional Integrated Circuit Design*. Morgan Kaufmann, 2009.

[3] E. Salman, "Noise Coupling Due to Through Silicon Vias (TSVs) in 3-D Integrated Circuits," in *Proc. of the IEEE Int. Symp. on Circuits and Systems*, May 2011, pp. 1411–1414.

[4] N. H. Khan *et al.*, "Through-Silicon Via (TSV)-Induced Noise Characterization and Noise Mitigation Using Coaxial TSVs," in *Proc. of the IEEE Int. Conf. on 3D System Integration*, September 2009, pp. 1–7.

[5] J. Cho *et al.*, "Guard Ring Effect for Through Silicon Via (TSV) Noise Coupling Reduction," in *Proc. of the IEEE CPMT Symposium Japan*, Aug. 2010, pp. 1–4.

[6] P. Saguet, "The 3D Transmission-Line Matrix Method: Theory and Comparison of the Processes," *Int. Journal of Numerical Modelling: Electronic Networks, Devices and Fields*, vol. 2, no. 4, pp. 191–201, December 1989.

[7] J. Cho *et al.*, "Modeling and Analysis of Through-Silicon Via (TSV) Noise Coupling and Suppression Using a Guard Ring," *IEEE Trans. on Components, Packaging and Manufacturing Tech.*, vol. 1, no. 2, pp. 220–233, February 2011.

[8] S. Ramaswami *et al.*, "Process Integration Considerations for 300 mm TSV Manufacturing," *IEEE Trans. on Device and Materials Reliability*, vol. 9, no. 4, pp. 524–528, December 2009.

[9] I. Savidis and E. G. Friedman, "Closed-Form Expressions of 3-D Via Resistance, Inductance, and Capacitance," *IEEE Trans. on Electron Devices*, vol. 56, no. 9, pp. 1873–1881, September 2009.

[10] J. Kim, J. Cho, and J. Kim, "TSV Modeling and Noise Coupling in 3D IC," in *Proc. of the IEEE Electronic System-Integration Tech. Conf.*, September 2006, pp. 1–6.

[11] J. S. Pak *et al.*, "PDN Impedance Modeling and Analysis of 3D TSV IC by Using Proposed P/G TSV Array Model Based on Separated P/G TSV and Chip-PDN Models," *IEEE Trans. on Components, Packaging and Manufacturing Technology*, vol. 1, no. 2, pp. 208–219, February 2011.

A Portable High-Frequency Digitally Controlled Oscillator (DCO)

Muhammad E. S. Elrabaa
Computer Engineering Department
King Fahd University of Petroleum and Minerals
Dhahran 31261, Saudi Arabia
Tel.: +9665-08979203
elrabaa@kfupm.edu.sa

ABSTRACT

A novel digitally-controlled oscillator (DCO) is reported. Utilizing a new capacitive load, the new DCO is capable of producing much higher output frequencies than existing DCOs. All other components are fully digital and modular, allowing portability to any CMOS process and customization for different applications. At the heart of the DCO is a digital ring oscillator (DRO) that utilizes the new shunt-capacitive loads. Unprecedented higher frequencies are obtained through a novel idea of electrically removing the effect of un-enabled loads. Simple design conditions for achieving proper operation of the DRO are provided. Spice simulations verified the correct and superior operation of the DCO even with device mismatch. A custom layout of the DRO was generated using LFoundry's 150 nm technology. The total DRO area was found to be 418 μm^2.

Categories and Subject Descriptors

B.7.0 [**Integrated Circuits**]: General

Keywords

Digitally-Controlled Oscillator (DCO), Systems-on-Chip (SoCs), Digital Circuits

1. INTRODUCTION

Many applications implemented as systems-on-chip require the generation of high speed on-chip clocks with minimal area and power consumption. Analog Phase-locked loops (PLLs) can provide precise frequencies but contains analog circuits and filters that take up large chip area and can't be ported from one fabrication process to another. All-digital PLLs (ADPLLs) are more portable and have smaller area. They, however, require digital DCOs with monotonic behavior, fine frequency resolution and good period linearity.

Over the years, many DCOs have been proposed. Almost all of the reported DCOs use two stages for frequency tuning; a coarse tuning stage and a fine tuning stage. This allows the DCO to have a large frequency range with fine resolution while using a minimal number of control bits. This however, may also limit the maximum output frequency of the DCO. Most of these DCOs employ one or more of the techniques illustrated in Figure 1 to change the delay. These techniques are; current-starved inverters as delay stages [1-3,9,12], inverters with switched shunt MOS capacitors [4-5, 12-13], and multiplexors to select the number of delay stages (path selection) [1,6-11]. The current starvation and path selection methods are seldom used on their own and are usually combined with other techniques (e.g. [1,9]). This is because current starvation has very limited range and path selection has very limited resolution. This means the former would require large number of delay elements (resistors) and the latter would require a large number of delay stages (inverters). For DCO's with shunt capacitors delay elements, some researchers use MOS varactors with differential drive due to their excellent linearity [12-14]. This, however, also requires a large number of delay stages due to the small capacitance of varactors.

Although all the above mentioned techniques were effectively used to produce fine resolution and large frequency range with adequate linearity, they all suffer from a basic shortcoming; limited achievable maximum frequency. This is due to the fact that whatever technique is used to control the DCO's period, the elements that are used to control the delay (series resistances, shunt capacitors, or selection multiplexors) always exist in the circuit and can't be physically eliminated at the highest DCO frequency. Also, switches (NMOS, PMOS or transmission gates) controlling these elements introduce significant parasitic capacitances reducing the DCO's maximum attainable frequency further. These two issues lead to a basic trade off in all existing DCOs; in order to increase the resolution and/or frequency range, more delay elements have to be added which reduces the DCO's intrinsic (maximum) frequency. Also, in order to increase the range, the range of values of the binary-weighted resistors or capacitors (used as delay control elements in the DCO) must be increased. This causes matching problems and can lead to non-monotonic DCO frequency characteristics at some control code words. This also forces designers to use the highly non-linear MOS capacitors to be able to get large capacitance values in reasonable silicon area.

The developed DCO circuit attempts to solve these problems by electrically removing delay elements (shunt capacitors) that are not enabled by the digital control word. This allows obtaining a maximum intrinsic frequency (when the control word is all 0s) close to that of an unloaded oscillator. It also allows obtaining the desired frequency range with much smaller values of shunt capacitors. In section 2 the general architecture of the proposed DCO is introduced followed by a detailed description of the digital

GLSVLSI'13, May 2–3, 2013, Paris, France.

ring oscillator (DRO) which is the core of the DCO in section 3. Simulation results showing the superior operation of the operation of the DRO are shown in section 4. Finally, a custom layout of the new DRO is shown in section 5 to illustrate the small area of the new DRO.

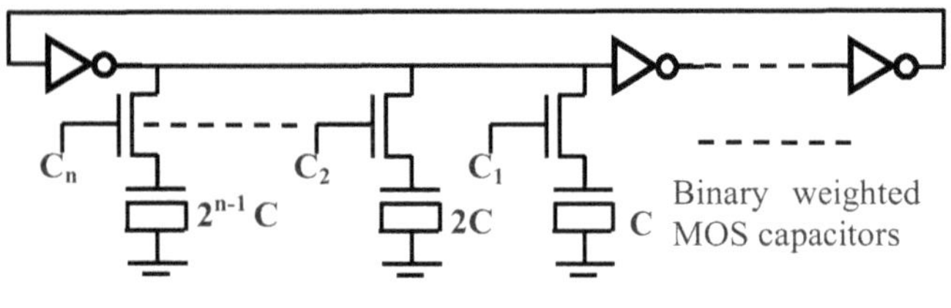

(a) A Digital Oscillator based on shunt MOS capacitors.

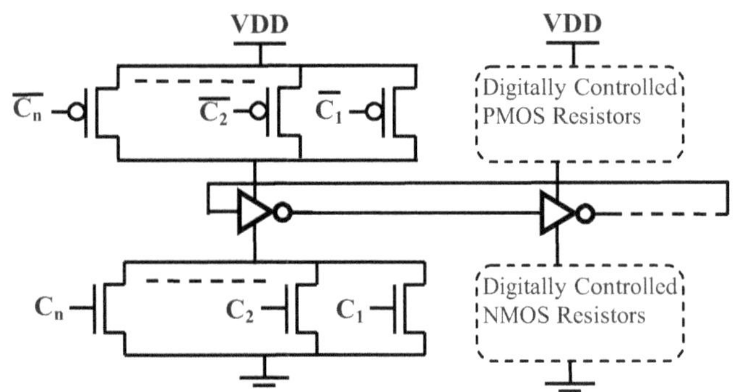

(b) A Digital Oscillator based on current starving.

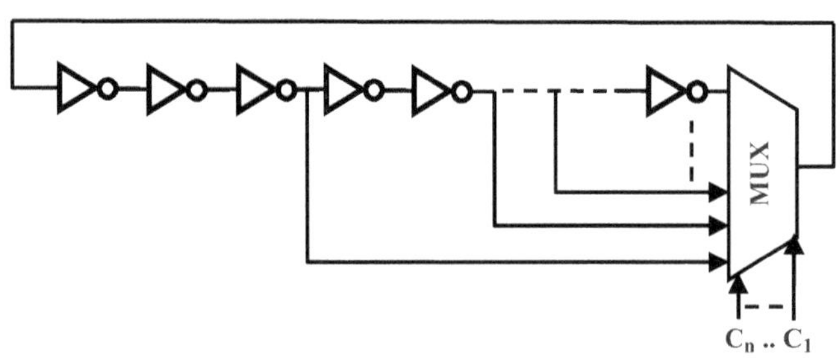

(c) A Digital Oscillator based on path selection.

Figure 1. Conventional methods for implementing DCOs.

2. DCO ARCHITECTURE

Figure 2 below shows the general architecture of the proposed DCO. It is made of a new digitally-controlled digital ring oscillator (DRO), a frequency divider (counter) and a range selection MUX. The DRO generates the basic high frequency range, the counter generates lower frequency ranges (through division by powers of 2), and the MUX provide range selection. Hence, for proper operation, the DRO frequency range must be: $f_{max} - 0.5f_{max}$, where f_{max} is the maximum frequency. This ensures continuous frequency range with the successive division by 2. As such, the required range of the DRO is relatively small, allowing for finer resolution with few bits of control word. It also means that the maximum shunt capacitance is significantly reduced, preserving DRO's signal slopes and improving linearity. The reduced range of capacitance values also makes it easier to ensure monotonicity of the oscillator with process variations.

3. THE DRO

3.1 DRO Circuitry

The DRO is made of two digital oscillators with identical inverters and a merging NAND (MNAND) gate, Figure 3. The two oscillators oscillate at the same frequency and phase. The merging NAND gate, Figure 3(b), resets the oscillators to start at the same exact moment and 'merges' the two oscillators' last-stage signals ensuring phase and frequency synchronization between the two oscillators under process variations. Newly developed, binary weighted, digitally controlled, capacitive loads are connected between each delay stage in one oscillator and the corresponding stage in the other oscillator, Fig. 3(c). As this Figure shows, a capacitive load cell is made of two identical capacitors connected in series and an NMOS switch that conditionally connects the node in between them to ground. These switches are controlled by the DCO's fine frequency control word.

3.2 Concept of Operation

Figure 4 illustrates the basic concept of the new capacitive load. When the NMOS switch is off, only its small (relative to C) drain junction capacitance (Cj) is connected between the common node and ground. Since the capacitor combination is driven from both ends with identical phase, the two inverters in the two oscillators see very small capacitances (< 0.5 Cj). When the switch is on the common node is connected to ground and each inverter sees a load of C which increases its delay while identical phase is still maintained. Hence, unlike conventional shunt capacitance schemes, this new scheme results in a large capacitance difference between the two states since there are no parasitic capacitance at the inverters' outputs. This means that the required frequency range and resolution could be obtained with small values of C and few load cells. It should be noted here that the shunt capacitances must be implemented with metal layers not MOS capacitors to avoid excessive parasitic capacitances. This however is not a problem since the required values are very small. For example, for a 3-stage DRO implemented in a 0.13 μm technology with 4-bit/stage control, the required value of C is 0.5 fF. This yields an astonishing DRO frequency range of 2.5 – 5 GHZ. The on-chip area of such a capacitor would be in the order of few square micrometers.

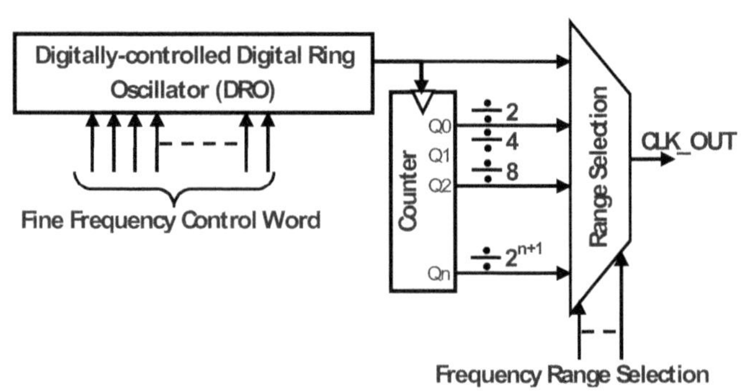

Figure 2. The general architecture of the new DCO.

3.3 Design Constraints

For proper portability of the DRO to any fabrication process, it has to be ensured that $\mathbf{f_{max}} \geq 2\mathbf{f_{min}}$ (i.e. the frequency range is one octave). That means the maximum delay through a DRO stage must be at least twice its minimum delay, i.e. $\mathbf{TD_{max}} \geq 2\mathbf{TD_{min}}$. Also, to improve linearity of the DRO, the sizes of the NMOS switches in the capacitance cells have to be increased at the same ratio of the capacitance (i.e. binary-weighted fashion).

Now the minimum value of C (the unit load capacitance, Fig. 2(c)) that is needed for proper DRO operation can be estimated as follows; first, the following two equations give approximate values of $\mathbf{TD_{max}}$ and $2\mathbf{TD_{min}}$ for a 3-stage DRO based on simple RC delay model:

$$\mathbf{TD_{min} = R_{eq} * [2C_{in} + 0.5*(2^n - 1)\, C_j]} \quad (1)$$
$$\mathbf{TD_{max} = R_{eq} * [2C_{in} + (2^n - 1)\, C]} \quad (2)$$

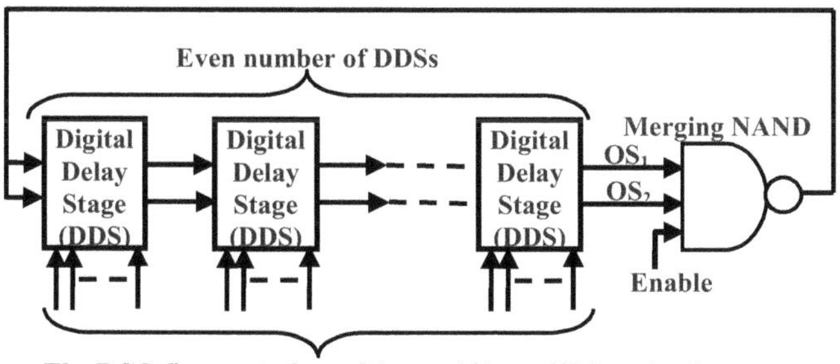

(a) Basic structure of the DRO

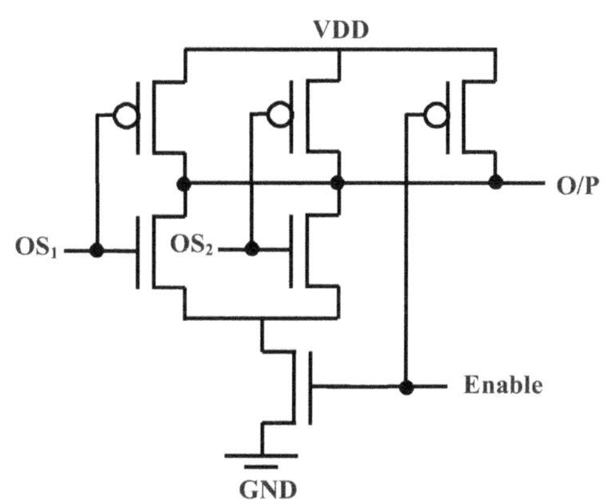

(b) The Merging NAND gate.

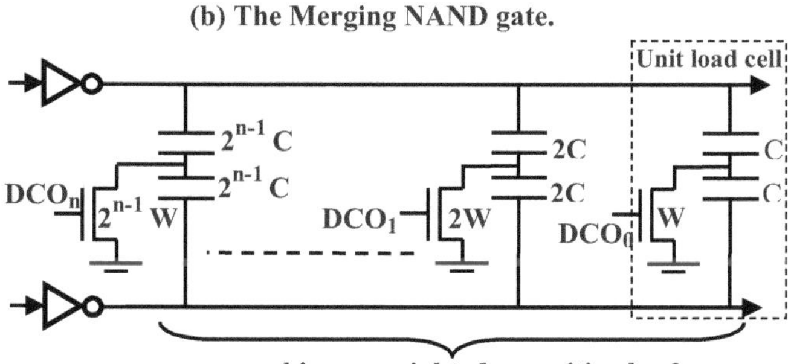

***n* binary-weighted capacitive loads**

(c) A Digital Delay Stage (DDS)

Figure 3. The DRO's circuitry.

Where R_{eq} is the equivalent resistance of the CMOS inverter in the DDS, C_{in} is its input capacitance, C_j is the unit drain junction capacitance of the NMOS switch in the capacitance cell (its value is split between the two inverters), and n is the number of control bits/DDS. In (1) and (2) above, the delay of the MNAND gate (which is relatively constant) is divided among the two DDSs, hence the factor of 2 in front of C_{in}. The intrinsic delay of the MNAND is approximately twice that of the inverters in the DDSs since it has twice the FanOut. So from the above equations, the condition for proper operation of the DRO is:

$$C \geq \frac{2C_{in} + \left(2^n - 1\right)C_j}{\left(2^n - 1\right)} = \frac{2C_{in}}{2^n - 1} + C_j$$

For the 0.13µm, 1.2V technology used in this work, the values of C_{in} and C_j were 1.6 fF and 0.27 fF, respectively. According the condition obtained above and for 4-bit control/DDS, the value of C should be 0.48fF. The actual value found from simulations was actually 0.5fF, an excellent agreement with the predicted value.

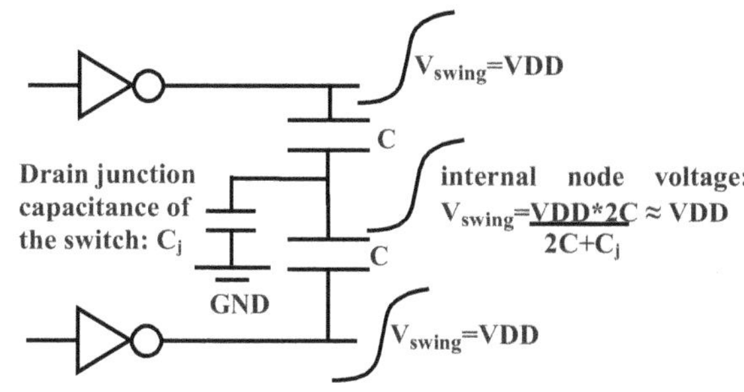

(a) When the switch is off, each inverter sees a load capacitance ≈ 0.5(C C_J)/(C+ C_J) < 0.5 C_J

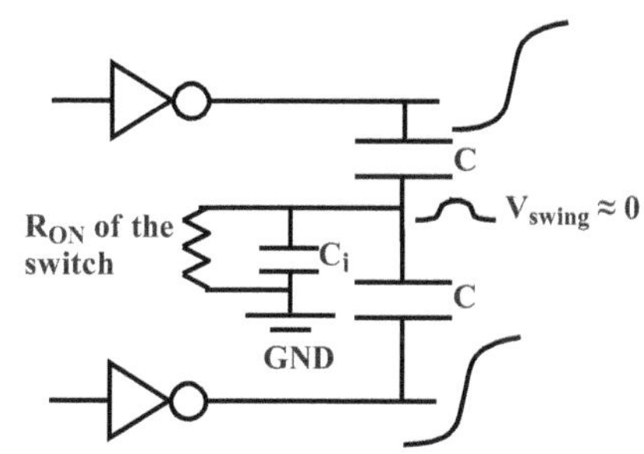

(b) When the switch is ON, each inverter sees a load ≈ C

Figure 4. Concept of operation of the new capacitive load cell.

3.4 DRO Control Schemes

Since each DDS has its own control word, there are many ways to control the DRO. All DDSs' words can't be treated as one big word because that will result in many redundant codes (producing equal frequencies) which in turn will cause non-monotonic RDO characteristics. As such, there are two main ways to properly control the DRO to ensure non-redundant codes; one way is to control the DDSs in a binary round-robin fashion (one DDs at a time) and the other is to control all the DDSs' loads together in a thermometer coding fashion. In the first method, starting with the first DDS, its control word is incremented till it reaches its maximum value, then the control word for the 2nd DDS is incremented till it reaches its maximum value, then the control word for the 3rd DDS is incremented and so on as illustrated in Figure 5(a). This is equivalent to dividing the frequency range into a number of sub-ranges equal to the number of DDSs. In the second DRO control method, Figure 5(b), the load capacitances are enabled in a thermometer coding fashion; 1st ***C*** (the smallest capacitance) of the 1st DDS is enabled, then that of the 2nd DDS, then that of the 3rd DDS till the last DDS's ***C*** is enabled. Next the ***2C*** of the 1st DDS is enabled and its ***C*** is disabled, then the ***2C*** of the 2nd DDS is enabled and its ***C*** is disabled, and so on till all the

*2C*s are enabled. Then another round of enabling the *C*s, followed by a round of enabling the *4C*s and disabling the *C* and *2C* of each DDS. This will be followed by another round of enabling the *C*s, and then the *2C*s, followed by a round of enabling the *8C*s and disabling the *C*, *2C* and *4C* of each DDS and so on till all loads in all DDSs are enabled (minimum frequency). This type of control achieves better linearity but is more complex to implement. Both methods yield the same number of total distinct codes; for *m* DDS stages and *n*-bits/stage control word, the total number of distinct codes is $m*(2^n - 1) + 1$.

4. SIMULATION RESULTS

4.1 Validation of the Basic Operation

To verify the validity of the new DCO design a 3-stage DRO (two DDSs) and a 3-bit counter were simulated with Spice using a 0.13 m, 1.2V CMOS technology. All NMOS transistor widths were set to a minimum of 1.5*L (L is the minimum channel length) and that of the PMOS transistors to twice that (i.e. 3*L). C was selected to achieve the required frequency criteria as explained in 3.3 ($\mathbf{f_{max} \geq 2f_{min}}$). Fig. 6 shows the voltage waveforms at the outputs of the first stage of both oscillators of the DRO (VOUT1 & VOUT2), and the internal node of the unit capacitance load cell (the smallest capacitance) for two conditions; (a) when all control words are 0s (maximum frequency), and (b) when the 1st control word is 0011 (three steps above minimum period). This figure illustrates how the concept of the new capacitive load cell works. When the switch is OFF, the internal node follows the inverters' outputs very closely, resulting in very small 'effective' capacitive loads for these inverters, thus the attained extremely high frequency. The figure also shows how the two oscillators remain in-phase all the time (with or without the capacitive cells enabled).

4.2 Resolution and Linearity

Figure 7 shows the period and frequency of the DRO's output with the two types of control methods. As this figure shows, the thermometer coding method produces a very linear response. It also shows that with only 4-bits of control word/stage the period step is ~ 6ps, a remarkable performance. Lower frequency ranges, obtained through frequency division, will have larger period steps. Also, when using the binary round-robin control, the period step slightly increases as we move from the 1st DDS to the second DDS due to the effects of increased signal slopes. Hence there will be a set of sub-ranges within the DRO's intrinsic range equal to the number of DDSs within the DRO when using this type of control. The DRO's response is still very acceptable.

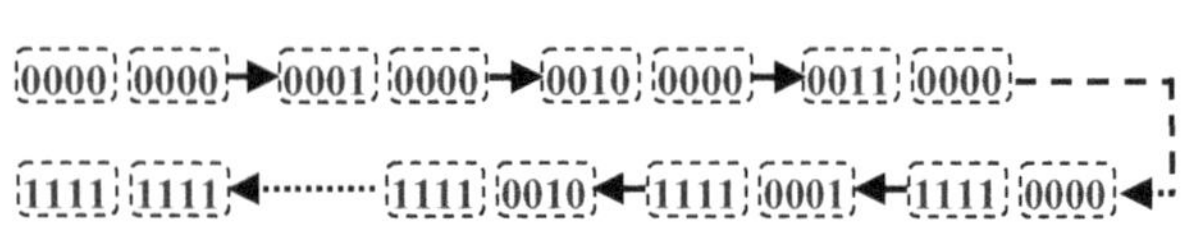

(a) Controlling the DDSs in a binary round-robin fashion.

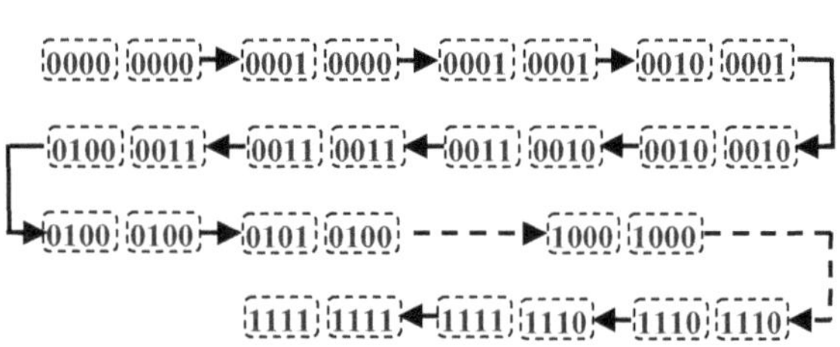

(b) Controlling the DDSs in a thermometer coding fashion.

Figure 5. The two main methods for controlling the DRO (illustrated for 2 DDSs with 4-bit/stage control words).

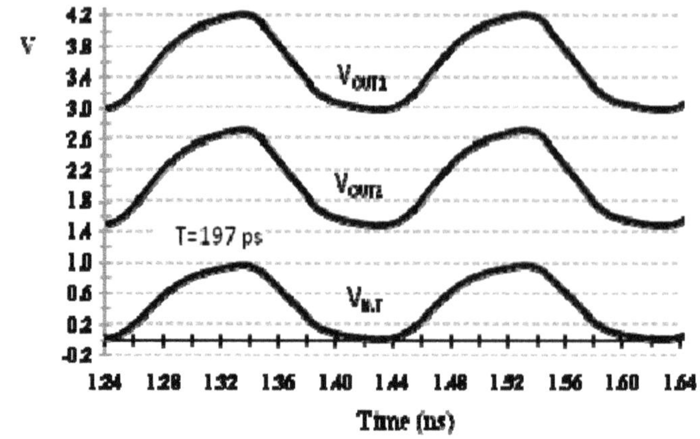

(a) Voltage waveforms with the NMOS switch in the capacitance cell off; the internal voltage tracks the inverters' output hence the inverters practically see no capacitance. Waveforms (V_{out1} & V_{out2}) have been shifted up for clarity.

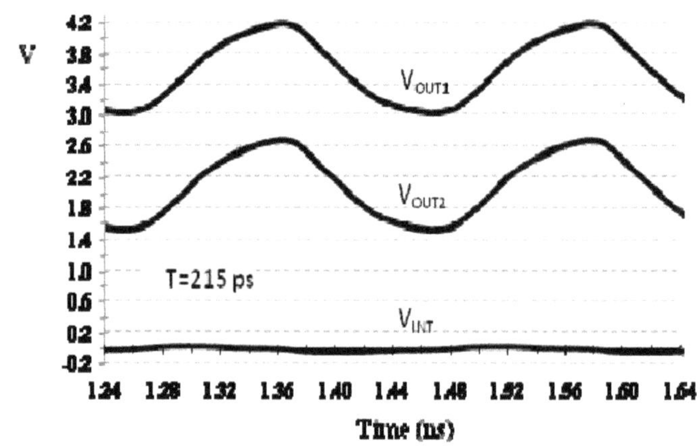

(b) The waveforms with the NMOS Switch ON; the internal voltage is held close to 0 V (i.e. the inverters in the DRO see the full load capacitance.

Figure 6. The voltage waveforms of the 1st stage of the DRO and internal node of the unit capacitance load cell.

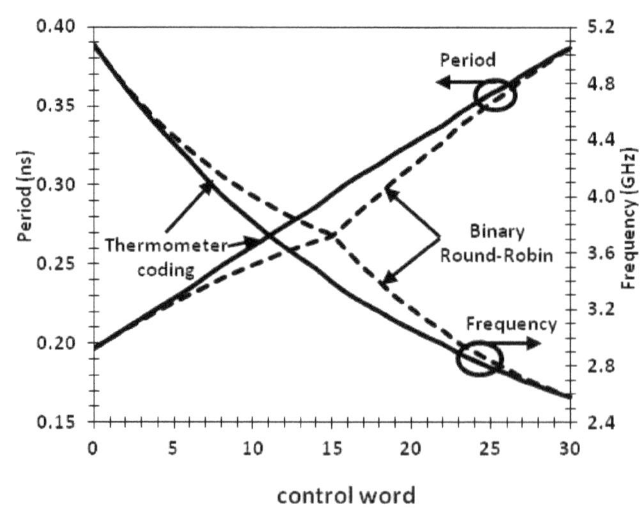

Figure 7. The DRO's Characteristics (period/frequency versus control word) using the two control methods.

It should be noted that, when lower frequency ranges are obtained through successive divisions by 2, non-montonic transitions at the borders between adjacent ranges can occur due to $\mathbf{f_{max}}$ being > $\mathbf{2f_{min}}$. This, however, won't cause any problems since the transition from one range to another is accomplished using the selection MUX while the DRO has the same frequency (i.e. the DCO would go from a frequency **f** to 0.5**f**, 2**f**, 0.25**f**, or 4**f** ...etc.). So any search algorithm used in a feedback loop (such as in a digital PLL) won't be stuck in a non-monotonic region.

4.3 Comparison with Conventional DCO

A 3-stage, 4-control bits/stage conventional DCO based on shunt MOS capacitive loads (such as in Figure 1 (a)) was designed and simulated. Figure 8 shows the period and frequency of this DCO versus the control word. Also reproduced on the same Figure are the results for the proposed DCO. Both DCO's had the same inverters' sizes but the loads were adjusted such that the ratio between $\mathbf{f_{max}/f_{min}} \sim 2$. As this figure shows, the new DCO can achieve double the maximum frequency of the conventional DCO due to the new capacitive cell. It also has a much better linearity than the conventional DCO.

4.4 Effects of Devices Mismatch

Since the new DCO is utilizing two oscillators that are suppose to oscillate in perfect synch for the new load cell to operate properly, one concern might arise; what would be the effect of devices mismatch between the two oscillators? The concern here is that the device mismatch would cause phase mismatch between the two oscillators. With such phase mismatch, capacitances in cells that are off which are suppose to 'appear' as open circuited would have an actual value that depends on the mismatch. To check this concern the DRO was simulated with the channel lengths of all MOS transistors in one of oscillator being 10% larger than the minimum. This goes beyond any reasonable process mismatch. Figure 9 shows the output of the DRO at maximum frequency (the most sensitive point to phase mismatch) without the mismatch (Vout1) and with the mismatch (Vout2). As this figure clearly shows, the capacitive cell (and the DRO) still operate exactly as they should. The MNAND gate actually 'mixes' the two oscillators frequency and the resulting frequency is an interpolation between the two. The 10% increase in the channel length of one of the oscillators increased its period by ~7%. The net result, as shown in Figure 9, is that the DRO's period actually increased by ~3.5%.

5. DRO AREA

A custom layout of the DRO was generated using LFoundry's 150 nm technology and is shown in Figure 10 below. M1-M2 capacitances were used for the load cells (at the top part of the layout). The total DRO's area came out to be 44 µm X 9.5 µm (i.e. 418 µm^2), a remarkably small area even at this technology node.

6. CONCLUSIONS

A new fully-digital DCO has been developed. It utilizes a new type of shunt capacitive load for delay control. The new load significantly reduces the effects of parasitic capacitance resulting in a large increase of the maximum achievable frequency. Two methods for controlling the DCO have been devised. Also a simple yet very accurate equation was obtained for selecting the value of the unit capacitance load to ensure proper operation of the DCO.

Spice simulations were carried out to verify the operation of the new DCO and evaluate its performance using a 0.13 µm, 1.2V technology. The DCO achieved a maximum frequency of 5.08 GHz at this technology node, an un-precedent performance. With only 4-bit control word/stage, the DCO's resolution at the maximum frequency was 6ps/step. Comparison results show the superior performance of the new DCO compared to conventional shunt capacitive load based DCOs. Also, the custom layout of the DRO shows its remarkable small area.

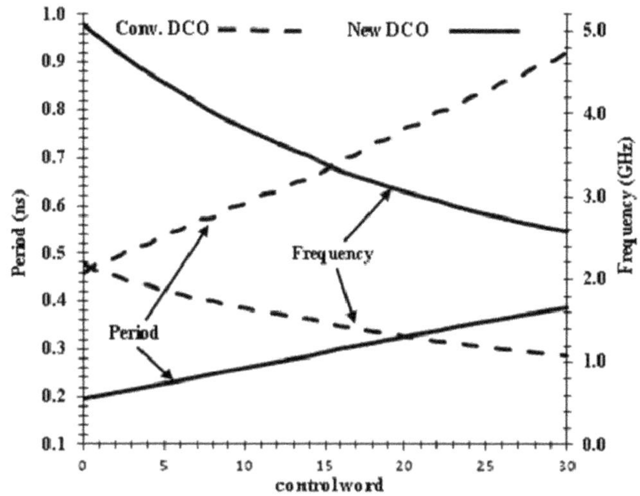

Figure 8. Comparison results for the new DCO and a conventional DCO based on MOS shunt capacitive loads.

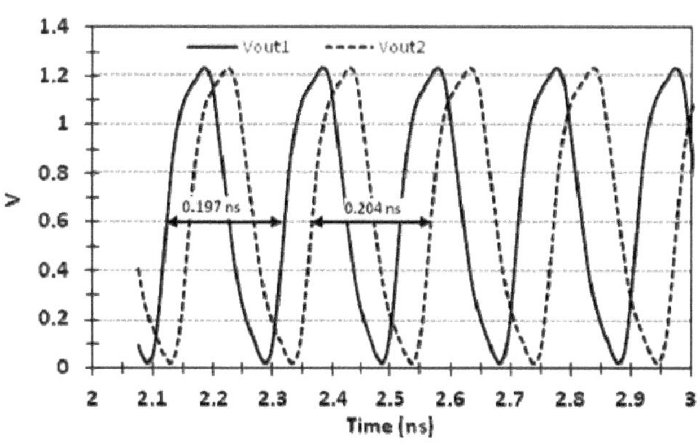

Figure 9. Simulation results of the DRO's output with and without device mismatch.

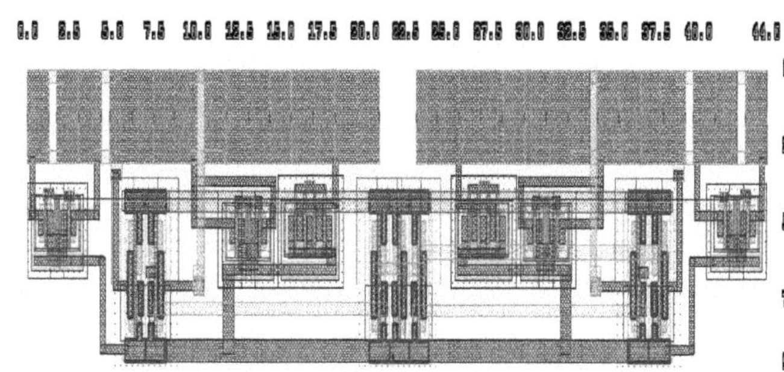

Figure 10. Custom layout of the DRO using LFoundry's 150 nm technology.

7. ACKNOWLEDGMENTS

This work has been supported by King Abdel-Aziz City for Science and Technology through development grant 4-14/2010. KFUPM's facilities support is highly appreciated by the author.

8. REFERENCES

[1] Jen-Shim Chiang and Kuang-Yuan Chen, "The design of an all-digital phase-locked loop with small DCO hardware and fast phase lock," IEEE Transactions on Circuits and Systems II: Analog and Digital Signal Processing, Vol. 46, No 7, pp. 945-950, 1999.

[2] M. Saint-Laurent and G. P. Muyshondt, "A digitally controlled oscillator constructed using adjustable resistors," in *Proceedings of IEEE Southwest Symposium on Mixed-Signal Design*, pp. 80–82, 2001.

[3] R. B. Staszewski and P. T. Balsara, "Phase-domain all-digital phase-locked loop," IEEE Transactions on Circuits and Systems II, vol. 52, no. 3, pp. 159–163, 2005.

[4] P. Raha, S. Randall, R. Jennings, B. Helmick, A. Amerasekera, and B. Haroun, "A robust digital delay line architecture in a 0.13µm CMOS technology node for reduced design and process sensitivities," in *Proceedings of the International Symposium on Quality Electronic Design (ISQED '02)*, pp. 148–153, 2002.

[5] Hiroyoshi Tomita, "Delay Circuit Having a Capacitor and Having Reduced Power Supply Voltage Dependency," US Patent 7,352,223 B2, April 1, 2008.

[6] Chia-Tsun Wu, Wei Wang, I-Chyn Wey, and An-Yeu (Andy) Wu, "A Scalable DCO Design for Portable ADPLL Designs," in *Proceedings of* ISCAS 2005, pp. 5449 – 5452, Vol. 6.

[7] Duo Sheng, Ching-Che Chung, and Chen-Yi Lee, "An All-Digital Phase-Locked Loop with High-Resolution for SoC Applications," in *Proceedings* Int. Sym. on VLSI Design, Automation and Test, pp. 1-4, 2006.

[8] Duo Sheng, Ching-Che Chung, and Chen-Yi Lee, "An Ultra-Low-Power and Portable Digitally Controlled Oscillator for SoC Applications," IEEE Transactions on Circuits and Systems II: Express Briefs, Vol. 54, No. 11, pp. 954-958, Nov. 2007.

[9] Shu-Yu Hsu, Jui-Yuan Yu, and Chen-Yi Lee, "A Sub-10-μW Digitally Controlled Oscillator Based on Hysteresis Delay Cell Topologies for WBAN Applications," IEEE Transactions on Circuits and Systems II: Express Briefs, Vol. 57, No. 12, pp. 951-955, Dec. 2010.

[10] Kouichi Ishimi, "Digital Delay Line," US Patent 6,366,150 B1, April 2, 2002.

[11] Robert G. Warren, "Programmable Digital Delay," US Patent 7,952,411 B2, May 31, 2011.

[12] Hong-Yi Huang, Jen-Chieh Liu, and Kuo-Hsing Cheng, "All-Digital PLL Using Pulse-Based DCO," in *Proceedings of* 14th IEEE Int. Conf. on Digital Electronics, Circuits and Systems (ICECS'07), pp. 1268-1271, 2007.

[13] Heesoo Song, et al., "A 1.0–4.0-Gb/s All-Digital CDR With 1.0-ps Period Resolution DCO and Adaptive Proportional Gain Control," IEEE J. of Solid-State Circuits, Vol. 46, No. 2, pp. 424-434, Feb. 2011.

[14] S. Yoo, et al., "A 5.8-GHz High-Frequency Resolution Digitally Controlled Oscillator Using the Difference Between Inversion and Accumulation Mode Capacitance of pMOS Varactors," IEEE Transactions on Microwave Theory and Techniques, Vol.59, no.2, pp.375-382, Feb. 2011.

A 250mV Sub-threshold Asynchronous 8051 Microcontroller with a Novel 16T SRAM Cell for Improved Reliability in 40nm CMOS

Jaeyoung Kim
University of Michigan
Ann Arbor, MI
hesed@umich.edu

Kwen-Siong Chong
Nanyang Technological University
Nanyang Avenue, Singapore
KSChong@ntu.edu.sg

Joseph Sylvester Chang
Nanyang Technological University
Nanyang Avenue, Singapore
EJSCHANG@ntu.edu.sg

Pinaki Mazumder
University of Michigan
Ann Arbor, MI
mazum@umich.edu

ABSTRACT

Asynchronous approach for digital systems is a way to resolve increased timing uncertainty with technology scaling since timing issue is eliminated in asynchronous systems. This paper presents a sub-threshold operating asynchronous 8051 microcontroller (A8051) with a novel 16T SRAM cell for improved reliability in asynchronous systems. This A8051, adopting a 4-phase dual-rail protocol, can operate up to 250 mV. A8051 has 67.53 μs as a critical path delay with 91.6 nW power consumption at 250 mV, which is equivalent to 12.88 kHz in synchronous systems. At 1.0 V, the delay of a critical path of A8051 microcontroller is 5.74 ns, which is equivalent to 151.55 MHz, with 8.98 mW power consumption. The proposed 16T SRAM cell is applied in memory blocks. The 16T SRAM structure eliminates charge contentions between devices during read and write operations so that SRAM can be operated fully in static mode, bringing about improved write margin (WM). The WM of this 16T SRAM cell is 1.81 times greater than the conventional 6T SRAM cell and 1.58 times better than 8T SRAM cell. At 250 mV, the SNM of SRAM cell is 12.5 mV under process and mismatch variations. Write delay of the asynchronous SRAM block is 4.02 μs (equivalent to 248.5 kHz) with 5.44 pJ energy dissipation, while read delay is 12.61 μs (equivalent to 79.3 kHz) with 9.08 pJ energy dissipation.

Categories and Subject Descriptors

B.7.1 [**Hardware**]: Types and Design Styles—*VLSI*

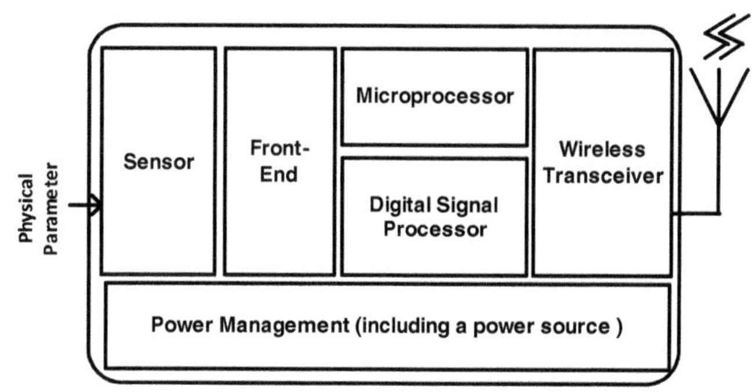

Figure 1: Wireless Sensor Network.

General Terms

Performance, Design, Reliabiliy, Experimentation

Keywords

Asynchronous system, Microcontroller, Microprocessor, Sub-threshold operation, SRAM, Reliable memory structure

1. INTRODUCTION

Wireless sensor networks (WSNs) are increasingly ubiquitous, in part, due to their ultra-low power, high reliability operation, and a small form factor. Figure 1 depicts a typical WSN, comprising six main modules: a sensor, a front-end (e.g. an analog-to-digital converter), a microprocessor, a digital signal processor, a wireless transceiver, and a power management unit including a power source [12]. As this WSN is typically designed for a long operational lifespan, power is carefully budgeted where pertinent, and it is energized only when required so that the overall average power is typically 10μW to 100μW. In a typical case, the power source in the power management unit allocates 20% of power to each module; the actual power breakdown will vary depending on specific applications. The microprocessor module with ultra-low power dissipation is highly desirable as it often remains active for parameter monitoring, and it, in part, enables various power-efficient techniques (e.g. with a very low duty cycle for wireless data transmission) [2]. In this paper, an ultra-low power microprocessor for WSNs is of specific interest.

For realization of a ultra-low power microprocessor, many design approaches have been proposed, including adoptation of limited instruction sets at the cost of reduced programmability and versatility [13], usage of smaller feature size devices and their corresponding lower supply voltages, clever design techniques with a similar objective to reduce switched capacitances and switching activities [3], and adaptive circuits and systems for lowering power budgets, such as Dynamic Voltage and Frequency Scaling (DVFS). Many approaches to this adaptive circuits have been suggested: so-called "always correct" such as look-up tables [14] and canary circuits [10, 11, 7, 16], and "fail and correct" such as RazorII [6]. These approaches, however, could have some issues. For example, look-up tables have fixed data in a static memory block, so these data should be determined by intensive simulation results under Process-Voltage Temperature (PVT) variations, so it is not fully fitted optimized. Canary circuits should be designed as the worst case in performance for monitoring, so there is a timing margin between this canary's delay and the critical path of a real fabricated chip. Razor has an assumption that secondary latch is always correct, meaning that this latch should be designed as a slower one than its primary latch under PVT variations. Furthermore, Razor requires additional cycles to correct erroneous data, bringing about performance degradation. The issues of the above mentioned approaches to the adaptive circuits and systems mainly relate to timing uncertainty. This is because inappropriate, insufficient budget for clock period or cycle time in digital circuits and systems could lead to incorrect functioning. This timing uncertainty is due to many components, such as Phase-Lock-Loop jitter, clock skew and jitter, power supply noise, PVT variations, and etc. As a result, either supply voltage is increased or a cycle time is longer as a guard band for ensuring correct operation, resulting in more energy consumption [8]. Among these components, PVT variations have been a significant issue with technology enhancement, switching the design paradigm from deterministic to statistical [4]. Due to increased timing uncertainty, the required guard band has also been increased to ensure correct operation under this increased timing uncertainty. PVT variations are even compounded for operation at the sub-threshold voltage region, further increasing this guard band. Several earlier studies even reported that the timing uncertainty could be more than 200 times in the sub-threshold region due to PVT variations [9].

In this paper, we present an adaptive sub-threshold 8051 microcontroller by means of asynchronous-logic; we denote it as 'A8051' for brevity. The main motivation of adopting asynchronous-logic is to remove the clock, hence resolving the timing uncertainty issue. Our proposed A8051 consists of an A8051 core, a 1,024×8 bit ROM, a 128×8 bit RAM, and a 1,024×8 bit XRAM. The overall architecture of the A8051 in this work is close to that of the design [5], but the latter employs a memory intellectual property provided by the manufacturer. Hence, the latter is not able to operate robustly in the sub-threshold regime largely due to memory failures and in part due to PVT variations for the memory interfaces, and it only features limited-range DVS from nominal to near-threshold voltage. In this work, we address these

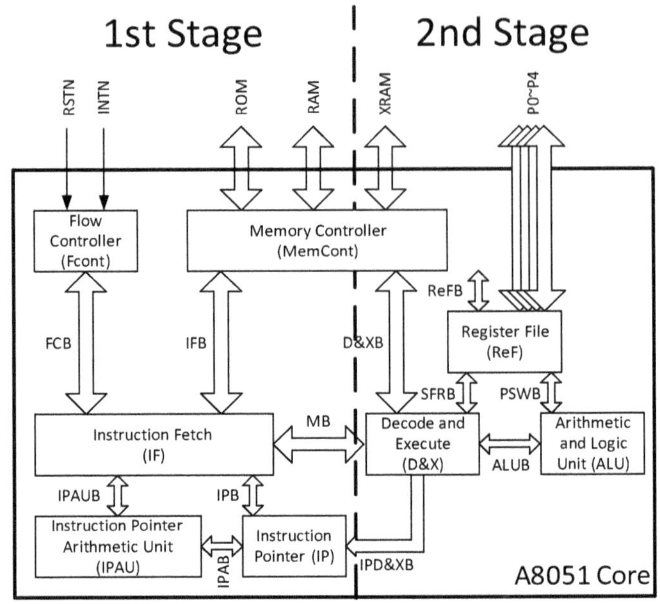

Figure 2: Block diagram of the A8051 μC core.

problems, and our proposed A8051 features several interesting attributes and novelties. First, our proposed A8051 is able to operate robustly in the sub-threshold region (e.g. at 250mV). Second, our proposed A8051 can innately accommodate PVT variations at any prevailing operating conditions, hence enabling a full-range DVS from nominal voltage through near-threshold to sub-threshold voltage uninterruptedly. Third, we propose a novel 16-Transistor (16T) SRAM cell for the ROM, RAM and XRAM memories. This 16T SRAM cell eliminates charge contentions between the cross-couple devices therein, hence operating fully in static mode and bringing about improved write margin. The 16T SRAM cell also features low bit line leakage, enabling the ROM, RAM and XRAM memories to operate at low voltage. Fourth, the A8051 is designed using combinational-logic gates (i.e. no sequential-logic gates such as clocked latch or flip-flop), hence potentially having lower error rates as delineated in the ITRS-2011 report where the sequential-logic has significant higher error rates than combinational logic for contemporary and future CMOS fabrication processes [1].

The remaining parts of this paper are as follows. Section 2 presents the overall architecture of the A8051. Section 3 describes memory design for the A8051. Simulation results for A8051 and memory blocks are presented in Section 4, and finally Section 5 provides conclusions.

2. A8051 ARCHITECTURE

A8051 consists of a 1,024×8 bit read-only memory (ROM) for programing instructions, a 128×8 bit random-access memory (RAM) for storing data, a 1,024×8 bit external RAM (XRAM), A8051 core, and an interface block for controlling program ports and general purpose I/O ports. The A8051 core mainly consists of two pipeline stages: Instruction Fetch (IF) and Decode and Execute (D&X) (see Figure 2). IF, Flow Controller (FCont), Instruction Pointer Arithmetic Unit (IPAU), Instruction Pointer (IP), and Memory

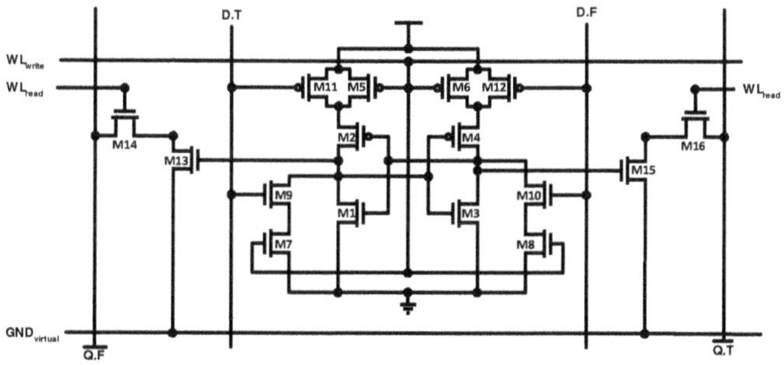

Figure 3: The proposed 16T SRAM cell structure.

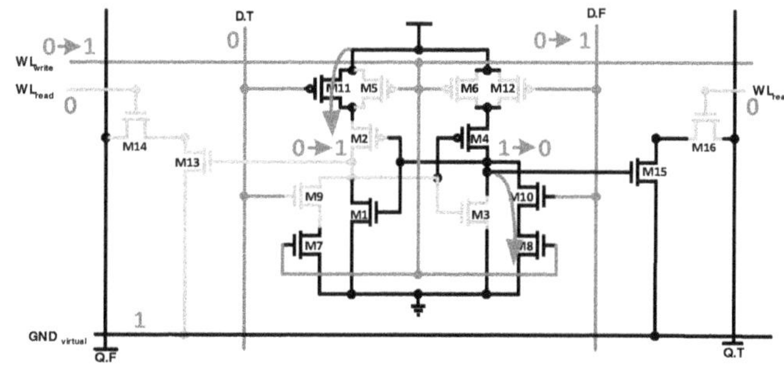

(a) WRITE operation.

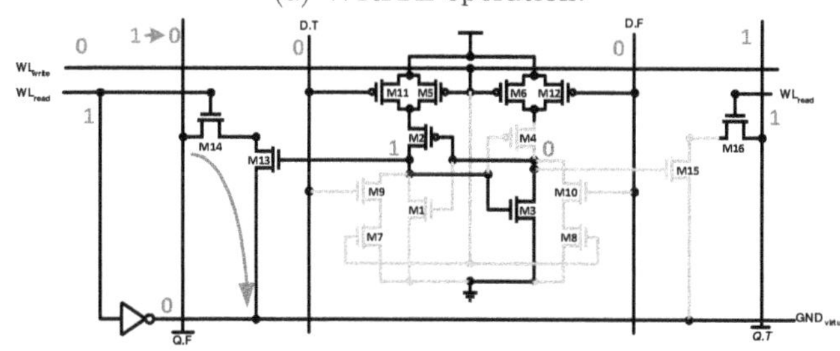

(b) READ operation.

Figure 4: WRITE/READ operation principle of the proposed 16T SRAM. (a) During WRITE, only one path exists at each storage node, so there is no short circuit current as well as charge contentions. (b) When READ, access transistors M13 and M14 evaluates the stored state as conventional 8T SRAM.

Controller (MemCont) participate in the first pipeline stage in which instruction fetch and exceptions including interrupts and branching are processed. Instructions are fetched by one byte at each cycle in this IF stage since majority of instructions have 1-byte length. In the second pipeline stage, operands fetch, operation execution, and write back are conducted by D&X, Arithmetic and Logic Unit (ALU), Register File (ReF), and MemCont, managing operands fetch, executing operation, and writing back. Each pipeline stage is almost independent from each other so that pipeline stalls can be minimized. Although a deeper pipeline stage can increase the overall throughput of A8051, it requires significant area and energy overhead for a pipeline controller to deal with data and control dependency. In addition, pipeline registers might be needed for stalling in the increased pipeline stages, which, in turn, dilutes the strength of asynchronous circuits: no register at all. Notice that each bus between blocks has handshake channels along with data channels so that each block knows when the resulting data should be delivered to the next block.

3. MEMORY DESIGN

In this section, a novel memory structure is presented. A 16T SRAM cell structure is presented in Section 3.1, and its operation principle is described in Section 3.2. Sizing constraints are provided in Section 3.3, and Section 3.4 describes a memory block.

3.1 SRAM Cell Structure

Reliability in memory system is one of the most challenging issues since A8051 is targeted at sub-threshold operation. One of the advantages of asynchronous systems is no timing issue, meaning that no flip-flop is required. Thus, memory for asynchronous systems should have such a property so that the entire system can be more robust in terms of signal timing; otherwise, memory blocks should be a bottleneck in reliability. In order to have this property, a memory cell should have freedom from latch-like structure (i.e. back-to-back inverters). Latch structure, however, is inevitable when a memory cell is composed only of transistors, not of a capacitor or any other element; otherwise, data cannot be retained. Accordingly, this latch structure is necessary for holding a state , but for read operation (READ) or write operation (WRITE), it should be circumvented in some ways. One possible solution is static operation mode.

Figure 3 shows our proposed 16T SRAM cell. Device M1 to M4 are back-to-back inverters, and additional four PMOS devices (M5, M6, M11, and M12) are attached, two of which are connected to the source node of each PMOS of the inverters in parallel. At each output node of each inverter, two additional NMOS devices (M7 to M10) are connected to the ground node in series. These attached eight devices (M5 to M12) are for eliminating contentions between devices during WRITE so that the SRAM can be operated in static mode during WRITE. In addition, two additional NMOS devices are connected to each storage node for decoupling the output node from the storage node as in the case of conventional 8T SRAM cells [17]. In particular, one of these transistors is connected to virtual power rail for reducing leakage current to the output bit lines. Since the system needs timing valanced dual-rail output, those two additional devices (M13 to M16) are attached to both storage nodes. Accordingly, either M13 and M14 or M15 and M16 can be detached for the other systems, such as synchronous systems.

3.2 Operation Principle

16T SRAM is operated in fully static mode during READ and WRITE (see Figure 4).

3.2.1 Write Operation

Devices M5 to M12 are involved in WRITE. When a word line for WRITE (WL_{WRITE}) is enabled, M5 and M6 are opened, so no current is flown to inverters from the supply, while M7 and M8 are on, so stored data can be discharged through these transistors if M9 or M10 is enabled. When a desired state is asserted to bit lines, D.T and D.F, one path either from the supply node or to the ground node is formed at each storage node, so when different states are asserted these bit lines, a storage node has a different path from the other. At this moment, charging and discharging take place sequentially without any charge contention, meaning static logical operation. After discharging one of the internal nodes, the other node is applied with new data. For this sequential operation, WL_{WRITE} should be asserted after applying new data to bit lines, D.T and D.F. Once WRITE is completed, WL_{WRITE} is disabled so that no charge can be discharged through M7 to M10, while keeping power supply connected to inverters through M5 and M6.

3.2.2 Read Operation

Devices M13 to M16 are involved in READ. M13 and M15 decouple output nodes from storage nodes as conventional 8T SRAM. M14 and M16 are access transistors, and an inverter is connected to virtual ground node ($GND_{virtual}$). When idle mode, access transistors are inactive, and $GND_{virtual}$ is driven to VDD. Since READ evaluation occurs by discharging, this VDD driven $GND_{virtual}$ prevents read bit lines, $\overline{Q.T}$ and $\overline{Q.F}$ from misreading due to leakage through M14 and M16. Notice that $\overline{Q.T}$ and $\overline{Q.F}$ are precharged to VDD even if memory blocks are not used. When READ, M14 and M16 are turned on, and $GND_{virtual}$ is forced to GND, so evaluation occurs through M13 to M16.

3.3 Memory Cell Sizing

The proposed SRAM structure has an advantage in sizing. Every transistor can be minimum since there is no contention of charge between devices, bringing about reduction in engineering effort to design a SRAM cell.

3.4 Memory Block

In A8051, three memory blocks are included: two 1,024×8 bit block for ROM and XRAM and a 128×8 bit blocks for RAM. ROM is identical to RAM since A8051 is designed for flexible applications due to reprogram. The 1,024×8 bit block consists of eight banks, one of which is exactly same as the 128×8 bit block. 128 rows of SRAM cells share a bit line. During idle mode, bit lines are precharged as VDD, and a balancer, which is a p-type pass transistor, evenly distributes charges stored at both bit lines. When READ, these prechargers and a balancer are disabled, and charges on either side of the bit lines begin to be discharged through enabled read access transistors. Notice that little leakage flows to these bit lines since the virtual ground nodes of the other cells on the bit lines is driven to VDD, meaning that voltage between the node and the bit line is zero. In addition, since fully static logic is more robust at sub-threshold regime, an inverter is used as a sense amplifier even though a commonly used cross-coupled sense amplifier is faster. Our approach is to eliminate any feedback loop during WRITE although a feedback loop is inevitable in a SRAM cell. In addition, dual-rail outputs are needed since 4-phase dual-rail protocol is adopted in A8051, so an inverter can be a good candidate as a sense amplifier.

4. SIMULATION RESULTS

Since A8051 has no flip-flop, a critical path should carefully be considered for the evaluation of its performance. Figure 5 shows how each block communicates with a 4-phase dual-rail handshake protocol. When the channel is available (i.e., acknowledge signal is logically low), logic block A transmits data to logic block B. When the block B receives data from block A, it asserts acknowledge signal. Then, block A transmits empty data, and block B acknowledges by de-

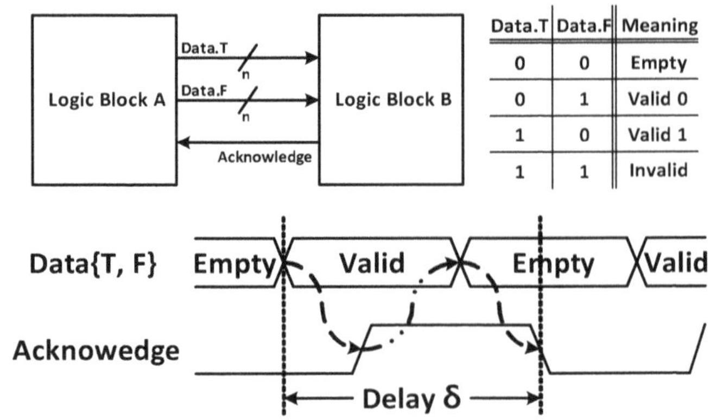

Figure 5: 4-phase dual-rail protocol.

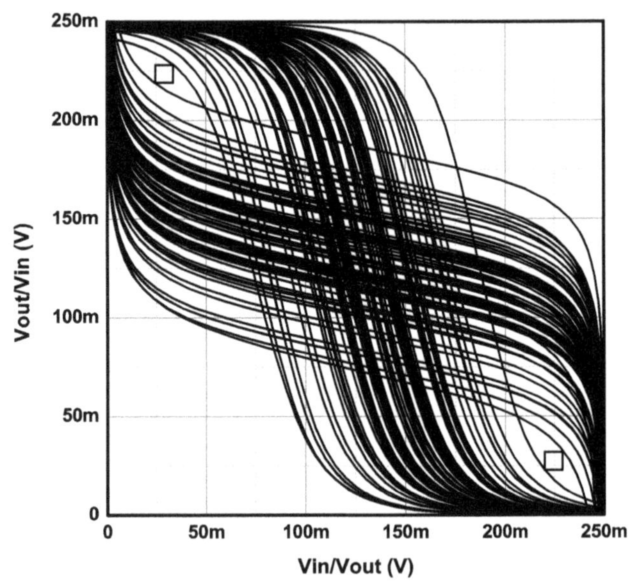

Figure 6: SNM of SRAM cell. Under process and mismatch variations, SNM=12.5 mV.

asserting acknowledge signal. These are the complete handshake process between two asynchronous logic blocks. Similarly, the performance of A8051 core can be evaluated with the same manner. When the completion time of handshake between the two pipeline stages is acquired, this time can be assumed as a critical delay for A8051 core since acknowledge signal for each pipelining stage is only asserted when it is completely done. The performance of asynchronous memory blocks can also be evaluated as A8051 core.

4.1 SRAM

Figure 6 shows the static noise margin (SNM) simulation results of a SRAM cell under process and mismatch variations. The SNM of this proposed design is 12.5 mV at 250 mV power supply, which can overcome 5% of VDD peak-to-peak ripple under random-dopant fluctuations.

The delay and energy simulation results of SRAM block is shown in Table 1. All results are from the simulations at 250mV. In addition, delay and energy were evaluated by referring a request signal to the related acknowledge signal. Layout cell size is 1750 λ^2, which is almost twice as conven-

Table 1: SRAM Simulation Results(@250mV)

		1,024×8bit 16T SRAM block
Type		16T SRAM
Cell Size		1.54×1.82 μm^2 (1750 λ^2)
SNM		12.5 mV
Delay	Read	12.61 μs (*79.3 kHz)
	Write	4.02 μs (*248.5 kHz)
Energy	Read	9.08 pJ
	Write	5.44 pJ

* This figure is equivalent value to synchronous systems

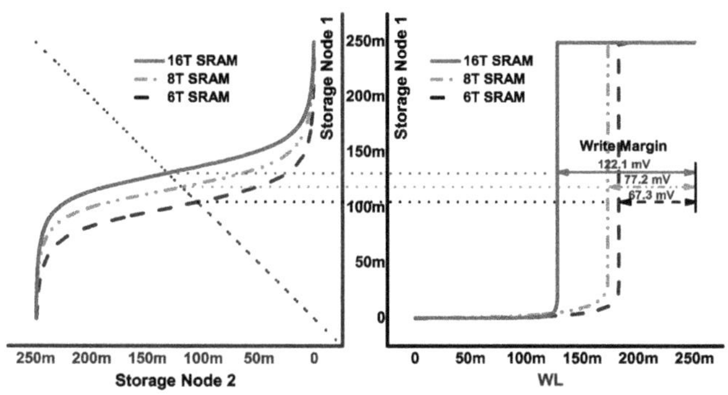

Figure 7: SRAM cell write margin simulation results at V_{DD}=250mV. The WM of 6T, 8T, and 16T SRAM are 67.3 mV, 77.2 mV, and 122.1 mV, respectively

tional 8T SRAM. It can operate at 250mV under process and mismatch variations, but, in typical, it can even operate at 200 mV. Write performance is higher than READ performance due to the fact that there is no charge contention during WRITE, and inverters are used as a sense amplifier for READ.

Figure 7 shows write margin (WM) simulation results of each SRAM cell (6T, 8T, and 16T) at V_{DD}=250 mV. The definition in [15] is used for the WM simulations. WM can be acuiqred from DC analysis when '1' is written to storage node 1 storing '0' value. Left curve is the voltage transfer characteristics (VTC) of the inverter of each SRAM cell, showing where the switching threshold is. Since WRITE occurs at the switching threshold, WM can be from the point to VDD. The WM of 6T, 8T, and 16T SRAM cells are 67.3 mV, 77.2 mV, and 122.1 mV, respectively. Accordingly, the proposed SRAM cell has 1.81 times more WM than the conventional 6T SRAM cell and 1.58 times WM than 8T SRAM cell.

4.2 Asynchronous 8051 microcontroller

Regarding the performance and power consumption of A8051, transistor level simulations were performed, and power and delay of three blocks including A8051 core, ROM, and RAM were acquired from these simulations. A benchmark program, Dhrystone [18], was converted to HEX code for programming ROM. Regarding the delay of A8051 core, a request signal and its corresponding acknowledge signal to a pipeline stage was referred. For ROM and RAM, request signals from A8051 core to ROM and RAM were referred, indicating how fast A8051 core passes these request signals to ROM and RAM so that the processing performance of

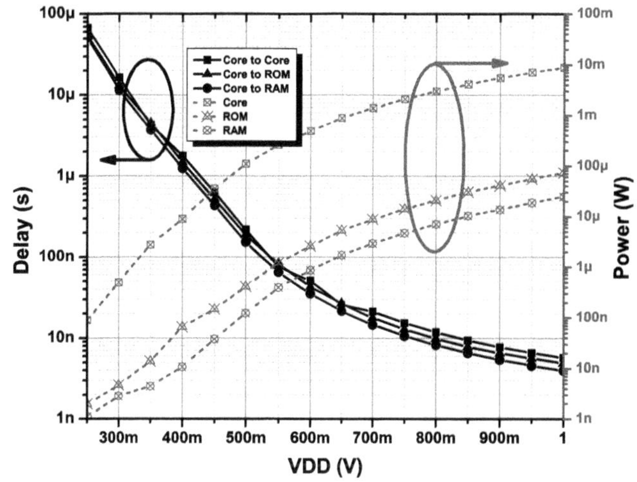

Figure 8: Delay and power simulation results for Core, ROM, and RAM.

Table 2: Asynchronous 8051 μC features

Technology	40nm CMOS technology
Die Size	1×1 mm^2 (254×545 μm^2 without IO PADs)
Operating supply voltage	250 mV ~ 1.0V
Number of Devices	795,249 (Total) 455,926 (Core) 158,904 (ROM) 158,904 (XRAM) 20,174 (RAM) 1,342 (Buffers, MUXes, and etc.)
Critical path delay	67.53 μs @250mV (*12.88 kHz) 5.74 ns @1V (*151.55 MHz)
MIPS	0.0074 MIPS (@250mV) 87.11 MIPS (@1V)
Total Average Power	91.6 nW (@250mV) 8.98 mW (@1V)

* This figure is equivalent value to synchronous systems. The inverse value of a critical path delay is multiplied by 1.15 as a guard band.

ROM and RAM to A8051 core can be evaluated. Figure 8 shows the curve of these delay values at different VDD from 250mV to 1V. Power consumption of each block was acquired during the execution of the benchmark program, which is also shown in Figure 8.

Table 2 shows the features of A8051. At 250 mV, the delay of a critical path in A8051 core is 67.53 μs, which is equivalent to 12.88 kHz when a guard band is considered as 1.15×. It performs 0.0074 MIPS when CPI is regarded as 2 since the majority of instructions in 8051 microcontroller are 1 byte, and A8051 is 2 stage pipelined. Accordingly, the MIPS of A8051 would be lowered in reality. Total average power consumption is 91.6 nW. When operating at nominal voltage, the critical path delay is 5.74 ns, which is equivalent to 151.55 MHz. It performs 87.11 MIPS, and total average power is 8.98 mW. The die size of A8051 is 1×1 mm^2, and the chip layout is shown in Figure 9.

Figure 9: A8051 chip Layout. Die size is 1×1 mm^2.

5. CONCLUSIONS

In this paper, a sub-threshold asynchronous 8051 microcontroller is demonstrated along with a novel sub-threshold 16T SRAM cell for asynchronous systems. Asynchronous approach eliminates timing uncertainty issue for robust subthreshold operation. The proposed 16T SRAM cell eliminates charge contentions between devices, so this structure improves write margin. A8051 chip is implemented in 40nm CMOS technology. The simulation results demonstrate A8051 operates up to 250 mV. At 250 mV, A8051 has 67.53 μs as a critical delay with 91.6 nW power consumption, while it has a critical delay of 5.74 ns with 8.98 mW power consumption at nominal voltage (i.e. 1 V). The READ operation for 1,024×8 bit SRAM block requires 12.61 μs with 9.08 pJ, while WRITE needs 4.02 μs with 5.44 pJ at 250mV.

6. ACKNOWLEDGMENTS

The authors thank Dr. Kok-Leong Chang for his technical support. This research is sponsored in part by DARPA/AFOSR (FA9550-12-1-0033).

7. REFERENCES

[1] International technology roadmap for semiconductors. Technical report.

[2] G. Asada, M. Dong, T. Lin, F. Newberg, G. Pottie, W. Kaiser, and H. Marcy. Wireless integrated network sensors: Low power systems on a chip. In *Solid-State Circuits Conference, 1998. ESSCIRC '98. Proceedings of the 24th European*, pages 9 – 16, sept. 1998.

[3] L. Benini, G. De Micheli, E. Macii, D. Sciuto, and C. Silvano. Asymptotic zero-transition activity encoding for address busses in low-power microprocessor-based systems. In *VLSI, 1997. Proceedings. Seventh Great Lakes Symposium on*, pages 77 –82, mar 1997.

[4] S. Borkar, T. Karnik, S. Narendra, J. Tschanz, A. Keshavarzi, and V. De. Parameter variations and impact on circuits and microarchitecture. In *Proceedings of the 40th annual Design Automation Conference*, DAC '03, pages 338–342, New York, NY, USA, 2003. ACM.

[5] K.-L. Chang and B.-H. Gwee. A low-energy low-voltage asynchronous 8051 microcontroller core. In *Circuits and Systems, 2006. ISCAS 2006. Proceedings. 2006 IEEE International Symposium on*, pages 4 pp. –3184, may 2006.

[6] S. Das, C. Tokunaga, S. Pant, W.-H. Ma, S. Kalaiselvan, K. Lai, D. Bull, and D. Blaauw. Razorii: In situ error detection and correction for pvt and ser tolerance. *Solid-State Circuits, IEEE Journal of*, 44(1):32 –48, jan. 2009.

[7] A. Drake, R. Senger, H. Deogun, G. Carpenter, S. Ghiasi, T. Nguyen, N. James, M. Floyd, and V. Pokala. A distributed critical-path timing monitor for a 65nm high-performance microprocessor. In *Solid-State Circuits Conference, 2007. ISSCC 2007. Digest of Technical Papers. IEEE International*, pages 398 –399, feb. 2007.

[8] R. Franch, P. Restle, N. James, W. Huott, J. Friedrich, R. Dixon, S. Weitzel, K. Van Goor, and G. Salem. On-chip timing uncertainty measurements on ibm microprocessors. In *Test Conference, 2007. ITC 2007. IEEE International*, pages 1 –7, oct. 2007.

[9] R. Jorgenson, L. Sorensen, D. Leet, M. Hagedorn, D. Lamb, T. Friddell, and W. Snapp. Ultralow-power operation in subthreshold regimes applying clockless logic. *Proceedings of the IEEE*, 98(2):299 –314, feb. 2010.

[10] T. Kuroda, K. Suzuki, S. Mita, T. Fujita, F. Yamane, F. Sano, A. Chiba, Y. Watanabe, K. Matsuda, T. Maeda, T. Sakurai, and T. Furuyama. Variable supply-voltage scheme for low-power high-speed cmos digital design. *Solid-State Circuits, IEEE Journal of*, 33(3):454 –462, mar 1998.

[11] M. Nakai, S. Akui, K. Seno, T. Meguro, T. Seki, T. Kondo, A. Hashiguchi, H. Kawahara, K. Kumano, and M. Shimura. Dynamic voltage and frequency management for a low-power embedded microprocessor. *Solid-State Circuits, IEEE Journal of*, 40(1):28 – 35, jan. 2005.

[12] G. J. Pottie and W. J. Kaiser. Wireless integrated network sensors. *Commun. ACM*, 43(5):51–58, May 2000.

[13] M. Seok, S. Hanson, Y.-S. Lin, Z. Foo, D. Kim, Y. Lee, N. Liu, D. Sylvester, and D. Blaauw. The phoenix processor: A 30pw platform for sensor applications. In *VLSI Circuits, 2008 IEEE Symposium on*, pages 188 –189, june 2008.

[14] B. Stackhouse, S. Bhimji, C. Bostak, D. Bradley, B. Cherkauer, J. Desai, E. Francom, M. Gowan, P. Gronowski, D. Krueger, C. Morganti, and S. Troyer. A 65 nm 2-billion transistor quad-core itanium processor. *Solid-State Circuits, IEEE Journal of*, 44(1):18 –31, jan. 2009.

[15] K. Takeda, H. Ikeda, Y. Hagihara, M. Nomura, and H. Kobatake. Redefinition of write margin for next-generation sram and write-margin monitoring circuit. In *Solid-State Circuits Conference, 2006. ISSCC 2006. Digest of Technical Papers. IEEE International*, pages 2602 –2611, feb. 2006.

[16] J. Tschanz, K. Bowman, S.-L. Lu, P. Aseron, M. Khellah, A. Raychowdhury, B. Geuskens, C. Tokunaga, C. Wilkerson, T. Karnik, and V. De. A 45nm resilient and adaptive microprocessor core for dynamic variation tolerance. In *Solid-State Circuits Conference Digest of Technical Papers (ISSCC), 2010 IEEE International*, pages 282 –283, feb. 2010.

[17] N. Verma and A. Chandrakasan. A 256 kb 65 nm 8t subthreshold sram employing sense-amplifier redundancy. *Solid-State Circuits, IEEE Journal of*, 43(1):141 –149, jan. 2008.

[18] R. P. Weicker. Dhrystone: a synthetic systems programming benchmark. *Commun. ACM*, 27(10):1013–1030, Oct. 1984.

GPU Implementation of a Scalable Non-Linear Congruential Generator for Cryptography Applications

Aditya Belsare
Department of Electrical Engineering
Texas A&M University
College Station, Texas 77843
asbelsare@tamu.edu

Steve Liu
Department of Computer Science and Engineering
Texas A&M University
College Station, Texas 77843
liu@cse.tamu.edu

Sunil P. Khatri
Department of Electrical Engineering
Texas A&M University
College Station, Texas 77843
sunilkhatri@ece.tamu.edu

ABSTRACT

Fast key generation algorithms which can generate random sequences of varying atomic lengths and throughput are important for secure data communication. In this paper, we present a non-linear congruential method for generating high quality random numbers at flexible throughput rates of upto 66 Gbps, on a GPU platform. Each random number can have up to 4096 key bits. The method can be easily extended for implementation on hardware platforms like FPGAs and ASICs as well. Our key generator is comprised of N Linear Congruential Generators (LCGs) running in parallel; we have chosen $N = 4096$ for the GPU implementation. The outputs of the LCGs are combined using N encoded majority functions. The encoded majority function used for any bit is changed in every generation iteration. In our GPU implementation of the Non-Linear Congruential Generator (GPU-NLCG), it is possible to alter the LCG functions on the fly by changing the primes periodically without interrupting the generation. Our GPU-NLCG can be used for high speed cryptographic key generation for rates up to 66 Gbps and can be easily integrated into multi-threaded applications in cryptography and Monte Carlo methods. The GPU-NLCG passes the NIST, Diehard and Dieharder battery of tests of randomness, which ensure the quality of our ciphers.

Categories and Subject Descriptors

I.0 [**Computing Methodologies**]: General; E.3 [**Data Encryption**]: Metrics—*Public Key Cryptosystems*

Keywords

Cryptography, Graphics Processing Units, Block Cipher, pseudo random sequence

General Terms

Security, Random number generation

1. INTRODUCTION

Random numbers find varied applications in all spheres of security, particularly for securing high speed data communication systems. High speed secure communication for communication standards like OC-768 on fiber optic channels (which operate at 40 Gbps) consume random numbers at equally high rates. Random numbers are also used for solving Monte Carlo based computational algorithms, which rely on repeated random sampling to compute their results. Such algorithms require high quality random numbers at extremely high rates.

For secured communication systems, the message to be sent (called *plaintext*) is usually combined with a pseudo-random sequence (called a *key*). The transformed message, called the *ciphertext* is then transmitted to the recipient over the communication channel. The *cipher*, which is the algorithm used to combine plaintext and the key can operate in the following two ways -

- On the stream of plaintext one bit at a time, called a *stream cipher*. Example - A bitwise XOR of key bits and plaintext bits.
- On a block of plaintext at a time, called a *block cipher*. The block cipher encrypts a message by breaking it down into fixed or variable sized blocks and encrypting data in each block. Examples - DES [13] (64-bit block size), AES [14] (128-bit block size), Diffie-Hellman [15] and RSA [16] (variable block sizes).

At the receiving end, the original message is decrypted using the same cipher and key that was used during encryption. Block ciphers are preferred over stream ciphers for many reasons. Stream ciphers are generally slower than block ciphers. If the throughput of a stream cipher is to be increased, very high clock rates need to be employed. Block ciphers are better at resynchronization. In a synchronous stream cipher, if a cipher text character is lost during transmission, the sender and receiver must resynchronize their key generators before transmission can be resumed. In a context of a block cipher, the block can be simply discarded or error-corrected using redundant bits in the block. For these reasons, we chose the GPU-NLCG to be a block cipher with adjustable block length, which is actually equal to the number of generators used in our CUDA implementation. The maximum block length can be upto 4096 bits in our realization.

Linear generators like Linear Feedback Shift Registers (LFSR) and Linear Congruential Generators (LCG) are a popular choice for pseudo-random number generation, mainly due to their simple hardware implementation and fast performance. For cryptographic applications however, linear generators like LFSRs and LCGs are not suitable. In an n-bit LFSR for example, the feedback polynomial can be inferred by observing $2n$ consecutive bits of its sequence [1].

A Linear Congruential Generator (LCG) represents one of the oldest and best known pseudo-random number generator algorithms. A general LCG function can be represented by the following recurrence relation:

$$X_n = [m \times X_{n-1} + p] \; mod \, \alpha \quad (1)$$

where m is called the *multiplier*, p the *increment*, α the *modulus* and $0 \leq X_0 < \alpha$ the *initial seed*. The period of this LCG would be at most α, and for some choices of m and p much less than that. The LCG will have a full period if and only if the following conditions are satisfied [2]:

1. α and p are co-primes (i.e. have no common factors other than 1.)
2. $m - 1$ is divisible by all prime factors of α.
3. If 4 divides α, then 4 divides $m - 1$.

To satisfy these conditions, we choose $\alpha = 2^{32}$, m as a fixed prime number such that 4 divides $m - 1$ and p as a variable prime number. Hence, α and p now become co-primes and hence first condition is satisfied. Since $m - 1$ is divisible by 4, it is also divisible by the prime factor of 2^{32}, i.e. 2. Hence, the second condition is satisfied. Also, the third condition gets automatically satisfied based on our choice of m and α. For the LCG generators used in the first stage of our implementation, we fix m to be a large prime and vary p with different values of primes across generators.

LCGs are fast, require minimal memory and are valuable for simulating multiple independent streams. However, when used on their own for cryptographic applications, they have certain severe shortcomings. If an LCG is seeded with a character and then iterated once, the result is a simple classical cipher called an *affine cipher*; this cipher is easily broken by standard frequency analysis [3].

At a high level, the NLCG is comprised of two computation stages. The first stage constitutes of 4096 32-bit LCGs operating in parallel. A separate primality test kernel operates in parallel with the generating LCGs. After every 10000 random sequence generation by each congruential generator, the primality test kernel supplies the LCG kernel with a new set of 4096 prime numbers for p as required by Eq. 1. For the second stage, the 32-bit outputs from the LCGs are arranged as 32 rows of 4096-bits each. A 12-bit LFSR and the 12-bit index position is then used to select a unique encoded majority function for each of the 4096 bit positions, where the encoded majority function takes in 32 inputs per bit position and outputs a binary value. Any encoded majority function is a 32-input and 1-output function, which is a '1' for 2^{31} random input combinations. Thus, the NLCG outputs a block of 4096 bits per iteration.

In GPU-NLCG, we have exploited parallelism by launching a large number of independent generators, before being permuted in a non-linear fashion by the encoded majority functions. Each generator is mapped to one CUDA thread. Hence, the throughput can be simply adjusted by changing the number of LCG CUDA threads, which makes the NLCG a suitable candidate for different communication rates and standards. Moreover, it is secure because we change the primes used during generation after every 10000 iterations and permute the generated bits using a pool of encoded majority functions, which are selected randomly and dynamically.

The primality test kernel in the first stage of generator is a parallelized implementation of the Miller-Rabin Primality Test, a fast yet probabilistic method for determining primality of a number. In order to supply prime numbers to the congruential functions, the primality test kernel begins the search starting at a random offset number and returns 4096 primes after it has scanned an appropriately large range around the offset. This kernel can be used as a standalone application, yielding a GPU implementation of a fast primality checking algorithm for a block of numbers. Primality testing, which itself finds crucial applications for RSA algorithms where a product of two large primes is used as a key, can make good use of our GPU based fast primality test for finding a pool of candidate primes quickly.

To evaluate the quality of random numbers generated by the NLCG, we tested the results with 3 well known batteries of tests of randomness, namely the NIST [10], Diehard [11] and Dieharder [12]. The results passed for all the the variants of these three test suites. These test suites are stringent and exhaustive, and frequently used for analyzing the quality of ciphers generated by various algorithms.

The remainder of this paper is organized as follows. Section II surveys the existing work done on LCGs, particularly on the GPUs. Section III describes our NLCG design, while Section IV reports experimental results, including results from NIST and Dieharder test suites. Finally in Section V, we make concluding comments and discuss further work that needs to be done in this area.

2. PREVIOUS WORK

We use a non-linear function which acts on the LCGs, since LCGs have shortcomings when used on their own, as described earlier. A non-linear generator utilizes a non-linear function for the next state feedback or for the output. In a theoretical study [5], a non-linear congruential pseudo-random number generator has been shown to be better than a conventional LCG from a statistical view point. However, no implementation was proposed. Our algorithm for such a non-linear function is inspired by the Non-Linear Feedback Shift Register (NLFSR) proposed in [4], with the important difference of using a large array of dynamically seeded LCGs with periodically changing generating functions instead of the 3 LFSRs used in [4]. The advantage of our approach is that since it is software based, it is not restricted to fixed bit-widths as is the case for a hardware approach like [4]. Additionally, we keep the array length of the dynamic LCGs equal to the size of the block cipher desired, and select functions from an ordered array of majority functions instead of using a multiplexer with functions as inputs.

There have been several efforts to exploit LCGs for non-cryptographic purposes. In [6], for example, a GPU implementation for LCG is presented where a large number of LCGs are launched in parallel to obtain an immense speedup on a Fermi GPU. No mention is however, made about the quality of the resulting random numbers. As discussed before, using LCGs for pseudorandom number generation is not suitable for cryptographic applications. We tested the random sequences generated by using LCGs on the GPU, using the NIST and Diehard tests and found that it failed many tests. In particular, the *Runs Statistical Test* on NIST (with a *p-value* of 0.003), and the *BlockFrequency Test* (with a *p-value* of 0.005) performed very poorly. The failure in *BlockFrequency Test* can be attributed to the short period of an LCG, which can be no more than 2^{32} for a 32-bit LCG.

The *cuRand* library in CUDA provides two Random Number Generators namely Mersenne Twister and XORWOW algorithms. There are two problems with the Mersenne Twister implementation - the first being the fact that it requires a very large state (624 words), which makes it unwieldy if a random number sequence is desired per CUDA thread, with tens of thousands of threads. Thus, the size of the state rules out thread level parallelization. Second, the Mersenne Twister is not suitable for cryptography. Observing a sufficient number of iterations (624 in case of MT19937) allows one to predict all future iterations [7]. The XORWOW algorithm introduced by Marsaglia [8] belongs to the *Xorshift* class of generators. These generate the next number in their sequence by repeatedly taking the XOR of a number with a bit shifted version of itself. They are a subclass of LFSRs, and hence they are fast. However since they are based on LFSRs, they are not suitable for cryptographic applications. Moreover, the parameters have to be chosen carefully in order to achieve a long period [9]. Thus, the GPU-NLCG can be better suited for cryptographic applications than CUDA's own implementation using Mersenne Twister and Xorshift generators.

3. TECHNICAL APPROACH

In this section, we present our NLCG design and its implementation on CUDA. We first describe the operation of the LCGs used in first stage of generation, then the second stage which permutes the results of LCGs into a N-bit output using encoded majority functions. For our CUDA implementation, N is chosen to be 4096.

Stage 1: The LCGs

We use Linear Congruential Generators for the first stage of random number generation. A 32-bit LCG function can be represented by Eq. 1. Parallelization of this function can be achieved by varying either m or p or both at the same time. For our implementation, we choose to fix the multiplier m for all generators, while varying p with values obtained from the primality test. We have an implementation with the N generators, and the data structure for the entire LCG generator is as shown:

$$
\begin{aligned}
&typedef\ struct\ LCG\ \{\\
&\qquad\qquad unsigned\ int\ state[N];\\
&\qquad\qquad unsigned\ int\ p[N];\\
&\}\ LCG\_RNG;
\end{aligned}
$$

The values of p are refreshed after every 10000 iterations by a separate *primality kernel*, which implements the Miller-Rabin Primality test for a block of numbers assessed for primality. Each GPU thread assesses one number for primality, implementing the Miller-Rabin algorithm:[1]

Inputs:

1. $n > 3$, an odd integer to be tested for primality.
2. k, a parameter that determines the accuracy of the test.

Output: *Probably prime* if n is prime, else composite.

Write $n-1$ as $2^s.d$ with d odd by factoring powers of 2 from $n-1$

Loop: Repeat k times:
  Pick a random integer a in the range $[2, n-2]$
  $x \leftarrow a^d mod\, n$
  if $x = 1$ or $x = n-1$ then do next Loop
  for $r = 1 \ldots s-1$
    $x \leftarrow x^2 mod\, n$
    if $x = 1$ then return composite
    if $x = n-1$ then do next Loop
  return composite
return probably prime

The test completes in $\widetilde{O}(k \log^3 n)$, where k is number of test iterations. The worst-case error bound of the Miller-Rabin primality test is 4^{-k}. We choose $k = 1000$, so the probability that a composite number is incorrectly determined as a prime by this method is $4^{-1000} \approx 0$.

As shown in Miller Rabin Algorithm, the primality test requires a random integer a in the range $[2, n-2]$. Each primality thread uses its own LFSR for this purpose, which runs k times. The feedback polynomial for this 32-bit LFSR is given by:

$$x^{31} + x^{21} + x^1 + x^0$$

Each LFSR uses the same feedback polynomial but different starting seed, which is provided along with the number to be tested for primality. Thus, at the end of every iteration in Stage 1, we obtain N 32-bit random numbers, one from each of the LCGs, with the LCG functions themselves changing every 10000 iterations. The choice of 10000 iterations is experimentally determined - using fewer sequences may not yield 4096 primes, while using more sequences would allow the LCG coefficients to be susceptible to being deciphered. When algorithm **??** returns a *composite*, we ignore the result and continue the search till all the N primes have been accumulated. The output of stage 1 is array of N primes in the CUDA global memory; this array is used by Stage 2 of the computation.

Stage 2: The Non-Linear Permutation Block for LCGs

Our proposed GPU-NLCG utilizes N LCGs, whose outputs are combined in a non-linear manner. The outputs of the LCGs are represented as M rows of N bits each. We have chosen $N = 4096$ and $M = 32$ for our CUDA implementation, as shown in Figure 1. We refer to each of these M rows as *transformed LCGs*. The ith bit $(0 \leq i < N)$ of each row j $(0 \leq j < M)$ is connected to an encoded majority function, and the output of this encoded majority function is the ith bit of the resulting NLCG. Thus, output of NLCG is N bits per iteration. The encoded majority function is non-linear, hence the combined state of the N majority outputs results in an NLCG.

We now describe the encoding majority functions and their selection in greater detail. The encoded majority function has the following characteristics -

[1] `http://en.wikipedia.org/wiki/Miller-Rabin_primality_test`

Figure 1: 4096-bit NLCG with Encoded Majority Function Blocks

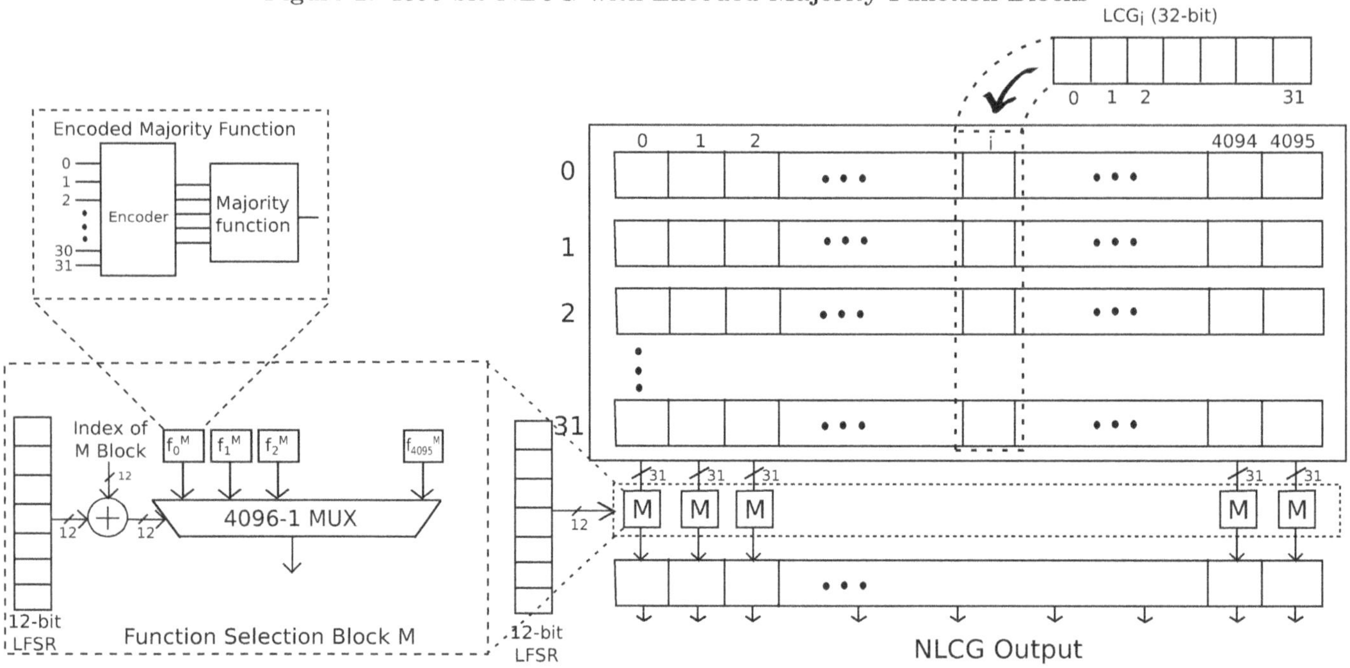

- The encoded majority function is different for each bit position in any generation iteration (or clock cycle).
- The encoded majority function for any bit position changes at each iteration (or clock cycle).
- Each encoded majority function has 32 inputs and one output. The output is a '1' for 2^{31} randomly arranged inputs.

We have a pool of 4096 encoded majority functions, each of which is mapped to a single CUDA thread. An encoded majority function in our approach is a two-block function. The first is an encoding block which receives 32 binary inputs and outputs 5 binary bits, as shown on the top left of Figure 1. Each encoded majority function is differentiated from another by the sequence (bit-position) of the 32 inputs. The order of these inputs determine the 5-bit output, which is next provided to a majority function, defined as below:

$$M(p_1, p_2, \ldots, p_5) = \left\lfloor \frac{1}{2} + \frac{\sum_{i=1}^{5} p_i - 1/2}{5} \right\rfloor$$

Assuming an NLCG output length of $N = 4096$ and 32-bit LCGs, we need 4096 encoded majority functions ($f_0^M - f_{4095}^M$). Each encoded majority function has 32 inputs, one from each transformed LCG (see figure 1). The number of such unique encoded majority functions possible is ${}^{2^{32}}C_{2^{31}}$. We choose 4096 of them at random for our implementation.

Now we describe how the encoded majority functions are selected. The ith bit of each LCG is used as an input to the ith *function selection block* M_i (See bottom left, Figure 1). In each function selection block, the inputs are connected to all 4096 encoded majority functions and a 4096-1 multiplexer selects one of these functions for the ith bit NLCG output. For a software based implementation like in a GPU, we do not need a multiplexer per se, so we simply select the appropriate function from an ordered array of 4096 functions. The selection signal is controlled by XOR-ing the 12-bit bit index of the function selection block M_i with the output of the 12-bit LFSR (leftmost part of Figure 1). The 12-bit LFSR outputs a new value at every LCG generation iteration, and the output is shared by all 4096 function selection blocks. Figure 1 shows only one such block. Thus, all bit positions select a unique and different encoded majority function every cycle.

Length Sequence of the NLCG

We now assert and prove that the NLCG of length $M \times N$ has an asymptotically maximal length sequence. Assume that in any iteration j, the state of the *transformed LCG* L_i is V_i^j $(0 \leq i \leq 31)$. The NLCG operation is denoted as $N^j = M(V_0^j, V_2^j, \ldots, V_{31}^j)$, where M is as shown in Figure 1. We show that the probability that the NLCG output does not repeat within the composite sequence of 32 transformed LCGs, the length of which is $N \times M = 32 \times 4096$, can be made asymptotically low. In other words, we need to show that for two different iterations j and k in the maximal sequence, the following holds true for an appropriately large n:

$$p(M(V_0^j, V_2^j, \ldots, V_{31}^j) = M(V_0^k, V_2^k, \ldots, V_{31}^k)) \rightarrow 0$$

To show this, we note that for the encoded majority function computing the output of the NLCG, the output will be the same across two iterations j and k if all 32 bits that are input to the function selection block M_i (for all i) are identical. For any i, this probability is $\left(\frac{1}{2}\right)^M$. Since the transformed LCGs have N columns the probability of the output of the NLCG being same for two iterations j and k is $\left[\left(\frac{1}{2}\right)^M\right]^N$. With a large M and N, this probability can be made arbitrarily small. This means that the composite sequence (of length $M \times N$) is maximal. For our implementation with $M = 32$ and $N = 4096$, this probability is $\left[\left(\frac{1}{2}\right)^{32}\right]^{4096} \approx 0$.

4. EXPERIMENTAL RESULTS

We report our results both in terms of quantity of numbers generated (throughput) and the quality of random numbers (results of NIST, Diehard and Dieharder benchmarks).

4.1 Quantity of Random Numbers

We implemented the program for primality testing independently, comparing results for CPU and GPU versions of the program. We then integrated the primality tester with the LCG kernel and observed the time taken for generating a fixed quantity of random numbers.

The CPU used for comparison had a 64-bit Intel Xeon Processor with 8 cores and 2 cores per thread. It had 16 GB of memory, 8 MB of L3 cache and was clocked at 2.9 GHz. The CPU used Ubuntu 11.10 distribution of the Linux operating system. For the GPU, we used a NVIDIA Kepler GeForce GTX 680 card with 8 Multiprocessors (MP), 192 CUDA cores/MP, and 2 GB of global memory. Table 1 reports the comparison of CPU and GPU versions of the NLCG generators with 4096-bit output. Note that we have assumed that the cryptography application, Monte Carlo method or any random number consuming application resides in the same kernel as the NLCG generator itself. This makes the GPU-NLCG integrated with the consuming function, and removes the overhead of transferring a large set of random numbers generated every second to the CPU. The results reported in Table 1 correspond to $k = 1000$ for the Miller-Rabin algorithm and 4096-bit random numbers.

Table 1: Results for CPU and GPU versions of NLCG Program

Program	# Random	Time	Speedup
NLCG_CPU	10^9	3.01 h	-
NLCG_GPU	10^9	61.9 s	175x
NLCG_CPU	$2*10^9$	6.22 h	-
NLCG_GPU	$2*10^9$	124 s	180x

The GPU throughput calculated from Table 1 comes to be 66.1 Gbps for the GPU version and 337 Mbps for the CPU version of the program. Thus, the GPU-NLCG can sustain data rates for communication standards like OC-768, which operate at 40 Gbps.

4.2 Quality of Random Numbers

We used 3 benchmark tools for assessing the quality of random numbers generated by our NLCG - *Marsaglia's Diehard battery of tests of randomness* [11], *Dieharder* (an open source random number/uniform deviate generator and tester [12]), and *NIST*, which is a statistical test suite for the validation of RNGs and PRNGs for cryptographic applications [10]. The Dieharder test suite incorporates many of the tests included in NIST, and is available as a standard software package in Ubuntu's software repository, and can test either P-bit bitstrings (with P being user specified) or double precision floating point numbers in the range [0.0, 1.0). For the Diehard and NIST suites, we had to split each 4096-bit random number into chunks of 64-bit numbers for testing. As required by each of the three test suites, we generated binary output files from the NLCG with 8 million samples each.

The results of all three test suites are reported as *p-values*. The *p-value* for a particular test is compared with the following two parameters to decide the outcome of that test:

- *Weak threshold* - The test is deemed as *weakly passed* if *p-value* is found less than the weak threshold. Default is 0.005 for Dieharder tests. NIST does not use this parameter.
- *Fail threshold* - The test is deemed as *certainly failed* if the *p-value* is found lesser than the fail threshold. Dieharder recommends this to be 0.000001, while NIST refers to this parameter as α, with a recommendation of 0.01 be used for all tests.

An α of 0.001 indicates that one would expect one sequence in 1000 sequences to be rejected by the test if the sequence was random. For a *p-value*$\geq$0.001, a sequence would be considered to be random with a confidence of 99.9%. For a *p-value*$\leq$0.001, a sequence would be considered to be non-random with a confidence of 99.9%.

We used default sample sizes recommended by the individual tests. For the GPU-NLCG, all the 15 tests in Diehard, 12 major tests and their 162 variants in NIST, and the 25 major tests and their 209 variants in Dieharder passed. Table 2 lists the results of important tests from all the 3 test suites. In this table, any test with a prefix "NIST" comes from the NIST suite, while a prefix "DHr" and "DH" refer to the Dieharder and Diehard suites respectively. Column 2 lists the t-samples which is the total random samples consumed by the test under consideration.

Table 2: Results from Statistical Test Suites. (All tests pass.)

Test-name	t-samples	*p-value*
NIST_Runs	100000	0.54890228
NIST_Serial	100000	0.21857258
NIST_BlockFrequency	100000	0.13538794
NIST_Monobit	100000	0.13499881
DHr_Bitstream	2097152	0.87918116
DHr_Squeeze	100000	0.36519217
DHr_Sums	100	0.35587054
DHr_rgb_lagged_sum	1000000	0.24969625
DHr_rgb_permutations	100000	0.17348102
DHr_min_distance	10000	0.53072256
DHr_3d_Sphere	4000	0.71973711
DH_rank_32x32	40000	0.00505979
DH_dna	2097152	0.23968055

V. Conclusion

We present a GPU-based parallelized implementation of a simple linear congruential generator and extend it to serve as a non-linear generator. The resulting implementation, which we call GPU-NLCG, is an inexpensive method for fast random number generation with variable bit-size on any GPU based platform. The GPU-NLCG is perfectly suited for fast key generation for cryptography applications. The NLCG uses 4096 independent LCGs, the outputs of which are permuted using 4096 encoded majority functions selected randomly per bit per iteration. The throughput of our system is limited not by the generator but by the bandwidth of the CPU-GPU PCI-e bus. Hence, an FPGA or ASIC based implementation of our NLCG may yield yet higher throughputs.

5. REFERENCES

[1] E. Zenner, "Cryptanalysis of LFSR-based pseudorandom generators - a survey," 2004.

[2] Severance, Frank (2001). "System Modeling and Simulation," John Wiley & Sons, Ltd. pp. 86. ISBN 0-471-49694-4.

[3] Joan Boyar (1989). "Inferring sequences produced by pseudo-random number generators," Journal of the ACM 36 (1): 129–141.

[4] Kent Lin, Sunil Khatri (2010). "VLSI Implementation of a Non-Linear Feedback Shift Register for High-Speed Cryptography Applications," Proceedings of the 20th symposium on Great lakes symposium on VLSI, Pages 381-384.

[5] J. Eichenauer, J. Lehn, A non-linear congruential pseudo random number generator, Statist. Papers 27 (1986) 315-326.

[6] Shuang Gao and Gregory D. Peterson, GASPRNG: "GPU Accelerated Scalable Parallel Random Number Generator Library," Computer Physics Communications.

[7] "Reversing the Mersenne Twister RNG Temper Function". http://b10l.com/?p=24

[8] G. Marsaglia. Xorshift RNGs. J. of Statistical Software, 8(14):1-6, 2003.

[9] Panneton, FranÃğois (October 2005). "On the xorshift random number generators," ACM Transactions on Modeling and Computer Simulation (TOMACS) Vol. 15 (Issue 4).

[10] NIST computer security resource center. http://csrc.nist.gov/groups/ST/toolkit/rng/index.html

[11] The Marsaglia Random Number CDROM including the Diehard Battery of Tests of Randomness. http://www.stat.fsu.edu/pub/diehard/

[12] Dieharder: A testing and benchmarking tool for random number generators. Dieharder page for Ubuntu

[13] Data Encryption Standard (DES), U.S. Department of Commerce/National Institute of Standards and Technology. http://csrc.nist.gov/groups/ST/toolkit/rng/index.html

[14] 'Announcing the Advanced Encryption Standard (AES)'. Federal Information Processing Standards Publication 197. United States National Institute of Standards and Technology (NIST). November 26, 2001.

[15] 'New Directions in Cryptography'. W. Diffie and M. E. Hellman, IEEE Transactions on Information Theory, vol. IT-22, Nov. 1976, pp: 644-654.

[16] Rivest, R.; A. Shamir; L. Adleman (1978). "A Method for Obtaining Digital Signatures and Public-Key Cryptosystems". Communications of the ACM 21 (2): 120-126.

Fast and Memory-Efficient GPU Implementations of Krylov Subspace Methods for Efficient Power Grid Analysis

Takumi Morishita Hiroshi Tsutsui
Hiroyuki Ochi Takashi Sato
Department of Communications and Computer Engineering, Kyoto University
Yoshida-hon-machi, Sakyo, Kyoto 606-8501, Japan
paper@easter.kuee.kyoto-u.ac.jp

ABSTRACT

Power grid analysis for modern LSI is computationally challenging in terms of both runtime and memory usage. In this paper, we implement Krylov subspace based linear circuit solvers on a graphics processing unit (GPU) to realize fast power grid analysis. Efficiencies of memory space and access performance are pursued by improving a data structure that stores elements of large sparse matrices. Experimental results on benchmark circuits show that the proposed data structures are more suitable than widely used compressed sparse row (CSR) format and our GPU implementations can achieve up to 17× speedup over CPU implementations.

Categories and Subject Descriptors

B.7.2 [**Integrated Circuits**]: Design Aids—*simulation*; D.1.3 [**Programming Techniques**]: Concurrent Programming—*Parallel Programming*

General Terms

Performance

Keywords

Power grid analysis, Krylov subspace method, GPGPU

1. INTRODUCTION

Because of relentless trend toward higher density of semiconductor devices and lower power supply voltages, supply voltage drop (IR drop) in power distribution network has been giving even greater influence on the performance of LSI systems than before. It is reported that a 10 % voltage drop causes both 6 % operation frequency degradation and 18 % power consumption increase [1,2]. For this reason, IR drop in a power grid should be carefully considered in all phases of chip design.

IR drop analysis is widely adopted to verify power distribution networks, to prevent from IR-drop related timing failure, as well as to take preventive efforts early in the design stages. Full-chip analysis is required to confirm that sufficient voltages are distributed to all devices on the chip. Power grid of an LSI is generally modeled using linear circuit components, i.e., resistors, capacitors, voltage sources, and current sources. The total number of components in equivalent circuits is now very large as it grows continuously with the increase of the number of devices in LSIs. In terms both of runtime and memory usage, it is a challenging task to simulate a large scale power grid. Fast and low-memory-usage methods for power grid analysis are strongly demanded.

IR-drop simulation requires solving a very large linear system

$$A\boldsymbol{x} = \boldsymbol{b}, \tag{1}$$

where A is the conductance matrix formulated through nodal or modified nodal analysis, $\boldsymbol{x}$ is the node voltage vector and $\boldsymbol{b}$ is the independent current source vector. Various methods have been developed to solve Eq. (1) efficiently. The methods are classified into two broad categories: direct methods and iterative methods.

Direct methods (e.g., LU factorization) factorize the conductance matrix into a simpler triangular form and then solve the matrix by row-wise substitutions. Through direct methods, exact solution is available within a fixed number of arithmetic operations. Direct methods are also considered as robust. With these reasons, the direct methods have been the first choice in IR drop analyses. On the other hand, the computational complexity of direct methods is super-linear. Even in the case of power grid network, in which the conductance matrix becomes sparse, the computational complexity of its LU factorization is approximately $O(N^{1.2})$ to $O(N^{1.8})$. Also, it requires a large memory space due to fill-ins when solving large problems such as those found in modern power grid analysis.

Iterative methods (e.g., Jacobi method) successively approximate the solution of Eq. (1). The methods use less memory than direct methods, while their runtime depends on the number of iterations required until the convergences are achieved.

In power grid analysis, iterative methods are gaining popularity as the size of the problems becomes larger. In particular, the idea of projection into Krylov subspaces is a very popular approach. The conjugate gradient (CG) method [3], which belongs to the class of Krylov subspace methods, is frequently used. The CG method can be applied to problems in which matrix A is symmetric positive definite (SPD). This condition is satisfied in IR-drop analysis. Another Krylov subspace method, generalized minimal residual (GMRES) method [4], is also used. In GMRES, SPD is not a prerequisite.

As a computing platform, graphics processing unit (GPU) is paid much attention in various fields. GPU, which was originally developed for accelerating operations of computer graphics, is now more broadly applied because of their highly parallel architecture. This technique is called general-purpose computing on GPUs (GPGPU). GPU acceleration for power grid analysis has been also studied as Poisson solver for regular power grid [5] and as geometric multigrid solver [6,7]. In this paper, we accelerate power grid analysis by implementing two Krylov subspace methods, CG and GMRES, on GPU. We further study sparse matrix storage format suitable for improving memory access time on GPU.

This paper is organized as follows. In Section 2, we describe the outline of GPGPU and GPU architecture. In Section 3, we explain the two Krylov subspace methods: CG and GMRES. In Section 4, we consider GPU implementation of the Krylov subspace methods especially in terms of memory access. In Section 5, we show experimental results and evaluate out implementations, and finally we summarize this paper in Section 6.

2. GPGPU

In this section, we describe the outline of the Fermi GPU architecture supporting CUDA [8], a GPGPU environment on which we implement IR-drop analysis.

A Fermi-based GPU has 16 streaming multiprocessors. Each streaming multiprocessor is a single-instruction multiple-thread (SIMT) processor which includes 32 streaming processors. Up to 32 threads can be executed simultaneously on a streaming multiprocessor.

Memory hierarchy in the GPU is shown in Fig. 1. "Global memory" is the main memory (up to several GB of DRAM) and can be accessed by all streaming multiprocessors. Access to the Global memory from the streaming multiprocessors requires several hundred clocks, which is extremely slow compared to the high operating capability of GPU processors. Each streaming multiprocessor has "shared memory". Shared memory is 64 kB of SRAM which can be shared among the streaming processors within a multiprocessor. Two cache memories are provided for global memory; L1 cache is the part of shared memory for each streaming processor, and L2 cache (768 kB of SRAM) is shared among all streaming multiprocessors.

When some data stored in global memory are required by a thread, the streaming multiprocessor receives 384 bits of sequential data including requested ones at once, and delivers them to the thread. Until all requested data received, global memory access continues and the client thread sleeps. As mentioned above, global memory access is extremely slow, so it is significantly important to devise data structure and access patterns so that the alignment will be achieved. Reducing the number of access counts to global memory significantly improves program performance. For example, when two threads issue data requests concurrently and if both requests are aligned in the same 384 bits block, the streaming multiprocessor needs to access global memory only once. Otherwise, two accesses are required and thread parallelism is significantly deteriorated.

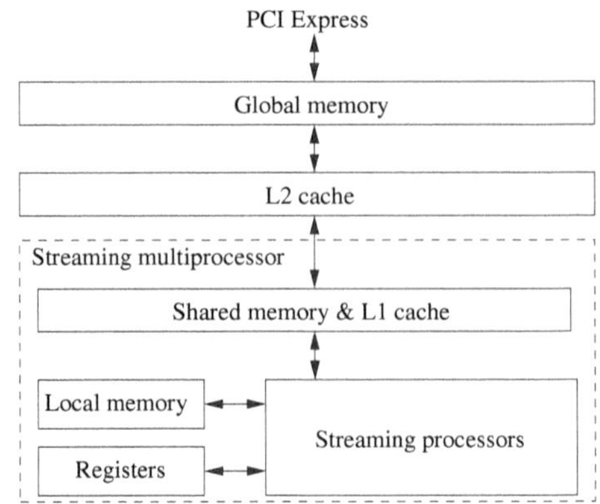

Figure 1: Memory architecture of GPU.

Algorithm 1 Conjugate gradient method

$r_0 = b - Ax_0,\ p_0 = r_0$
for $k = 0, 1, 2, \cdots$ **do**
 $\alpha_k = \dfrac{(r_k, p_k)}{(p_k, Ap_k)}$
 $x_{k+1} = x_k + \alpha_k p_k,\ r_{k+1} = r_k - \alpha_k Ap_k$
 $\beta_k = -\dfrac{(r_{k+1}, Ap_k)}{(p_k, Ap_k)}$
 $p_{k+1} = r_{k+1} + \beta_k p_k$
 Break loop if $\|r_k\|_2 < \varepsilon$
end for

3. KRYLOV SUBSPACE METHOD

Krylov subspace method is a class of iterative methods. It finds the approximate solution x_k that minimizes the residual $r_k = b - Ax_k$ on the subspace $x_0 + \mathcal{K}_k(A, r_0)$ iteratively, where

$$\mathcal{K}_k(A, r_0) = \text{span}(r_0, Ar_0, A^2 r_0, \cdots, A^{k-1} r_0)$$

is k-dimensional Krylov subspace. To minimize the residual, Rits-Galerkin approach and minimal residual approach are widely used. CG uses Rits-Galerkin approach while GMRES uses minimal residual approach.

Since $r_0, Ar_0, A^2 r_0, \cdots, A^{k-1} r_0$ are almost linearly dependent and thus are unsuitable to minimize residual, CG and GMRES generate the orthogonal basis $v_1, v_2, \cdots, v_k$ of $\mathcal{K}_k(A, r_0)$ by using Arnoldi process or Lanczos process.

3.1 Conjugate Gradient Method

Conjugate gradient method (CG) determines the k-th residual r_k and the k-th approximate solution x_k so that

$$r_k \perp \mathcal{K}_k(A, r_0).$$

This approach is called Rits-Galerkin approach. An efficient algorithm to calculate r_k and x_k in each iteration has been developed. The outline of CG is presented in Algorithm 1.

3.2 Generalized Minimal Residual Method

Instead of Rits-Galerkin approach in CG, GMRES uses minimal residual approach to determine approximate solutions. The residual minimization is written as

$$x_k = \underset{x \in x_0 + \mathcal{K}_k(A, r_0)}{\text{argmin}} \|b - Ax\|_2. \qquad (2)$$

Algorithm 2 Generalized minimal residual method

1: Generate V_k and $\tilde{H}_k$ by using Arnoldi process.
2: Calculate $\boldsymbol{y}_k \in \mathcal{K}_k(A, \boldsymbol{r}_0)$ which minimizes $\|\boldsymbol{b} - A(\boldsymbol{x}_0 + V_k\boldsymbol{y}_k)\|_2$.
3: Output the approximate solution $\boldsymbol{x}_k = \boldsymbol{x}_0 + V_k\boldsymbol{y}_k$.

Algorithm 3 Generalized minimal residual method (sequential update)

1: $\boldsymbol{r}_0 = \boldsymbol{b} - A\boldsymbol{x}_0$, $\boldsymbol{v}_1 = \boldsymbol{r}_0/\|\boldsymbol{r}_0\|_2$
2: **for** $k = 1, 2, \cdots, m$ **do**
3: $\quad \boldsymbol{w} \leftarrow A\boldsymbol{v}_k$
4: $\quad$ **for** $i = 1, 2, \cdots, k$ **do**
5: $\quad\quad h_{ik} \leftarrow (\boldsymbol{w}, \boldsymbol{v}_i)$
6: $\quad\quad \boldsymbol{w} \leftarrow \boldsymbol{w} - h_{ik}\boldsymbol{v}_i$
7: $\quad$ **end for**
8: $\quad h_{k+1,k} \leftarrow \|\boldsymbol{w}\|_2$, $\boldsymbol{v}_{k+1} \leftarrow \boldsymbol{w}/\|\boldsymbol{w}\|_2$
9: $\quad$ **for** $j = 1, 2, \cdots, k-1$ **do**
10: $\quad\quad \begin{bmatrix} h_{kj} \\ h_{k,j+1} \end{bmatrix} \leftarrow \begin{bmatrix} c_j & s_j \\ -s_j & c_j \end{bmatrix} \begin{bmatrix} h_{kj} \\ h_{k,j+1} \end{bmatrix}$
11: $\quad$ **end for**
12: $\quad c_k \leftarrow \dfrac{h_{kk}}{\sqrt{h_{kk}^2 + h_{k+1,k}^2}}$, $s_k \leftarrow \dfrac{h_{k+1,k}}{\sqrt{h_{kk}^2 + h_{k+1,k}^2}}$
13: $\quad \begin{bmatrix} g_k & h_{kk} \\ g_{k+1} & h_{k+1,k} \end{bmatrix} \leftarrow \begin{bmatrix} c_k & s_k \\ -s_k & c_k \end{bmatrix} \begin{bmatrix} g_k & h_{kk} \\ g_{k+1} & h_{k+1,k} \end{bmatrix}$
14: $\quad$ Break loop if $|g_{k+1}| < \varepsilon$
15: **end for**
16: **for** $i = k, k-1, \cdots, 1$ **do**
17: $\quad y_i \leftarrow \dfrac{1}{h_{ii}}\left(g_i - \sum_{j=i+1}^{k} h_{ij}y_j\right)$
18: **end for**
19: $\boldsymbol{x} \leftarrow \boldsymbol{x}_0 + V_m\boldsymbol{y}$

Using the orthogonal basis of $\mathcal{K}_k(A, \boldsymbol{r}_0)$ generated by Arnoldi process, Eq. (2) is reformed into

$$\begin{cases} \boldsymbol{x}_k = \boldsymbol{x}_0 + V_k\boldsymbol{y}_k\,, \\ \boldsymbol{y}_k = \underset{\boldsymbol{y} \in \mathcal{K}_k(A, \boldsymbol{r}_0)}{\operatorname{argmin}} \|\boldsymbol{b} - A(\boldsymbol{x}_0 + V_k\boldsymbol{y})\|_2\,, \end{cases}$$

where $V_k = [\boldsymbol{v}_1, \boldsymbol{v}_2, \cdots, \boldsymbol{v}_k]$. By using Givens rotation, $\boldsymbol{y}_k$ can be calculated. The outline of GMRES is shown in Algorithm 2.

Since Algorithm 2 computes $\boldsymbol{x}_k$ after the calculations of k vector(s) $\boldsymbol{v}_1, \boldsymbol{v}_2, \cdots, \boldsymbol{v}_k$, it cannot update $\boldsymbol{x}_i$ and $\boldsymbol{r}_i$ incrementally in each step to calculate $\boldsymbol{v}_i$ for $i = 1, 2, \cdots, k$ and cannot break off the calculation even after the residual $\boldsymbol{r}_i$ become small enough, as many other iterative methods do. So in practical use, the derived version like Algorithm 3 which update the approximate solution sequentially is used. Lines 3 to 8 are Arnoldi process and lines 9 to 13 are Givens rotation.

In Algorithms 2 and 3, the number of elements of the basis increases as k increases. This is not feasible in terms of memory usage. Hence, "GMRES(m)" or "restarted GMRES" is used, in which GMRES is restarted using the final approximate solution obtained after the m iterations being the initial solution. In this paper, we use GMRES(m).

Both of CG and GMRES consists of matrix-vector multiplication, inner product and addition of vectors, and these calculations can be parallelized for each elements of the vector. So we implement data-level parallelism of CG and GMRES on GPU.

3.3 Preconditioning

In Krylov subspace methods, the number of iterations can be drastically reduced by preconditioning. Therefore, to accelerate these methods, we need to select and apply appropriate preconditioning.

There are several types of preconditioners. For example, the left preconditioning multiplies a preconditioning matrix P to the both sides of Eq. (1). In this case, linear solvers solve the preconditioned system

$$PA\boldsymbol{x} = P\boldsymbol{b}.$$

To apply the left preconditioning to CG or GMRES, we have only to replace A with PA and $\boldsymbol{b}$ with $P\boldsymbol{b}$ in Algorithms 1 and 3.

4. IMPLEMENTATION

4.1 Preconditioner

Jacobi, SSOR and incomplete Cholesky factorization are popular and commonly used preconditioners. However, SSOR and incomplete Cholesky factorization are not suitable for parallel processing on GPU because they include forward and backward substitution for each iteration. In Jacobi preconditioner, on the other hand, though we only need to compute PA and Pb where $P = \mathrm{diag}(A)^{-1}$ before iteration, the effect on convergence speed is low.

In this paper, we apply the left preconditioner introduced in [9] to CG and GMRES. It preconditions a linear system with the matrix

$$P = F^{-1}(2I - AD^{-1}), \tag{3}$$

where

$$\begin{aligned} D &= \mathrm{diag}(A), \\ F &= \mathrm{diag}((2I - AD^{-1})A). \end{aligned}$$

This preconditioning matrix is designed assuming the linear system to be generated from a regular grid. This assumption is not necessarily satisfied in general power grids. However, it is proven that the matrix serves as a good preconditioner for CG [9]. Since this preconditioning is based on the structure of the target graph, we consider it can be widely applied to other solvers. In this paper, we apply the preconditioner in Eq. (3) to our GPU implementation of GMRES to further improve its performance.

4.2 Vector and Matrix Storages for Coalesced Access

As stated earlier, with the lack of large hierarchical cache memory, data alignment and access orders for global memory is the critical factor that determines acceleration performance on GPU. Both in CG and GMRES, calculations of

- inner product of vectors and
- matrix-vector product

are the performance-dominant operations. In the following, we discuss efficient storage format of vectors and matrices in GPU implementation.

$$\begin{bmatrix} 5 & 1 & 3 & \\ 1 & 2 & & 1 \\ 3 & & 4 & \\ & 1 & & 2 \end{bmatrix}$$

(a)

value			column		
5	1	3	1	2	3
1	2	1	1	2	4
3	4	0	1	3	0
1	2	0	2	4	0

(b)

value {5 1 3 1 1 2 4 2 3 1 0 0}
column {1 1 1 2 2 2 3 4 3 4 0 0}

(c)

Figure 2: Example illustration of ELL format.

A vector is stored as a single array and the elements processed by a thread are determined according to their indices. With this assignment, adjacent threads access consecutive addresses (coalesced access). Data utilization of a load from global memory is improved, and the number of load counts to the global memory is reduced. For the performance improvement of inner product calculation, intermediate data stored in the global memory is arranged so that the next load addresses become contiguous.

A matrix data should be also aligned so that adjacent threads access continuous addresses. Compressed sparse row (CSR) format [10] is the most widely used sparse matrix representation. For example, CUSPARSE, a basic linear algebra library for sparse matrix released by NVIDIA, adopts CSR. CSR achieves efficient memory utilization because only non-zero elements of the matrix are stored. However, considering the matrix-vector operation of Krylov subspace methods, its irregular access pattern often causes the increase of global memory access.

In this paper, we proposed to apply ELL[1] format [10] to enhance performance. Figure 2 illustrates the structure of ELL by an example. Assume the matrix in Fig. 2(a) is given, we store non-zero elements and their column indices in the matrix as respective two compressed matrices. Packing of the non-zero elements and their column indices are shown in Fig. 2(b) for the example of Fig. 2(a). Those will be stored as two arrays shown in Fig. 2(c). When each thread processes one row to calculate a matrix-vector multiplication, ELL allows global memory accesses to be coalesced.

Although ELL is more efficient in terms of coalesced access on GPU than CSR is, the memory utilization of the ELL is less efficient than CSR. Trailing zeroes are stored unless the number of non-zero elements of the row is equal for all rows. Figure 3 shows the memory efficiency of the ELL, in which black bars represent non-zero elements and white spaces represent zero paddings, respectively. As shown in Fig. 3(a), ELL requires more memory than the actual number of non-zero elements to achieve alignment. The amount of padding zeros depends on the variance of the number of non-zero elements in each row.

When solving very large circuits, the memory inefficiency limits the size of power grid circuits. To remedy this issue, we propose a memory efficient implementation of ELL, called *tight ELL*

Figure 3(b) shows an implementation of the tight ELL, where the matrix row is segmented to reduce the maximum number of non-zero elements in a segment and reduce the

[1]"ELL" comes from "ELLPACK", a software for solving elliptic problems, which uses that format.

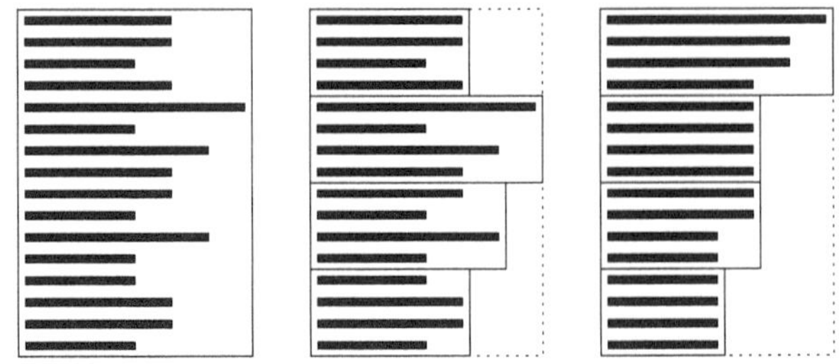

(a) (b) (c)

Figure 3: Memory efficient ELL formats. (a) conventional ELL, (b) tight ELL (segmented), and (c) tight ELL (sorted and segmented).

number of padded zeroes. When the tight ELL is utilized, the column lengths in the value-column arrays in Fig. 2(b) varies by rows depending on which segment the row in interest belongs to. Storing a set of boundaries and column lengths for each segment enables the tight ELL. The memory overhead of the boundary table is negligible compared with the storage saved by segmentation. The accesses through the boundary table have a slight impact on performance, which will be numerically evaluated in the next section.

Figure 3(c) shows a further compact implementation of the tight ELL, where the matrix rows are first sorted and then segmented. These procedures further squeeze the memory usage of the ELL with an additional overhead due to row number translations.

In the case of CG and GMRES, the translations can be realized as matrix reordering, because the results of them are independent of matrices and vector ordering and they use only one matrix[2]. Here, the reordering requires storing the order of matrix rows, which can be also applicable for vectors orders and the column orders. In general, the row translation table can be realized as one dimensional array with the length of row-numbers. Hence, in this case, if more than one padding zero par row is saved on average, the tight ELLs can successfully reduces the memory usage.

4.3 Kernels

The code executed in streaming multiprocessors of GPU is called "kernel". By specifying a kernel and a number of threads in runtime, we can execute the code on GPUs. Data communications with other kernels are made only through the global memory. Therefore, issuing many kernel calls increases the number of global memory accesses. However, the only way to synchronize more than two streaming multiprocessors is to divide a kernel into several kernels, since all streaming multiprocessors are synchronized at the time of each kernel call. In order to pursue efficiency of the code, it is important to detect the points where the synchronization is really required. Reducing the number of kernel calls and the number of writing back the intermediate data to the global memory are critical for enhancing performance.

We show an example of synchronization using the calculation of inner product $(\boldsymbol{p}, A\boldsymbol{p})$, where $\boldsymbol{p}$ is a vector and A is a matrix. When general-purpose interfaces such as BLAS are

[2]To be precise, they use two matrices A and P. However, since P in Eq. (3) has the same non-zero pattern as A, we can apply one reordering matrix to both of them.

Algorithm 4 CG (implemented version)

1: $\boldsymbol{v} \leftarrow \boldsymbol{b} - A\boldsymbol{x}_0$
2: $\boldsymbol{r} \leftarrow P\boldsymbol{v}$, $\boldsymbol{p} \leftarrow \boldsymbol{r}$
3: **for** $k = 0, 1, 2, \cdots$ **do**
4: $\quad \boldsymbol{v} \leftarrow A\boldsymbol{p}$
5: $\quad \boldsymbol{v} \leftarrow P\boldsymbol{v}$, first step of $(\boldsymbol{r}, \boldsymbol{p})$ and $(\boldsymbol{p}, \boldsymbol{v})$
6: $\quad c \leftarrow 1/(\boldsymbol{p}, \boldsymbol{v})$, $e \leftarrow (\boldsymbol{r}, \boldsymbol{p})$, $\alpha \leftarrow ce$
7: $\quad \boldsymbol{x} \leftarrow \boldsymbol{x} + \alpha\boldsymbol{p}$, $\boldsymbol{r} \leftarrow \boldsymbol{r} - \alpha\boldsymbol{w}$, first step of $(\boldsymbol{r}, \boldsymbol{w})$
8: $\quad \beta \leftarrow -c(\boldsymbol{r}, \boldsymbol{w})$
9: $\quad \boldsymbol{p} \leftarrow \boldsymbol{r} + \beta\boldsymbol{p}$
10: $\quad$ Break loop if $e < \varepsilon$
11: **end for**

Table 1: The detail of benchmark circuits [2].

Circuit	# of nodes	Dimension of linear system	# of non-zero elements
#1	30,638	16,327	75,827
#2	127,238	126,905	542,895
#3	851,581	285,344	1,508,802
#4	953,583	952,618	4,054,056
#5	1,079,310	540,220	2,691,958
#6	1,670,494	834,252	4,125,504

used, the above operation would be typically implemented by series of two basic operations: $\boldsymbol{v} \leftarrow A\boldsymbol{p}$ and $c \leftarrow (\boldsymbol{p}, \boldsymbol{v})$. This calls for synchronization between the two operations. This synchronization can be eliminated by defining the kernel to calculate $\boldsymbol{v} \leftarrow A\boldsymbol{p}$, $w_i \leftarrow p_i v_i$ and the first step of reduction for $\boldsymbol{w}$, which saves two global memory accesses for $\boldsymbol{v}$.

Our CG implementation is shown as Algorithm 4. Each line is implemented as one kernel except the sum reduction for vector inner product (lines 6 and 8) which has to be divided into some kernels for synchronization. In lines 5 and 7, as mentioned, first step of inner product is processed in the same kernel to reduce the number of global memory accesses.

This optimization cannot be well applied to GMRES. So we implemented the same kernel interface to BLAS except some special operations. In Algorithm 3, Givens rotation (lines 9 to 13) and backward substitution (lines 17 to 19) are sequential and hard to parallelize. We implemented them as single thread kernels. Though single thread processing on GPU is not effective, they are expected to have little influence on the runtime because the total runtime is dominated by Arnoldi process.

5. EXPERIMENTAL RESULTS

5.1 GPU vs. CPU and CG vs. GMRES

We implemented CG and GMRES to run on both CPU (single thread) and GPU to compare runtime. IBM power grid benchmark [2] suite is used as test circuits. Table 1 shows their details. The supply voltage is 1.8 V for all benchmark circuits. Numerical experiments were carried out on a Linux PC with Intel Core i7 970 (3.20 GHz) CPU and NVIDIA GeForce GTX 580 GPU.

Table 2 shows the experimental results: runtime, the number of iterations, and the error based on the comparison with the solutions by a commercial circuit simulator. GMRES did not reach convergence for circuit #2. In this experiment, we used the original ELL format because the GPU has enough memory to store these matrices.

As shown in Table 2, GPU implementations of both CG and GMRES achieve speedup of approximately 10× for most cases, and 17× speedup in the best case. In this experiment with the preconditioner mentioned in Section 4, CG is better than GMRES in terms of both runtime and convergent speed. The reason why the number of iterations and the error of GPU implementation are not equal to those of CPU implementation seems to be the difference of floating-point arithmetic units.

The theoretical performance and main memory bandwidth of the GPU are 790 GFLOPS and 192 GB/s, respectively, while those of the CPU are 12.8 GFLOPS and 10.7 GB/s, respectively. So the GPU has 62× FLOPS and 18× memory bandwidth. Since main memory access is the bottleneck for both CG and GMRES on GPU, the upper bound of the runtime improvement is 18×. Our implementations achieve 17× speedup at the maximum, which is considered to be very close to the limit.

5.2 Matrix Storage

For the four different matrix formats described in the previous section, i.e., CSR, ELL, tight ELL, and sorted tight ELL, we counted the number of elements required to store the matrices of benchmark circuits and measured runtime of matrix-vector multiplication during the iterations in the CG algorithm. The results for the block size of 256 are shown in Tables 3 and 4, and Fig. 4 and 5 compare each version of ELL with CSR, normalizing the results by that of CSR.

According to Table 3 and Fig. 4, the number of elements of the original ELL is about 150 % of that of CSR at the worst case. We expect that this large memory usage can be alleviated by the use of the proposed tight ELL formats. As expected, sorted and segmented ELL formats stored matrices as almost equally memory efficiently as the CSR.

Considering the runtime, according to Table 4 and Fig. 5, ELL is about 1.5 times faster than CSR and segmentation (tight ELL) has an insignificant impact on the processing speed. For this reason, tight ELL is superior to the original ELL. On the other hand, in the case of sorted tight ELL, the runtime slightly increases from that of the original and the tight ELL. Considering runtimes of the original ELL and tight ELL are equal, the reason seems to be that, in calculation of matrix-vector multiplication, access pattern for vector data is altered by the reordering associated to realize sort, which lowers cache hit ratio.

From these experiments, we conclude that the best choice is to use the tight ELL. When a matrix is too large for an available memory size, then, the sorted tight ELL is recommended. Most of the matrices that has been traditionally stored using CSR format can also be stored and solved more efficiently with tight ELL with our GPU implementation.

6. SUMMARY

In this paper, we have presented efficient power grid analyses on GPUs that are based on Krylov subspace methods. Compared with single-threaded CPU implementations of those methods, our GPU implementation achieved 17x performance acceleration at the maximum and 10x performance acceleration on average for analyzing IBM benchmark circuits. Our implementation utilizes a newly devised tight ELL format to store sparse matrices, with which memory consumption and access pattern have been improved over more widely used CSR format.

Table 2: Experimental results of power grid analysis.

	CG								GMRES							
	CPU implementation				GPU implementation				CPU implementation				GPU implementation			
	Time	# of	Error [mV]		Time	# of	Error [mV]		Time	# of	Error [mV]		Time	# of	Error [mV]	
Circuit	[s]	Iter.	Max.	Avg.	[s]	Iter.	Max.	Avg.	[s]	Iter.	Max.	Avg.	[s]	Iter.	Max.	Avg.
#1	0.05	140	0.20	0.034	0.01	138	0.20	0.034	0.11	183	0.18	0.023	0.18	183	0.22	0.023
#2	1.02	343	0.10	0.019	0.11	341	0.10	0.019	N/A	N/A	N/A	N/A	N/A	N/A	N/A	N/A
#3	3.16	375	0.32	0.025	0.24	373	0.32	0.025	8.16	724	0.30	0.023	0.67	726	0.30	0.023
#4	12.65	444	0.11	0.015	0.74	442	0.11	0.015	40.91	1043	0.13	0.025	2.29	1043	0.13	0.025
#5	6.88	514	0.22	0.020	0.51	512	0.22	0.020	18.48	950	0.18	0.036	1.17	950	0.18	0.036
#6	12.37	557	0.21	0.028	0.85	555	0.21	0.028	34.07	1087	0.35	0.046	2.19	1084	0.35	0.046

Table 3: Required element numbers for each storage format.

Circuit	CSR	ELL	Tight ELL	Sorted tight ELL
#1	75,827	81,920	81,920	76,288
#2	542,895	634,560	586,752	543,488
#3	1,508,802	2,283,008	1,734,144	1,509,888
#4	4,054,056	4,763,200	4,292,352	4,054,784
#5	2,691,958	3,781,568	2,715,136	2,693,376
#6	4,125,504	5,840,128	4,185,008	4,126,464

Table 4: Required time to execute the iteration part of CG algorithm.

Circuit	CSR	ELL	Tight ELL	Sorted tight ELL
#1	0.007	0.005	0.005	0.007
#2	0.083	0.059	0.061	0.074
#3	0.300	0.144	0.145	0.305
#4	0.723	0.504	0.508	0.911
#5	0.487	0.299	0.301	0.330
#6	0.796	0.525	0.533	0.709

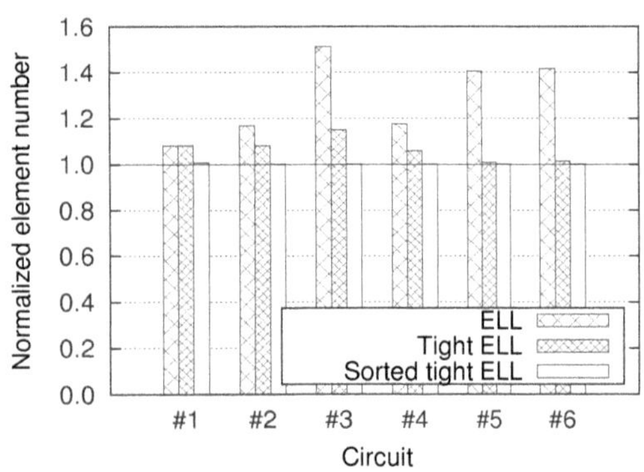

Figure 4: Required element numbers for each ELL format (normalized by that of CSR format).

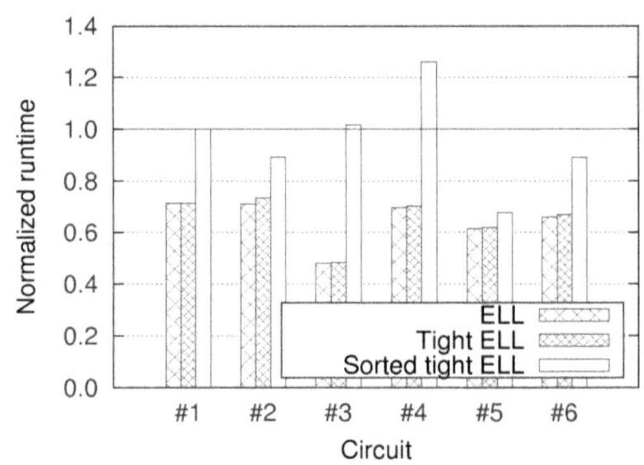

Figure 5: Required time to execute the iteration part of CG algorithm (normalized by that of CSR format).

7. REFERENCES

[1] J. Friedrich, B. McCredie, N. James, B. Huott, B. Curran, E. Fluhr, G. Mittal, E. Chan, Y. Chan, D. Plass, S. Chu, H. Le, L. Clark, J. Ripley, S. Taylor, J. Dilullo, and M. Lanzerotti, "Design of the Power6 microprocessor," in *ISSCC Dig. Tech. Papers*, Feb. 2007, pp. 96–97.

[2] S. R. Nassif, "Power grid analysis benchmarks," in *Proc. ASPDAC*, Jan. 2008, pp. 376–381.

[3] M. R. Hestenes and E. Stiefel, "Methods of conjugate gradients for solving linear systems," *J. Res. Natl. Bur. Stand.*, vol. 49, no. 6, pp. 409–436, Dec. 1952.

[4] Y. Saad and M. H. Schultz, "GMRES: A generalized minimal residual algorithm for solving nonsymmetric linear systems," *SIAM J. Sci. and Stat. Comput.*, vol. 7, pp. 856–869, Jul. 1986.

[5] J. Shi, Y. Cai, W. Hou, L. Ma, S.-D. Tan, P.-H. Ho, and X. Wang, "GPU friendly fast Poisson solver for structured power grid network analysis," in *Proc. DAC*, Jul. 2009, pp. 178–183.

[6] Z. Feng and P. Li, "Multigrid on GPU: Tackling power grid analysis on parallel SIMT platforms," in *Proc. ICCAD*, Nov. 2008, pp. 647–654.

[7] Z. Feng, X. Zhao, and Z. Zeng, "Robust parallel preconditioned power grid simulation on GPU with adaptive runtime performance modeling and optimization," *IEEE Trans. CAD*, vol. 30, no. 4, pp. 562–573, Apr. 2011.

[8] NVIDIA, "CUDA Zone," http://www.nvidia.com/object/cuda_home_new.html.

[9] C.-H. Chou, N.-Y. Tsai, H. Yu, C.-R. Lee, Y. Shi, and S.-C. Chang, "On the preconditioner of conjugate gradient method – a power grid simulation perspective," in *Proc. ICCAD*, Nov. 2011, pp. 494–497.

[10] N. Bell and M. Garland, "Efficient sparse matrix-vector multiplication on CUDA," NVIDIA Corporation, NVIDIA Technical Report NVR-2008-004, Dec. 2008.

Translation Validation of Scheduling in High Level Synthesis

Tun Li, Yang Guo, Wanwei Liu, Mingsheng Tang
School of Computer, National University of Defense Technology,
ChangSha, HuNan, P. R. China, 410073
{tunli, guoyang, wwliu, mstang}@nudt.edu.cn

ABSTRACT
The growing design-productivity gap has made designers shift toward using high-level synthesis (HLS) techniques to generate register transfer level design from high-level languages. Unfortunately, this translation process is very complex and may introduce bugs into the generated design, which can create a mismatch between what a designer intends and what is actually implemented in the circuit. In this paper, we present an equivalence checking method to validate the result of HLS scheduling against the initial high-level program. Finite state machine with data path (FSMD) models were used to represent designs before and after scheduling. The proposed method uses a bisimulation relation approach to prove equivalence. The automatically established bisimulation relation guarantees that for each execution sequence in the design before scheduling, a related and equivalent execution sequence exists in the design after scheduling and vice versa. Our method provides a unified way to deal with various scheduling optimizations. We have implemented our validation technique and compared it with a state-of-the-art HLS scheduling verification method. The promising results show the effectiveness and efficiency of our method.

Categories and Subject Descriptors
B.7.2 [**Integrated Circuits**]: Design Aids – *verification.*

General Terms
Algorithms, Design, Verification.

Keywords
Translation Validation; Scheduling; High Level Synthesis; Non-Structure-Preserving; Bisimulation Relation.

1. Introduction
High-level synthesis (HLS) is generally a process of generating a register-transfer level (RTL) design from a high-level behavioral description expressed in languages like C/C++. HLS is seen as a solution to bridge the design productivity gap and typically consists of several inter dependent subtasks such as specification, compilation, scheduling, allocation and binding [1]. Scheduling generally assigns operations of a behavior description with specific clock cycles based on data dependencies and given constraints of area and/or delay. Consequently, the results of scheduling do not have a one-to-one mapping to their original behavior descriptions. This aspect raises the most tough verification challenge in HLS. The objective of this paper is to check the computational equivalence of design descriptions before and after scheduling.

In order to use a minimal number of time steps to schedule the operations, scheduling generally modifies input behavior in some ways. For example, the basic block (BB)-based scheduling algorithms, such as list scheduling [2] and force-directed scheduling [3], perform code transformations within a BB. Global code transformation techniques such as code motion and common subexpression elimination (CSE) [4] usually lead to moving operations across BB boundaries to improve the performance and the resource utilization. The commonly used code motion techniques include duplicating down (DD)/duplicating up (DU), merging up (MU)/merging down (MD), useful move (UM), speculation (Sp) and reverse speculation (RSp). All these type of optimizations are typical structure-preserving transformation techniques. Also, path-based scheduling (PBS) [5] constantly merges certain qualified consecutive paths to minimize the number of control states, which may modify the control structure of original high-level description and is so called non-structure-preserving transformations. Employing aforementioned scheduling methods is the major reason why the scheduling result cannot be one-to-one mapped to its original high-level description.

There are a few related works addressing the equivalence checking issue of scheduling. A theorem-proving approach for verifying the As Soon As Possible scheduling algorithm is reported in [6]. A Petri net model based model checking technique for checking the correctness of BB-based algorithms is presented in [7]. In [8], by inserting invariant properties as program assertions inside the implementation of the force-directed list scheduler (FDLS) algorithm, the correctness of the FDLS algorithm is proved using the prototype verification system (PVS) [9]. In [10], scheduling results are verified based on the precondition-based correctness and completeness of register transfer split [11]. Eveking et al. represent the prescheduled and postscheduled behaviors in the language of labeled segments (LLS), and develop basic transformations to prove the computational equivalence of LLS [12]. PVS is also used in [13] to prove the equivalence between descriptions before and after scheduling by introducing the critical paths. In [14], by defining the functional equivalence between two FSMDs [1] representing the descriptions before and after scheduling, and introducing break-points to decompose them, the equivalence of the two FSMDs is proved using PVS. The authors extend their work in [15] to further handle MU and DD code motions by concatenating critical paths.

Recently, Karfa et al. proposed an equivalence checking method by transforming the equivalence checking between FSMDs to the equivalence checking of paths [16]. The algorithm runs in an extension-and-try manner, therefore, due to the "blindly" path extension, the complexity of the method is very high. As a result, to prove a path cover, many iterations are needed if the structure of the description cannot be one-to-one mapped. Lee et al. enhance Karfa's algorithm to verify scheduling generated by code transformations employing speculation and global CSE [17].

This work is supported by National Natural Science Foundation of China under grant No. 61272335, 61133007 and 61103012

GLSVLSI'13, May 2–3, 2013, Paris, France.

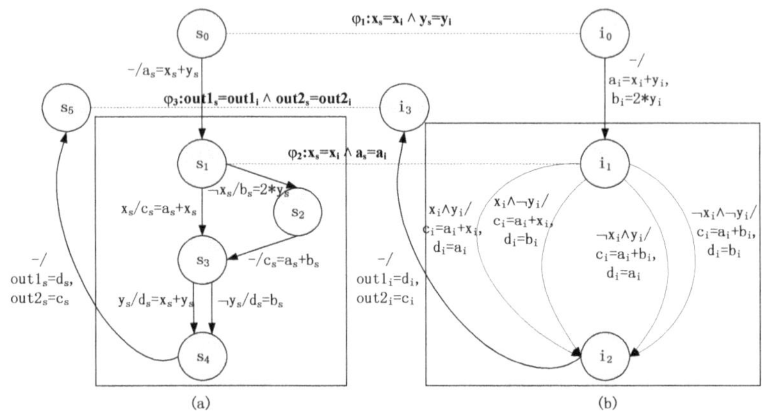

Figure 1. A simple example (a) FSMD before scheduling (b) FSMD after scheduling

However, their method cannot deal with loop shifting. In [18], a method checking the equivalence of two descriptions before and after optimization carried out by Spark [19] using translation validation technique is proposed. The proposed method is suitable for handling structure-preserving transformations; while in HLS the structure usually may not be preserved after scheduling. Tun et al. [19] improved the method proposed in [18] by redefining the bisimulation relations to dramatically reduce the number of query of automatic theorem prover (ATP), such as Z3 [22].

In this paper, we propose a translation validation technique [18] based equivalence checking method for scheduling verification in HLS. In our approach, descriptions before and after scheduling are represented by FSMDs. We construct hierarchical graphs based on FSMD to automatically capture the non-structure-preserving transformations. Then the bisimulation relations for both types of transformations are built automatically. Finally, our method uses a weakest precondition computation approach on the bisimulation relations to prove equivalence between the descriptions before and after scheduling. The automatically established bisimulation relation states what points in the description before scheduling are related to what points in the description after scheduling. Our method guarantees that for each execution sequence in the description before scheduling, a related and equivalent execution sequence exists in the description after scheduling and vice versa.

One contribution of our method is to show how translation validation can effectively be applied on a previously unexplored structure: FSMD. The other contribution is a unified method to validate most of the scheduling optimizations except loop pipelining. The scheduling optimization techniques covered both structure-preserving and non-structure-preserving transformations. To the best of our knowledge, verifying all the aforementioned scheduling techniques have not been properly addressed in a unified framework.

The rest of the paper is organized as follows. Section 2 presents some basic definitions to introduce our method. Section 3 presents the necessary background of our method, such as definition of equivalence, bisimulation relations and the correctness of our method. Section 4 provides the full details of the translation validation algorithm. Section 5 gives the experimental results on several benchmarks. Finally, Section 6 presents our conclusions and future directions of research.

2. Preliminaries

In this paper, the description before (specification) and after (implementation) scheduling scheduling, are modeled using FSMD.

We extend the definition of traditional FSMD by defining data state associated with location. An FSMD M is defined as a 8-tuple $<Q, q0, I, O, V, f, h, \Sigma>$, where

1) Q is the finite set of locations,

2) $q0$ is the reset location,

3) I is the finite set of inputs,

4) O is the finite set of outputs,

5) V is the finite set of storage variables,

6) f: $Q \times S \rightarrow Q$ is the *state transition function*, where S is the set of *status expressions* consisting of arithmetic predicates over $I \cup V$,

7) h: $Q \times S \rightarrow U$ is the *update function* of the outputs and storage variables, where $U = \{x{=}e \mid x \in O \cup V$ and e is an arithmetic predicate or expression over $I \cup V\}$ is a set of storage variables or output assignment statements.

8) Σ is defined as the union of I, V and O, which is to represent the data state of each locations. For location $q_i \in Q$, the data state is denote by σ_i and $\sigma_i \in \Sigma$. Therefore, when transferring from one location to another one, the data will be changed by the update function along the transition.

An FSMD M, is inherently deterministic; for any state q, if the status expressions $s1 = s2$, it implies that $f(q, s1) = f(q, s2)$ and $h(q, s1) = h(q, s2)$. Parallel edges between two states capture the disjunction of status expressions. For any location $q_i \in Q$, a *configuration* is defined to be a pair $<q_i, \sigma_i>$. An example used through this paper is depicted in Figure 1. Figure 1 (a) and (b) depict the FSMD model of behavioral specification before and after scheduling, where the label on each transition is of the form f/h, such as $\mathbf{x_s/c_s{=}a_s{+}x_s}$ on the transition from s1 to s3. In this paper, to simplify the bisimulation relation building process, we make a copy of the reset location to be a final location, such as state s_5 and i_3. Therefore, all the incoming edges of the reset location are altered to point to the final location.

PBS usually merges certain qualified consecutive paths. Generally, the merge pattern can be classified into two types as shown in Figure 2. Figure 2 (a) depicts the merge of a loop node with adjacent condition nodes to reduce the scheduling steps from 4 to 3. Figure 2(b) shows the merge of adjacent condition nodes to reduce scheduling steps from 5 to 4.

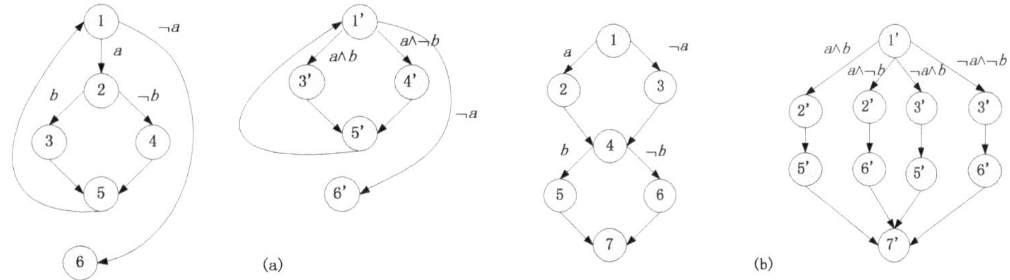

Figure 2. Two types of non-structure-preserving transformations (a) loop and condition (b) adjacent conditions

To capture the non-structure-preserving transformations, especially path merging presented in Figure 2, we further classify the node of FSMD into three types: *basic node*, *condition node* and *loop node*.The condition node and loop node are also called compound node. Intuitively, we define the three types of node according to the number of incoming and outgoing edges. The basic node has only one incoming and one outgoing edge. The condition node has only one incoming edge and may have more than one outgoing edges, but along any outgoing edge, it will not visit itself again and will join at one node. The loop node has more than one incoming and outgoing edges. However, except one incoming and outgoing edge, the other edges will form a circle.

Therefore, a hierarchical graph (HG) is defined to capture the control flow between the basic nodes and the compound nodes. The HG is built on the FSMD and only associates some information on FSMD to capture non-structure-preserving transformations. A HG is a directed acyclic graph that consists of two types of nodes: the basic node and the compound node. Compound nodes are hierarchical in nature, i.e., they contain other HG nodes. Generally, a compound node consists of a head node and a tail node which are basic nodes and a body which is a compound node. One type of compound node is the condition node, which is used to represent structures like if-then-else structures, switch-case blocks and adjacent branch structures. For example, the nodes 1, 2, 3, 4, 5, 6, 7 and the nodes 1', 2', 3', 5', 6', 7' in Figure 2 (b) form a condition node, respectively, where node 1 and 1' are the head nodes, node 7 and 7' are the tail nodes, and the other nodes form the bodies. The other type of compound node is the loop node, which is used to represent various types of loops (for, while-do, do-while). For example, the nodes 1 to 6 and the nodes 1' to 6' in Figure 2 (a) form a loop node respectively, where node 1 and 1' are loop heads, node 6 and 6' are loop tails and the other nodes form the loop body, respectively. Obviously, the nodes enclosed in the rectangle in Figure 1 form two condition nodes, which are s_1, s_2, s_3, s_4 and i_1, i_2, respectively.

An execution path η in FSMD is formulated as $q_i \underset{c_i}{\rightarrow} q_{i+1} \underset{c_{i+1}}{\longrightarrow} \dots \underset{c_{i+n-1}}{\longrightarrow} q_n = q_j$, where $q_k \in Q$ for all k, $i \leq k \leq i+n$, and there exists $c_k \in S$ as well as function f_k such that $f_k(q_k, c_k)=q_{k+1}$ for all k, $i \leq k \leq i+n$, and q_i is the starting state and q_j is the ending state. We denote by $\mathcal{N}$ the set of all execution paths in an FSMD.

In the compound node, the execution is started from the head node to the tail node. To map the execution path from the specification to the implementation, we use the path condition (PC) to identify paths that potentially can be matched. PC is a concept of symbolic execution, which is a conjunction of all the conditions along the execution path. For example, the PC of path 1-2-4-5-7 in Figure 2 (b) is $a \wedge b$, while the PC of path 1'-2'-5'-7' is also $a \wedge b$, which means that the two paths have the same PCs and may be matched.

3. Equivalence of FSMDs

Let M_S=<Q_S, $q0_S$, I_S, O_S, V_S, f_S, h_S, Σ_S> and M_I=<Q_I, $q0_I$, I_I, O_I, V_I, f_I, h_I, Σ_I> be two FSMDs before and after scheduling, We define the equivalence of M_S and M_I as: for the given reset locations $q0_S \in Q_S$ and $q0_I \in Q_I$, M_S and M_I are said to be equivalent if for every execution path of M_S starting and ending in $q0_S$ there is an equivalent execution path of M_I starting and ending in $q0_I$ and vice-versa. As mentioned in section 2, we make a copy of the reset location to be a final location. Therefore:

M_S and M_I are said to be equivalent if for every execution path of M_S starting in $q0_S$ and ending in final location, there is an equivalent execution path of M_I starting in $q0_I$ and ending in final location and vice-versa.

Here we use cutpoint which is identical to Floyd-Hoare's method [11, 21] to decompose FSMD. However, for equivalence checking, the cutpoint in the FSMDs of both specification and implementation should be correlated. We use the *verification relation* to correlate these cutpoints in both descriptions.

A *verification relation* between two FSMDs M_1=<Q_1, $q0_1$, I_1, O_1, V_1, f_1, h_1, Σ_1> and M_2=<Q_2, $q0_2$, I_2, O_2, V_2, f_2, h_2, Σ_2> is a set of triples (q_1, q_2, φ), where $q_1 \in Q_1$, $q_2 \in Q_2$ and φ is a predicate over the variables live at locations q_1 and q_2. φ is defined to be a set of such predicates that $\varphi(\sigma_1, \sigma_2)=true$ is used to indicate that φ is satisfied in $(\sigma_1, \sigma_2) \in \Sigma_1 \times \Sigma_2$.

The *Simulation relations* and *bisimulation relations* are verification relations with a few additional properties. Supposing $<q_1, \sigma_1>$ and $<q'_1, \sigma'_1>$ is the starting and ending configurations of execution path η_1 of M_1, denoted by $<q_1, \sigma_1> \overset{\eta_1}{\rightarrow} <q'_1, \sigma'_1>$. Therefore, a simulation relation R for two FSMDs M_1 and M_2 is a verification relation such that:

$$
\begin{aligned}
& R(q0_1, q0_2, true): \\
& \forall (q_1, q_2, q'_1, \sigma_1, \sigma_2, \sigma'_1, \varphi, \eta_2) \in Q_1 \times Q_2 \times Q_1 \times \Sigma_1 \times \Sigma_2 \times \Sigma_1 \times \Phi \times \mathcal{N}_2. \\
& \left[<q_1, \sigma_1> \overset{\eta_1}{\rightarrow} <q'_1, \sigma'_1> \wedge R(q_1, q_2, \varphi) \wedge \varphi(\sigma_1, \sigma_2) = true \right] \Rightarrow \\
& \exists (q'_2, \sigma'_2, \varphi', \eta_2) \in Q_2 \times \Sigma_2 \times \Phi \times \mathcal{N}_2 \\
& \left[<q_2, \sigma_2> \overset{\eta_2}{\rightarrow} <q'_2, \sigma'_2> \wedge R(q'_1, q'_2, \varphi') \wedge \varphi'(\sigma'_1, \sigma'_2) = true \wedge \eta_1 \equiv \eta_2 \right] \qquad (1)
\end{aligned}
$$

Intuitively, these conditions respectively state that:

- the reset location of M_1 and M_2 must be related;
- if M_1 and M_2 are in a pair of related configurations, and M_1 can proceed several steps producing an execution path η_1, then M_2 must also be able to proceed several steps, producing an execution path η_2 that is equivalent to η_1, and the two resulting configurations must be related.

Therefore, a verification relation R is a bisimulation relation for M_1 and M_2 *iff* R is a simulation relation for M_1 and M_2 and R^{-1}={(q_2, q_1, φ) | $R(q_1, q_2, \varphi)$} is a simulation relation for M_2 and M_1. Intuitively, for two FSMDs M_1 and M_2, the bisimulation relation consists of a set of entries of the form (q_1, q_2, φ), where q_1 and q_2 are locations in M_1 and M_2 respectively, and φ is a predicate over variables of M_1 and M_2. The pair (q_1, q_2) represents how the control state of M_1 and M_2 is related, whereas φ captures how the data is related.

4. Translation Validation Algorithms

```
TV(MS, MI)
{
  worklist=InferBR(MS, MI);
  BuildPC(MS, MI);
  while not all Ψbr in worklist reach its fixpoint do
     br=SelectBR(worklist);
     cons=BuildConstraints(br);
     Strongcons=cons;
     if ATP(cons) ≠ valid
       if br=(q0S, q0I)
          ErrorandExit("Not Equivalent")
       else
          Strongcons=Strengthen(cons);
       endif
     endif
     if Strongcons≠cons
          worklist.update(cons, Strongcons);
     end if
  end while
  Exit("Equivalent");
}
```

Figure 3. Equivalence checking algorithm

The proposed equivalence checking algorithm is presented in Figure 3. The algorithm takes two FSMDs M_S and M_I with HG information as inputs representing the specification and implementation, respectively. Supposing the scheduling transformation will not change the names of inputs, outputs and storage variables.

We will use a simple example to explain our algorithm. The example is presented in Figure 1. Figure 1 (a) and (b) are the original description and the result of running the PBS HLS algorithm, respectively. This design takes x and y as inputs, and computes and outputs the values of d and c. During PBS, several transformations are also performed, such as speculation and common subexpression elimination (CSE). Subscript s and i are used respectively to denote variables in the specification which is the original design description before scheduling and the implementation which is the description after scheduling. After scheduling, the adjacent branch started respectively at s1 and s3 are merged into one branch started at i1. Assignment "$b_i=2*y_i$" is moved speculatively out of the branch it originally located. The assignment to variable "d_i" is safely replaced by "a_i" by CSE. The reset states s1 and i1 are duplicated as final states and are renamed as s5 and i3 respectively.

4.1 Inferring Potential Bisimulation Relations

The procedure **InferBR** is used to automatically infer all the potential bisimulation relations. If a branch is encountered, **InferBR** uses the structure of the branch predicate and the structure of the instructions on each side of the branch to guess a correlation. **InferBR** use syntax to check if the branches encountered in both descriptions are with the same predicates.

The reset location pair is naturally a bisimulation relation and is added to *worklist*. For the reset location pair, the inputs in set $I_1 \cap I_2$ are correlated to form predicate φ. For the given example in Figure 1, the bisimulation relation for the reset location pair is (s0, i0, $x_s=x_i \wedge y_s=y_i$). From reset location pair, InferBR will proceed forward in parallel in both FSMDs of the specification and the implementation. It stops when either a final location or a location with divergent outgoing edges is encountered. For final locations such as s5 and i3 in Figure 1, **InferBR** will build the predicate φ to ensure that all the variables in set $O_1 \cap O_2$ to be output at here have the same values. Therefore constraints"$out1_s=out1_i \wedge out1_s=out2_i$" is built for bisimulation (s5, i3). Note that the two FSMDs are all associated with HG information. Therefore, for locations with divergence outgoing edges, we can easily identify them as a loop node or a condition node, and can be handled accordingly:

(1) For loop node, **InferBR** just checks the loop conditions in M_S and M_I, if both conditions can be matched, the head nodes of both the loop nodes may be bisimulation relation. For example, in Figure 2 (a), although the node 2 is merged with the node 1, **InferBR** will easily identify that the node 1 and the node 1' are all loop node and their loop exit conditions are identical. Therefore, **InferBR** will guess that the node 1 and 1' as potential bisimulation relations and build predicate φ to correlate them.

(2) For adjacent branch nodes, as shown in Figure 4, the left and right flows are FSMDs before and after scheduling, respectively. There may be n adjacent conditions and after scheduling k conditions may be merged, where $1 \le k \le n$. As the example shown in Figure 2(b), the condition of branch node i ($1 \le i \le k$) will be a part of the condition of the node 1. Therefore, **InferBR** will compare the predicate of the nth condition with the predicate at location 1', if matched, it means the nth condition is merged; otherwise, **InferBR** will iterate from the $(n-1)^{th}$ condition back to the first condition to decide k. The iteration will be stopped when a match is found, supposing the kth condition, which means the condition nodes from 1 to k are merged. Then **InferBR** will shrink the range of adjacent conditions to only enclose the first k conditions, and build bisimulation between node 1 and 1', k+1 and k'+1,..., n and n', just as depicted in Figure 4.

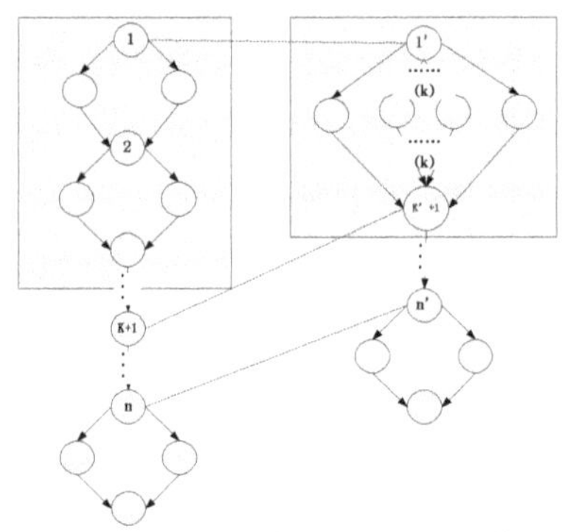

Figure 4. Merge of adjacent conditions

For both loop node and condition node, the predicates of the bisimulation relation are conjunction of equality of variables used in conditions. Therefore, for our example, after executing **InferBR**, the *worklist* will have three elements as shown in the 2nd column of Table 1.

The second step **BuildPC** only applies to the compound nodes in M_S and M_I. For all the bisimulation relations corresponding to the compound nodes, it will utilize symbolic execution on all the execution paths from the head node to the tail node to compute PC for each path. After computing PCs, the paths in M_S and M_I with matched predicates are grouped as path pair. For example in Figure 1(a), there are four paths from s1 to s4, whose path conditions are $(x_s \wedge y_s)$, $(x_s \wedge \neg y_s)$, $(\neg x_s \wedge y_s)$ and $(\neg x_s \wedge \neg y_s)$, respectively. There are also four paths in Figure 1(b) from i1 to i2, whose path conditions can be matched from left to right to the corresponding four paths in Figure 1(a). The path pairs are added to *worklist* as properties of corresponding bisimulation relations at compound nodes.

Table 1. Computation of bisimulation relation

$\varphi(q_S, q_I)$	1st iteration	2nd iteration	3rd iteration
$\varphi_{(s0, i0)}$	$x_s=x_i \wedge y_s=y_i$	$x_s=x_i \wedge y_s=y_i$	$x_s=x_i \wedge y_s=y_i$
$\varphi_{(s1, i1)}$	$x_s=x_i \wedge a_s=a_i$	$x_s=x_i \wedge a_s=a_i \wedge x_s+y_s=a_i \wedge a_s+x_s=a_i+x_i \wedge a_s+2*y_s=a_i+b_i \wedge x_s+y_s=a_i \wedge c_s=c_i$	$x_s=x_i \wedge a_s=a_i \wedge x_s+y_s=a_i \wedge a_s+x_s=a_i+x_i \wedge a_s+2*y_s=a_i+b_i \wedge x_s+y_s=a_i \wedge c_s=c_i$
$\varphi_{(s5, i3)}$	$out1_s=out1_i \wedge out1_s=out2_i$	$out1_s=out1_i \wedge out1_s=out2_i$	$out1_s=out1_i \wedge out1_s=out2_i$

4.2 Checking Bisimulation Relations

The core of our algorithm is iterations over all the bisimulation relations in *worklist*. At first, **SelectBR** selects a bisimulation relation from *worklist*, which is denoted by *br*. It is directed heuristically to select bisimulation relations near the final location and from there to the reset location in a backward way. In our practice, we found that this heuristic selection will accelerate the checking of bisimulation relations. Next, all the bisimulation relations that can reach *br* without passing other bisimulation relations in both FSMDs are picked out and form a relation set, denoted by *brs*. The relations in *brs* are used to build constraints to state the relationship between *br* and other pairs of related locations. According to formula (1), if starting at (q_S, q_I) in location satisfying $\varphi_{(q_S,q_I)}$, the specification and implementation can execute in parallel to reach another related pair of locations (q'_S, q'_I), then $\varphi_{(q'_S,q'_I)}$ must hold in the resulting locations. This kind of constraints guarantees that the bisimulation relation inferred by **InferBR** is really a bisimulation. Here, (q'_S, q'_I) is corresponding to *br* and (q_S, q_I) to the relation in *brs*.

This kind of constraints are be stated over the variables $\varphi_{(q_S,q_I)}$ and $\varphi_{(q'_S,q'_I)}$ using the weakest precondition operator (*wp*) [21], which is computed as $wp(w_1, wp(w_2, \varphi_{(q'_S,q'_I)}))$, where w_1 and w_2 are update functions along the paths from q_S to q'_S and from q_I to q'_I, respectively. The formula computes the weakest precondition with respect to w_2 and then with respect to w_1. When computing *wp* with respect to onesequence, we treat all variables from the other sequence as constants. Obvious, the order in which we process the two sequences does not matter. If there are several path pairs between (q_S, q_I) and (q'_S, q'_I), the $wp(w_1, wp(w_2, \varphi_{(q'_S,q'_I)}))$ is a conjunctions of *wp* of each path pair. Finally, procedure **BuildConstraints** generates a new constrain $\varphi_{(q_S,q_I)} \Rightarrow wp(w_1, wp(w_2, \varphi_{(q'_S,q'_I)}))$, denoted by *cons*. *Cons* is checked by querying an ATP if it is valid. If not and (q_S, q_I) is reset location pair, the algorithm will stop and unequivalence is reported. Otherwise, procedure **Strengthen** will strengthen $\varphi_{(q_S,q_I)}$ using conjunction of $wp(w_1, wp(w_2, \varphi_{(q'_S,q'_I)}))$. The strengthened constraint is compared with $\varphi_{(q_S,q_I)}$ to check if they are the same. If so, the corresponding bisimulation relation is marked as reaching its fixpoint. Otherwise, $\varphi_{(q_S,q_I)}$ is replaced with the strengthened constraint in *worklist*. If all the bisimulation relations reach their fixpoints, the algorithm will stop and report equivalence.

For the given example in Figure 1, the process of checking bisimulation relation is presented in Table 1. Bisimulation relation (s5, i3) is selected first. We can find that the relation that can reach (s5, i3) is (s1, i1). According to the path matching information from **BuildPC**, we can compute *wp* for this two relations and get "$x_s+y_s=a_i \wedge a_s+x_s=a_i+x_i \wedge a_s+2*y_s=a_i+b_i \wedge x_s+y_s=a_i \wedge c_s=c_i$". The constraint "$x_s=x_i \wedge a_s=a_i \Rightarrow x_s+y_s=a_i \wedge a_s+x_s=a_i+x_i \wedge a_s+2*y_s=a_i+b_i \wedge x_s+y_s=a_i \wedge c_s=c_i$" is checked by an ATP and is not valid. As a result, the predicate of (s1, i1) is strengthened by conjunction of the results of *wp* computation. We find that after 3 iterations, all the bisimulation relations reach their fixpoints and stop checking process with equivalent results returned.

5. Implementation and Experimental Results

Our equivalence checking algorithm has been implemented in C++ and all experiments have been conducted on a workstation with a 2.67 GHz Intel Core i5 CPU and 4 GB RAM. The ATP used is the Z3 theorem prover [22].

Some of the benchmarks are from Spark public available example suite, which are *sumto10*, *idct*, *findmin8*, and *minsort*. The rests are from high level synthesis benchmarks [23, 24], such as *gcd*, *barcode*, *lru* and *parker1986*. The benchmarks selected not only include control intensive designs such as *minsort* and *gcd*, but include data intensive designs, such as *idct*. For all the test cases, the structure-preserving transformations, such as DD/DU, MU/MD/, UM, Sp/RSp, and loop shifting, are applied using Spark synthesizer [19], and then non-structure-preserving transformations are applied manually according to PBS [5].

We compared our method with the method in [18] to show the ability of the proposed method. The experimental results are shown in Table 2. The first column is the benchmarks used in our experiments and the 2nd column shows the number of line in C language of each design. The 3rd column shows the number of the bisimulation relations found in each benchmark. The 4th column shows the number of ATP querying of our method. The 5th and 7th columns show the validation time cost of our method and the method in [18], respectively. The time consists of bisimulation relation inferring and checking costs. The 6th and 8th columns shows the ability of handling scheduling optimization techniques of our method and the methods proposed in [18], respectively. A "P" means a validation result is returned and an "F" means validation is failed with no result returned. In the 8th column, when applying PBS to merge paths, the method in [18] was not able to build bisimulation relations and would fail to check equivalence.

We also give the time costs of the method in [18] when no PBS is applied in the 9th column which also includes bisimulation relation inferring and checking costs. When comparing with the results shown in the 5th column. We find that although our method spent more time on inferring bisimulation relations to capture non-structure-preserving transformations than that of the method in [18], the total costs are almost the same as the method in [18] due to the heuristics in **SelectBR**. However, our method can deal with more scheduling optimization techniques than the [18]'s method.

The last column of Table 2 shows the validation result of each case, "Yes" means the transformation is equivalent, while "No" means not equivalent. The results confirm that our proposed method can correctly identify the equivalence transformation of Spark, and also uncover the bug revealed in [18].

Table 2 confirms that our proposed method can correctly address various structure-preserving and non-structure-preserving transformations in a unified framework.

6. Conclusions

In this paper, we present a novel approach for scheduling verification in HLS. The innovation in our work lies in two factors. First is applying translation validation techniques for equivalence checking of FSMDs. Second is except loop pipelining transformations, our method can deal with all the structure-preserving and non-structure-preserving scheduling optimization techniques. The experimental results on various scheduling transformations indicate the effectiveness of our method. In the near future, we intend to extend

Table 2. Experimental results

Benchmarks	C #Line	# BRs	# ATP Querying	Our Method		The Method in [18]		If Passed, Time of the Method in [18] (ms)	Equivalent
				Time(ms)	Passed	Time(ms)	Passed		
gcd	20	4	19	187.2	P	-	F	182.1	Yes
sumto10	23	3	22	233.9	P	-	F	227.5	Yes
idct	47	4	13	163.7	P	-	F	163.2	Yes
barcode	39	7	34	381.8	P	380.9	P	380.9	Yes
lru	43	7	17	179.1	P	-	F	170.4	Yes
parker1986	31	6	30	429.5	P	417.3	P	417.3	Yes
findmin8	44	9	29	354.3	P	-	F	341.6	Yes
minsort	15	5	33	647.6	P	-	F	644.7	No

our method not only for the remaining HLS phases: parsing, binding and code generation, but also for loop pipelining optimization.

7. References

[1] Daniel D. Gajski, Nikil D. Dutt, Allen C.-H. Wu, Steve Y.-L. Lin. 1992. *High-Level Synthesis: Introduction to Chip and System Design.* Kluwer Academic Publishers, Norwell, MA, USA.

[2] Charles Y. Hitchcock, III and Donald E. Thomas. 1983. A method of automatic data path synthesis. In *Proceedings of the 20th Design Automation Conference* (DAC '83). IEEE Press, Piscataway, NJ, USA, 484-489.

[3] P. G. Paulin, J. P. Knight.1989. Force-directed scheduling forthe behavioral synthesis of ASIC's. *IEEE Trans. on Computer-Aided Design of Integrated Circuits and Systems.* 8, 6 (Jun. 1989), 661-679. DOI=http://dx.doi.org/10.1109/43.31522.

[4] Minjoong Rim, Yaw Fann, Rajiv Jain. 1995. Global scheduling with code-motions for high-level synthesis applications. *IEEE Trans. Very Large Scale Integr. Syst.* 3, 3 (Sep. 1995), 379-392. DOI=http://dx.doi.org/10.1109/92.406996.

[5] R. Camposano. 1991. Path-based scheduling for synthesis. *IEEE Trans. on CAD of Integrated Circuits and Systems*. 10, 1(Jan. 1991), 85-93. DOI=http://dx.doi.org/10.1109/43.62794.

[6] D. P. Anderson J. Ainscough. 1994. The verification of scheduling algorithms. In *Proceedings of IEE Colloq. Structured Methods Hardware Syst. Des.*. 7/1-7/5.

[7] Tsung-Hsi Chiang, Lan-Rong Dung. 2007. Verification method of dataflow algorithms in high-level synthesis. *J. Syst. Softw.* 80, 8 (Aug. 2007), 1256-1270. DOI=http://dx.doi.org/10.1016/ j.jss.2006.12.547.

[8] Naren Narasimhan, Elena Teica, Rajesh Radhakrishnan, Sriram Govindarajan, Ranga Vemuri. 2001. Theorem Proving Guided Development of Formal Assertions in a Resource-Constrained Scheduler for High-Level Synthesis. *Form. Methods Syst. Des.* 19, 3 (Oct. 2001), 237-273. DOI=http://dx.doi.org/10.1023/ A:1011250531814.

[9] Sam Owre, John M. Rushby, Natarajan Shankar. 1992. PVS: A Prototype Verification System. In *Proceedings of the 11th International Conference on Automated Deduction: Automated Deduction* (CADE-11), Deepak Kapur (Ed.). Springer-Verlag, London, UK, 748-752. DOI=http://dx.doi.org/10.1007/3-540-55602-8_217.

[10] Rajesh Radhakrishnan, Elena Teica, Ranga Vemuri. 2001. Verification of Basic Block Schedules Using RTL Transformations. In *Proceedings of the 11th IFIP WG 10.5 Advanced Research Working Conference on Correct Hardware Design and Verification Methods* (CHARME '01), Tiziana Margaria, Thomas F. Melham (Eds.). Springer-Verlag, London, UK, 173-178. DOI=http://dx.doi.org/10.1007/3-540-44798-9_16.

[11] R.W. Floyd. 1967. Assigning meaning to programs. In *Proceedings of 19th Symp. Appl.Math.* J. T. Schwartz, Eds., American Mathematical Society. 19-32.

[12] Hans Eveking, Holger Hinrichsen, Gerd Ritter. 1999. Automatic verification of scheduling results in high-level synthesis. In *Proceedings of the conference on Design, automation and test in Europe* (DATE '99). ACM, New York, NY, USA, Article 12. DOI=http://doi.acm.org/ 10.1145/307418.307449.

[13] Nazanin Mansouri and Ranga Vemuri. 1998. A Methodology for Automated Verification of Synthesized RTL Designs and Its Integration with a High-Level Synthesis Tool. In *Proceedings of the Second International Conference on Formal Methods in Computer-Aided Design* (FMCAD '98), Ganesh Gopalakrishnan, Phillip J. Windley (Eds.). Springer-Verlag, London, UK, 204-221. DOI=http://dx.doi.org/10.1007/3-540-49519-3_15.

[14] Youngsik Kim, Shekhar Kopuri, and Nazanin Mansouri. 2004. Automated Formal Verification of Scheduling Process Using Finite State Machines with Datapath (FSMD). In *Proceedings of the 5th International Symposium on Quality Electronic Design* (ISQED '04). IEEE Computer Society, Washington, DC, USA, 110-115. DOI=http://dx.doi.org/10.1109/ISQED.2004.1283659.

[15] Youngsik Kim and Nazanin Mansouri. 2008. Automated formal verification of scheduling with speculative code motions. In *Proceedings of the 18th ACM Great Lakes symposium on VLSI* (GLSVLSI '08). ACM, New York, NY, USA, 95-100. DOI=http://doi.acm.org/10.1145/1366110.1366134.

[16] Chandan Karfa, D. Sarkar, C. Mandal, and P. Kumar. 2008. An Equivalence-Checking Method for Scheduling Verification in High-Level Synthesis. *Trans. Comp.-Aided Des. Integ. Cir. Sys.* 27, 3 (March 2008), 556-569. DOI=http://dx.doi.org/10.1109/ TCAD.2007.913390.

[17] Chi-Hui Lee, Che-Hua Shih, Juinn-Dar Huang, and Jing-Yang Jou. 2011. Equivalence checking of scheduling with speculative code transformations in high-level synthesis. In *Proceedings of the 16th Asia and South Pacific Design Automation Conference* (ASPDAC '11). IEEE Press, Piscataway, NJ, USA, 497-502. DOI=http://dx.doi.org/10.1109/ASPDAC.2011.5722241.

[18] Sudipta Kundu, Sorin Lerner, and Rajesh K. Gupta. 2010. Translation validation of high-level synthesis. *IEEE Trans. Comp.-Aided Des. Integ. Cir. Sys.* 29, 4 (April 2010), 566-579. DOI=http://dx.doi.org/10.1109/TCAD.2010.2042889.

[19] S. Gupta, N. Dutt, R. Gupta, A. Nicolau. 2003. Spark: A high-level synthesis framework for applying parallelizing compiler transformations. In *Proceedings of the 16th International Conference on VLSI Design* (VLSID 2003). IEEE Press, Piscataway, NJ, USA, 461-466. DOI=http://dx.doi.org/ 10.1109/ ICVD.2003.1183177.

[20] Tun Li, Yang Guo, Wanwei Liu, Chiyuan Ma. 2013. Efficient Translation Validation of High-Level Synthesis". Accepted to be appeared in proceedings of ISQED 2013.

[21] C. A. R. Hoare. 1969. An axiomatic basis for computer programming. *Commun. ACM* 12, 10 (Oct. 1969), 576-580. DOI=http://doi.acm.org/10.1145/363235.363259.

[22] Leonardo De Moura and Nikolaj Bjørner. 2008. Z3: an efficient SMT solver. In *Proceedings of the Theory and practice of software, 14th international conference on Tools and algorithms for the construction and analysis of systems* (TACAS'08/ETAPS'08), C. R. Ramakrishnan, Jakob Rehof (Eds.). Springer-Verlag, Berlin, Heidelberg, 337-340. DOI=http://dx.doi.org/10.1007/978-3-540-78800-3_24.

[23] Preeti R. Panda and Nikil D. Dutt. 1995. 1995 high level synthesis design repository. In *Proceedings of the 8th international symposium on System synthesis* (ISSS '95). ACM, New York, NY, USA, 170-174. DOI=http://doi.acm.org/ 10.1145/224486.224537.

[24] http://computing.ece.vt.edu/~mhsiao/hlsyn.html

LASER – Layout-aware Analog Synthesis Environment on Laker

Yu-Ching Liao Yen-Lung Chen Xian-Ting Cai Chien-Nan Jimmy Liu Tai-Chen Chen
Dept. of Electrical Engineering, National Central University, Jung-Li City, Taiwan, ROC
{ 100521024, 995401004, 100521026 }@cc.ncu.edu.tw; { jimmy, tcchen }@ee.ncu.edu.tw

ABSTRACT

In modern technology, layout effects have more and more impacts on circuit performance. However, most of the existing analog automation tools consider the circuit sizing and layout generation in two separate steps, which often result in time-consuming sizing-layout iterations. In this paper, a layout-aware analog synthesis tool is presented to generate the required designs from specifications to layout through a user-friendly GUI. In order to provide a strong link between sizing and layout steps, a parasitic-aware circuit sizing flow is proposed based on a flexible layout template to prevent the performance from failing to meet the specifications after layout. Routability-aware analog placement is then performed with a simple routing algorithm to generate the corresponding layout with minimized cost. As demonstrated in the experimental results, this analog synthesis tool is able to generate the required circuits in seconds with high quality layouts and effectively guarantees the post-layout performance with less over-design.

Categories and Subject Descriptors: B.7.2

[**Integrated Circuits**]: Design Aids – *layout, and placement and routing.*

Keywords: Analog, Synthesis, Parasitic-aware sizing, Routability-aware Placement.

1. INTRODUCTION

Typical analog design procedure often requires simultaneous optimization between different specifications. Manually designing the circuits to meet the requirements is often considered as a difficult task, which often takes a lot of time. While the device size is continuously shrinking in deep submicron process, more non-ideal effects also bring more challenges to sensitive analog circuits. Therefore, various design automation approaches have been proposed for analog circuits to solve this design bottleneck [1]. Besides the benefit of fast design procedure, good design automation algorithms also provide proper considerations between multiple objectives.

Analog design automation has two important steps: circuit sizing and layout generation. In the literature, the automatic circuit sizing approaches can be roughly classified into three categories. Knowledge-based optimization [2-3] generalizes designers' experience and obtains a design solution based on some pre-defined design flows and database. Although this method is quite fast, the optimality of the solution cannot be guaranteed. The general solution for the sizing problem is adopting circuit simulators in the sizing iterations until finding the relatively better design [4-5]. This approach can obtain more accurate results, but the required computation time is relatively longer among various kinds of approaches. Equation-based approaches use the mathematical solvers to solve the circuit equations and obtain the global optimal solutions, such as the geometric programming (GP) based approaches in [6-7]. These approaches are extremely fast, but the accuracy is often limited due to the simplified circuit equations.

After the device parameters are determined, the next design step is to generate the corresponding layout. In the literature, layout automation techniques for analog circuits are also a popular research direction [8-14]. In recent years, many researches [8-10] focus on studying the analog placement problem to consider the extra layout constraints such as matching, proximity, etc. The succeeding routing issue is often not considered due to the small number of nets in analog circuits. However, for sensitive analog circuits, improper routing paths may incur unexpected electrical effects. Unmatched routing paths may break the perfect matching in placement. Over-the-cell routing may incur extra parasitic capacitors and crosstalk that affect the circuit performance. In order to reduce the parasitic capacitance and crosstalk effect, the routing nets above the active area of the devices are often avoided in the routing step [11]. Previous works [8-10] do not consider the routing resource in the placement stage and will incur compact placements which are not practical in analog designs. Previous works [12-14] perform congestion-driven adjustments during placement. However, the estimated routing congestion may not faithfully represent the required routing resource. Besides, the adjustments after a placement is generated may deviate the placement.

In modern technology, layout effects have more and more impacts on circuit performance. Due to the parasitic effects, the performance between pre-layout simulation and post-layout simulation might be quite different, which may make the circuit performance fail to meet the specifications after layout. In such cases, designers have to redesign the circuits and layouts, which is also a time-consuming loop. In conventional design flow, those layout-induced parasitic effects can only be estimated after layout completion. Therefore, several sizing-layout iterations are still required in previous layout-aware synthesis approaches [15-18]. Due to the tedious post-layout simulations, the sizing-layout iterations greatly slow down the whole synthesis procedures.

In the literature, several CAD tools [19-21] have been developed to automate the design procedure of analog circuits and the corresponding layouts. However, most of these tools consider the circuit sizing and layout generation in two separate steps. Because the layout-induced parasitic effects are not considered during the sizing step, the performance degradation after layout may lead to several time-consuming sizing-layout iterations. Giving the objectives higher than the required specifications is a popular way for designers to solve this issue. However, how many margins should be preserved is often decided by experience, which may incur too much over-design. In order to solve these issues, an integrated analog synthesis tool, named as Layout-aware Analog Synthesis Environment on Laker (LASER), is presented in this paper. The key contributions are listed as follows.

GLSVLSI'13, May 2–3, 2013, Paris, France.

This work was supported in part by R.O.C. National Science Council under Grant NSC101-2628-E-008-003-MY3. Their supports are greatly appreciated.

- A layout-aware analog synthesis tool that integrates both circuit sizing and layout generation is presented in this paper. A user-friendly GUI is also provided to help users complete the design flow successfully and efficiently. An automatic layout script is then generated to implement the corresponding layout with Laker [25] in seconds, which is convenient for users to connect existing EDA flow.
- In order to provide a strong link between sizing and layout steps, a layout-aware automatic sizing flow for analog circuits is proposed in this work to consider the layout-induced parasitic effects based on a flexible layout template. Therefore, the cases that the performances fail to meet the specifications after layout can be avoided. Since the possible performance degradation has been considered, no preserved design margins and re-design cycles are required in the proposed sizing flow, which significantly reduces the design overhead.
- A routability-aware analog placement algorithm is also provided in this tool. Besides the default layout template, several tool-generated layout placement results are also provided for users to choose their own preferred layout style with minimum cost. Since proper routing spaces have been preserved during the placement stage, a simple routing algorithm is enough to generate the corresponding layout without over-cell nets.

The rest of this paper is organized as follows. A brief introduction about the design flow and environment of LASER is given in Section 2. In Section 3, the proposed layout-aware sizing flow is presented with the consideration of parasitic effects. The layout generation of LASER is described in Section 4. Section 5 demonstrates the experimental results and shows the generated layout. Finally, a conclusion is drawn in Section 6.

2. LAYOUT-AWARE SYNTHESIS ENVIRONMENT

LASER is a layout-aware synthesis tool for analog integrated circuits, which integrates both circuit sizing and layout generation. This tool offers a user-friendly GUI to help designers complete the design flow successfully and efficiently. At the end, it will automatically generate the corresponding layout on Laker. The design flow of LASER is shown in Fig. 1. Given the technology file and the available supply voltages of the target process, the required transistor information is automatically extracted at the first step. After that, the circuit structure is selected by users to begin the following design procedure. Then, the best bias voltages and bias currents that make the circuit satisfy all specifications with optimized cost are identified by a linear programming (LP) solver with the pre-built sizing tables and transistor parameters. Finally, given the layout design rules, the final circuit layout will be generated on Laker according to the layout template.

The GUI of the proposed tool is illustrated in Fig. 2. There are five steps in the left window to help users complete the design flow easily. In the middle window, there are two pictures showing the corresponding design information. Take the typical two-stage Operational Amplifier (OPA) as an example. The top picture is the schematic of this OPA, and the bottom one is the flexible layout template of this circuit. The right window shows the status reports and warning messages during the execution. The details of the five steps are introduced as follows.

1) *Environment Setup:* There is another window for users to setup the design environment including technology file, process file, circuit simulator and supply voltage.

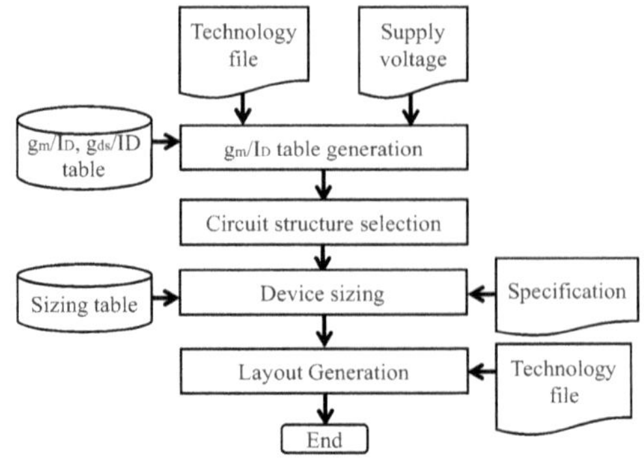

Figure 1. The flow chart of LASER

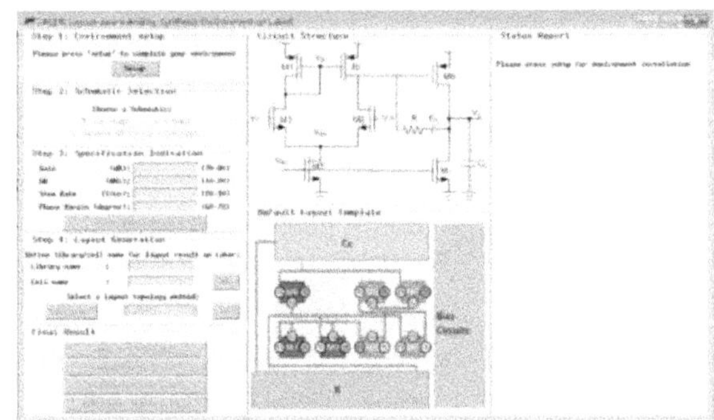

Figure 2. The Graphical User Interface (GUI) of LASER

If the process information has not been extracted, the simulator is used to extract the required information automatically. Otherwise, this extraction step will be ignored after the setup is completed.

2) *Schematic Selection*: Currently, four different OPA circuits are supported in LASER, which are A) Two-stage B) Folded C) Current mirror D) Telescopic, as shown in Fig. 3(a). Once the selection button is pressed, the corresponding schematic of the selected circuit is displayed as shown in Fig. 3(b).

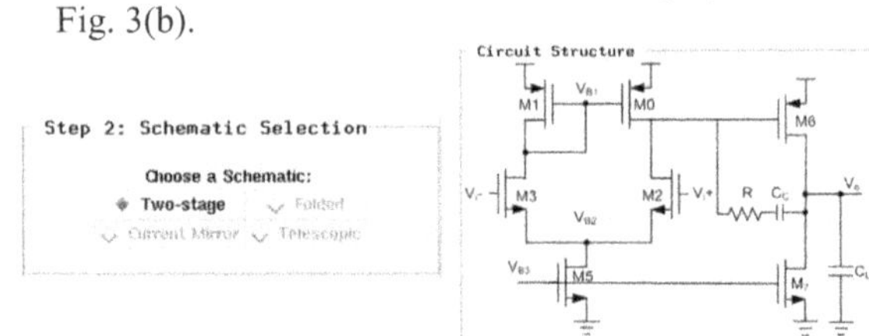

Figure 3. Step 2 (a) Schematic Selection (b) Circuit Structure

3) *Specification Indication*: It allows users to input the desired performance of the OPA circuit such as gain、bandwidth、slew rate and phase margin. Then, clicking the button below starts the device sizing procedure. The obtained device sizes are directly displayed on the circuit schematic as shown in Fig. 4(b) to provide more intuitive view of the results.

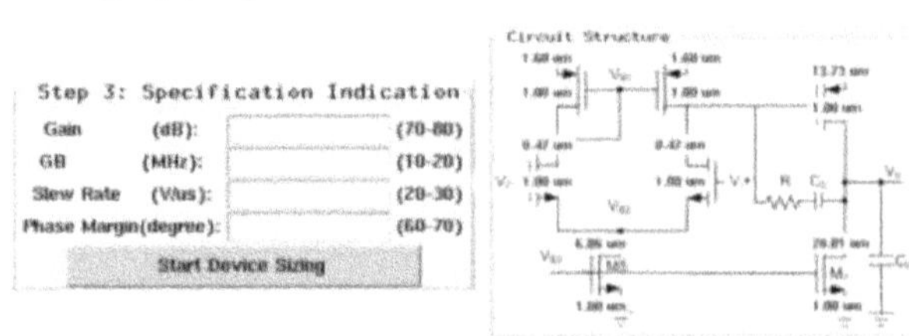

Figure 4. Step 3 (a) Specification Indication (b)Circuit Structure

4) *Layout Generation*: In this step, designers can input the desired library and cell names while drawing the layout on Laker. In addition, this tool provides several layout templates for users as illustrated in Fig.5(a) and Fig.5(b). Besides the default template, there are also several layout templates generated by the proposed placement optimization algorithm. After the layout template is determined, an automatic layout script is then generated to implement the corresponding layout with Laker. The generated layout will be verified by Caliber to check the design rules automatically.

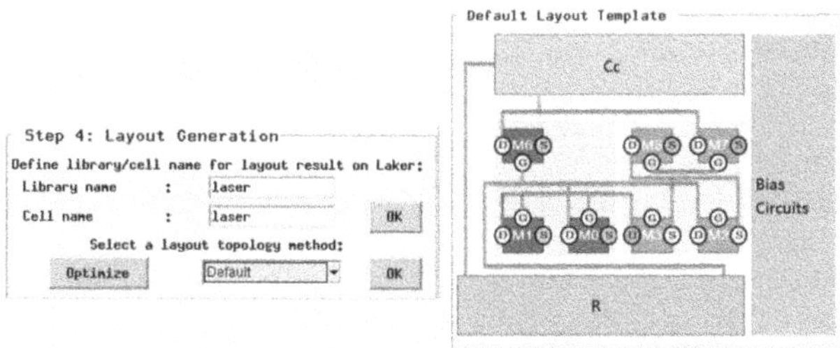

Figure 5. Step 4 (a) Layout Generation (b) Default Template

5) *Final Result*: After the above four steps, the device sizes in the circuit and the corresponding layout are all generated. The last step is presenting the final results for users. In LASER, the following information can be displayed on users' request, as shown in Fig.6. 1) Netlist File : It shows the SPICE netlist of the circuit with detail sizes. 2) Layout Script : It displays the final layout automatically on Laker. 3) Post-netlist File : It shows the netlist with the parasitic RC post-layout extracted by PEX. 4) Performance Report : It shows the circuit performance of sizing prediction, pre-layout simulation and post-layout simulation.

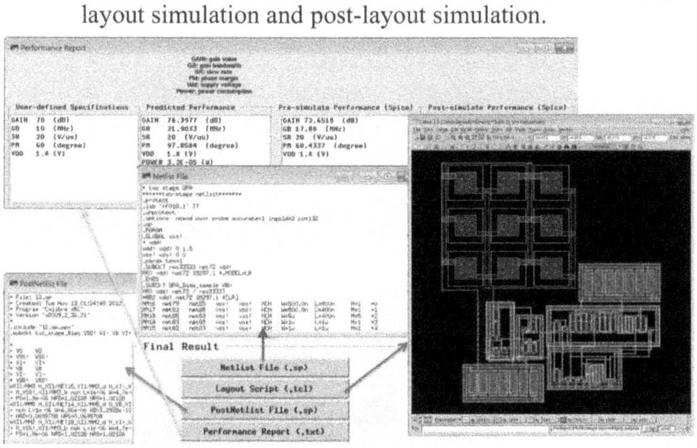

Figure 6. Step 5 of Final Result

3. LAYOUT-AWARE SIZING APPROACH

3.1 Parasites in the Layout Template

The parasitic effects can be classified into intra-device effects and inter-device effects. In post-layout simulation, extra parasitic capacitors will appear in each MOS as illustrated in Fig. 7. These parasitic capacitors will have significant impacts on both DC and AC performance, which are called intra-device parasitic effects in this paper. If the circuit performance is estimated based on the ideal device behavior that is characterized without parasitic capacitors, the performance after layout will be different to the prediction at the circuit sizing stage. Theoretically, those parasitic capacitors are proportional to the diffusion area (width * length) of the transistor. However, accurate parasitic values can only be extracted with detailed layout information. Therefore, in conventional design flow, the parasitic effects are analyzed in the post-layout simulation, which requires long simulation time.

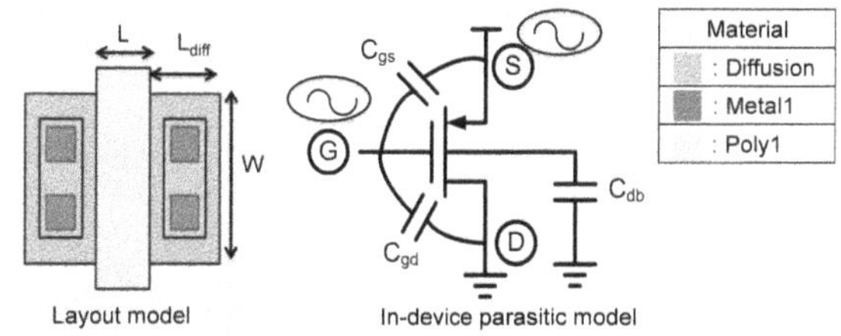

Figure 7. The intra-device parasitic model of a transistor

As to the inter-device parasitic effects, they are proportional to the length of interconnection wires. The wire length can be estimated from the given layout template. Then, the parasitic capacitors and resistors can be estimated from the wire length and technology parameters. By replacing each wire with the typical π-model, those parasitic RC can be added into the performance equations. As illustrated in Fig. 8, a simple single-input OPA is used as an example. If the input wire is replaced by its equivalent π-model, the small-signal model of this circuit will be modified as in the right of Fig. 8. If the performance equations are modified according to the new circuit model, the layout-induced parasitic effects can be considered in circuit sizing flow without tedious post-layout simulation.

$$C_{gs} = C_{gs1} + C_{wire} \tag{4}$$

$$C_{gd} = C_{gd1} + C_{wire} \tag{5}$$

$$R_L = r_{o1} \,||\, r_{o2} \tag{6}$$

$$C_L = C_{db1} + C_{db2} + C_{gd2} \tag{7}$$

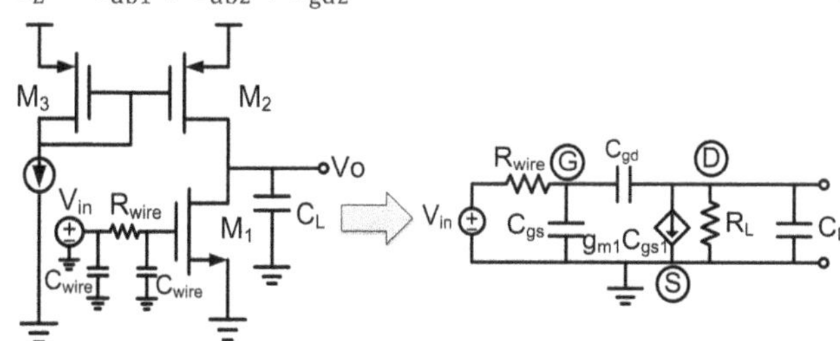

Figure 8. An example of the parasitic-aware performance model

In this paper, the inter-device parasitic effects are defined as the impacts from the parasitic devices on the interconnection wires between each device. Those parasitic devices are often modeled as resistors and capacitors, as illustrated in Fig. 9. Theoretically, the parasitic resistors and capacitors of a wire are proportional to its wire length, which is not available before layout completion. Therefore, in conventional design flow, these parasitic effects are analyzed in the post-layout simulation, too.

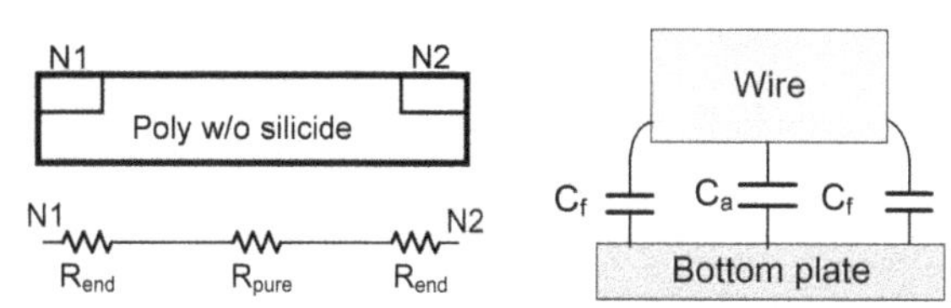

Figure 9. The parasitic resistance and capacitance of a wire

Most of the intra-device parasitic effects come from the internal capacitors of a transistor, such as C_{gs}, C_{gd}, etc. Theoretically, those internal capacitances of a transistor are proportional to its diffusion area, i.e. width(W) * length(L). The I_d current of a transistor is proportional to its W/L ratio. If L is the same value, both the internal capacitors and I_d current are proportional to W. In other words, the internal capacitors of a transistor are proportional to its I_d current. It implies that C/I_d is also a constant for the same L. Therefore, this work uses the similar table-based approach to construct the capacitor models for frequency-domain analysis, which are functions of (V_g, V_d, V_s), too.

Because the proposed circuit sizing flow, the layout template is assumed available at the circuit sizing stage. With the given layout template, it is possible to estimate the wire length between each device according to the device sizes and relative placement, as illustrated in Fig. 10. Since the wire length can be predicted, the parasitic resistors and capacitors can be predicted also. This enables the circuit sizing algorithm to estimate the parasitic effects on the circuit performance without post-layout simulation, which is one of the key ideas of the proposed approach in this paper.

$$L_H = 2 \cdot L_{dif} + L_m + \text{M\_M} + 2 \cdot \text{dif}_H \quad (1)$$

$$L_V = (W_{m5} + W_{m2}) + \text{M\_M} + 2 \cdot \text{dif}_V \quad (2)$$

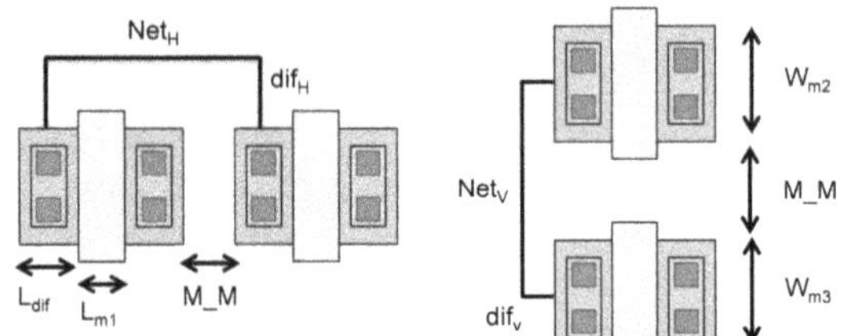

Figure 10. The estimated wire length of interconnection wires

3.2 Table-Assisted Circuit Sizing Flow

The proposed layout-aware circuit sizing flow is built on the bias-driven equation-based optimization approach [22]. The whole synthesis flow is shown in Fig. 11. In conventional equation-based synthesis approaches for analog circuits, significant prediction errors in some circuit performance often exist due to the errors of transistor parameters, such as g_m and g_{ds}. Instead of using complex models for g_m and g_{ds}, this work models the g_m/I_d and g_{ds}/I_d of transistors. Given the node voltages of a transistor, its g_m/I_d value is almost a constant, no matter what its size is. Then, the g_m and g_{ds} of each transistor and the corresponding circuit performance can be derived after multiplied by the bias current. By using this approach, the models for transistor parameters are a function of its node voltages (V_g, V_d, V_s) only, which can be captured easily with better accuracy. In order to avoid the fitting errors, those transistor parameters are modeled with lookup tables in this work instead of a single equation. Those accurate results significantly improve the accuracy of the predicted solution in the mathematical solver, as demonstrated in the experimental results.

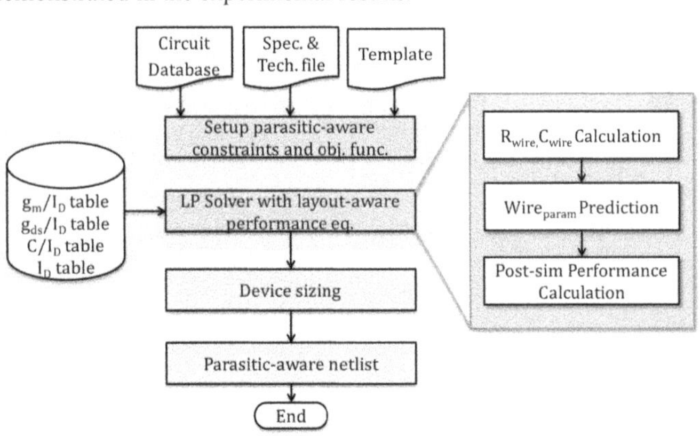

Figure 11. The proposed layout-aware synthesis flow

After the bias voltages and bias current for each transistor are identified by the mathematical solver, the size of each transistor can be determined also under a given technology. Actually, deriving the corresponding device size from the basic MOS current equation is not accurate due to the non-ideal effects. Therefore, this work adopts a table-based approach to find the corresponding transistor sizes. Under different combinations of node voltages, the corresponding I_d of a unit-size transistor is recorded in a pre-built table, which includes the non-ideal effects automatically. Then, multiplied by its bias current can obtain the approximate transistor size for this case. Using this approach, the device sizes can be obtained quickly and accurately. Since building the sizing table is a one-time effort that can be built in advance, this approach greatly improves the efficiency of circuit sizing and optimization.

4. LAYOUT GENERATION

4.1 MOS Generator

The MOS generator extracts the transistor information from the SPICE netlist generated in the circuit sizing step and transforms each transistor to the corresponding Tcl syntax defined in Laker. The layout design rules with multi-finger supports, as shown in Fig.12, are considered to generate DRC-clean layouts. Those layout design rules will be automatically extracted from the given technology file at the setup step.

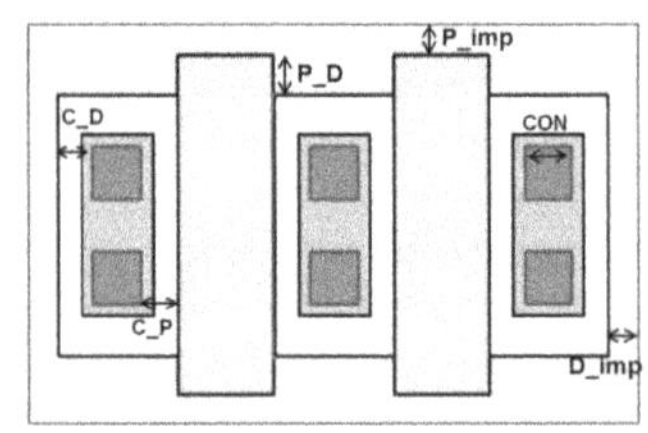

Design Rules	
C_D	Contact to Diffusion
C_P	Contact to Poly
P_D	Poly to Diffusion
CON	Contact Width
P_imp	Poly to Imp
D_imp	Diffusion to Imp

Figure 12. The design rules of MOS

4.2 Resistor and Capacitor

The whole resistor can be connected by a series of unit-sized resistors, as illustrated in Fig.13. Dummy resistors are often added at the two sides to avoid the unexpected etching that changes the desired resistance. In order to avoid the noise interference, a guard ring is added outside of the resistor. Fig.13(b) shows a real layout example of resistors on Laker. Similarly, the whole capacitor can be connected by a series of unit-sized capacitors. The common-centroid structure is often adopted for better element matching, as illustrated in Fig.14(a). Fig.14(b) shows an actual layout example of capacitors on Laker.

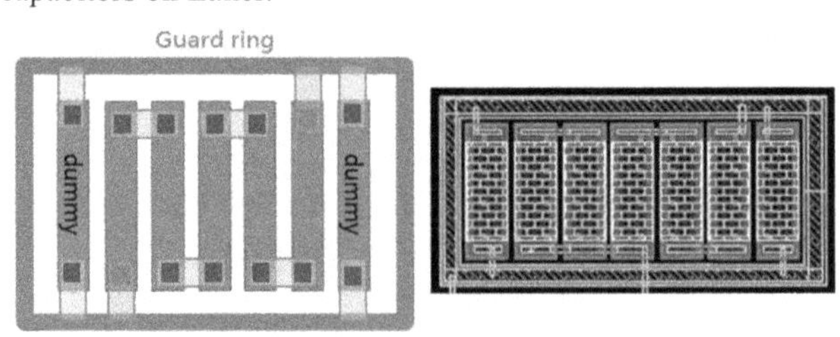

Figure 13. Resistor (a) template (b) layout

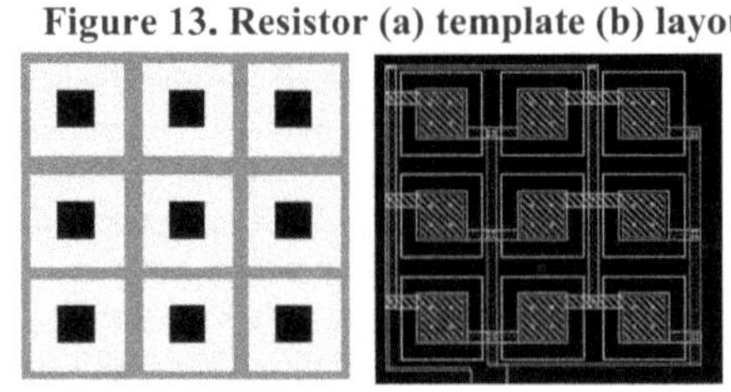

Figure 14. Capacitor (a) template (b) layout

4.3 Layout Template

In LASER, the layout placement is based on a flexible layout template. The key information in the layout template includes the relative placement of devices and the global routing directions. For users' convenience, a default template is provided for each circuit. According to the different device sizes and different design rules, the layout generation algorithm automatically adjusts the device positions on Laker to avoid DRC violations. The routing space is also preserved during the placement based on the global routing information in the layout template. Therefore, a simple routing algorithm is enough to connect all the devices. Finally, the generated layout is also verified with Caliber and post-layout simulations to ensure that the final performances reach the required specifications.

Another type of layout templates is the flexible template which is generated by the proposed routability-aware analog placer. Previous works did not consider the required routing resource in the placement stage [8-10] or used an unfaithful congestion to adjust the generated placement [12-14]. To accurately predict the required routing resource during placement and generate non-stochastic placements, a routability-aware analog placement flow is proposed. Based on the slicing-tree representation [23] and the deferred decision making technique [24], the proposed placement flow can reserve the required routing resource during placement and consider the topological constraints (symmetry, proximity, etc.). Besides, the proposed placement flow can handle soft blocks (variable width/height) and generate multiple solutions for designers. At the end, multiple solutions are provided for users to choose the preferred layout style with minimized cost. According to the selected layout template, the device sizes will be automatically adjusted to reflect the different parasitic effects. Therefore, the post-layout performance can still reach the desired specifications. Fig.15 shows an example of the optimal template for two-stage Operational Amplifier (OPA).

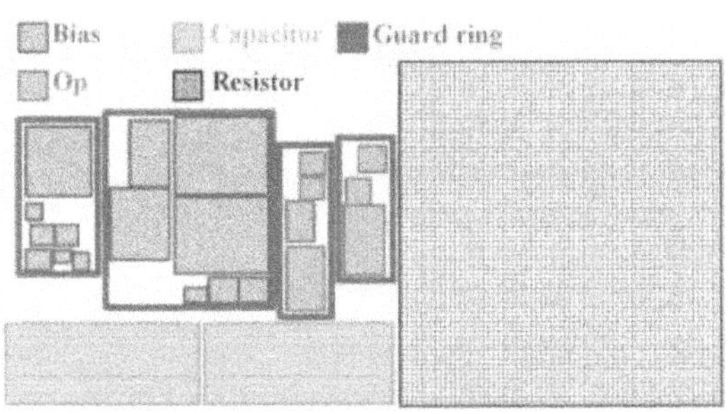

Figure 15. Optimal template

5. EXPERIMENTAL RESULTS

The LASER tool is programmed in several languages. The GUI is implemented with Tcl/TK as shown in Fig. 1. The circuit sizing flow is implemented with C++ and CPLEX [26], which is the adopted mathematical solver to solve those convex equations. The layout script generation is also implemented with C++ to generate the Tcl script for Laker [25]. The corresponding layout is then automatically generated in Laker environment. The circuit simulation results are obtained from HSPICE. All the results are measured on the same machine with Intel Xeon quad-core processor at 2.67GHz and 4GB memory.

The experiment demonstrates the results of a two-stage OPA circuit with TSMC 0.18um process. Its circuit structure and default layout template are shown in Fig. 16(a) and Fig. 16(b) respectively. The performance reports of different design approaches are shown

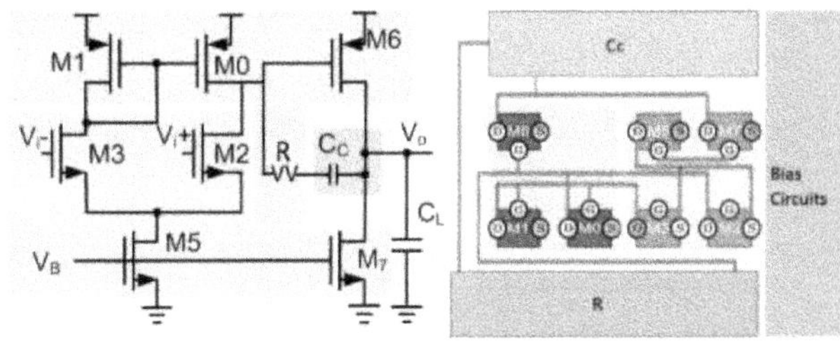

(a) (b)

Figure 16. A two-stage OPA (a) schematic (b) layout template

in Table I. In Table I, the column "Perf." shows the compared OPA performance, open-loop gain (Gain), gain bandwidth (GB), output slew rate (SR), and phase margin (PM), and the given specifications. The column "Conventional" reports the achieved performance without layout consideration. With the same circuit sizing approach, the column "5% margin" reports the results with 5% margin on each specification. The achieved circuit performance of the proposed layout-aware approach is reported in the column "Parasitic-aware". The power consumption, the total transistor area and the total computation time for the entire automation procedure are also reported in the bottom three rows of Table I.

According to the simulation results shown in Table I, the GB of conventional approach fails to meet the required performance in post-layout simulation (post-sim). Although the performance has reached the required specification in pre-layout simulation (pre-sim), the post-layout may still fail because the layout-induced parasitic effects are not considered. The straight-forward solution for this issue is giving higher specification to preserve some design margins for possible degradation after layout, as demonstrated in the column "5% margin". Although this approach reaches all specifications in post-layout simulation, the power and area overhead are quite large (63% in power, 8.3% in area). By using the proposed layout-aware sizing approach, the generated circuit also meets all specifications even in post-layout simulation. Only 0.5% overhead in power consumption is required. These results have demonstrated that the proposed approach is able to effectively avoid the time-consuming redesign cycles and unnecessary preserved design margin.

TABLE I. Final performance of OPA (0.18um)

Perf.	Conventional	5% margin	Parasitic-aware
Gain ≥ 75 (dB)	75.5^{+} 75.1^{++}	78.9^{+} 78.5^{++}	75.9^{+} 75.7^{++}
GB ≥ 20 (kHz)	20.1^{+} **19.2^{++}**	22.5^{+} 21.9^{++}	20.5_{+} 20.2_{++}
SR ≥ 2 (V/ms)	2.6^{+} 2.2^{++}	3.3^{+} 3.0^{++}	2.8^{+} 2.6^{++}
PM ≥ 60 (°)	68.5^{+} 67.9^{++}	69.8^{+} 67.9^{++}	68.7^{+} 68.3^{++}
Power(mW)	2.17	3.54	2.18
Area (μm^2)	4479	4850	**3344**
Time (sec.)	< 1	< 1	< 1

$^{+}$: pre-sim. $^{++}$: post-sim.

The final layouts generated by the proposed automation tool, including the capacitor, resistor, bias circuits and routing, are also shown in Fig. 17 and Fig. 18 for readers' reference. Fig. 17 shows the layout with the default template. Fig. 18 shows the layout with the optimized template. The empty space can be effectively reduced with the proposed layout optimization algorithm while preserving the analog layout constraints.

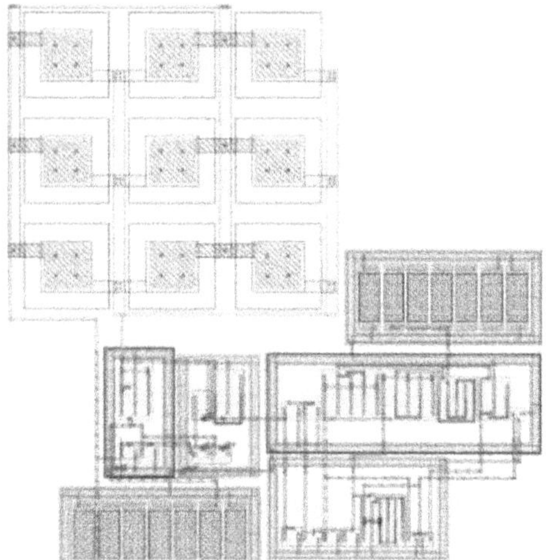

Figure 17. Layout with default template

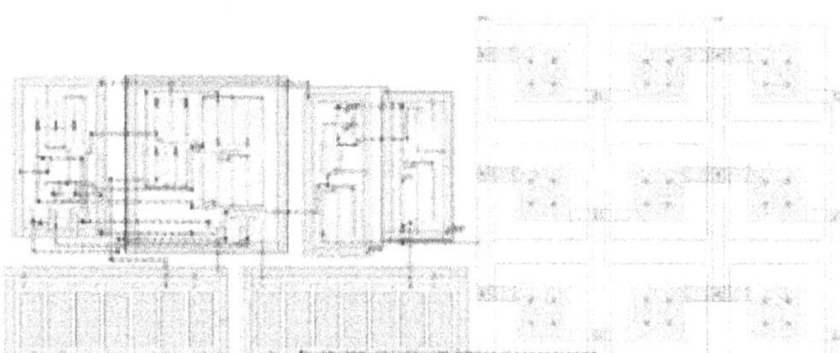

Figure 18. Layout with optimal template

6. CONCLUSIONS

In this paper, a layout-aware analog synthesis tool is presented to generate the required designs from specifications to layout through a user-friendly GUI. In order to prevent the circuit performance from failing to meet the specifications after layout, a layout-aware automatic sizing flow for analog circuits is proposed in this work to consider the layout-induced parasitic effects based on a flexible layout template. Besides the default layout template, a routability-aware analog placement algorithm is performed to generate several templates for users to choose their own preferred layout style with minimum cost. As shown in the experimental results, this analog synthesis tool efficiently generates the required circuits in seconds with high quality layouts. The failed performance in post-layout simulations can be avoided with only 2% overhead. These results have demonstrated that the proposed approach is able to effectively avoid the time-consuming redesign cycles and unnecessary preserved design margins.

7. REFERENCES

[1] R. A. Rutenbar, G. G. E. Gielen, and J. Roychowdhury, "Hierarchical Modeling, Optimization, and Synthesis for System-Level Analog and RF Designs," *in Proc. of the IEEE*, vol. 95.,no. 3, pp. 640-669, Mar. 2007.

[2] F. El-Turky and E.E. Perry, "BLADES: An Artificial Intelligence Approach to Analog Circuit Design", *IEEE Trans. on Computer-Aided Design*, vol. 8, no. 6, pp. 680-692, Jun. 1989.

[3] J. Mahattanakul and J. Chutichatuporn, "Design Procedure for Two-Stage CMOS Op-Amp with Flexible Noise-Power Balancing Scheme", *IEEE Trans. on Circuits and Systems-I: Regular Papers*, vol. 52, no. 8, pp. 1508-1514, Aug. 2005.

[4] R. Phelps, M. Krasnicki, R. A. Rutenbar, L.R. Carley and J.R. Hellums, "Anaconda: Simulation-Based Synthesis of Analog Circuits via Stochastic Pattern Search", *IEEE Trans. on Computer-Aided Design*, vol. 19, no. 6, pp. 703-717, Jun. 2000.

[5] C.-W. Lin, P.-D. Sue, Y.-T. Shyu and S.-J. Chang, "A Bias-Driven Approach for Automated Design of Operational Amplifiers", in *Proc. Int'l Symp. on VLSI Design, Automation, and Test*, pp. 119-121, April 2009.

[6] M. del M. Hershenson, S. P. Boyd, and T. H. Lee, "Optimal Design of a CMOS Op-Amp via Geometric Programming", *IEEE Trans. on Computer-Aided Design*, vol. 20, no. 1, pp. 1-21, Jan. 2001.

[7] W. Gao and R. Hornsey, "A Power Optimization Method for CMOS Op-Amps Using Sub-Space Based Geometric Programming", *in Proc. DATE*, pp. 508-513, Mar. 2010.

[8] P.-H. Lin, Y.-W. Chang and S.-C. Lin, "Analog Placement Based on Symmetry-Island Formulation," *IEEE Trans. on Computer-Aided Design*, Vol. 28, No. 6, pp. 791-804, Jun. 2009.

[9] Y.-P. Weng, H.-M. Chen, T.-C. Chen, P.-C. Pan, C.-H. Chen and W.-Z. Chen, "Fast Analog Layout Prototyping for Nanometer Design Migration," *in Proc. ICCAD*, pp. 517 – 522, Nov. 2011.

[10] P.-Y. Chou, H.-C. Ou, Y.-W. Chang, "Heterogeneous B*-trees for Analog Placement with Symmetry and Regularity Considerations," *in Proc. ICCAD*, pp. 512-516,. 2011.

[11] H.-C. Ou; H.-C. Chang Chien and Y.-W. Chang, "Non-uniform multilevel analog routing with matching constraints," *in Proc. DAC*, pp. 549–554, Jun. 2012.

[12] L. Xiao, E.F.Y. Young, X. He, K.P. Pun, "Practical Placement and Routing Techniques for Analog Circuit Designs," *in Proc. ICCAD*, pp. 675-679, 2010.

[13] M.-K. Hsu, S. Chou, T.-H. Lin and Y.-W. Chang, "Routability-driven analytical placement for mixed-size circuit designs," *in Proc. ICCAD*, pp. 517 – 522, 2011.

[14] C.-W. Lin, C.-C. Lu, J.-M. Lin, and S.-J. Chang, "Routability-driven Placement Algorithm for Analog Integrated Circuits," *in Proc. ISPD*, pp. 71 – 78, 2012.

[15] M. Ranjan, W. Verhaegen, A. Agarwal, H. Sampath, R. Vemuri, G. Gielen, "Fast, Layout-Inclusive Analog Circuit Synthesis using Pre-Compiled Parasitic-Aware Symbolic Performance Models", *in Proc. DATE*, pp. 604-609, Feb. 2004.

[16] G. Zhang, A. Dengi, R. A. Rohrer, R. A. Rutenbar, L. R. Carley, "A Synthesis Flow Toward Fast Parasitic Closure For Radio-Frequency Integrated Circuits," *in Proc. DAC*, pp. 155–158, July 2004.

[17] R. Castro-López, O. Guerra, E. Roca, and F. V. Fernández, "An Integrated Layout-Synthesis Approach for Analog ICs", *IEEE Trans. on Computer-Aided Design*, vol. 27, no. 7, pp. 1179-1189, Jun. 2008.

[18] H. Habal and H. Graeb, "Constraint-Based Layout-Driven Sizing of Analog Circuits", *IEEE Trans. on Computer-Aided Design*, vol. 30, no. 8, pp. 1089-1102, Jun. 2011.

[19] K. Lampaert, D. Garrod, R. Rutenbar, and L. Carley, "Koan/Anagram II: New Tools for Device-Level Analog Placement and Routing," *IEEE J. Solid-State Circuits*, vol. 26, no. 3, pp. 330-342, Mar. 1991.

[20] L. Zhang, U. Kleine and Y. Jiang, "An automated design tool for analog layouts", *IEEE Trans. on VLSI System*, vol. 14, no. 8, pp. 881-894, Aug. 2006.

[21] E. Yilmaz, Günhan Dündar "Analog Layout Generator for CMOS Circuits", *IEEE Trans. on Computer-Aided Design*, vol. 28, no. 1, pp. 32-45, Jan. 2009.

[22] Y.-F. Cheng, L.-Y. Chan, Y.-L. Chen, Y.-C. Liao, and C.-N. Jimmy Liu, "A Bias-Driven Approach to Improve the Efficiency of Automatic Design Optimization for CMOS OP-Amps", in Proc. Asia Symposium on Quality Electronic Design, pp. 59-63, 2012.

[23] D. F. Wong and C. L. Liu, "A New Algorithm for Floorplan Design," *in Proc. DAC*, pp. 101-107, 1986.

[24] J. Z. Yan and Chris Chu, "DeFer: Deferred Decision Making Enabled Fixed-Outline Floorplanning Algorithm," *IEEE Trans. on Computer-Aided Design*, vol. 29, no. 3, pp. 367-381, Mar. 2010.

[25] Laker^{TM} from Synopsys, http://www.springsoft.com

[26] ILOG CPLEX^{TM} from IBM, http://www.ilog.com/products/cplex/

Hardware Acceleration of Retinal Blood Vasculature Segmentation

Dimitris Koukounis, Christos Ttofis, Theocharis Theocharides
KIOS Research Center
ECE Department, University of Cyprus
123 Kyreneias Ave, Nicosia, Cyprus
{koukounis.dimitris, ttofis.christos, ttheocharides}@ucy.ac.cy

ABSTRACT

Retinal vessel tree extraction is a complex and computationally intensive task used in several medical and biometric applications. The emergence of portable biometric authentication applications, as well as on-site biomedical diagnostics, raises the need for hardware-accelerated, power-efficient architectures that can satisfy the performance and accuracy requirements of retinal vessel tree extraction. As such, this paper presents a VLSI implementation of a retina vessel segmentation system, in an attempt to illustrate the advantages and performance benefits that result from a dedicated VLSI solution. The proposed design implements an unsupervised, vessel segmentation algorithm, which utilizes match filtering with signed integers to enhance the difference between the blood vessels and the rest of the retina. The design simplifies the process of obtaining a binary map of the vessel tree by using parallel processing and efficient resource sharing, thus offering real-time performance. FPGA-based simulation results indicate significant performance improvements (up to 90x) when compared to existing hardware and software implementations.

Categories and Subject Descriptors

B.7.1 [Integrated Circuits]: Types and Design Styles – Algorithms implemented in hardware.

General Terms

Algorithms, Performance, Design.

Keywords

Retinal Vessel Segmentation, Portable Biometrics, On-site medical diagnostics, Reconfigurable Parallel Architectures, Hardware Acceleration.

1. INTRODUCTION

Separating the blood vessel tree from the retina, is an important and daunting process in biometric and biomedical applications. As such, several algorithms have been implemented, mostly on software, that perform retinal blood vasculature segmentation, and most of them target medical applications where speed is extremely important [2]. These algorithms are typically used by physicians to diagnose and investigate different diseases that cause changes to the blood vasculature. Many of those algorithms manage to provide satisfactory results in terms of accuracy. However, the use of the retina in identification or authentication systems traditionally was rarely used due to the high cost of the image capturing equipment, but most importantly due to the complex computational demands [1].

Emerging research suggests hardware acceleration [3-5] that aims to speed up the processing time of vessel segmentation algorithms, thus realizing their use in biometric systems. However, to take advantage of the emerging need for portable biometric authentication devices, as well as on-site diagnostic requirements, further performance improvement is still necessary, with the power consumption an additional significant constraint. Portable biometric authentication systems and on-site medical diagnostic equipment require high performance that achieves real-time response, low resource requirements and an acceptable accuracy able to authenticate/identify an individual or detect accurately a disease. Existing hardware attempts have various insufficiencies in regard to the processing time and their scalability to high-resolution images, thus have partially satisfied these requirements so far.

This paper therefore presents a VLSI implementation of an unsupervised vessel segmentation algorithm. The proposed hardware architecture adopts the main algorithmic idea from [12] that was implemented in software, but integrates hardware optimization techniques in an attempt to make the algorithm hardware-friendly while achieving high performance and accuracy. A matched filtering technique is used to augment the vessels from retinal images, which helps the extraction of the binary map of the vessel tree. The binary image can be used at a later stage in a segmentation-driven registration system in order to make the registration process more accurate. The paper presents several architectural optimizations stemming from the targeted hardware architecture, and thus is able to achieve significant performance increase (~90x) when compared to other hardware and software implementations, with acceptable accuracy (within ~4% of the highest reported accuracy achieved with hardware acceleration). The proposed architecture was implemented and evaluated on a Spartan 6 FPGA, in order to derive synthesis and performance evaluation results, yielding *0.03139s* per *768x584* pixels image.

The paper is organized as follows. Section II presents background information and describes the algorithm used to segment the blood vessel tree. Section III summarizes and reviews the related work. Then the proposed hardware architecture is presented in Section IV and Section V presents the simulation results. Finally, this paper concludes in Section VI where conclusions and future directions are presented.

2. Background

2.1 Vessel Segmentation

A large number of algorithms relating to the segmentation of the retinal blood vessels have been published so far, and can be classified into two broad categories: supervised and unsupervised [2]. Furthermore, [2] classifies the blood vessels segmentation algorithms in seven main categories based on the methodology followed: (1) pattern recognition techniques, (2) matched filtering (MF), (3) vessel tracking/tracing, (4) mathematical morphology, (5) multiscale approaches, (6) model based approaches and (7) parallel/hardware based approaches.

Among the aforementioned categories, matched filtering approaches give the more accurate results [7], thus the proposed architecture is inspired by a matched filter approach, and specifically, the algorithm presented in [12]. This algorithm integrates matched filtering and a threshold probing technique to achieve the desirable results. As proposed in [8], which presented the design of a hardware image processing system based on matched filtering, this approach has another significant advantage; the filter is implemented using integers, rather than floating point values, making the algorithm hardware friendly.

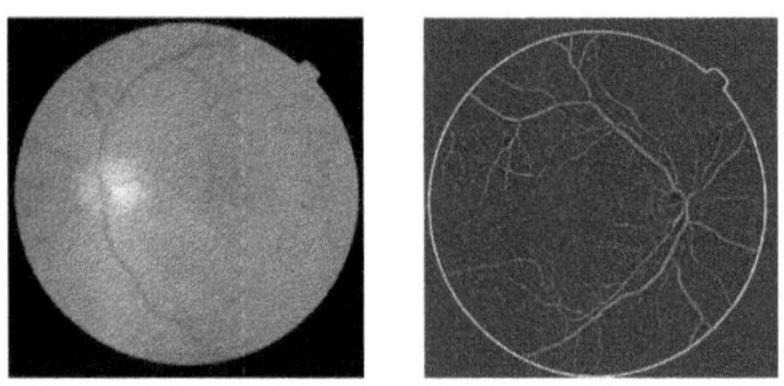

Fig. 1.(a) Retinal Image. (b) Matched Filter Response.

2.2 Matched filtering-based algorithm

The matched filter describes the expected appearance of the blood vessels in retinal images. The design of the matched filter in [8] is based on three main characteristics of the blood vessels: the first characteristic is that the vessels have a small curvature so they can be approximated by piecewise linear segments, the second characteristic is that the vessels have a lower reflectance compared to the rest of the retinal and as a result vessels appear darker in comparison with the background, and lastly, the third feature of the vessels is that their width lies between 2-10 pixels.

Utilizing these characteristics, [8] proposes a Gaussian function as a model for a blood vessel profile. In order to extend the model to two dimensions (X and Y), it is assumed that a vessel has fixed width and direction, for a short length. The matched filter is then rotated from 0° to 180° with a 15° step. The twelve resulting filter kernels are convolved with the image. The matched filter response (MFR) is the highest scoring value at each pixel. The MFR of a retinal image can be seen in Figure 1b.

Once the convolution with the matched filter finishes, the vessels are well distinguished from the rest of the retina. The MFR is thresholded by using a threshold value calculated automatically through iterative thresholding, and selected based on the number of pixels that have the most common pixel value on the image.

The next step involves threshold probing, where the segments created by the thresholding are used to locate a set of starting points to initialize the probe queue. Probing is iterative and is used to determine the appropriate threshold for the area being probed.

3. Related Work

The majority of the algorithms presented in literature, have been implemented in software running on general-purpose processors or clusters. An ad-hoc parallel implementation for the segmentation of high-resolution images is presented in [6]. In this approach, the image is divided into sub-images that have overlapping regions and each sub-image is distributed across computers for feature extraction and region growing; the segmentation results are then combined. The proposed algorithm processes 7 megapixel images in less than two minutes, facing memory shortage issues stemming from the use of such high-definition images. A similar parallel approach to segment high-resolution images is proposed in [10]. This approach reports a processing time of 3.54s for segmenting a 10-megapixel image, at 92.5% accuracy (the accuracy is measured by the ratio of the total number of correctly classified pixels to the total number of pixels in the original image that actually belong to the vasculature). This approach spends 20% of the total execution time for communication and I/O purposes. In addition, [10] indicates that the use of multiprocessing computing platforms can enable the segmentation of high definition images in reasonable time. However, while easy to implement and extremely flexible, these approaches require state-of-the-art and expensive hardware platforms to execute, and consume excessive amount of energy. Therefore, the use of such systems can only be limited to certain medical applications, where power and cost are not an issue.

A similar approach to the one described in this paper is shown in [21], and segments veins from medical video processing. It reports a performance of 215 frames per second (0.005s per image), however the performance value stated in [21] is theoretical and not verified experimentally. It relies only on the listed FPGA operating frequency, without considering issues that may affect performance such as I/O handling and memory accesses.

The emergence of portable biometrics and on-site medical diagnosis, raised the need for dedicated hardware approaches, which have emerged recently in an attempt to overcome the constraints found in software-based solutions [2]. In [5], an FPGA-based implementation of vessel extraction is presented. The specific algorithm implements a single-instruction multiple data (SIMD) architecture on a Spartan-3 FPGA. It requires a total of 1.4s to extract the vessels from a 768x584 retinal image. Alonso-Montes et al. [4] present another hardware approach where the vessels are segmented by using local dynamic convolutions and morphological operations together with arithmetic and logical operations. These are implemented and tested in a fine-grained SIMD processor array. [4] reports an average accuracy of 91.8% and a processing time of 0.19s per image.

While the hardware implementations in [3-5] achieve significant improvements in comparison with the software approaches, they partially satisfy the high performance requirement of portable systems that impose a satisfactory accuracy and most importantly vessel extraction in real-time response. Furthermore, most of the approaches follow a straight algorithm-to-hardware mapping, without any algorithmic optimizations or considerations for hardware-friendly design optimization techniques, especially focused on the reconfigurable fabric. As the segmentation process is the most computationally intensive part of the entire authentication or diagnosis procedure that involves retina examination, further performance improvement is necessary. As such, the work presented in this paper targets a hardware-based

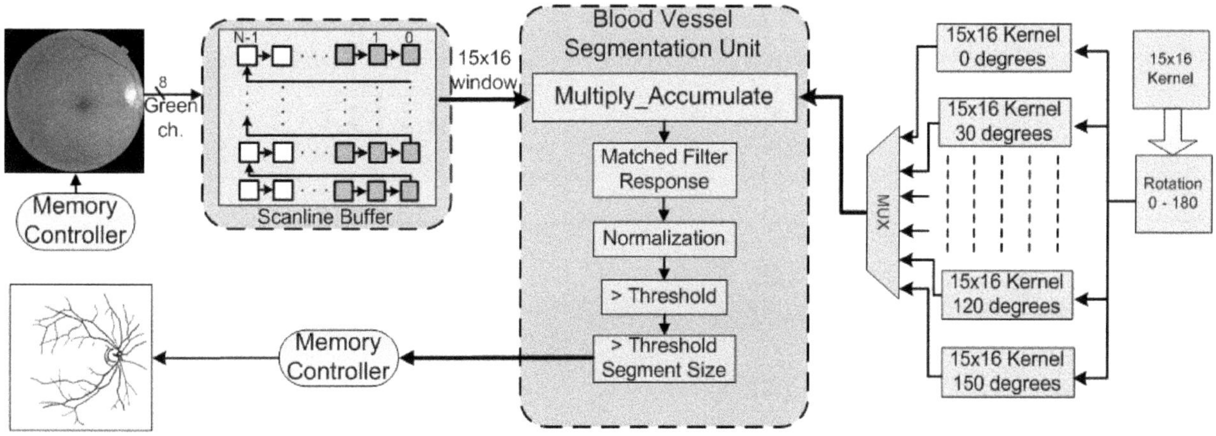

Fig. 2. Proposed Architecture Overview

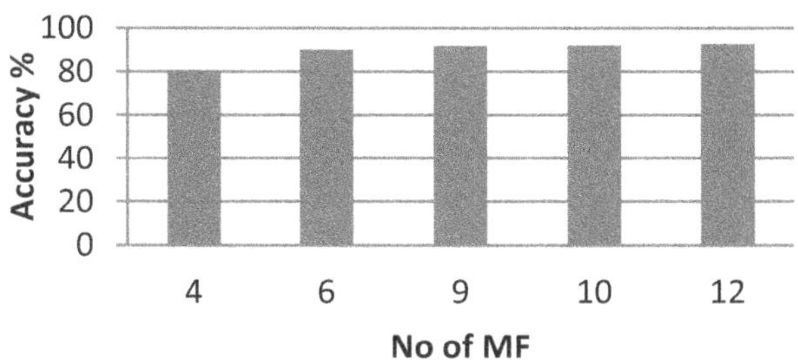

Fig. 4. Accuracy vs Number of Matched Filters

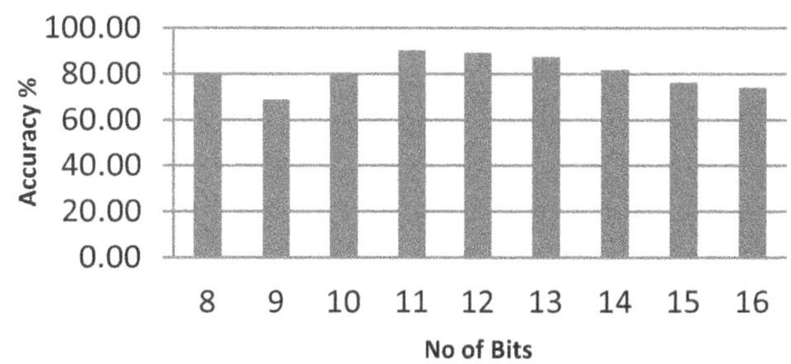

Fig. 6. Accuracy vs Number of Bits per pixel

vessel segmentation approach that manages to achieve extremely high performance by reducing certain computationally intensive steps, sacrificing minimal (but acceptable) accuracy for performance speedup and power reduction. The hardware-targeted optimizations enable easy integration of the architecture in existing portable biomedical and biometric systems.

4. Proposed Hardware Architecture

The proposed architecture consists of three major hardware units; the I/O unit, the Match Filter unit and the Threshold unit. These units are pipelined, thus are able to operate concurrently, and the interconnection between them is accomplished using on-chip buffers. Particularly, the system uses one scanline buffer to temporarily store the image data, and six smaller on-chip buffers to store the rotated matched filters. The complete system architecture and the communication flow between the units is shown in Fig. 2. It is assumed that input images are captured by a specialized retina camera, and are loaded through an external memory controller on the targeted hardware in a raster scan format.

4.1 I/O Unit

The memory controller communicates with on-chip Block RAM memory and fetches the green channel values (8 bits) to the scanline buffer. The red and blue channels are not used as the green channel utilizes a better difference between the vessels and the retinal background [15]. The use of the scanline buffer provides manageable parallelism in order to continuously feed the processing units with the data needed without any delays. It allows to sequentially read the image from the BRAM, and uses it in parallel in order to exploit the available parallelism on the hardware platform. The only introduced delay by using this mechanism is an initial delay until the first N rows of the image are loaded (where N is the width of the kernel window). Subsequently, this mechanism can provide data needed by the match filter unit every cycle.

4.2 Match Filter Unit (MFU)

In contrast to the original software-based algorithm in [12], the number of the matched filters were reduced to 6 by increasing the rotation step from 15° to 30°, in an attempt to reduce the computational complexity and memory requirements. Simulations indicate a less than 3% reduction in the accuracy when reducing the number of rotations. Figure 4 illustrates how the number of rotations used are affecting the accuracy of this vessel segmentation approach. When the rotation step is increased to 45° (4 matched filters) a significant reduction in accuracy is observed(less than 80%). At 30° the reduction in the accuracy is not significant (~3%) and the resulting accuracy of 90% is more than enough for the aimed applications. In addition, if more matched filters were used the hardware area overhead would increased significantly and thus increase the power consumption. The MFU performs the convolution between the image and the six resulting kernels (Fig.4a). Fig. 3 shows the pipeline stages followed during the convolution process. The first N lines of the image are being loaded to the scanline buffer, while the matched

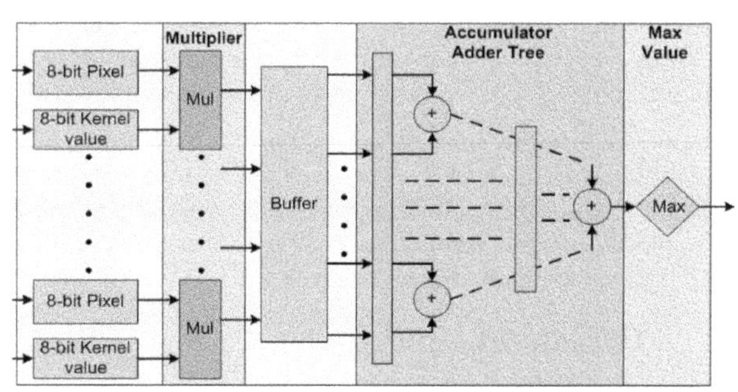

Fig. 3. Pipeline Stages followed in Match Filter Unit

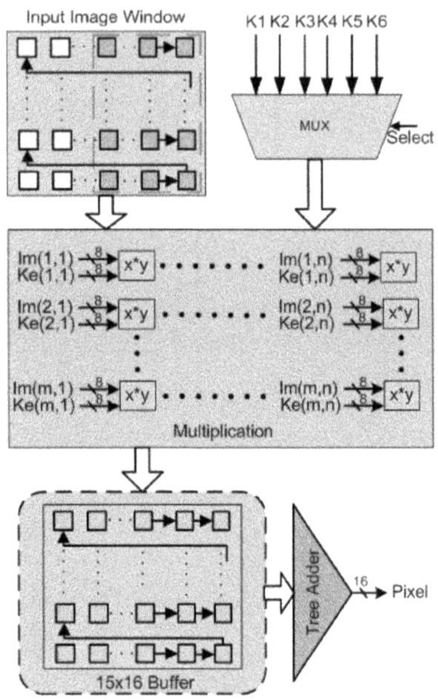

a.

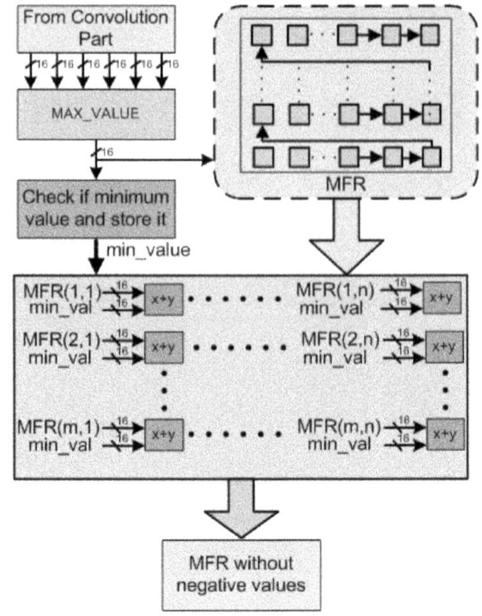

b.

Fig. 5. (a) Architecture of the Convolution Unit. (b) Architecture of the scaling done between the Convolution and Threshold Unit

filter is rotated and stored in six different buffers.

As soon as the scanline buffer fills up, the multiplexer selects the first kernel for pixel-wise multiplication. The number of parallel multiplications equals to the number of elements of the scanline buffer output window. Multiplication results are stored in a 16-bit buffer in order to have simultaneous access to all the elements, which are then all forwarded to the accumulation stage in the following clock cycle. At this point, the multiplexer selects the next kernel, and the operation iterates. Fig. 4a ("Tree Adder") shows the accumulation as the last step of the Convolution Unit, implemented as a tree adder, which enables a cost-effective implementation of the accumulator. The hardware layout resulting from this type of addition results in the highest performance and frequency achievable.

The resulting accumulated 16-bit value is stored in a register and every time a new value is computed, a comparison between the previous value and the new value is done. Six values are compared, one for each kernel, and the maximum value remains as the final pixel value. The final pixel values are stored in another scanline buffer called the "Matched Filter Response". The MFR scanline buffer can be seen in the blood vessels segmentation unit, below the multiply-accumulate unit in Fig. 2.

This process is repeated six times for every image window that the starting scan line buffer fetches, in order to convolve the image window with the six kernels (Fig. 5a). The pipelined architecture of the convolution unit allows it to sustain a valid pixel value every six clock cycles (given that convolution is done for all six kernels). This process is done for all windows in the image.

The convolution unit can be easily modified to process more than one kernel in a single cycle, thus computing a valid pixel in less than 6 clock cycles, depending on the amount of parallelism desired, and the available hardware resources.

4.3 Threshold Unit

Prior to thresholding, it is necessary to normalize the accumulation result of the pixels computed at the previous stage in order to simplify and reduce storage requirements. The authors in [12] normalize the resulting value by dividing each pixel with a constant value. To implement a hardware friendly normalization, we used only the 11 most significant bits of each pixel value (Fig. 2 – Normalization Stage). Figure 6 displays the accuracy against the number of bits per pixel used. The accuracy reaches 90,07% which the highest achievable accuracy of the presented approach at 11 bits per pixel. At a lower number of bits, the accuracy is reduced considerably since important features of the vessels are ignored. As the number of bits is increased, the accuracy is reduced, as a result of the unwanted details included at such high numbers of bits per pixel. Moreover, the MFR consists of both positive and negative integer values; hence, all values are subsequently positively biased by the *absolute* value of the *minimum* pixel value, to create only positive integers and reduce the complexity of the threshold operation. This is described in Fig. 5b.

The largest number from all the resulting pixels that have similar value (i.e. all characterize similar intensity in the image, which is usually the background and not the vasculature) is then used to compute the threshold value. As said, this value should represent the background of the retinal image. As shown in Fig. 1b, the background of the MFR is well distinguished from the vessels. The thresholding operation is done in parallel. All pixels above the threshold are replaced with '1' (white) and the rest of the pixels are replaced with '0' (black).

5. Experimental Evaluation and Results

The proposed retinal blood vessel architecture was implemented on a Spartan 6 XC65LX150T FPGA, and was evaluated using images from publicly available databases such as the STARE database [12], the DRIVE database [11] and the MESSIDOR database [13]. The input retinal images were loaded into the BlockRAM of the FPGA and used as input to the system shown in Figure 2. The implemented system was evaluated subsequently in terms of performance, power consumption, accuracy and hardware requirements. Results of the obtained segmented images are shown in Figure 7, along with the corresponding software implementation results.

TABLE III. Processing speed and Accuracy Comparison for Various Systems

	Database	Accuracy	Frequency (MHz)	Execution time (s)	Platform	Algorithm
Chaudhuri et al [9]	N/A	0.8850	N/A	2,0000	Special Image processing board	MFR
Hoover et al [7]	STARE	0.9267	N/A	3,0000	Sun SPARCstation 20	MFR+Threshold Probing
Alonso Montes et al [4]	DRIVE	0.9185	N/A	0,1925	Cellular Neural Networks	Pixel Level Snakes
Cinsidiki and Aydun [8]	DRIVE	0.9293	N/A	3,0000	Cluster Computer	MFR+ANT algorithm
A. Nieto et al [5]	N/A	0,9100	53	1,4000	FPGA	Active Contour
Proposed Architecture	DRIVE	0,9007	100	0, 0313	FPGA	MFR+Thresholding
	STARE	0,9019	100	0,0318	FPGA	MFR+Thresholding

TABLE I. Image size and Execution time

Image Size	Number of Windows	Total Execution Time (s)
768x584	448512	0,03139
700x605	455000	0,03185

TABLE II. Complete System Hardware Overheads

Platform	Slice LUTs (92152)	Slice Registers (183304)	DSP48A 1s (180)	BRAMs (268)	Freq. (MHz)
Spartan 6 XC65LX150T	66%	14%	100%	55%	100

5.1 Performance

The processing speed of the proposed hardware implementation is measured by the time (seconds) needed to extract the blood vessel tree of a retinal image. We used 20 images from the DRIVE [11] and STARE [12] database to compute an average processing time for the proposed architecture and the original software-based algorithm. Experimental results indicate that the proposed system offers a total performance acceleration of up to 90 times relative to the corresponding software implementation of the algorithm.

Table I illustrates how the processing time of the proposed architecture is affected by the image size. A promising observation is that the processing time remains relatively small for a small image size (of course this is expected), but increases linearly as the image size increases, which is attributed to the I/O handling, since once the original set of pixels are loaded, processing is then dependent clearly on the amount of available FPGA resources. Given that the overall input image size plays a linear role in the initial scanline buffer filling up, the delay is then directly proportional to the overall number of input pixels per line, which explains the linear delay increase. This is attributed to the fact that the number of the windows being processed is equal to the number of pixels in the image (see Table I).

Table III presents a comparison between existing implementations and the proposed architecture, showing the algorithm adopted in each work as well as the implementation platform. The proposed system is faster than all implementations, as it needs only 0.03139s to extract the vessels from an image size of 768x584 while the processing time of the majority of the works listed in the table is comparatively lower (comparisons are based on the same image databases that were used in existing implementations).

5.2 Accuracy

The accuracy of the system is defined as the ratio of the total number of correctly classified pixels (sum of true positives and true negatives) to the number of pixels in the image field of view. We compare the accuracy of the proposed architecture with the software-base implementation of the algorithm. The original algorithm has an accuracy of 92.67%, while the accuracy of the proposed architecture is 90.07%. This accounts for a quality drop of about 2.5%, and is attributed to the hardware optimizations (kernel reduction and thresholding). While this might be a considerable drop for a high-end medical system, the accuracy drop-off is not as important in portable biometrics and biomedical diagnostics; furthermore, other algorithms can be used instead of thresholding in order to segment the blood vessels which can increase the accuracy. This, however, is left as future work. It is worth noting that only the work in [5] has been implemented on an FPGA platform, has an accuracy of 91%, which is roughly the same as our reported accuracy but with a total speed up in performance of up to 50x.

5.3 Area and Power Overheads

The proposed FPGA-based vessel segmentation architecture yields promising processing speed. To determine its requirements in terms of the targeted FPGA, we implemented the experimental architecture on a Spartan 6 XC65LX150T FPGA. Table II gives synthesis results for the entire system implemented on the FPGA. The table lists area results for slice LUTs, slice registers and DSP components in order to give a complete picture of the required hardware overheads associated with the system. It can be observed that the entire system utilizes 66% of the FPGA LUTs and 14% of the FPGA slice registers. The slice LUTs are dominated by the convolution unit, which consumes 60%, while the slice registers are dominated by the scanline buffer. Additionally, the system takes full advantage of the DSPs available on the FPGA platform to perform most of the multiplication operations of the convolution unit. We used the Xilinx X-Power Analyzer tool and the image data as input to obtain the average dynamic consumption figures. The total power needed for one frame is just 2.146W, making the design suitable for embedded portable applications [19,20]. We anticipate that an ASIC implementation will further increase the performance while reducing the power as well.

Conclusively, synthesis and performance results indicate that vessel segmentation from retinal images can be effectively implemented on reconfigurable hardware platforms, with high potential for biometric applications that require real-time processing speed. The proposed system consumes a relatively reasonable number of resources when implemented on a typical low-end FPGA such as the Spartan-6, leaving room for the host application, let that be biometric or biomedical, to expand the

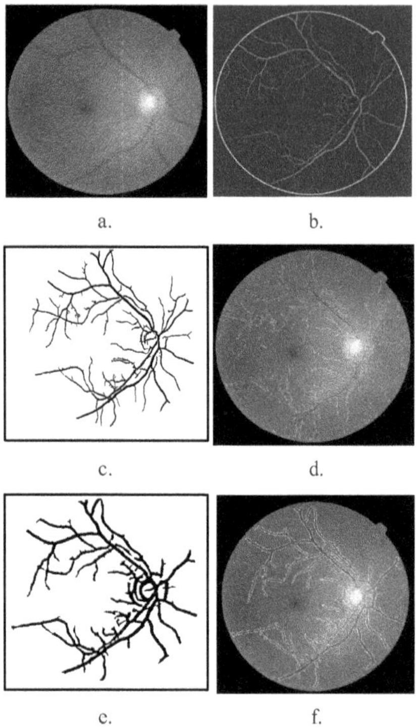

Fig.7. (a) Original retinal image used as input to the system. (b) Matched Filter Response. (c) B/W image with vessels segmented from the software version of the algorithm. (d) Image outlining the vessels based on the segmentation result. (e) B/W image with vessels segmented from the hardware version of the algorithm. (f) Image outlining the vessels based on the segmentation result.

capabilities of the system by integrating registration and pattern recognition modules.

6. Conclusion and Future Work

This paper presented a high performance retinal blood vessel segmentation accelerator architecture. The architecture features a highly-parallel match filtering unit, enabling real-time blood vessel tree extraction. FPGA-based evaluation suggests that the proposed architecture outperforms existing works, demonstrating the capabilities of reconfigurable hardware for the acceleration of such algorithms.

Our on-going work focuses on incorporating pre-processing and post-processing retinal processing methods into the existing system, in order to further improve accuracy. The objective is to integrate registration and recognition algorithms and their efficient implementation on reconfigurable hardware, for portable biometric and biomedical applications.

7. ACKNOWLEDGMENTS

This work was co-funded by the European Regional Development Fund and the Republic of Cyprus through the Research Promotion Foundation (Project NEA ΥΠΟΔΟΜΗ/ΣΤΡΑΤΗ/0308/26).

8. REFERENCES

[1] Usher, D. Y. Tosa, and M. Friedman, 2008. Ocular biometrics: simultaneous capture and analysis of the retina and iris, *Advances in Biometrics: Sensors, Algorithms and Systems*, Springer Publishers, London, UK, 133-155.

[2] Fraz, M.M. P. Remagnino, A. Hoppe, B. Uyyanonvara, A.R. Rudnicka, C.G. Owen, S.A. Barman, 2012, Blood vessel segmentation methodologies in retinal images – A survey, *Computer Methods and Programs in Biomedicine*, Volume 108, Issue 1, 407-433.

[3] Perfetti, R, E. Ricci, D. Casali, G. Costantini, 2007, Cellular neural networks with virtual template expansion for retinal vessel segmentation, *IEEE Trans. Circuits Syst.*, 54 (2), 141–146.

[4] Alonso-Montes, C, M Ortega, M.G. Penedo, D.L. Vilarino, 2008 *Pixel parallel vessel tree extraction for a personal authentication system*, Circuits and Systems,. ISCAS 2008. IEEE International Symposium on , vol., no., pp.1596-1599, 18-21.

[5] Nieto, A, V.M. Brea, D.L. Vilarino, 2009,*FPGA-accelerated retinal vessel-tree extraction*, Field Programmable Logic and Applications, 2009. FPL 2009. International Conference on , 485-488.

[6] Palomera-Perez, M.A., M.E. Martinez-Perez, H. Benitez-Perez, J.L. Ortega-Arjona, 2010, Parallel Multiscale Feature Extraction and Region Growing: Application in Retinal Blood Vessel Detection, *IEEE Transactions on Information Technology in Biomedicine*, vol.14, no.2, 500-506.

[7] Muhammed Gökhan Cinsdikici, Doğan Aydın, 2009, Detection of blood vessels in ophthalmoscope images using MF/ant (matched filter/ant colony) algorithm, *Computer Methods and Programs in Biomedicine*, Volume 96, Issue 2, November 2009, 85-95.

[8] Chaudhuri, S, S. Chatterjee, N. Katz, M. Nelson,; Goldbaum, M, 1989, Detection of blood vessels in retinal images using two-dimensional matched filters, *IEEE Transactions on Medical Imaging*, 8 (3). pp. 263-269. ISSN 0278-0062.

[9] Raja J.B., C.G. Ravichandran, November 2001, Blood Vessel Segmentation For High Resolution Retinal Images, *Inter. J. of Comp. Science Issues*,vol.8, Issue 6, no 2.

[10] Hijazi, M. H. A, F. Coenen, and Y. Zheng, 2010, Retinal image classification using a histogram based approach. In *Proc. International Joint Conference on Neural Networks*, pages 3501– 3507. IEEE..

[11] Niemeijer M. , J.J. Staal, B.v. Ginneken, M. Loog, M.D. Abramoff, *DRIVE: digital retinal images for vessel extraction*, http://www.isi.uu.nl/Research/Databases/DRIVE,2004.

[12] Hoover, A.D. Kouznetsova, V. Goldbaum, M, March 2000, Locating blood vessels in retinal images by piecewise threshold probing of a matched filter response, *IEEE Transactions on Medical Imaging*, vol.19, no.3, pp.203-210.

[13] *MESSIDOR: Methods for Evaluating Segmentation and Indexing Techniques Dedicated to Retinal Ophthalmology*, http://messidor.crihan.fr/index-en.php/2004.

[14] Jain Anil, Lin Hong, and Sharath Pankanti. 2000. Biometric identification. *Commun. ACM* 43, 2 (February 2000), 90-98. DOI=10.1145/328236.328110.

[15] Feng P., Y. Pan, B. Wei, W. Jin, and D. Mi, 2007, Enhancing retinal image by the contourlet transform. *Pattern Recognition Letters*, 4(28):516-522.

[16] Kanski J.J., *Clinical Opthalmology*, 6th ed., Elsevier Health Sciences, London, UK, 2007.

[17] Teng T., M. Lefley, D. Claremont, 2002, Progress towards automated diabetic occular screening: a review of image analysis and intelligent systems for diabetic retinopathy, *Medical and Biological Engineering and Computing* 40 (2002) 2-13.

[18] Heneghan C., J. Flynn, M. O'Keefe, M. Cahill, 2002, Characterization of changes in blood vessel width and tortuosity in retinopathy of prematurity using image analysis, *Medical Image Analysis* 6, 407-429.

[19] Fowers, J., Brown, G., Cooke, P., and Stitt, G. A Performance Energy Comparison of FPGAs, GPUs, and Multicores for Sliding-Window, Applications. *Proceedings of the ACM/SIGDA international symposium on Field Programmable Gate Arrays* (FPGA'12), New York, NY, USA, 47-56.

[20] Maclean, W. J. 2005. An Evaluation of the Suitability of FPGAs for Embedded Vision Systems. *Proceedings of the 2005 IEEE Computer Society Conference on Computer vision and pattern recognition*, San Diego,CA, 131.

[21] Thisius, R.S., A, Kjaer-Nielsen, A, S, Sorensen, 2011, Real-time medical video processing, enabled by hardware accelerated correlations, *Journal of Real-Time Image Processing*, September 2011, Volume 6, Issue 3, pp 187-197.

A 191µW BPSK Demodulator for Data and Power Telemetry in Biomedical Implants

Li-Lan Wang, Chia-Hsiang Yang, Herming Chiueh
National Chiao Tung University
1001 University Road
Hsinchu 30010, Taiwan
+886-3-57121211 ext 54597
{liliness, chy, chiueh}@soclab.org

ABSTRACT

This paper presents a fully digital binary-phase-shift keying (BPSK) demodulator for data and power telemetry. This demodulator recovers BPSK signals by detecting the symbol edge of the digitized received carrier. Parameters of the coupling coils, rectifier DC output, and data rate are taken into consideration in the early design stage. Given a limited coil size and quality factor, the demodulator achieves a data rate of 678kb/s with BER < 10^{-9} at a carrier frequency of 13.56MHz. Fabricated in a 0.18µm CMOS process, the chip area is 0.445mm^2. The chip core dissipates 191µW at 13.56MHz. A system prototype was developed to transmit data and power simultaneously through a pair of coils.

Categories and Subject Descriptors

B.7.m [**Integrated Circuits**]: System and Integrated Circuits for Biomedical Implants.

General Terms

Design, Reliability, Experimentation

Keywords

Binary-Phase-Shift Keying, Biomedical Implants, Mixed-Signal Integrated Circuits

1. INTRODUCTION

Power and data telemetry is the key to battery-less biomedical implants. Wireless power transmission and wireless data telemetry have been extensively studied in [1][2] and [6]-[18], respectively. In order to reduce the number of implanted coils, systems that transmit both data and power were proposed in [6][7]. Amplitude-shift keying (ASK) or on-off keying (OOK) is frequently used for data transmission due to its simplicity. However, ASK and OOK modulations suffer from low data rate and instability for power transmission. In contrast, frequency-shift keying (FSK) modulation requires a LC resonator with a low quality factor, but this leads to low power efficiency. Constant-envelope, fixed-carrier frequency modulations conveys stable power regardless of data pattern, which is a better solution for simultaneous data and power transmission. Binary phase-shift keying (BPSK) is preferred in order to reduce the power consumption of demodulator. Another advantage of BPSK is that it requires less transmitted power to achieve the same bit-error rate (BER) when compared to FSK or ASK.

For BPSK demodulation, phase-locked loop (PLL) was proposed and it provides stable data and power transmission [6][7]. However, the high power consumption of PLL-based demodulator is not suitable for implantable devices since PLL requires phase synchronization for coherent detection. Extracting clock from received data can be used to eliminate the use of PLL, as shown in Fig. 1(a). This clock-recovery technique was adopted for data transmission [9][10][18], but applying this technique to simultaneous data and power transmission has not been explored.

In this work, a clock-recovery-based BPSK demodulator for data and power telemetry targeting for epilepsy seizure detection is proposed. Since power consumption is an important issue for implantable devices, a symbol edge detector is used instead of a power-hungry oscillator, as shown in Fig. 1 (b). This replacement further reduces power consumption. Fig. 2 shows the overall data and power transmission system. The power is transmitted through coils and received inside the human body. The data demodulator is powered by the rectified power.

The paper is organized as follows. Section II provides theoretical analysis on power efficiency and data rate. Section III describes the circuit design of the BPSK demodulator. Chip implementation, test setup, and experimental results are show in Section IV. Finally, Section IV concludes the paper.

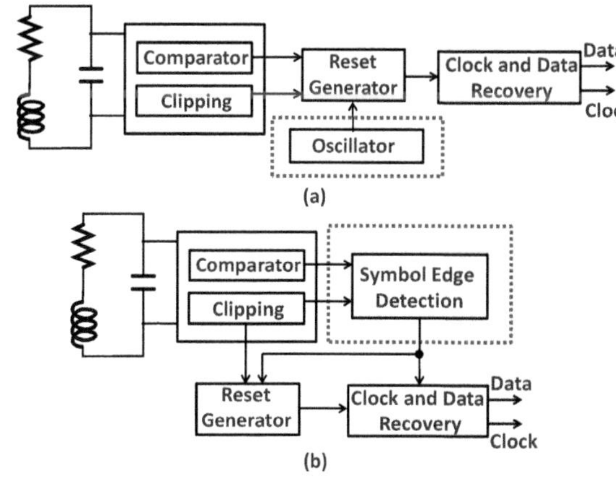

Figure 1. (a) Conventional clock-recovery-based demodulator and (b) proposed oscillator-less demodulator

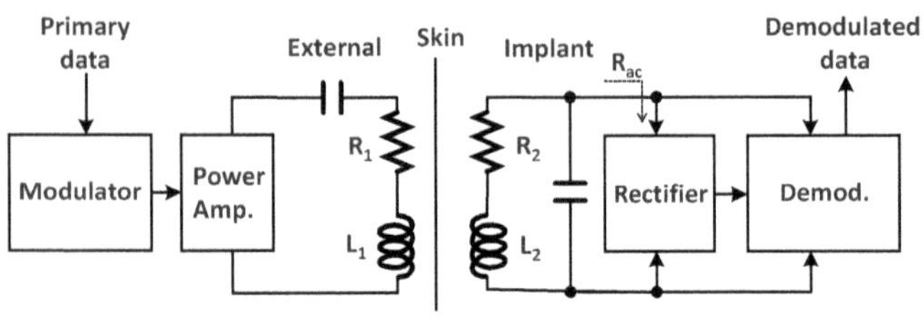

Figure 2. Data and power transmission system

2. PERFORMANCE ANALYSIS

An inductive power link is composed of a transmitter's and receiver's coils (L_1 and L_2), as shown in Fig. 2. The coupling coefficient of two coils k ($0 < k < 1$) depends on the portion of the total flux lines that cuts both primary and secondary coils. An AC voltage is applied across the transmitter's coil, and this induces an AC voltage on the receiver's coil. A bridge rectifier s used to for convert AC-to- DC conversion. The power efficiency η between transmitted and received powers (P_0 and P_1) is represented by

$$\eta = \frac{P_o}{P_1} = \frac{k^2 Q_1 Q_2^3 R_2 R_{ac}}{\left(R_{ac} + Q_2^2 R_2\right)\left[\left(1 + k^2 Q_1 Q_2\right) R_{ac} + Q_2^2 R_2\right]} \quad (1)$$

As shown in Fig. 2, R_2 is the resistance of internal coil and R_{ac} is the input resistance of the rectifier. Q_1 (Q_2) is the quality factor of coil L_1 (L_2). This relationship provides the maximum power efficiency given these circuit parameters.

The received power also affects the data transmission. To evaluate the maximum data rate, channel capacity is evaluated. Channel capacity C is impacted by the received signal power P_R, LC tank bandwidth B, and noise power N.

$$C = B\log\left(1 + \frac{P_R}{N}\right) \quad (2)$$

Considering only thermal noise, the noise power N is given by the Johnson's noise formula:

$$N = BKT, \quad (3)$$

where T is the temperature in Kelvin and K is the Boltzmann constant (1.38×10^{-23}). In our simulation, we assume that the system operate at the room temperature of 27°C. The achievable bandwidth B is a function of the Q-factor of transmitter's and receiver's coils [5]. The ratio between signal bandwidth B and carrier frequency f_0 is described below.

$$\frac{B}{f_0} = \frac{\sqrt{-(Q_1^2 + Q_2^2) + \sqrt{(Q_1^2 + Q_2^2)^2 + 4Q_1^2 Q_2^2}}}{\sqrt{2}Q_1 Q_2} \quad (4)$$

The carrier resonance frequency is set to 13.56MHz. The parameters of coils are summarized in Table 1.

3. CIRCUIT DESIGN

A clock-recovery-based demodulator architecture is adopted to reduce the complexity induced by PLL. BPSK modulation converts baseband data 0 and 1 into two in-phase waveforms. It converts analog signals to digital signals through comparator and clipping circuit. After digital signals are extracted, there is an edge detection circuit to detect rising or falling edges. The right symbol edge should be detected in order to recover the data. We propose an area-efficient circuit implementation for data-and-clock recovery (DCR). It consists of two D flip-flops whose clocks come from the differential outputs of the comparator and data input is connected to high. Fig. 3 shows the functionality of the start circuit. An external power on signal is switched to enable the detection process. One of the differential outputs of the clipping circuit (clip+ or clip-) is used to extract the clock frequency and trigger the start signal. Timing diagram of the key signals are shown in Fig. 4. When two falling edges are detected, the reset signal is pulled to high. The reset signal goes down if a rising edge is detected. Once the reset signal goes high, an edge detector generates right signals to DCR.

Table 1. COIL SPECIFICATIONS

	External coil	Internal coil
L (μH)	7	0.382
C (pF)	18.8	374.25
R (Ω)	3.6	1.05
Q	165.66	30.99

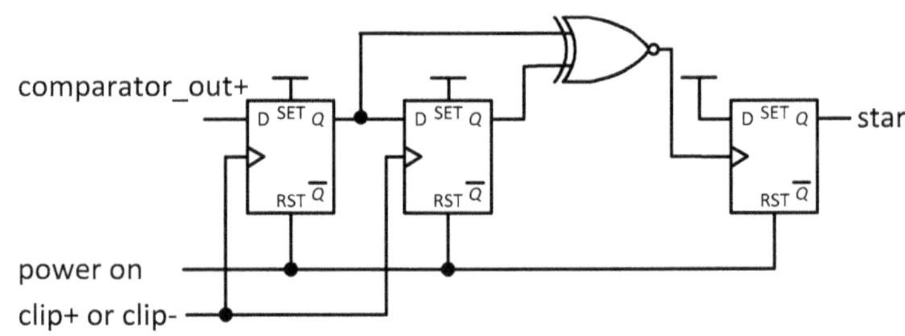

Figure 3. Schematic of start circuit for edge detection

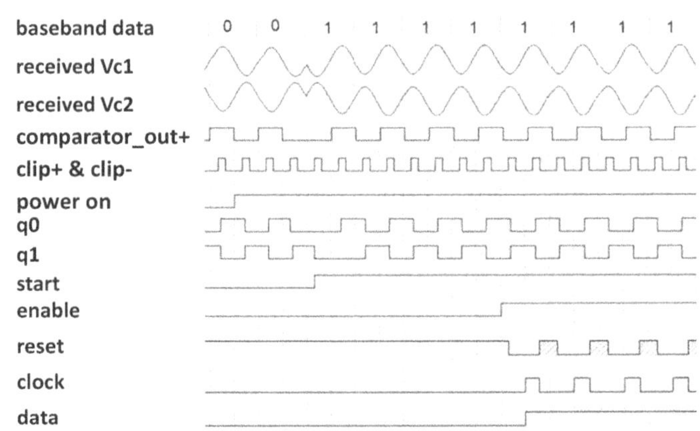

Figure 4. Timing diagram of system signals

Since data and power are transmitted simultaneously, the input waveform distortion caused by rectifier is taken into consideration. According to the specs of the rectifier, the input voltage is set to $5.8V_{pp}$ (peak-peak voltage) at a data rate of 678kbps. The rectifier output supplies a 1.8V DC power to the BPSK demodulator circuit.

4. TEST SETUP AND EXPERIMENTAL RESULTS

A Schottky diode bridge rectifier connects to internal coil. The DC output of the rectifier supplies stable 1.8V to BPSK demodulator circuit. The test setup, measurement environment, and chip implementation are shown in Fig. 5, the diameter and thickness of the external (internal) coil are 4.2cm (1.8cm) and 1.1cm (0.2cm), respectively. The spacing between two coils is 1cm. The die size of the chip is 0.668x0.666 mm^2 and the core area is 0.139×0.07 mm^2 in a 0.18μm CMOS process.

Fig. 6 shows a screen shot of the measurement result of data transmission. The signal is attenuated by 10 times for detecting larger-amplitude signals. Therefore, the measured coil input voltage of 580mV_{pp} is equivalent to received voltage of 5.8V_{pp} on channel 2. The 1.8V DC voltage from rectifier is measured on channel 3.

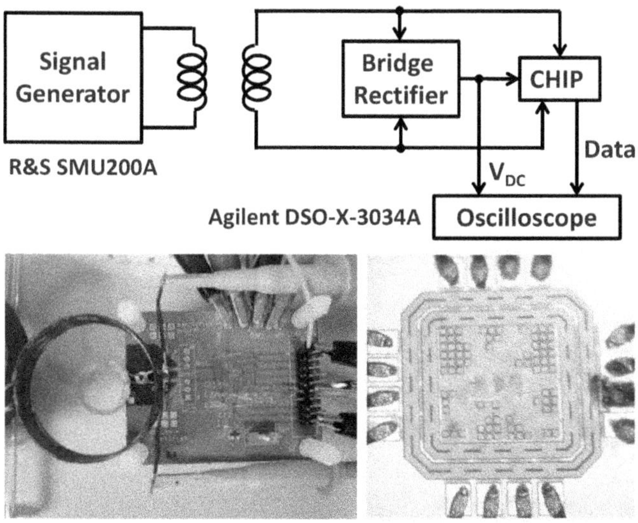

Figure 5. Test environment of BPSK demodulator

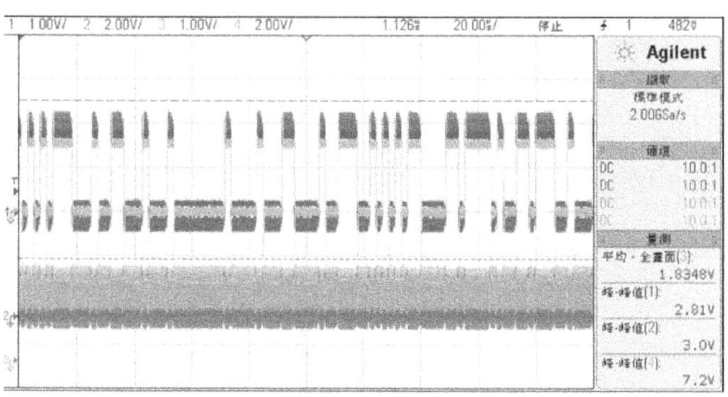

Figure 6. Measured waveforms of BPSK demodulator and bridge rectifier

The received signals are modulated at a carrier frequency of 13.56MHz carrier frequency, resulting in a non-single tone wave. Fig. 7 shows that the rectifier drop-out voltage varies according to the supply voltage and modulated data rate. The drop-out voltage increase with the increased supply voltage and it increases with increased data rate. The degradation of drop-out voltage is used to adjust the transmit power to compensate the power loss. To supply a stable 1.8V DC power for different data rates, appropriate primary (external) input power must be chosen. To maintain a constant received power, more primary power is needed for higher data rate, resulting in lower power efficiency. Fig. 8 shows the required primary power and corresponding power efficiency for the change of data rate.

Once the receive power at different data rates is known, the channel capacity can be derived from (2). When data rate achieves 1.13Mbps, the received power decreased, leading to degradation of the channel capacity. This causes unreliable data transmission. Therefore, data rate must be lower than 1.13Mbps according to the measurement result. In theory, higher SNR contributes to higher BER performance, but it is impossible to achieve the theoretical bound because additional noise sources are not taken into consideration. The unexpected noise causes extra transmission errors. Fig. 9 shows the relation of channel capacity and data rate. The maximum data with near-errorless performance (BER < 10^{-9}) can be achieved at data rate of 678kbps. The BER = 10^{-9} bound is limited by the test infrastructure.

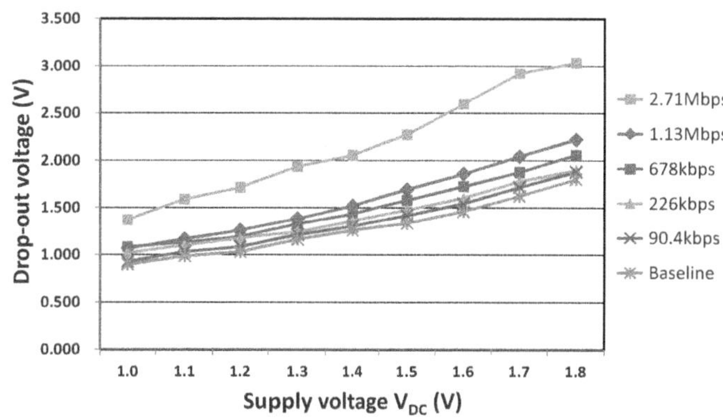

Figure 7. Drop-out voltage versus different data rate

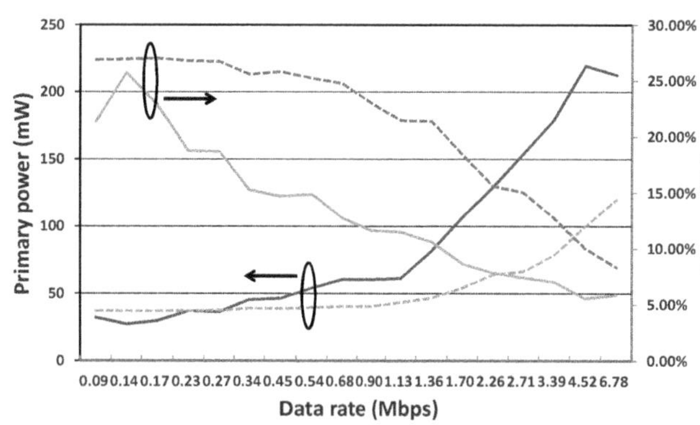

Figure 8. Primary power and power efficiency versus data rate

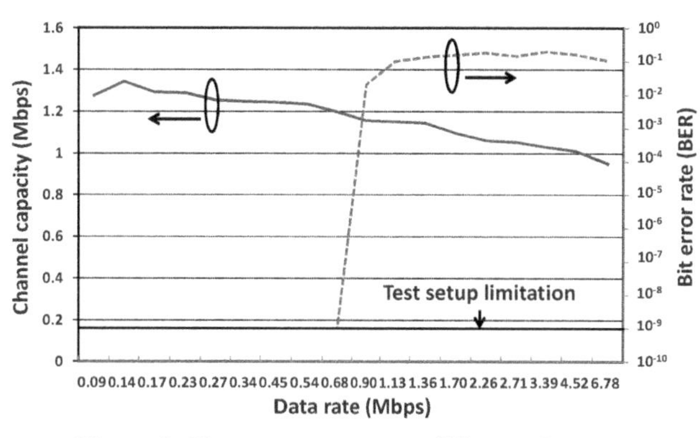

Figure 9. Bit error rate versus different data rate

A performance comparison with prior clock-recovery-based BPSK demodulators is listed in Table 2. Except this work, other designs are only verified in simulation and coils are not included in the system. The data rate to carrier frequency ratio of this work achieves 100% with comparable power consumption. Table 3 shows a performance comparison of this work with other demodulators for biomedical implants. Most of the designs are designed for data transmission only. Compared with designs with capability of data and power transmission [6-7], this work dissipates less power and the core area is smaller. When compared with other BPSK demodulators dedicated to data transmission, this design still has the lowest power consumption normalized by carrier frequency.

Table 2. BPSK DEMODULATOR COMPARISON

	Data transmission		Power consumption (µW)	CMOS Tech. (µm)	
	Data rate (Mbps) Mod: BPSK	Carrier freq. (MHz)			
This work	13.56 (without coil) 0.678 (with coil)	13.56	(chip) 294 @ 1.8V (core) 191 @ 1.8V	0.18	measured
[10]	0.8/4/8/20	0.8/4/8/20	46/93/148/310 @ 1.8V	0.18	simulation
[12]	10	10	119 @ 1.8V	0.18	simulation
[18]	10	10	232 @ 1.8V	0.18	simulation

Table 3. DATA AND POWER TRANSMISSION COMPARISON

	Data transmission		Power transmission			CMOS Tech. (µm)	Area (mm^2)
	Data rate (Mbps) Mod: BPSK	Carrier freq. (MHz)	Received power (mW)	Coil distance (cm)	Power consumption (mW)		
This work	0.678	13.56	5	1	(chip) 294 @ 1.8V (core) 191 @ 1.8V	0.18	(chip) 0.445 (core) 0.0097
[6]	1.69	13.56	22.5	0.22	5 @ 3.3V	0.5	0.1
[7]	1.69	13.56	22.5	0.22	2.3 @ 3.3V	0.5	0.293
[8]	0.02	13.56	-	-	3 @ 3.3V	0.5	1
[9]	0.8	4	-	-	0.059 @ 1.8V	0.18	0.0043
[15]	1-2.5	10	-	-	0.485 @ 3.8V	0.35	0.161
[16]	1.12	13.56	-	-	0.414 @ 1.8V	0.18	0.148

5. CONCLUSTION

A clock-recovery-based BPSK demodulator for data and power telemetry is designed and implemented for biomedical implantable devices. A symbol detector is used instead of oscillator circuitry for power and area reduction. The measurement results show that the rectifier output supplies a stable 1.8V power supply for the BPSK demodulator. The maximum data rate for reliable data transmission is up to 678 kbps with BER < 10^{-9}. The power consumption of the chip is 191µW from a 1.8V supply. Compared to prior work, this work supports both data and power transmissions and it dissipates the lowest power.

6. ACKNOWLEDGMENTS

This research is supported by National Science Council, Taiwan under contract number NSC 101-2220-E-009-019, NSC 101-2220-E-009-064, and MoE ATU Program of Biomedical Electronics Translational Research Center (BETRC) at National Chiao Tung University, Taiwan. The author would like to acknowledge the prototype fabrication and design environment provided by National Chip Implementation Center (CIC,) Taiwan.

7. REFERENCES

[1] G. A. Kendir, et al., "An efficient inductive power link design for retinal prosthesis," in Proc. IEEE Int. Symp. Circuits and Systems, vol. 4, pp. IV 41-44, May 2004.

[2] W. H. Ko, et al., "Design of radio-frequency powered coils for implant instruments," Med.Bio.Eng.Comput., vol. 15, pp. 634-640, 1977.

[3] T. H. Lee, "The Design of CMOS Radio-Frequency Integrated Circuits.Cambridge," *UK: Cambridge Univ. Press,* 1998.

[4] H. Yamu, M. Sawan, "A fully integrated low-power BPSK demodulator for implantable medical devices," in *Proc. IEEE Trans. Circuits and Systems-I*, vol. 52, no. 12, pp. 2552-2562, Dec. 2005.

[5] J. I. Agbinya and M. Masihpour, "Power equations and capacity performance of Magnetic Induction body area network nodes," in *Conf. on Fifth Int. Broadband and Biomedical Communications (IB2Com)*, pp. 15-17, Dec. 2010.

[6] S. Sonkusale and Z. Luo, "A complete data and power telemetry system utilizing BPSK and LSK signaling for biomedical implants," in *30th Annul IEEE Engineering in Medicine and Biology Society (EMBS)*, pp. 20-25, Aug. 2008.

[7] X. Wangren, et al., "Fully Digital BPSK Demodulator and Multilevel LSK Back Telemetry for Biomedical Implant Transceivers," in *IEEE Trans. Circuits and Systems-II*, vol. 56, no. 9, pp. 714-718, Sept. 2009.

[8] L. Zhenying and S. Sonkusale, "A Novel BPSK Demodulator for Biological Implants," in *IEEE Trans. Circuits and Systems-I*, vol. 55, no. 6, pp. 1478-1484, July 2008.

[9] C. S. A. Gong, et al., "Low-power and area-efficient PSK demodulator for wirelessly powered implantable command receivers," in *Electronics Letters*, vol. 44, no. 14, pp. 841-842, July 2008.

[10] F. Asgarian and A. M. Sodagar, "A high-data-rate low-power BPSK demodulator and clock recovery circuit for implantable biomedical devices," in *Proc. 4th Int. IEEE/EMBS Conf. Neural Engineering*, pp. 407-410, May 2009.

[11] L. Zhijun and M. Sawan, "An 8 Mbps data rate transmission by inductive link dedicated to implantable devices," in *Proc. IEEE Int. Symp. Circuits and Systems (ISCAS)*, pp. 18-21, May 2008.

[12] F. Asgarian and A. M. Sodagar, "A carrier-frequency-independent BPSK demodulator with 100% data-rate-to-carrier-frequency ratio," in *Proc. IEEE Biomedical Circuits and Systems Conference (BioCAS)*, pp. 3-5, Nov. 2010.

[13] F. Asgarian and A. M. Sodagar, "A low-power noncoherent BPSK demodulator and clock recovery circuit for high-data-rate biomedical applications," in *Proc. IEEE Int. Conf. Engineering in Medicine and Biology Society (EMBC)*, pp. 3-6, Sept. 2009.

[14] C. P. Li, et al., "A new integrated low-power BPSK demodulator for wireless implantable neural recording system," In *Proc. IEEE Int. Conf. Electron Devices and Solid-State Circuits (EDSSC)*, pp. 15-17, Dec. 2010.

[15] H. Cheng-Han, et al., "Low-power 13.56 MHz RF front-end circuit for body sensor network," in International Symp. The Bioelectronics and Bioinformatics (ISBB), pp. 3-5, Nov. 2011.

[16] M. Sawan, "Wireless Smart Implants Dedicated to Multichannel Monitoring and Microstimulation," in *Proc. IEEE/ACS Int. Conf. Pervasive Services (ICPS)*, pp. 19-23, July 2004.

[17] Z. H. Wu, et al., "A new BPSK demodulation circuit for command transmission in wireless implantable neural recording system," in *Proc. IEEE Asia Pacific Conf. on the Circuits and Systems (APCCAS)*, Dec. 2008.

[18] C. P. Li, et al., "A new integrated low-power BPSK demodulator for wireless implantable neural recording system," in *Proc. IEEE Int. Conf. Electron Devices and Solid-State Circuits (EDSSC)*, pp. 15-17, Dec. 2010.

Custom Memory Architecture for Multi-Core Implementation of Face Detection Algorithm

David Watson, Ali Ahmadinia, Gordon Morison, Tom Buggy
School of Engineering and Built Environment
Glasgow Caledonian University
Glasgow, Scotland
{david.watson2, ali.ahmadinia, gordon.morison, t.buggy}@gcu.ac.uk

ABSTRACT

Here we present our multi-core architectures for object detection. We move away from the traditional architecture of Multi-Processors (MPs) by using cacheable accesses to main memory to create atomic cores and utilising local memory for all program data. Main memory is partitioned through software into dedicated data regions to allow atomic accesses by cores, without the need for synchronisation primitives. In doing this, we demonstrate how multi-threading techniques such as Interleaved Task Reordering (ITR) can be utilised to balance the processing loads on available cores. We implement and test up to 7 soft-cores with the Viola Jones face detection algorithm and achieve a performance increase of up to 9.14x with a 100% detection rate: surpassing the theoretical performance increase of multi-core processors for all designs and test images. Furthermore, we surpass the performance increases of multi-core implementations from the literature, thus proving our custom designs to be a more viable solution for multi-core object detection applications. Finally, resource and power consumption estimates indicate our designs to be suitable for embedded systems deployment.

Keywords

Custom memory; multi-core; Microblaze; Viola Jones

Categories and Subject Descriptors

C.1.2 [**Multiple Data Stream Architectures (Multiprocessors)**]: Parallel processors; D.4.1 [**Process Management**]: Multiprocessing/multiprogramming/multitasking

1. INTRODUCTION

Multi-core processing has become an integral part of modern day electronics and is used in products ranging from desktop Central Processing Units (CPUs), to tablets and mobile phones. Incorporating multiple cores into one processor increases the computational power of the processor, without increasing the clock speed. This increase in throughput makes multi-core processors an attractive platform for implementing object detection algorithms. However, the architecture of Symmetrical Multi-Processors (SMPs) and the memory architecture of Multi-Processors (MPs) can inhibit performance and scalability. Performance decreases can be experienced due to poor load balancing on cores, cache incoherencies, synchronisation primitives, etc. The Viola Jones algorithm [10] is well known to accurately detect faces in images using Haar-like features (2-D wavelets) to classify image regions. However, it involves large amounts of computations coupled with large quantities of data that can make implementations problematic.

In this work, we propose and evaluate multi-core architectures of face detection using the Microblaze soft-core processor and custom memory architectures. We aim to overcome the problems of multi-core processing by creating custom processing platforms that are capable of operating atomically. The memory architecture of the designs is addressed in two key areas: data accesses to main and program memory. Software is used to partition main memory into dedicated data buffers and program data is organised locally to each core through the use of Local Memory Extension (LME). We implement multi-threaded load balancing techniques such as Interleaved Task Reordering (ITR) through fine-grained threading over all available cores. Thread affinity is maximised by executing each core independently, which also eliminates synchronisation primitives. Cache coherency and false-sharing issues are eliminated by omitting level-2 caches and customising the execution path of threads before runtime.

GLSVLSI'13, May 2–3, 2013, Paris, France.

2. RELATED WORK

Lai et al [8, 7] optimise the locality of data for a shared-memory SMP to promote data reuse during feature evaluations. However, the data reuse promoted is for the feature data contained within the cascade and not integral data. This is achieved through loop restructuring and grouping the execution of cascade stages. Chiang et al [4] simulate a concurrent implementation of Viola Jones with up to 64 ARM v5 cores. The inherent parallelism of the algorithm is exploited at the resizing and classifying stages. The Viola Jones algorithm is implemented by Chen et al [2] using the IBM cell of a PlayStation 3. Task interleaving, re-ordering, and splitting are used to achieve optimal load balances on the cores by processing image rows in parallel and interleaving large and small search windows: a method known as Interleaved Task Re-ordering (ITR).

Xin et al [11] implement the Scale Invariant Feature Transform (SIFT) algorithm on multi-core DSPs. To overcome the poor expandability and shared memory bottleneck of the SMP, the authors create a distributed shared memory parallel machine, which achieves a speedup of 5x using 6 cores, but fail to continue a linear speedup trend. Hefenbrock et al [6] use GPUs to implement the Viola Jones algorithm. Search windows are executed in parallel using SIMD/SIMT instructions, with threads from a thread block contributing to shared memory accesses. Shekhar and Varaganti [5] implement a dual-core face detection engine which uses Sobel filtering and feature extraction to detect faces. False sharing is reduced through data partitioning complete image scales to individual cores and recording face detections is ensured atomicity using a serial buffer.

Chen et al [3] implement a parallelised version of a body tracking algorithm for multi-core processors. Data and functional parallelism is used to describe fine- and coarse-grained parallelism respectively, where the former is used for parallelising the algorithm. Ranjan and Malik [9] present a multi-core implementation of a face detection and tracking system which uses the Viola Jones algorithm. The algorithm is coarse-grained parallelised by allocating no. scales/no. cores scales for processing. Finally Ach et al [1] implement a multi-core object detection algorithm for the application of detecting road signs which uses the same Haar-features as the Viola Jones algorithm. A circular buffer is used for the integral image which is then classified: 10 worker threads are used to copy and process data in the circular buffer.

The multi-core implementations mentioned above rely on the promotion of data reuse, concurrency exploitation, and the reduction of synchronisation primitives. Furthermore, proper load balancing, false-sharing, and thread affinity have been highlighted as issues affecting the performance of multi-core processors. However, the majority of these issues can be addressed by one main concept: the memory architecture. The greatest consideration for multi-core programming should not be for thread granularity or data reuse concepts; it should be for how memory is organised, accessed, and utilised. These considerations are required for both main memory accesses for data, but also for program data. Object detection algorithms rely on the use of predefined libraries in order to classify images. These libraries offer a better reuse of data than image data, as not all pixels of an image will be used during classification. How we organise main memory can also impact how programs are threaded. By firstly considering how memory will be accessed, it can be organised and partitioned to facilitate the data accesses of a multi-threaded implementation. It is for these reasons we investigate the use of multi-soft-core processing architectures for the application of object detection. The memory architecture of the multi-processor is tailored to suit multi-threading needs through the use of custom buffer assignments and the localisation of all program data. To our best knowledge there has not been any such investigation.

3. VIOLA JONES FACE DETECTION

The Viola Jones algorithm searches an image at multiple scales to detect faces. The original algorithm [10] stipulates image scales generated at 80% yields the best balance between computational load and detection rate. A sub-window of 20x20 pixels is moved horizontally and vertically across these image scales at one or two pixel increments, in which a *Binary Decision Tree*, known as the Attentional Cascade (AC), is used to detect facial features. The AC consists of 22 stages, each of which contains a different number of Haar-features, which are used to detect facial features. Haar-features (also known as Haar-like features) are 2-D wavelets that consist of 2 or 3 rectangles. Figure 1 illustrates stage 1 of the AC and a 2 rectangle Haar-feature. A sub-window is deemed to contain a face if it passes all 22 stages of the AC. To facilitate the fast computation of features, integral images are used. The integral image at location (x,y) contains the sum of the pixels above and to the left of (x,y) inclusive:

$$ii(x,y) = \sum_{x' \leq x, y' \leq y} i(x', y') \tag{1}$$

where $ii(x,y)$ and $i(x,y)$ are the integral image and original image respectively. Each Haar-feature is composed of 2-3 rectangles, each of which consists of 4 coordinates (corners) used to reference pixels in the integral image, and a weight. Haar-features also contain a *Feature Threshold*, as well as *Left* and *Right* values. The 4 pixels of the integral image that correspond to the rectangles corners are used to calculate the rectangle sum using Equation 2:

$$Rect\ Sum = (A - B - C + D) * W \tag{2}$$

where A, B, C, D are the corners of the rectangles of a Haar-feature and the weight (W) is defined as -1, 2, or 3 in the cascade. Rectangle sums are totalled for a given feature to give a *Feature Sum*. E.g. the sum of *Haar_Feature_0* would equal the sum of *Rect_0* and *Rect_1*. This sum is compared to the *Feature Threshold*, where, if the *Feature Sum* is less than the *Feature Threshold*, the *Left* value is taken; otherwise the *Right* value is taken. The cumulative sum of the *Left* and *Right* values taken for all features of a stage is compared to the *Stage Threshold*. If the cumulative sum is less than the *Stage Threshold*, the sub-window does not contain a face and the next sub-window is processed. Otherwise it may contain a face and the next stage is processed. If a sub-window passes all 22 stages, it contains a face and is ear-marked as such.

The first stage of the algorithm is the pre-processing stage, where image scales are generated and their integral images calculated. Following this, the classification of the image scales is carried out according to the rules detailed above. The classification stage is responsible for approximately 90% of the run-time of the algorithm and is the main focus of all concurrent implementations. E.g. the pre-processing stage for a VGA image (640x480) generates 15 integral images, each of which must be scanned using the AC.

4. IMPLEMENTATION

In order to fully test the feasibility of multi-soft-core processing architectures, we implement a standard single-core design which is used as a benchmark for all our multi-core designs. The standard implementation uses a performance optimised Microblaze core with 64KB local memory, 32KB I- and D-cache, uses dynamic memory allocations to manage integral image buffers, and stores all program data in main memory. Our multi-core implementations utilise up to 7 performance optimised cores with 32KB D-cache, no I-cache, and 176KB of local memory. All Microblaze cores are clocked at 100MHz and use the AXI4 bus interface for main memory accesses: the AXI4 bus interface ensures clock

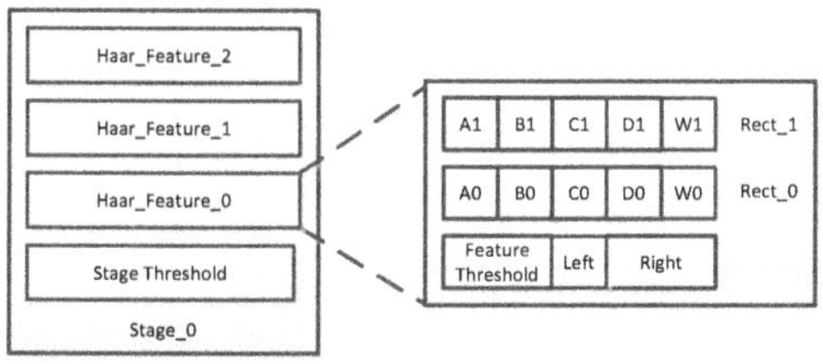

Figure 1: Viola Jones Haar-based computation at pixel level. First stage of the AC (left), structure of a 2-rectangle Haar-feature (right)

Version	Description
Standard	1 core, dynamic memory, no LME, coarse-grained
Single-core	1 core, MO, LME, coarse-grained
Dual-core	2 cores, MO, LME, ITR
Tri-core	3 cores, MO, LME, ITR
Quad-core	4 cores, MO, LME, ITR
Pent-core	5 cores, MO, LME, ITR
Hex-core	6 cores, MO, LME, ITR
Hept-core	7 cores, MO, LME, ITR

Table 1: Descriptions of all Implementations

frequency has no impact on design scalability. Our software uses the same AC from the OpenCV v2.0 Viola Jones Haar-feature implementation and is normalised to 32-bit fixed-point notation, with the scan-window moved 2 pixels at a time. Multi-core implementations decompose the classification stage of the algorithm into multiple steps to allow for concurrent processing of image scales. The classification stage is multi-threaded and the pre-processing stage of a scale is carried out serially by one core. Prior to the pre-processing and classification stages, initialisations are performed and cores in a multi-core design perform a simple hand-shaking protocol with each other. All designs are created using the Xilinx Embedded Development Kit (EDK) and implemented on a Xilinx Virtex 6 development board with a XC6VLX240T FPGA. A summary of the designs implemented can be found in Table 1.

4.1 Hardware Optimisations - Program and Main Memory

All multi-core implementations use resource optimised Local Memory Extension (LME) to allow local access to all program data. As mentioned in Section 2, object detection algorithms rely on predefined libraries to classify image regions, therefore by keeping libraries local to a core, the performance of the algorithm can be improved. Furthermore, using the Local Memory Bus (LMB) of the Microblaze gives a latency of 1 clock cycle for program data accesses: giving the same result as a core having all instructions in I-cache. In doing this we no longer require I-cache for the Microblaze and can reduce the power consumption of the core and its resource footprint. Level-2 cache is not used for any design as the structure of our software decouples cores from this burden, eliminates false sharing, and an unnecessary level of synchronicity.

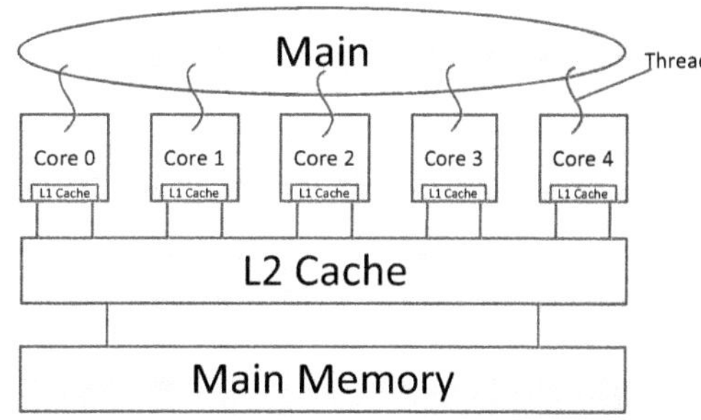

Figure 2: Typical architecture of a multi-processor, where multiple levels of cache can inhibit performance and developers have limited control over thread execution

Figure 2 depicts the typical structure of a multi-core processor, where a hierarchy of caches is used to facilitate the execution of threads. Threads (cores) rely on the assumption that another thread has accessed the piece of data they require and go through the cache hierarchy depicted in Fig. 2 to find this data. However, depending on how an object detection algorithm is threaded, threads may not require the image data of another thread. Furthermore, the algorithm structure dictates the data reuse possible within the implementation. Performance can also be hampered by the size of the shared cache, where cache-thrashing can occur when threads have little execution flow in common. We therefore present the multi-core processing structure depicted in Fig. 3. Here it can be seen how we have removed level-2 caches, as well as I-caches: I-cache is now shared with local memory and is accessed as such. Microblaze cores utilise Xilinx Cache Links (XCL) for fast cacheable accesses to image data and are considered atomic processing agents for the object detection algorithm. These memory considerations address the reuse of program data from object detection libraries instead of the reuse of image data, which, for the Viola Jones algorithm, can be rarely used more than once.

4.2 Software Optimisations - Main Memory

We remove dynamic memory management and replace it with Memory Optimised (MO) buffers for each scale which can be accessed by distributed pointers. In doing this, we no longer think of memory as a contiguous piece of storage, but as separate and independent data buffers which are accessed and managed by the threads of execution. When we compare Figs. 2 and 3 we can see how we view main memory and how a typical multi-core processor views it. By partitioning main memory such as we have, we can obtain greater control over memory accesses for the threads in execution, since we have this knowledge before runtime. In doing this, we consider the threads of our multi-core processor independent of all other threads that may be in execution, and we eliminate synchronisation primitives often associated with multi-core processors. Finally, memory partitions are aligned on 4-byte boundaries to facilitate burst memory accesses and for general housekeeping.

4.3 Threading

We achieve thread affinity by executing one main routine per core, as shown in Fig. 3. Here each statement of a serial 'C' program can be visualised as part of the same thread.

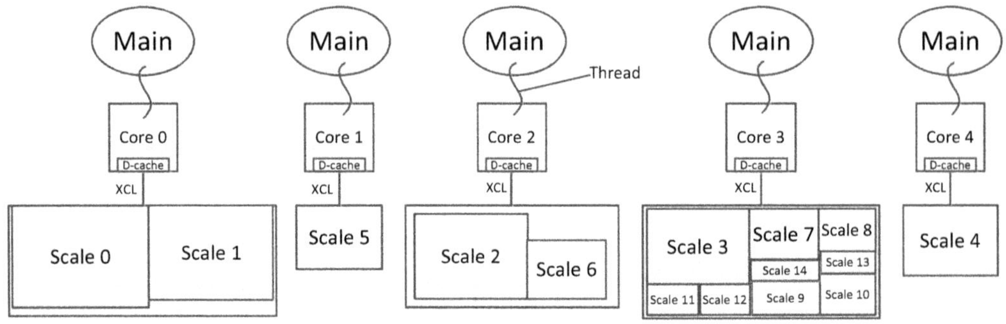

Figure 3: Proposed multi-core architecture, where memory is conceptualised as independent data buffers and cores are considered threads

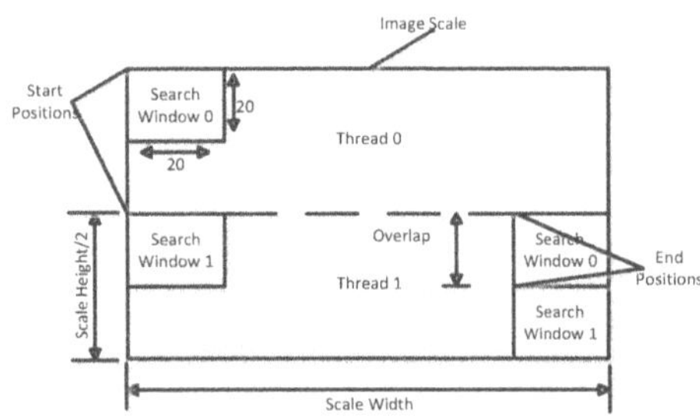

Figure 4: Multi-threading image scales for the dual-core implementation

Version	Lena (512x512)	VGA (640x480)	QVGA (320x240)
Standard	1.00x	1.00x	1.00x
Single-core	1.13x	1.19x	1.26x
Dual-core	2.21x	2.32x	2.48x
Tri-core	3.27x	3.56x	3.79x
Quad-core	4.41x	4.67x	5.03x
Pent-core	5.49x	5.84x	6.70x
Hex-core	6.55x	6.95x	8.38x
Hept-core	7.81x	8.02x	9.14x

Table 2: Performance improvement factors of all implementations for all test images

Fine-grained threading is implemented by decomposing image scales into n rows of equal size, where n is the number of cores, and processing them concurrently. Figure 4 illustrates this concept for the dual-core implementation, where 2 different threads process an image scale concurrently. As can be seen, there is an overlapping of search windows (20 pixels), which is easily managed through software as threads always start at the top of their respective row. When the space between rows becomes smaller than the 20 pixels, we risk contention and therefore do not implement the fine-threads. We also implement Interleaved Task Reordering (ITR) in a bid to obtain the optimal load balance on each core, as well as to eliminate contention. This is achieved by interleaving the threads of larger and smaller image scales. We also implement ITR at the system level, where core tasks are separated from other similar core tasks. In doing this, we all but eliminate contention for shared resources.

5. RESULTS

To allow a good comparison with the literature, we present and critique our results in terms of performance increase and detection rate. Implementations are tested using VGA and QVGA images containing 15 faces, as well as the well-known Lena image and achieve a 100% detection rate throughout. Only processing times are measured for each core, with the time taken for initialisations and timer calls omitted. Execution times are obtained via on-board AXI timers. For the multi-core implementations, we use the average execution time of the cores. Table 2 summarises the results for each implementation when tested with each image.

5.1 Test Images Performance

Performance improvement results for the Lena test image surpass the theoretical performance increases of multi-core processors. We first see how the proposed hardware and software architectures provide an increase in performance of 1.13x when using one Microblaze core; this highlights how the reuse of the detection library can benefit even single-core designs. As we increase the number of cores to 2 and 3, the performance increases surpass the theoretical multi-core improvement with a modest 2.21x and 3.37x respectively. The use of 4 cores not only achieve greater than 4x improvement, but we also see how the performance increase is almost double the performance increase of the dual-core design at 4.41x. This is an interesting trend, as multi-core performance increases are traditionally quoted from 1 core, but here we can demonstrate how our proposed methods maintain this trend when using multiples of core instantiations. Using 5 and 6 cores gives performance increases with similar trend to the quad-core improvement. Finally, the use of 7 cores gives an impressive performance increase of 7.81x when used with our proposed hardware and software based memory optimisations. Furthermore, if compared to the expected performance increases when moving from 2-6 cores to 7 cores, we can see how the performance increase of the hept-core design surpasses even these expectations. We therefore do not only surpass the theoretical performance increases for multi-core designs, but also for our own multi-core designs.

From Table 2 we see our single-core implementation improvement exceeds that of the Lena image by achieving an improvement factor of 1.19x. Dual- and tri-core improvements excel those of the Lena image at 2.32x and 3.56x. These improvements can be attributed to the local access of the detection library, as the VGA image contains a larger search space and is more complex. These two attributes will result in more sub-windows, which may be more complex, that have to be searched using the Attentional Cascade. The use of 4 Microblaze cores gave a performance increase of 4.67x, far exceeding the expected performance increase of a multi-core design. The performance increase attributed to the use of 6 and 7 cores is very impressive and exceeds the theoretical increase of 7 cores by an order of magnitude. These results clearly show the benefits of customising the memory architecture of multi-core designs, especially for designs that are used for object detection applications.

For the QVGA images, using 1-3 cores surpassed the performance increases of the Lena and VGA test images, due to the QVGA images' smaller search space. However, when 4 cores are instantiated, we see a performance increase of 5.03x: far exceeding the theoretical performance increase by an order of magnitude. This large increase in performance is due to the finer threading of the algorithm, allowing more sub-windows to be evaluated and also due to the ability of cores to readily access the detection library of the Viola Jones algorithm. The pent- and hex-core design achieved performance increases of 6.7x and 8.38x respectively, again far surpassing the theoretical multi-core performance increase and reaffirming the benefits of our proposed architecture. However, we can see from Table 2 that the hex-core implementation achieved an impressive performance increase of 9.14x: an extra 2x better than the theoretical. This massive increase in performance can be attributed to our customised memory architectures and also to the smaller search space of the QVGA image.

5.2 Core Loads

We also offer an analysis of the average core loads of our designs when tested with all test images, the results of which can be found in Fig. 5. We can see how the use of ITR allows a good load balance over all cores for all designs. The biggest deviation between core loads is in the order of 30ms for the hex- and hept-core designs. These discrepancies can be attributed to the fine-grained threading employed. Despite the fine-grained threading allowing more cores to process an image scale, the complexity of the image scale is not distributed evenly across all available cores, due to the nature of the Viola Jones algorithm. If we consider the case where each sub-window of a scale could be evaluated concurrently by a core, we would still view load imbalances over the available cores since some sub-windows would contain a face, and some would not; as sub-windows that contain faces require more computation and therefore more time. This issue could be addressed through restructuring the algorithm, however we are more interested in the improvements that can be achieved through hardware and software design, as opposed to algorithmic design which has been considered by other works.

5.3 Literature Comparison

Figure 6 and Table 3 summarise the performance increases from the relevant literature, as well as our proposed solutions. Results from the literature are normalised to performance increases relative to 1 core. From Fig. 6 we see core instantiations of 3-6 cores outperform all performance

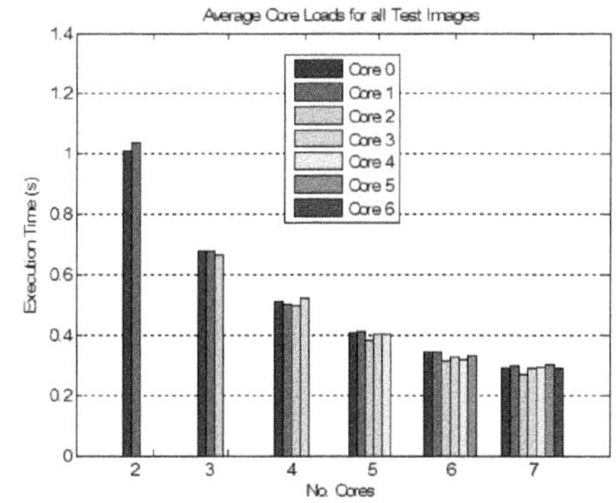

Figure 5: Average core loads of all designs with all test images

increases of [2] when tested with VGA and QVGA images. These performance increases are more impressive when we consider our test images are more computationally complex than the images tested in [2]. Our proposed dual-core solution achieves a higher performance increase than [8, 7, 4] over the use of 1 core. We again see from Table 3 our quad-core design outperforms those of [8, 7, 4] in terms of performance increase. However, when we compare the frame-rates obtained from [8, 7] to our own, we achieve superior frames-rates of 1.09FPS and 2.17FPS, compared to 0.24FPS and 0.45FPS for dual- and quad-core implementations respectively. When compared to [4], we achieve frame-rate improvements of 1.63x and 1.62x for dual- and quad-core designs respectively. Multi-core processors often suffer from performance increase issues when scaled, but here we have shown that by customising the memory architecture of a multi-core processor, maintaining core atomicity, and custom fine-threading that this problem can be overcome.

5.4 Power Consumption

Power consumption details for the proposed designs can be found in Table 4, as well as the average performance increase of the designs over all test images. From Table 4 we can see how static power consumption almost remains

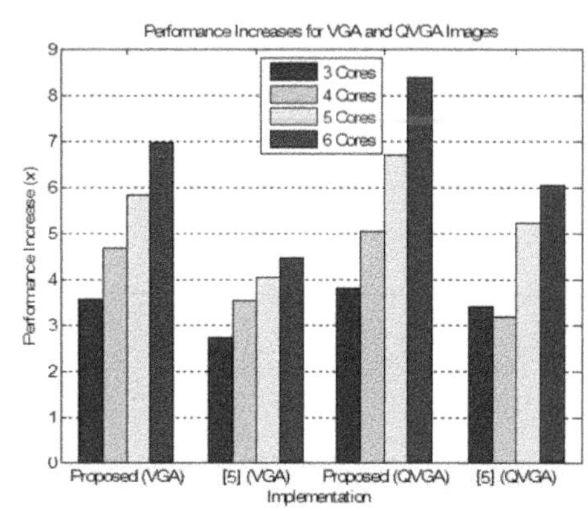

Figure 6: Performance improvements comparison for VGA and QVGA test images

Work	No. Cores	Improvement (x)	Frame-rate (FPS)
[8, 7]	2	2	0.24
	4	4	0.45
[4]	2	2.03	0.67
	4	4.05	1.34
Proposed	2	2.21	1.09
	4	4.41	2.17

Table 3: Performance increase comparisons for Lena test image

constant for all designs, with any deviations in the order of milliwatts. Dynamic power consumption gradually increases as the number of cores used increases due to the instantiation of more BRAM primitives. We can see the LME of each core only leads to an additional dynamic power consumption of approximately 200mW. However, we see from the average performance increase obtained for the designs that the total estimated power consumption increases only slightly when compared to the average performance increases obtained. The increased power consumption of designs using more cacheable memory is also a common theme for typical multi-core processors. We can therefore conclude our multi-core designs are low-powered and offer superior performance increase than current commercial embedded multi-core processors.

6. CONCLUSIONS AND FUTURE WORK

We have presented the findings of our initial investigation into the utilisation of soft-core processors for face detection using the Viola Jones algorithm. We customise the memory architecture of our designs via software and hardware optimisations. First, the memory architecture for program data is extended to accommodate the detection library of the Viola Jones algorithm. Secondly, main memory is partitioned into data buffers which are accessed via global pointers available to all cores. By considering data accesses before run-time, core accesses to main memory are controlled and contention all but eliminated. Using 7 Microblaze cores, we achieve a performance increase of up to 9.14x with a complex QVGA test image. We demonstrate how the typical shared L2 cache of MPs is not required to obtain impressive performance increases for multi-core object detection. Furthermore, when compared to the literature we obtain larger performance increases than embedded multi-core implementations and when compared to implementations that utilise ARM V5 cores, we achieve frame-rate improvements of up to 4.82x when using 4 Microblaze cores: thus proving the validity of our concepts and designs. Our implementations also consume low levels of dynamic power, with a near constant leakage for all designs. Future work will focus on the memory architecture of this initial design, where we are interested in developing distributed program and main memories that fully complement concurrent object detection applications. In doing this we aim to increase the level of parallelism available to the applications executing on these designs and the performance increases possible.

Version	Dynamic (W)	Static (W)	Total (W)	Average Performance Increase (x)
Standard	0.557	6.149	6.707	1.00
Single	0.702	6.155	6.857	1.22
Dual	0.929	6.163	7.091	2.35
Tri	1.181	6.172	7.353	3.54
Quad	1.394	6.18	7.574	4.78
Pent	1.596	6.187	7.784	6.01
Hex	1.824	6.195	8.02	7.29
Hept	2.038	6.203	8.241	8.32

Table 4: Power consumption details for all implementations and average performance increases of all test images

7. REFERENCES

[1] R. Ach, N. Luth, and A. Techmer. Real-time detection of traffic signs on a multi-core processor. In *Intelligent Vehicles Symposium*, pages 307 –312, june 2008.

[2] S.-K. Chen, T.-J. Lin, and C.-W. Liu. Parallel object detection on multicore platforms. In *Signal Processing Systems, SiPS. IEEE Workshop on*, pages 75–80, 2009.

[3] T. Chen, D. Budnikov, C. Hughes, and Y.-K. Chen. Computer vision on multi-core processors: Articulated body tracking. In *Multimedia and Expo, International Conference on*, pages 1862 –1865, july 2007.

[4] C.-H. Chiang, C.-H. Kao, G.-R. Li, and B.-C. Lai. Multi-level parallelism analysis of face detection on a shared memory multi-core system. In *VLSI Design, Automation and Test (VLSI-DAT), 2011 International Symposium on*, pages 1 –4, april 2011.

[5] T. Deepak Shekhar and K. Varaganti. Parallelization of face detection engine. In *Parallel Processing Workshops (ICPPW), 2010 39th International Conference on*, pages 113 –117, sept. 2010.

[6] D. Hefenbrock, J. Oberg, N. Thanh, R. Kastner, and S. Baden. Accelerating viola-jones face detection to fpga-level using gpus. In *Field-Programmable Custom Computing Machines (FCCM), 18th International Symposium on*, pages 11 –18, may 2010.

[7] B.-C. Lai, C.-H. Chiang, and G.-R. Li. Classifier grouping to enhance data locality for a multi-threaded object detection algorithm. In *Parallel and Distributed Systems (ICPADS), 17th International Conference on*, pages 268 –275, dec. 2011.

[8] B.-C. Lai, C.-H. Chiang, and G.-R. Li. Data locality optimization for a parallel object detection on embedded multi-core systems. In *Software Engineering and Service Science (ICSESS), International Conference on*, pages 576 –579, july 2011.

[9] A. Ranjan and S. Malik. Parallelizing a face detection and tracking system for multi-core processors. In *Computer and Robot Vision (CRV), 2012 Ninth Conference on*, pages 290 –297, may 2012.

[10] P. Viola and M. Jones. Rapid object detection using a boosted cascade of simple features. In *Computer Vision and Pattern Recognition, CVPR. Proceedings of IEEE Conference on*, pages 511–518 vol.1, 2001.

[11] L. Xin, C. Wenjie, M. Tao, and X. Lishuang. Real-time algorithm for sift based on distributed shared memory architecture with homogeneous multi-core dsp. In *Intelligent Control and Information Processing, International Conference on*, pages 839–843, july 2011.

A Novel and Improved Design of a Ternary CNTFET-based Cell

Geunho Cho and Fabrizio Lombardi
Northeastern University
Department of Electrical and Computer Engineering
360 Huntington Avenue, Boston, MA
gcho@ece.neu.edu and lombardi@ece.neu.edu

ABSTRACT
A novel ternary CNTFET-based SRAM cell is proposed in this paper; the operation of this CNTFET SRAM is nearly independent of the ternary values, therefore it is said to be balanced. Different from previous ternary cells, the proposed cell does not require a read buffer for changing the voltage level of the read bit line, because it uses additional CNTFETs to sink the bit lines to ground. By using four additional CNTFETs for ternary operation, a conventional (two-valued) sense amplifier is then used for output response. The contribution of this paper is not restricted to the design of the SRAM cell, but also the systematic modifications of the CNTFETs to substantially improve specific performance metrics. CNTFET features (such as sizing) and performance metrics (such as SNM, power delay product (PDP) and write/read times) are considered and assessed in detail. Extensive simulation results are provided to show that the proposed ternary SRAM cell has better performance compared to a previous CNTFET-based ternary cell as well as binary-based cells (implemented by either MOSFET or CNTFET).

Categories and Subject Descriptors
B.7.1 [**Integrated Circuits**]: Types and Design Styles – *Advanced Technologies, Memory Technologies, VLSI (Very Large Scale Integration)*

Keywords
CNTFET, Ternary SRAM design, Process variation

1. INTRODUCTION
Recently the number of application requiring memory intensive operations has considerably increased [1]; high memory capacity and low power dissipation are two stringent requirements in today's computing and communication systems. Ternary logic allows more information to be processed and stored for many computationally demanding applications [2]; ternary logic can reduce design complexity (such as area and interconnect) and reduce the computational requirements of many arithmetic functions for efficient parallel processing.

GLSVLSI'13, May 2–3, 2013, Paris, France.

The Carbon NanoTube Field Effect Transistor (CNTFET) is one of the most promising emerging technologies to extend and complement the silicon-based MOSFET [3]. Based on a previously proposed Standard Ternary Inverter (STI) [4], a ternary CNTFET-based cell has been proposed in [5] (and denoted as TC SRAM0 in this paper). As the threshold voltage of the CNTFET is function of the geometry (i.e. the chirality and diameter), a multi-diameter (multi-threshold voltage) CNTFET-based ternary memory cell design has been initially pursued in [5]. In the memory cell of [5], the read/write operations are executed by introducing in the design a transmission gate and a buffer to make these operations separate. Simulation results have shown [5] that TC SRAM0 performs the correct function during the read and write operations, even though its operations are dependent on the logic value (the worst performance occurs for a stored "1"). This unbalanced operation is caused by design features when handling the three states (in particular "1"); therefore, it raises significant concerns in terms of average performance metrics and stability. Also, voltage control of the three supply lines is still required and has design implications such as the requirement of a ternary-to-Boolean converter in the output response circuit.

A ternary CNTFET SRAM cell (denoted as TC SRAM1) with a balanced operational profile (i.e. not dependent on the stored value) is proposed in this paper; features such as sizing, SNM, power delay product (PDP) and write/read times are considered. The contribution of this paper is not restricted to the design of TC SRAM1; systematic modifications to substantially improve performance metrics are also proposed. The novel features of the proposed SRAM cell is that it uses additional CNTFETs to sink the bit lines to ground; this cell uses four additional CNTFETs for ternary operation. Different from TC SRAM0, TC SRAM1 does not require a read buffer for changing the voltage level of RBL; moreover, it requires a conventional (two-valued) sense amplifier for output response. Extensive simulation results are provided to show that the proposed ternary SRAM cell has better average performance for all considered metrics compared to the TC SRAM0 cell of [5] as well as binary-based cells (implemented by either MOSFET or CNTFET).

2. REVIEW
In this section, different features related to CNTFET technology and previous works on ternary design are reviewed. A CNTFET is usually manufactured by depositing CNTs to a substrate and is made from a sheet of graphene rolled into tube; a CNT can be either metallic or semiconducting depending on the chirality vector (i.e., the direction in which the graphene sheet is rolled). The chirality vector is denoted by (n,m) [3]; a CNT is metallic if $n=m$ or $n\text{-}m=3i$ (i is an integer), otherwise, CNT is semiconducting. In this paper, only single-walled semiconducting

CNTs are considered and when the chirality vector is changed, it is assumed that m is always zero. The diameter of a CNT (D_{CNT}) is determined based on the chirality vector (n,m) and is an important parameter when determining the current of a CNTFET. The threshold voltage also changes depending on D_{CNT} because the threshold voltage of the intrinsic CNT channel can be approximated to the first order as the half bandgap and an inverse function of the diameter [6][7]. D_{CNT} also changes the amount of current of a CNTFET, because the CNT is used as conducting channel between the drain and the source. So, when the chirality of a CNTFET has a small value, a small current flows only when a high voltage is applied.

Based on the previously proposed STI inverter [4], this paper proposes a ternary CNTFET SRAM cell. The performance of this new CNTFET SRAM cell is also compared with existing binary MOSFET and CNTFET SRAM cells, denoted as BM and BC SRAM respectively. For fair comparison, the gate length of the MOSFET and the CNTFET is the same, i.e. 32 nm (corresponding also to the minimum gate width) at a 0.9V supply. When the gate width is determined in a CNTFET circuit, general width/length ratios (such as 0.5, 1, and 2) are used as also applicable to CMOS circuit design. The number of CNTs in a CNTFET is determined by considering also the pitch (given by 32nm). The well-known Predictive Technology Model (PTM) for conventional 32nm MOSFET technology [10] and Stanford University CNFET model [11] are utilized in this simulation.

In the TC SRAM cells, the threshold voltages of the CNTFETs are 0.290V, 0.550V, and 0.423V based on the chirality vectors (19,0), (10,0) and (13,0) respectively [4][5]. Due to the high threshold voltages (0.423V and 0.550V) and the configuration of the STI inverter, it takes longer to read the stored value than for a BM or BC SRAM cell; so, additional transistors are required to reliably and quickly read the stored value. Figure 1 shows TC SRAM0 [5]; it uses single-ended access mechanisms for the write and read operations. The data on the Write Bit Line (WBL) is written into the memory cell through the write transmission gates (MN1 and MP1) as data is forced. The read buffer consists of a P-CNTFET and a N-CNTFET (with threshold voltages of -0.559V and 0.559V respectively). Before reading the stored value, RBL is charged to 0.45V. Therefore, when the cell is storing a '0' ('2'), MN3 and MP3 are turned on and off (off and on). The two values can be read through the read transmission gates (MN2 and MP2). When the cell stores a'1', both MN3 and MP3 are turned off; so, RBL remains at 0.45V. Therefore, to read the stored value in TC SRAM0, a sense amplifier that can detect three voltage levels, is required.

3. PROPOSED CELL

Figure 2 shows the proposed TC SRAM cell (TC SRAM1); it uses additional CNTFETs to sink the bit line charge to ground, so TC SRAM1 does not require a read buffer for changing the voltage level of RBL. The chirality vector of MN11 and MN12 is (19,0); so, MN11 and MN12 are turned on when 0.45V or 0.9V are stored at q or nq. Different from TC SRAM0, TC SRAM1 requires a (conventional) sense amplifier with two output signals. i.e. the output signals' values are '0.9V and 0V', '0.9V and 0.9V', and '0.9V and 0V', corresponding to the '0', '1', and '2' values stored in the cell. However, this design needs proper modifications to operate for best performance.

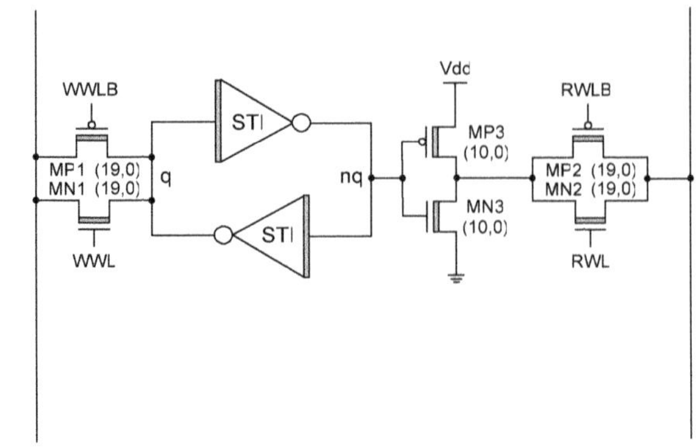

Figure 1. Ternary CNTFET SRAM cell of [5] (TC SRAM0)

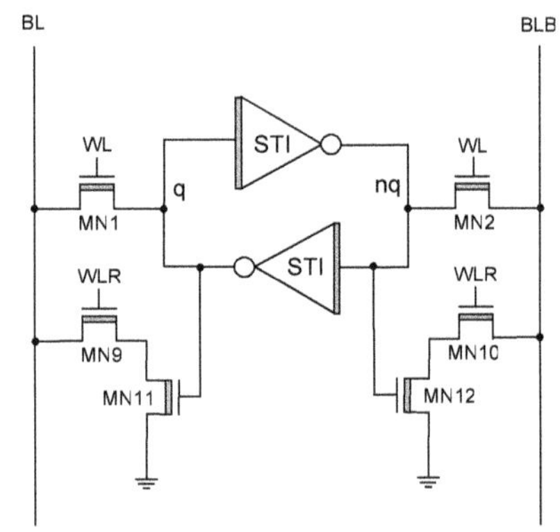

Figure 2. Proposed ternary CNTFET SRAM cell (TC SRAM1)

Consider first sizing; when sizing a 6T SRAM cell, the MN4/MN2 and MP5/MN1 ratios must be optimized [10][11]. An appropriate sizing ratio between MN4 and MN2 is required, such that the voltage at node 'nq' is lower than the threshold voltage (V_{th}) and the stored logic value does not change during the read operation. During the write operation, the pass gate MN1 must be significantly more conductive than MP5: so, the voltage at node 'q' should be greater than V_{th}.

Sizing of TC SRAM1 is based on CNTFET grouping. Consider the memory cell and corresponding inverters shown in Figure 3. Each group of three CNTFETs (i.e. MN3, MN4, MP5, and MP6) is defined by its CNTFET label; the member of each group is then numbered accordingly (for example, the members of MN3 are MN31, MP32, and MP33). Using the grouping of Figure 3, the gate widths of the three CNTFETs in each group are changed at the same time. The increase in voltage at node 'nq' versus the MN4/MN2 ratio (at node 'q' versus the MP5/MN1 ratio) is shown in Figure 4 (Figure 5) [10][11]. Figures 4 and 5 show that the voltages at 'nq' and 'q' of TC SRAM1 are similar to a BM SRAM cell (the V_{th} of a MOSFET is 0.160V and the smallest V_{th} of a CNTFET in TC SRAM1 is 0.289V). The MN4/MN2 and MP5/MN1 ratios that are used in BM SRAM, BC SRAM, and TC SRAM1 and the gate widths, are given in Table 1.

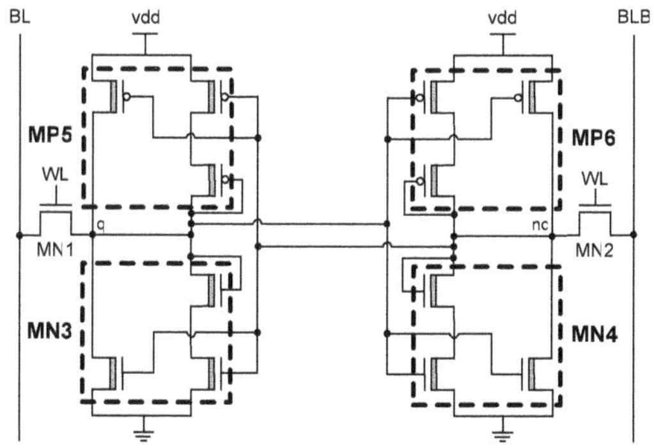

Figure 3. Grouping CNTFETs for sizing

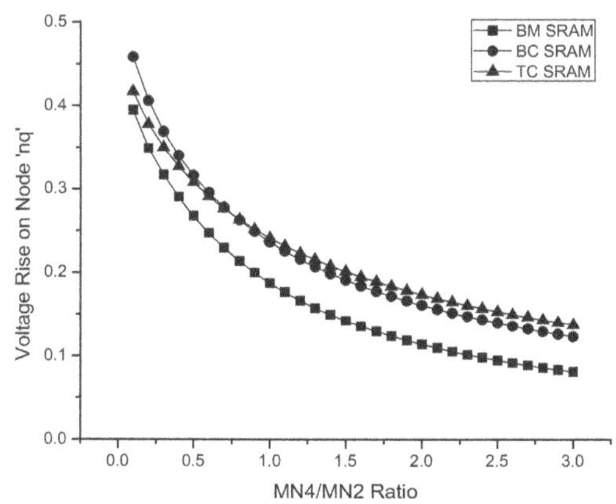

Figure 4. MN4/MN2 ratio versus voltage increase at nq

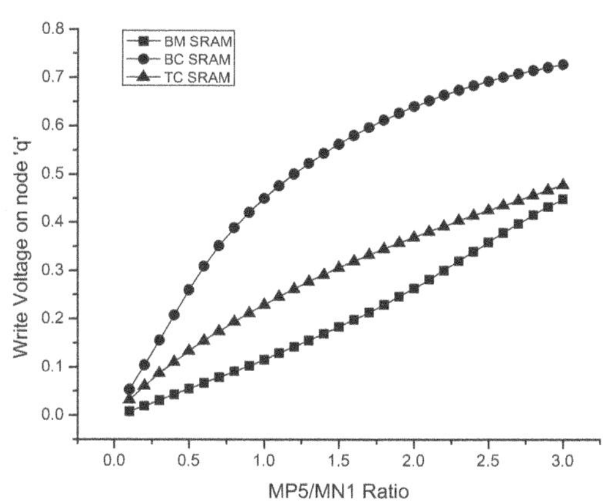

Figure 5. MP5/MN1 ratio versus voltage increase at q

A concern of TC SRAM0 that is addressed by TC SRAM1, is the feature of balanced operation. The static noise margin (SNM) must be considered; the SNM is an important measure for stability of a SRAM design. The SNM is defined as the maximum value of DC noise voltage that can be tolerated by the SRAM cell without changing the value of the stored bit [12]. Figure 6 shows the Voltage Transfer Characteristics (VTC) diagram of the STI of [4]; its SNM has two different values (Figure 7), so the SRAM cell with this STI has different static stabilities depending on the stored value, i.e. it is unbalanced. A new STI with static stability of nearly equal values (balanced) is therefore needed. In the modified STI, the middle voltage levels of the VTC (corresponding to the smallest value in SNM) are increased by changing the chirality vector from (10,0) to (8,0) for MN32, MN42, MP52, and MP62 of Figure 3. The VTC diagram of the modified STI is plotted in Figure 6, while Figure 7 shows the SNM; the SNM values are now close regardless of the ternary voltage levels, i.e. it is balanced.

Table 1. Transistor Ratio and Gate Width

	Ratio		Gate Width (nm)		
	MN4 /MN2	MP5 /MN1	MN1 MN2	MN3 MN4	MP5 MP6
BM SRAM	1.5	0.5	64	96	32
		1.0	96	64	64
	2.0	0.5	64	128	32
		1.0	32	64	32
	2.5	0.5	64	160	32
		1.0	32	80	32
	3.0	0.5	64	192	32
		1.0	32	96	32
BC SRAM	1.0	0.5	64	64	32
	1.5		64	96	32
	2.0		64	128	32
	2.5		64	160	32
	3.0		64	192	32
TC SRAM1	0.5	0.5	64	32	32
		1.0	64	32	64
	1.0	0.5	64	64	32
		1.0	32	32	32
	1.5	0.5	64	96	32
		1.0	64	96	64
	2.0	0.5	64	128	32
		1.0	32	64	32
	2.5	0.5	64	160	32
		1.0	64	160	64
	3.0	0.5	64	192	32
		1.0	32	96	32

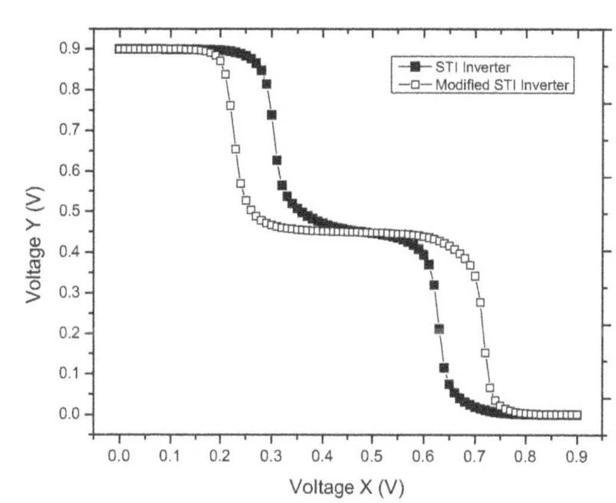

Figure 6. VTC diagram of STI and Modified STI inverters

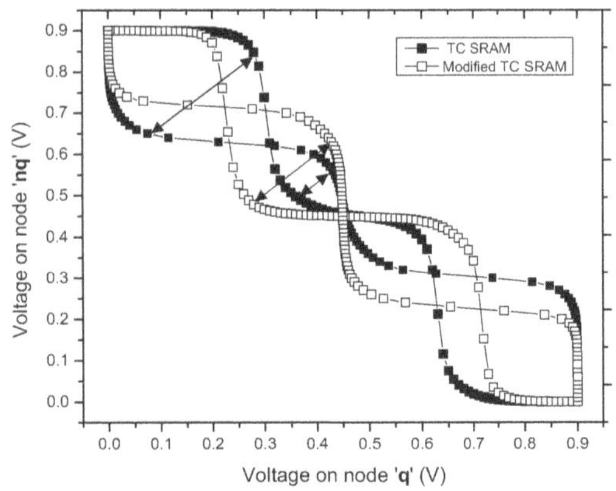

Figure 7. VTC diagram of TC SRAM cells

Next, the improvement in PDP of TC SRAM1 is analyzed by changing the chirality of individual CNTFETs (so, differently from the improvement in SNM in which the chirality of multiple CNTFETs is changed at the same time as previously). In TC SRAM1, the chiralities of MN33, MN43, MP53, and MP63 are very important, because these CNTFETs operate effectively as resistors (so, the VTC is not significantly affected). However if the chirality vectors of other CNTFETs are changed, the VTC diagram is adversely modified, because the voltage levels change.

It is important to note also (as verified by simulation) that when the chirality vector '*n*' of MN33, MN43, MP53 and MP63 is changed from 10 to 14, there is no significant change in the SNM of TC SRAM1. As discussed previously, the delay and PDP values are affected by the stored value in the cell. So in Figure 8 when the chirality is increased from 13 to 14, the average delay (PDP) is decreased (increased) by 2.217% (14.046%). However, when the chiralities of these CNTFETs are decreased to 10, the delay (PDP) is increased (decreased) by 2.183% (28.269%). In a CMOS inverter, the finite slope of the input signal causes a direct current path (short-circuit current) between V_{DD} and GND for a short period of time during switching (the NMOS and PMOS transistors are conducting simultaneously); this is the main cause of dynamic power consumption in an inverter [13]. By decreasing the chirality of a CNTFET used as a resistor, the short-circuit currents are decreased. Therefore, the chiralities of the CNTFETs MN33, MN43, MP53 and MP63 are reduced from 13 to 10 for achieving a better PDP.

In a binary SRAM cell, the number of voltage values possible at 'q' and 'nq' is given by two, i.e. 0.9V and 0V. So, performance metrics (such as delay, power, and PDP) mostly reflect a nearly equal change in these two values for 'q' and 'nq' due to the symmetric structure of the cell itself. This condition is not generally applicable to a ternary SRAM cell, i.e. there are six combinations of values due to the three possible voltages (0V, 0.45V, and 0.9V). However, among these six combinations, three of them have same performance values, because the proposed SRAM cell has symmetry in its structure and balanced operation. As mentioned previously, 0, 1, and 2 are given by 0V, 0.45V, and 0.9V respectively. Therefore, a binary SRAM cell has a single type of value change (such as 0 -> 2 (2 -> 0)), while a ternary SRAM cell has three types of value change (such as '0 -> 1 (2 -> 1)', '0 -> 2 (2 -> 0)', and '1 -> 0 (1 -> 2)'). The MN4/MN2 and MP5/MN1 ratios that are utilized in BM SRAM, BC SRAM and TC SRAM are reported in Table 2. The write operation of TC SRAM1 is similar to the BM SRAM cell; after setting up the desired voltage level for BL and BLB, MN1 and MN2 are turned on. As mentioned previously, MN9 and MN10 are turned on with a high voltage at WLR for the read operation of TC SRAM1. Having established the gate width and chirality of each CNTFET, Figures 9 and 10 show the write and read operations of TC SRAM1 for all three values (0, 1 and 2). BL and BLB are discharged faster when 0 or 2 are stored (Figure 10), because a smaller voltage is applied to the gates of MN9 and MN10 when a 1 is stored.

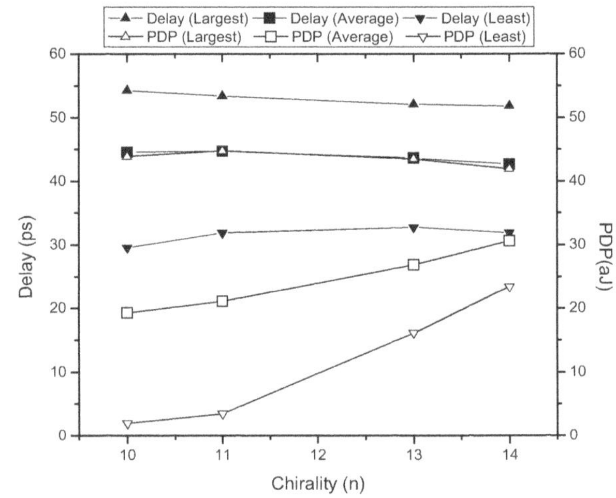

Figure 8. Delay and PDP when chirality vectors in MN33, MN43, MP53, and MP63 are changed

Table 2. Transistor Ratio and Performance

	Change of Values	Ratio		Performance				SNM (mV)
				Write			Read	
		MN4 /MN2	MP5 /MN1	Delay (ps)	Power (uW)	PDP (aJ)	Delay (ps)	
BM SRAM	0 -> 2 (2 -> 0)	2.0	1.0	49.009	1.648	80.762	58.061/ N/A	364.962
BC SRAM	0 -> 2 (2 -> 0)	1.0	0.5	25.922	1.504	38.984	73.211 / N/A	456.388
TC SRAM1	0 -> 1 (2 -> 1)	0.5	0.5	54.259	0.036	1.966	129.450 / 281.930	221.539 /181.573
	0 -> 2 (2 -> 0)	0.5	0.5	49.851	0.880	43.844		
	1 -> 0 (1 -> 2)	0.5	0.5	29.558	0.405	11.972		
	Avg.	0.5	0.5	44.556	0.440	19.261	205.690	201.556
	Std. Dev.	0	0	13.174	0.423	21.870	107.820	28.260

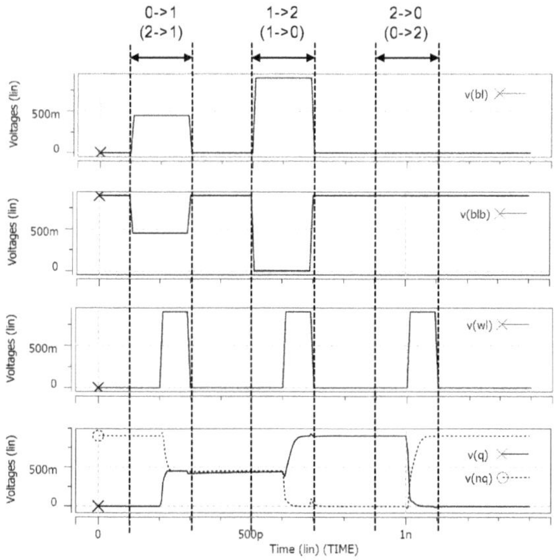

Figure 9. Write operation of TC SRAM1

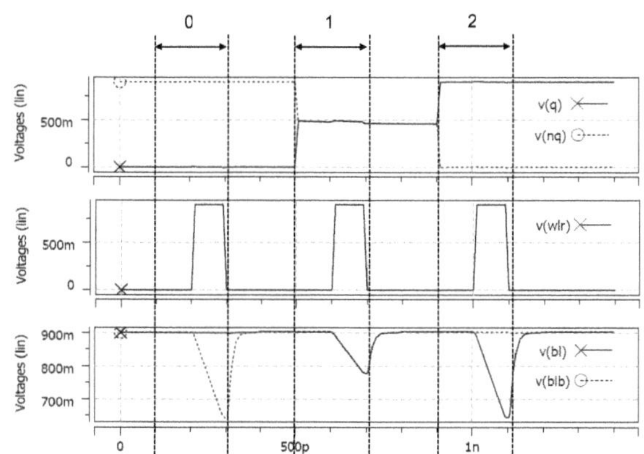

Figure 10. Read operation of TC SRAM1

However TC SRAM1 is expected to be slower than a BC SRAM for an average write/read delay. So a modification to TC SRAM1 is required to reduce the delay by considering only those CNTFETs involved in the read/write circuits. Consider the MN4/MN2 and MN5/MN1 ratios in Table 2; MN1 and MN2 in the write circuits should be more conductive than other CNTFETs, i.e. it is worth to increase the chirality in the CNTFETs of the write circuit to have a larger current flow through MN1 and MN2. To increase the current, the gate width and the number of CNTs in each CNTFET can be increased at the same time. However, this would result also in an area penalty; so only the chirality must be increased for increasing the current, while decreasing the write/read delays.

4. EVALUATION

Table 2 shows also different figures of merit such as delay, power consumption, PDP and SNM for BM SRAM, BC SRAM and TC SRAM1 by considering the various input vectors (as given by the column of 'Change of Values' in Table 2) and discussed in a previous section. The following features can be observed by analyzing the results of Table 2. (1) TC SRAM1 has a substantially lower PDP than the BM and BC SRAM cells. At the same supply voltage, a smaller SNM is unavoidable in a ternary cell, because more voltage levels are utilized compared to a binary cell. (2) On average, TC SRAM1 accomplishes significant improvements in PDP and power dissipation.

TC SRAM1 is also compared with TC SRAM0 of [5]; the simulation results for TC SRAM0 are given in Table 3. Simulation has been performed under the same operating conditions such as clock cycle, rise/fall time, delay points and capacitance in bit lines. The simulation results in Table 3 are different from those reported in [5], because MP1 and MP2 of Figure 1 are now included in the measurements. In Table 3, TC SRAM0 has six sets of performance metrics (write/read delay, power, and PDP) due to its unbalanced circuit nature.

Compared to TC SRAM1, TC SRAM0 has a larger delay, power consumption and PDP (5.700%, 34.980%, and 58.328% on average); for the write operation. Moreover, these values are more deviated (larger standard deviation values) depending on the figure of merit (104.964%, 13.871%, and 58.133% in delay, power and PDP on average respectively). For the read operation, TC SRAM0 incurs in a smaller delay (by 17.736%) on average and additional power consumption is due to the read buffer and the charging process of RBL (as reported in Table 3 also for the read operation). Moreover, the power to charge RBL is determined by its capacitance (a value of 10fF is used in this paper [13]) and increases as function of the size of the SRAM memory. In theory when a 1 is read, the voltage value of RBL should not change (MP3 and MN3 should be turned off). However, as mentioned previously the threshold voltage of MN3 (MP3) in Figure 1 is 0.559V (-0.559V) and the voltage of nq is 0.45V when a 1 is stored; so, MP3 and MN3 are slightly turned on at the same time. So as found by HSPICE simulation, the power consumption is 0.621uW and the voltage level of RBL decreases (more than 0.1V) at the frequency of operation considered in this paper.

The average PDP of TC SRAM1 has been reduced compared with TC SRAM0 due to the significantly lower power consumption, albeit a modest decrease in average delay (for both operations) is also encountered. The significant advantage of TC SRAM1 is the significant reduction in deviation of the SNM whose value is independent of the stored value. As mentioned previously, TC SRAM1 has a better PDP than the BM and BC SRAM cells, but as expected TC SRAM1 is slower than a BC SRAM for the average write/read delay.

Table 3. Read and Write Delay of TC SRAM0 Cell [5]

	Write				Read				SNM (mV)
	Change of Values	Delay (ps)	Power (uW)	PDP (aJ)	Values	Delay (ps)	Power (uW)	PDP (aJ)	
TC SRAM0	2 -> 1	76.363	1.223	93.369	0	159.040	0.026	4.126	299.993 /99.339
	0 -> 1	81.554	0.282	22.984					
	0 -> 2	45.714	0.122	5.578	1	63.409	0.621	39.345	
	2 -> 0	40.364	1.154	46.560					
	1 -> 0	17.293	0.549	9.500	2	282.110	2.400	677.064	
	1 -> 2	21.286	0.234	4.983					
	Avg.	47.096	0.594	30.495	**Avg.**	168.186	1.015	240.179	199.666
	Std. Dev.	27.002	0.482	34.584	**Std. Dev.**	109.637	1.235	378.763	141.883

As discussed in a previous section, the modification to TC SRAM1 to reduce the delay by increasing the chirality in those CNTFETs involved in the read/write circuits has been evaluated. Figure 11 shows that as chirality increases, the delay and the PDP decrease at the same time, because more current flows through the write circuit, while the power consumption remains nearly unchanged. At some point (when the chirality is 20 and 22), the PDP increases, because more current flows (more power is consumed) than the reduced transient time. Using a similar process, the read time can also be decreased by increasing the chirality in the read circuit (consisting of MN9, MN10, MN11, and MN12); Figure 12 shows that the read delay decreases as chirality increases.

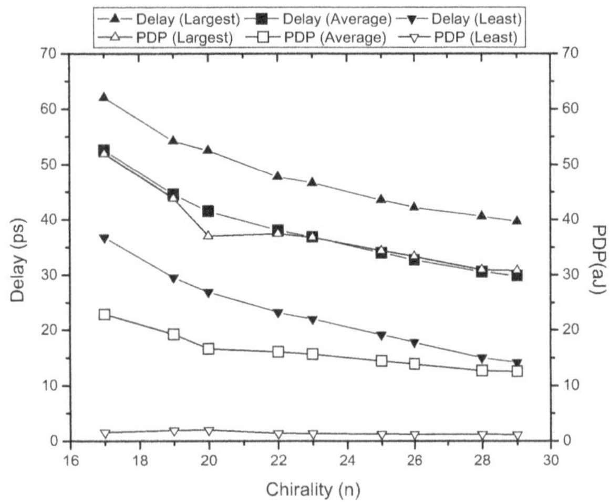

Figure 11. Largest, average, and least write delay and PDP values when chirality in CNTFETs of write circuits (MN1 and MN2) is changed in TC SRAM1

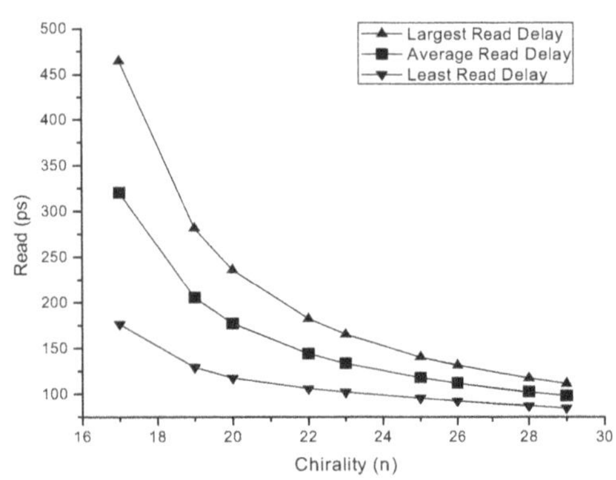

Figure 12. Largest, average, and least read delay when chirality in CNTFETs of read circuit (MN9, MN10, MN11, and MN12) is changed in TC SRAM1

5. CONCLUSION

This paper has presented a ternary CNTFET SRAM based on a previously suggested standard ternary inverter. This cell denoted as TC SRAM1 offers balanced operation (not dependent on the stored value in the memory). Different from a previous ternary cell [5], the proposed cell does not require a read buffer for changing the voltage level of the read bit line, because it uses additional CNTFETs to sink the bit lines to ground. By using four additional CNTFETs for ternary operation, a conventional (two-valued) sense amplifier is used for output response. Its operation has been analyzed in detail; appropriate sizing of the CNTFETs has been employed. Sizing has been performed by grouping the CNTFETs of the ternary cells; chirality of the CNTFETs has been changed to improve the SNM and to reduce power dissipation while retaining balanced operation. To reduce the write/read times in a ternary CNTFET SRAM, chirality of the CNTFETs in the write and read circuits has been increased, yielding significant improvements in PDP for the proposed SRAM cell.

6. REFERENCES

[1] M. Qazi, M.E. Sinangil, A.P. Chandrakasan, "Challenges and Directions for Low-Voltage SRAM," *IEEE Design & Test of Computers*, vol.28, pp.32-43, 2011.

[2] P.C. Balla, A. Antoniou, "Low power dissipation MOS ternary logic family," *IEEE Journal of Solid-State Circuits*, vol. 19, no. 5, pp. 739-749, 1984.

[3] A. Raychowdhury, K. Roy, "Carbon-Nanotube-Based Voltage-Mode Multiple-Valued Logic Design," *IEEE Transaction on Nanotechnology*, vol. 4, pp. 168-179, 2005.

[4] S. Lin, Y. Kim, F. Lombardi, "CNTFET-Based Design of Ternary Logic Gates and Arithmetic Circuits," *IEEE Transactions on Nanotechnology*, vol. 10, pp. 217-225, 2009.

[5] S. Lin, Y. Kim, F. Lombardi, "Design of a Ternary Memory cell using CNTFETs," *IEEE Transactions on Nanotechnology*, vol. 11, pp. 1019-1025, 2012.

[6] J. Deng, H.-S. P. Wong, "A Compact SPICE Model for Carbon-Nanotube Field-Effect Transistors Including Nonidealities and Its Application - Part I: Model of the Intrinsic Channel Region," *IEEE Transactions on Electron Devices*, vol 54, pp. 3186-3194, 2007.

[7] J. Deng, H.-S. P.Wong, "A compact SPICE model for carbon nanotube field effect transistors including non-idealities and its application—Part II: Full device model and circuit performance benchmarking," *IEEE Transactions on Electron Devices*, vol. 54, pp. 3195–3205, 2007.

[8] Stanford University Nanoelectronics Group. CNFET Models. http://nano.stanford.edu/models.php.

[9] Arizona State University. Nanoscale Integration and Modeling (NIMO) Group. PTM Models. http://ptm.asu.edu

[10] S. Lin, Y-B. Kim, and F. Lombardi, "Design of a CNTFET-Based SRAM Cell by Dual-Chirality Selection," *IEEE Transactions on Nanotechnology*, vol. 9, pp.30-37, 2010

[11] A. Chandrakasan,W. J. Bowhill, and F. Fox, "Design of High-Performance Microprocessor Circuits." Piscataway, IEEE Press, 2000.

[12] E. Seevinck, F. J. List, and J. Lohstroh, "Static-noise margin analysis of MOS SRAM cells," *IEEE J. Solid-State Circuits*, vol. 22, no. 5, pp. 748–754, Oct. 1987.

[13] J. M. Rabaey, A. Chandrakasan, and B. Nikolic, "Digital Integrated Circuits: A Design Perspective (2nd Edition)," Prentice Hall, 2003.

Variability Evaluation of Feedback Circuits Used in Nanoelectronic Memristive/CMOS Circuits

Arne Heittmann and Tobias G. Noll
Chair of Electrical Engineering and Computer Systems
RWTH Aachen University
Schinkelstr. 2, D-52062 Aachen, Germany

ABSTRACT

One significant challenge in the design of nanoelectronic circuits which comprise resistive switches (RS) beyond 10 nm feature size is to control the variability and the irreproducibility of RS device conductance. For RS, which are based on the electrochemical metallization effect (ECM), this variability is – amongst others - caused by random discrete electrochemical reactions during the filament growth. Based on a physical model for ECM cells a 1D Monte Carlo circuit simulation model was derived which models the occurrence of individual electrochemical reactions by a Poisson process. By circuit simulation it was possible to show that the conductance variation during the filament growth can be controlled by a regulating device, which establishes a negative feedback on the electrochemical reaction rate. In particular, for three different load types (current source, linear resistor, and source coupled NMOS transistor) the regulating properties in regard to variability reduction (and partly observed in experiments) were shown by Monte Carlo simulation. It was especially possible to show that the obtained conductance variation appears to be the consequence of the respective dynamic joint interaction between RS and the regulating device and is –in turn- not a predetermined property of the RS alone.

Categories and Subject Descriptors

B.3.3 [Memory Structures]: performance analysis and design aids;

Keywords

nanoelectronics, resistive switches, hybrid circuit, electrochemical metallization, monte carlo simulation, variability, irreproducibility, scaling

1.INTRODUCTION

In recent years resistive switches (RS) have gained considerable attention as these devices provide non-volatile and multilevel storage capabilities. In particular, these devices are commonly seen as an area-efficient successor to well-established storage devices since their comparatively simple setup may help to overcome the technological as well as economical limits anticipated for scaling CMOS beyond the 10nm node [1]. Due to their multilevel capabilities a broad spectrum of potential application are foreseen for RS such as memories, logic which is based on look-up tables (LUTs) and programmable wiring junctions [2], and very specialized applications such as filters [3] and neuromorphic circuits [4]. However, scaling beyond 10nm feature size - especially for resistive switches - has serious consequences in respect to reliability and variability. By now, the most crucial challenge for the utilization of the multilevel capabilities is to cope with the variability of device conductance which has been observed experimentally [5,6,7]. The variability seriously harms the overall circuit performance [5] by the degradation of signal swing, signal margin, and signal to noise ratio. Even if the multilevel capability is not needed (for instance in simple memories), the variability of RS devices requires special write schemes in passive nanoelectronic crossbars [8].

Both, the evaluation of circuit performance as well as the introduction of effective means in order to actively limit the variability can only be accomplished in a systematic way by (i) identifying fundamental sources of randomness in the dynamics of the switching process, and (ii) providing simulation models, which reproduce a realistic device behavior. Although many device models are found in literature today [9-14] which are suitable for circuit simulation, variability is covered only by the ability to vary model parameters which could be drawn from pre-defined distribution functions. However, these models are unable to capture the experienced cycle-to-cycle variations of RS conductance, which is frequently observed in almost all experimental setups [5]. Furthermore, it is not possible to meaningfully capture scaling effects and effects of a joint interaction between the programming circuit and an RS.

In order to incorporate variability and irreproducibility into a circuit simulation flow the sources of variability have to be modeled. In this paper, modeling is done for a particular class of resistive switches which are based on the Electrochemical Metallization effect (ECM) [14-16]. For this particular device class a closed physical device model has been found in literature [14] which is based on the description of physical processes. Based on this model a simple model extension was proposed, which covers the statistics of random ion deposition on a filament. Derived from this model an equivalent electrical circuit of the statistical device model is presented. The equivalent circuit is implemented in such a way that it can be executed on standard circuit simulation platforms such as SPICE or SPECTRE. As an application, four different circuit setups for conductance programming comprising an RS and a particular load circuit were examined in regard to explore the variability of device conductance.

The examined load circuits are partly known from literature and are well known for their variation-limiting properties by experience. By extensive Monte Carlo simulations using the proposed device model it was possible to demonstrate the variation-limiting properties of the load circuits. In particular, the standard deviations of the device conductance were quantified which allows for comparison of the variation-limiting quality of particular programming circuits. Finally, the variation-limiting properties of the examined circuits were theoretically examined and reasoned.

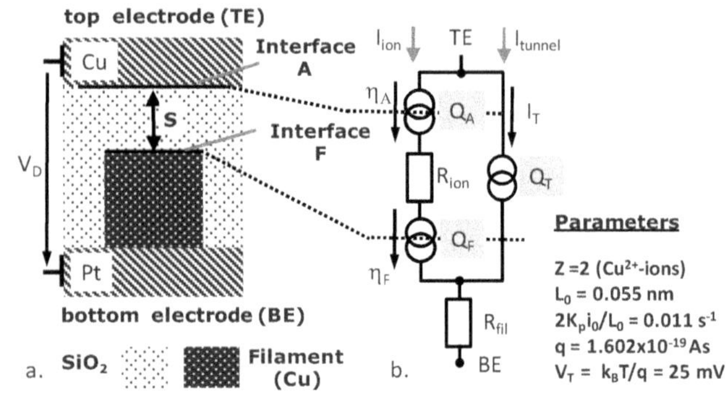

Figure 1: a. Cross section of an ECM device based on $Cu/SiO_2/Pt$. b. equivalent circuit which models ionic and tunnel current over a gap of size S.

2. THE EXTENDED ECM DEVICE MODEL

The basic device model is derived from the results presented in [14]. In Fig. 1a a cross section of an ECM device based on Cu-SiO_2-Pt is illustrated. The electrodes are separated by an electronically insulating material which additionally acts as an ion conducting layer (electrolyte). If a voltage V_D is applied, electrochemical active ions dissolve into the ion conducting layer and drift towards the opposite electrode. This counterelectrode is separated from the top electrode by a gap of size S. The deposition of ions on the counterelectrode results in the continuous growth of a filament while the gap continuously shrinks.

The current density for the charge transfer across the electrolyte-electrode interface F during the cathodic reduction is described by the Butler-Volmer-equation [16]

$$I_{ion} = A_{fil} \cdot i_0 \cdot \left(e^{\alpha \cdot z \cdot \eta_F / V_T} - e^{(\alpha - 1) \cdot z \cdot \eta_F / V_T} \right). \tag{1}$$

In (1) A_{fil} denotes the area of the filament surface, i_0 is the reaction rate at zero voltage, V_T is the temperature voltage, z the charge count of the involved ions, η_F the potential difference across the interface F (outer Helmholtz layer) and α a symmetry factor [17]. A similar equation holds for the interface A. Since the ionic charge transport is associated with material transport towards the filament, the filament growth is proportional to the ion current density and is reflected by a shrink of the gap size S

$$\frac{dS}{dt} = -K_p \cdot \frac{I_{ion}}{A_{fil}}. \tag{2}$$

In (2) K_p models the relation between electronic charge transport and matter deposition. For small gap sizes S the voltage across both interfaces (A and F) is approximately half of the device voltage V_D. As soon as S becomes smaller than approximately 0.8 nm the electronic current becomes dominated by a tunneling current which depends on the gap size S [14]. As soon as the tunnel current dominates the overall device conductance can be approximated by

$$G \approx \frac{q^2}{4 \cdot \pi \cdot h} \cdot e^{-S/L_0} \cdot \frac{A_{fil}}{S \cdot L_0} \tag{3}$$

using the definition

$$L_0 = \frac{1}{\sqrt{2m_e \cdot q \cdot \phi_0}} \cdot \frac{h}{4 \cdot \pi}. \tag{4}$$

From (3) the sensitivity of G with regard to the gaps size S is given by

$$\frac{\partial G}{\partial S} \approx -\frac{1}{L_0} \cdot G(S), \tag{5}$$

which constitutes an almost exponential dependency of G from S:

$$G(S + \Delta S) = G(S) \cdot e^{\Delta S / L_0}. \tag{6}$$

Consequently, the deposition of few atoms on the active filament surface has to be controlled if a particular conductance has to be programmed assuming a required precision.

The ECM model described by equations (1)-(4) establishes a macroscopic view of the device behavior which can be efficiently used for circuit simulation. However, in this model variability of the device behavior can only be constituted by variation of individual model parameters. Such approach has been applied for instance in [18] and is useful if the active device area is larger than roughly $100nm^2$ [8]. Such approach involves the property of repeatability of the I-V characteristics; under identical conditions (parameters, voltages, currents, temperature, initial conditions) the simulated device *perfectly* repeats its characteristics. By considering devices which are scaled below $100nm^2$ the switching characteristics of a bipolar resistive switch depends more and more on the (random) movement (or deposition) of individual particles (e.g. ions, atoms, vacancies). Due to the stochastic nature of the particle movement the device characteristics appears to be irreproducible at the nanoscale. In [8] the impact of irreproducibility on circuit design is discussed for a particular write scheme used in passive nanoelectronic crossbars.

In order to obtain a device model which includes the property of irreproducibility one has to model the microscopic device physics. Based on this microscopic device model a macroscopic model can be derived which keeps the most significant (stochastic) properties of the microscopic level. With respect to ECM devices, we will concentrate on the process of filament growth. At the nanoscale the growth of a filament is a discrete process since only integer numbers of atoms are participating in the matter deposition on the filament. The deposition sites as well as the time of the depo-sition events are to some extent random. An appropriate stochastic model for the discrete ion deposition is given by a Poisson process. Poisson models have been successfully used for kinetic Monte Carlo simulations describing epitaxial growth [19] and in particular for the description of the growth of filaments [6].

In a Poisson process subsequent events are exponentially distributed. In the extended ECM device model events are associated with electrochemical reactions (i.e. charge transfer and deposition). The particular average rate of such reactions is given by

$$r = \frac{I_{ion}}{qz} \tag{7}$$

using (1) as the ionic current, which is the result of an applied electrical stimulus delivered from a given load device.

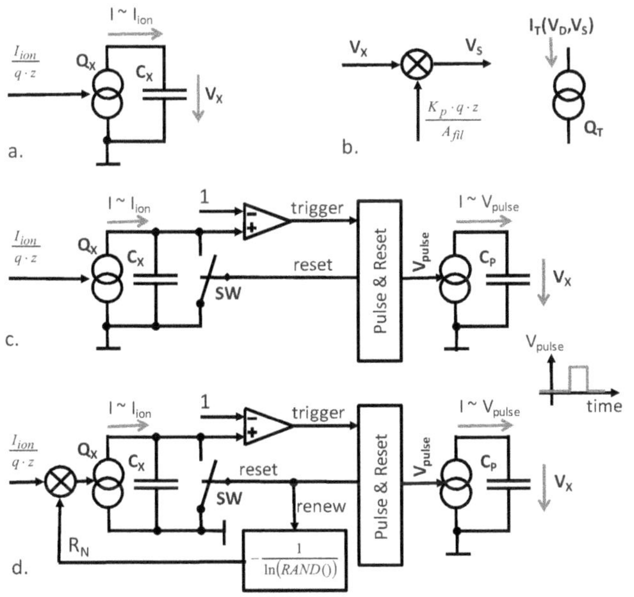

Figure 2: a) continuous integration of the ionic current (2). V_X represents the accumulated number of events. b. transformation of events to gap size S and current source which models the tunnel current, which depends on $S(V_S)$ and V_D. c. pulsed implementation of a. Each event is represented by a pulse V_{pulse} which are integrated on C_P. d. modulation of event rate by random numbers. The pulses on V_{pulse} are exponentially distributed.

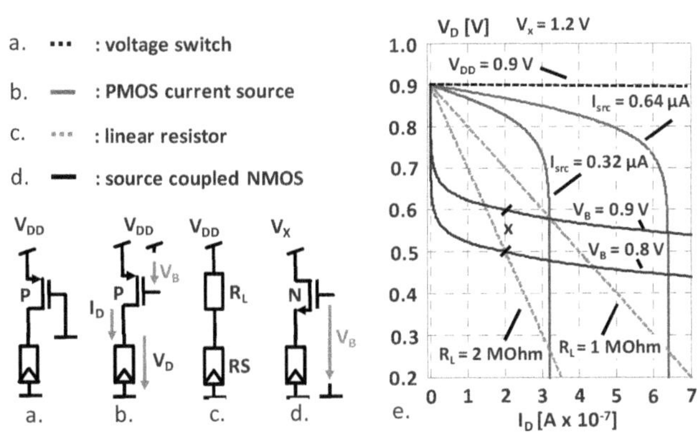

Figure 3: Four different load circuits for programming the low resistive state (LRS) of a resistive switch. e. I-V characteristics for exemplary parameters of the load circuits. blue curves: PMOS current source; red curves: linear resistor; black curves: source coupled NMOS.

For exponentially distributed events the probability density function (PDF) is given by

$$f(r_i,t)=\begin{cases} r_i \cdot e^{-r_i \cdot t} & t \ge 0 \\ 0 & otherwise \end{cases} . \tag{8}$$

In (8) r_i specifies the average event rate and t the time. Then, the function $f(r_i,t)$ specifies the relative frequency of events observed at time t. However, for m subsequent and independent events the time T proceeds, and at

$$T = \sum_{i=1}^{m} t_i \tag{9}$$

the *m-th* event is observed. For any event (i.e. an electrochemical reaction) the gap size changes by

$$\Delta S = \frac{K_p \cdot q \cdot z}{A_{fil}} \tag{10}$$

as modeled by (2). If it is assumed that the stimulus to the RS is applied with known initial conductance G_0 the obtained conductance G_m after m consecutive events is

$$G_m = G_0 \cdot e^{m \cdot \Delta S / L_0} . \tag{11}$$

As for most stimuli the time, for which the given stimulus lasts, is fixed to T the expected number of events as well as the variation of obtained events until T is of interest. For the following theoretical examinations it is assumed that the electrical conditions are only determined by the mutual interaction between a load device and the RS. Specifically, the device voltage V_D as well as the device current I_D change only, if the conductance G changes due to (10). Then, by starting at an initial conductance G_0 the sequence r_i of reaction rates is known and deterministic for a deterministic load. Here, it is possible to calculate the PDF for m subsequent events as the m-fold convolution of (8):

$$f_m(T) = \left\{ \mathop{*}_{i=1}^{m} f(r_i,t) \right\}(T) . \tag{12}$$

The function $f_m(T)$ predicts the relative frequency of the *m-th* event observed at time T. The PDF for obtaining *exactly m* events at time T is then given by

$$g(m,T) = e^{-T \cdot r_{m+1}} \int_0^T f_m(T) \cdot e^{r_{m+1} \cdot \tau} d\tau . \tag{13}$$

If $\sigma(T)$ denotes the standard deviation of $g(m,T)$, the variance of the gap S is given by

$$\sigma_S^2(T) = \Delta S^2 \cdot \sigma^2(T) \tag{14}$$

using (10). The standard deviation of the conductance G is given by

$$\sigma_G(T) = -\frac{\partial G}{\partial S} \cdot \sigma_S(T) = \frac{G}{L_0} \cdot \sigma_S(T) . \tag{15}$$

3. CIRCUIT SIMULATION MODEL

In order to provide a simulation model which can be executed on a standard circuit simulator (SPICE, SPECTRE, Verilog AMS), equivalent circuits have to be provided that model the device dynamics by electrical quantities as a substitute for arbitrary dynamic variables. In particular, the gap S is modeled by a voltage V_S (cf. Fig. 2). In turn, V_S is derived from a voltage V_X which accumulates all preceding events. In Fig.2a the continuous model (i.e. the standard model) is shown which realizes the integration accordingly to (2).

Fig. 2c shows the pulsed implementation of (2). The input rate is continuously integrated (as in Fig. 2a) but a change of V_X is accomplished only if the accumulated rate reaches integer multiples of a given threshold: the accumulated rate is continuously monitored by a threshold element which generates a short output pulse (event) on V_{pulse} in case that the accumulated rate reaches the given threshold. Each pulse charges C_P and increases V_X by a constant amount. After each pulse event C_X is reset to an initial level by the switch SW and the integration of the input rate starts again.

In the circuit shown in Fig. 2d the input rate is modulated by R_N. R_N is an exponentially distributed random number. Then, at constant input rate the PDF of the pulses on V_{pulse} is an exponential distribution (which is the desired property). For each subsequent event a new random number is drawn and logarithmically transformed into R_N. RAND() specifies a uniformly distributed random number. Note, that this circuit is also able to correctly process non-stationary input rates.

The circuits shown in Fig. 2d and Fig. 1b represent the implementation of the proposed extended ECM device model. Each pulse on V_{pulse} represents the deposition (or dissolution) of an individual Cu atom on the filament. Given V_S as well as the device voltage V_D the overall device current $I_D=I_{Tunnel}+I_{ion}$ is obtained by solving the equilibrium state of the voltage-controlled current sources Q_F, Q_A and Q_T which is done automatically by the circuit simulator, cf. Fig. 1b.

4. THEORY OF GAP VARIANCE

In the following section an approach is presented to evaluate the variance-limiting properties of particular load circuits which are commonly used to program resistive switches (multivalued or bi-valued). It is assumed that the resistive switch is initially in a high resistive state (HRS). The initial conductance G_0 of the RS is chosen in such a way that shortly after turning on of the load device the device voltages is close to the supply voltage V_{DD}. Then, the load device is turned on for a time interval T. After T has been elapsed the conductance G of the RS is assessed. After several (infinite) repetitions of this experiment the variance of G is determined.

The mutual and exclusive interaction between the load device and the RS results in a deterministic sequence of rates r_i. From these rates exponentially distributed event times are derived which establish the randomness of the filament growth. If the individual rates r_i appear to be successively diminishing such as given by an exponential law

$$r_i = r_0 \cdot e^{-i \cdot \beta} , \qquad (16)$$

the variance of $g(m,T)$ (13) becomes independent from T for sufficiently large T and converges towards

$$\sigma^2(T) = \frac{1}{2\beta} , \qquad (17)$$

i.e. the rates are successively dying out. For sufficiently large T almost no new events are generated. Note, that (17) is also independent from the initial rate r_0. Combined with (14) as well as (15) this particular property (17) constitutes the basis for conductance regulation.

The particular value for β can be derived from the electrical interaction between the regulating load device and the resistive switch. By using (1) as well as (7) the value of β can be estimated to

$$\beta = \frac{\alpha \cdot (1-\alpha) \cdot z}{V_T} \cdot \left(V_{D,i} - V_{D,i+1}\right) = \frac{\alpha \cdot (1-\alpha) \cdot z}{V_T} \cdot \Delta V_i . \qquad (18)$$

$V_{D,i}$ and $V_{D,i+1}$ are the (deterministic) device voltages after the deposition event i and the deposition event $i+1$, respectively. ΔV_i is the observed voltage step after the $i+1$-th deposition. Note, that β is not necessarily constant. Typically, for a specific operating point β gets maximized which constitutes the minimal variance (17) which is possible at all.

4.1 The Voltage Switch

In Fig. 3a the typical circuit for the implementation of a voltage switch is shown. In the circuitry the internal resistance of the driving PMOS transistor is sufficiently low. Hence, until time T, the device voltage V_D is equal to V_{DD}. For this particular setup the rates r_i (16) are also equal (β=0), consequently a Poisson distribution can be assumed here. For this setup, the variance of S is given by

$$\sigma_S^2(T) = K_P^2 \frac{q \cdot z \cdot T \cdot j_0}{A_{fil}} \cdot e^{\alpha \cdot z \cdot V_{DD} / 2V_T} . \qquad (19)$$

Clearly, (19) has no variance limiting property as σ_S significantly grows with time T.

4.2 The Resistive Load

The resistive load is the simplest circuit element which features current-limiting properties. In Fig.3c the principal circuit setup comprising the load resistance R_L is shown. The voltage step between two consecutive ion depositions on the RS's filament is given by

$$\Delta V_i = V_{D,i} - V_{D,i+1} = V_{DD} \cdot \frac{1}{R_L} \frac{G_{i+1} - G_i}{\left(1 + R_L G_i\right) \cdot \left(1 + R_L G_{i+1}\right)} . \qquad (20)$$

The voltage step gets maximized at a particular step m for which the *matching* of load R_L and RS conductance is fulfilled:

$$R_L \cdot G_m \approx 1 \;\Rightarrow\; \Delta V_{D,\max} = \frac{V_{DD}}{4} \cdot \left(e^{\Delta S / L_0} - 1\right) . \qquad (21)$$

Using (18) the value of β can be estimated to

$$\beta_R \approx \frac{\alpha \cdot (1-\alpha) \cdot z}{V_T} \cdot \frac{V_{DD}}{4} \cdot \left(e^{\Delta S / L_0} - 1\right) \approx \frac{\alpha \cdot (1-\alpha) \cdot z}{V_T} \cdot \frac{V_{DD}}{4} \cdot \frac{\Delta S}{L_0} . \qquad (22)$$

The expected variance of the gap size S results to

$$\sigma_S^2(T) = \frac{4 \cdot L_0 \cdot V_T \cdot q \cdot K_P}{A_{fil} \cdot \alpha \cdot (1-\alpha) \cdot V_{DD}} . \qquad (23)$$

4.3 The PMOS Current Source

A current source (CS) is a typical element to limit the write current for multivalued programming of conductive states for RS. In integrated circuits the simplest way to implement a CS is to use (for instance) a biased PMOS transistor [20]. Fig. 3b shows the associated circuit. By fulfilling the required preconditions, the PMOS current source initially operates in the linear regime

$$I_D(V_D) = \mathrm{B} \cdot \left(V_B - \left|V_{Thp}\right| - \frac{V_{DD} - V_D}{2}\right) \cdot \left(V_{DD} - V_D\right) . \qquad (24)$$

In (24) V_{THP} represents the threshold voltage of the PMOS transistor and B a prefactor. For a given device voltage $V_{D,0}$ the current I_D can be locally approximated in a linear way by

$$I_D(V_D) \approx \left(V_{SRC}(V_{D,0}) - V_D\right) \cdot g_D(V_{D,0}) . \qquad (25)$$

In (24) the definition

$$g_D(V_{D,0}) = -\frac{\partial I_D}{\partial V_D}(V_{D,0}) \qquad (26)$$

was used to model a conductance. Additionally, the definition

$$V_{SRC}(V_{D,0}) = \frac{I_D(V_{D,0})}{g_D(V_{D,0})} + V_{D,0} \qquad (27)$$

was used to describe a (virtual) voltage source driving g_D. Similar to the analysis of the resistive load the maximum voltage step is obtained by maximizing (20). The load resistance R_L has to be replaced by the transistor conductance g_D and the supply voltage V_{DD} has to be replaced by the virtual voltage source V_{SRC}. Note that both, the virtual voltage source as well as the involved conductances depend on the operating point given by V_D. By variation of V_D the expression (20) is maximized for

$$V_{D,\max} = V_{DD} - (V_B - |V_{Thp}|) . \qquad (28)$$

At this point, the PMOS transistor operates at the transition between linear mode and saturation mode and delivers almost the full current. The associated voltage step is given by

$$\Delta V_i(V_{D,\max}) = \left(V_{DD} - V_B + |V_{Thp}|\right) \cdot \left(e^{\Delta S / L_0} - 1\right) \qquad (29)$$

The variance of the gap size S is given by

$$\sigma_S^2(T) = \frac{L_0 \cdot V_T \cdot q \cdot K_P}{A_{fil} \cdot 2 \cdot \alpha \cdot (1-\alpha) \cdot (V_{DD} - V_B + |V_{Thp}|)} . \qquad (30)$$

4.4 The Source Coupled NMOS

The circuit comprising a source coupled NMOS transistor as a load circuit for the RS is shown in Fig.3d. This particular circuit is of interest since it allows a change of the RS device conductance by a smooth power law and is therefore well-suited for incremental multilevel programming. In the following section the dynamics of this particular circuit is evaluated by using the continuous equations (1-3). Based on these equations the expected variance of the gap S is derived. In the initial phase the transistor operates in subthreshold mode [20] and delivers a current which is given by

$$I_D = I_0 \cdot \left(e^{(V_B - V_D)/k \cdot V_T} - 1\right) . \qquad (31)$$

In (30) k, V_T, I_0 are the non-ideality factor, temperature voltage, and reverse current, respectively. In Fig.3e (the black curves) the I-V characteristics is plotted for exemplary parameters V_B=*0.9V (0.85V)* I_0=$2x10^{-11}$*A*, and *k*=1.3. For large device voltage V_D the dynamics of the RS conductance is (continuously) described by

$$\frac{dG}{dt} = \frac{1}{L_0} \cdot G(t) \cdot \frac{j_0 \cdot K_P}{A_{fil}} \cdot e^{\alpha \cdot z \cdot V_D / 2V_T} . \qquad (32)$$

From (31) the device voltage is obtained by

$$e^{V_D / V_T} = \left(\frac{I_0}{I_D}\right)^k = \left(\frac{I_0}{G(t) \cdot V_D}\right)^k . \qquad (33)$$

By combination of (32) and (33), and taking the marginal slope of the device voltage V_D into account the dynamics of G is expressed by

$$\frac{dG}{dt} \approx C \cdot G^{(1 - \frac{k \cdot z \cdot \alpha}{2})} \qquad C = \frac{j_0 \cdot K_P}{A_{fil} \cdot L_0} \cdot \left(\frac{I_0}{V_{D,0}}\right)^{\frac{k \cdot z \cdot \alpha}{2}} . \qquad (34)$$

By evaluation of (34) the characteristics of G is given by a power law

$$G(t) \propto t^{\left(\frac{2}{k \cdot z \cdot \alpha}\right)} . \qquad (35)$$

The maximum voltage drop can be determined using (20) again. The (locally linear) load conductance of the NMOS transistor is given by

$$g_D(V_{D,0}) = -\frac{\partial I_D}{\partial V_D}(V_{D,0}) = -\frac{I_D}{k \cdot V_T} . \qquad (36)$$

The virtual voltage source is given by

$$V_{SRC}(V_{D,0}) = \frac{I_D(V_{D,0})}{g_D(V_{D,0})} + V_{D,0} = k \cdot V_T + V_{D,0} . \qquad (37)$$

ΔV_i (20) is maximized by maximizing the device voltage V_D. It follows that the maximum voltage step is given by

$$\Delta V_i(V_{D,\max}) = k \cdot V_T \cdot \left(e^{\Delta S / L_0} - 1\right) , \qquad (38)$$

which turns out to be almost independent from the operating point. The variance of the gap size is then described by

$$\sigma_S^2(T) = \frac{L_0 \cdot q \cdot K_P}{A_{fil} \cdot \alpha \cdot (1-\alpha) \cdot k} . \qquad (39)$$

5. RESULTS AND CONCLUSION

For each load circuits with variance limiting properties (Fig. 2b-d) $2x10^3$ Monte Carlo simulations were executed on a circuit simulation environment in order to quantify the variance of *S* (Fig. 4) as well as of *G* (Fig. 5) dependent on *T*. For the three circuits parameters were specified in such a way that after starting with an identical initial gap of *S=5nm* the expected device conductance at *T=2ms* (point A in Fig. 5a) and *T=1.3ms* (point B in Fig. 5a), respectively, are almost equal. In Fig. 4 the development of the standard variation of the gap size is shown. Initially, until *T=0.2ms*, σ_S rises proportional to the square root of *T* which indicates a variance in agreement to (19); the circuits deliver the full supply voltage to the RS and act as voltage sources. If *G* is becoming sufficiently large, the feedback of the load devices becomes effective. This is seen by the rapid reduction of σ_S by more than one order of magnitude within less than 100µs (Fig. 4).

After this short transition phase the variance settles to an almost constant level which strongly depends on the circuit type but which is less dependent on the specific circuit parameters and time *T*. These observations confirm, that the assumption (16) was valid for the theoretical examination. Especially, the results (23), (30), (39), and (19) are in good agreement with the simulation results. On one hand, the effect of device scaling is covered in the theoretical deduction by the specific dependence of the variance from the filament area A_{fil}. On the other hand, (23), (30), (39), and (19) show that the obtained variance of the conductance G depends on parameters which are *not* ECM device parameters but parameters of the regulating device: V_{DD}, V_B, V_{Thp}, and k. Hence, as initially stated, the variance is the consequence of a *joint interaction* between RS device and the load circuit. The results emphasize the necessity of statistical simulation using a Monte Carlo approach for the design of future nanoelectronic systems and architectures which are based on scaled RS devices as well as scaled CMOS devices. With the presented device model the proposed statistical evaluations are possible.

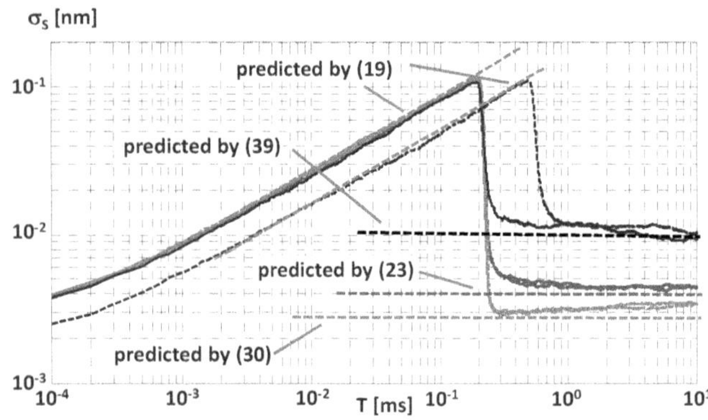

Figure 4: Variance of the gap S vs. time T for different load types. Red curve: PMOS current source 250nA, dashed: 16nA, blue curve: resistor 22MOhm, dashed: 1.6MOhm, black curve: source coupled NMOS V_B=0.9V, dashed: 0.85V. A_{fil}= 12.6nm^2

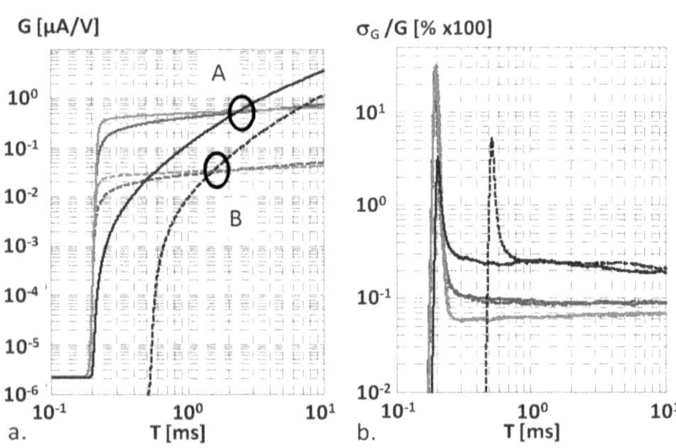

Figure 5: a. Advance of G vs. time T. for different load types. Red curve: PMOS current source 250nA, dashed: 16nA, blue curve: resistor 22MOhm, dashed: 1.6MOhm, black curve: source coupled NMOS V_B=0.9V, dashed: 0.85V. b. Variance of G vs. time T for different load circuits.

6. ACKNOWLEDGMENTS

The authors like to thank Rainer Waser, Stephan Menzel and Eike Linn from RWTH IWE II for providing the fundamental ECM device model as well as representative parameters for simulation

7. REFERENCES

[1] ITRS roadmap, http://www.itrs.net

[2] S. Tanachutiwat, M.Liu, W. Wang, „FPGA based on Integration of CMOS and RRAM", IEEE Trans. On Very large Scale Integration (VLSI) Systems, Vol.19, No.11, pp.2023-2032, Nov. 2011

[3] T.Driscaoll, J.Quinn, S.Klein, H.T.Kim, B.J.Kim, Y.V.Pershin, M.DiVentra, D.N.Basov, „Memristive Adaptive Filters", Applied Physics Letters 97, 093502, 2010

[4] S.Yu, Y.Wu, R.Jeyasingh, D.Kuzum, H.-S. P. Wong, „An Electronic Synapse Device Based on Metal Oxide Resistive Switching Memory for Neuromorphic Computation", IEEE Transactions on Electron Devices, Vol. 58, No.8, pp 2729-2737, Aug. 2011

[5] An Chen; Ming-Ren Lin; , „Variability of resistive switching memories and its impact on crossbar array performance", *Reliability Physics Symposium (IRPS), 2011 IEEE International*, vol., no., pp.MY.7.1-MY.7.4, 10-14 Apr. 2011

[6] R.Soni, P.Meuffels, G.Staikov, R.Weng, C.Kügeler, A.Petraru, M.Hambe, R.Waser, H.Kohlstedt, „On the stochastic nature of resistive switching in Cu doped Ge0.3Se0.7 based memory devices", Journal of Applied Physics 110, 054509, 2011

[7] C.Schindler, M.Meier, R.Waser, M.N. Kozicki, „Resistive Switching in Ag-Ge-Se with extremely Low Write Current", NVMTS,pp.82-85, 2007

[8] A.Heittmann, T.G.Noll, „A Monte Carlo Analysis of a Write Method used in Passive Nanoelectronic Crossbars", Procs. of ACM/IEEE Nanoarch, 2012, Amsterdam, The Netherlands, 2012

[9] S. Shin, K. Kim, and S.-M. Kang, „Compact Models for Memristors Based on Charge-Flux Constitutive Relationships", IEEE Transactions on Computer-Aided Design of Integrated Circuits and Systems, vol. 29, no. 4, pp. 590–598, 2010.

[10] M.D.Pickett, D.B.Strukov, J.L.Borghetti, J.J.Yang, G.S. Snider, D.R.Stewart, R.S.Williams, „Switching dynamics in titanium dioxide memristive devices", Journal of Applied Physics 106, 074508, 2009

[11] H.Abdalla, M.D.Pickett, "SPICE Modeling of Memristors", IEEE International Symposium on Circuits and Systems (ISCAS 2011), pp 1832-1835, May 2011

[12] X.Guan, S.yu, H.-S.P. Wong, „On the Switching Parameter Variation of Meta-Oxide RRAM – Part I: Physical Modeling and Simulation Methodology", IEEE Transactions on Electron Devices, Vol. 59, No.4, pp. 1172-1182, April 2012

[13] S.Yu, X. Guan, H.-S.P. Wong, „On the Switching Parameter Variation of Metal Oxide RRAM- Part II: Model Corrobation and Device Design Strategy", IEEE Transactions on Electron Devices, Vol. 59, No.4, pp. 1183-1188,Apr. 2012

[14] S.Menzel, U.Böttger, and R.Waser, „Simulation of Multilevel Switching in Electrochemical Metallization Memory Cells", Journal of Applied Physics Vol. 111, pp.014501-014501-5, Jan. 2012

[15] R.Waser, „Electrochemical and Thermochemical Memories", IEDM, pp 1-4, 2008.

[16] Waser, R., Dittmann, R., Staikov, G. and Szot, K., „Redox-Based Resistive Switching Memories – Nanoionic Mechanisms, Prospects, and Challenges", Adv. Mater., 21: 2632–2663, 2009

[17] Bockris, John O'M., Reddy, Amulya K.N., Gamboa-Aldeco, Maria E., „Modern Electrochemitry 2A", 2nd ed., Springer,2000

[18] D. Querlioz et al., „Bioinspired Networks with Nanoscale Memristive Devices that combine the unsupervised and supervised learning approaches ", Procs. of ACM/IEEE Nanoarch, 2012, Amsterdam, The Netherlands, 2012

[19] B.Meng, W.H. Weinberg, „Dynamical Monte Carlo studies of molecular beam epitaxial growth models: interfacial scaling and morphology", Surface Science 364, pp. 151-163, 1996

[20] S.C Liu, J.Kramer, G.Indiveri, T.Delbrück, R.Douglas, „Analog VLSI:Circuits and Principles", Cambridge, MA: MIT Press,2002

Asymmetric-access Aware Optimization for STT-RAM Caches with Process Variations*

Yi Zhou
Beijing Institute of Technology
zhouyi.initial@gmail.com

Chao Zhang
Peking University
zhang.chao@pku.edu.cn

Guangyu Sun
Peking University
gsun@pku.edu.cn

Kun Wang
IBM Research - China
wangkun@cn.ibm.com

Yu Zhang
IBM Research - China
zhyu@cn.ibm.com

ABSTRACT

STT-RAM (Spin Transfer Torque Random Access Memory) has been extensively researched as a potential replacement of SRAM (Static RAM) as on-chip caches. Prior work has shown that STT-RAM caches can improve performance and reduce power consumption because of its advantages of high density, fast read speed, low standby power, etc. However, under the impact of process variations, using worst-case design can induce significant performance and power overhead in STT-RAM caches. In order to overcome the problem of process variations, we propose to apply the variable-latency access method to STT-RAM caches by introducing a variation-aware LRU (Least Recently Used) policy. Moreover, we show that simply applying traditional variable-latency access method is inefficient due to the read/write asymmetry. First, we demonstrate that a write-oriented data migration is preferred. Second, a block remapping is necessary to prevent some cache sets from being significantly affected by process variations. After using our techniques, the experimental results show that the performance can be improved by 13.8% and power consumption can be reduced by 14.1% compared to a prior approach [3].

Categories and Subject Descriptors

B.3.2 [**Memory Structures**]: Design Styles–*Cache memories*

Keywords

STT-RAM caches, process variations, asymmetric-access

1. INTRODUCTION

As the mainstream of computer system moves to the regime of multi-/many-core designs, the number of processing cores integrated on the same chip has been increasing. It also results in an increasing requirement for on-chip memory capacity to bridge the gap between processors and off-chip main memory. The traditional SRAM technology, however, cannot satisfy the requirement due to its problems of low cell density, high standby power, and vulnerable to soft errors [4], which lead to the well-known problem "memory wall" [14]. In order to attack this problem, STT-RAM has been extensively proposed to replace SRAM as on-chip memory because of its advantages of high cell density and immunity to soft errors [12, 13, 11].

Prior research has shown that replacing SRAM with STT-RAM in various levels of on-chip caches can help improve performance, reduce power consumption, and achieve a better reliability. Yet the main focus of prior research of STT-RAM is on how to leverage the density, power consumption advantages and try to mitigate the well-known issue of asymmetric-access. The impact of process variations on STT-RAM caches and corresponding optimization techniques, however, are not well studied at the architectural level. In fact, since STT-RAM is normally employed as lower level caches (e.g. L2) with large capacity, process variations have a significant impact STT-RAM caches. Moreover, it becomes more complicated when the character of asymmetric read/write access is considered.

For the popular one-transistor-one-MTJ (Magnetic Tunneling Junction) based STT-RAM cell, various design parameters, such as MTJ shape, MgO thickness, access transistor's gate length/width, etc., can be affected by process variations. Consequently, the characteristics of STT-RAM cells in an STT-RAM cache can be quite different from each other. In the simple worst-case design employed for STT-RAM caches, a large overhead can be induced for access latency and power consumption due to process variations. Sun *et al* proposed a process-variation aware data management for non-uniform access STT-RAM caches [12]. In their work, the write latency variation is comparable to that consumed on data movement among different banks. Thus, the data migration policy is tailored to adapt the variable write latency caused by process variations. However, their approach is not feasible for uniform access STT-RAM cache, especially when the read-write asymmetric access is considered.

In this work, we apply the variable-latency access method to uniform access STT-RAM caches by introducing a novel variation-aware LRU policy. Moreover, we show that simply applying traditional variable-latency access method is inefficient due to the read/write asymmetry. First, we demonstrate a write-oriented data migration is preferred. Second, a block remapping is necessary to prevent some cache sets from being significantly affected by process variations. After using our techniques, the experimental results show that the performance can be improved by 13.8% and power consumption can be reduced by 14.1%.

The rest of the paper is organized as follows. In the next section, we will give a brief review of STT-RAM and the impact of

* *This work is supported by the National Natural Science Foundation of China 61202072.*

process variations on STT-RAM cache cells as well as the basic idea of variable-latency access method. Then, we will present our variation-aware LRU policy that can achieve a variable-latency access STT-RAM cache in Section 3. In addition, we will discuss the impact of asymmetric access and show that the remapping technique is necessary. The evaluation results are shown in Section 4, followed by a conclusion.

2. PRELIMINARIES

In this section, we first give a review of STT-RAM technology and how to model the effect of process variations on STT-RAM. Then, we introduce the method of variable-latency access to caches under process variations.

2.1 STT-RAM Basics

Different from traditional CMOS based SRAM, STT-RAM uses MTJ devices to store the information. A MTJ has two ferromagnetic layers (FL) and one oxide barrier layer (BL). The resistance of MTJ, which is used to represent information stored, is decided by the relative magnetization directions (MDs) of the two FLs. When the MDs are parallel or anti-parallel, the MTJ is in low (bit '0') or high resistance state (bit '1'). R_h and R_l are usually used to denote the high and the low MTJ resistance, respectively. In a MTJ, the MD of one FL (reference layer) is pinned while the one of the other FL (free layer) can be flipped by applying a polarized write current though the MTJ. In the "1T1J" cell structure, the MTJ write current is supplied by a NMOS transistor [11, 10].

2.2 Modeling of Process Variations of STT-RAM

In the popular "1T1J" STT-RAM cell, the transistor is used to control the access to the MTJ while the MTJ is used to story data. Both access transistor and MTJ can be affected by process variations. Due to different access mechanisms in read and write operations, the process variations of these two components have different impact on read and write to STT-RAM cells. Note that in this subsection, the read/write latency is just for a single STT-RAM cell, which should be differentiated from the cache access latency for a whole STT-RAM cache. In addition, we find that variation of a STT-RAM cell is significantly larger than that of the peripheral circuitry. Thus, we focus on the variation of the STT-RAM cell in this work.

The read variation is mostly determined by variations on parameters of the access transistor, which include effective channel length/width, threshold voltage, oxide thickness, etc. Similar to prior research, variations of these parameters of transistors can be modeled with normal distributions using a well-know method [1, 9, 2]. Having distributions of these parameters, we can run Monte-Carlo based simulation to model the characters of all access transistors in a STT-RAM cache. Then, these numbers can be used to calculate the access latency and energy consumption of a read operation to cells in a STT-RAM cache. Obviously, these cells will demonstrate different access latency and energy, which are projected in the variable-latency access method in the next subsection.

Compared to read, the write operation of STT-RAM can have even lager variances of latency and energy consumption due to process variations of both the access transistor and MTJ. For the MTJ, the process variations have effects on several design parameters, such as MTJ shape, MgO thickness, and the localized fluctuation of magnetic anisotropy [11, 13]. All these parameters are related to the programming current, which is key factor for access latency and energy of a write operation. Wen *et al* and Sun *et al* investigated the typical distributions in various STT-RAM cell designs. A dual-exponential function about programming current is used to provide an excellent accuracy, which can be described in equation (1) [13]. Due to page limitation, the detailed relationship among those design parameters and characters of write operation is not presented in this paper. For those readers who are interested in the model details, please refer to Wen's and Sun' work for reference.

$$f(I) = \begin{cases} a_1 e^{b_1(I-\mu)} & I \leq \mu \\ a_2 e^{b_2(\mu-I)} & I > \mu \end{cases} \quad (1)$$

2.3 Variable-latency Cache Access

Variable-latency access is an effective method to mitigate the impact of process variations on caches [3, 8, 11, 7]. In a cache using variable-latency access technique, the cache access latency is no longer a fixed value as in a traditional worst-case cache design. Instead, the access latency has a variance depending on the affect of process variations on STT-RAM cells. In other words, the access to some parts of the caches can be faster than the rest parts due to different effects of process variations on the STT-RAM cells. Since the access granularity is a cache line, the access latency of a cache line is decided by the slowest STT-RAM cell belonging to the cache line. It should be addressed that the energy consumption to a cache line also varies in a cache because the energy consumption is related to both access latency and current. Note that the definition of variable-latency access cache is different from that of a Non-uniform cache architecture (NUCA). For the later one, the non-uniform access latency is caused by the specific architecture design (e.g. network-on-chip) and also exists even when there are no any process variations.

Extensive research has been done on the variable-latency cache architecture. One variable-latency cache architecture uses a delay storage to record the delay time for each cache line [3]. In another variable-latency cache design, they enabled the variable access cache by modifying the function units and adding special queues to store the dependent instructions [8]. Similarly, a LA (Latency Aware)-LRU policy is put forward to mitigate the process variation [7]. They also uses the delay-storage architecture, and dynamically swapping data using the LA-LRU policy. The optimization targets in these works are traditional SRAM caches so that the asymmetric read/write access to STT-RAM caches are not considered. Sun *et al* proposed a variation aware data management for non-uniform cache architectures. They compensate write time variations via dynamic data migration [11]. Their data migration method, however, is not feasible for uniform caches. Considering the limitation of prior works, we propose a novel LRU policy with consideration of data migrations for asymmetric access STT-RAM caches, which will be introduced in next section.

3. VARIABLE-LATENCY ACCESS STT-RAM CACHE DESIGN

In this section, we first introduce the basic structure of our STT-RAM cache that can support variable-latency access. Then, we propose a novel LRU policy tailored to enable the data migrations in the variable-latency access cache. Finally, we will discuss the impact of cache line remapping and analyze design overhead.

3.1 Basic Structure

With the process variations, the caches lines in a STT-RAM cache demonstrate various access latency, in respect of read and write operations. For example, the distribution of write latency for a STT-RAM cache with 512 cache lines are shown in Figure 1. In our variable latency cache, we partition these cache lines into two parts, which are called the fast-access part (FAP) and the slow-access part (SAP). The partitioning is based on the distribution of the access latency to different cache lines. Since the STT-RAM cache has asymmetric read/write access latency, we can have different partitioning policies based on read latency, write latency, or a combination of them. The impact of partitioning policies is discussed later in subsection 3.3.

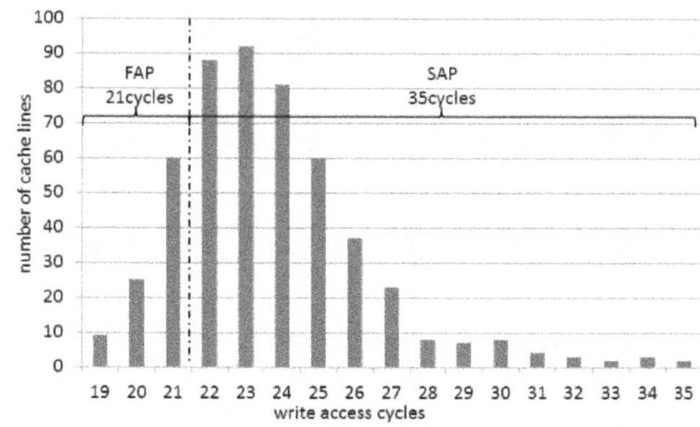

Figure 1: An example of distribution for cache write latency.

In the example of Figure 1, we partition the STT-RAM cache based on write latency and set 21 *clock cycles* as the threshold to differentiate SAP from FAP. In other words, for the cache lines that can be written within 21 *cycles*, they are grouped together as the FAP of the STT-RAM cache. The rest caches lines, which have the longer write latency than 21 *cycles*, are organized as the SAP of the STT-RAM. The discussion of selecting a proper threshold will be introduced in subsection 3.3. For each part of the STT-RAM cache, the write latency is fixed and determined by the longest access latency of the cache line in the part. For the same example in Figure 1, the write latency to SAP and FAP are 21 *cycles* and 35 *cycles*, respectively.

The organization of FAP and SAP for STT-RAM cache is different from traditional variable-access latency cache structure in prior work [3, 8, 7]. In prior work, the access latency is identified with a fine granularity, which is a single cache line. It means that the access to each cache line can be finished as soon as possible. For our variable-latency access STT-RAM cache, however, the cache is partitioned with a coarse granularity (e.g. SAP and FAP), and the access latency to a cache line is decided by the slowest cache line in the same part.

It is easy to find that the traditional fine granularity based structure can achieve a better performance than ours for the *static data allocation*. Static data allocation means that a data are placed in fixed positions (cache blocks) before they are evicted. On the contrary, a *dynamic data allocation* means that data can be moved from SAP to FAP if they are frequently accessed so that the performance is improved. It should be addressed that a dynamic data allocation is necessary for variable-latency access cache due to temporal data locality. Otherwise, if most frequently accessed data are always allocated in the cache lines with slow access latency, we may gain little benefits of using variable latency access cache.

When the dynamic data allocation is employed, our variable-latency access cache structure can outperform those in prior work for three reasons [3, 8, 7]. First, data migration is feasible to be realized when there are only SAP and FAP. On the contrary, in the traditional structure, there will be many different levels of access latency, which can significantly increase the design complexity since data migration can happen between any two cache lines with different access latency. Moreover, when the cache remapping technique is applied (introduced later in the paper), design complexity of a fine granularity policy is further significantly increased. In fact, the dynamic allocation policy is not introduced in most of traditional structures [3, 8]. Second, a coarse partitioning granularity can help reduce the overhead of tracing access latency of different cache lines. In traditional structure, a table is needed to record the access latency of each individual cache line. Such a table can induce non-trivial overhead, especially for STT-RAM cache which normally has the large capacity and significant variance of write latency. Third, the design complexity for variable-access latency control can be reduced. For example, as introduced in Bennaser's work [3], a buffer is needed to enable the variable-latency access. The design complexity of such a buffer and related control logic increases with the total number of different access latency. Since we only have two different access latency for SAP and FAP, the design can be really simple.

3.2 A Novel LRU Policy for Data Migration

As discussed in the last subsection, a dynamic data allocation is critical to exploit the benefits of using variable-latency access cache. In Sun's work, dynamic data allocation is enabled through data migration among different cache banks in the NUCA [11]. This data migration method, however, is not feasible for the uniform cache architecture. First, the data migration technique in NUCA can be enabled without the existence of process variations. Sometimes, the migration may even depends on the infrastructure of the NUCA (e.g. network-on-chip). When we apply the data migration in uniform cache, the extra design overhead must be considered. Second, the data locality is not considered in their approach, which may destroy the LRU replacement policy. Third, using the DPVA-NUVA-2 policy, both read and write operations can trigger the data migration. We will discuss in next subsection that such a combination of migration technique can harm the performance in uniform cache. Thus, we propose a new LRU policy, which can also enables the data migration at the same time.

The new LRU policy is illustrated in Figure 2 with a set of cache lines belonging to an eight-set-associative STT-RAM cache. As shown in the figure, the cache lines of the FAP are represented using blocks with shade. In this example, there are cache lines in FAP and five cache lines in SAP. In order to simplify the discussion, we assume that the positions of these cache lines in FAP are physically next to each other. The number labeled on top of each cache line is used to represented the position of the cache line in the LRU stack. The cache line with number 7 is the one that is just accessed. The cache line with number 0 is always evicted when a cache replacement happens.

Figure 2 (a)-(c) demonstrates the initial filling process for an empty cache set. As shown in Figure 2 (b), the data are always allocated in the FAP when there are slots in it. Figure 2 (c) shows the status when half of the cache set is filled. Note that the most recently loaded data are allocated in the SAP (the fifth cache line from left) at this time. Our experiments show that there is little enhancement if we carry out data swapping policy in the filling process. It is because high frequent data reuse is rare at the initialization stage. Moreover, these data may generate more swap operations when the data swap policy is applied. Considering the extra power consumption and latency for data swapping, we do not swap data in the filling process.

Figure 2 (d) and (e) shows the process when a cache replacement happens. When new data are loaded and the cache set is full, the cache line labeled with 0 is evicted. At the same time, we need to check whether the evicted cache line is in the SAP. If it is true, a data migration is triggered. The cache block

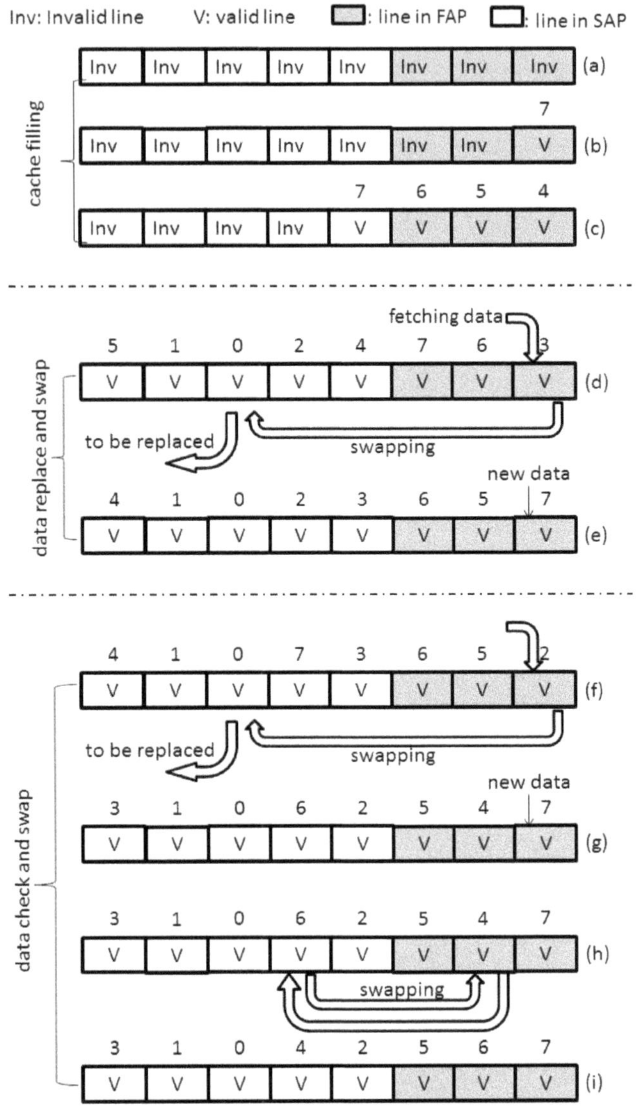

Figure 2: The illustration of LRU with the consideration of data migrations.

with smallest label number (e.g. the least recently used cache line in FAP), which is 3 in this example is, moved to the evicted cache line. Then, the new data is allocated in FAP (e.g. the last cache block in Figure 2 (e)). Such a replacement can promise that the most frequently loaded data are always allocated in the FAP.

As shown in Figure 2 (f), it is possible that the most recently accessed cache line resides in the SAP. Such scenarios usually happen during the initial filling process and also exist when there are some frequently accessed data that have a long reuse distance. In order to mitigate such problems, an extra check-and-swap step is adapted during the replacement. As shown in Figure 2 (g) to (i), after the replacement, if the cache labeled with number 6 is detected in the SAP, another data swapping is employed to move the SAP. With such a LRU policy, most data that are frequently accessed can be kept in the FAP to improve the performance.

Note that the data migration is only triggered during the cache replacement to avoid some useless data migration caused by cache hits, which are used in previous work [11]. In addition, we only demonstrates our data migration method with a perfect LRU policy to simplify the discussion. In fact, such a method can also be applied to some fake LRU polices in real processors. Due to page limitation, we will not discuss the details in this work.

3.3 Impact of Management Policies

The efficiency of variable-latency access is related to different cache management policies. In this subsection, we will discuss the impact of two important policies including partitioning policy and the cache line remapping.

3.3.1 Partitioning Policy

Obviously, the STT-RAM cache partitioning for SAP and FAP is one of the most important key factor for a variable-latency access cache design. First, it is critical to choose what type of access latency as the basis of latency partitioning. Second, the threshold for separating SAP from FAP also has an important impact on the efficiency of cache design.

As discussed in subsection 2.1, the variance of read and write latency is decided by different design parameters in STT-RAM. Thus, read latency and write latency have different distributions, although there is a correlation between them. To this end, we have to choose from read latency, write latency, or the combination of them for the cache partitioning. The comparison among these three partitioning methods is discussed as follows. In prior work for traditional SRAM caches, the read latency is normally used because the read latency is on critical path of execution time and the access latency is similar for read and write operations.

The case for STT-RAM caches, however, is quite different from that in SRAM caches. The reason is in three-folds: 1) The latency of a write operation is much longer than that of a read latency so that read operations may be blocked by write operations for a long period [10, 12]; 2) STT-RAM is usually used in lower cache levels like L2 caches. If a L1 cache is write-through based, the intensity of write operations can be significantly higher than that of the read operations in the L2 cache. Thus, the access time of write operations may dominate the total cache access time; 3) For each data migration, extra write operations are induced. Since the variance of read latency is also quite smaller than that of the write operation, a large number of subsequent accesses to the same data is needed to amortize the overhead of data migration. Later in section 4, we will show that the read-based partitioning cannot effectively improve the performance and may even degrade the performance sometime when a frequently accessed data is migrated to a cache line with short read latency but long write latency. Moreover, when the energy consumption is consider at the same time, a partitioning based on write operation is more attractive because cache lines in FAP also consume less energy in a write operation.

The third choice is to consider both read and write at the same time. This method, however, has a worse performance compared to the write-based partitioning. It is because a competition of same data may happen between a cache line with short read latency and a cache line with short write latency. Such a competition can result in a "ping-pong" style data migration and induce huge overhead. Consequently, only the pure read-based and write-based partitioning is considered in this work.

Another important issue in partitioning policy is the ratio between SAP and FAP. On one hand, if the SAP is too large, we cannot gain much benefits because most cache lines have the access latency of SAP, which is equal to that of the worst case in all cache lines. In addition, there will be intensive competition for the cache lines in the FAP, which may induce more data migrations. On the other hand, if the FAP is too large, more cache lines with long access latency will be included in the FAP so that the access latency to FAP will be increased. Thus, the benefits of data migration will also be reduced because the gap of access latency between SAP and FAP is shrunk. The optimized ratio between SAP and FAP is related to the distribution of process variations on STT-RAM cells and the data access pattern. For the STT-RAM cache and workloads in section 4, the results show that the optimized ratio is in the range of 4 to 7. Thus, the threshold 21 *cycles* in subsection 3.1 is plausible, for the ratio between SAP and FAP is nearly 4.

3.3.2 Cache Line Remapping

Due to the correlation of process variations, the physical position allocation of cache lines in FAP or SAP can be quite non-uniform. It means that the ratio between SAP and FAP can vary a lot all through the whole cache. Consequently, the efficiency of using variable-latency access cache will be decreased. In order to mitigate this problem, a cache remapping method introduced in prior research need to be adopted to achieve a uniform distribution of ratio between SAP and FAP [6]. Figure 3 shows an example of such a remapping method. Note that the number labeled on the cache lines represent the index of cache set that those cache lines belong to. The cache lines with shade color represent the cache lines in FAP. We can find that, after remapping, the ratio between SAP and FAP become uniform among different cache sets.

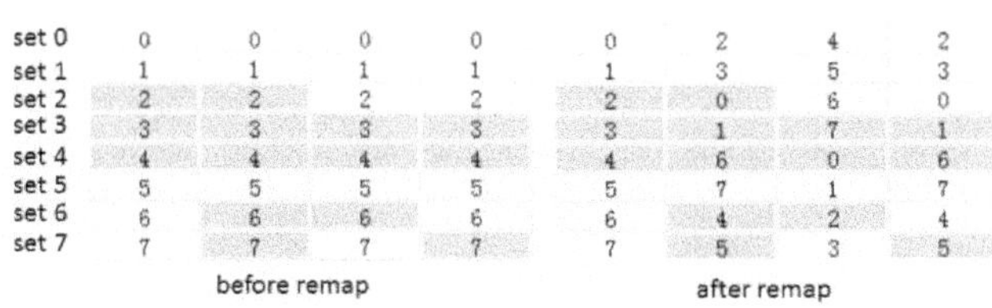

Figure 3: An example of cache line remapping.

4. EVALUATION

In this section, we first evaluate our variable-latency access cache with different management policies. Then, we compare our optimization design to the worst case design and an approach from prior work [3], in respect of performance and energy consumption.

4.1 Experiment Setup

For system level simulation, we use SIMICS simulator to evaluate performance. It is configured to model an eight-core processor. Each core is Ultra-SPARC-like with a 2GHz frequency. There are three levels of caches. The IL1/DL1 caches are SRAM based and the capacities are set to 64KB/64KB. The L2 cache is a 16-way 4MB STT-RAM cache. The simulator captures data addresses from all loads, stores, and prefetch operations. We use the information to calculate the memory access intensity, and use that to compute the energy consumption of the cache hierarchy. Our workloads are sets of multi-threaded benchmarks from PARSEC.

In order to estimate the access latency and energy consumption of different caches, we extend the widely used tool NVSim [5] to support STT-RAM technology with process variations considered. The parameters used in this work are listed in Table 1 and Table 2. Figure 4 shows the probability of programming current in write operations, which are fitted based on Monte-Carlo simulation for different STT-RAM cells. Note that in our simulations, extra latency and power consumption are included for the data migration. For each data migration, we add extra write access latency (30 *cycles*) and corresponding power consumption.

Table 1: Device Parameters

Device	Parameters	Mean
Transistor	Channel Length L	$45nm$
	Channel Width W	design dependent
	Threshold Voltage V_{th}	$0.466V$
MTJ	MgO Thickness τ	$2.2nm$
	Cross Section A	$45 \times 90nm^2$

Table 2: 3σ parameters of the transistor.

Width	*Length*	V_{th}
$-15\%\mu$	$+15\%\mu$	$+15\%\mu$

4.2 Experimental Results

First, we compare the performance of different benchmarks with variable-latency access caches using different partitioning policies in Figure 5. The y-axis represent the normalized execution time of different benchmarks. The first set of results labeled with *Worst* are for the baseline case, in which worst case design method is used in STT-RAM cache design. The second set of results labeled with *VAL-R* are for the cache using variable-latency access caches with a read-based partitioning policy. Similarly, the third set of results labeled with *VAL-W* are for the cache using variable-latency access caches with a write-based partitioning policy. The results show that the read-based partitioning cannot help improve the performance even with data migration supported. It is because the variance of a read latency is not significant and read operations only contribute a small fraction to the total access numbers. On the contrary, using write-based partitioning can significantly improve performance. Thus, the write-based partitioning should be used for variable-latency access STT-RAM caches. For the rest of the section, all variable-latency access caches will use write-based partitioning.

In order to study the impact of ratio between SAP and FAP, we assume a perfect allocations of cache blocks in SAP and FAP. In this perfect allocation, each cache set has an identical number of cache lines belonging to FAP. Then, we vary this number from 1 to 8 and compare the normalized reduction of execution time for a typical workload (*bodytrack*) in Figure 6. Note that the results for other workloads show the similar trend and we hide them due to page limitation. The results show that the best performance can be achieved for most benchmarks when the cache line numbers of FAP are in the range of 2 to 4. Consequently, we estimate an sub-optimized ratio between SAP and FAP as $(16-3)/3$, or 4.3.

The impact of remapping on performance is shown in Figure 7. *VAL-NoRemap* represents the normalized execution time without using cache remapping. *VAL-Remap* represents the normalized execution time for the case of using cache remapping. the ratio between SAP and FAP of the whole cache is still set to about 4.3 for both cases. It can be found that the performance can be significantly improved with remap method. The reason is that, without cache remapping, the ratio between SAP and FAP in many caches are quite different from that of the whole cache. Thus, the benefits of using variable-latency access caches are decreased significantly.

We compare our method to a prior approach using the same

$$f(I) = \begin{cases} 0.0802e^{0.1604(I-84.77)} & I \leq 84.77 \\ 0.0802e^{0.1604(84.77-I)} & I > 84.77 \end{cases}$$

Figure 4: Fitting equation of programming current

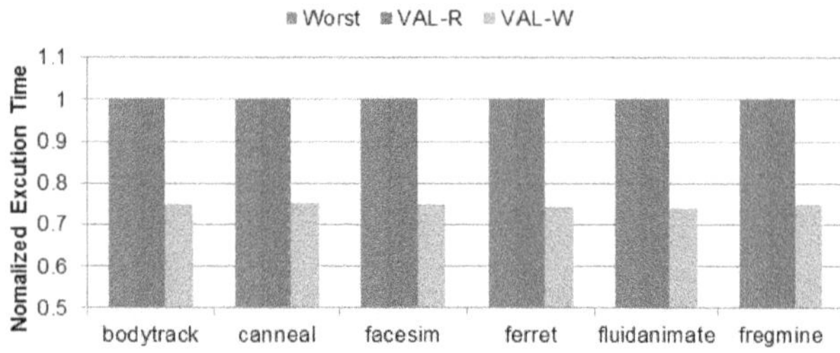

Figure 5: Comparison of performance with different partitioning policies.

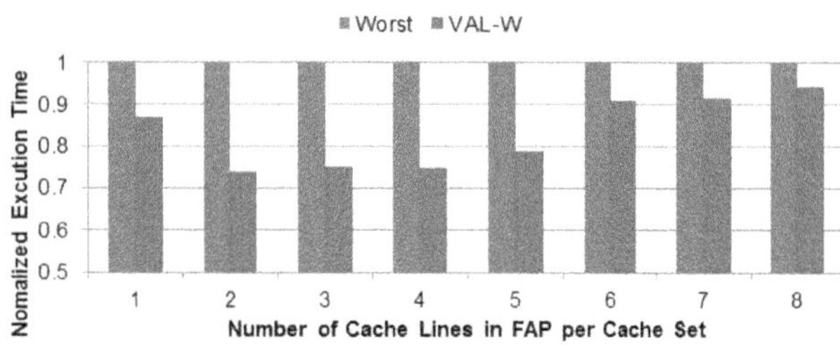

Figure 6: Comparison of performance with different ratios between SAP and FAP.

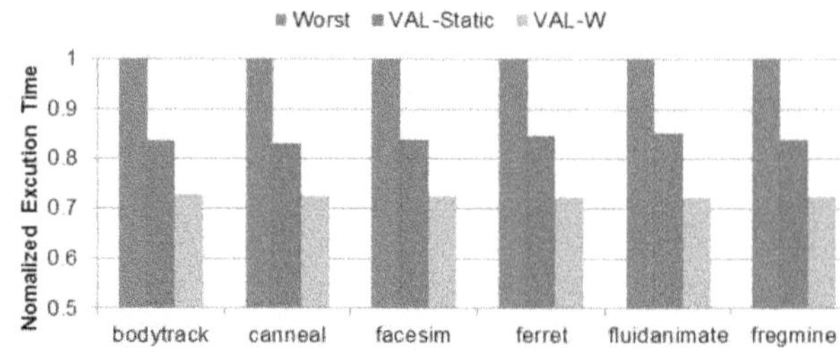

Figure 8: Comparison of performance for different variable-latency access methods.

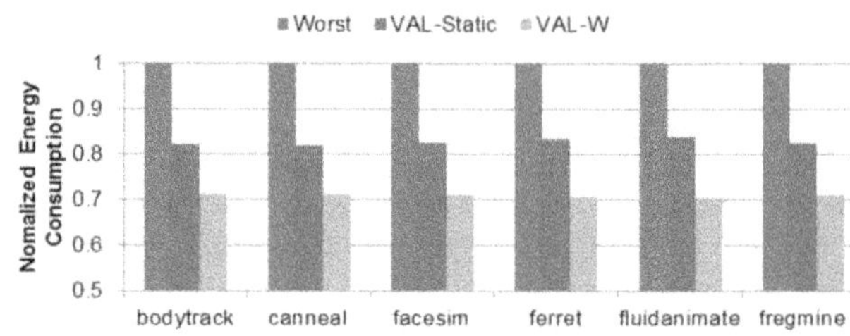

Figure 9: Comparison of energy for different variable-latency access methods.

experimental setup, in respect of performance and energy consumption [3]. The results of normalized execution time are shown in Figure 8. The label *VAL-Static* represents the set of results using the variable-latency access cache design in Bennaser's work [3]. On average, the performance can be improved by about 13.8%, compared to the prior approach [3]. Also, we compare the results of normalized energy consumption in Figure 9. We can draw a similar conclusion for the results of energy consumption. It is because the write energy consumption of a STT-RAM cell is roughly proportional to the product of programming latency and current. Since the write energy dominates in the total energy consumption of STT-RAM caches, a similar conclusion can be drawn in the comparison. On average, the energy consumption can be reduced by about 14.1%. Note that the VAL-Static architecture in Bennaser's work uses fine granularity, but without cache line remapping. The result shows that our coarse granularity policy with cache line remapping and dynamic data migration works better than their approach. Theoretically speaking, fine granularity with remapping and data migration can perform better. However, as discussed in subsection 3.1, the design complexity of such a fine-granularity architecture is significantly higher.

5. REFERENCES

[1] A. Agarwal, D. Blaauw, and V. Zolotov. Statistical timing analysis for intra-die process variations with spatial correlations. In *Computer Aided Design, 2003. ICCAD-2003. International Conference on*, pages 900 – 907, nov. 2003.

[2] A. Agarwal, D. Blaauw, V. Zolotov, S. Sundareswaran, M. Zhao, K. Gala, and R. Panda. Statistical delay computation considering spatial correlations. In *Design Automation Conference, 2003. Proceedings of the ASP-DAC 2003. Asia and South Pacific*, pages 271 – 276, jan. 2003.

[3] M. Bennaser, Y. Guo, and C. Moritz. Data memory subsystem resilient to process variations. *Very Large Scale Integration (VLSI) Systems, IEEE Transactions on*, 16(12):1631 –1638, dec. 2008.

[4] Y. G. Choi, S. Yoo, S. Lee, and J. H. Ahn. Matching cache access behavior and bit error pattern for high performance low vcc l1 cache. In *Design Automation Conference (DAC), 2011 48th ACM/EDAC/IEEE*, pages 978 –983, june 2011.

[5] X. Dong, C. Xu, Y. Xie, and N. Jouppi. Nvsim: A circuit-level performance, energy, and area model for emerging nonvolatile memory. *Computer-Aided Design of Integrated Circuits and Systems, IEEE Transactions on*, 31(7):994 –1007, july 2012.

[6] M. Hussain and M. Mutyam. Block remap with turnoff: A variation-tolerant cache design technique. In *Design Automation Conference, 2008. ASPDAC 2008. Asia and South Pacific*, pages 783 –788, march 2008.

[7] A. Jain, A. Shrivastava, and C. Chakrabarti. La-lru: A latency-aware replacement policy for variation tolerant caches. In *VLSI Design (VLSI Design), 2011 24th International Conference on*, pages 298 –303, jan. 2011.

[8] S. Ozdemir, A. Mallik, J. C. Ku, G. Memik, and Y. Ismail. Variable latency caches for nanoscale processor. In *Supercomputing, 2007. SC '07. Proceedings of the 2007 ACM/IEEE Conference on*, pages 1 –10, nov. 2007.

[9] G. Sun, C. Xu, and Y. Xie. Modeling and design exploration of fbdram as on-chip memory. In *Design, Automation Test in Europe Conference Exhibition (DATE), 2012*, pages 1507 –1512, march 2012.

[10] G. Sun, Y. Zhang, Y. Wang, and Y. Chen. Improving energy efficiency of write-asymmetric memories by log style write. In *Proceedings of the 2012 ACM/IEEE international symposium on Low power electronics and design*, ISLPED '12, pages 173–178, New York, NY, USA, 2012. ACM.

[11] Z. Sun, X. Bi, and H. Li. Process variation aware data management for stt-ram cache design. In *Proceedings of the 2012 ACM/IEEE international symposium on Low power electronics and design*, ISLPED '12, pages 179–184, New York, NY, USA, 2012. ACM.

[12] Z. Sun, X. Bi, H. H. Li, W.-F. Wong, Z.-L. Ong, X. Zhu, and W. Wu. Multi retention level stt-ram cache designs with a dynamic refresh scheme. In *Proceedings of the 44th Annual IEEE/ACM International Symposium on Microarchitecture*, MICRO-44 '11, pages 329–338, New York, NY, USA, 2011. ACM.

[13] W. Wen, Y. Zhang, Y. Chen, Y. Wang, and Y. Xie. Ps3-ram: A fast portable and scalable statistical stt-ram reliability analysis method. In *Design Automation Conference (DAC), 2012 49th ACM/EDAC/IEEE*, pages 1187 –1192, june 2012.

[14] M. Yamaoka and H. Onodera. A detailed vth-variation analysis for sub-100-nm embedded sram design. In *SOC Conference, 2006 IEEE International*, pages 315 –318, sept. 2006.

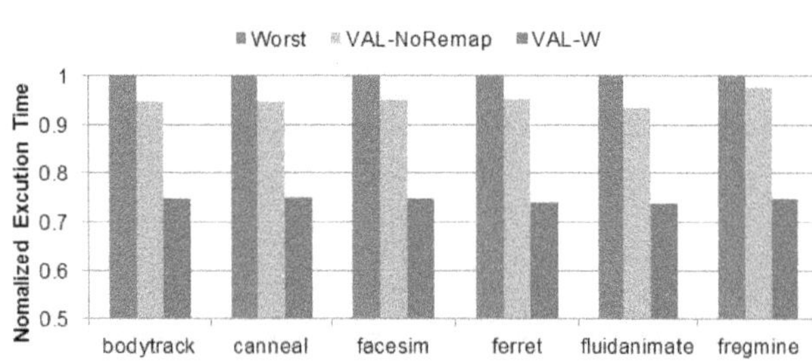

Figure 7: Comparison of performance for cases using and without using remapping.

A New Extension Method of Retention Time for Memory Cell on Dynamic Random Access Memory

Yoshiro Riho
Department of Electrical Engineering and Computer Science,
Graduate School of Engineering,
Nagoya University,
Nagoya, 464-8603, Japan

Kazuo Nakazato
Department of Electrical Engineering and Computer Science,
Graduate School of Engineering,
Nagoya University,
Nagoya, 464-8603, Japan

ABSTRACT

Demands have been placed on a dynamic random access memory (DRAM) to not only have increased memory capacity and data transfer speed but also have reduced operating and standby currents. When a system uses a DRAM, a refresh operation is necessary because of its data retention time restriction: each bit of the DRAM is stored as an amount of electrical charge in a storage capacitor. Power consumption for the refresh operation increases in proportion to the memory capacity. A new method is proposed to reduce refresh power consumption dynamically when the full memory capacity is not required, by effectively extending the memory cell retention time. Conversion from 1 cell/bit to 2^N cells/bit reduces the variation of retention time among memory cells. This method reduces the frequency of disturbance and its power consumption by two orders of magnitude. The conversion itself can be realized very simply from the structure of the DRAM array circuit, maintaining all conventional functions and operations in case of the full array access mode.

Categories and Subject Descriptors

B.3 [**Memory Structures**]: Dynamic memory (DRAM)

Keywords

partial array self refresh (PASR), composed memory cell, partial access mode (PAM)

1. INTRODUCTION

A dynamic random access memory (DRAM) has been playing the main role of storage in computer systems including high performance systems, personal computers, and mobile phones. The memory capacity of a DRAM has been increased to meet system demands, and this has been supported by the development of a semiconductor process technology. In fact, needs from the system side have included not only memory capacity but also data transfer speed, operation current reduction, and standby current reduction. Since a refresh operation is necessary in a DRAM, the increase in memory capacity has directly caused two problems: disturbance and power consumption.

In particular, standby power consumption has become one of the most serious problems for using a DRAM in mobile applications. Following the constant-electric-field scaling theory, the voltage should be lowered. However, the rewrite voltage is saturated by the difficulty of real operation. This

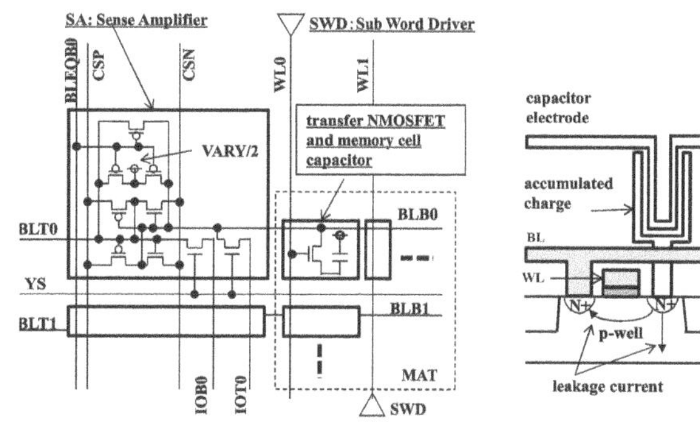

(a) Array Circuit (b) Memory Cell

Figure 1: DRAM array, (a) array circuit and (b) schematic cross section of memory cell.

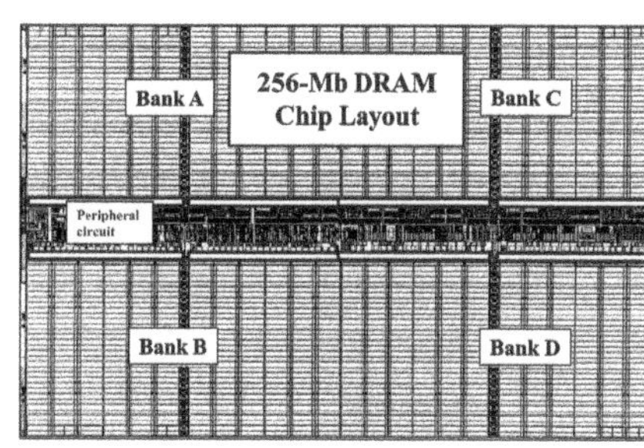

Figure 2: 256-Mb DRAM chip layout (CAD data) which is implemented and evaluated, bank structure = 4 banks, X address = X0-X12, Y address = Y0-Y9, Data = X8, refresh times = 8-K, page size = 8-K.

GLSVLSI'13, May 2–3, 2013, Paris, France.

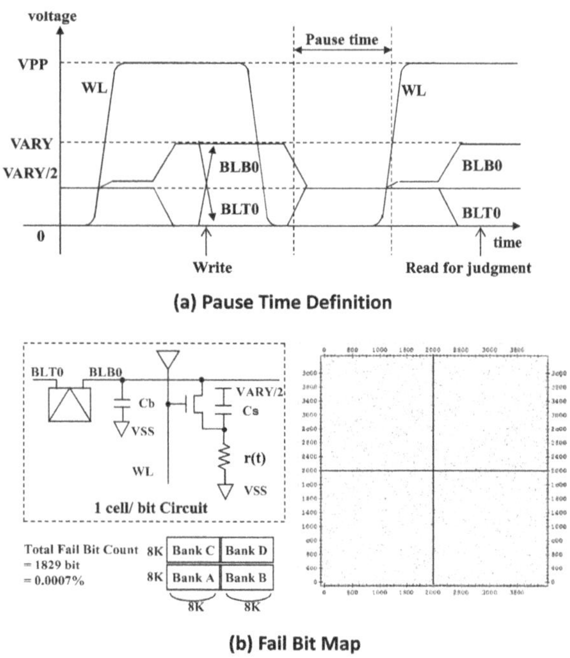

Figure 3: (a) Pause time definition and (b) random fail bit map.

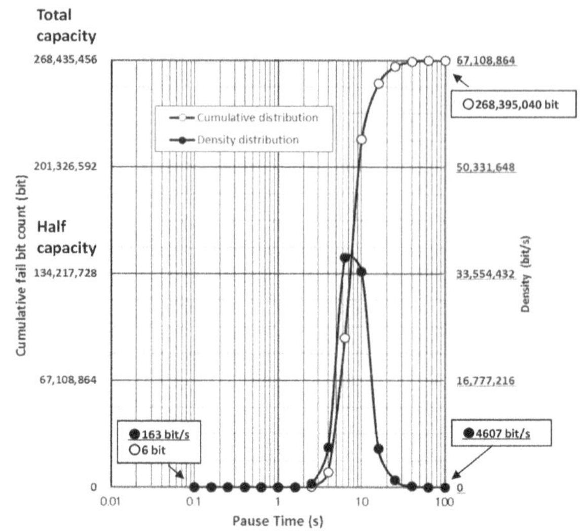

Figure 4: Cumulative fail bit count and the density distribution of 256-Mb DRAM. Temperature is 85 °C.

means that the standby power increases with the memory-chip capacity.

This paper proposes a new method to reduce refresh current when memory capacity is not fully required. The proposed method, partial access mode (PAM), is based on the conversion from 1 cell/bit to 2^N cells/bit, which can reduce the frequency of disturbance and its power consumption by two orders of magnitude. We will prove the effect of PAM by the estimation from the DRAM measured data, comparing to the conventional method, partial array self refresh (PASR).

2. REFRESH OPERATION AND RETENTION TIME

2.1 Refresh Operation

A DRAM stores a single bit in a memory cell as an amount of electrical charge on a storage capacitor. The leakage current collapses charged data as shown in Fig. 1(b). This means that a DRAM needs a rewrite operation before the memory cell loses its storage charge. This rewrite operation is called *refresh* [9]. A refresh is performed by sense amplifier (SA) in Fig. 1(a). An external auto-refresh command (AREF) activates SA. Since refresh is a kind of disturbance for system where normal operation, i.e., read or write, is forbidden, the frequency of AREF command should be minimized. In case of 256-Mb DRAM with 4 banks in Fig. 2, the refreshment of all memory cells is completed by issuing 2^{13} = 8,192 AREF commands, called an 8-K AREF operation, which means one AREF operation causes one wordline (WL) refresh in 8-K WLs per bank [10]. Each data retention time of cell is generally expressed by t_{ret}. The minimum retention time of memory cell, $t_{ret,min}$ during which all memory cells keep own charges, should be longer than 64 ms, standardized by cell refresh time t_{ref}. Thus, the maximum time interval of an AREF command, t_{REF}, is 64 ms/2^{13} = 7.8 μs in an 8-K AREF operation.

2.2 Retention Time

The AREF interval time, t_{REF}, should be longer for system requirement. A user usually sets this time to 7.8 μs that is the maximum value defined from the specification, t_{ref} of 64 ms. The retention time t_{ret} depends on the state of bit line (BL) since the leakage current is influenced by the voltages of BL, H (VARY), L (VSS=0V), and the half voltage (= VARY/2). This causes the variety of t_{ret} among some conditions [1-2]. One is that BL voltage is maintained at VARY/2 and bank is in pre-charge standby state as shown in Fig. 3(a). We call this state *pause*. The other is that BL voltage is VARY or VSS, and bank is in active standby state. We call this state *disturb* [9–10].

The t_{ret} as measured data in *pause* is obtained from the time when the fail bit judgment of the read command appears after a pause time in the pre-charge mode from the write operation of memory node voltage H, as shown in Fig. 3(a). There is no destruction from L to H, since the leakage current flows only into body (p-well) which is at L level. Cumulative distribution in Fig. 4 shows the fail bit counts of 256-Mb DRAM in each pause time after H write operation to all memory cells at 85°C, the worst condition for the leakage current in DRAM specifications. Fail bit appears at pause time = 100ms but error ratio (= fail bit count/total capacity) begins to increase at pause time = 4 s, which is 40 times larger than $t_{ret,min}$. Density distribution in Fig. 4 is defined from the difference of fail bit counts and of pause time (ΔCumulative distribution/Δpause time), which clarifies the distribution feature. The variation of t_{ret} is larger than 3 orders and the average value of t_{ret} is around 10 s. This indicates that the $t_{ret,min}$ of 256-Mb DRAM is 100 ms although t_{ret}s of almost all memory cells in 256-Mb is longer than 4 s. This is caused by the t_{ret} variation among memory cells [1-3,5].

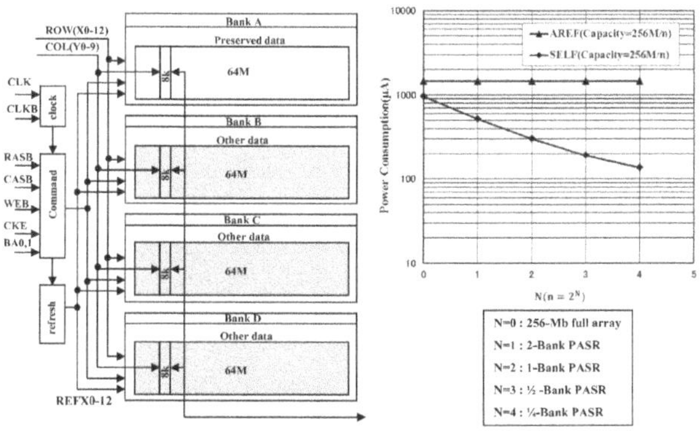

(a) 64M bits PASR (1-Bank-PASR) (b)Measured current value

Figure 5: (a) Block diagram of conventional 256-Mb DRAM with PASR. (b) Measured current consumption of PASR

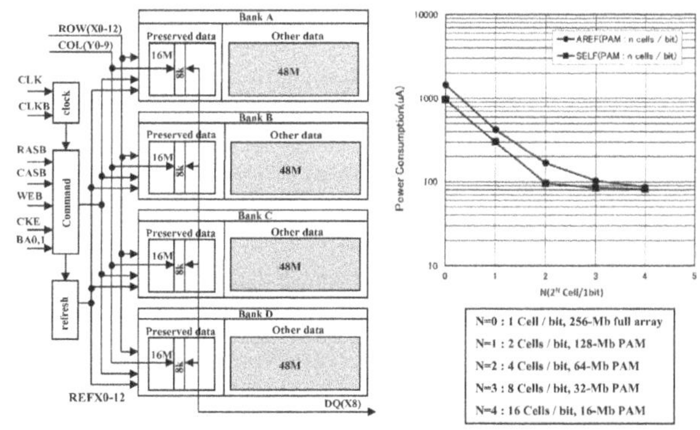

(a) 64M bits PAM (b) Estimated current value

Figure 6: (a) Block diagram 256-Mb DRAM with PAM. (b) Estimated current consumption of PAM

3. POWER REDUCTION METHOD

3.1 Self Refresh Mode

A DRAM has a specification of low power mode (SELF). The DRAM does refresh operation by itself after receiving a SELF command. If it is known in advance that there is no access to DRAMs in certain interval, a system issues SELF command and the DRAM reduces the power consumption by setting t_{REF} to the longest value of *pause* during which storage charge is not lost.

3.2 Conventional Method

A system usually demands much greater power reduction than when it is in the SELF mode. Partial array self refresh (PASR) is added to DRAM specification. In case of 1-Bank PASR in Fig. 5(a) the refresh operation is abandoned for 2 or 3 banks in 4 banks. Both memory capacity and refresh current decrease by a factor of 2 or 4 in the SELF mode. The power consumption in Fig. 5(b) is given by the measured standby current of 256-Mb DRAM. The memory capacity in SELF is limited to 256-M/n where $n = 2^N$. The constant power consumption in AREF is estimated from the maximum AREF interval t_{REF}. PASR has no regulation in the normal mode. PASR has some problems from the system-side viewpoint[11].

(1) There is mismatch in the number of banks between the AREF operation in the normal mode and PASR. Fig. 5(a) shows the block diagram for a one-bank-PASR. At the first refresh after SELF EXIT, an AREF command is issued to select WLs of all 4 banks, although 3 of those do not keep data. This means that power consumption is four times larger than needed, since a refresh operation for 3 banks is not necessary.

(2) The system usually activates the SA for all banks and accesses 4 pages (8-K $\times$ 4 = 32-K size) of 4 banks, because the frequency of the disturbance factor, i.e., pre-charge, should be less. However, the PASR mode cannot preserve all 32-K data in 4 banks, so 4 pages must be activated separately with 4 pre-charge operations before entering the SELF mode, to preserve the 32-K accessed data.

3.3 Partial Access Mode

We propose partial access mode (PAM) which eliminates

Table 1: Specification of PAM

	A2-A0	Access Array	N	n	*Capacity*	t_{ref}	*times*	t_{REF}
PAM	000	All Arrays	0	1	256M	64ms	8Kref	7.8μs
	001	128M(X0=0)	1	2	128M	128ms	4Kref	31.3μs
	010	64M(X0=X1=0)	2	4	64M	256ms	2Kref	125μs
	011	restriction	–	–	–	–	–	–
	100	restriction	–	–	–	–	–	–
	101	32M(X0=X1=X2=0)	3	8	32M	512ms	1Kref	500μs
	110	16M(X0=X1=X2=X3=0)	4	16	16M	1024ms	512ref	2000μs
	111	restriction	–	–	–	–	–	–

these mismatches and uses all memory cells efficiently, as shown in Fig. 6(a).

(1) One characteristic of PAM is that it holds data using $n = 2^N$ cells per bit to extend retention time, as shown in Table 2. Parameter n has both meanings of the composed cell number and the division number of memory capacity.

(2) Another characteristic is the control method between 1 cell/bit and 2^Ncells/bit that is located higher in hierarchy than the control of the normal and SELF modes, as shown.

These two characteristics realize a reduction of refresh operation frequency in cases of both the normal and the SELF modes. As a result, the AREF command frequency in the normal mode and power consumption in the SELF mode are reduced at once. In contrast, PASR reduces power consumption in the SELF mode only.

PAM ENTRY operation indicates a conversion from 1 cell/bit to 2^N cells/bit. This is simply a copy operation from a memory cell connected with 1 WL to 2^N- 1 memory cells connected with 2^N- 1 WLs in the same mat. Table 2 shows how this operation is simply achieved by only a delayed WL selection. PAM ENTRY operation is completed through 8-K/2^Ntimes copy operations applied to all memory cells in 256-Mb DRAM [9–10].

In the PAM, 2^N memory cells compensate for each other's storage charge by selecting 2^N WLs at the same time in both the normal and the SELF modes.

PAM EXIT operation indicates reconversion from 2^Ncells/bit to 1 cell/bit which is necessary because the storage charge in 1 memory cell is not enough to read for 1 bit. 2^N times the charge in the PAM is used for 1 bit. PAM EXIT operation is just continuous 8-K/2^N times refresh operations to all memory cells, as shown in Table 2.

PAM enables extension of the AREF command interval, t_{REF} by dual effects, the retention time extension and the

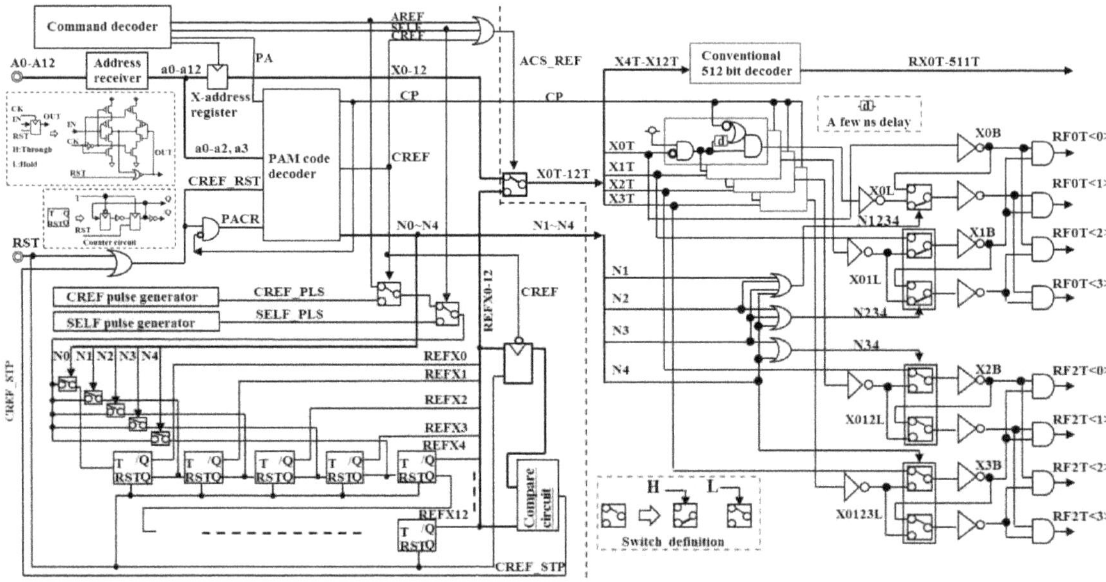

Figure 7: The circuit modification for PAM.

Table 2: Partial Access Conversion ($n = 2^N$=4)

PAM ENTRY	Partial Access	PAM EXIT
Continuous copies 1 cell / bit → 4 cells / bit	A simultaneous select 4 cells / bit	Continuous refreshes 4 cells / bit → 1 cell / bit

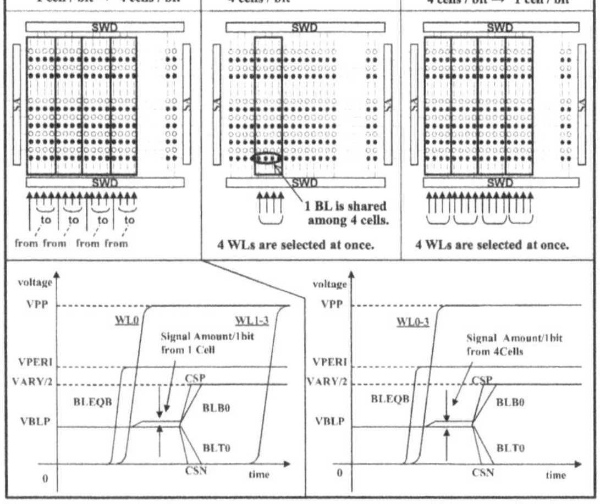

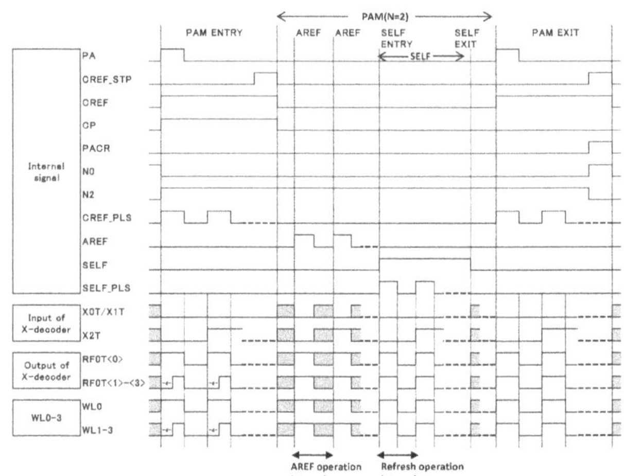

Figure 8: Functional waveforms of PAM

reduction of memory capacity. This is summarized in Table 1. In this example of 256-Mb DRAM, the accessed data capacity is 256-M/2^N and t_{REF} = 7.8 $\times 2^N \times 2^N \mu$s. On the other hand the t_{REF} in the SELF mode is set from the DRAM measured data complied during the manufacturing process of the silicon wafer. These effects just accompany with $(2^N - 1) \times 10\%$ charging current increase per SA amplification and no area overhead.

3.4 Implemented circuit and functional waveform

In this section we show that a few simple modifications to conventional DRAM enable PAM. The left panel of Fig. 7 is refresh control including the modification for PAM ENTRY and PAM EXIT. When the DRAM receives PAM ENTRY, command decoder generates PA, a one shot pulse, which enables receipt of a PAM code. A0-A2 for PAM code enters PAM code decoder, which outputs N0-N4 signals to keep composed cell number n until the DRAM receives the next PAM EXIT command. PAM code decoder outputs CREF = H and CP = H (set by A3) those decide the continuous copies operation. CREF = H selects CREF_PLS as input signal into refresh counter, which sends the pulses for a continuous copies operation in 100ns interval repeatedly until compare circuit outputs CREF_STP = H which stops continuous copies at the point of the full array copy completion. The reset operation changes CREF to L and CP to L. Fig. 8 shows these signal transitions. In PAM ENTRY operation N0-N4 signal selects the lowest bit of refresh counter among X0T-X4T.

In PAM, AREF command interval is defined in Table 1 and the refresh operation interval in SELF is determined by measured t_{ret} data in wafer condition which is estimated in the next section. In both cases output signals REFX0-12 in Fig. 7 enables refresh operation selected by ACS_REF as X0T-12T. If PAM is selected, for example, when a mobile phone system is in a call waiting state, both refresh operation intervals of normal and SELF can be extended.

When the DRAM receives PAM EXIT, the circuit operation about CREF is same. But PAM code decoder keeps CP to L by A3. The A0-A2 code in PAM code decoder is reseted at the point of the continuous refreshes completion by the signal, CREF_STP = PACR = H. PACR = H selects N0 through PAM code decoder.

CP and N1-N4 become input signals of X-decoder circuit, two 4-bit decoders for X0T/X1T and X2T/X3T in the right panel of Fig. 7. This circuit enables delayed $n-1$ wordlines selections in case of CP = H, PAM ENTRY in Table 2, and n simultaneous selection in case of CP = L, PAM EXIT in Table 2. N1-N4 decide composed cell number n and CP decides whether operations become continuous copies or continuous refreshes [7]. This address specification X0T = X1T = L in case of N2 (4 cells/bit mode) makes a delay of a few nanoseconds of X0L = X01L = L to the next inverters. This operation generates the delayed selections of RF0T<1-3> from RF0T<0>, which correspond to the delayed selections of WL<1-3> from WL<0>, for RF0T<0-3> and RF2T<0-3> are input signals of the 16-bit decoder for 16 WLs.

4. EFFECT OF IMPROVEMENT

4.1 Mechanism of Data Losing (*pause*)

We examine the pause time dependence of the signal amount to estimate t_{ret} for the refresh operation in SELF of the PAM. The destruction of memory node voltage H restored in the DRAM memory cell is caused by the discharge of the storage capacitor through leakage current. Destruction itself means that BL voltage is amplified to a different side from the written side because of storage charge reduction. When the pause time is 0, owing to charge conservation and sharing between the storage capacitor and the BL, the signal amount ΔV is expressed by

$$\Delta V = \frac{\text{VARY}}{2}\frac{C_s}{C_b + C_s}, \tag{1}$$

where C_b and C_s are the capacitances of the BL and the storage capacitor, respectively. Figs. 3(a) and (b) explain this situation. Figs. 9(a) and (b) show how the signal amount depends on the pause time. These data show that the signal amount decreases in proportion to the pause time t. The retention time is defined by the time when the signal amount is zero, i.e., $\Delta V(t_{ret}) = 0$. Thus, the signal amount $\Delta V(t)$ is given by the following equation:

$$\Delta V(t) = \frac{\text{VARY}}{2}\frac{C_s}{C_b + C_s}\left(1 - \frac{t}{t_{ret}}\right), \tag{2}$$

which means that the storage charge $Q(t)$ is given by

$$Q(t) = \frac{\text{VARY}}{2}C_s\left(1 - \frac{t}{t_{ret}}\right), \text{ for } t \le 2t_{ret}. \tag{3}$$

From the viewpoint of $Q(t)$, Figs. 9(a) and (b) express that $Q(t)$ decreases in proportion to the pause time from VARY$\cdot C_s/2$ at $t = 0$ to $-$VARY$\cdot C_s/2$ at $t = 2t_{ret}$. For $t > 2t_{ret}$, $Q(t)$ is constant at $-$VARY$\cdot C_s/2$ since the memory node voltage has settled to the lowest voltage VSS [8].

4.2 Data Retention Time in Case of 2^N cells/bit ($N \ge 1$)

The signal amount in case of 2^Ncells/bit ($N \ge 1$) can be achieved by the same method of the charge conservation law, eq. (1). And eq. (4) is the format rewritten with $Q(t)$ in eq. (3).

$$\Delta V(j_1, \cdots, j_n, t) = \frac{1}{C_b + nC_s}\sum_{i=1}^{n} Q(j_i, t), \tag{4}$$

n is the composed cell number ($n = 2^N$) and j means a fixed combination of n memory cells, respectively. When $\Delta V(j_1, \cdots, j_n, t) > 0$ the composed cell j is judged as pass. When $\Delta V(j_1, \cdots, j_n, t) \le 0$ the composed cell j is judged as fail.

The expected value of fail bit count of $256/n$-Mb DRAM is expressed in eq. (5), where t is pause time, θ is the function defined in eq. (6), and $p_{j1}, \cdots p_{jn}$ is the probability that each cell has a certain pause time.

$$E^{(n)}(t) = \frac{256M}{n}\sum_{j1}\cdots\sum_{jn}\theta(j_1, \cdots, j_n, t)p_{j1}\cdots p_{jn}, \tag{5}$$

$$\theta(j_1, \cdots, j_n, t) = \begin{cases} 1 & (\Delta V(j_1, \cdots, j_n, t) \le 0) \\ 0 & (\Delta V(j_1, \cdots, j_n, t) > 0), \end{cases} \tag{6}$$

Error ratios by eq. (5) are plotted in Fig. 10 for $n = 2, 4, 8$, and 16. They are compared to the measured data for $n = 1$.

The data for $n = 1$ is same as cumulative distribution in Fig. 4. But the vertical axis is converted to error ratio by using an inverse cumulative distribution function in order to compare fail counts in different pause times and check whether the distribution coincides with a standard normal distribution. If the distribution takes a standard normal distribution, the line becomes straight. Fig. 10 shows that the measured data for $n = 1$ has a kink, indicating the existence of two kinds of standard normal distributions, tail and main distributions. The t_{ret} variation which is larger than three orders contains one order of variation in the tail distribution area whose error ratio is less than 0.1%. This means that $t_{ret,min}$ of 256-Mb DRAM is 100 ms although t_{ret}s of 99.9% are included in main distribution area which is longer than 3 s. A DRAM supplier can replace the worst 20 or 30 memory cells with other good ones prepared in advance. Fig. 10 shows the effect of 20-bit replacement. The DRAM has a circuit to realize such a replacement [1–5].

This results of $n = 2, 4, 8$, and 16 are obtained from two assumptions. One is the random formation of the fail bit distribution. This is proved by the measured fail bit map in Fig. 3(b). The other is an approximation, the majority decision rule, that a composed cell with 2^N cells/bit fails if half the memory cells in 2^Ncells fail, based on the mechanism of data loss. This approximation means that memory node voltage of each cell takes either H or L and that each cell charge compensates for charges of other memory cells [6-7].

This result shows that t_{ret} improves from $t_{ret} = 100$ ms (1 cell/bit) to $t_{ret} = 1$ s (4 cells/bit). Increasing n brings a reduction in the variation of t_{ret} and eliminates the tail distribution. Therefore, the error ratio in Fig. 10 increases steeply at the pause time = 7 s (50% error ratio) with increasing n.

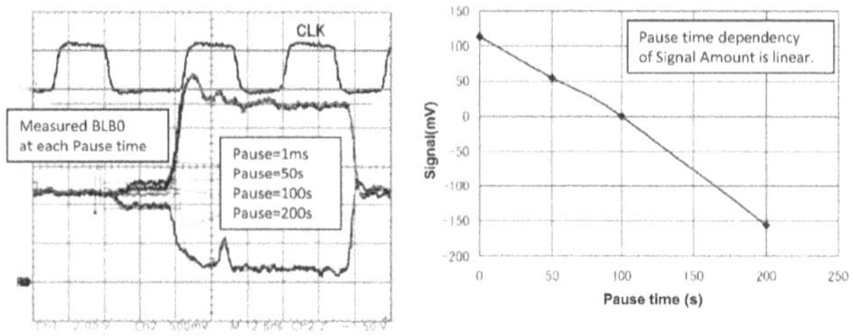

(a) Real Waveform (b) Measured Signal Amount

Figure 9: Signal amount, (a) waveform and (b) pause time dependency at room temperature.

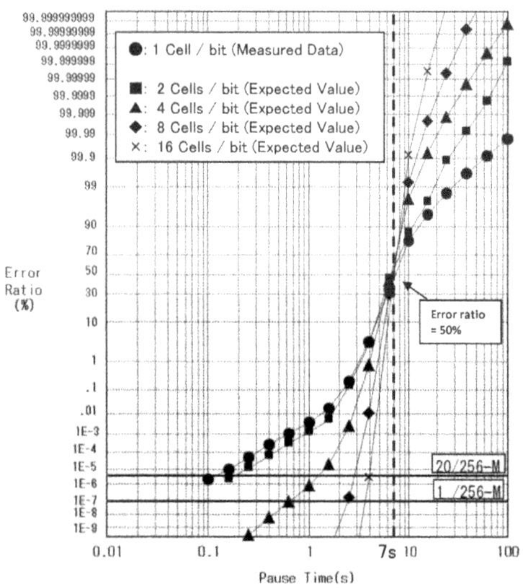

Figure 10: Error ratio of measured data and expected values. Temperature is 85 °C. Chip is assembled with ball grid array package.

4.3 Summary of current consumption

The summarized current consumption is in Fig. 6(b). The refresh operation current and the DC current compose total current consumption in the SELF mode. The DC current is the sum of the all transistor OFF current, the operating current for internal voltage generators and the current for refresh period oscillators. The DC current of 256-Mb DRAM is measured as 81 μA. Fig. 6(b) is the current consumption of PAM, which is different from that of PASR in Fig. 5(b). The traditional 256-Mb DRAM reduces the refresh operation current by a factor of 2^N in SELF by PASR. This is caused by the preserved density reduction. There is no current reduction in AREF. By contrast PAM reduces current consumption by a factor of $2^N \times 2^N$ in both cases of SELF and AREF. The current consumption of $N = 2$ in SELF and that of $N = 3$ in AREF are almost equal to the DC current, 100 μA.

If a mobile phone system selects the full access mode ($N = 0$) in normal operating state and PAM ($N \geq 2$) in a call waiting state, total power consumption is extraordinary decreased (Fig. 8), since almost of all time sharing in a mobile phone system is occupied by waiting call state.

5. CONCLUSION

The DRAM including PAM derives the t_{REF} extension effect of $2^N \times 2^N$ times. This indicates that the refresh operation frequency and the power consumption are both reduced by a factor of $2^N \times 2^N$. This new function changes the standby current of DRAMs to its DC current, even though DRAMs keeps holding written data in the fixed memory space defined by X0-X3.

6. REFERENCES

[1] S. Amakawa and K. Nakazato, "A new approach to failure analysis and yield enhancement of very large-scale integrated systems," ESSDERC, pp.147-150, 2002.

[2] O. K. B. Lui and P. Migliorato, "A new generation-recombination model for device simulation including the Poole-Frenkel effect and phonon-assisted tunneling," Solid-State Electron., 41, pp.575-583, 1997.

[3] W. Shockley and W.T.READ, Jr., "Statistics of the recombinations of holes and electrons," Phys. Rev., 87, pp.835-842, 1952.

[4] A. Hiraiwa, M. Ogasawara, N. Natsuaki, Y. Itoh, and H. Iwai, "Field-effect trap-level-distribution model of dynamic random access memory data retention characteristics," J. Appl. Phys., Volume 81, pp.7053-7060, 1997.

[5] K. Yamaguchi, "Temperature dependence of anomalous currents in worst-bit cells in dynamic random-access memories," Journal of Applied Physics, Volume 87, 11, pp.8064-8069, 2000.

[6] R. Takemura, K. Itoh, T. Sekiguchi, S. Akiyama, S. Hanzawa, K. Kajigaya, and T. Kawahara, "Long-Retention-Time, High-Speed DRAM Array with 12-F2 Twin Cell for Sub 1-V Operation," IEICE TRANS. ELECTRON., VOL.E90-C, 4, pp.758-764, 2007.

[7] T. Iwai, M. Kaku, T. Miyazaki, H. Iwai, H. Takenaka, A. Suzuki, S. Miyano, and M. Hamada, "Low power embedded DRAM using 0.6V super retention mode with word line data mirroring," A-SSCC 2009. pp.209-212, 2009.

[8] Y. Mori, S. Kamohara, M. Moniwa, K. Ohyu, T. Yamanaka, and R. Yamada, "Direct observation of worst-bit leakage currents of DRAM," IEEE Transactions on Electron Devices, Volume 53, 2, pp.398-400, 2006.

[9] N. C. C. Lu and H. H. Chao, "Half-V/SUB DD/ bit-line sensing scheme in CMOS DRAMs," IEEE Journal of Solid-State Circuits, vol.SC-19, No.4, pp.451-454, Aug. 1984.

[10] R. C. Foss, G. Allan, P. Gillingham, F. Larochelle, V. Lines, G. Shimokura, "Application of a high-voltage pumped supply for low-power DRAM," Digest of Technical papers, VLSI Circuits, pp.106-107, 1992.

[11] S. Phadke, Ann Arbor, S Narayanasamy. "MLP aware heterogeneous memory system," Design Automation and Test in Europe Conference and Exhibition, 2011.

A Hardware-Oriented Dynamically Adaptive Disparity Estimation Algorithm and its Real-Time Hardware

Abdulkadir Akin, Ipek Baz
STI, EPFL, 1015
Lausanne, Switzerland
{abdulkadir.akin, ipek.baz}@epfl.ch

Baris Atakan
EEE, METU,
Ankara, Turkey
el174054@metu.edu.tr

Irem Boybat
EE, SU,
Istanbul, Turkey
iremboybat@sabanciuniv.edu

Alexandre Schmid, Yusuf Leblebici
STI, EPFL, 1015,
Lausanne, Switzerland
{alexandre.schmid, yusuf.leblebici}@epfl.ch

ABSTRACT
The computational complexity of disparity estimation algorithms and the need of large size and bandwidth for the external and internal memory make the real-time processing of disparity estimation challenging, especially for High Resolution (HR) images. This paper proposes a hardware-oriented adaptive window size disparity estimation (AWDE) algorithm and its real-time reconfigurable hardware implementation that targets HR video with high quality disparity results. The proposed algorithm is a hybrid solution involving the Sum of Absolute Differences and the Census cost computation methods to vote and select the best suitable disparity candidates. It utilizes a pixel intensity based refinement step to remove faulty disparity computations. The AWDE algorithm dynamically adapts the window size considering the local texture of the image to increase the disparity estimation quality. The proposed reconfigurable hardware of the AWDE algorithm enables handling 60 frames per second on Virtex-5 FPGA at a 1024×768 XGA video resolution for a 120 pixel disparity range.[1]

Categories and Subject Descriptors
C.3 [**Special-Purpose and Application Based Systems**]: Real-time and embedded systems, signal processing systems

General Terms
Algorithms, Design.

Keywords
Disparity Estimation, Census Transform, Sum of Absolute Differences, High Resolution, Real-Time Implementation, FPGA.

1. INTRODUCTION
Depth estimation is an algorithmic step in a variety of applications such as autonomous navigation, robot and driving systems, 3D geographic information systems, object detection and tracking, medical imaging, computer games, 3D television, stereoscopic video compression, and disparity-based rendering.

[1]This research has been partly conducted with the support of the Swiss NSF under grant number 200021-125651.

GLVLSI'13, May 2–3, 2013, Paris, France.

Many Disparity Estimation (DE) algorithms have been developed with the goal to provide high-quality depth map results. These are ranked with respect to their performance in the evaluation tool for the Middlebury benchmarks [1]. Although top performer algorithms provide impressive visual and quantitative results [2-3], their implementations in real-time High Resolution (HR) stereo video are challenging due to their complex multi-step refinement processes or their global processing requirements that demand huge memory size and bandwidth.

Various hardware architectures that are presented in literature provide real-time DE [4-9]. Some implemented hardware architectures only target CIF or VGA video [4-6]. The hardware proposed in [4] only claims real time for CIF video. It uses the Census transform [10] and currently provides the highest quality disparity results compared to real time hardware implementations in ASICs and FPGAs. The hardware presented in [4] uses low complexity Mini Census method to determine the matching cost, and aggregates the Hamming costs following the method in [2]. Due to high complexity cost aggregation, the hardware proposed in [4] requires high memory bandwidth and intense hardware resource utilization, even for Low Resolution (LR) video.

Real-time DE for HR images offers some crucial advantages compared to low resolution DE. Processing HR stereo images increases the disparity map resolution which improves the quality of the object definition. In addition, DE for HR stereo images is able to define the disparity with sub-pixel efficiency compared to the DE for LR image. Therefore, the DE for HR provides more precise depth measurement than the DE for LR. However, the use of HR stereo images brings some challenges. Pixel-wise stereo matching operations cause a sharp increase in computational complexity when DE for HR is targeted. Moreover, DE for HR stereo images requires stereo matching checks with larger number of candidate pixels than the disparity estimation for LR images.

The systems proposed in [7-9] claim to reach real time for HR video. Still, their quality results in terms of the HR benchmarks given in [1] are not provided. [7] claims to reach 550 fps for 80 pixel disparity range at a 800×600 video resolution, but it requires extremely large hardware resources. A simple edge-directed method presented in [8] reaches 50 fps at a 1280×1024 video resolution and 120 pixel disparity range, but does not provide satisfactory DE results due to a low-complexity architecture. In [9], a hierarchical structure with respect to image resolution is presented to reach 30 fps at a 1920×1080 video resolution and 256 pixel disparity range, but it does not provide high quality DE for HR.

In this paper, we present a hardware-oriented adaptive window size disparity estimation (AWDE) algorithm and its real-time reconfigurable hardware implementation to process HR stereo video with high-quality disparity estimation results. The proposed algorithm combines the strengths of the Binary Window SAD (BW-SAD) [11] and Census Transform methods thus enables an efficient hybrid solution for the hardware implementation.

The benefit of using different window sizes for different texture features on the image is observed from the DE results in [11]. The hardware presented in [11] is not able to dynamically change the window size, since it requires to re-synthesize the hardware for using different window sizes. The hardware presented in this paper provides dynamic configurability to have satisfactory disparity estimation quality for the images with different contents. It provides dynamic reconfigurability to switch between window sizes of 7×7, 13×13 and 25×25 pixels in run-time to adapt to the texture of the image.

The proposed dynamic reconfigurability provides better DE results than existing real-time DE hardware implementations for HR images [7-9] for the tested HR benchmarks. The proposed hardware can reach 60 frames per second on Virtex-5 FPGA at a 1024×768 XGA video resolution and 120 pixel disparity range.

2. HARDWARE-ORIENTED AWDE ALGORITHM

The main focus of the AWDE algorithm is its compatibility with real-time hardware implementation while providing high quality DE results for HR. The algorithm consists of three main parts: window size determination, disparity voting, and disparity refinement.

As a terminology, we use the term "block" to define the 49 pixels in the left image that are processed in parallel. The term "window" is used to define the 49 sampled neighboring pixels of any pixel in the right or left images with variable sizes of 7×7, 13×13 or 25×25. The pixels in the window are used to calculate the Census and BW-SAD cost metrics during the search process. The parameters that are used in the AWDE algorithm are given in Section 4.

2.1 Window Size Determination

The window size of the 49 pixels in each block is adaptively determined according to the Mean Absolute Deviation (MAD) of the pixel in the center of the block with its neighbors. The formula of the MAD is presented in (1), where **c** is the center pixel of the block and **q** is the pixel in the neighborhood, Nc, of **c**. The center of the block is the pixel located at block(4, 4) in Fig. 1. (a). Three different window sizes are used. As expressed in (2), a 7×7 window is used if the MAD of the center pixel is high. A very small deviation is the sign of a region with low texture content, and a 25×25 window is used for these regions of the image.

$$MAD(\boldsymbol{c}) = \frac{1}{48} \times \sum_{q \in N_c} |I_L(\boldsymbol{q}) - I_L(\boldsymbol{c})| \quad (1)$$

$$window\ size = \begin{cases} 7 \times 7 & if\ \text{MAD}(\mathbf{c}) > tr_{7\times7} \\ 13 \times 13 & else\ if\ \text{MAD}(\mathbf{c}) > tr_{13\times13} \\ 25 \times 25 & else \end{cases} \quad (2)$$

As a general rule, increasing the window size increases the algorithm and hardware complexity [11]. As shown in Fig. 1. (b), in our proposed algorithm, in order to provide constant hardware complexity over the three different window sizes, 49 neighbors are constantly sampled for different window sizes. "1", "2" and "3" indicate the 49 pixels used for the different window sizes 7×7, 13×13 and 25×25, respectively.

2.2 Disparity voting

In this work, the BW-SAD is used as cost metrics instead of SAD. The use of BW-SAD provides better results than using the SAD when there is disparity discontinuity since it combines shape information with the SAD [11]. However, the computational complexity of the BW-SAD is high, thus result of this metric is provided for nine of the 49 pixels in a block and they are linearly interpolated to find the BW-SAD values for the remaining 40 pixels in a block. The selected nine pixels for the computation of BW-SAD are shown in Fig. 1 (a). The low complexity Census metric is computed for all of the 49 pixels of a block.

The formula expressing the BW-SAD for a pixel **p** is shown in (3) and (4). The BW-SAD is calculated over all pixels **q** of a neighborhood Np, where the notation d is used to denote the disparity. The binary window, **w**, is used to accumulate absolute differences of the pixels, if they have an intensity value which is similar to the intensity value of the center of the window. The multiplication with **w** in (4) is implemented as reset signal for the resulting absolute differences (AD). In the rest of the paper, the term, "*Shape*" is indicated by **w**.

Depending on the texture of the image, a hybrid selection method is used to combine Census and the BW-SAD. As shown in (5) and (6), an adaptive penalty (ap) that depends on the texture observed in the image is applied to the cost of the Hamming differences between the Census values. Subsequently, the disparity with the minimum *Hybrid Cost* (HC) is selected as the disparity of a searched pixel. 2's order penalty values are used to turn the multiplication operation into a shift operation. If there is a texture on the block, the BW-SAD difference between the candidate disparities needs to be more convincing to change the decision of Census, thus a higher penalty value is applied. If there is no texture on the block, a small penalty value is applied since the BW-SAD metric is more reliable than the decision of Census.

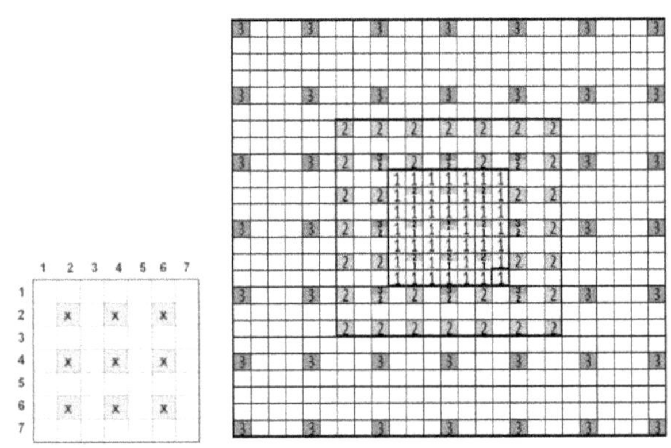

Figure 1. (a) 9 selected pixels in a block for BW-SAD calculation. (b) 49 selected pixels of adaptive windows (yellow (1): 7×7, green (2): 13×13 and blue (3):25×25).

$$\boldsymbol{w} = \begin{cases} 0 & if\ |I_L(\boldsymbol{q}) - I_L(\boldsymbol{p})| > threshold_w,\ \boldsymbol{q} \in N_p \\ 1 & else \end{cases} \quad (3)$$

$$BW\text{-}SAD(\boldsymbol{p}, d) = \sum_{q \in N_p} |I_L(\boldsymbol{q}) - I_R(\boldsymbol{q} - d)| * \boldsymbol{w} \quad (4)$$

$$HC(\boldsymbol{p}, d) = BW\text{-}SAD(\boldsymbol{p}, d) + hamming(\boldsymbol{p}, d) \times ap \quad (5)$$

$$ap = \begin{cases} ap_{7x7} & if\ \ window\ size(\boldsymbol{p}) == 7 \times 7 \\ ap_{13x13} & else\ if\ window\ size(\boldsymbol{p}) == 13 \times 13 \\ ap_{25x25} & else\ if\ window\ size(\boldsymbol{p}) == 25 \times 25 \end{cases} \quad (6)$$

2.3 Disparity Refinement

The proposed Disparity Refinement (DR) process assumes that neighboring pixels within the same *Shape* needs to have an identical disparity value, since they may belong to one unique object. In order to remove the faulty computations, the most frequent disparity value within the *Shape* is used.

The DR process of each pixel is complemented with the disparities of 16 neighbor pixels and its own disparity value. Finally, the most frequent disparity in the selected 17 contributors is replaced with the disparity of that pixel. The disparity of the processed pixel and the disparity of its four adjacent pixels always contribute to the selection of the most frequent disparity.

In Fig. 2, examples of the selection of contributing pixel locations are shown for three different window sizes. Since the proposed hardware processes seven rows in parallel during the search process of a block, the DR process only takes the disparity of pixels in the processed seven rows. Considering the proposed contributor selection scheme, the pixels in the same row with the same window size have identical masks. The masks for the seven rows of a block and three window sizes are different. Therefore, 21 different masks are applied in the refinement process. These masks turn out to simple wiring in hardware.

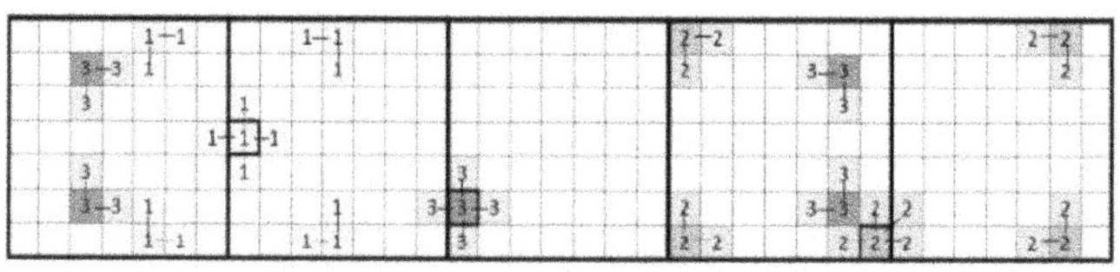

Figure 2. Examples for selecting 17 contributing pixels for 7×7, 13×13 and 25×25 window sizes during the disparity refinement process (yellow (1): 7×7, green (2): 13×13 and blue (3): 25×25).

3. HARDWARE IMPLEMENTATION

3.1 System Overview

The top-level block diagram of the proposed reconfigurable disparity estimation hardware and the required embedded system components for the realization of the full system are shown in Fig. 3. Since the main improvement of the proposed system relates to the Reconfigurable Disparity Map Estimation module, it is further explained in detail.

The Reconfigurable Disparity Map Estimation module involves 5 sub-modules and 62 dual port BRAMs. These five sub-modules are the Control Unit, Reconfigurable Data Allocation, Reconfigurable Computation of Metrics (RCM), Adaptive Disparity Selection (ADS) and Disparity Refinement. 31 of the 62 BRAMs are used to store 31 consecutive rows of the right image, and the remaining 31 BRAMs are used to store 31 rows of the left image.

External memory bandwidth is an important limitation for disparity estimation of HR images. Our proposed memory organization and data allocation scheme require reading each pixel only one time from the external memory during the search process.

3.2 Data Allocation and Disparity Voting

The block diagram of the Reconfigurable Data Allocation module is shown in Fig. 4. The data allocation module reads pixels from BRAMs, and depending on the processed rows, it rotates the rows using the Vertical Rotator to maintain the consecutive order.

The search process starts with reading the 31×31 size window of searched block from the BRAMs of the left image. Therefore, the Control Unit sends the *image select* signal to the multiplexers that are shown in Fig. 4 to select the BRAMs of the left image. While the window of searched block are loaded to the D flip flop (DFF) Array, the RCM computes and stores the 49 Census transforms, 49 *Shapes* and 9 windows pertaining to the pixels in the block for the computation of BW-SAD.

The Census transforms and windows of the candidate pixels in the right image are also needed for the matching process. After the computation of metrics for the 7×7 block, the Control Unit selects the pixels in the right image by changing the *image select* signal, and starts to read the pixels in the right image from the highest level of disparity by sending the address signals of the candidate pixels to the BRAMs. The disparity range can be configured by the user depending on the expected distance to the objects.

The detailed block diagram of the DFF Array and the Weaver are shown in Fig. 5. They are the units of the system that provide the configurability of the adaptive window size. As a terminology, we used the term "weaving" to mean "selecting 49 contributor pixels in different window sizes 7×7, 13×13 and 25×25 by skipping 1, 2 and 4 pixels respectively". Seven rows and one column are processed in parallel, and the processed pixels flow inside the DFF Array from the left to the right. Additionally, the weaving process is applied to the location (15, 8) of the DFF Array at the beginning of the search process only, to select the window size by computing the deviation of the center of the block from its neighbors for 7×7 and 13×13 windows.

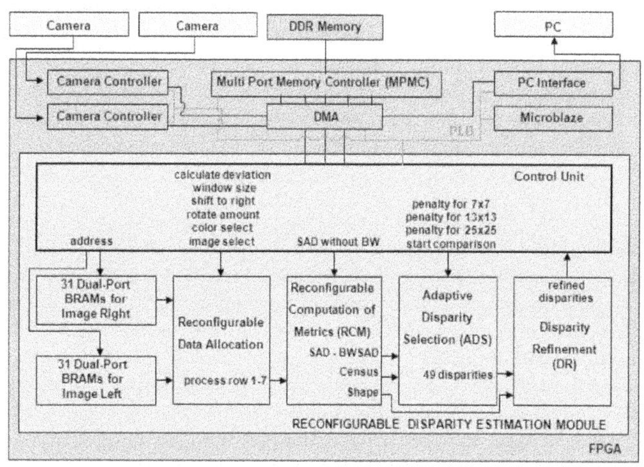

Figure 3. Top-Level Block Diagram of the System Architecture.

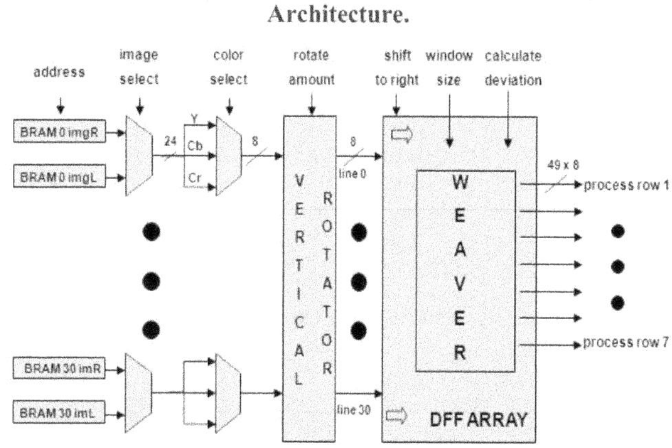

Figure 4. Reconfigurable Data Allocation Module.

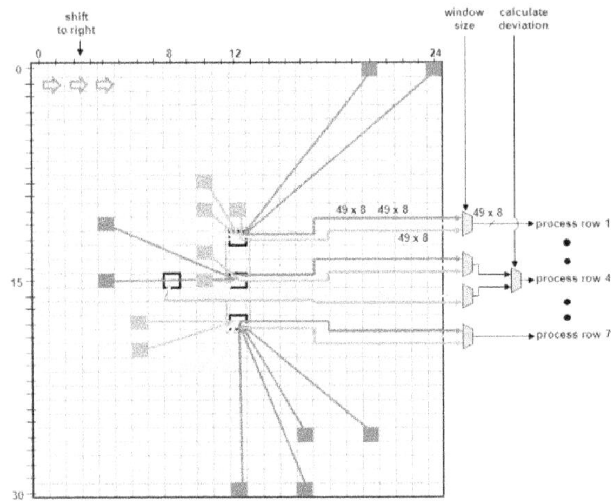

Figure 5. DFF Array and the Weaver (yellow: 7×7, green: 13×13 and blue: 25×25).

The DFF Array is a 31×25 array of 8-bit registers as shown in Fig. 5. While the pixels are shifting to the right, the Weaver is able to select the 49 components of the 7×7, 13×13 and 25×25 window sizes from the DFF Array with simple wiring and multiplexing architecture. Some of the contributor pixels of the windows for different window sizes are shown in Fig. 5 in different colors. The Weaver sends seven windows to be processed by RCM as *process row 1* – *process row 7*, and each *process row* consists of 49 selected pixels.

A large window size normally involves high amount of pixels and thus requires more hardware resource and computational cost to support the matching process [11]. By using the proposed weaving architecture, even if the window size is changed, always 49 pixels are selected for the window. Therefore, the proposed hardware architecture is able to reach the largest window size (25×25) among the hardware architectures implemented for DE [4-9].

During the weaving process of the 49 pixels in the block and the candidate pixels in the right image, the RCM computes the Census and *Shape* of these pixels in a pipeline architecture. The block diagram of the RCM is shown in Fig. 6. In Fig. 6, the registers are named as "Shape $_{row\_column}$" and "Census $_{row\_column}$". Since the BW-SAD is only applied for 9 of the 49 pixels, the BW-SAD computation sub-modules are only implemented in process rows 2, 4 and 6. The computation of the Hamming distance requires significantly less hardware area than the BW-SAD. Therefore, the Hamming computation is used for all of the 49 pixels in a block.

As shown in Fig. 7, the proposed hardware searches 49 pixels in a block in parallel. While the proposed architecture computes the Hamming distance for the left-most pixels of the block, the Hamming for disparity **d**, rightmost pixels of the block computes their Hamming for disparity **d**+6. Therefore, the resulting Hamming costs are delayed in the ADS to synchronize the costs. This delay is also an issue of the BW-SAD results and they are also synchronized in the ADS.

The ADS module shifts the Hamming results of the candidate pixels depending on the 2's order adaptive penalty for the multiplication process as shown in formula (5). The ADS module adds the resulting Hamming penalty on the BW-SADs to compute *Hybrid Costs*. 49 comparators are used to select 49 disparity results that point minimum *Hybrid Costs*.

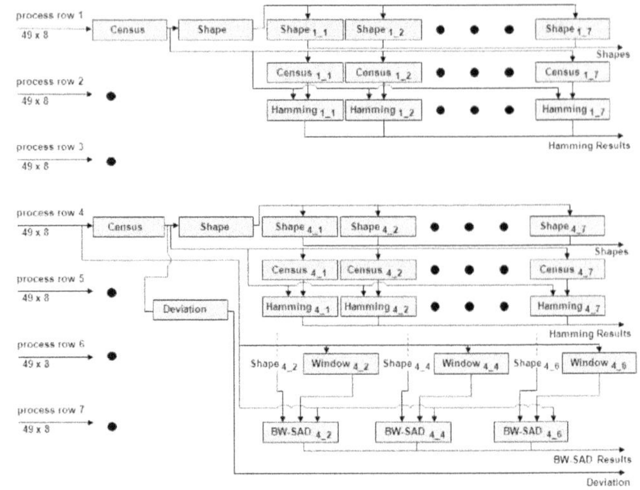

Figure 6. Reconfigurable Computation of Metrics.

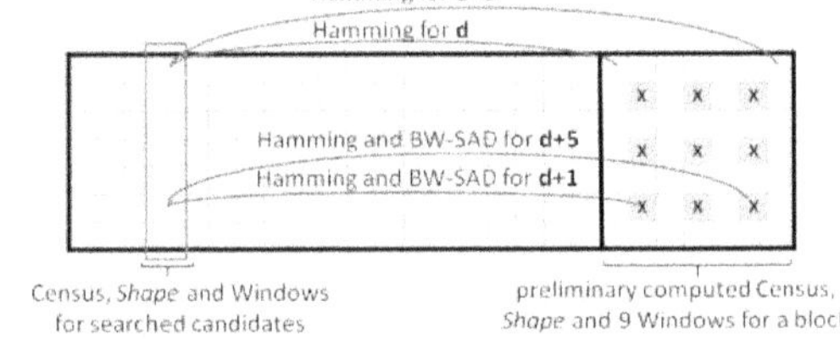

Figure 7. Processing Scheme ("x" indicates 9 selected pixels in a block for BW-SAD calculations).

3.3 Disparity Refinement

The DR module receives the 49 disparity results from the ADS and the *Shapes* of the 49 pixels of a block from the RCM and determines the final refined disparity values.

As presented in Fig. 8, after the ADS module has computed 49 disparity values in parallel, it loads this data with the respective *Shape* information in to the DFF Array of DR module (DR-Array). The DR-Array has a size of five blocks for the refinement process. DR-Array is designed to shift the disparity and *Shape* values from right to left to allocate data for the refinement processes.

The DR module involves seven identical Processing Elements (DR-PE). As presented in Fig. 8, DR-PEs are positioned to refine seven disparities in 15th column of DR Array in parallel while the disparity and shape values shift through the DR-Array. The hardware architecture of a single DR-PE is presented in Fig. 9.

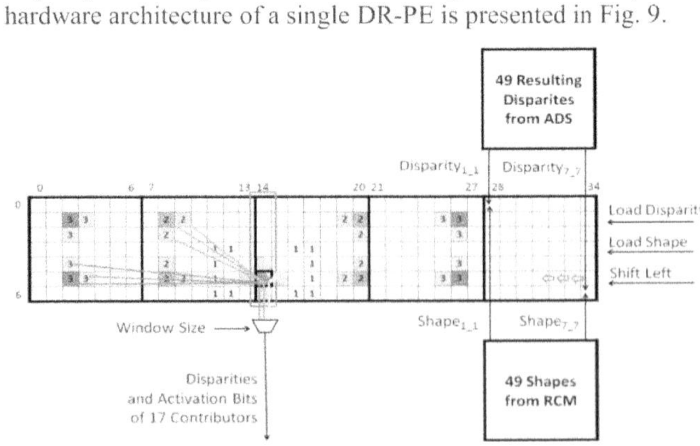

Figure 8. DR-Array of the Disparity Refinement Module (yellow (1): 7×7, green (2): 13×13 and blue (3):25×25).

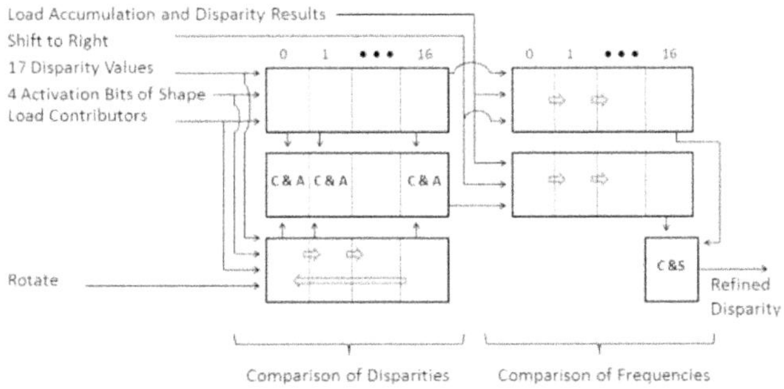

Figure 9. Processing Element of the Disparity Refinement Module.

In Fig. 8, while 17 disparity values are selected by the multiplexers, the *Shape* information corresponding to the four corners are also selected from the 48-bit shape information of the processed pixel. The selected 4-bits inform the DR-PE which of these 17 disparity values will be used while computing the highest frequency disparity. These 4 bits of the *Shape* are called *activation bits* in Fig. 9. Each activation bit activates itself together with its two adjacent disparities. The DR-PE uses shift arrays, 17 Compare and Accumulate (C&A) and 17 Compare and Select (C&S) sub-modules to select the disparity with the highest frequency as refined disparity.

4. IMPLEMENTATION RESULTS

The reconfigurable hardware architecture of the proposed AWDE algorithm is implemented using Verilog HDL, verified using Modelsim 6.6c. The Verilog RTL models are mapped to a Virtex-5 XCUVP-110T FPGA comprising 69k Look-Up Tables (LUT), 69k DFFs and 144 Block RAMs (BRAM). The proposed hardware consumes 59% of the LUTs, 51% of the DFF resources and 42% of the BRAM resources of the Virtex-5 FPGA. The proposed hardware operates at 190 MHz after place & route and computes the disparities of 49 pixels in 195 clock cycles for 120 pixel disparity range. Therefore, it can process 60 fps at a 768×1024 XGA video resolution.

The parameters of the AWDE algorithm are shown in Table 1. Parameters are selected by sweeping to obtain high quality DE of HR images considering different features.

Table 2 and Table 3 compare the disparity estimation performance and hardware implementation results of the AWDE architecture with other existing hardware implementations that targets HR [7-9] and currently the highest quality DE hardware that targets LR [4]. These papers do not provide the disparity estimation quality results for the HR benchmarks of the Middlebury data-set. Thus, we implemented [4], [7], and [9] in software, and the software implementation of [8] is obtained from the authors. The DE results for the Census and the BW-SAD metrics for different window sizes are also presented in Table 2. The comparisons of the resulting disparities with the ground-truths are done as prescribed by the Middlebury evaluation module. If the estimated disparity value is not within a ∓1 range of the ground truth, the disparity estimation of the respective pixel is considered as erroneous. 18 pixels located on the borders are neglected in the evaluation of LR benchmarks, and a disparity range of 30 is applied for all algorithms. 30 pixels located on the borders are neglected in the evaluation of HR benchmarks, and a disparity range of 120 is applied for all algorithms.

The Census and BW-SAD results that are shown in Table 2 are provided by sampling 49 pixels in a window. Although the Census and the BW-SAD algorithms do not provide individually very efficient results, the combination of these algorithms into a reconfigurable hardware provides an efficient hybrid solution, as demonstrated from the AWDE results. If the sampling is not applied and all the pixels in a window are used during the matching process, the complexity of the AWDE algorithm increases by 12 times. The result of the high complexity version of the AWDE algorithm (AWDE-HC) is also provided in Table 2 for comparison. The AWDE-HC provides almost same quality results as the AWDE. Considering the hardware overhead, the low complexity version of the algorithm, AWDE, is selected for hardware implementation, and its efficient reconfigurable hardware is presented.

Table 1. Parameters of the AWDE

tr_{7x7}	tr_{13x13}	ap_{7x7}	ap_{13x13}	ap_{25x25}	$threshold_w$
5	2	32	16	4	8

Table 2. Disparity estimation performance comparisons

Algorithm	Error Rate (%)				
	Tsukuba 288 x384	**Venus 383 x484**	**Aloe 1110 x1282**	**Art 1110 x1390**	**Clothes 1110 x1300**
Chang [4]	4.15	0.56	3.75	12.80	2.97
Ttofis [8]	13.21	4.56	8.88	32.18	7.67
Greisen [9]	12.42	4.14	8.65	23.46	5.30
Georgoulas [7]	12.38	15.20	6.97	23.75	9.15
Census7	26.05	30.80	20.36	45.39	21.80
Census13	18.19	18.83	11.21	31.65	9.36
Census25	15.94	15.38	10.41	29.66	7.16
BWSAD7	12.19	19.45	8.31	34.03	13.33
BWSAD13	11.23	15.16	7.13	28.57	9.27
BWSAD25	10.43	11.12	6.74	24.74	6.28
AWDE	7.64	5.33	4.94	16.33	2.89
AWDE-HC	7.47	4.73	4.92	16.17	2.95

Table 3. Hardware performance comparisons

Hardware	Technology	Image Resolution	Disparity Range	fps	Clock Speed (MHz)
Chang[4]	ASIC-90nm	352×288	64	42	95
Ttofis[8]	Virtex-5	1280×1024	120	50	100
Greis.[9]	Stratix-III	1920×1080	256	30	130
Geor.[7]	Stratix-IV	800×600	80	550	511
Proposed (AWDE)	Virtex-5	1024×768	120	60	190
		640×480	60	224	
		352×288	60	680	

The algorithm presented in [4] uses the Census algorithm with the cost aggregation method, and provides the best results for both LR and HR stereo images except the HR benchmark Clothes. As shown in Table 3, due to the high-complexity of cost aggregation, it only reaches 42 fps for CIF images, thereby consuming a large amount of hardware resource. If the performance of [4] is scaled to 1024×768 for disparity range of 120, less than 3 fps can be achieved.

None of the compared algorithms that have a real-time HR hardware implementation [7-9] is able to exceed the DE quality of AWDE for HR images. The overall best results following the results of AWDE are obtained from [9]. The hardware presented in [9] consumes 20% of the 270k Adaptive LUT (ALUT) resources of a Stratix-III. It provides high disparity range due to its hierarchical structure. However, this structure easily causes faulty computations when the disparity selection finds wrong matches in low resolution.

The hardware implementation of [7] provides the highest speed performance in our comparison. However this hardware applies 480 SAD computations for a 7×7 window in parallel. The hardware presented in [7] consumes %60 of the 244k ALUT resources of a Stratix-IV FPGA. In our hardware implementation we only use 9 SAD computations in parallel for the same size window and this module consumes 16% of the resources of Virtex-5 FPGA on its own. Therefore, the hardware proposed in [7] may not fit in to 3 Virtex-5 FPGAs.

The visual results of the AWDE algorithm for the HR benchmarks Clothes, Art and Aloe are shown in Fig. 10 (a-f). The 1024×768 resolution disparity map result of the AWDE algorithm for the pictures taken by our stereo camera system is shown in Fig. 10 (g-h). Our hardware architecture provides both quantitative and visual satisfactory results and reaches real-time for HR.

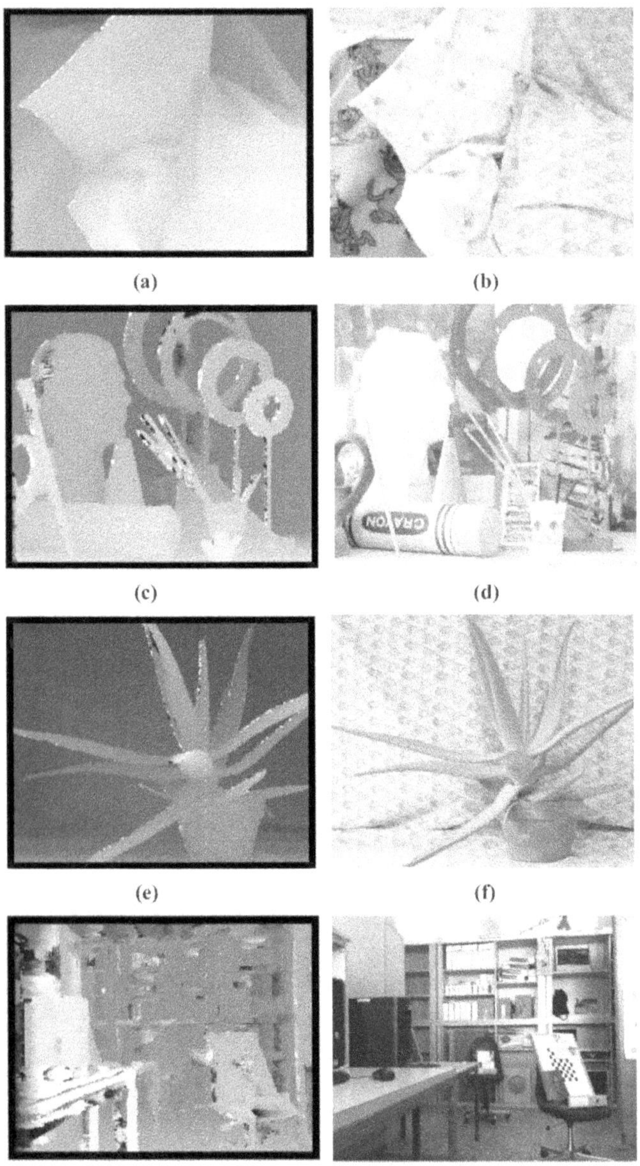

(a) (b) (c) (d) (e) (f) (g) (h)

Figure 10. Visual disparity estimation results of AWDE algorithm for HR benchmarks (a-b) Clothes, (c-d) Art, (e-f) Aloe, and (g-h) LSM lab.

5. CONCLUSION

In this paper, a hardware-oriented adaptive window size disparity estimation algorithm and its real-time reconfigurable hardware implementation are presented. The proposed AWDE algorithm dynamically adapts the window size considering the local texture of the image to increase the disparity estimation quality. Currently, the AWDE algorithm and its real-time hardware implementation reach the highest DE quality compared to existing real-time DE hardware implementations for HR images. The proposed reconfigurable hardware can process 60 fps on Virtex-5 FPGA at a 1024×768 XGA video resolution for 120 pixel disparity range. The AWDE algorithm and its reconfigurable hardware can be used in consumer electronic products where high-quality real-time disparity estimation is needed for HR video.

6. ACKNOWLEDGMENTS

The authors of [8] are acknowledged for kindly sharing their software implementation. Authors thank Omer Cogal for his support on hardware development.

7. REFERENCES

[1] D. Scharstein, and R. Szeliski. 2002. A Taxonomy and evaluation of dense two-frame stereo correspondence algorithms. *Int. J. Comput. Vision*, (Apr. 2002), 7–42.

[2] X. Mei, X. Sun, M. Zhou, S. Jiao, H. Wang, and X. Zhang. 2011. On building an accurate stereo matching system on graphics hardware. *In Proc. of GPUCV*, (Nov. 2011).

[3] A. Klaus, M. Sormann, and K. Kamer. 2006. Segment-based stereo matching using belief propagation and a self-adapting dissimilarity measure. *In Proc. ICPR*, 3 (2006).

[4] N. C. Chang, T. H. Tsai, B. H. Hsu, Y. C. Chen, and T. S. Chang. 2010. Algorithm and architecture of disparity estimation with mini-census adaptive support weight. *IEEE Trans. on Circuits and Systems for Video Technology*, 20, (Jun. 2010), 792-805.

[5] S. Jin, J. Cho, X. D. Pham, K. M. Lee, S. K. Park, M. Kim, and J. W. Jeon. 2010. FPGA design and implementation of a real-time stereo vision system. *IEEE Trans. on CSVT*, (Jan. 2010), 15 - 26.

[6] S. H. Lee, and S. Sharma. 2011. Real-time disparity estimation algorithm for stereo camera systems. *IEEE Trans. on Consumer Electronics*, 57, (Aug. 2011).

[7] C. Georgoulas, and I. Andreadis. 2009. A Real-Time Occlusion Aware Hardware Structure for Disparity Map Computation. *ICIAP*, (2009), 721-730.

[8] C. Ttofis, S. Hadjitheophanous, A. S. Georghiades, and T. Theocharides. 2012. Edge-directed hardware architecture for realtime disparity map computation. *In Proceedings of the IEEE Transactions on Computers*, (Jan. 2012).

[9] P. Greisen, S. Heinzle, M. Gross, and A. P. Burg. 2011. An FPGA-based Processing Pipeline for High-Definition Stereo Video. *EURASIP Journal on Image and Video Processing*, (Nov. 2011).

[10] R. Zabih, and J. Woodfill. 1994. Non-parametric local transforms for computing visual correspondence. *In Proc. 3rd Eur. Conf. Comput. Vision*, 2, (1994), 151–158.

[11] A. Motten, and L. Claesen. 2010. A binary adaptable window SoC architecture for a stereo based depth field processor. *In Proceedings of the IEEE VLSISOC*, (Sep. 2010), 25 - 30.

A Source-synchronous Htree-based Network-on-Chip

Ayan Mandal, Sunil P. Khatri and Rabi N. Mahapatra
Texas A&M University
College Station, TX, USA
ayan1731@tamu.edu, sunilkhatri@tamu.edu and rabi@neo.tamu.edu

ABSTRACT

Most existing Network-on-Chip (NoC) designs operate at the same or lower clock speed as the processing elements (PEs). Recently, a new source-synchronous ring-based NoC architecture has been proposed, which runs significantly faster than the PEs and offers a significantly higher bandwidth and lower communication latency. However, the ring-based design assumes a separate clock distribution scheme for the NoC and the PEs, and uses a standard mesh topology for the NoC. In this work, we present a source synchronous ring-based NoC, laid out in an H-tree topology, with each data link being routed parallel to a clock ring. The clock is generated and distributed by multiple standing wave oscillator (SWO) rings, which are also laid out in an H-tree topology. Our design allows the PEs to extract a low jitter clock directly from the high speed ring-based SWO clock by division. Moreover, since the PEs are synchronous with the ring clock, they do not need synchronizers while communicating with the NoC. We also show that by recursively duplicating links in the H-tree based source synchronous NoC (*Hnoc*), we can obtain new hybrid NoC structures. In the limit, this recursive duplication causes the H-tree based NoC to morph into the mesh-based source synchronous NoC (*Mnoc*). The performance of each such intermediate hybrid NoC structure is quantified in terms of area, link utilization and contention free latency. We also enhance the performance of the hybrid NoCs by widening congested links, and quantify the tradeoffs. Experimental results show that the hybrid NoC designs can provide significantly lower latency (upto 5× lower) and are able to sustain a higher injection rate (upto 6.8× higher) compared to a state of the art mesh. Moreover, these hybrid NoC designs use fewer buffers (upto 19.4% less) and lower wire length (upto 19.7% lower) compared to a mesh. Based on the performance and the area tradeoffs, an NoC designer can select any hybrid NoC structure among the presented.

Categories and Subject Descriptors: B.4.3 [Interconnections (subsystems)]: Asynchronous/synchronous operation; Topology (e.g., bus, point-to-point)

Keywords: H-tree; NoC; Source-synchronous

1. INTRODUCTION

Traditionally the H-tree has been used as a popular synchronous clock distribution approach. An n-level H-tree has 2^n leaves (endpoints), which are uniformly spaced, and cover the entire IC area. A clock source is present at the center and the H-tree is typically constructed with the help of wire segments (often driven by inverters) which are arranged in a recursive "H" pattern. The local clock regenerators (at the leaves of the H-tree) buffer the clock signal to drive the clock locally, with a sufficiently high slew-rate. Due to the symmetry of the H-tree, the resulting clock has low skew. Recently, a standing wave oscillator (SWO) [1, 2] has been demonstrated as an alternate high-speed on-chip synchronous clock distribution and generation scheme. These SWOs rely on the passive inductive and capacitive parasitics of on-chip wiring rings to create a resonant structure and a high clock frequency of oscillation, and also exhibit a very low jitter. Oscillations are sustained by using a single cross coupled inverter pair which provides the negative resistance essential for oscillation. Charge is recirculated in these resonant oscillators, and therefore they exhibit a low power consumption (which arises due to resistive losses in the ring, as well as the power consumed by the inverter pair).

With the increasing size of Chip Multi-Processors (CMPs), designers are opting for a separate synchronous communication network between the regions on the die, while each region is clocked in a synchronous manner. This is referred to as multi-synchronous or mixed clock communication approach. Most of the NoCs are implemented using the multi-synchronous paradigm. Recently, the authors in [3] proposed a high-speed, source-synchronous NoC architecture, where the PEs communicate with the NoC using the multi-synchronous paradigm. However, a multi-synchronous design suffers from the following disadvantages : i) it requires more area for the asynchronous FIFOs present at clock crossing boundaries, ii) it incurs an extra latency due to synchronization, and iii) the synchronizers present at the clock crossing boundaries have a non-zero probability of failure due to metastability issues. Hence, a common synchronous clock for the NoC and the PEs is desirable, and utilized in this paper. We present a source synchronous NoC (laid out in an H-tree topology) with each link being routed in parallel with an SWO clock ring. The clock is implemented as a standing wave oscillator (SWO). Multiple clock rings are used (one for each H-tree segment), and each clock ring is injection locked with other clock rings that intersect it. Hence the entire NoC is synchronous. The PEs can extract a low jitter clock from the high speed ring clock by clock division, making them synchronous with the ring clock. We also show that by recursively duplicating links in the H-tree based source synchronous NoC (*Hnoc*) we can obtain new hybrid NoC designs. In the limit, in this manner, the H-tree based NoC morphs into the mesh-based source synchronous NoC (*Mnoc*). The performance of each such intermediate hybrid NoC structure is quantified. The *Hnoc* has the lowest area but worst performance. The *Mnoc* has largest area and best performance. Based on the performance requirements and the area budget, an NoC designer can select any intermediate hybrid NoC structure. All our designs are compared with a state-of-the art mesh based NoC [4].

The key contributions of this paper are:

- To the best of the authors' knowledge, this paper is the first to propose a common ring-based resonant clock distribution scheme for the PEs and the NoC.
- Our design allows the PEs to extract a low jitter clock directly from the high speed ring-based resonant NoC clock.
- The PEs are synchronous with the ring clock and hence we avoid the need for synchronizers between the PEs and the NoC.
- Architectural simulations show that our proposed NoC designs

GLSVLSI'13, May 2–3, 2013, Paris, France.

can provide significantly lower latency (upto 5×) and are able to sustain a higher injection rate (upto 6.8×) compared to a state of the art mesh.

- Our proposed NoC designs use fewer buffers (by upto 19.4%) and lower wire length (by upto 19.7%) compared to a state of the art mesh.
- We provide a quantitative comparison of the various variants of the H-tree NoC based designs in terms of area and performance.

The rest of the paper is organized as follows. Section 2 describes previous approaches in this area. Section 3 presents our approach, while Section 4 describes the experimental results which we performed to validate our approach. We conclude in Section 5.

2. PREVIOUS WORK

The goal of an NoC is to provide a high performance, modular connectivity between PEs in a CMP. The NoC topology defines the way in which the PEs are connected, and directly impacts the area requirement (logic and wire) and performance of the NoC. In terms of topology, the mesh [5] and torus [6] have received greatest attention due to their regular and modular structure, making them easy to implement in an IC. Topologies like the 2D flattened butterfly [7] and fat tree [8] provide a dense connectivity compared to a mesh, with the downside of the requirement for large-radix routers and increased link area. A topology of concentric rings (similar to ring road in city) has been proposed in [9], which reduces the risk of congestion in the central parts of the network. In [10], the authors implement a simple yet effective reconfigurable source-synchronous NoC, which can sustain a peak throughput of one word per cycle. In [11], the authors have implemented a low-overhead asynchronous NoC, which has significantly lower latency and competitive throughput for mid-range injection rates, but suffers degradation at higher injection rates.

In all the above implementations, design decisions are made based on the fact that the NoC runs at the same or lower frequency as the PEs. In [3], the authors present a fast source synchronous ring-based NoC architecture which runs significantly faster (by a factor of 9×) than the PEs' clocks. Data is driven in a source-synchronous manner along with a high speed resonant clock, providing significantly higher bisection bandwidth with narrower links (yielding a lower area and power for the same bisection bandwidth). The authors have validated their approach by means of thorough circuit simulations. In [12], the authors focus on the architectural aspects of the ring-based NoC, providing deadlock free routing using link ordering and virtual channels. Architectural results obtained on synthetic traffic demonstrate that the ring-based NoC has up to 3.5× lower latency and up to 2.9× higher maximum sustained injection rate compared with a state of the art mesh-based NoC. However, in [12], the authors assume a mesh topology, where the ring clocks are synchronous amongst each other, and the PE clocks are also synchronous amongst each other. The ring and the PE clocks are asynchronous, thereby requiring asynchronous FIFOs for the communication between the PEs and the NoC. In this paper, in contrast, we propose an unified synchronous clock distribution for the NoC and the PEs. In our design, the PEs can extract a low jitter clock directly from the high speed ring clock by division. Moreover, since the PE clocks are synchronous with the ring clocks, we eliminate the need for synchronizers and asynchronous FIFOs while communicating between the PEs and the NoC.

Our approach is based on the use of a very fast resonant clock, and data that is source-synchronous with this clock. This clock is obtained using a standing wave resonant oscillator (SWO) [1, 2]. SWOs are a promising technique to generate a high-frequency on-chip clock signal with low power. In [2], the authors present a tiled SWO based resonant grid for high frequency clock distribution. In contrast to the tile-based topology of [2], we focus our attention on a H-tree based resonant clock distribution. We route the NoC datapath parallel to the resonant clock grid, both of which are arranged in an H-tree pattern.

3. OUR APPROACH

Figure 1 shows the baseline architecture of our proposed resonant clock rings (arranged in an H-tree pattern) and the overlaid datapath, for a three level H-tree serving 8 *PEs*. Each of the clock rings (dotted lines) oscillates at the same frequency, and they are injection locked with intersecting clock rings as shown in Figure 1. This ensures that all rings have the same phase and frequency. A bidirectional link (solid line) is responsible for carrying the data for the NoC, and is routed parallel to the corresponding clock ring. The PEs of the CMP connect to the ring at the leaves of the H-tree through *Insertion-Extraction Stations* (IES'), which are labeled "×" in Figure 1. *Junction Stations* (JS') are present where the links intersect (JS' are labeled as "○" in Figure 1) and are responsible for routing the flits. The IES' and JS' extract a high speed, low jitter clock directly from the clock rings. The *PEs* extract their clock by division from the ring clock. The design thus eliminates the need of synchronizers between the *PE* and the IES, which improves the communication latency. Moreover, we avoid the need of extra hardware to maintain the read and write pointers of the FIFOs (responsible for data transfer between the *PE* and the IES) in two different clock domains. This is because the FIFOs are synchronous.

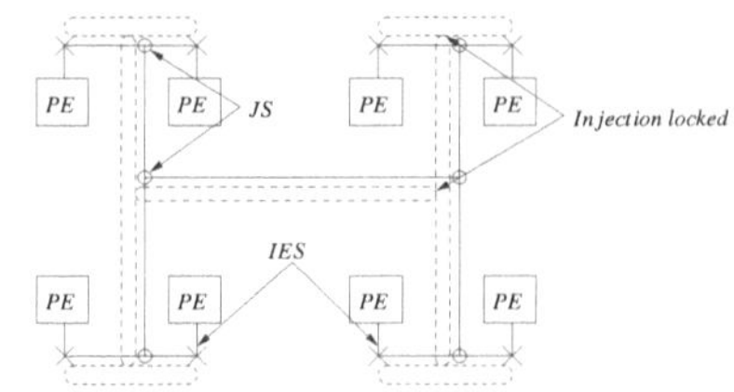

Figure 1: Clock Rings of *Hnoc* (dotted) with Datapath Overlaid (solid)

3.1 Baseline Hnoc Clocking

In this section, we discuss our approach towards the resonant clock distribution used in our design. Figure 2 a) shows a standing wave oscillator (SWO) [1]. A long wiring ring (connected in a mobious fashion) is used, and oscillations are sustained in this resonant ring by a single inverter pair. The clock signal at any point in the ring is sinusoidal with the same phase, but varying amplitude at all points along the ring. The amplitude of the signal reduces as the distance from the inverter pair increases, reaching at a virtual "zero" at the center of the ring. A clock recovery circuit (implemented as a differential amplifier) is used to recover a full-amplitude clock anywhere along the ring.

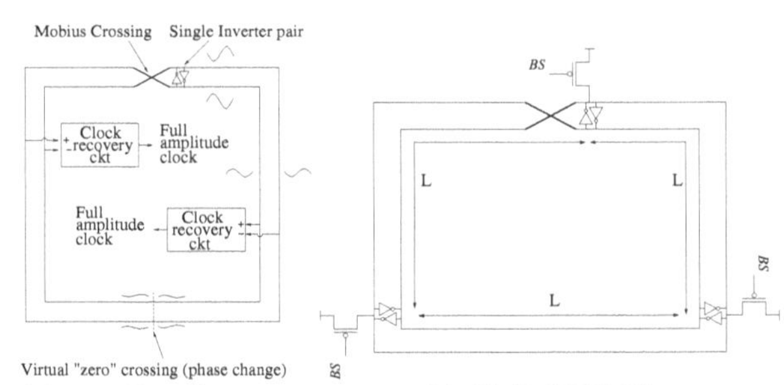

(a) Standing-wave Resonant Clock (b) $3\lambda/2$ SWO Ring

Figure 2: Standing Wave Resonant Clocking Concept

For an SWO with perimeter p, the total phase change is $\lambda/2$ while traversing the ring once. Since the different links of the H-tree have different lengths, we need to implement SWOs with different perimeter, but same frequency. Hence we increase the perimeter of the ring to $k \cdot p$ (where k is odd), so that the total phase change over a single traversal of the ring is $k \cdot \lambda/2$. Such an implementation will require p equally spaced inverter pairs, and an odd number (typically one) of mobius connections. Such a ring with perimeter $k \cdot p$ (where k is odd) oscillates at the same frequency as a ring with perimeter p. The circuit configuration for a $3 \cdot \lambda/2$ ring is shown in Figure 2 b). Note that it in order to ensure that the resonant structure bootstraps in a standing wave configuration, the signals at all inverter pairs are initialized using a global bootstrap signal (labeled *BS* in Figure 2) b). We choose the k value for each ring of Figure 1 so that the ring has the desired length of the H-tree segment. Intersecting SWOs (of the same frequency) use injection locking (as in [2]) to ensure identical phase across all the rings. Hence, we achieve a high-speed, synchronous clock distribution across the die, laid out in an H-tree topology. Note that from Figure 1, we need a clock ring of perimeter $2x$ to serve a bidirectional link of length x. Based on HSPICE [13] simulations in a 22nm PTM [14] process (discussed later), we obtain a 14 *GHz* clock frequency for the NoC. The required length of the metal wire is $2.75mm$ (corresponding to $\lambda/2$) which can serve a bidirectional link of length $1.375mm$. For links with length greater than $1.375mm$, we use a ring perimeter of $k \cdot \lambda/2$, (where k is odd, and appropriately selected). The PEs (present at the leaves of the H-tree) can extract their clock by division of the SWO clocks. In general, fractional clock dividers can be used, allowing for the ability to scale PE performance. The above design suffers from a drawback that the clock provided to the PEs are not phase locked to the external reference. However, this is reconciled by observing that only IO PEs need a phase locked loop (PLL) in order to establish off-chip communication. Hence the IO PEs can have a separate PLL, and would require a single synchronizer while communicating with the NoC. The rest of PEs (which are not IO PEs) can derive their clock from the SWOs by division, without requiring synchronizers.

3.2 Baseline Hnoc Data Network

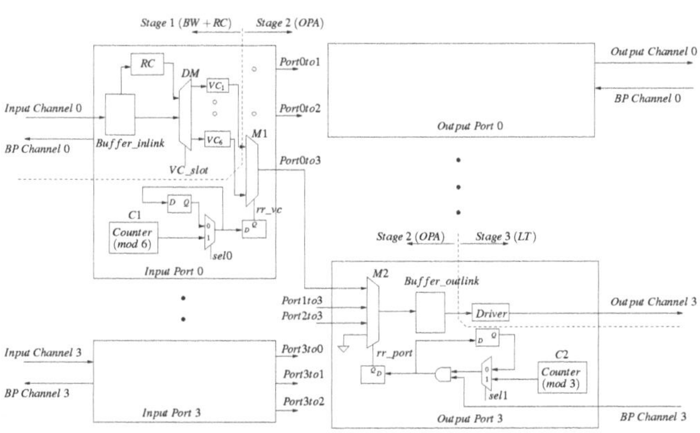

Figure 3: Router Architecture

The ring-based NoC uses two different types of routers called IES and JS. An IES has 3 input and 3 output ports (including a pair of ingress/egress ports connected to the local PE). A JS has a maximum of 5 input and 5 output ports. Figure 3 shows the major components of a 3 stage pipelined virtual-channel router with 4 input and 4 output ports. This design is used both for IES and JS.

The 3 pipeline stages of the router are described below:

- **Buffer Write (BW) + Route Compute (RC)** : A flit arriving at the input port is stored at the register *Buffer_inlink*. The address of the flit is sent to the route computation (*RC*) block to calculate the output port. The flit, along with its output port, is written into one of the free virtual channel buffers pointed by the *VC_slot*. A one bit back-pressure signal (*BP*) is send by each input port to its upstream router, signifying the availability of virtual channel buffers.
- **Output Port Allocation (OPA)** : Individual flits waiting at the input virtual channel buffers (VC_1 - VC_6) arbitrate for access to physical channels at the output port. Arbitration is performed in two stages. The first stage of arbitration is performed between 6 input virtual-channels. This requires a 6 input arbiter for each input-output pair (for a total of 12 arbiters since there are 4 input ports and 3 output ports per input port). Let us consider the arbiter ($M1$) from input port zero to output port 3 as shown in Figure 3. The *rr_vc* signal chooses one of the 6 virtual-channels. The *rr_vc* is generated with the help of a modulus 6 counter ($C1$). However the *rr_vc* signal is updated with the new counter value only when $sel0$ is true. The signal $sel0$ is a conjunction of two events : the virtual channel buffer pointed at by the new counter value is not empty and the output port value of the flit pointed at by the new counter is 3. The second stage of arbitration occurs between the input ports. Once a flit is selected from the input port for a certain output port, the output port performs an arbitration between the three input ports. Let us consider the arbiter ($M2$) at output port 3, whose control input is the *rr_port* signal. The *rr_port* signal is generated with the help of a modulus 3 counter ($C2$). However the *rr_port* signal is updated with the new counter value only when $sel1$ is true. This happens when the port pointed by the new counter value is not empty. Note that the *rr_port* signal is driven to zero when the corresponding back-pressure (*BP*) signal is low, signifying the unavailability of buffers at the downstream router. The flit which passes through both arbiters ($M1$ and $M2$) gets written into the *Buffer_outlink*.
- **Link Traversal (LT)** : Flits from *Buffer_outlink* are driven out through the output link in the third pipeline stage.

3.3 NoC Topologies Considered

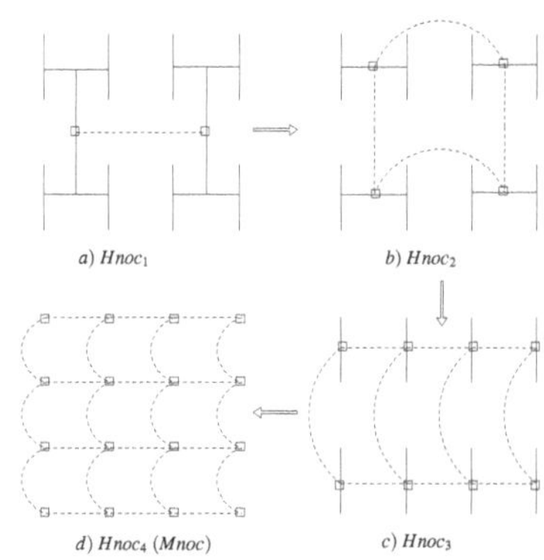

Figure 4: Morphing of Hnoc into a Mnoc

In this section, we discuss how the H-tree based source synchronous NoC (*Hnoc*) is modified by link duplication, resulting in various hybrid intermediate NoC structures, yielding a *Mnoc* in the limit. We consider a m level H-tree, serving $n = 2^m$ PEs. Figure 4 shows an example of the generation of hybrid NoC structures starting with $Hnoc_1$, for $m = 4$. In our baseline architecture, we create two $m-1$ level H-tree based source synchronous NoCs to serve the n PEs (where $m = log_2(n)$). These two H-tree based source synchronous NoCs are connected (using dotted lines) to realize $Hnoc_1$, as shown in Figure 4 a). The dotted connections are realized using a mesh based source synchronous NoC topology. In the second variant, we create four $m-2$ level H-tree based source synchronous NoCs. These four H-tree based source synchronous NoCs are connected in form of a mesh (dotted lines) to realize the $Hnoc_2$ (Figure 4 b)). In the third variant, we create eight 1-level H-tree based source synchronous NoCs and connect these eight nodes in the form of a mesh to realize $Hnoc_3$ (Figure 4 c)). Finally we create sixteen 0-level H-tree based source synchronous NoCs and connect these sixteen nodes in the form of a mesh to realize $Hnoc_4$. Note that $Hnoc_4$ is same as *Mnoc* in our example. In general, at i-th step, we create 2^i $(m-i)$ level H-tree based source synchronous NoC. We connect the centers of these 2^i H-trees in the form of a mesh based source synchronous NoC, to realize $Hnoc_i$. Note that $Hnoc_m$ is identical to a *Mnoc*.

We compare the different hybrid NoC designs as constructed above, in terms of capacity and contention-free latency. Capacity is defined as the injection bandwidth offered to each of the PEs for uniform random traffic, where injection bandwidth is defined as the maximum injection rate times the flit width. For a given NoC frequency, capacity is dependent on the bisection channel count (which we report) and provides an indicator of global wiring requirements. We also perform experimental simulations on synthetic traffic (uniform, tornado and bit-complement) to quantify latency and maximum number of flits delivered (as a function of injection rate) for the various hybrid NoC topologies. In addition, we analyze the average link utilization of all the links in the NoC. This provides an insight into the bottlenecks that limit the maximum sustainable injection rate of any NoC topology. Based on this study, we propose and quantify additional NoC variants, as described in the sequel.

3.4 Deadlock Freedom

In [15], the authors prove that the necessary and sufficient condition for deadlock-free routing is the absence of cycles in the channel dependency graph. An H-tree by construction is acyclic. However, as we add additional links to get hybrid NoC designs, we introduce cycles in the channel dependency graph. We visualize each hybrid NoC design as a combination of a mesh, with an H-tree rooted at each of the mesh nodes. The routing in the H-tree part of the topology is deadlock free (since H-tree is acyclic). We implement a deadlock free routing in the mesh by employing a dimension-ordered routing (DOR)-XY. This achieves deadlock free routing for all the hybrid NoC designs.

4. EXPERIMENTAL RESULTS

4.1 Circuit Results

Our design was implemented in the 22 nm PTM [14] technology, with $VDD = 0.8V$. All circuit simulations were conducted in HSPICE [13]. RLC parasitics for all the wires were extracted using Raphael [16]. Each PE tile is assumed to be $1.229mm \times 1.229mm$, as reported in [3]. The routers (IES' and JS') are designed to drive a maximum link length of $0.615mm$. For links greater than $0.615mm$, we insert repeater(s) to ensure that largest link driven is $0.615mm$. The delay of the first stage pipeline (BW + RC, as described in Section 3.2) was found to be 44.2 ps. The delay of the second stage pipeline (OPA) was found to be 42.8 ps. For the last pipeline stage (LT), the delay was found to be 41.7 ps. Worst-case cross talk patterns were assumed to occur among any 3 adjacent wires in a link when computing any wiring delays. The clock-to-Q delay and the setup time of the registers are added to each of the above pipeline stage delays to determine the clock frequency. PVT or on-chip variations (IR drop, local hot-spots, etc) can reduce the operating frequency of the ring-based NoC. We address this issue by adding a nominal guard-band, which results in a operating speed of 14 GHz. Compared to alternate clock distribution schemes, our SWO has significantly lower jitter alleviating PVT variation issues.

Figure 5 shows the cross section of the wiring used in implementing the SWO rings. The clock wires are implemented in METAL9, and are shielded on either side, as well as above and below, by grounded wires.

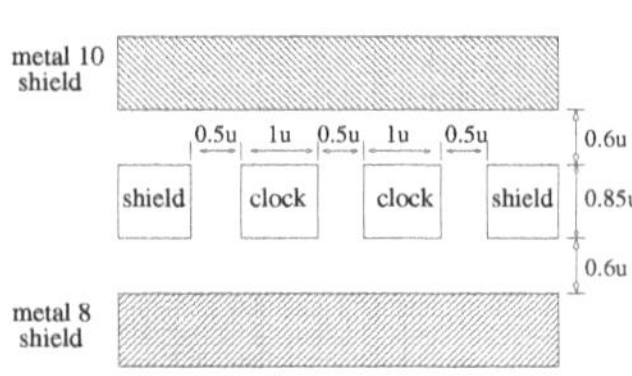

Figure 5: Cross section of Wires in any H-tree Segment

4.2 Evaluation Platform

We use a modified version of GEM5 [17] for cycle-accurate micro-architectural NoC simulations. We simulate a network with 64 PEs. In other words, $m = 6$, and the baseline $Hnoc_1$ has 6 levels. The link width in any direction is assumed to be 18 bytes. We assume a single flit packet of 18 bytes. Note that, the above assumption simplifies the routing logic with the overhead of transmitting the header information in each packet. Since our NoCs have 64 PEs, the source and destination addresses require 12 bits. This amounts to an overhead of $\sim$1% for a packet of 18 bytes. Each of the routers (IES' and JS') support 6 virtual channels per port and perform routing as discussed in Section 3.4. We compare our architectural results with a state of the art mesh with virtual channel buffered flow control [4]. The mesh operates at 2 GHz. Although $Hnoc_6$ is topologically same as the mesh [4], it uses high speed source synchronous links (operating at 14 GHz). The routers in the mesh of [4] are 3-stage pipelined supporting 6 virtual channels and perform a dimension-ordered (DOR)-XY routing. The routers in the mesh are capable of supporting multi-flit packet (unlike our work) by the virtue of wormhole routing, which requires complex logic. Moreover, the lack of availability of a faster on-chip clock restricts the mesh design [4] to operate at the same frequency (2 GHz) as the PEs.

4.3 Design Tradeoffs

4.3.1 Area

Table 1 reports the number of links, wire length and the total number of buffers across all the routers for different hybrid NoC structures. The second column is intended to provide a coarse measure of connectivity. Note that the links reported in Column 2 are not all of the same length. Hence, we report the total wire length of all the links in Column 3. Column 4 indicates the wiring length as a fraction of the mesh topology.

The buffers were shown to occupy 75% of the total on-chip network area in the TRIPS chip [18] and hence we use them as an indicator of the global logical area requirement. Note that each router supports 6 virtual channels and hence the number of buffers is six times the number of ports. The number of buffers used is indicated in Column 5, while Column 6 presents the ratio of the number of buffers to that of a mesh design.

Topology	Links	Wiring (mm)	Wiring (%)	Buffers	Buffers (%)
$Hnoc_1$	126	206.64	75	1500	86.8
$Hnoc_2$	124	216.48	78.6	1488	86.1
$Hnoc_3$	122	226.32	82.1	1464	84.7
$Hnoc_4$	120	236.16	85.7	1440	83.3
$Hnoc_5$	116	265.68	96.4	1392	80.6
$Hnoc_6$ ($Mnoc$)	112	275.52	100	1728	100
Mesh [4]	112	275.52	100	1728	100

Table 1: Area Comparison for Hybrid NoC Topologies

Clearly $Hnoc_6$ (which has the same topology as a mesh) has the longest wire length and the highest number of buffers. Wire length is lowest for $Hnoc_1$ and increases as we duplicate links to create other hybrid NoC designs. In particular, $Hnoc_1$ and $Hnoc_5$ respectively have 25% and 3.6% lower wire length compared to a mesh. In contrast, the number of buffers reduces from $Hnoc_1$ to $Hnoc_5$. The reason behind this trend is that the number of routers decreases as we transform $Hnoc_i$ to $Hnoc_{i+1}$ (for $i \leq 4$), since JS routers are reused in this transformation, by adding an extra port. When transforming from $Hnoc_{m-1}$ to $Hnoc_m$, each of the IES requires an extra pair of ports, resulting in a steep increase in the number of required ports. In particular, $Hnoc_1$ and $Hnoc_5$ respectively have 13.2% and 19.4% fewer buffers compared to a mesh.

4.3.2 Contention Free Latency

Table 2 compares the analytical network performance of various hybrid NoC structures for a 64 PE CMP. B_C corresponds to the bisection channel count and $\frac{B_C}{T_{clk}^{NoC}}$ defines the aggregate bandwidth (in flits/sec) available. We also report the average contention-free latency (T_0) incurred by a flit from source s to destination d which comprises: 1) the average hop count (H) from s to d, 2) router traversal latency (t_r) and 3) the average channel traversal latency (T_c), which is the latency induced by repeaters in long links. The average contention-free latency T_0 is calculated as :

$T_0(s,d) = H(s,d) * t_r + T_c(s,d)$

Note that we report results in Table 2 in terms of PE clock cycles. The bisection channel count increases from $Hnoc_1$ to $Hnoc_6$ since we duplicate links to morph the H-tree into a $Mnoc$. In particular, $Hnoc_1$ and $Hnoc_4$ respectively have 8× and 2× lower bisection channel count compared to a mesh. The clock frequency for the mesh was assumed to be same (2 GHz) as the PEs. Our source synchronous hybrid NoC designs ($Hnoc_1$ through $Hnoc_6$) run 7× faster than the PEs. This helps us achieve a higher available aggregate bandwidth for $Hnoc_2$ through $Hnoc_5$ compared to a regular mesh, despite a lower bisection channel count. In particular, $Hnoc_2$ and $Hnoc_5$ respectively have 1.75× and 5.25× better available aggregate bandwidth compared to a mesh, with a 4× and 1.33× lower bisection channel count respectively. $Hnoc_6$, which has the same bisection channel count as the mesh, has a 7× higher available aggregate bandwidth.

The router traversal latency (t_r) is 3 *PE* clock cycles for the mesh and 3/7 *PE* clock cycles for the hybrid NoC designs (since the NoC clock runs 7× faster than the *PE*s). The average hop count decreases as we duplicate links in a H-tree to get hybrid structures. For link lengths greater the $0.615mm$, we introduce repeaters for both mesh and source synchronous NoC designs and this contributes to an increase in the average channel traversal latency (T_c). Note that only the links are buffered, not the SWO itself. Note that for both the mesh and the hybrid NoC designs, we report T_c in terms of *PE* clock cycles. We improve the average contention free latency (T_0) by 4.1× for $Hnoc_1$ and by 7× for $Hnoc_6$.

Topology	B_C	$\frac{B_C}{T_{clk}^{NoC}}$	H	$T_c(cycles)$	$T_0(cycles)$
$Hnoc_1$	2	14	9.68	1.66	5.81
$Hnoc_2$	4	28	8.92	1.48	5.30
$Hnoc_3$	6	42	7.90	1.27	4.66
$Hnoc_4$	8	56	7.22	0.89	3.98
$Hnoc_5$	12	84	6.75	0.82	3.71
$Hnoc_6$ (*Mnoc*)	16	112	6.25	0.75	3.42
Mesh [4]	16	16	6.25	5.25	24

Table 2: Network Performance for Hybrid NoC Topologies

4.3.3 *Link Utilization*

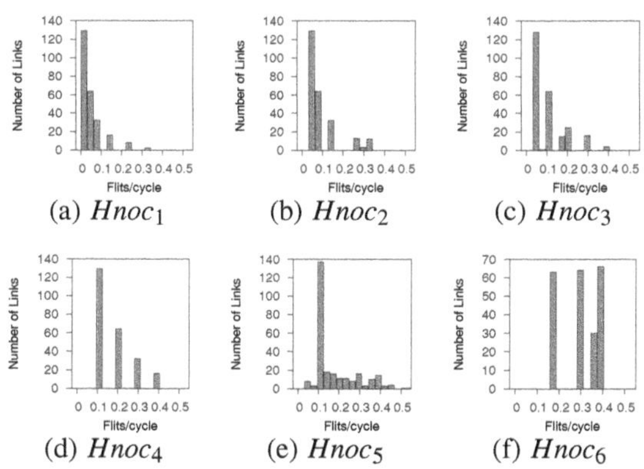

Figure 6: Link Utilization for Hybrid NoC Topologies

We next analyze the utilization of the various links for different hybrid NoC designs. Figure 6 shows the histogram of the link utilization for various hybrid NoC designs, obtained through experiments using uniform traffic. All links are assumed to have the same width. Note that $Hnoc_1$ and $Hnoc_6$ are identical to an H-tree and a mesh respectively, in terms of topology. The X-axis represents the link utilization in units of flits/cycle and Y-axis represents the number of links with a particular value of link utilization. The *PEs* inject traffic uniformly and the injection rate is chosen to ensure that the corresponding network does not saturate. Clearly for $Hnoc_1$, we observe a small number of links showing a very high link utilization. They correspond to the lowest level of H-tree (the level that is farthest from the leaves). We also observe a higher number of links with a very low utilization. They correspond to links which are present towards the leaves of the H-tree, servicing fewer *PEs* and hence exhibiting a lower utilization. Next, we construct $Hnoc_2$ by removing the lowest level of H-tree. We observe more links with higher link utilization (corresponding to the mesh portion of the hybrid NoC design). As we keep removing levels of the H-tree and growing the size of the mesh portion of the hybrid NoC, the link utilization is spread in the mesh core. The peaks in the hybrid NoC designs ($Hnoc_2$ through $Hnoc_5$) correspond to the links which are present towards the leaves of the H-tree. They serve fewer *PEs* and hence exhibit a lower utilization. As we morph from $Hnoc_5$ to $Hnoc_6$ (which is topologically same as the mesh), we observe a major change in pattern of the link utilization. This happens because we remove links that were responsible for serving a single PE in $Hnoc_5$ and instead create a grid of links. An ideal link utilization histogram needs to be unimodal with a very low variance. The bandwidth of a NoC design is limited by the highest utilized link. When there are a very few links in the network with a high utilization, most of the links are being under-utilized when the network saturates. Such a design will not be able to sustain a high injection rate.

The purpose of the above experiment is to provide a systematic way to debug link utilization. Based on this experiment, we analyze the overhead of selectively widening congested links. We widen the two lower level links (the ones farthest from the leaves) for $Hnoc_1$ and the mesh portion for the rest ($Hnoc_2$ to $Hnoc_6$) of the hybrid NoC designs by a factor of two. The wiring and buffer overheads are reported in Table 3. Each variant with widened links is referred to with a "fat" subscript. We observe that $Hnoc_{4,fat}$, $Hnoc_{5,fat}$ and $Hnoc_{6,fat}$ have significantly higher overheads compared to the baseline mesh topology. So we present the results for $Hnoc_{1,fat}$, $Hnoc_{2,fat}$, $Hnoc_{3,fat}$, $Hnoc_4$, $Hnoc_5$ and $Hnoc_6$ in the next section.

Topology	Links	Wiring (mm)	Wiring (%)	Buffers	Buffers (%)
$Hnoc_{1,fat}$	132	221.40	80.3	1560	90.3
$Hnoc_{2,fat}$	128	226.32	82.1	1536	88.9
$Hnoc_{3,fat}$	132	295.20	107.1	1560	90.2
$Hnoc_{4,fat}$	144	354.68	128.7	1728	100
$Hnoc_{5,fat}$	168	521.52	189.3	2064	119
$Hnoc_{6,fat}$	224	551.04	200	3456	200
Mesh [4]	112	275.52	100	1728	100

Table 3: Area Comparison for Hybrid NoC Topologies with Wide Links

4.4 Architectural Simulations

We compare the performance of different hybrid NoC structures described in Section 3.3 by running synthetic traffic (uniform, tornado and bit-complement). Figure 7 a), c) and e) presents the latency (in terms of *PE* cycles) while Figure 7 b), d) and f) presents the number of flits delivered over 10K *PE* cycles as a function of flit injection rate for the three synthetic traffic patterns. Note that the legends in Figure 7 b), d) and f) corresponds to the legends in Figure 7 a), b) and e) respectively, and are omitted for clarity.

We summarize the data of Figure 7 (normalized to the mesh [4]) in Table 4 and Table 5 respectively. Based on Table 4, all design variants achieve a better latency than the mesh of [4], for all three traffic types. From Table 5, we note that $Hnoc_{3,fat}$, $Hnoc_4$, $Hnoc_5$ and $Hnoc_6$ achieve a higher number of flits than the mesh of [4] for all traffic types. For tornado traffic, $Hnoc_{2,fat}$ improves over the mesh, while for bit-complement traffic, all variants beat the mesh in terms of flit delivered. The fact that the hybrid NoC designs run significantly faster than the mesh (by 7×) contributes to these improvements.

For uniform traffic, we observe that $Hnoc_{1,fat}$ and $Hnoc_{2,fat}$ provide 1.9× and 2× lower latency compared to the mesh of [4]. However, we observe that they can only sustain an injection rate of 0.31 and 0.32 in comparison to 0.38 for the mesh. $Hnoc_{3,fat}$, $Hnoc_4$, $Hnoc_5$ and $Hnoc_6$ provide 2.3×, 3.5×, 3.5× and 4.6× lower latency compared to a mesh and are able to sustain an injection rate of 0.63, 0.7, 1 and 2.6 respectively.

For tornado traffic, we observe that $Hnoc_{1,fat}$ and $Hnoc_{2,fat}$ provide on average 1.4× and 1.6× lower latency compared to the mesh of [4] and can only sustain an injection rate of 0.23 and 0.24 in comparison to 0.35 for the mesh. $Hnoc_{3,fat}$, $Hnoc_4$, $Hnoc_5$ and $Hnoc_6$ provide a 1.9×, 3.1×, 2.9× and 4.2× lower latency compared to a mesh and are able to sustain better injection rates than the mesh of [4] (0.45, 0.6, 0.8 and 2.4 respectively). The maximum number of flits delivered over 10K *PE* cycles for $Hnoc_{2,fat}$, $Hnoc_{3,fat}$, $Hnoc_4$, $Hnoc_5$ and $Hnoc_6$ are between 1.2× and 7× higher compared to the mesh of [4], as reported in Table 5. For bit-complement traffic, we observe that $Hnoc_{1,fat}$ and $Hnoc_{2,fat}$ provide 2.3× and 2.4× lower latency compared to the mesh of [4] but can only sustain an injection rate of 0.18 and 0.19 respectively in comparison to 0.25 for the mesh. $Hnoc_{3,fat}$, $Hnoc_4$, $Hnoc_5$ and $Hnoc_6$ pro-

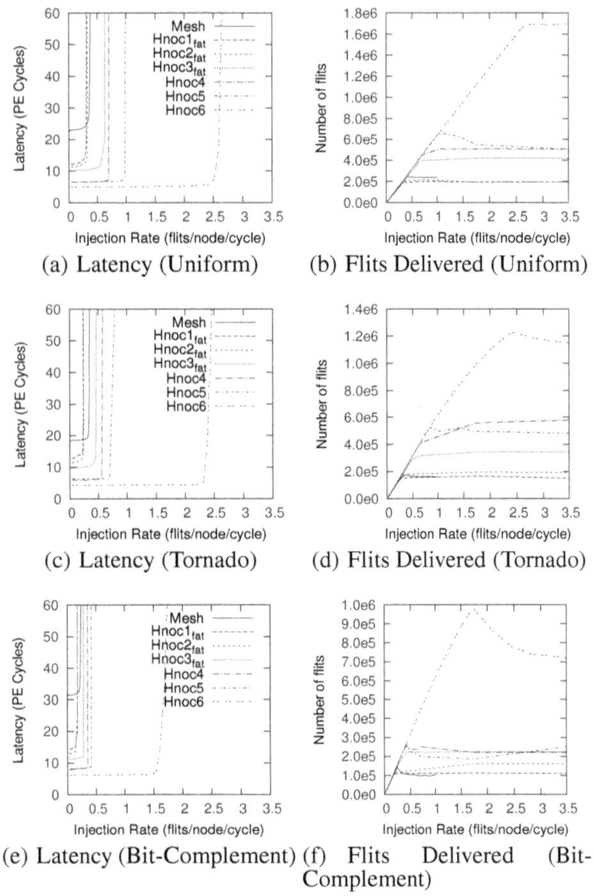

(a) Latency (Uniform) (b) Flits Delivered (Uniform)

(c) Latency (Tornado) (d) Flits Delivered (Tornado)

(e) Latency (Bit-Complement) (f) Flits Delivered (Bit-Complement)

Figure 7: Network Performance Comparison

vide a 2.8×, 3.9×, 3.8× and 5× lower latency compared to the mesh of [4] and are able to sustain a higher injection rate than the mesh (0.3, 0.37, 0.45 and 1.8 respectively). The maximum number of flits delivered over 10K *PE* cycles for all *Hnoc* variants are between 1.1× and 7× higher than the mesh of [4], as reported in Table 5.

Hence we conclude that $Hnoc_{1,fat}$, $Hnoc_{2,fat}$ have significantly lower latency and comparable maximum sustained injection rate in comparison to the state of the art mesh. Moreover, $Hnoc_{3,fat}$, $Hnoc_4$, $Hnoc_5$ and $Hnoc_6$ provide a better latency and are able to sustain a higher injection rate for all synthetic patterns in comparison to a state of the art mesh. Among them, $Hnoc_5$ has the lowest number of buffers and $Hnoc_4$ has the smallest wire length. By taking this tradeoff into account, a designer can choose a hybrid NoC design depending on the area-performance characteristics they wish to achieve.

Topology	Uniform	Tornado	Bit-Complement
$Hnoc_{1,fat}$	1/1.9	1/1.4	1/2.3
$Hnoc_{2,fat}$	1/2.0	1/1.6	1/2.4
$Hnoc_{3,fat}$	1/2.3	1/1.9	1/2.8
$Hnoc_4$	1/3.5	1/3.1	1/3.9
$Hnoc_5$	1/3.5	1/2.9	1/3.8
$Hnoc_6$	1/4.6	1/4.2	1/5.0
Mesh [4]	1	1	1

Table 4: Latency Comparison

Topology	Uniform	Tornado	Bit-Complement
$Hnoc_{1,fat}$	0.8	0.9	1.1
$Hnoc_{2,fat}$	0.8	1.2	1.6
$Hnoc_{3,fat}$	1.8	2.1	2.2
$Hnoc_4$	2.1	3.6	2.2
$Hnoc_5$	2.1	3.0	2.5
$Hnoc_6$	7.0	7.0	7.0
Mesh [4]	1	1	1

Table 5: Maximum Flits Delivered Comparison

5. CONCLUSIONS

In this work, we present a family of high-speed source synchronous NoCs organized in an H-tree topology, with each data link being routed parallel to a clock ring. The clock is generated with the help of injection locked standing wave oscillators of varying length as required by the topology. *PEs* directly extract a low jitter clock from the high speed ring clock by division. We have not compared the power of our scheme relative to the mesh, since we do not have high level power estimates for the hybrid NoCs. However, our past work [3] on the circuit level implementations of a fast source-synchronous NoC demonstrates a 35% power improvement over a mesh. We compare different hybrid NoC designs with a traditional mesh [4] in terms of area, link utilization and contention-free latency. We also perform experimental simulations on synthetic traffic (uniform, tornado and bit-complement) to quantify latency and maximum number of flits delivered as a function of injection rate for various hybrid NoC structures. Our results demonstrate significant improvements over a traditional mesh. The fact that the hybrid NoC designs run significantly faster than the traditional mesh (by 7×) contributes to these improvements. Using our results, a designer can choose any hybrid NoC design depending on the area-performance characteristics desired.

6. REFERENCES

[1] V. H. Cordero and S. P. Khatri, "Clock Distribution Scheme Using Coplanar Transmission Lines," in *Proceedings of the Conference on Design, Automation and Test in Europe*, DATE '08, (New York, NY, USA), pp. 985–990, ACM, 2008.

[2] A. Mandal, V. Karkala, S. Khatri, and R. Mahapatra, "Interconnected Tile Standing Wave Resonant Oscillator Based Clock Distribution Circuits," in *VLSI Design (VLSI Design), 2011 24th International Conference on*, pp. 82 –87, Jan. 2011.

[3] A. Mandal, S. P. Khatri, and R. N. Mahapatra, "A fast, source-synchronous ring-based network-on-chip design," in *DATE*, pp. 1489–1494, 2012.

[4] J. Balfour and W. J. Dally, "Design Tradeoffs for tiled CMP On-chip Networks," in *Proceedings of the 20th annual international conference on Supercomputing*, ICS '06, (New York, NY, USA), pp. 187–198, ACM, 2006.

[5] W. Dally and B. Towles, "Route packets, not wires: on-chip interconnection networks," in *Design Automation Conference, 2001. Proceedings*, pp. 684 – 689, 2001.

[6] J. Duato, S. Yalamanchili, and N. Lionel, *Interconnection Networks: An Engineering Approach*. San Francisco, CA, USA: Morgan Kaufmann Publishers Inc., 2002.

[7] J. Kim, J. Balfour, and W. Dally, "Flattened butterfly topology for on-chip networks," *Computer Architecture Letters*, vol. 6, pp. 37 –40, Feb. 2007.

[8] C. E. Leiserson, "Fat-trees: universal networks for hardware-efficient supercomputing," *IEEE Trans. Comput.*, vol. 34, pp. 892–901, October 1985.

[9] H. Samuelsson and S. Kumar, "Ring road NoC architecture," in *Norchip Conference, 2004. Proceedings*, pp. 16 – 19, Nov. 2004.

[10] A. T. Tran, D. N. Truong, and B. Baas, "A Reconfigurable Source-Synchronous On-Chip Network for GALS Many-Core Platforms," *Computer-Aided Design of Integrated Circuits and Systems, IEEE Transactions on*, vol. 29, pp. 897 –910, June 2010.

[11] M. Horak, S. Nowick, M. Carlberg, and U. Vishkin, "A Low-Overhead Asynchronous Interconnection Network for GALS Chip Multiprocessors," in *Networks-on-Chip (NOCS), 2010 Fourth ACM/IEEE International Symposium on*, pp. 43 –50, May 2010.

[12] A. Mandal, S. P. Khatri, and R. N. Mahapatra, "Architectural Simulations of a Fast, Source-Synchronous Ring-based Network-on-Chip Design," in *ICCD*, Oct 2012.

[13] I. Meta-Software, "HSPICE user's manual," Campbell, CA.

[14] "PTM website." `http://www.eas.asu.edu/~ptm`.

[15] W. J. Dally and C. L. Seitz, "Deadlock-free message routing in multiprocessor interconnection networks," *IEEE Trans. Comput.*, vol. 36, pp. 547–553, May 1987.

[16] "Raphael Interconnect Analysis Tool: User's Guide."

[17] N. Binkert, B. Beckmann, G. Black, S. K. Reinhardt, A. Saidi, A. Basu, J. Hestness, D. R. Hower, T. Krishna, S. Sardashti, R. Sen, K. Sewell, M. Shoaib, N. Vaish, M. D. Hill, and D. A. Wood, "The GEM5 simulator," *SIGARCH Comput. Archit. News*, vol. 39, pp. 1–7, Aug. 2011.

[18] P. Gratz, C. Kim, R. McDonald, S. Keckler, and D. Burger, "Implementation and Evaluation of On-Chip Network Architectures," in *Computer Design, 2006. ICCD 2006. International Conference on*, pp. 477 –484, Oct 2006.

A Parallel VLSI Architecture for Markov Chain Monte Carlo based MIMO Detection

Uwe Deidersen, Dominik Auras, Gerd Ascheid
Institute for Communication Technologies and Embedded Systems
RWTH Aachen University, 52056 Aachen, Germany
{deidersen,auras,ascheid}@ice.rwth-aachen.de

ABSTRACT

Multiple-input multiple-output (MIMO) wireless transmission together with iterative decoding at the receiver is a key technique to achieve high spectral efficiency. However, particularly the required soft-input soft-output (SISO) MIMO detector entails a very high complexity, which motivates the investigation of suboptimal detectors with reduced complexity. In this paper, we present—to the best of our knowledge—the first implementation of a parallel VLSI architecture for a SISO detector based on Markov chain Monte Carlo (MCMC) methods. The proposed architecture is scalable and allows to exploit the parallelism inherent in the considered MCMC algorithm. We investigate the implementation costs and show that this architecture covers a wide range of trade-offs between throughput and silicon area.

Categories and Subject Descriptors

C.3 [**Special-Purpose and Application-Based Systems**]: Signal processing systems

General Terms

Design

Keywords

Multiple-input multiple-output (MIMO) wireless transmission; soft-input soft-output (SISO) detection; Markov chain Monte Carlo (MCMC); VLSI architecture

1. INTRODUCTION

Multiple-input multiple-output (MIMO) transmission schemes with spatial multiplexing have been adopted by most state-of-the-art wireless communication standards, due to the increased spectral efficiency compared to single antenna systems. In addition, receivers operating according to the bit-interleaved coded modulation with iterative decoding (BICM-ID) principle, where soft-information is exchanged iteratively between the detector and the channel decoder, offer a significant improvement of the communication performance over non-iterative receivers.

Unfortunately, these benefits come at the cost of a considerable complexity. In particular, the exponential complexity of the optimal soft-input soft-output (SISO) detector has led to the investigation of numerous suboptimal detectors with lower complexity. Among these are detectors based on linear filtering as well as detectors based on tree-search algorithms such as the sphere detector.

Another family of suboptimal detectors that employ Markov chain Monte Carlo (MCMC) probabilistic methods has been proposed in [4]. These detectors randomly draw a set of candidate symbol vectors with high a-posteriori probability used to estimate the posterior distribution of the transmitted bits. While MCMC-based detectors exhibit good communication performance [4][7], even in low SNR regimes, and offer other advantages such as a fixed run time, only little work has so far been dedicated to the investigation of their implementation complexity.

An efficient VLSI implementation should allow to exploit the parallelism inherent in MCMC-based detection algorithms. While the architecture proposed in [5] does not provide the means for this, the architecture for a soft-output detector proposed in [9] does, although only as part of a hybrid architecture in conjunction with a sphere detector, that provides the initial values required for a parallel implementation. Feedback from the channel decoder is not considered by that architecture, nor by the architecture proposed in [1]. Moreover, in both [9] and [1] results are only given in terms of operation counts.

Contributions:

In this paper, we present—to the best of our knowledge—the first implementation of a parallel VLSI architecture for MCMC-based SISO detection in iterative MIMO receivers. The proposed architecture is scalable and allows to exploit the inherent parallelism. To this end, we introduce a novel initialization scheme incurring no additional hardware costs. Implementation-friendly algorithmic transformations are applied throughout the derivation of the architecture. We investigate the implementation costs in terms of silicon area and execution time.

Outline:

In the remainder of this section, the assumed system model is described. Section 2 introduces the considered MCMC algorithm for SISO MIMO detection. The proposed VLSI architecture is then derived in Section 3. Implementation results are discussed in Section 4.

GLSVLSI'13, May 2–3, 2013, Paris, France.

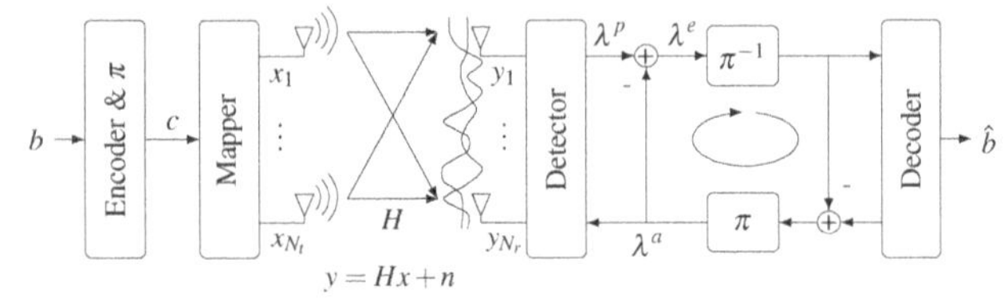

Figure 1: Block diagram of the MIMO BICM-ID system.

1.1 System Model

A $N_t \times N_r$ spatial multiplexing MIMO system following the BICM-ID principle, as shown in Fig. 1, is assumed. At the transmitter, a block $\boldsymbol{b} \in \{0,1\}^{rKN_tQ}$ of bits is encoded by a channel encoder with rate r and then interleaved (π). The resulting encoded bits are split into K vectors $\mathbf{c}$ of length N_tQ that are transmitted in sequence. To this end, $\mathbf{c}$ is mapped to the symbol vector $\boldsymbol{x} = [x_1, \ldots, x_{N_t}]^T \in \mathcal{A}^{N_t}$, where each element x_i ($i = 1, \ldots, N_t$) is one of the 2^Q symbols defined by the modulation alphabet $\mathcal{A} \subset \mathbb{C}$. Each of the N_t transmit antennas then sends one element of $\boldsymbol{x}$. The symbol vector $\boldsymbol{y} \in \mathbb{C}^{N_r}$ received by the $N_r \geq N_t$ receive antennas is given by $\boldsymbol{y} = \boldsymbol{Hx} + \boldsymbol{n}$, where $\boldsymbol{H} \in \mathbb{C}^{N_r \times N_t}$ is the channel matrix and $\boldsymbol{n} \in \mathbb{C}^{N_r}$ is a white Gaussian noise vector with element-wise variance N_0. At the receiver, detector and channel decoder iteratively exchange extrinsic information $\boldsymbol{\lambda}^e = \boldsymbol{\lambda}^p - \boldsymbol{\lambda}^a$ about the coded bits, where $\boldsymbol{\lambda}^a$ are the prior L-values fed back from the decoder and $\boldsymbol{\lambda}^p$ are the posterior L-values from the detector.

2. MCMC-BASED MIMO DETECTION ALGORITHM

We consider a variant of the Rao-Blackwellized MCMC algorithm proposed in [7], that is based on uniform sampling[1], as this promises the lowest implementation complexity. This yields an approximation of the exact posterior L-value of a particular bit c_k given by

$$\lambda_k^p = \log \frac{P(c_k = 0 \mid \boldsymbol{y}, \boldsymbol{H}, \boldsymbol{\lambda}^a)}{P(c_k = 1 \mid \boldsymbol{y}, \boldsymbol{H}, \boldsymbol{\lambda}^a)} \approx \log \frac{\sum_{s=1}^{N_s} P(c_k = 0, \mathbf{c}_{\setminus k}^{(s)} \mid \boldsymbol{y}, \boldsymbol{H}, \boldsymbol{\lambda}^a)}{\sum_{s=1}^{N_s} P(c_k = 1, \mathbf{c}_{\setminus k}^{(s)} \mid \boldsymbol{y}, \boldsymbol{H}, \boldsymbol{\lambda}^a)} \quad (1)$$

where $\mathbf{c}^{(s)}$ ($s = 1, \ldots, N_s$) denotes the random samples, i.e. the candidate vectors that were sent with high probability, and $\mathbf{c}_{\setminus k} = (c_1, \ldots, c_{k-1}, c_{k+1}, \ldots, c_{N_tQ})$. Assuming statistically independent bits, each term in (1) is given by

$$\begin{aligned} &P(c_k = \beta, \mathbf{c}_{\setminus k}^{(s)} \mid \boldsymbol{y}, \boldsymbol{H}, \boldsymbol{\lambda}^a) \\ &= P(\boldsymbol{y} \mid c_k = \beta, \mathbf{c}_{\setminus k}^{(s)}, \boldsymbol{H}) P(c_k = \beta, \mathbf{c}_{\setminus k}^{(s)} \mid \boldsymbol{\lambda}^a) [P(\boldsymbol{y})]^{-1} \quad (2) \\ &\propto \exp\left(-\frac{1}{N_0} \left\| \boldsymbol{y} - \boldsymbol{H}\boldsymbol{x}_{k\beta}^{(s)} \right\|^2\right) \prod_{l=1}^{N_tQ} \exp(-c_{k\beta,l}^{(s)} \lambda_l^a) \quad (3) \end{aligned}$$

with $\mathbf{c}_{k\beta} = (c_1, \ldots, c_k = \beta, \ldots, c_{N_tQ})$ and $\beta \in \{0,1\}$. $\boldsymbol{x}_{k\beta}$ is the symbol vector obtained by mapping $\mathbf{c}_{k\beta}$. In (3) all factors that eventually cancel out in (1) are omitted. Using (3), the application of the max-log approximation [6] to (1) then results in

$$\lambda_k^p \approx \max_s \mu(\mathbf{c}_{k0}^{(s)}) - \max_s \mu(\mathbf{c}_{k1}^{(s)}) \quad (4)$$

$$\mu(\mathbf{c}_{k\beta}^{(s)}) = -\frac{1}{N_0} \left\| \boldsymbol{y} - \boldsymbol{H}\boldsymbol{x}_{k\beta}^{(s)} \right\|^2 - \sum_{l=1}^{N_tQ} c_{k\beta,l}^{(s)} \lambda_l^a. \quad (5)$$

The summation problem in (1) has thus been turned into an optimization problem that involves determining the two maximum values in (4). To this end, the metric in (5) has to be evaluated twice per bit for each candidate vector $\mathbf{c}^{(s)}$.

The candidate vectors $\mathbf{c}^{(s)}$ are drawn by a Gibbs sampler realizing a Markov chain that starts from an initial state and then iterates over the individual bits c_k updating them by drawing a new value from their marginal distribution, conditioned on the current values of the other bits $\mathbf{c}_{\setminus k}$:

Initialize $\mathbf{c}^{(0)}$
for $s = 1, \ldots, N_s$ **do**
  for $k = 1, \ldots, N_tQ$ **do**
    $c_k^{(s)} \sim P(c_k | c_1^{(s)}, \ldots, c_{k-1}^{(s)}, c_{k+1}^{(s-1)}, \ldots, c_{N_tQ}^{(s-1)})$
  end
end

The uniform distribution of the random candidate vectors is achieved by discarding repetitions of samples drawn from the Gibbs sampler. This is done implicitly in (4), since the maximum search remains unaffected by repeated occurences of the same value. Due to the max-log approximation we also do not need a burn-in period, usually suggested for the Gibbs sampler to converge to the target distribution.

As shown in [5], it is easier to calculate $P(c_k \mid \mathbf{c}_{\setminus k}^{(s)})$ in the log-domain in form of the L-value

$$\begin{aligned} \gamma_k^{(s)} &= \log \frac{P(c_k = 0 \mid \mathbf{c}_{\setminus k}^{(s)})}{P(c_k = 1 \mid \mathbf{c}_{\setminus k}^{(s)})} \\ &= \frac{1}{\eta N_0} \left(\left\| \boldsymbol{y} - \boldsymbol{H}\boldsymbol{x}_{k1}^{(s)} \right\|^2 - \left\| \boldsymbol{y} - \boldsymbol{H}\boldsymbol{x}_{k0}^{(s)} \right\|^2 \right) + \lambda_k^a. \quad (6) \end{aligned}$$

The parameter η constitutes a system temperature that allows to avoid lock-in effects in high SNR regimes. Here, we always assume $\eta = 2^\alpha$ with $\alpha \in \mathbb{N}$.

The communication performance can be improved by drawing the random candidate vectors with N_{gs} independent Gibbs samplers that realize uncorrelated Markov chains with different intitial states $\mathbf{c}^{(0)}$. These initial states in turn can either be random or obtained from an initial hard-output detection. The use of independent Gibbs samplers introduces the parallelism that is exploited by the proposed VLSI architecture.

3. VLSI ARCHITECTURE

An overview of the proposed VLSI architecture is given in Fig. 2. A circuit for front-end processing (FEP) first transforms the channel observations into a representation more suitable for subsequent processing. At the core of the architecture is the Gibbs sampler (GS) circuit that iteratively

[1]Similar to the MCRB-U detector proposed in [4].

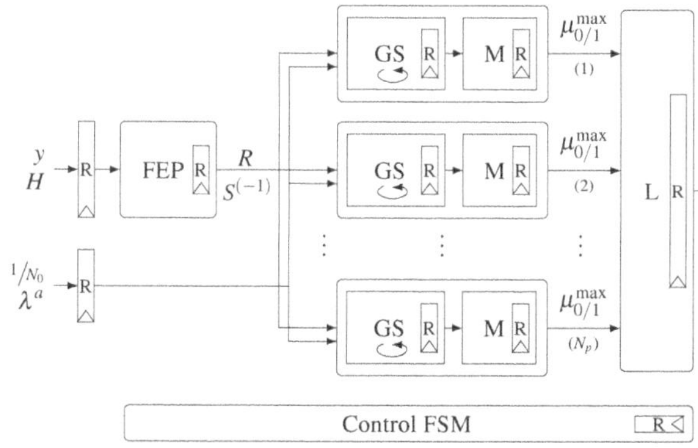

Figure 2: Block diagram of the proposed VLSI architecture consisting of a front-end processing (FEP) circuit, N_p parallel instances of the Gibbs sampler (GS) and metric calculation (M) circuits as well as a circuit for posterior L-value calculation (L).

draws the candidate vectors in a bit-wise fashion, thus taking N_tQ clock cycles per candidate vector. The metrics required for posterior L-value calculation are evaluated by a metric calculation (M) circuit. We exploit the inherent parallelism of the algorithm by instantiating N_p pairs of GS/M-circuits. Each of these realizes N_{gs}/N_p independent Gibbs samplers sequentially. Before the N_s/N_{gs} candidate vectors of one independent Gibbs sampler are drawn, the GS-circuit transitions bit-wise from a predefined state $\boldsymbol{c}^{(-1)}$ to the respective initial state $\boldsymbol{c}^{(0)}$ in an initialization phase, that allows us to employ efficient update schemes, even in the presence of independent GS (Section 3.5). At the end of drawing all candidate vectors, the L-value calculation (L) circuit computes the posterior L-values based on the N_p maximum metrics determined for each bit by the M-circuits.

In the following, the individual circuits as well as details regarding the initialization phase are described.

3.1 Front-end Processing

In order to simplify the calculation of $\gamma_k^{(s)}$ in the GS-circuit, the channel observations $\boldsymbol{y}$ and $\boldsymbol{H}$ are processed once. To this end, we define $\boldsymbol{y}^{mf} = \boldsymbol{H}^H\boldsymbol{y} \in \mathbb{C}^{N_t}$ and $\boldsymbol{R} = \boldsymbol{H}^H\boldsymbol{H} \in \mathbb{C}^{N_t \times N_t}$. $\boldsymbol{R}$ is hermitian and thus fully specified by its upper-triangular matrix. We also define $\boldsymbol{S}^{(s)} = \boldsymbol{y}^{mf} - \tilde{\boldsymbol{R}}\boldsymbol{x}^{(s)} \in \mathbb{C}^{N_t}$, where $\tilde{\boldsymbol{R}}$ is $\boldsymbol{R}$ with the diagonal elements set to 0 and $\boldsymbol{x}^{(s)}$ is the complex symbol vector obtained by mapping the current Gibbs sampler state $\boldsymbol{c}^{(s)}$. The FEP-circuit computes $\boldsymbol{R}$ and $\boldsymbol{S}^{(-1)}$, i.e. $\boldsymbol{S}^{(s)}$ evaluated for the predefined state $\boldsymbol{c}^{(-1)}$ which maps to $\boldsymbol{x}^{(-1)}$ with $x_i^{(-1)} = 1 + j1$ for $i = 1, \ldots, N_t$. Thus, multiplication by $x_i^{(-1)}$ only requires two adders. The FEP-circuit features a complex multiply-and-accumulate unit that alternatingly computes one row of the upper-triangular matrix of $\boldsymbol{R}$ and one element of $\boldsymbol{y}^{mf}$, requiring N_r clock cycles per element. A second accumulator directly adds the elements $r_{l,i}$ multiplied by $-x_i^{(-1)}$ ($i = 1, \ldots, N_t$, $i \neq l$) and y_l^{mf}, yielding the l-th element ($l = 1, \ldots, N_t$) of $\boldsymbol{S}^{(-1)}$. A compact processing schedule is ensured by overlapping the accumulation of the lower-triangular elements of $\boldsymbol{R}$ with the calculation of y_l^{mf}. Registers store $\boldsymbol{S}^{(-1)}$ and the upper-triangular matrix of $\boldsymbol{R}$.

3.2 Gibbs Sampler Circuit

The GS-circuit shown in Fig. 3 draws one bit $c_k^{(s)}$ per clock cycle, updating a register that holds the Gibbs sampler state $\boldsymbol{c}^{(s)}$. For this purpose, the L-value $\gamma_k^{(s)}$ from (6) is computed. Since $\boldsymbol{c}_{k0}^{(s)}$ and $\boldsymbol{c}_{k1}^{(s)}$ differ only in the k-th bit, the symbol vectors $\boldsymbol{x}_{k0}^{(s)}$ and $\boldsymbol{x}_{k1}^{(s)}$ differ only at the n-th position $x_{k\beta,n}^{(s)}$ ($\beta \in \{0, 1\}$). Thus, most terms in (6) cancel out, yielding

$$\gamma_k^{(s)} = \frac{1}{\eta N_0}\left(\Re\{r_{n,n}\}\,\|\Delta\|_k^{2(s)} - 2\Re\left\{S_{k-1,n}^{(s)}\Delta_k^{(s)*}\right\}\right) + \lambda_k^a \tag{7}$$

as shown in [5][3], where $\|\Delta\|_k^{2(s)} = \|x_{k1,n}^{(s)}\|^2 - \|x_{k0,n}^{(s)}\|^2$ and $\Delta_k^{(s)} = x_{k1,n}^{(s)} - x_{k0,n}^{(s)}$. Furthermore, $r_{n,n}$ is the n-th element of the diagonal of $\boldsymbol{R}$, while $S_{k-1,n}^{(s)}$ denotes the n-th element of $\boldsymbol{S}^{(s)}$ after updating the $(k-1)$-th bit of $\boldsymbol{c}^{(s)}$.

Since $x_{k0,n}^{(s)}$ and $x_{k1,n}^{(s)}$ belong to a small set of symbols defined by $\mathcal{A}$ and we assume a Gray-coded signaling scheme, the differences $\|\Delta\|_k^{2(s)}$ and $\Delta_k^{(s)}$ can only take on a few values, either real or purely imaginary. Multiplication by these values in (7) can therefore be realized efficiently with a few constant shifts and additions. We exploit this in a similar way to [9], by instantiating simplified multiplier circuits. These consist of a datapath and a circuit generating control signals based on $\boldsymbol{c}^{(s)}$ and the index k.

Instead of computing $S_{k-1,n}^{(s)}$ directly, which would incur a dot-product in the critical path of the GS-circuit, we employ a novel recursive update scheme. This allows us to compute $S_{k-1,n}^{(s)}$ one clock cycle earlier, i.e. in parallel to drawing

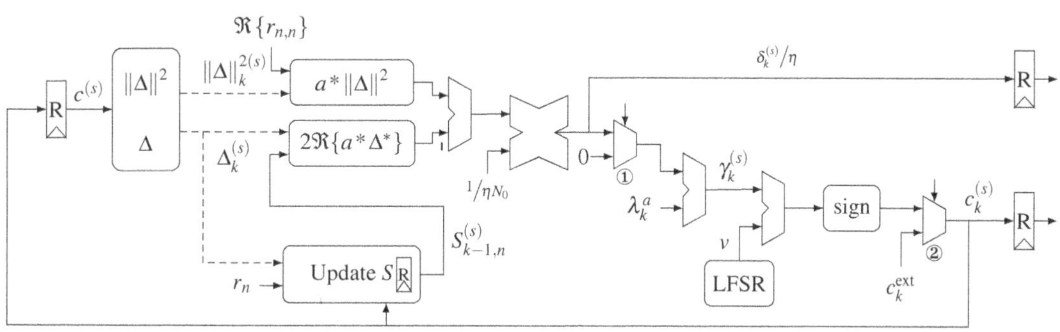

Figure 3: Block diagram of the GS-circuit (r_n denotes the n-th column of matrix $\boldsymbol{R}$. a refers to the respective data inputs of simplified multipliers). The multiplexers ① and ② are used for initialization.

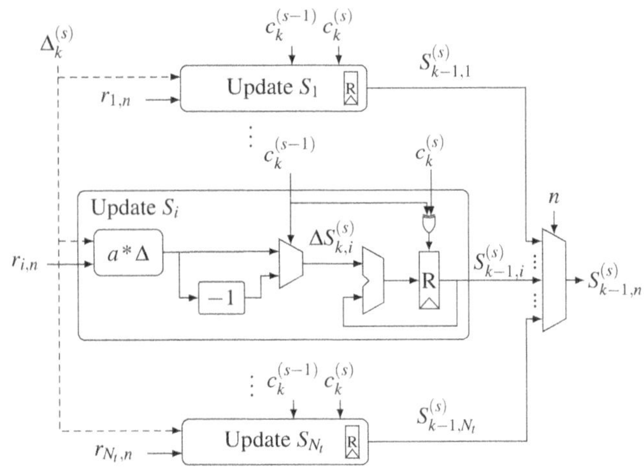

Figure 4: Block diagram of the Update S circuit consisting of N_t identical circuits. Details are depicted for the circuit updating element S_i.

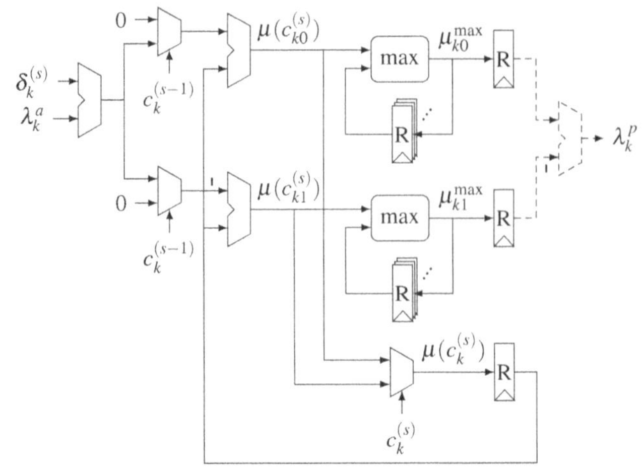

Figure 5: Block diagram of the M-circuit. The dashed part indicates the L-circuit for the case of $N_p = 1$.

$c_{k-1}^{(s)}$. In addition to reducing the critical path delay, this update scheme also allows us to use simplified multiplier circuits, thereby reducing the required area. To this end, we define the recursion $\boldsymbol{S}_k^{(s)} = \boldsymbol{S}_{k-1}^{(s)} + \Delta\boldsymbol{S}_k^{(s)}$, that is initialized with $\boldsymbol{S}^{(-1)}$ computed by the FEP-circuit. After drawing $c_k^{(s)}$, each element of $\boldsymbol{S}_k^{(s)}$ is updated by adding the increment $\Delta S_{k,i}^{(s)} = S_{k,i}^{(s)} - S_{k-1,i}^{(s)}$, which is given by

$$\Delta S_{k,i}^{(s)} = \begin{cases} -r_{i,n}\Delta_k^{(s)} & c_k^{(s-1)} = 0, c_k^{(s)} = 1 \\ r_{i,n}\Delta_k^{(s)} & c_k^{(s-1)} = 1, c_k^{(s)} = 0 \\ 0 & \text{else} \end{cases} \tag{8}$$

for $i = 1, \ldots, N_t$ and $i \neq n$. The increment $\Delta S_{k,n}^{(s)} = 0$, since diagonal elements of $\boldsymbol{R}$ are omitted in the definition of $\boldsymbol{S}^{(s)}$. The circuit implementing this update scheme is shown in Fig. 4. From (8) it follows, that the increment value in case of $\Delta S_{k,i}^{(s)} \neq 0$ can be computed before $c_k^{(s)}$ has been drawn. The newly drawn $c_k^{(s)}$ is then only used to generate a write-enable signal for the register in order to store the new value if $c_k^{(s)} \neq c_k^{(s-1)}$. Thus, the impact of the proposed update scheme on the critical path delay is almost negligible.

The remaining computation of $\gamma_k^{(s)}$ is implemented in a straightforward manner. The multiplication by $1/\eta N_0$ incurs the only full multiplier in the GS-circuit (Fig. 3).

In order to draw $c_k^{(s)}$, the probability $P(c_k = 1 \mid \boldsymbol{c}_{\setminus k}^{(s)})$ has to be retrieved from the log-domain representation $\gamma_k^{(s)}$. The costs of exact computation can be avoided by employing $P(c_k = 1 \mid \boldsymbol{c}_{\setminus k}^{(s)}) = (1 + \exp(-\gamma_k^{(s)}))^{-1} \approx 1/2 + 1/8(-\gamma_k^{(s)})$ [5]. While the computation so far has been completely deterministic, $c_k^{(s)}$ is drawn randomly according to $P(c_k = 1 \mid \boldsymbol{c}_{\setminus k}^{(s)})$. To this end, a pseudo-random number $u \sim U(0,1)$ uniformly distributed in the unity interval can be generated by a pseudo-random number generator based on a linear-feedback shift-register (LFSR). The bit $c_k^{(s)}$ is then set to 1 if $P(c_k = 1 \mid \boldsymbol{c}_{\setminus k}^{(s)}) - u > 0$, which is equivalent to $\gamma_k^{(s)} + v < 0$ with $v \sim U(-4, 4)$. This is implemented as shown in Fig. 3.

3.3 Metric Calculation Circuit

After drawing bit $c_k^{(s)}$ with the GS-circuit, the corresponding M-circuit, shown in Fig. 5, evaluates the metrics $\mu(\boldsymbol{c}_{k0}^{(s)})$ and $\mu(\boldsymbol{c}_{k1}^{(s)})$ required for the calculation of (4). In order to avoid the costs incurred by the computation of the Euclidean distance in (5), we adapt a recursive update scheme proposed in [5] to our needs. To this end, we note that both required Euclidean distances can be retrieved from the term $\delta_k^{(s)} = \frac{1}{N_0}(\|\boldsymbol{y} - \boldsymbol{H}\boldsymbol{x}_{k1}^{(s)}\|^2 - \|\boldsymbol{y} - \boldsymbol{H}\boldsymbol{x}_{k0}^{(s)}\|^2)$ which is calculated in (6), albeit scaled by η. Thus, $\delta_k^{(s)}$ is obtained by a left-shift by α bits. The metrics are then given by

$$\mu(\boldsymbol{c}_{k0}^{(s)}) = \begin{cases} \mu(\boldsymbol{c}_{k-1}^{(s)}) & c_k^{(s-1)} = 0 \\ \mu(\boldsymbol{c}_{k-1}^{(s)}) + \delta_k^{(s)} + \lambda_k^a & c_k^{(s-1)} = 1 \end{cases} \tag{9}$$

$$\mu(\boldsymbol{c}_{k1}^{(s)}) = \begin{cases} \mu(\boldsymbol{c}_{k-1}^{(s)}) - \delta_k^{(s)} - \lambda_k^a & c_k^{(s-1)} = 0 \\ \mu(\boldsymbol{c}_{k-1}^{(s)}) & c_k^{(s-1)} = 1 \end{cases} \tag{10}$$

$$\mu(\boldsymbol{c}_{k}^{(s)}) = \begin{cases} \mu(\boldsymbol{c}_{k0}^{(s)}) & c_k^{(s)} = 0 \\ \mu(\boldsymbol{c}_{k1}^{(s)}) & c_k^{(s)} = 1 \end{cases} \tag{11}$$

where $\mu(\boldsymbol{c}_k^{(s)})$ keeps track of the current metric, i.e. (5) evaluated for the Gibbs sampler state $\boldsymbol{c}^{(s)}$ after updating the k-th bit. For the sake of a simplified implementation, $\mu(\boldsymbol{c}_k^{(s)})$ is initialized with 0. The compensation of the error introduced by this is discussed in Section 3.5.

In order to determine the maximum metrics required for (4), the metrics obtained from (9) and (10) are compared to the respective current maximum metrics $\mu_{k0}^{\max}$ and $\mu_{k1}^{\max}$ held in register files, possibly yielding new maximas.

3.4 L-Value Calculation Circuit

Once bit c_k has been drawn by the parallel GS-circuits in their last iteration and the corresponding metrics are evaluated, the posterior L-value λ_k^p can be computed. To this end, the L-circuit has to determine the maximum metrics across the N_p M-circuits, as these only determine the local maximum metrics $\mu_{k0}^{\max}$ and $\mu_{k1}^{\max}$ across the N_s/N_p candidate vectors drawn by the N_{gs}/N_p independent Markov chains of the respective GS/M-circuit pair (Fig. 2). Therefore, two compare-select tree structures of depth $\lceil \log_2 N_p \rceil$ are instantiated, that determine the global maximum metrics. Subsequent subtraction then yields λ_k^p. In case $N_p = 1$, the L-circuit is reduced to a subtractor, as indicated in Fig. 5.

3.5 Initialization of Gibbs Samplers

During the initialization phase of each independent Gibbs sampler, the respective GS-circuit is put into its initial state $\boldsymbol{c}^{(0)}$ in a bit-wise fashion. Random initial states are drawn from the prior distribution given by $\boldsymbol{\lambda}^a$. This is achieved by omitting the channel-based term in (6). For this purpose, the GS-circuit features a multiplexer (① in Fig. 3) that ensures $\gamma_k^{(0)} = \lambda_k^a$. An initial state $\boldsymbol{c}^{\text{ext}}$ obtained from hard-output detection in turn is enforced by setting $c_k^{(0)} = c_k^{\text{ext}}$ with a second multiplexer (②).

A problem that arises from drawing the candidate vectors with several independent Gibbs samplers stems from the fact that the initial value of $\mu(\boldsymbol{c}_k^{(s)})$ in (11) is set to 0. While this was motivated by avoiding the direct calculation of the Euclidean distance in (5), it entails that the metrics obtained by (9) and (10) differ from their exact values in the actual initial value $\mu(\boldsymbol{c}^{(0)})$. In the context of a single Gibbs sampler, this difference can be ignored as it cancels out in (4) eventually [2]. However, independent Gibbs samplers with different initial states $\boldsymbol{c}^{(0)}$ lead to different initial values $\mu(\boldsymbol{c}^{(0)})$ that do not cancel out and thus cannot be neglected, as this would distort the search for maximum metrics in (4).

We resolve this issue without resorting to the direct calculation of the Euclidean distance. To this end, the M-circuit is activated during the initialization phase. As (9) and (10) require a previous state ($\boldsymbol{c}^{(s-1)}$), we introduce the predefined state $\boldsymbol{c}^{(-1)}$ common to all Gibbs samplers. While the GS-circuit is initialized with the individual state $\boldsymbol{c}^{(0)}$, it computes $\delta_k^{(0)}$ which is used by the M-circuit for the metric update scheme. At the end of the initialization phase, the $\mu(\boldsymbol{c}_k^{(s)})$ register holds the value $\mu(\boldsymbol{c}^{(0)}) - \mu(\boldsymbol{c}^{(-1)})$. As $\mu(\boldsymbol{c}^{(-1)})$ is equal for all Gibbs samplers, it eventually cancels out in (4) and can therefore be neglected.

Since $\boldsymbol{c}^{(-1)}$ is defined at design time, an arbitrary value can be chosen. We do this as described in Section 3.1, in order to simplify the computation of $\boldsymbol{S}^{(-1)}$ during FEP.

4. IMPLEMENTATION RESULTS

The proposed VLSI architecture presented in Section 3 was implemented in VHDL. It features design time parameters including the number of parallel GS/M-circuit pairs (N_p) as well as the wordlengths. Moreover, the number of antennas (N_t, N_r) and the modulation order (2^Q) as well as the MCMC parameters (N_s, N_{gs}) are run time configurable within a range specified at design time.

4.1 Throughput and Execution Time

The number of clock cycles required by the FEP-circuit as well as the GS-circuit are given by

$$n_{fep} = \frac{1}{2} N_t (N_t + 3) N_r + 3 \tag{12}$$

$$n_{gs} = \frac{N_{gs}}{N_p} * (\frac{N_s}{N_{gs}} + 1) * N_t Q + 3. \tag{13}$$

Both (12) and (13) account for the respective pipeline delay of 3 clock cycles, which in case of n_{gs} comprises the latencies of the M- and L-circuits.

The overall cycle count is thus given by $n_{total} = n_{fep} + n_{gs}$. However, the detection of received symbol vectors can be scheduled according to a macro-pipelining scheme, where the FEP and the main processing are overlapped for successive symbol vectors. Moreover, FEP only has to be performed during the first iteration between the detector and the channel decoder. Therefore, in the following, we only consider n_{gs} for the throughput and execution time.

The throughput of the proposed VLSI architecture can then be determined by

$$\Theta = \frac{N_t Q}{n_{gs}} f_{clk} \quad \text{[bit/s]}, \tag{14}$$

while the execution time in turn is given by $T_{exec} = n_{gs} t_{clk}$.

4.2 ASIC Synthesis Results

Layout-aware gate-level synthesis has been performed with Synopsys Design Compiler G-2012.06 in topographical mode using a 90 nm, 1.0 V standard-performance standard cell library. We investigate four instances of the scalable VLSI architecture that differ in $N_p \in \{1, 2, 4, 8\}$, i.e. the number of parallel GS/M-circuit pairs. All instances support up to 4×4 antennas as well as modulation up to 64-QAM. Fixed-point wordlengths[3] have been selected to allow 4-,16-,64-QAM modulation with negligible performance loss in terms of the coded frame error rate (FER) compared to floating-point simulations. Throughout the following discussion, we assume 4×4 antennas and 64-QAM modulation. Moreover, $N_s = 64$ and $N_{gs} = 8$ have been selected as suggested in [7].

Implementation results are shown in Fig. 6 in form of the AT_{exec}-diagram, that depicts the trade-offs between the area[4] A and the execution time T_{exec} for the scenario described above and the maximum achievable clock frequency. The proposed architecture covers a wide range of trade-offs, where the fastest variants of the least and most parallelized ($N_p = 1$ and $N_p = 8$) instances differ roughly by 208 kGE and 4.4 µs respectively. The design with the shortest T_{exec} achieves an AT_{exec}-product of 185.5 kGEµs close to the overall smallest AT_{exec}-product (181.7 kGEµs). It further exhibits a maximum clock frequency $f_{max} = 312$ MHz, an execution time $T_{exec} = 0.7$ µs and an area $A = 265$ kGE. The FEP-circuit accounts for 11 kGE, while each GS/M-circuit pair accounts for 29.1 kGE, where the circuit for updating $\boldsymbol{S}^{(s)}$, the remainder of the GS-circuit and the M-circuit account for 7.7 kGE, 9.2 kGE and 12.2 kGE respectively. The L-circuit for $N_p = 8$ accounts for 3.3 kGE. The remainder of the area is mainly incurred by registers, including additional registers required to buffer $\boldsymbol{R}$ and $\boldsymbol{S}^{(-1)}$ from the FEP-circuit for the macro-pipelining scheme.

[2]This reasoning is presented in [5], indicating that only one GS is considered, which does not introduce any parallelism.

[3]Wordlengths [integer.fractional] of quantized signals: y_i[7.9], $h_{i,j}$[3.8], λ_i^a[13.4], $1/N_0$[6.11], $r_{i,j}$[6.8], $S_{k,i}$[9.10], δ/η[17.8] and $\mu_{k\beta}$[19.6]. Real and imaginary parts of complex signals have identical wordlengths.

[4]The unit for area is gate equivalent (GE). One GE corresponds to the area of a 2-input drive-1 NAND gate.

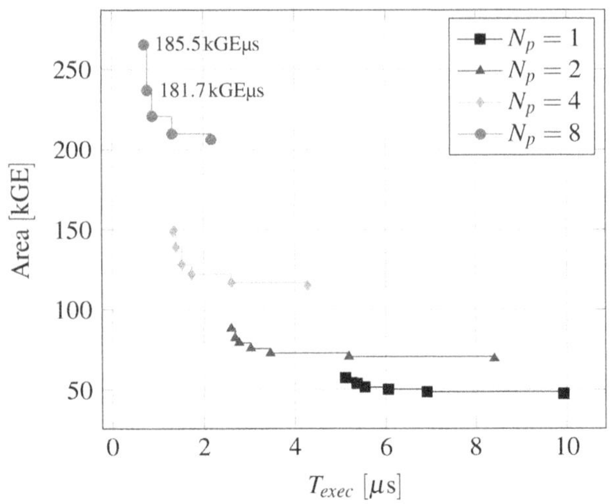

Figure 6: Area vs. execution time trade-off for a 4×4 system with 64-QAM ($N_t = 4$, $Q = 6$) and 64 random candidate vectors drawn by 8 independent Gibbs samplers ($N_s = 64$, $N_{gs} = 8$)

	This work	[2]	[8]
Number of antennas	$\leq 4 \times 4$	$\leq 4 \times 4$	$\leq 4 \times 4$
QAM order	≤ 64	≤ 64	≤ 64
Area [kGE]	265[a]	212[b]	410
Avg. throughput [Mbit/s]	34.2	46.3[c]	757
Area eff. [Mbit/s/kGE]	0.13	0.22	1.85

[a] Area for optional initial detection not included
[b] Area for required QRD not included [2]
[c] We assume 100 visited nodes per symbol vector [8].

Table 1: Comparison of implementation results with other SISO detectors. 4×4 antennas, 64-QAM and $N_s = 64$, $N_{gs} = 8$ are assumed.

Some of the implementation results for the variant with the shortest T_{exec} are summarized in Table 1. This variant achieves the overall highest throughput of 34.2 Mbit/s for 4×4 antennas, 64-QAM and $N_s = 64$, $N_{gs} = 8$.

Table 1 also shows the implementation results reported for VLSI architectures of other SISO detectors. Since MCMC-based detectors are known to perform better in low SNR regimes [4] and the SISO STS SD in [2] exhibits a variable run time depending on the SNR, we follow [8] and assume 100 visited nodes per symbol vector. Under these practically relevant channel conditions, our architecture achieves a throughput in the same range as [2]. Although Table 1 might suggest a better area efficiency of [2], it has to be noted that the area does not include the required QR-decomposition (QRD), whereas an initial detection prior to the MCMC-based detector is only optional. A fair comparison of our architecture with the SISO MMSE PIC in [8] requires a thorough analysis of the communication performance of both algorithms, as [8] might need a higher number of iterations between detector and channel decoder in order to achieve the targeted FER, particularly in low SNR regimes.

It is worth to note, that according to (14) the throughput of the proposed VLSI architecture is decisively determined by the parameters N_s, N_{gs} as well as the parallelism degree N_p. While throughout this section we always assumed the values for N_s, N_{gs} suggested in [7], these were determined purely by algorithmic considerations. An analysis considering both the targeted communication performance as well as the impact on implementation costs and throughput might lead to more appropriate values.

5. CONCLUSIONS

In this paper we presented the first implementation of a parallel VLSI architecture for MCMC-based SISO detection in iterative MIMO receivers. The proposed architecture is scalable and allows to exploit the inherent parallelism. The analysis of the implementation costs shows that the proposed architecture covers a wide range of trade-offs between area and execution time. Moreover, we have shown that our architecture can achieve a throughput and area efficiency in the range of the SISO STS SD, which makes it a viable alternative in low SNR regimes.

Future research will focus on further architectural optimizations of the proposed architecture. Moreover, suitable parametrizations will be investigated with the aim to increase throughput and area efficiency, while considering both communication performance and implementation costs.

Acknowledgements

This work has been supported by the UMIC Research Centre, RWTH Aachen University. The authors would like to thank Martin Senst for the invaluable prior work.

6. REFERENCES

[1] K. Amiri et al. Architecture and algorithm for a stochastic soft-output MIMO detector. In *Proc. of IEEE ACSSC*, pages 1034 –1038, Nov. 2007.

[2] F. Borlenghi et al. A 772 Mbit/s 8.81 bit/nJ 90 nm CMOS soft-input soft-output sphere decoder. In *Proc. of IEEE A-SSCC*, pages 297 –300, Nov. 2011.

[3] R.-R. Chen et al. Approaching MIMO capacity using bitwise Markov chain Monte Carlo detection. *IEEE Trans. Commun.*, 58(2):423 –428, Feb. 2010.

[4] B. Farhang-Boroujeny et al. Markov chain Monte Carlo algorithms for CDMA and MIMO communication systems. *IEEE Trans. Signal Process.*, 54(5):1896 – 1909, May 2006.

[5] S. Laraway et al. Implementation of a Markov Chain Monte Carlo based multiuser/MIMO detector. *IEEE Trans. Circuits Syst. I*, 56(1):246 –255, Jan. 2009.

[6] P. Robertson et al. A comparison of optimal and sub-optimal MAP decoding algorithms operating in the log domain. In *Proc. of IEEE ICC*, volume 2, pages 1009 –1013, Jun. 1995.

[7] M. Senst et al. A Rao-Blackwellized Markov chain Monte Carlo algorithm for efficient MIMO detection. In *Proc. of IEEE ICC*, pages 1 –6, Jun. 2011.

[8] C. Studer et al. ASIC implementation of soft-input soft-output MIMO detection using MMSE parallel interference cancellation. *IEEE J. Solid-State Circuits*, 46(7):1754–1765, Jul. 2011.

[9] F.-L. Yuan et al. A hardware-efficient VLSI architecture for hybrid sphere-MCMC detection. In *Proc. of IEEE GLOBECOM*, pages 1 –6, Dec. 2011.

Scaling RTL Property Checking Using Feasible Path Analysis and Decomposition

Lingyi Liu and Shobha Vasudevan
Dept. of Electrical and Computer Engineering,
University of Illinois at Urbana-Champaign, Urbana, IL, USA
liu187@illinois.edu, shobhav@illinois.edu

ABSTRACT

Property checking at the Register Transfer Level (RTL) is a critical problem for verifying complex digital design. In this paper, we present a scalable solution for property checking at RTL. We check the properties of the form $G(A => X^{=t}B)$, which means that once A is valid, B should be valid t cycles later. We introduce a decomposition strategy to scale high level bounded property checking. This decomposition strategy partitions the monolithic SMT based BMC problem into multiple smaller, independent subproblems. Every path in the RTL program is analyzed for *feasibility/relevance* using (a) a hybrid of concrete and symbolic execution and (b) property based pruning using the antecedent condition A. The partitions of the RTL source code that correspond to the feasible paths are then checked with respect to the property of interest using an SMT solver. We manage to prune a large percentage of the RTL design paths using feasibility check, such that the decomposed subproblems are small and easily verifiable.

Categories and Subject Descriptors

B.7.2 [**Integrated Circuits**]: Design Aids; Verification

Keywords

RTL, property checking, symbolic execution

1. INTRODUCTION

Due to a wide deployment of multi-processors and embedded memories, System-on-a-Chip (SoC) design is becoming increasingly complex and challenging. In the meantime, the decreased time-to-market and plummeting costs result in a pressure to shorten the entire design cycle. Design verification is still a primary source of bottlenecks in the entire design flow. Simulation based verification and formal verification are widely used in industry. Static formal verification (property checking) aims to mathematically prove the functional correctness of a design with respect to a property. In state-of-the-art, this property checking method can only be used to verify small to medium size designs because of the state explosion problem.

In this paper, we present a scalable approach to check properties at RTL. Our approach automatically decomposes the RTL design on the basis of property based path pruning. The result of the decomposition is multiple sub-programs (sets of program statements) that are reachable and relevant to the property that is being checked. The decomposition approach uses a combination of concrete and symbolic execution to select a program path. The constraints for the program path are provided in conjunction with the antecedent constraints in the property to an Satisfiability Modulo Theory (SMT) solver [1]. If the SMT solver presents a satisfiable assignment, it is mutated to obtain the next path constraint. Each such iteration produces an *antecedent pruned program* that can then be verified with respect to the property. We use an SMT solver to verify if the antecedent pruned program satisfies the property up to a given bound. We unroll the RTL for a bounded number of cycles starting from initial states and then check property on the paths within unrolled RTL. The SMT based bounded model checking (BMC) [2] is used for checking property on each antecedent pruned program.

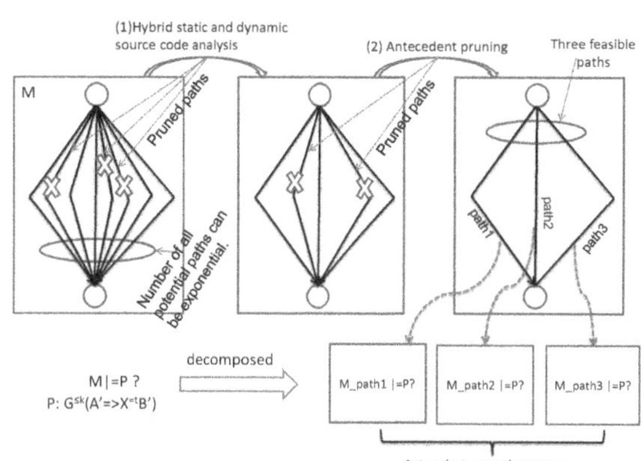

Figure 1: Figure showing our RTL decomposition using feasible path analysis including (1)hybrid static and dynamic analysis and (2)antecedent pruning. M_path_1, M_path_2 and M_path_3 are antecedent pruned programs on feasible $path_1$, $path_2$ and $path_3$. The RTL property checking is decomposed into property checking on each antecedent pruned program. k in P denotes the bounded of SMT based BMC.

Unlike previous work in decomposition where the partitioning is based on clusters/sets of paths in the CFG [3], we have an analysis phase before decomposition. The analysis phase determines that (a) every statement along a path is reachable, and (b) that the path with reachable statements is relevant to the property being checked. This means that our analysis checks to see if the path will be executed when the antecedent condition of the property holds. We use the term *relevance* of a path interchangeably with the term *feasibility* of a path in this paper. Our decomposition, then identifies only such feasible paths and presents the statements along this path for verification. This antecedent pruned program is verified using SMT based BMC [2]. The SMT based BMC instance, instead of verifying the entire monolithic RTL design, is now decomposed disjunctively into smaller and independent subproblems. The independence of each of the subproblems is ensured by mutating the symbolic path constraint in each iteration of the decomposition algorithm.

In a sense, our method does not strictly decompose the design. It also analyzes the verifiability of the design. It applies the feasibility analysis for two purposes. The first purpose is to eliminate the paths (and the statements along those paths) that do not need to be analyzed in the model, to verify a given property. The second purpose is to ensure that the paths that are being analyzed for feasibility are mutually exclusive, thereby making each of the smaller verification subproblems independent. The independence assists in the proof that the decomposition is sound and complete. Figure 1 visualizes the decomposition process.

Our analysis proceeds completely at a higher abstraction (RTL level), and is easily extensible to the system level, since we treat RTL design as software program in this paper. All generated constraints, as well as the verification operates at a higher level of abstraction as compared to bit level. Reasoning at higher levels of abstraction provides a more macroscopic view of the design and is more time efficient. However, any bit-level verification mechanism like model checking can be used to verify our decomposed antecedent pruned programs.

We apply our methodology to an open source RISC processor using our automatic decomposition technique. We are able to show that the feasibility analysis reduces the number of paths significantly. The antecedent pruning can reduce the number of paths up to 22.7%.

The principle contribution of this work are the following: 1)We present an automatic decomposition technique that depends on hybrid dynamic and static analysis to scale high level bounded verification. 2)We use the antecedent condition of the RTL property to limit the program paths, as well as the statements along that path that need to be verified.

2. PRELIMINARIES

We provide here, definitions that will be used in the explanation of our approach. We define everything in terms of the RT-level, but the technique of high level verification can be applied to the system level in principle. We particularly define Verilog RTL semantics that are useful to us. We consider Verilog as a "program" for the purpose of this work [4].

A **Control Flow Graph** (CFG) of a Veriog progam is a tuple: $< N, E, n_i, n_o, P, Q >$ where N is a set of nodes, $E \subseteq N \times N$ is a set of directed edges. $n_i \in N$ is an enter node and $n_o \in N$ is an exit node. Each edge $e \in E$ is labeled with a formula by $P(e)$. Each node $n \in N \setminus \{n_o\}$ is labeled with a Verilog statement by $Q(n)$.

A Verilog design typically has multiple CFGs, each of which corresponds to a concurrently executing process. Verilog statements can be classified into assignment statement and branch-control statement. If it is branch-control statement, the CFG node has multiple out-edges and $P(e)$ specifies each corresponding branch condition expression(formula). If it is assignment statement, $P(e)$ labels the node's out-edge as true.

A Verilog **Use-Definition Chain** (UD Chain) is a data structure defined using nodes of the CFG. It consists of a use of a variable and the transitive closure of all definitions to that variable. The definitions should be able to reach the use without any intervening definitions.

A Verilog UD chain captures the data dependency among the program variables. The Verilog definition statement is different from that in software. There are two kinds of assignments in Verilog: blocking assignment and non-blocking assignment. In a clock-triggered process, the variable defined by a non-blocking assignment is used in next cycle. We assume that the RTL code uses a synthesizable subset of Verilog RTL to ease our implementation.

A Verilog **single-cycle path** in the CFG is a path which is executed in a single cycle and it starts at node n_i and ends at node n_o. If there are multiple CFGs, a Verilog path consists of the paths in all CFGs. A Verilog **multiple-cycle path** refers to a path that is executed across multiple cycles. In order to account for the sequential behavior of RTL, the Verilog program is sequentially unrolled and the variables in each cycle are annotated with the corresponding cycle number.

Given a Verilog path, each used variable has exact one definition on the path. A Verilog **dynamic UD chain** consists of one used variable and its corresponding definition reaching that use on the path.

Given a Verilog path, **symbolic execution** is the process of assigning symbolic values to inputs (instead of concrete values) to execute the Verilog statements along the path, and representing every variable on the path as a symbolic expression in terms of the input symbolic variables.

Given a Verilog path, its **path constraint** is a formula in terms of symbolic inputs by accumulating $P(e)$ on every edge of the path. If the variable used in $P(e)$ is not input variable, Verilog symbolic execution can replace it with symbolic expressions over input variables. A path is feasible if its path constraint is satisfiable. Otherwise, it is an infeasible path.

Given a Verilog path, all statements appearing on that path constitute the **path program**. Path program is a decomposition of the original Verilog program. A path program is different from path constraints in that path constraints only include branch condition expressions in terms of symbolic inputs while the path program includes all statements along the path expressed by the path constraint.

Antecedent Pruned Program is a path program such that the conjunction of its path constraint and the antecedent condition A of a property of the form $G^{\leq k}(A => X^{=t}B)$ is satisfiable.

It should be noted that the antecedent should annotated with cycle information from 0 to k when the Verilog deisgn is unrolled for k cycles.

3. ALGORITHM FLOW

In this section, we explain the algorithm flow for checking

properties at RTL using automatic path-based decomposition. The algorithm flow is shown in Figure 2. It consists of two major parts: (1)automatic generation of RTL feasible path for property checking (Step 1-8); and (2)verification of the property in the antecedent pruned program (Step 9). The bounded depth k are assumed to be given beforehand.

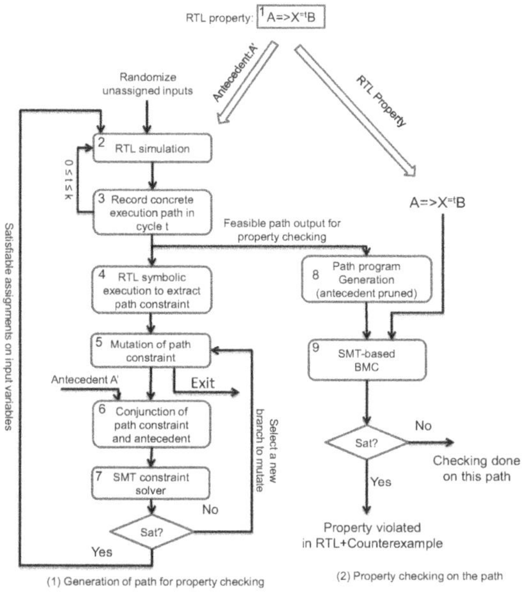

Figure 2: The algorithm flow for checking RTL property in the form of $G(A \Rightarrow X^{=t}B)$. A bounded depth k is also given beforehand. This property is then checked at RTL using path-based decomposition.

3.1 RTL concrete execution and recoding of concrete execution path (Step 2&3)

Given an input vector, it will execute a concrete path in the RTL CFG. It should be noted that not all primary inputs from cycle 0 to cycle k get concrete assignments. In this case, we need to randomize the unassigned input variables. In the initial stage, all input variables from cycle 0 to k use randomized values. The advantage of using randomized concrete values is that the input vector will execute a feasible path in the RTL CFG since we only randomize the primary input variables.

The assignments on input variables are iteratively generated from the SMT constraint solver in Step 7. The executed path is recorded in the form of taken branches. We will explain how to record concrete execution path in the form of taken branches in detail in Section V.

3.2 Symbolic execution to extract path constraint (Step 4)

Symbolic execution works on the CFG to extract a path constraint from cycle 0 to cycle k in this step. Path constraint consists of a conjunction of branch condition expressions on concrete path. However, not all variables used in condition expression are primary input. Symbolic execution is responsible for replacing these variables with expressions in terms of primary input. It should be noted that there are multiple possible paths from primary inputs to variables used in condition expression. Symbolic execution only considers the concrete execution path to extract expressions for those variables.

In addition, according to traversal order, the branch condition expressions are pushed into a stack. We extract path constraint for a concrete execution path because we want to mutate one branch condition expression in current path constraint to generate a new path. The symbolic execution implementation will be presented in Section V.

3.3 Constraint mutation to explore new path (Step 5)

After the generation of one feasible path for property checking, the algorithm tries to generate another new path until all feasible paths are generated for property checking. This is done by mutating one branch condition expression of the current path constraint. In order to systematically try all possible paths, we use a constraint stack to sustain an order to select a branch condition expression. It can guarantee that the same path will not be selected repetitively. The detail is also shown in Section V. The terminating condition for feasible path generation in the flow is in this step. When there is no candidate branch condition expression in constraint stack for mutation, all possible paths have been explored and the algorithm flow will exit.

3.4 Using antecedent to prune paths (Step 6)

We consider properties of the form of $G(A' \Rightarrow X^{=t}B')$. Antecedent A' specifies the condition on which consequent B' satisfies. If antecedent A' is false, it is unnecessary to check the consequent B'. On the other hand, we need to check properties on all k-cycle paths in RTL design. In general, the number of paths will increase exponentially as we increase the bounded depth k. The mutation of any branch condition expression in any cycle will lead to a new path.

If the path constraint conflicts with an antecedent A' of the property being checked, the corresponding path program will also be at conflict with antecedent A'. In other words, A' is always false on the path program, which implies that the property will be true on this path program since our property is in the form of $G(A' \Rightarrow X^{=t}B')$. Therefore, it is unnecessary to check the property on this path. In this sense, the antecedent in property can be used to prune paths. The number of feasible paths on which we need to check property, can therefore be reduced.

3.5 SMT to determine path feasibility (Step 7)

In this step, we use an SMT solver to decide the constraints generated from Step 5. The conjunction of a path constraint and an antecedent condition is input to the SMT solver. If it is satisfiable, SMT solver is able to give concrete assignments to all variables used in the constraints. The concrete assignments are then used for simulation in Step 2. If SMT solver reports unsatisfiable, the current path is infeasible and should be excluded from further property checking. In this situation, the flow goes to Step 5 and continues to search for another candidate branch condition expression in the constraint stack to mutate.

Let $Const\_path_j^{(0,k)}$ be path constraint of k-cycle path j and $Ante(\phi^{(i)})$ be the antecedent condition of property ϕ. Only the paths satisfying formulas (1) are reported as feasible paths and used for further property checking. We use the disjunction of all antecedent conditions in every cycle, because the path should be retained for property checking as long as the path constraint does not conflict with the antecedent in anyone cycle between 0 and k.

$$(Const\_path_j^{(0,k)}) \wedge (\bigvee_{0 \le i \le k} Ante(\phi^{(i)})) \qquad (1)$$

3.6 Generation of antecedent pruned program (Step 8)

After the generation of feasible paths, we extract the corresponding path program for property checking. Our analysis phase has determined that all statements along the feasible path are reachable if the antecedent of the given property holds true.

Conceptually, the RTL design is unrolled for k cycles. The RTL variables in every cycle are annotated with cycle number. However, the algorithm can work on RTL CFGs to extract path program for a k-cycle concrete path starting from initial states. It traverses RTL CFGs cycle by cycle. In each cycle, it follows exactly the concrete path. At every branch node in CFG, it inquires the recorded branch taken information. All statements on the path are retained in path program and variables in each statement are annotated with corresponding cycle number.

3.7 Property checking on the antecedent pruned program (Step 9)

In this step, we check given property on the extracted antecedent pruned program using SMT based BMC. If the property is violated, the SMT based BMC is able to report a counterexample to invalid the property being checked. Or else, the algorithm continues to check for the validity of the property in the next feasible path. The detailed analysis of property checking using SMT based BMC is explained in section VI.

3.8 Optimizations using backward slicing for property checking

We employ bounded model checking to falsify a given property on each path program. However, not all statements in the path program are related to the property being checked. For instance, an assignment statement defines a variable which is never used by any other statements on the path. We use property preserving slicing to remove those statements that have no impact on the property being checked [5] [6]. In other words, the reduced path program still preserves its behavior with respect to the property begin checked.

The backward slicing works on dynamic UD chain and CFGs and the algorithm is similar to path constraint generation. For every variable used in the property being checked, backward slicing algorithm transitively tracks its definition along dynamic UD chain until all used variables are inputs. The definition statements appearing on the chain are retained. It should be noted that the entire path constraint should be included in path program, because path constraint determines the control dependency for every executed statements on the path.

4. IMPLEMENTATION OF THE HYBRID STATIC AND DYNAMIC ANALYSIS

The static analysis phase of the hybrid analysis corresponds to symbolically executing the path of interest, while the dynamic analysis corresponds to concrete execution of paths in the Verilog CFG. Our hybrid symbolic and concrete execution engine plays an important role in automatic decomposition of RTL into paths. In this section, we present its working through an example. The Verilog program is instrumented to help record concrete execution paths. The CFG is constructed using a Verilog parser and conditional expressions are also annotated to edges. Symbolic execution traverses each CFG node along recorded concrete path and extracts path constraints.

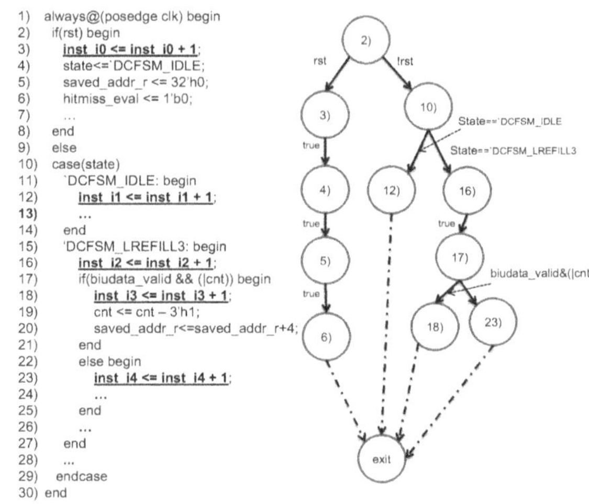

Figure 3: A Verilog example from OpenRisc [7] design which shows the instrumentation and CFG. The instrumented codes are underscored. In each node, it shows the line number of corresponding RTL code.

4.1 RTL code instrumentation

We only allow the symbolic execution to extract constraints on feasible paths, which are executed by concrete execution. RTL source code needs to be instrumented to record these concrete paths. The instrumentation serves as communication channel between concrete execution and symbolic execution. As the concrete execution proceeds in every cycle, the taken branch information is passed to symbolic execution and symbolic execution can go along taken branch to extract path constraints that caused the concrete execution path to occur.

Figure 3 shows a Verilog RTL example design with instrumented code. The $inst\_iX(X = 0, 1, ...)$ statements, which are underscored in the figure, are instrumented code. At the end of each cycle, the value of variable $inst\_iX$ is compared with its value in the last cycle. If the value is updated, it implies corresponding branch is taken in this cycle during concrete execution. The taken branch is recorded in corresponding branch node of CFG. In the case of multiple-cycle path, it is not necessary to replicate CFG. A *cycle indexed branch array* is sustained for each branch variable. When the concrete execution is done, the symbolic execution can exactly follow the taken branch in every cycle by looking up the indexed branch array.

4.2 Static analysis and UD chain construction

Symbolic execution represents every variable in RTL as symbolic expression in terms of input variables. However, the variables used in branch condition statement may not be in terms of primary input variables. Dynamic UD chain can be used to identify definitions to variables which are used in branch condition expression. In addition, dynamic UD chain is used to do backward slice with respect to property.

UD chain describes data dependency between variables in RTL and can be constructed using static analysis on RTL code. Static analysis program traverses all CFGs twice to construct UD chain. In the first time, each statement in each CFG is analyzed and the defined variable is identified and recorded. In the second time, for each used variable in each statement, all corresponding definitions are annotated to that variable's use. The reason why CFG needs to be traversed twice is that RTL code is not sequential and a used variable in one CFG may be defined in another CFG. Dynamic UD chain can be constructed in the same way by only retaining the definitions on executed path. It should be noted that the definitions of used register variables are from previous cycles when considering unrolling the RTL for multiple cycles.

In the example shown in Figure 3, variable *saved_addr_r* is defined in statements 5) and 20). If it is used in branch condition expression, these two definitions are included in UD chain. If the RTL path is $2)-10)-16)-17)-18)-...$, the dynamic UD chain only includes the definition in 20).

4.3 Path constraint generation

A path constraint is a formula and applying any of its satisfiable assignments as an input vector on the design will follow the same path. It can be calculated using symbolic execution. The symbolic execution engine traverses the CFGs one by one at every cycle. At each branch node in each cycle, the engine decides the taken path by inquiring the cycle indexed branch array. The branch condition formula labeled by P(e) on taken branch is collected. If the variable used in formula is not input variable, symbolic execution engine should transitively collect variable's definition along the path until all used variables are inputs.

Consider the example in Figure 3. In the example, we assume that the concrete execution path is $2)-10)-16)-17)-18)-...$ and the current cycle number is i. The extracted path constraint is $(\neg rst[i]) \wedge (state[i] == `DCFSM\_LREFILL3)$ $\wedge (biudata\_valid[i]\ \&\&\ (|cnt[i])) \wedge\ (cnt[i] == cnt[i-1]-1) \wedge (state[i]...)$. It can be observed that variable $cnt[i]$ is not input variable and the previous cycle's definition is therefore included in the constraint. We assume that statement 19) is executed in cycle $i-1$.

4.4 Constraint mutation for new path and constraint stack

After extracting one path constraint, we want to mutate one branch condition expression to explore another path. The mutated path constraint is sent to SMT solver. If feasible, SMT solver is able to give an assignment to variables used in the constraint. This assignment can be considered as input vector and the concrete execution of it will execute a new path.

In order to systematically explore all possible paths, we use a ***constraint stack*** to store all extracted constraints according to CFG traversal order. It should be noted that a branch condition expression will not be considered as a candidate for mutation if this branch condition expression is mutated before. A mutate flag is used to indicate whether a branch condition expression has been mutated before. When a branch condition expression is popped out from constraint stack, its mutate flag is reset which means the branch condition expression is not mutated before. When a branch condition expression is mutated, its mutate flag is set. In general, one branch condition expression can possibly be popped out or pushed in constraint stack multiple times in the whole process.

Using constraint stack to systematically explore all feasible paths can detect constraint conflict earlier and avoid the enumeration of infeasible paths. It is different from simply enumerating all possible combinations of condition branches in RTL design. After mutation of one condition branch, the conflict may result from two branches's dependency. Taking of one branch implies that another branch should not be taken. For instance, two branches in cycle $r(0 \le r \le k)$ are conflict with each other, using constraint stack can detect it when all constraints between cycle $r+1$ and k are popped out from stack. Therefore, it will no longer enumerate all path following these two conflict branches from cycle $r+1$ to cycle k.

5. PATH BASED DECOMPOSITION OF RTL DESIGN FOR SMT BASED BMC

In this section, we formally analyze our path-based decomposition for property checking. We argue that our decomposition is sound and complete. Each feasible path is exclusive.

BMC unrolls a design for a fixed number of cycles k and determines if an LTL property ϕ can be falsified in k or fewer cycles [8]. Let $M^{(i)}$ be the RTL model annotated with cycle number i, $\phi^{(i)}$ be the property annotated with cycle number i and $i \in [0...k]$. Both of them can be translated to formulas. SMT based BMC can be regarded as checking the satisfiability of the following formulas.

$$SMT\_BMC^k \triangleq (\bigwedge_{0 \le i \le k} M^{(i)}) \wedge (\bigvee_{0 \le i \le k} \neg \phi^{(i)}) \quad (2)$$

In this work, we decompose the RTL source code into N feasible paths, each of which is executed k continuous cycles from initial states. N is the number of all feasible RTL paths within k cycles. It should be noted that the feasible path should take into account antecedent of property.

Let $M\_path_j^{(0,k)}$ be the k-cycle antecedent pruned program corresponding to path j and $j \in [1...N]$. The property checking process on each feasible path is shown in equation (3). Considering that path constraint is one part of path program, equation (4) should be satisfiable according to equation(1).

$$SMT\_BMC^k_{path_j} \triangleq (M\_path_j^{(0,k)}) \wedge (\bigvee_{0 \le i \le k} \neg \phi^{(i)}) \quad (3)$$

$$(M\_path_j^{(0,k)}) \wedge (\bigvee_{0 \le i \le k} Ante(\phi^{(i)})) \quad (4)$$

$$\forall j \in [1, N]\ \ M\_path_j^{(0,k)} \models \phi \Leftrightarrow \bigwedge_{0 \le i \le k} M^{(i)} \models \phi \quad (5)$$

After decomposition RTL into feasible path with respect to property's antecedent, we can use SMT based BMC to verify property on each corresponding antecedent pruned program.

We argue that our path-based decomposition is **sound and complete**. In other words, formula (5) should be valid. We can prove them by contradiction. For soundness, if the property is not true in M, we can find a counterexample.

Properties	Unrolled cycles K	#Potential paths	#Tested paths	#Feasible paths	#Pruned paths(%)	Running time(s)
P1	5	2^{25*5}	3648	3612	32(0.9%)	59.93
P2	5	2^{25*5}	3648	3612	32(0.9%)	55.73
P3	5	2^{20*5}	14925	11536	3389(22.7%)	81.10
P4	5	2^{20*5}	14925	11536	3389(22.7%)	83.24
P5	5	2^{25*5}	3648	3612	32(0.9%)	56.50
P6	5	2^{20*5}	24375	21631	2744(11.3%)	148.08
P7	5	2^{20*5}	32580	3083	2197(6.7%)	216.88

Table 1: The detail for checking RTL temporal property using path-based decomposition. K is equal to bounded depth from initial states. Runtime is measured in seconds. The number of different kinds of path are shown. We unroll the RTL design for 5 cycles because the sequential depth in these two units are not very large. The percentage of pruned paths in tested paths is also shown in parenthesis.

Applying counterexample for simulation can trigger a feasible path. This contradicts with the left side of formula (5). For completeness, if the property is not true in one feasible path, we can use it as a counterexample for M. This contradicts with the right side.

$$\forall p, q \in [1, N] \; and \; p \neq q \quad M\_path_p^{(0,k)} \bigwedge M\_path_q^{(0,k)} \tag{6}$$

We argue that our path-based decomposition is **exclusive**. In other words, any two feasible paths are exclusive. Formally, the formula (6) should be unsatisfiable. This exclusiveness can be proved using contradiction. Supposing formulas (6) is satisfiable, we can get satisfiable assignments. Applying the satisfiable assignments as vectors for simulation, two paths will be executed.

6. EXPERIMENTAL EVALUATIONS

We implement the algorithm flow with C++ using direct programming interface(DPI) and Yices [1] SMT solver with its C Library Interface. There are about 6000 lines of C++ code in total. All the following experiments are performed on a four Intel i5 2.67GHz processor cores machine with 16GB of memory running Linux. To evaluate the effectiveness of our entire flow, we employ the RTL designs of instruction *decoder* unit and *execution* unit from OpenRisc [7]. We manually write RTL properties (P1-P7) according to the design specification. We verify the refined properties on our implemented RTL designs using path-based decomposition method.

We check the RTL temporal properties on each feasible RTL path. Antecedent condition of the property being checked is used to prune feasible paths in RTL design. We report the experimental detail in Table I. We demonstrate the effectiveness of our approach by showing the number of different kinds of paths. The potential paths means any combination of RTL branches and it is exponential. The tested path means that the corresponding path constraint is checked for feasibility by using SMT solver. The feasible path means that property should be checked on the path. The pruned path means the path does not satisfy antecedent condition and path constraint.

From Table I, it can be observed that our hybrid static and dynamic analysis can help to remove a lot of infeasible potential paths. Although the number of potential paths is exponential, the number of tested paths is largely reduced. For example, there are 25 branches in *decoder* unit, the number of paths sent to SMT solver for feasibility testing is only 3648. The effectiveness of path pruning using antecedent depends on the antecedent constraints. If there are more constraints in antecedent, more paths can be pruned. It can be observed that 22.7% of tested paths get pruned in the *execution* unit when verifying property P3 and P4.

7. RELATED WORK AND CONCLUSION

We discuss work that is related to different aspects of our approach. Compositional model checking reduces the verification of a large system to several small and tractable modules [9]. One relation can be assumed to be true while verifying another. However, this technique is a domain specific strategy and is difficult to generalize it to different kinds of designs. The assumption generation from one module's outputs is also a challenge problem.

Symbolic execution for software testing is extensively studied due to the development of constraint solver technology. The combination of concrete execution and symbolic execution is used to generate test to fail the program [10] [11]. In hardware, symbolic execution is also used to generate high coverage tests [12].

Using SMT solver for high level bounded model checking is not a completely new idea in software verification [2] [13] [3]. In [3], they use path-based decomposition and each subproblem consists of a set of paths(tunnels) instead one path. However, the unfeasible paths are also included in the constraints for property checking in this method. Our approach is different from theirs in that we use symbolic execution and concrete execution to choose only feasible paths for property checking and we also take into account antecedent for path prunning.

We have shown a novel approach to verify property at RTL using path-based decomposition. The hybrid of static and dynamic analysis and property's antecedent are used to decompose RTL design into feasible paths. Each property is checked on antecedent pruned program.

8. REFERENCES

[1] "Yices SMT solver *http://yices.csl.sri.com/*."

[2] A. Armando, J. Mantovani, and L. Platania, "Bounded model checking of software using smt solvers instead of sat solvers," *Int. J. STTT*, vol. 11, no. 1, pp. 69–83, 2009.

[3] M. Ganai and A. Gupta, "Tunneling and slicing: Towards scalable bmc," in *Proc. of DAC*, 2008, pp. 137 –142.

[4] S. Vasudevan, "High level static analysis of system descriptions for taming verification complexity," Ph.D. dissertation, The University of Texas at Austin, 2007.

[5] E. M. Clarke, M. Fujita, S. P. Rajan, T. W. Reps, S. Shankar, and T. Teitelbaum, "Program slicing of hardware description languages," in *Proc. of CHARME*, 1999, pp. 298–312.

[6] V. M. Vedula, J. A. Abraham, J. Bhadra, and R. Tupuri, "A hierarchical test generation approach using program slicing techniques on hardware description languages," *J. Electron. Test.*, vol. 19, no. 2, pp. 149–160, 2003.

[7] "Openrisc web page: *http://www.opencores.org*."

[8] E. M. Clarke, A. Biere, R. Raimi, and Y. Zhu, "Bounded model checking using satisfiability solving," *Formal Methods in System Design*, vol. 19, no. 1, pp. 7–34, 2001.

[9] K. L. McMillan, "A methodology for hardware verification using compositional model checking," *Sci. Comput. Program.*, vol. 37, pp. 279–309, May 2000.

[10] P. Godefroid, N. Klarlund, and K. Sen, "Dart: directed automated random testing." in *Proc. of PLDI*, 2005.

[11] K. Sen, D. Marinov, and G. Agha, "Cute: a concolic unit testing engine for c," in *Proc. of ESEC/FSE-13*, 2005, pp. 263–272.

[12] L. Liu and S. Vasudevan, "Efficient validation input generation in RTL by hybridized source code analysis," in *Proc. of DATE*, 2011, pp. 1596–1601.

[13] M. K. Ganai and A. Gupta, "Accelerating high-level bounded model checking," in *Proc. of ICCAD*, 2006, pp. 794–801.

SIREN: A Depth-First Search Algorithm for the Filter Design Optimization Problem

Levent Aksoy
INESC-ID
Lisboa, Portugal
levent@algos.inesc-id.pt

Paulo Flores
INESC-ID/IST TU Lisbon
Lisboa, Portugal
pff@inesc-id.pt

José Monteiro
INESC-ID/IST TU Lisbon
Lisboa, Portugal
jcm@inesc-id.pt

ABSTRACT

This paper addresses the filter design optimization (FDO) problem that is to find a set of filter coefficients which yields the least design complexity while meeting the required filter constraints. The design complexity of a filter is defined in terms of the total number of adders/subtracters, assuming that the multiplication of coefficients by the filter input is realized under a shift-adds architecture. Existing algorithms use efficient search methods, but none of them can guarantee the minimum design complexity. Hence, we propose an exact algorithm, called SIREN, that finds an optimum solution of the FDO problem under the minimum quantization value. It is based on a depth-first search method equipped with an exact technique, that finds the minimum number of adders/subtracters in the multiplier block of the filter, and search pruning techniques that enable it to be applicable to practical instances. Experimental results show that SIREN can still find better solutions than efficient FDO algorithms and its solutions lead to filters with significantly less area when compared to a straightforward filter design technique.

Categories and Subject Descriptors

G.1.6 [**Mathematics of Computing**]: Optimization

General Terms

Algorithms, Design

Keywords

Filter design optimization, depth-first search, linear programming, multiplierless filter design

1. INTRODUCTION

Finite impulse response (FIR) filters are widely used in digital signal processing (DSP) applications due to their output stability and phase linearity. The output of an N-tap FIR filter $y(n)$ is computed as $\sum_{i=0}^{N-1} h_i \cdot x(n-i)$, where h_i is the ith filter coefficient and $x(n-i)$ is the ith previous filter input. The realization of this expression in the transposed form is shown in Figure 1, where the multiplication of coefficients by the filter input dominates the design complexity.

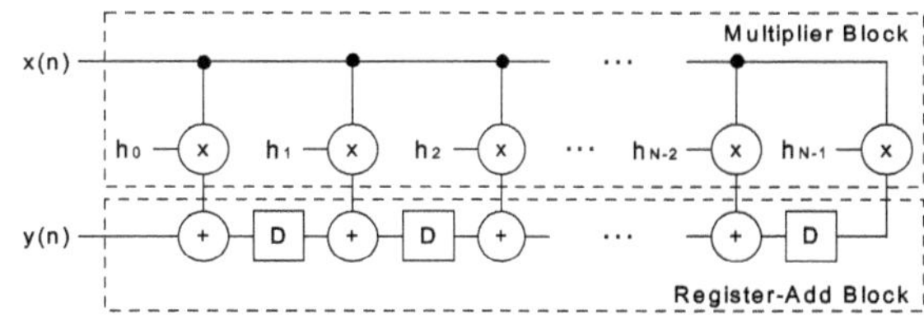

Figure 1: Transposed form FIR filter design.

However, the multiplier block of the FIR filter can be realized using only shifts and adders/subtracters, since the filter coefficients are determined to be constant [10]. Note that shifts can be implemented using only wires without representing any hardware cost. Thus, a fundamental optimization problem, called the multiple constant multiplications (MCM) [13], is defined as: given a set of constants to be multiplied by an input variable, find the minimum number of adders/subtracters that realize the constant multiplications. Existing MCM algorithms [1, 2, 8, 13] aim to maximize the sharing of partial products among the constant multiplications. In this manner, a straightforward way to design a multiplierless FIR filter requires two steps. First, a set of coefficients, that respects the filter constraints, is obtained, and second, the multiplier block of the filter is designed under a shift-adds architecture using an MCM algorithm.

In turn, the FDO problem [12] is defined as: given the filter specifications (*fspec*) denoted as a 5-tuple (filter length N, pass-band w_p and stop-band w_s frequencies, and maximum allowed pass-band δ_p and stop-band δ_s ripples), find a set of coefficients that leads to a filter design with minimum number of adders/subtracters and satisfies the filter constraints. The design complexity of the transposed form FIR filter is computed based on the number of adders/subtracters both in its multiplier and register-add blocks, assuming that the constant multiplications in its multiplier block are realized under a shift-adds architecture. Existing FDO algorithms include efficient techniques that minimize the number of nonzero digits in coefficients [3, 4, 11], maximize the common nonzero digits in coefficients [7, 14], and dynamically expand a set of partial products which are shared among the coefficient multiplications [12, 16]. However, to the best of our knowledge, there is no algorithm that guarantees a filter with the minimum design complexity.

In this paper, we first present the fundamental concepts on multiplierless design of constant multiplications and give an overview on prominent FDO algorithms. Then, we intro-

GLSVLSI'13, May 2–3, 2013, Paris, France.

duce the SIREN algorithm which can find a set of fixed-point filter coefficients that satisfies the filter constraints and leads to a filter design requiring minimum number of adders/subtraters under the smallest quantization value (Q). In SIREN, first, the floating-point lower and upper bounds of coefficients are computed. In an iterative loop, these lower and upper bounds of coefficients are quantized using Q, that is initially set to 0 and is increased by 1 in each iteration. Then, a depth-first search (DFS) method is applied to find a solution that yields a filter design with minimum complexity, satisfying the filter constraints or to prove that there exists no solution to be selected from the quantized lower and upper bounds of coefficients. The most important properties of the DFS method of SIREN are: i) it constructs a search tree based on a predetermined ordering of coefficients; ii) as the values of coefficients are determined, it updates the lower and upper bounds of coefficients in the search tree; iii) it computes the lower bound on the filter complexity and backtracks if this lower bound is greater than or equal to the best solution found so far; iv) it uses the exact MCM technique of [2] to find the minimum number of operations in the multiplier block of the filter. This paper also presents two variations of SIREN that are applied to slightly different FDO problems where search space is highly restricted. Finally, we introduce their time-complexity on randomly generated FIR filters, compare their solutions with prominent FDO algorithms, and give the gate-level results of FIR filters synthesized based on the solutions of SIREN. Experimental results indicate that they can be applied to real-size FIR filters and can find a minimum solution where prominent FDO algorithms cannot obtain a solution with this value or cannot ensure that their solutions are the minimum solutions.

2. BACKGROUND

2.1 Multiplierless Constant Multiplications

A straightforward method to realize constant multiplications under a shift-adds architecture, called the digit-based recoding (DBR) [5], initially defines the constants in multiplications under a number representation, *e.g.*, canonical signed digit (CSD)[1] [8] or binary. Then, for the nonzero digits in the representation of the constant, it shifts the input variable according to the digit positions and adds/subtracts the shifted variable with respect to the digit values. As a simple example, consider the constant multiplications $39x$ and $83x$. Their decompositions in CSD are listed as:

$$39x = (10100\bar{1})_{CSD}x = x \ll 5 + x \ll 3 - x$$
$$83x = (101010\bar{1})_{CSD}x = x \ll 6 + x \ll 4 + x \ll 2 - x$$

and require 5 operations as shown in Figure 2(a).

The complexity of an MCM design can be further reduced by sharing the partial products among the constant multiplications. Existing MCM methods can be grouped in two classes: common subexpression elimination (CSE) [1, 8] and graph-based (GB) [2, 13] techniques. The CSE algorithms initially define the constants under a particular number representation. Then, considering possible subexpressions that can be extracted from the nonzero digits in representations of constants, the "best" subexpression, generally, the most

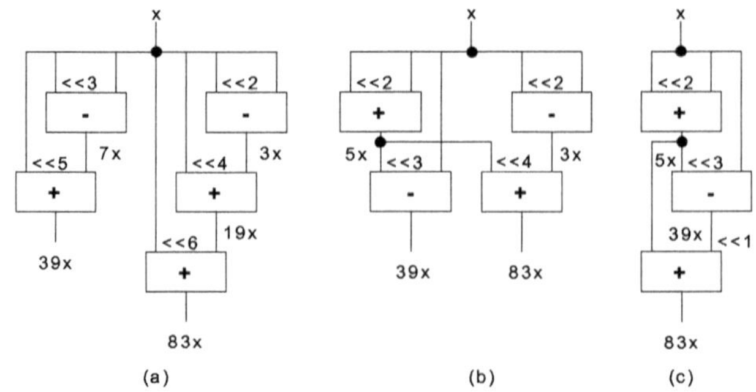

Figure 2: Shift-adds realizations of $39x$ and $83x$: (a) the DBR technique [5]; (b) the exact CSE algorithm of [1]; (c) the exact GB algorithm of [2].

common, is chosen to be shared among the constant multiplications. The GB methods are not restricted to any particular number representation and find the intermediate partial products that enable to realize the constant multiplications with minimum number of operations. They consider a larger number of realizations of a constant and may obtain better solutions than CSE methods [2, 13].

Returning to our example, the exact CSE algorithm [1] obtains a solution with 4 operations when constants are defined under CSD (Figure 2b) and the exact GB method [2] finds a minimum solution with 3 operations (Figure 2c).

2.2 Filter Design Optimization Algorithms

The zero-phase frequency response of a symmetrical FIR filter is given as[2]:

$$G(w) = \sum_{j=0}^{\lfloor M \rfloor} d_j h_j cos(w(M-j))$$

where $M = (N-1)/2$ and $d_j = 2 - K_{j,M}$ with $K_{j,M}$ is the Kronecker delta[3], $h_j \in \mathbb{R}$ with $-1 \leq h_j \leq 1$, and $w \in \mathbb{R}$ is the frequency in radians. Assuming that the desired pass-band and stop-band gains are equal to 1 and 0, respectively, the filter must satisfy the following constraints [3]:

$$\begin{array}{ll} 1-\delta_p \leq G(w) \leq 1+\delta_p, & w \in [0, w_p] \\ -\delta_s \leq G(w) \leq \delta_s, & w \in [w_s, \pi] \end{array} \quad (1)$$

The pass-band gain is not relevant for many DSP applications. Thus, a scaling factor (s) can be added into the filter constraints as a continuous variable as follows, where s^l and s^u are respectively the lower and upper bounds of s [7, 15].

$$\begin{array}{ll} s(1-\delta_p) \leq G(w) \leq s(1+\delta_p), & w \in [0, w_p] \\ s(-\delta_s) \leq G(w) \leq s(\delta_s), & w \in [w_s, \pi] \\ s^l \leq s \leq s^u & \end{array} \quad (2)$$

In some DSP applications, it is also desirable to minimize the normalized peak ripple (NPR) [12, 16], which is defined as δ/b, where δ and b are the peak ripple and pass-band gain, respectively. Other considered parameters are the peak weighted ripple [11] and the NPR magnitude [14].

FDO algorithms generally reduce the design complexity in two different ways. The algorithms of [3, 4, 11] try to find a set of coefficients with a minimum number of nonzero digits, since a constant represented with fewer number of

[1]In CSD, a constant is written as $\sum_{i=0}^{n-1} b_i 2^i$ using n digits, where $b_i \in \{1, 0, \bar{1}\}$ and $\bar{1}$ denotes -1. In its CSD representation, nonzero digits are not adjacent and minimum number of nonzero digits is used.

[2]Frequency response of an asymmetrical filter can be found in [9].

[3]$K_{a,b}$ function is 1 when a is equal to b. Otherwise, it is 0.

```
SIREN(fspec)
  Q = 0, sol_found = 0, sol = [ ]
  (h^l, h^u, s^l, s^u) = CLUBFCSF(fspec)
  O = OrderCoefs(h^l, h^u)
  repeat
     Q = Q + 1, H^l = ⌈(h^l · 2^Q)⌉, H^u = ⌊(h^u · 2^Q)⌋
     if are_lubfc_valid(H^l, H^u) then
        (sol_found, sol) = DFS(fspec, O, Q, H^l, H^u, s^l, s^u)
  until sol_found
  return sol
```

Figure 3: The SIREN algorithm.

nonzero digits requires fewer number of operations in general. However, the reduction of the number of nonzero digits does not always lead to a design with minimum number of operations, since the sharing of partial products among the constant multiplications is not considered (Figure 2). In order to exploit the sharing of partial products, the algorithms of [7, 14] search for the coefficients that have the most common nonzero digits. Also, in the methods of [12, 16], the common partial products are expanded dynamically as the filter coefficients are determined. However, these techniques also cannot ensure the minimum number of operations in the filter, since they are not equipped with exact methods.

FDO algorithms explore the filter coefficients using exhaustive [3, 7, 12, 15, 16] and local search [4, 11, 14] methods. The exhaustive search techniques use mixed-integer linear programming [7], branch-and-bound [3, 15], and DFS [12, 16] methods. While they can only be applied to low-order filters due to an exponential increase in run-time [3, 12], the local search methods can be applied to high-order filters, but cannot ensure the minimum design complexity, since all possible values of coefficients are not considered.

Meanwhile, SIREN is designed to handle the filter constraints in Eqn. 2, where the search space is increased due to s, yielding better solutions [7]. Its variations, that can handle the same filter constraints with a restricted s and those in Eqn. 1, are also described. It is based on the exhaustive DFS method and uses the exact MCM algorithm [2].

3. THE SIREN ALGORITHM

As shown in its pseudo-code given in Figure 3, SIREN takes the 5-tuple *fspec* denoting the filter specifications as input and returns a set of fixed-point filter coefficients *sol*. First, it computes the lower and upper bounds of filter coefficients and s in its *CLUBFCSF* function, where the lower bounds of coefficients are found by solving the following linear programming (LP) problem for each coefficient.

$$\begin{aligned} &minimize: f = h_i \\ &subject\ to: s(1-\delta_p) \leq G(w) \leq s(1+\delta_p), \quad w \in [0, w_p] \\ &\qquad -s(\delta_s) \leq G(w) \leq s(\delta_s), \quad w \in [w_s, \pi] \\ &\qquad s^l \leq s \leq s^u \end{aligned}$$

The filter coefficients and s are *continuous* variables of this LP problem and the initial s^l and s^u were set to 0.01 and 100, respectively. Thus, the value of h_i in the LP solution corresponds to its lower bound h_i^l and is stored in h^l. In a similar way, the upper bound of each coefficient h_i^u is found when the cost function is changed to $f = -h_i$ and is stored in h^u. The sets h^l and h^u consist of the floating-point lower and upper bound of each filter coefficient, respectively. The values of s^l and s^u are also updated similarly.

Then, its *OrderCoefs* function determines an ordering of filter coefficients to be used by the DFS method described in the following subsection while constructing the search tree. The main idea on finding an ordering of coefficients is based on a fact observed in experiments that if the coefficients with narrower upper and lower bound intervals are placed in lower depths of a search tree, less number of decisions are made and conflicts occur earlier. Thus, the runtime of SIREN can be reduced significantly, still exploring all possible values of coefficients. This function sorts the coefficients in ascending order according to their $h_i^u - h_i^l$ values and stores their indices i in this order in O.

In the iterative loop of SIREN, starting with the quantization value Q equal to 1, the floating-point lower (upper) bound of each coefficient is multiplied by 2^Q, rounded to the smallest following (the largest previous) integer, and is stored in H^l (H^u). The validity of these sets is tested by the *are_lubfc_valid* function by simply checking each coefficient if H_i^l is less than or equal to H_i^u. If they are not valid, Q is increased by one, and H^l and H^u are updated. Otherwise, the DFS method, that explores all possible values of each coefficient in between H_i^l and H_i^u, is applied to find a set of filter coefficients which respects the filter constraints and yields the minimum design complexity, or to prove that there exists no such a set of filter coefficients. If the latter occurs, Q is increased by one, H^l and H^u are updated, and the DFS method is applied again. Hence, SIREN ensures that its solution is a set of fixed-point filter coefficients obtained using the smallest Q. Note that Q is an important parameter in the filter design complexity, because as it increases, the bitwidths of coefficients are increased, requiring larger sizes of registers and structural adders in the register-add block (Figure 1), and most probably, requiring more number of operations in the multiplier block since larger integer coefficients are used. Similarly, in [3, 12, 16], the effective wordlength (EWL) of a set of coefficients, computed as $max\{\lceil log_2|h_i|\rceil\}$ when fixed-point coefficients are considered, was used to evaluate the quality of a solution in addition to the total number of adders/subtracters.

3.1 The Depth-First Search Method

The search tree is constructed based on the ordering of coefficients O, where a vertex at depth d, V_d, denotes the filter coefficient whose index is the dth element of O, *i.e.*, $h_{O(d)}$. An edge at depth d of the search tree, *i.e.*, a fanout of V_d, stands for an assignment to the vertex V_d from $[V_d^l, V_d^u]$ where V_d^l (V_d^u) denotes the lower (upper) bound of V_d. Note that the values of the vertex at depth d are assigned incrementally starting from V_d^l to V_d^u.

The DFS method first assigns $H_{O(1)}^l$ and $H_{O(1)}^u$ respectively to the lower and upper bounds of the vertex at depth 1, *i.e.*, the initial depth, and sets the value of the vertex V_1 to $H_{O(1)}^l$. However, at any depth greater than 1, $d > 1$, although the lower and upper bounds of a vertex can be taken from H^l and H^u, respectively, we can use tighter lower and upper bounds of a vertex, since the values of $d-1$ coefficients are determined and fixed at this stage. Thus, the lower bound of the vertex V_d is computed by solving the following LP problem, where the non-determined coefficients and s are the *continuous* variables of the LP problem.

$$\begin{aligned} &minimize: f = h_{O(d)} \\ &subject\ to: s(1-\delta_p) \leq G(w)/2^Q \leq s(1+\delta_p), \quad w \in [0, w_p] \\ &\qquad -s(\delta_s) \leq G(w)/2^Q \leq s(\delta_s), \quad w \in [w_s, \pi] \\ &h_{O(1)} \ldots h_{O(d-1)}: determined \qquad s^l \leq s \leq s^u \end{aligned}$$

In this LP problem, the lower and upper bounds of all non-determined filter coefficients are taken from H^l and H^u, respectively. The upper bound of V_d is also computed by solving a similar LP problem when the cost function is $f = -h_{O(d)}$. If there exist feasible solutions for both LP problems, the lower (upper) bound is rounded to the smallest following (the largest previous) integer. Also, if this integer lower bound is less than or equal to this integer upper bound, these values are determined to be the lower and upper bounds of V_d. Whenever there are no feasible lower or upper bounds for V_d, the search is backtracked chronologically to the previous vertex until there exists a value to be assigned among its lower and upper bounds.

When the values of all coefficients are determined, the implementation cost of the filter based on this coefficient set is computed as $TA = MA+SA$, where TA is the total number of adders/subtracters in the filter, and MA and SA are the number of adders/subtracters in the multiplier block and the number of structural adders in the register-add block, respectively. While MA is found using the exact MCM algorithm [2], SA is computed as $N-ncz-1$, where ncz is the number of coefficients equal to zero. Naturally, no adder is required for a coefficient set to zero in the register-add block. This coefficient set is stored in *sol* and *sol_found* is set to 1, if its TA value is smaller than that of the best one found so far, which was set to infinity in the beginning of search.

In order to prune the search tree, the implementation cost of the filter (TA) is estimated at depth d greater than $2M/3$ ($2N/3$) for symmetric (asymmetric) filters. This value is found empirically not to waste an effort for computing an estimate which usually does not yield a backtrack. The lower bound on MA is found using the determined coefficients as described in [6]. The lower bound on SA is computed after all non-determined coefficients are set to a value. To do so, the upper and lower bound interval of each non-determined coefficient is checked if 0 is included. If so, this non-determined coefficient is set to 0. Otherwise, it is assumed to be a constant different from 0.

The DFS method terminates when all possible values of coefficients have been explored. If *sol_found* is 0, it ensures that there exists no solution which respects the filter constraints under the given quantized lower and upper bounds of coefficients. Otherwise, its solution is a set of integer coefficients found using minimum Q that leads to a filter with minimum design complexity, satisfying the filter constraints.

As a simple example, consider a symmetric filter with *fspec* (8, 0.2π, 0.7π, 0.01, 0.01). The floating-point lower and upper bounds of coefficients are respectively computed as $h^l = \{h_0^l, h_1^l, h_2^l, h_3^l\} = \{-0.0966, -0.0915, 0.0015, 0.0039\}$ and $h^u = \{h_0^u, h_1^u, h_2^u, h_3^u\} = \{-0.0003, -0.0002, 0.4144, 1\}$. The ordering of coefficients is found as $O = \{1, 0, 2, 3\}$. The quantized lower and upper bounds of coefficients are determined as $H^l = \{-3, -2, 1, 1\}$ and $H^u = \{-1, -1, 13, 32\}$, respectively when Q is 5 (no solution was found with $Q < 5$).

The search tree constructed by the DFS method is shown in Figure 4, where a and b in $[a\ b]$ given next to each vertex stand respectively for its lower and upper bounds which are dynamically computed as coefficients are determined. In this figure, the actual traverse of the DFS method on filter coefficients can be followed from top to bottom and from left to right. Here, **Conflict** denotes that given determined coefficients, there exists no feasible lower/upper bound for the current depth vertex. Also, **Success** presents that the set

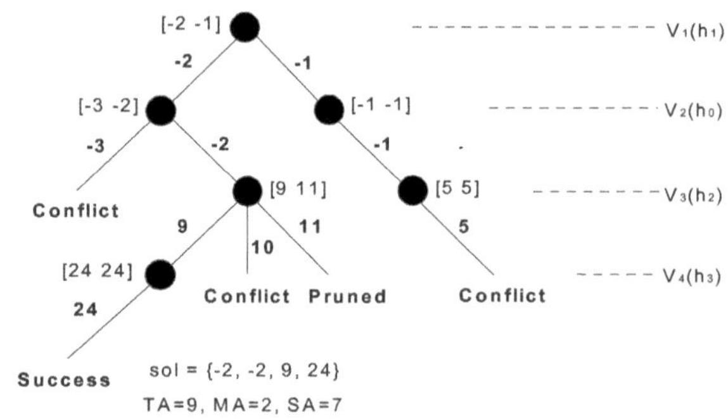

Figure 4: Search tree formed by the DFS method.

of determined coefficients leads to a solution satisfying the filter constraints and **Pruned** indicates that the set of determined coefficients cannot lead to a better solution than the best one found so far. Observe that as values are assigned to coefficients, the intervals between the lower and upper bounds of coefficients are significantly reduced when compared to those in the original H^l and H^u.

3.2 Modifications for Similar FDO Problems

Since SIREN is a general algorithm, it can easily be modified to handle filter constraints given in Eqn. 1 or those in Eqn. 2 with a restricted s. For the former, the lower and upper bounds of s are simply set to 1 and this type of the SIREN algorithm is denoted as SIREN(1). For the latter, its lower and upper bounds are set to given values. In another type of the SIREN algorithm called SIREN(3), they were respectively set to 0.7 and 1.4, $\mp$3dB gain tolerance.

4. EXPERIMENTAL RESULTS

In this section, we present the results of SIREN and its variations, and compare them with those of prominent FDO algorithms. We also introduce the results of a straightforward filter design technique (SFDT), where given *fspec*, the *firgr* function of MATLAB was used to obtain the filter coefficients which were quantized to integers with minimum Q, avoiding the effect of quantization errors on the filter constraints. Then, the exact MCM algorithm [2] was applied to these filter coefficients to realize the multiplier block of the filter using minimum number of adders/subtracters. Note that SIREN was written in MATLAB and was run on a PC with Intel Xeon at 2.33GHz and 4GB memory. As an LP solver, it uses *lp_solve 5.5.2.0*.

As the first experiment set, we used 21 randomly generated symmetric low-pass FIR filters whose N values range between 20 and 40. Figure 5 presents the results of algorithms in terms of Q, TA, and CPU time in seconds.

Recall that SIREN and its variations iteratively increase Q until they find a solution that satisfies the filter constraints. Observe from Figure 5(a) that SIREN always finds a solution on these instances using a smaller Q than its variations. The difference between the average Q value in its solutions and those in the solutions of SIREN(3), SIREN(1), and SFDT is 1.5, 2.5, and 4, respectively. In this case, the EWL values of filter coefficients in its solutions are also smaller than those in the solutions of these algorithms. This comes from the fact that since there is no restriction on s in SIREN, the filter constraints of Eqn. 2, can be satisfied with smaller coefficients, but with a larger s. Simply because of this fact and because no restriction on s increases the number of pos-

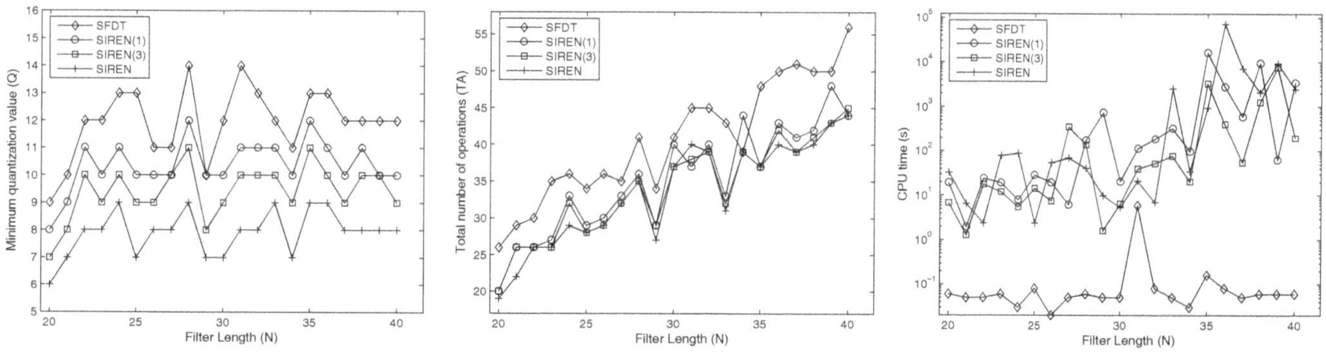

(a) (b) (c)

Figure 5: Results of algorithms on randomly generated filters (a) Q; (b) TA; (c) CPU.

Table 1: Specifications of symmetric FIR filters.

Filter	N	w_p	w_s	δ_p	δ_s
X1	15	0.2π	0.8π	0.0001	0.0001
G1	16	0.2π	0.5π	0.01	0.01
Y1	30	0.3π	0.5π	0.00316	0.00316
Y2	38	0.3π	0.5π	0.001	0.001
A	59	0.125π	0.225π	0.01	0.001
S2	60	0.042π	0.14π	0.012	0.001
L2	63	0.2π	0.28π	0.028	0.001

sible constants considered for a coefficient, SIREN generally obtains better solutions in terms of TA than these algorithms, as can be observed from Figure 5(b). On average, it finds 0.6, 1.7, and 7.2 less number of adders/subtracters than SIREN(3), SIREN(1), and SFDT, respectively. However, there is only one exception, the filter with an N value 31, where it obtains a worse solution than its variations. This is because the use of a larger Q in its variations may enable them to consider such filter coefficients that require no adder/subtracter, like 0 or shifted versions of 1, or that maximize the sharing of partial products[4]. On the other hand, SIREN requires the most CPU time on average, simply because no restriction on s increases the upper and lower bound intervals of coefficients, leading to a large number of branches (assignments) in the DFS method. Additionally, N and Q are other parameters that affect the runtime of SIREN and its variations. While an increase in N increases the number of vertices in the DFS method, an increase in Q increases the number of runs of the DFS method, the number of branches in the DFS method that is due to a larger upper and lower bound interval of a vertex, and the runtime of the exact MCM algorithm [2] since larger coefficients are considered. Hence, SIREN may obtain a solution using less CPU time than its variations due to a smaller Q.

As the second experiment set, we used the symmetric FIR filters given in Table 1. Table 2 presents the results of SIREN, SIREN(3), and previously proposed algorithms whose results were taken from [3, 12] as reported. The CPU time limit was set to 2 days for SIREN and SIREN(3). In this table, BST denotes the CPU time required to find the best solution and TT stands for the total CPU time. The algorithms were sorted according to their results on i) EWL, ii) TA, iii) TT, and iv) BST. Note that the solutions of [12] had been determined to be the best ones on these filters.

Observe from Table 2 that both SIREN and SIREN(3) can find a solution with an EWL value less or the same

[4]As SIREN can be initiated with any Q (Figure 3), it was applied to this instance with the same Q that SIREN(3) and SIREN(1) used and obtained better results than its variations.

Table 2: Summary of algorithms on FIR filters.

Filter	Method	EWL	MA	SA	TA	BST	TT
X1	[14]	13	7	8	15	–	–
	SIREN(3)	10	5	8	13	4s	6s
	SIREN	10	5	8	13	<1s	2s
	[12]	10	5	8	13	–	<1s
G1	[7]	7	2	13	15	–	–
	[12]	6	2	15	17	–	<1s
	SIREN(3)	6	2	15	17	<1s	<1s
	SIREN	6	2	15	17	<1s	<1s
Y1	[16]	10	6	23	29	–	21m30s
	SIREN(3)	9	7	29	36	2s	4s
	[12]	9	7	23	30	–	6s
	SIREN	9	6	23	29	2m17s	7m56s
Y2	[15]	12	–	–	39	–	–
	[12]	10	10	37	47	–	11s
	SIREN(3)	10	9	29	38	6m36s	7m29s
	SIREN	10	9	29	38	3m52s	4m29s
A	[3]	10	18	58	76	3h2m	4h14m
	[12]	10	14	54	68	–	2d2h
	SIREN(3)	*10*	*16*	*52*	*68*	*1d6h*	*2d**
	SIREN	*10*	*16*	*52*	*68*	*14h57m*	*2d**
S2	[3]	11	27	59	86	23m	27m
	[12]	10	17	59	76	–	16h42m
	SIREN	*9*	*14*	*57*	*71*	*16h4m*	*2d**
	SIREN(3)	*9*	*14*	*57*	*71*	*7h27m*	*2d**
L2	[3]	10	18	62	80	26m	54m
	SIREN	*10*	*16*	*60*	*76*	*1d23h*	*2d**
	SIREN(3)	*10*	*15*	*58*	*73*	*1d22h*	*2d**
	[12]	10	17	56	73	–	16h28m

* Time-limit is exceeded.

as prominent algorithms. The solutions of SIREN on small size filters X1, G1, Y1, and Y2 include the least or the same number of operations as those of the FDO algorithm [12]. In turn, on medium size filters A, S2, and L2, both SIREN and SIREN(3) cannot complete the search due to the CPU time limit. However, while SIREN(3) yields a solution with less or the same number of operations as the FDO algorithm [12], SIREN obtains a worse solution than SIREN(3) only on Filter L2. This is because it generally requires more CPU time than SIREN(3) to explore the whole search space.

We also designed Filter Y2 based on the solutions of SFDT and SIREN when N was in between 34 (the minimum N value found using the *firgr* function of MATLAB) and 38, and Filter G1 based on the solutions of SIREN when Q was 6 and 7. FIR filters were described in VHDL when the bitwidth of the filter input was 16, and they were synthesized using the Synopsys Design Compiler and UMCLogic 180nm Generic II library. Tables 3 and 4 present the results of FIR filters, where CA, NCA, and A denote the combinational area, non-combinational area, and total area, all in mm^2, respectively.

Table 3: Summary of results of algorithms on Filter Y2.

N	SFDT									SIREN									
	Q	EWL	MA	SA	TA	CPU	CA	NCA	A	Q	EWL	MA	SA	TA	CPU	CA	NCA	A	gain
34	15	14	18	31	49	5m6s	33.2	16.5	49.7	11	11	10	27	37	19m48s	23.2	15.1	38.3	23.0
35	15	14	18	34	52	7s	34.6	16.8	51.4	11	11	12	28	40	34m19s	25.0	15.0	40.0	22.2
36	14	13	15	29	44	1s	28.8	16.7	45.5	11	11	10	27	37	18m38s	23.3	15.3	38.6	15.3
37	14	13	17	36	53	15s	35.9	17.2	53.1	10	10	12	36	48	23s	30.0	15.6	45.6	14.1
38	14	13	15	35	50	15s	33.1	17.6	50.7	10	10	9	29	38	4m29s	23.5	16.1	39.6	21.9

Table 4: Summary of results of SIREN on Filter G1.

$Q=6$								$Q=7$							
EWL	MA	SA	TA	CPU	CA	NCA	A	EWL	MA	SA	TA	CPU	CA	NCA	A
6	2	15	17	<1s	9.1	5.6	14.7	7	2	13	15	25s	8.1	5.9	14.0

Also, *gain* in Table 3 is the area gain in percentage between the designs obtained by the solutions of SIREN and SFDT.

Observe from Table 3 that the solutions of SIREN with less number of adders/subtracters and with filter coefficients having smaller EWL values lead to FIR filters occupying smaller area with respect to the solutions of SFDT. Note that as N is increased, the EWL values of coefficients is decreased. Thus, the area values of Filter Y2 designed using the solutions of SIREN when N is 34 and 38 are close to each other, indicating the tradeoff between N and the EWL value of coefficients on the filter complexity. Hence, it is useful to design a filter with different N values, since different sets of coefficients yield filter designs with different complexity.

Also, observe from Table 4 that a filter with coefficients having a smaller EWL value does not always yield a filter design occupying smaller area. In this case, it is because of less number of structural adders in the register-add block. Thus, it is useful to run SIREN with also $Q+1$ to obtain such solutions. More importantly, these experiments suggest that to evaluate the quality of a solution of an FDO algorithm more accurately, the metrics, which take into account the gate-level implementation cost, should be derived.

5. CONCLUSIONS

This paper introduced the SIREN algorithm equipped with a DFS method, an exact MCM technique, and efficient search pruning techniques, and its variations applied to slightly different FDO problems. To our knowledge, they are the only algorithms that can guarantee the minimum filter design complexity defined as the total number of adders/subtracters using the minimum Q. Experimental results showed that they can be applied to medium size FIR filters and can still obtain better solutions than previously proposed prominent FDO algorithms. Also, the solutions of SIREN lead to less complex FIR filters with respect to the solutions of SFDT.

6. ACKNOWLEDGMENTS

This work was supported by national funds through FCT, Fundação para a Ciência e a Tecnologia.

7. REFERENCES

[1] L. Aksoy, E. Costa, P. Flores, and J. Monteiro. Exact and Approximate Algorithms for the Optimization of Area and Delay in Multiple Constant Multiplications. *IEEE TCAD*, 27(6):1013–1026, 2008.

[2] L. Aksoy, E. Gunes, and P. Flores. Search Algorithms for the Multiple Constant Multiplications Problem: Exact and Approximate. *Elsevier MICPRO*, 34(5):151–162, 2010.

[3] M. Aktan, A. Yurdakul, and G. Dündar. An Algorithm for the Design of Low-Power Hardware-Efficient FIR Filters. *IEEE TCAS-I*, 55(6):1536–1545, 2008.

[4] C.-L. Chen and A. N. Willson Jr. A Trellis Search Algorithm for the Design of FIR Filters With Signed-Powers-of-Two Coefficients. *IEEE TCAS-II*, 46(1):29–39, 1999.

[5] M. Ercegovac and T. Lang. *Digital Arithmetic.* Morgan Kaufmann, 2003.

[6] O. Gustafsson. Lower Bounds for Constant Multiplication Problems. *IEEE TCAS-II*, 54(11):974–978, 2007.

[7] O. Gustafsson and L. Wanhammar. Design of Linear-Phase FIR Filters Combining Subexpression Sharing With MILP. In *MWSCAS*, pages 9–12, 2002.

[8] R. Hartley. Subexpression Sharing in Filters Using Canonic Signed Digit Multipliers. *IEEE TCAS-II*, 43(10):677–688, 1996.

[9] Y. C. Lim. Design of Discrete-Coefficient-Value Linear Phase FIR Filters With Optimum Normalized Peak Ripple Magnitude. *IEEE TCAS*, 37(12):1480–1486, 1990.

[10] H. Nguyen and A. Chatterjee. Number-Splitting With Shift-and-Add Decomposition for Power and Hardware Optimization in Linear DSP Synthesis. *IEEE TVLSI*, 8(4):419–424, 2000.

[11] H. Samueli. An Improved Search Algorithm for the Design of Multiplierless FIR Filters with Power-of-Two Coefficients. *IEEE TCAS*, 36(7):1044–1047, 1989.

[12] D. Shi and Y. J. Yu. Design of Linear Phase FIR Filters With High Probability of Achieving Minimum Number of Adders. *IEEE TCAS-I*, 58(1):126–136, 2011.

[13] Y. Voronenko and M. Püschel. Multiplierless Multiple Constant Multiplication. *ACM Transactions on Algorithms*, 3(2), 2007.

[14] F. Xu, C. H. Chang, and C. C. Jong. Design of Low Complexity FIR Filters Based on Signed Powers of Two Coefficients With Reusable Common Subexpressions. *IEEE TCAD*, 26(10):1898–1907, 2007.

[15] J. Yli-Kaakinen and T. Saramaki. A Systematic Algorithm for the Design of Multiplierless FIR Filters. In *ISCAS*, pages 185–188, 2001.

[16] Y. J. Yu and Y. C. Lim. Optimization of Linear Phase FIR Filters in Dynamically Expanding Subexpression Space. *Circuits, Systems, and Signal Processing*, 29:65–80, 2010.

Generating Concise Assertions with Complete Coverage

Chen-Hsuan Lin
Department of Electrical and Computer Engineering
University of Illinois at Urbana-Champaign
Urbana, IL
clin54@illinois.edu

Lingyi Liu
Department of Electrical and Computer Engineering
University of Illinois at Urbana-Champaign
Urbana, IL
liu187@illinois.edu

Shobha Vasudevan
Department of Electrical and Computer Engineering
University of Illinois at Urbana-Champaign
Urbana, IL
shobhav@illinois.edu

ABSTRACT

Assertions are valuable and commonly applied to formal verification and simulation-based verification in IC design flow. Unfortunately, assertion generation is a time-consuming process that depends heavily on human efforts. Some dynamic methods based on simulation and static methods based on structure analysis are proposed to automate assertion generation process. However, dynamic methods cannot guarantee the quality of assertions due to incomplete simulation while static methods might have scalability limits. With the significant advances in Boolean satisfiability (SAT) solving, SAT solving becomes a promising technique to overcome these methods' weaknesses. In this paper, we successfully formulate assertion generation to a SAT problem and use unit assumption to generate concise assertions. Furthermore, we consider input constraints and word level features to generate meaningful and high-readability assertions. Experimental results on SpaceWire, Ethernet, and Floating Point designs show that the generated assertions can always achieve 100% input space coverage.

Categories and Subject Descriptors

B.7.2 [**Hardware**]: INTEGRATED CIRCUITS—*Design Aids*

Keywords

assertion generation; complete coverage; input constraints; word level features

1. INTRODUCTION

Verification is the bottleneck in IC design flow and this obstacle becomes more serious as the complexity of design grows. Assertion-based verification [16] is widely used in industry to facilitate checking of desired properties. Assertions are applied to formal verification for property checking and to simulation-based verification for monitoring design behavior and locating bugs [16].

In verification process, assertion generation is a tedious and time-consuming manual procedure. In order to generate high-quality assertions, this manual procedure would take several refining iterations and lots of manpower [9]. Therefore, methods to automatically generate assertions [15] [25] [7] [26] [12] [6] [13] [5] are proposed to relieve this hardship. Assertions, automatically generated from a Register Transfer Level (RTL) design, can be used in future regression test to guarantee the quality of design, and also help designers detect unexpected behaviors. Assertions are typically generated for a *target* RTL signal, and are of the form $A \to B$, which includes all propositional and temporal formulas in a standard temporal logic like LTL [19].

Dynamic methods generate assertions based on dynamic simulation or execution paths of RTL designs [5] [6] [15] [25] [7] [13]. Some of them also check the correctness of candidate assertions using formal techniques [25] [5]. These techniques examine simulation data and infer behavior, often using machine learning [25] [7] and pattern matching algorithms [13] for the same. The quality of assertions thus inferred heavily depends on the information provided by simulation data. The generated assertions are always biased towards the simulation data. This is contradictory to the intention of assertion generation process. If generated assertions have exactly the same behaviors as the simulation data from designers, no new bugs or design behaviors can be exposed or covered by the generated assertions.

Moreover, directed or even random simulation patterns cannot cover all behaviors of the design. Insufficient simulation may result in corner case or rare case scenarios not being covered. The generated assertions fail to cover the corner cases of the design. However, these *corner case assertions* are more valuable to the verification process than typical case scenarios. In addition, insufficient simulation will lead to poor guesses by the machine learning algorithms, thereby leading to overfitting, or many redundant propositions in the assertion [25] [7]. Candidate assertions generated from simulation data need extra time and effort to formally verify their correctness.

Static analysis of source code and models have been used in software for generating assertions [2] [21] and in deductive verification for generating invariants [22]. In hardware, the techniques in [12] [26] also generate assertions based on structural analysis of designs. Static analysis can generate accurate assertions without spending formal verification time and effort in checking them. It can also cover 100% of the design input space without reliance on "typical case scenarios".

The downside in static analysis is the limited scalability. In the context of assertion generation, however, scalability is not a critical issue, since assertions spanning across more than $2-3$ modules are generally unreadable for the users. For the scaling required by the assertion generation problem, sophisticated, scalable Boolean SAT based algorithms can be applied very effectively [8] [11] [10].

In this paper, we propose a **S**AT-**B**ased Automatic **A**ssertion **G**eneration method (**SAT-BAG**), which can avoid the drawbacks of dynamic methods while circumventing scalability issues faced by static methods. SAT-BAG converts the assertion generation problem into an ALL-SAT problem [4]. ALL-SAT problem is to generate all satisfying assignments for a SAT problem. The design information is completely stored in a SAT instance instead of simulation data. With this SAT instance, SAT-BAG can generate concise satisfying assignments for target signals. These satisfying assignments, together with the target signal value, can be regarded as assertions.

When directly applying SAT solvers to generate satisfying assignments for target signal, SAT solvers give concrete value to every variable in the design. As a result, the generated assertions that involve one proposition for each variable is very verbose. However, not all assignments are necessary for determining the target variable's value, and there are a lot of *redundant* propositions in the generated assertion. In this paper, we apply a SAT technique, *unit assumption* [8], to remove redundant propositions in assertions. Unit assumption is a SAT technique that allows users to make assumptions for variables during SAT solving and reports which assumptions contribute to the result. In our context, the propositions of an assertion can be viewed as assumptions for a target signal. We can examine which propositions contribute to the target, and include only those in our generated assertions.

Our SAT-BAG method stores the entire design as SAT instance and also considers design constraints on the input signals. Generating assertions without regards to the input constraints may lead to vacuous assertions, in which the antecedent conditions are never reached or satisfied in the design. Therefore, to avoid potentially generating assertions on unreachable states, SAT-BAG converts input constraints into SAT clauses that can filter out vacuous assertions generated for unreachable states, and also assist in further redundancy removal from the assertions. For example, a bus controller may specify the legal work mode of the bus, and our SAT-BAG method can then incorporate these constraints for shaping input signals in work mode.

To further optimize our proposed SAT-BAG method, we use *word level features* to improve the readability and input space coverage of generated assertions. Although SAT-BAG can generate concise and high coverage assertions, the generated assertions are at bit level and thus have low readability, and the number of bit level assertions might be large. Therefore, our SAT-BAG method uses word level features to elevate the assertions to word level. Take design $c = a[k:0] \geq b[k:0]$ as an example, SAT-BAG will generate one word level assertion instead of many bit level assertions for valid bit vector combinations of a and b. Word level features will be discovered from RTL design and added into SAT instances to let SAT-BAG generate word level assertions.

Our contributions in this paper are as follows. We propose a SAT-BAG method with unit assumption technique to generate concise assertions for given target signals. Our SAT-BAG method is not biased toward any simulation data and is able to generate assertions to fully cover both typical case scenarios and corner case scenarios. Our method can also filter out vacuous assertions by taking into account input design constraints. We also discover word level features from RTL design and apply these features to generate word level assertions, which have higher readability and expressiveness.

2. PRELIMINARIES

2.1 Assertion

Our assertions are of the form *antecedent* $\rightarrow$ *consequent*, where *antecedent* and *consequent* are propositional or temporal logic formulas. We use notation of SystemVerilog assertion [24] for expressing assertions in this paper. An example assertion is $(in_0 == 0$ && $in_1 == 1)$ ##1 $(in_1 == 1) \rightarrow (t == 1)$. It reads as follows: If both $(in_0 = 0)$ and $(in_1 = 1)$ in the current cycle, and $(in_1 = 1)$ in the next cycle, then the target signal "t" should evaluate to true.

In this paper, propositions can contain one of two features. A *bit level feature* is a bit signal evaluating to true or false. A *word level feature* is a first order formula in terms of word level variables, vectors of bit signals, that evaluates to true or false. In this paper, propositions in antecedents of assertions only consider *boundary input signals*, which are composed of primary inputs (PIs), and registers, also called pseudo primary inputs (PPIs). In our assertions, the consequent contains only one proposition – a target signal that evaluates to true/false (equals 1/0). *A word level assertion* is an assertion having at least one word level feature. An example of a bit level assertion is: $(in_0 == 1$ && $in_1 == 0) \rightarrow (t == 0)$ while an example of a word level assertion is: $(in_0[k:0] \geq in_1[k:0]) \rightarrow (t == 1)$.

Previous techniques generate assertions within a sliding time window [25] [7]. We use a similar concept here. A parameter *window size* (w) is used to restrict the duration of time cycles for temporal assertions we want to generate. For the example at the beginning of this section, w is set to 2, which means temporal assertions involving boundary input signals across two cycles will be generated.

A *vacuous assertion* is an assertion in which the antecedent cannot be satisfied. Consider the truth table of a target variable's function in terms of features. A table entry is *covered by a given assertion* if the concrete value of the entry can satisfy the antecedent of the given assertion. The *input space coverage* of an assertion denotes the percentage of truth table entries covered by the assertion.

2.2 Boolean satisfiability problem

A Boolean satisfiability (SAT) problem is a decision problem [4]. Given a finite set of Boolean variables $V = \{v_0, \ldots, v_n\}$, a *literal* l is either a Boolean variable v_i or its negated form $\neg v_i$ or $\sim v_i$. A *clause* is a disjunction of literals. A *SAT instance* is a conjunction of clauses, also named as *conjunction normal form (CNF)*. An *assignment* over V assigns each Boolean variable v_i either true or false value. A SAT instance is **satisfiable** if there exists an assignment, named *satisfying assignment*, such that the SAT instance evaluates to true; otherwise it is **unsatisfiable**.

An *ALL-SAT* problem is a variation of the SAT problem. While the traditional SAT problem only concerns one satisfying assignment for a satisfiable SAT instance, an ALL-SAT solution enumerates all satisfying assignments of a SAT instance. All satisfying assignments can be achieved by solving multiple SAT problems repeatedly. Each time a new satisfying assignment ψ is derived, the blocking clause of ψ will be added into the SAT instance to avoid deriving the same ψ again. The blocking clause of ψ is a disjunction of negated values of ψ. This collecting process will continue until no more satisfying assignments can be derived.

2.3 Conversion of circuit into SAT instance

An RTL design will be compiled to its corresponding circuit netlist. Given the circuit netlist, it can be converted to a SAT instance such that the functionality of design is completely preserved [23]. The conversion, named Tseitin transformation [23], can be done in linear time. For instance, a two input AND gate, $c = AND(a,b)$, can be converted into the following three clauses.

$$(v_a \vee \neg v_c) \wedge (v_b \vee \neg v_c) \wedge (\neg v_a \vee \neg v_b \vee v_c)$$

3. SAT-BASED AUTOMATIC ASSERTION GENERATION METHOD

This section will use a running example in Fig. 3 to introduce how to construct a SAT instance for assertion generation, and how to efficiently generate assertions with unit assumption technique and enhance the quality of assertions by applying input constraints. In addition, discovering word level features and generating word level assertions are also discussed in this section.

The overall flowchart of SAT-BAG is shown in Fig. 1. Our contributions are shown in dotted boxes. First of all, an RTL design is converted into a gate netlist with bit and word level features, which are discovered from static analysis of RTL's CDFG. Next, the gate netlist is converted to a SAT instance based on window size. The constraints for a target signal are set as well. If needed, the input constraints of the design will be converted to SAT clauses and added to the SAT instance. With this SAT instance, we begin to repeatedly generate satisfying assignments. Each assignment can determine features' values, and redundant features can be removed by unit assumption. After that, concise bit and word level assertions will be collected and reported. The algorithm terminates after no more satisfying assignments can be generated.

3.1 Formulation of an All-SAT problem for assertion generation

Given a sequential circuit synthesized from an RTL design, the transition functions (T) can be extracted as Fig. 2(a). PIs and POs are primary inputs and outputs, while PPIs and PPOs are register signals, pseudo primary inputs and outputs. Next, we unroll the circuit based on the parameter w, which determines how many cycles we generate temporal assertions for. As shown in Fig. 2(b), if window size $w = 3$, three T's will be duplicated. The internal register signals in these functions are connected. Signals in each T are annotated with superscripts, -2, -1, and 0, according to the corresponding time cycle. The superscript 0 represents the current cycle while -1 and -2 represent the previous two cycles.

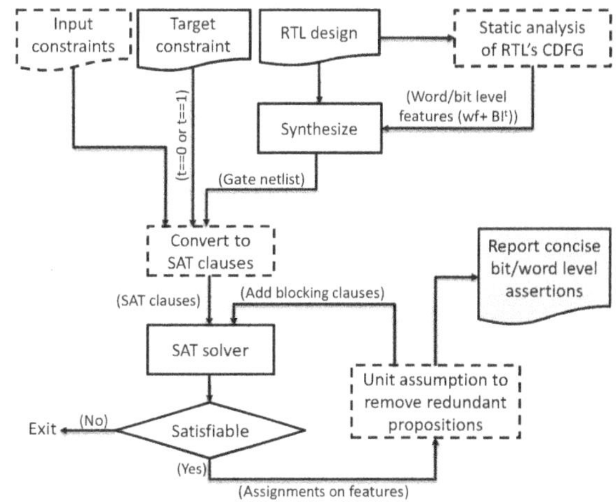

Figure 1: Flow chart of SAT-BAG.

A set of boundary input signals of the unrolled circuit based on window size w is $BI = PIs^0 \cup \ldots \cup PIs^{-(w-1)} \cup PPIs^{-(w-1)}$, where PIs^i and $PPIs^i$ denote the sets of PI and PPI signals at cycle i, respectively. For a target signal t^0, BI^{t^0} is a subset of BI only collecting boundary input signals that are in the logic cone of t^0, as the shaded region in Fig. 2(b). Therefore, the feature set F for generating assertions for the target signal t^0 will be set to BI^{t^0} because t^0's value is obviously determined by the boundary input signals in its logic cone.

DEFINITION 1. *Unit assumption [8] is a technique that allows users to assume specific conditions for each SAT solving through function interface $S.solve(assumps)$, denoted as $S^{assumps}$, where S is a SAT instance and* ***assumps*** *is the conjunction of literals like $a_0 \wedge a_1 \wedge \ldots \wedge a_n$. During each SAT solving, literals a_i are* ***temporarily*** *assigned to true. Therefore, if there exists an assignment that satisfies the SAT instance as well as assumptions, that satisfying assignment is returned. On the other hand, if the SAT problem is unsatisfiable under the given assumptions, the* ***subset*** *of those assumptions that collects assumptions contributing to the unsatisfiable result is returned. After each SAT solving, the unit assumptions are discharged.*

The unrolled circuit is converted to a SAT instance S as mentioned in Section II.C. To generate assertions for $t^0 = 1$, an assumption for the target signal $t^0 = 1$ would be added during solving S by function interface $S.solve(v_{t^0})$, where v_{t^0} is a Boolean variable representing signal t^0 in S. This assumed S is denoted as S^{t^0}. With S^{t^0} and F, ALL-SAT technique is applied to collect all satisfying assignments by repeatedly solving and blocking. Each satisfying assignment ψ for S^{t^0} can determine a unique vector of values for features in F which leads to $t^0 = 1$. Therefore, this vector of values and $t^0 = 1$ can be considered the antecedent and consequent of an assertion, respectively. With iterative collection of satisfying assignments, all assertions for $t^0 = 1$ can be generated.

Using the running example in Fig. 3 to demonstrate, we generate assertions for $o_2 = 0$ with window size $w = 1$. Here we omit cycle annotations of signals for better readability, since this is a single cycle assignment. The candidate feature set is $F = BI^{o_2} = \{a, b, ppi_1\}$. The design is converted to a SAT instance S and the corresponding Boolean variables in

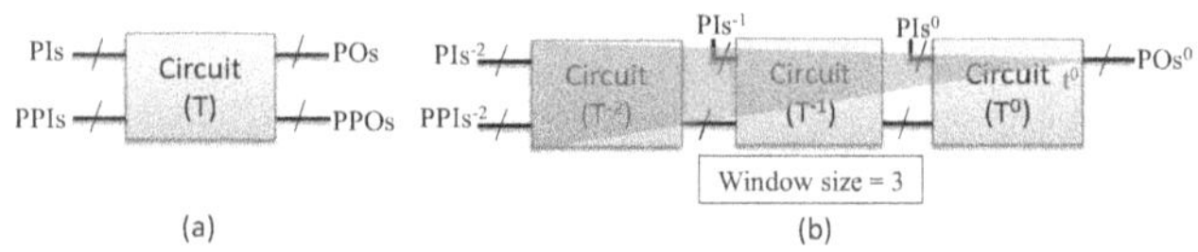

Figure 2: Unroll circuit based on parameter w and determine relevant boundary inputs by logic cone information.

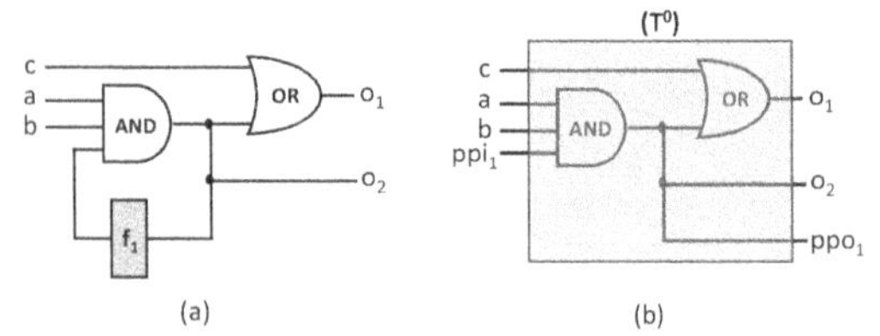

Figure 3: A running example.

S representing signals a, b, ppi_1 and o_2 are v_a, v_b, v_{ppi_1} and v_{o_2}, respectively. We then collect all satisfying assignments for $S^{\neg o_2}$. Each assignment can decide a vector of values for $[v_a,\ v_b,\ v_{ppi_1}]$. In this case, we will collect 7 vectors for $[v_a,\ v_b,\ v_{ppi_1}] = \{000, 001, \ldots, 110\}$, and each vector can be viewed as an assertion. For example, the vector $[0, 1, 0]$ stands for a valid assertion $(a == 0 \&\& b == 1 \&\& ppi_1 == 0) \rightarrow o_2 == 0$.

Obviously, these 7 assertions, converted from above 7 vectors, are valid assertions but not concise assertions due to including redundant propositions. Therefore, the method proposed to generate concise assertions using unit assumption is introduced in the next section.

3.2 Generation of concise assertions using unit assumption

When SAT-BAG computes a vector for features in F, it will apply unit assumption to check whether all these features are necessary to determine the value of target signal and remove redundant features from the vector. In addition, unit assumption also can accelerate the process of collecting vectors by blocking several satisfying assignments at the same time.

For the same example, if SAT-BAG computes a vector $[0, 1, 0]$ for $[v_a,\ v_b,\ v_{ppi_1}]$, unit assumption is applied to check whether all features $a = 0$, $b = 1$, and $ppi_1 = 0$ are necessary for determining $o_2 = 0$. First of all, we examine the necessity of $ppi_1 = 0$. The original assumption for S is changed from $o_2 = 0$ to $o_2 = 1$ and additional assumptions $a = 0$, $b = 1$ are added to S as well. Now we solve S under new assumptions, which is $S.solve(v_{o_2} \wedge \neg v_a \wedge v_b)$, and marked as $S^{o_2 \wedge \neg a \wedge b}$.

If the result of $S^{o_2 \wedge \neg a \wedge b}$ is **satisfiable**, that means signal o_2 can be 1 without the feature $ppi_1 = 0$. Hence, the feature $ppi_1 = 0$ is critical to determine $o_2 = 0$ and cannot be removed from the vector. And the process will continue examining the remaining features. In this case, the result of $S^{o_2 \wedge \neg a \wedge b}$ is **unsatisfiable**, which means these two features $a = 0$ and $b = 1$ are enough to determine signal $o_2 = 0$ and the feature $ppi_1 = 0$ is redundant. Therefore, the vector can be optimized from $[0, 1, 0]$ to $[0, 1, X]$, where X is a don't-care term.

Furthermore, many redundant features can be caught simultaneously, because unit assumption will also report the **subset** of assumptions that really contributes to the unsatisfiable result. In this case, $\{v_{o_2}, \neg v_a\}$, a subset of the original assumptions $\{v_{o_2}, \neg v_a, v_b\}$, is reported, and we can know that only feature $a = 0$ contributes to this unsatisfiable result and another redundant feature $b = 1$ is found. Therefore, the vector can be further optimized from $[0, 1, X]$ to $[0, X, X]$. Now we generate a concise assertion $(a == 0) \rightarrow o_2 == 0$ instead of $(a == 0 \&\& b == 1 \&\& ppi_1 == 0) \rightarrow o_2 == 0$. In addition, by blocking the optimized vector $[0, X, X]$, we can avoid generating vectors that are already covered by $[0, X, X]$. Therefore, the process of collecting vectors can be improved as well.

In other words, the original vectors can be regarded as minterms and the optimized vectors as cubes. Each time we will try to collect cubes instead of minterms. This not only can help SAT-BAG generate concise assertions but also can enhance the collecting process by blocking cubes instead of minterms.

3.3 Elimination of vacuous assertions and refinement using input constraints

There might exist constraints on boundary inputs of modules, called *input constraints*. These constraints should be taken into account while generating assertions. On one hand, the inputs of modules or PIs can be driven by other modules. Obviously these inputs are not real free variables, *i.e.* some combinations of the inputs are invalid. There is research [18] [10] on generating this kind of constraint on the connections between modules. On the other hand, register signals of modules, PPIs, are also not free variables, due to the presence of some unreachable states. Unreachable states are states that don't exist execution traces from the initial states of a given finite state machine, thus cannot be considered valid state values on register signals. Reachability analysis [20] [11] explores reachable states and this information can be regarded as constraints on register signals.

For the same example in Fig. 3, if the initial state of flip-flop f_1 is set to 0, the input ppi_1 will always be 0 during all clocks. Therefore, the input constraint $ppi_1 = 0$ should be considered while generating assertions. With this constraint, the assertion $(a == 1 \&\& b == 1 \&\& ppi_1 == 1) \rightarrow o_2 == 1$, though valid, will become a vacuous assertion because ppi_1 can never be 1. In addition, the assertion $(c == 0 \&\& ppi_1 == 0) \rightarrow o_1 == 0$ can be further refined to $(c == 0) \rightarrow o_1 == 0$.

However, previous work does not take into account input constraints. Their assertions might be vacuous and might have redundant propositions. Therefore, in this work, SAT-BAG will take input constraints into consideration by converting these constraints into SAT clauses [3]. For the same example, the constraint $ppi_1 = 0$ can be converted to a SAT clause $(\neg v_{ppi_1})$. With these clauses, SAT-BAG can generate more meaningful assertions and more accurate assertions.

3.4 Word level feature discovery and application

The evaluated values of word level variables are either true or false; therefore, these values can be displayed by bit signals wf_i as shown in Fig. 4. The signals wf_i are added to cycle copies (Ts) like primary outputs; thus, they can be considered features by SAT-BAG for generating assertions as well. The candidate feature set will be updated to $F = BI^{\Delta} \cup WF$, where $WF = \bigcup WPOs^i$. Furthermore, as to SAT-BAG, word level features in WF will have higher priority over bit level features in BI^{Δ} to be chosen as propositions in assertions.

Table 1: Comparison on assertion generation of SAT-BAG and the previous work, Goldmine [25].

Circuit (output)	SAT-BAG				Goldmine [25]			
	Asrt. Num.	Avg. Cov. %	Tot. Cov. %	Time	Asrt. Num.	Avg. Cov. %	Tot. Cov. %	Time
SWR-active_o	18	13.29	100	0.01	5	9.38	46.88	18.79
SWR-err_sqc	18	16.54	100	0.01	0	–	–	–
SWR-RST_tx_o	10	24.22	100	0.01	2	18.75	37.50	3.68
ETHNET-StartIdle	994	0.50	100	0.33	5	10.64	53.18	12.63
ETHNET-ByteCntMax	17	47.06	100	0.01	0	–	–	–
ETHNET-ExcessiveDefer	16	46.88	100	0.01	0	–	–	–
ETHNET-WB_ACK_O	8	17.58	100	0.10	9	11.11	100	7.86
FPU-expa_ff	9	44.49	100	0.01	119	0.39	46.81	18.49
FPU-infa_f_r	24	47.92	100	0.01	0	–	–	–
FPU-exp_in_80	9	44.49	100	0.01	9	11.11	100	3.67

* We list designs and target signal names, the number of assertions (Asrt. Num.), the average input space coverage per assertion (Avg. Cov.), and the total input space coverage (Tot. Cov.) for SAT-BAG and Goldmine.

* Avg. Cov. for SAT-BAG might not equal to 100% divided by the number of assertions due to coverage overlap between assertions.

To discover word level features, an RTL design will be compiled into the control data flow graph CDFG [27]. There are three typical nodes in the CDFG, branch, assignment and merge nodes. A branch node represents a branch statement in RTL; An assignment node represents an assignment statement in RTL; A merge node represents an end of branch. Word level features can be found in branch nodes (ex: $if\ (a \geq b)$) or assignment nodes (ex: $assign\ c = (a! = b)$). We will unroll the design CDFG based on the window size and trace target signal's fanin cone backward to the boundary inputs, PIs and PPIs. Word level features in terms of boundary input signals can be collected along the paths of fanin cone.

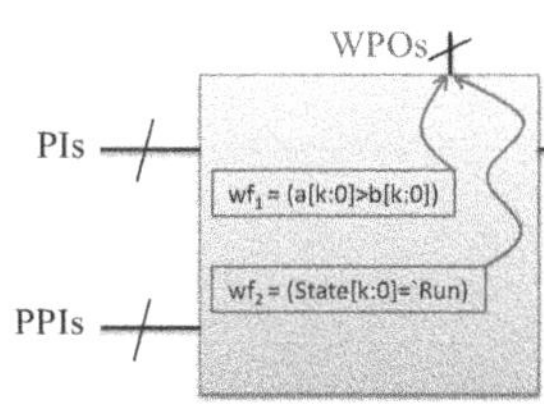

Figure 4: Use boundary signals wf_i to display the evaluated values of word level features.

4. EXPERIMENTAL RESULTS

We implemented our proposed method in ABC [1] with a build-in MiniSAT solver [8]. We present results on SpaceWire (SWR), Ethernet IP core (ETHNET) and Floating Point Unit (FPU) designs from OpenCores [17] and IWLS2005 suite [14] for our experiments. All experiments were run on a Linux machine with Intel Core 2 Quad 2.66GHZ CPU and 8GB RAM.

The experimental results are shown in two subsections. Section A shows comparison of generating bit level assertions with SAT-BAG and Goldmine [25], one of state-of-the-art dynamic works. Section B shows how optimization techniques, input constraints and word level features can benefit SAT-BAG.

4.1 Comparison between SAT-BAG and the previous work

Table 1 compares the performance of SAT-BAG with Goldmine [25] for generating bit level assertions. We set the window size w to 1 or 2 to get temporal assertions.

From this experiment, we observe that SAT-BAG generates high quality assertions that have higher input space coverage, and does not miss any corner case assertions. This is because all testcases can reach 100% input space coverage. In case StarIdle, SAT-BAG has lower average coverage per assertions than Goldmine because it generates more additional assertions that are difficult to be hit by simulation. With these assertions, SAT-BAG can reach 100% input space coverage while Goldmine can reach only 53.18%. Due to insufficient simulation, Goldmine also fails to find assertions in some cases. Furthermore, while Goldmine spends formal verification time and effort verifying "guesses" by data mining, SAT-BAG does not need this step.

Although Goldmine generates more assertions than SAT-BAG on cases WB_ACK_O: (Goldmine 9 vs. SAT-BAG 8) and expa_ff: (Goldmine 119 vs. SAT-BAG 9), some of Goldmine's assertions have redundant propositions in antecedents. This means more than one assertion will convey the same amount of information and have same input space coverage. Hence, high number of does not imply more information in Goldmine.

4.2 Improvements of SAT-BAG

This section will demonstrate the benefits of applying input constraints and word level features to SAT-BAG. Table 2 shows the results of SAT-BAG with and without input constraints. We use SpaceWire design, which has an one-hot encoding state machine, to show the effect of providing input constraints to SAT-BAG. The constraint for an one-hot encoded k bit state variable, $s[k-1:0]$, can be presented as: $\forall x, (k-1) \geq x \geq 0, s[x] = 1 \rightarrow \forall y, (k-1) \geq y \geq 0, y \neq x, s[y] = 0$. This constraint can be converted to SAT clauses by enumerating invalid values of the state variable. Therefore, SAT-BAG will avoid generating assertions from the invalid state values.

We observe that SAT-BAG can further filter out vacuous assertions and redundant propositions due to the usage of additional input constraints. Due to the input constraints, both the number of assertions and propositions decrease, and the average input space coverage increases. We don't show the total input space coverage here because SAT-BAG with constraints also achieves 100% input space coverage. Therefore, if we can provide input constraints to SAT-BAG, the generated assertions will be more meaningful and concise.

Table 3 shows the enhancement of SAT-BAG using word level features. The first column shows designs and target signals. The remaining columns show the same information as previous tables for SAT-BAG using two different kinds of features. SAT-BAG-bit and SAT-BAG-word represent

Table 2: Comparison on assertion generation of SAT-BAG with and without input constraints.

Circuit (output)	w/o constraints			w/i constraints		
	Asrt. Num.	Avg. prop.	Avg. cov. %	Asrt. Num.	Avg. prop.	Avg. cov. %
SWR-active_o	18	4.56	13.29	15	2.80	17.50
SWR-err_sqc	18	3.33	16.54	15	2.07	25.42
SWR-RST_tx_o	10	3.00	24.22	7	1.29	44.64

* Avg. prop. denotes the average number of propositions in antecedents of assertions.

Table 3: Comparison on bit and word level assertion generation of SAT-BAG.

Circuit (output)	SAT-BAG-bit				SAT-BAG-word			
	Asrt. Num	Avg. prop.	Avg. cov. %	Tot. Cov. %	Asrt. Num.	Avg. prop.	Avg. cov. %	Tot. Cov. %
SWR-active_o	18	4.56	13.29	100	12	3.08	14.01	100
SWR-err_sqc	18	3.33	16.54	100	7	2.14	17.97	100
SWR-RST_tx_o	10	3.00	24.22	100	4	1.50	25.00	100
ETHNET-StartIdle	994	8.80	0.50	100	6	2.33	26.39	100
ETHNET-ByteCntMax	17	1.88	47.06	100	2	1.00	50.00	100
ETHNET-ExcessiveDefer	16	1.88	46.88	100	3	1.33	49.99	100
ETHNET-WB_ACK_O	8	2.88	17.58	100	5	2.22	22.81	100
FPU-expa_ff	9	1.78	44.49	100	2	1.00	50.00	100
FPU-infa_f_r	24	1.92	47.92	100	2	1.00	50.00	100
FPU-exp_in_80	9	1.78	44.49	100	3	1.33	49.87	100

assertion generations using bit level features and word level features.

In this experiment, SAT-BAG-word generates few assertions but still achieves 100% input space coverage, which means it can generate more concise assertions, which have higher input space coverage, but also catch corner case assertions. Take **StartIdle** as an instance: the number of assertions can be reduced from 994 to 6 and the average numbers of propositions in antecedents decreases from 8.8 to 2.3. Furthermore, the average input space coverage increases from 0.5% to 26.39% drastically and the total input space coverage is still 100%. All these improvements result from using word level features, which can cover many bit level features. Thus, word level assertions can have high input space coverage and high readability.

5. CONCLUSION

SAT-BAG method was proposed to generate concise and high coverage assertions by using unit assumption technique as well as by considering input constraints and word level features.

6. REFERENCES

[1] Berkeley Logic Synthesis and Verification Group. *ABC: A system for sequential synthesis and verification.* http://www.eecs.berkeley.edu/~alanmi/abc/

[2] S. Bensalem, Y. Lakhnech, and H. Saidi, "Powerful Techniques for the Automatic Generation of Invariants," in *Proc. Computer Aided Verification Conf.*, pp. 323-335, 1996.

[3] C. W. Barrett, D. L. Dill, and A. Stump, "Checking Satisfiability of First-Order Formulas by Incremental Translation to SAT," in *Proc. Computer Aided Verification Conf.*, pp. 236-249, 2002.

[4] A. Biere, M. Heule, H. van Maaren, and T. Walsh (editors). "Handbook of Satisfiability," IOS Press, 2009.

[5] *BugScope*^{TM}: A Full-Chip Assertion Synthesis, http://www.nextopsoftware.com

[6] X. Cheng and M. S. Hsiao, "Simulation-directed invariant mining for software verification," in *Proc. of DATE.*, pp. 682ąV687, 2008.

[7] P.-H. Chang and L.-C. Wang, "Automatic assertion extraction via sequential data mining of simulation traces," in *Proc. Asian South Pacific Design Automation Conf.* pp. 607-612, 2010.

[8] N. Eén and N. Sörensson, "An Extensible SAT-solver," in *Proc. Int. Conf. on Theory and Applications of Satisfiability Testing*, pp. 502-518, 2003.

[9] H. Foster, D. Lacey, and A. Krolnik. "Assertion-Based Design," Norwell, MA, USA: Kluwer Academic Publishers, 2003.

[10] A. Gupta, K. L. McMillan and Z. Fu, "Automated Assumption Generation for Compositional Verification," in *Proc. Computer Aided Verification Conf.*, pp. 420-432, 2007.

[11] A. Gupta, Z. Yang, P. Ashar, and A. Gupta, "SAT-Based Image Computation with Application in Reachability Analysis," in *Proc. Formal Methods in Computer-Aided Design*, pp. 354-371, 2000.

[12] A. Hekmatpour and A. Salehi, "Block-based Schema-driven Assertion Generation for Functional Verification," in *Proc. Asian Test Symposium.*, pp. 34-39, 2005

[13] S. Hangal, N. Chandra, S. Narayanan, and S. Chakravorty, "IODINE: a tool to automatically infer dynamic invariants for hardware designs," in *Proc. Design Automation Conf.*, pp. 775ąV778, 2005.

[14] http://iwls.org/iwls2005/benchmarks.html

[15] W. Li, A. Forin, and S. A. Seshia, "Scalable specification mining for verification and diagnosis," in *Proc. Design Automation Conf.*, pp. 755ąV760, 2010.

[16] Mentor Graphics Inc. "Assertion-Based Verification," http://www.mentor.com/products/fv/methodologies/abv/

[17] OpenCores website: opencores.org/

[18] M. T. Oliveira and A. J. Hu, "High-Level specification and automatic generation of IP interface monitors," in *Proc. Design Automation Conf.*, pp. 129-134, 2002.

[19] A. Pnueli, "The temporal logic of programs,"in *Proc. of FOCS*, pp. 46-57, 1977.

[20] K. Ravi, F. Somenzi, "High-Density Reachability Analysis," in *Proc. Int. Conf. Computer-Aided Design*, pp. 154-158, 1995.

[21] S. Sankaranarayanan, H. Sipma, and Z. Manna, "Non-linear loop invariant generation using Gröbner bases," in *Proc. symposium on Principles of programming languages*, pp 318-329, 2004.

[22] A. Tiwari, H. Ruess, and N. Shankar, "A technique for invariant generation," in *Proc. of TACAS.*, pp. 113-127, 2001.

[23] G. Tseitin, "On the complexity of derivation in propositional calculus," *Studies in Constructive Mathematics and Mathematical Logic*, pp. 466-483, 1970.

[24] S. Vijayaraghavan and M. Ramanathan, "A Practical Guide for SystemVerilog Assertions," Springer Science+Business Media, Inc., New York, 2005.

[25] S. Vasudevan, D. Sheridan, S. Patel, D. Tcheng, B. Tuohy, and D. Johnson, "Goldmine: Automatic assertion generation using data mining and static analysis," in *Proc. of DATE.*, pp. 545-548, 2010.

[26] L.-C. Wang, M. S. Abadir, and N. Krishnamurthy, "Automatic generation of assertions for formal verification of powerpc microprocessor arrays using symbolic trajectory evaluation," in *Proc. Design Automation Conf.*, pp. 534ąV537, 1998.

[27] C. Wang, H. Kim, and A. Gupta, "Hybrid cegar: combining variable hiding and predicate abstraction," in *Proc. Int. Conf. Computer-Aided Design.* pp. 310-317, 2007.

Efficient Transistor-Level Design of CMOS Gates

[1]Vinicius N. Possani, [2]Vinicius Callegaro, [2]André I. Reis,
[2]Renato P. Ribas, [1]Felipe S. Marques, [1]Leomar S. da Rosa Junior
[1]Group of Architectures and Integrated Circuits, Federal University of Pelotas – UFPel, Pelotas, Brazil
[2]Institute of Informatics, Federal University of Rio Grande do Sul – UFRGS, Porto Alegre, Brazil
{vnpossani, felipem, leomarjr}@inf.ufpel.edu.br {vcallegaro, andreis, rpribas}@inf.ufrgs.br

ABSTRACT
The transistor arrangement optimization is an effective possibility to improve VLSI design, especially when generating CMOS logic gates to be inserted in standard cell libraries. This paper addresses this issue and presents a new methodology to generate efficient transistor networks. Starting from an input ISOP, the proposed method is capable to deliver series-parallel and non-series-parallel arrangements with reduced transistor count. By applying the proposed approach, it is possible to achieve optimized transistor arrangements since greedy choices are avoided during part of the generation process. The performed experiments have demonstrated the efficiency of this methodology when comparing to other available techniques.

Categories and Subject Descriptors
B.6.3 [**Logic Design**]: Design Aids – *Automatic synthesis, optimization, switching theory.*

General Terms
Algorithms, design, theory.

Keywords
Logic synthesis, switch network, transistor network, logic gate, digital design, VLSI design, CMOS, EDA.

1. INTRODUCTION
In current VLSI design, the total number of transistors necessary to implement logic gates is strongly related to the signal delay propagation, power consumption and area of integrated circuits (ICs) [1-3]. Transistor netlists are of special interest when designing standard cell libraries [4] or custom gates for IC design improvement [5]. To increase design quality in full-custom methodology, a handcraft generation of transistor netlists for each functional block may be performed. However, this is an extremely time-consuming task for larger ICs, making the adoption of such strategy prohibitive. Thus, it becomes crucial to have available efficient algorithms to automatically generate optimized transistor arrangements.

It is important to notice that CMOS gates are made of two logically complementary networks, one PMOS network for pull-up and one NMOS network for pull-down, as illustrated in Fig. 1. The remainder of this paper aims the generation of individual networks that can be used for either pull-up or pull down planes.

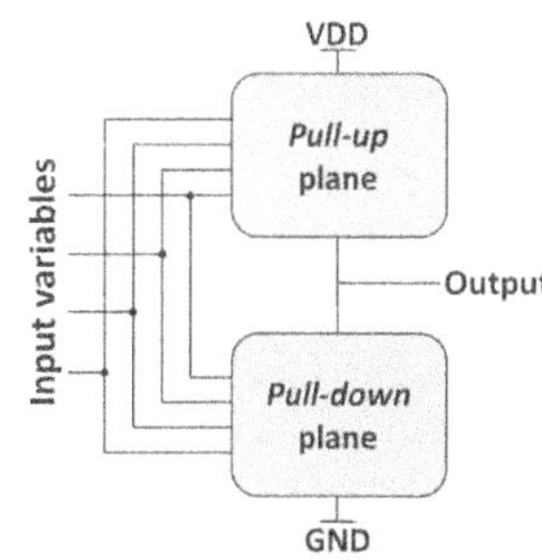

Figure 1. CMOS gate composed by two complementary networks.

In the last decades, several methods have been proposed to generate and optimize transistor networks. The most traditional solutions are based on algebraic and Boolean factorization [6-8]. In a Boolean expression, every instance of a variable is called *literal*, and a product of literals is formally called *cube*. The factorization process manipulates a Boolean expression in order to reduce the number of literals necessary to represent a function. Afterwards, the factored expression is directly translated to a transistor (switch) network. In this case, only series and parallel (SP) arrangements are obtained related, respectively, to AND and OR operations present in Boolean expressions.

Alternative methods to generate transistor networks are based on graph optimizations, where a Boolean expression is translated to a graph data representation. This graph can be optimized by edges sharing [9-11] or can be gradually composed from an input expression [12]. In some cases, these techniques are able to deliver better results than factorization based methods if non-series-parallel (NSP) arrangements are able to be found during the graph manipulation process, so reducing the total switch count. Such optimization obtained exploiting NSP topologies is due to the large sharing of logic paths that represent cubes of a function, overcoming SP arrangements.

It is worth to notice that existing techniques apply usually greedy strategies and deliver satisfactory results. However, to the best of our knowledge, no one of these methods can achieve the best results even for a set of simple functions when considering only the literal count metric [13]. We are taking into account functions that respect the minimum decision chain related to transistors in stack, being suitable for current CMOS technologies [14]. For instance, consider the following equation that represents a function from this set:

$$f = a.b + a.c + a.d + b.c.d \quad (1)$$

Available techniques [6-11] are not able to deliver the optimal topology in terms of switch count, comprising five switches, as illustrated in Fig. 2(a). The greedy strategies implemented in such techniques tend to share the literal ***a*** in the first step of the optimization process, becoming difficult to deliver the optimal solution. The methods presented in [6-11] find the solution depicted in Fig. 2(b), all of them resulting in a network with seven transistors.

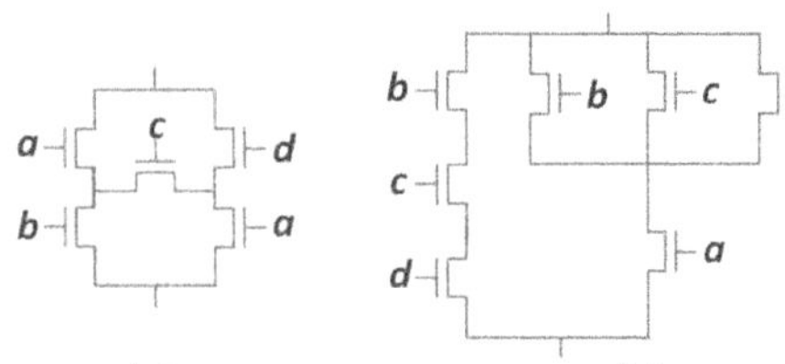

(a) (b)

Figure 2. Transistor networks corresponding to Equation (1): (a) optimal solution, (b) SP solution.

This paper proposes a new graph-based method able to generate optimized transistor networks. Our approach presents a structural algorithm to avoid greedy strategies during part of the generation process. NSP arrangements may be achieved, which leads to a reduction in the total switch count. Different from the approach presented in [11], this new method delivers the networks not only applying transistor sharing, but also considering topological information during the optimization. Finally, this paper improves the previous method described in [15], in which only the NSP kernel concept was effectively introduced.

The remaining of this paper is organized as follows. Section 2 provides some essential concepts for a better understanding of our approach. Section 3 introduces the novel method. Section 4 presents a detailed discussion about some case of studies and the obtained results. Finally, conclusions are outlined in Section 5.

2. PRELIMINARIES

In this section some basic concepts are briefly reviewed for a better understanding of this paper.

Series-parallel (SP) network – When every switch or cluster of the network is associated in series or in parallel with another switch or cluster. A network is SP if and only if there is no embedded network in a 'bridge' configuration [16]. Fig. 3(a) illustrates this kind of arrangement, obtained from the following equation:

$$f = a.b + a.c.e + d.e + b.c.d \qquad (2)$$

Non-series-parallel (NSP) network – A network with an embedded network containing 'bridges' (or NSP) associations [16]. The NSP network illustrated in Fig. 3(b) represents the logic function expressed by Equation (2).

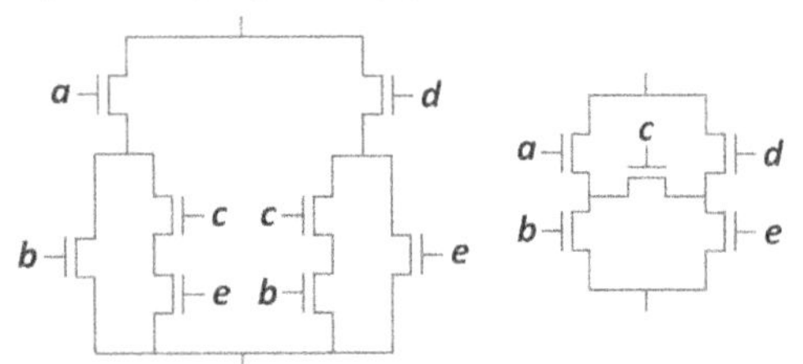

(a) (b)

Figure 3: Two possible transistor networks related to Equation (2): (a) SP network from exact factorization, and (b) NSP possible solution.

Irredundant sum-of-products (ISOP) expression – An ISOP is a sum-of-products (SOP) where each product is a prime implicant (PI) and no PI can be removed without changing the function represented by the Boolean expression [17].

3. SYNTHESIS METHODOLOGY

The proposed method starts from an ISOP, and tries to combine the cubes of a function to build NSP and SP kernels. These kernels are illustrated in Fig. 4(a) and in Fig. 6(a), respectively. Therefore, the method is divided in four well defined steps. The first three steps aim to build the kernels. The fourth one tries to merge the kernels found in order to deliver an optimized switch network. Thus, these steps are run in the following sequence:

(A) NSP Kernel Finder
(B) SP Kernel Finder
(C) Fake Cube Insertion
(D) Kernel Composition.

The cubes of the input ISOP are gradually taken into account according to the kernels that are found. Hence, the number of possible combinations to select the cubes and to build the kernels tends to decrease from the step *(A)* to the step *(D)*. On the other hand, if any of these routines does not find any kernel in the first steps, the following step receives all cubes still available at that moment. Depending on the Boolean function (expressed through an ISOP), the routines *(A)*, *(B)* and *(C)* may not find any kernel. Therefore, the switch network is generated during the step *(D)*, applying the edges sharing technique presented in [11]. This process can merge the kernels found and perform local optimizations on the arrangement.

3.1 NSP Kernel Finder

The NSP Kernel Finder algorithm proposed herein can be described as follows. For $|n=cubes(f)|$, four cubes are selected by combinations C_n^4. Afterwards, the algorithm builds a graph for each combination, as explained bellow.

We define a graph $G = (V,E)$ of a function H which is given by a SOP with exactly four cubes. The vertices in $V = \{v_1,v_2,v_3,v_4\}$ represent different cubes in H, and $|V|$ is the number of vertices in the set V. An edge $e = (v_i,v_j)$ in E exists if and only if at least one literal appears in both v_i and v_j. The operation $(v_i \cap v_j)$ represents common literals in both v_i and v_j vertices. Thus, an edge e formally exists if and only if:

$$(v_i \cap v_j) \neq \emptyset \qquad (3)$$

We define the label of e by using $label(e) = (lit(v_i) \cap lit(v_j))$, where $lit(v_i)$ are the literals in the vertex v_i. To ensure that the obtained graph is a valid NSP kernel two rules must be checked:

Rule 1 – Let E_{vi} be the set of edges that are connected to v_i. Each cube shares all its literals if the following equation is satisfied for all $v_i \in V$:

$$\prod_{i=1}^{|V|}\left(\bigcup_{j=1}^{|E_{vi}|} label(e_j) = v_i\right) \qquad (4)$$

Rule 2 – The obtained graph must be an isomorphic sub-graph to the graph template shown in Fig. 4(a). In this work, this structure is called *NSP kernel*.

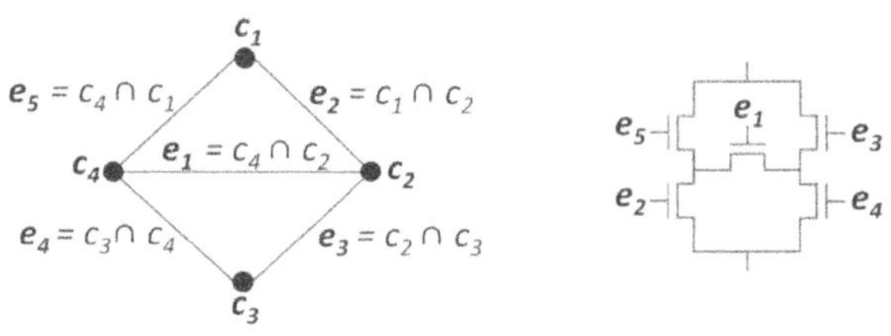

(a) (b)

Figure 4. Obtained network (b) from a NSP kernel (a).

The NSP kernels are mapped to switch networks applying a reordering in the label of the edges. Notice that in Fig. 4(a) the labels appear in a given order. However, in the switch network illustrated in Fig. 4(b) the labels are distributed in another order. It is done to represent the cubes of the Boolean function used as input, through the paths of the network. For instance, consider the NSP kernel depicted in Fig. 5(a). This kernel was obtained from Equation (2) and may be mapped directly to the transistor network illustrated in Fig. 5(b), through the edges reordering.

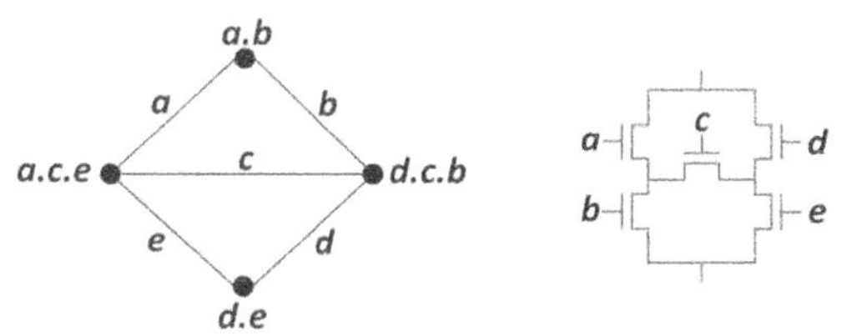

(a) (b)

Figure 5. NSP kernel (a) derived from Equation (2) and the resultant transistor network (b).

3.2 SP Kernel Finder

The SP Kernel Finder algorithm also selects the cubes by combinations C_n^4. Afterwards, for each combination, the algorithm builds a graph $G = (V,E)$ of a function H, which is given by a SOP with exactly four cubes, using the same definition and restrictions described in the NSP Kernel Finder routine. However, in this case, the second rule applied over the graph must be replaced by this other one:

Rule 3 – The obtained graph must be an isomorphic sub-graph to the graph template illustrated in Fig. 6(a). In this work this structure is called *SP kernel*.

Similarly to the step *(A)*, the algorithm of the step *(B)* must apply some transformations over the graph to map each found SP kernel to a switch network. Therefore, the first step consists in merging the rounded vertices of the template shown in Fig. 6(a) in a single vertex, as shown in Fig. 6(b). Afterwards, the edges reordering routine is applied over the graph illustrated in Fig. 6(b), resulting in the switch network illustrated in Fig. 6(c).

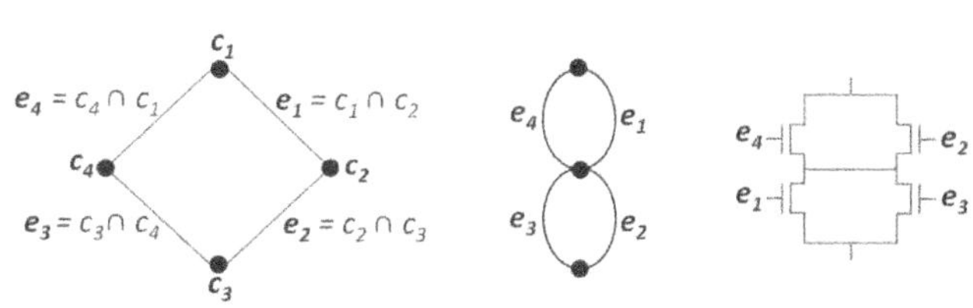

(a) (b) (c)

Figure 6. Vertices merging (b) and edge reordering process (c) on a SP kernel (a).

To a better understanding of such process, consider the SP kernel illustrated in Fig. 7(a), which was obtained from the following equation:

$$f = a.c + a.d + b.c + b.d \tag{5}$$

By applying the vertices merging illustrated in Fig. 7(b) and the edges reordering routine, this kernel may be mapped to the switch network illustrated in Fig. 7(c). Notice that the vertices merging and the edges reordering routines are necessary to implement the sharing between the paths of the network that represent the cubes from Equation (5).

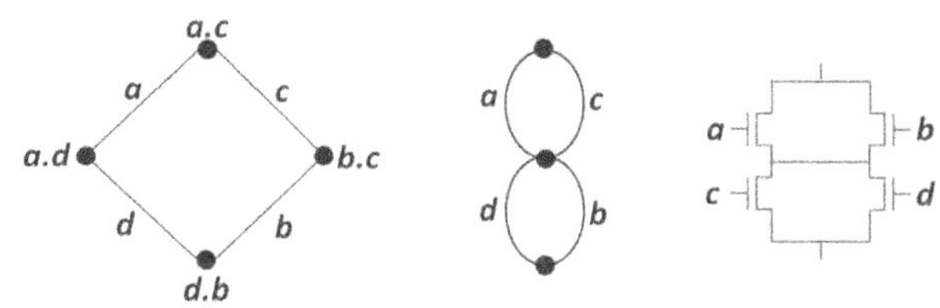

(a) (b) (c)

Figure 7. SP kernel (a) derived from Equation (5) and network (c) obtained after applying vertices merging (b) and edges reordering routine.

3.3 Fake Cube Insertion

In this step, the cubes are then selected by combinations C_n^3, and the algorithm tries to complement the graph using fake cubes. A fake cube is a cube that leads to redundant sub-functions. For instance, let f be the original function, and let f' be the function represented by a fake cube. The fake cube is valid only if $f' < f$ ($f' + f = f$). This means that the cube function f' is covered by another cube from the function f.

We define a graph $G=(V,E)$ of a function H, which is provided in a SOP form with exactly three cubes. The vertices in $V=\{v_1,v_2,v_3\}$ represent the different cubes in H, so $|V|=3$. The conditions to insert a labeled edge in the graph are the same of the NSP Kernel Finder routine. In order to complete the graph G, a new vertex v_z is inserted into the set V as follows. Let E_{vi} be the set of edges connected to the vertex v_i, and $|E_{vi}|$ the number of edges in this set. The literals of the new vertex are defined as:

$$lit(v_z) = \prod_{i=1}^{|V|} \left(lit(v_i) - \bigcup_{j=1}^{|E_{vi}|} label(e_j) \right) \tag{6}$$

Therefore, the Equation (3) is applied to the vertex v_z in order to insert the new edges E_{vz} in E. This process is illustrated in Fig. 8 and in Fig. 9 that show the graph obtained from the following equation, where some literals of the vertices were not shared:

$$f = !a.!d + !a.!b.c.d + !b.!c.!d \tag{7}$$

In Fig. 8, the rounded literals are not shared through any edge. This way, they can be merged into a single vertex, leading to the graph presented in Fig. 9. This graph is an NSP kernel with a non-sensitized cube, i.e., a cube that has at least one variable in both polarities, such as *!c.c.d*.

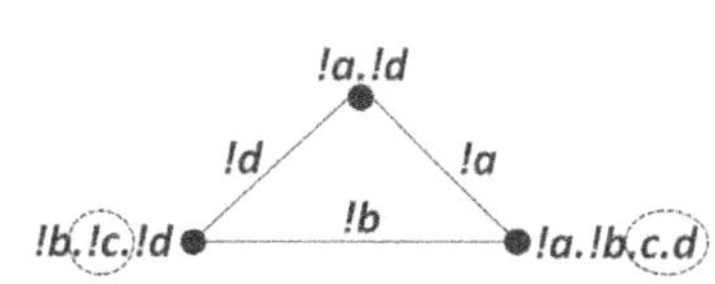

Figure 8. NSP kernel with a fake cube: intermediate graph to determine the fake cube.

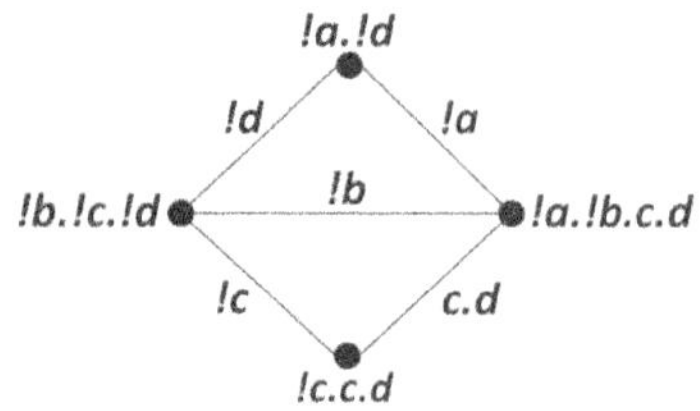

Figure 9. NSP kernel with a fake cube: the obtained NSP kernel.

The fake cube can be also introduced when the algorithm performs combinations C_n^4. It is easily found by the labels of the edges. An example of this situation is illustrated in Fig. 10 that shows a redundant switch ***a*** in the NSP kernel of the Equation (1). This kernel may be mapped, through the edges reordering, to the switch network illustrated in the Fig. 2(a), which other methods are not able to deliver.

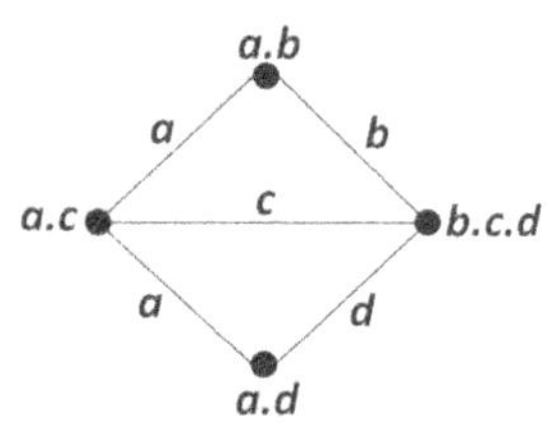

Figure 10. NSP kernel with a redundant switch.

3.4 Kernel Composition

It is important to notice that, depending on the input ISOP, multiple kernels can be found. Moreover, some cubes from the input ISOP cannot compose any kernel. Thus, five possible cases may occur when generating the transistor networks: (i) a network can be composed by just one kernel; (ii) a network can be composed by a kernel, and one or more cubes that are implemented as parallel transistor associations to this kernel; (iii) a network can be composed by multiple kernels in a parallel association; (iv) a network can be composed by multiple kernels, and one or more cubes that are implemented as parallel transistor associations; (v) there is no kernels and the network is implemented only with SP arrangements through the edges sharing algorithm. For each of these five cases, such topological composition is done gradually until achieving a network that is logically equivalent to the input Boolean function. During the composition process, the edges sharing procedure is applied to the network in order to eliminate redundant switches [11]. Such strategy allows a reduction in the total number of switches.

As an example of network generation with multiple kernels, let us consider the following equation as the input ISOP of our algorithm:

$$f = !a.!b.!c.!d + !a.b.d + !a.b.c + a.!b.d + \\ + a.!b.c + a.b.!c.!d + b.c.d \qquad (8)$$

Due to the characteristics of Equation (8), the step *(A)* does not find any NSP kernel. However, the SP kernel presented in Fig. 11(a) is found by applying the step *(B)*. After that, the step *(C)* can deliver a NSP kernel with a fake cube (rounded by dashed lines), as illustrated in Fig. 12(a). The switch networks obtained from each kernel are shown in Fig. 11(b) and in Fig. 12(b), respectively.

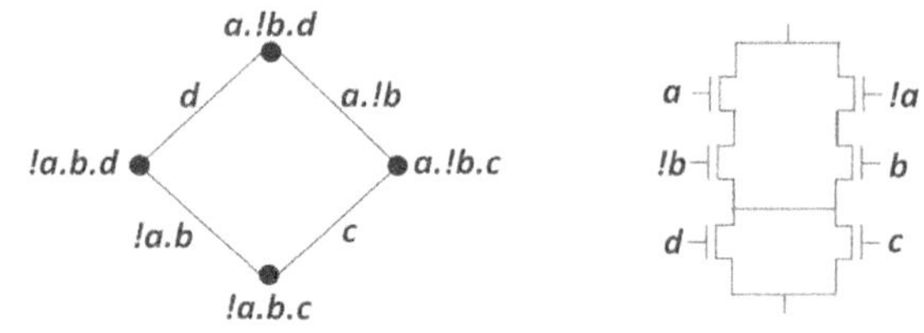

(a) (b)

Figure 11. SP kernel (a) and SP network (b) obtained from Equation (8).

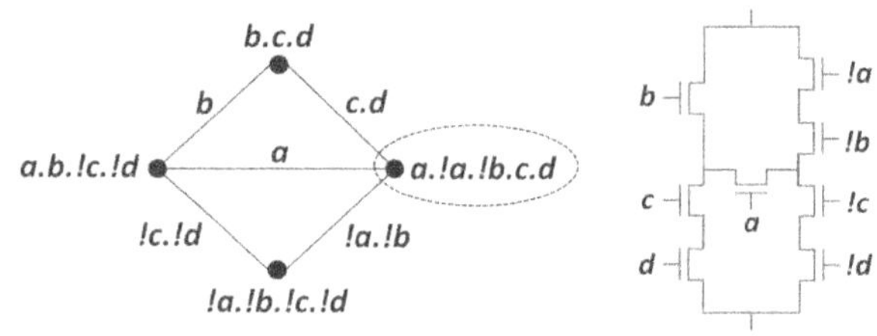

(a) (b)

Figure 12. NSP kernel (a) and NSP network (b) obtained from Equation (8).

In order to reach the target function, these networks are associated in parallel, as illustrated in Fig. 13(a). Notice that the resultant network has some redundant switches. Therefore, the edges sharing technique is applied to remove these redundancies, resulting in the network depicted in Fig. 13(b). This process is repeated until there are no more equivalent switches in the network. Finally, the minimal solution for the input ISOP can be implemented by the arrangement shown in Fig. 13(c).

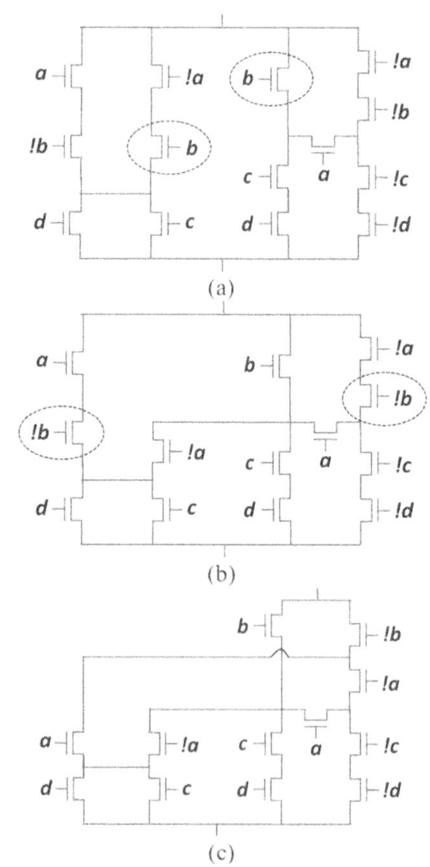

Figure 13. Parallel association of networks obtained from kernels (a), intermediate network (b) and final network (c).

4. EXPERIMENTAL RESULTS

We have implemented the proposed method in Java platform and comprehensively applied it over different set of logic functions. In order to provide a comparison of our methodology to other available solutions described in the literature, three different benchmarks were considered: (i) the set of 4-input P-class logic functions; (ii) a set of 53 logic functions that we have been manually generated in order to achieve the minimal solutions in terms of switch count; (iii) and a complex logic function with 11 variables that can be implemented with only 11 switches.

The first experiment was performed over the set of 4-input P-class logic functions, composed by 3982 functions. This set of functions was chosen because it contains simple functions that are more likely to be used as logic gates in real designs. We have generated gates for each function of this set, and compared them to other methods available in the literature. Table 1 shows the obtained results when considering the total switch count to compute the logic gates. These results also summarize the inverters needed to implement the gates. Inverters are needed to generate the complementary signal for input variables that appears in both polarities. As presented in Table 1, our method compares favorably to other existing approaches.

Table 1. Total switch count to compute logic gates for the set of 4-input p-class logic functions.

	[8]	[10]	[11]	[12]	Proposed Method
Total number of switches	102,668	103,049	96,804	97,174	**95,625**

The second experiment was performed over a set of 53 logic functions [13]. For each function of this set, we have generated a handmade optimal switch network and compared it to the networks generated by methods presented in [8,10-12] and our approach. The results are summarized in Table 2. The proposed method was able to decrease the transistor count, coming closer to the optimal solution. From 53 logic functions, only in one function our approach was not able to achieve the optimal solution.

Table 2. Total switch count for a set of 53 handmade optimal switch networks.

	Optimal	[8]	[10]	[11]	[12]	Proposed Method
Total number of switches	**356**	487	503	516	543	**359**

Fig. 14 shows the distribution of gains and losses of our approach compared to other solutions. As can be seen, the proposed method is able to reduce up to 8 transistors in some generated networks from the set of 53 logic functions. In general, the gains remain around 2 or 3 switches per network. Also, it is important to notice that only in one logic function our method delivers networks with an increasing in the switch count. In this case, the reason for that is the bad choice when merging edges during the multiple kernels merging process. An efficient heuristic to choose the edges that may be firstly merged would help to improve the results.

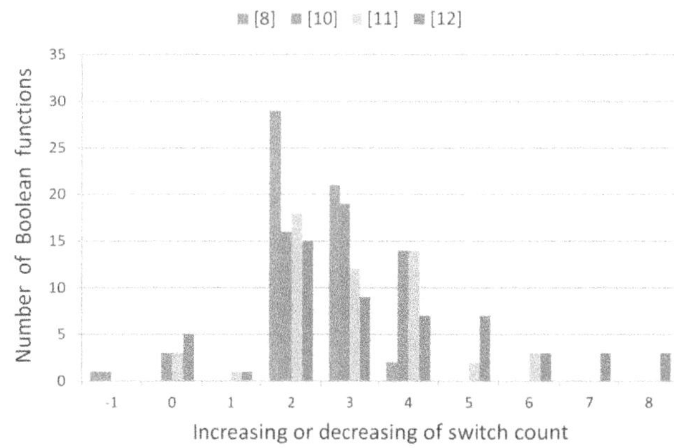

Figure 14. Number of functions with increasing or decreasing transistor count when comparing to other solutions.

Finally, we have also tested our method with more complex functions. As an example, consider the Boolean function expressed by the following ISOP composed by 99 literals:

$$z = a.i + c.k + b.d.i + b.f.k + a.g.j + b.e.j + c.h.j + c.f.d.i + b.e.g.i + c.h.g.i + a.d.f.k + a.g.h.k + b.e.h.k + b.d.g.j + a.d.e.j + c.f.e.j + b.f.h.j + c.h.e.d.i + c.f.e.g.i + b.f.h.g.i + a.g.e.f.k + b.d.g.h.k + a.d.e.h.k + c.f.d.g.j + a.d.f.h.j \quad (9)$$

For such logic function, our method delivers a network composed by multiple NSP kernels. After merging all kernels and applying the edges sharing routine, it finds the switch network presented in Fig. 15. This is the optimal solution, with only 11 switches. Table 3 shows the switch count obtained by other methods. As can be seen, our approach is able to deliver a network with at least 50% of reduction in transistor count.

Table 3. Number of switches, delivered by each method, to implement the logic function described in Equation (9).

	[8]	[10]	[11]	[12]	Proposed Method
Total number of switches	31	31	25	22	**11**

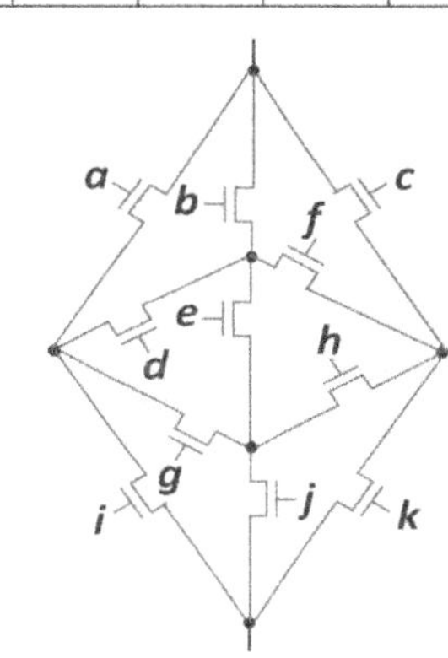

Figure 15. Optimal switch network provided by our method for Equation (9).

The total execution time to generate all the networks, described in the results, was less than 1.2 seconds when running in an Intel Core i5 at 2.8GHz with 4Gb of RAM. It demonstrates the feasibility of the proposed method to increase design quality when generating digital CMOS gates.

5. CONCLUSIONS

This paper described a new graph-based method to generate optimized switch networks. The method is able to generate arrangements more general than the usual series-parallel associations. It results in a minimization of switch count when compared to previously published techniques. It is known that a transistor count reduction in a logic gate may lead to better results in terms of signal delay propagation and power consumption. However, these associated gains were not explicitly investigated in this work, and they are left as a future work at gate, library and circuit design level.

6. ACKNOWLEDGMENT

Research partially supported by Brazilian funding agencies CAPES, CNPq and FAPERGS.

7. REFERENCES

[1] Lai, Y., Jiang, Y. and Chu, H. BDD Decomposition for Mixed CMOS/PTL Logic Circuit Synthesis, *In: IEEE Int. Symp. on Circuits and Systems (ISCAS 2005)*, p. 5649-5652.

[2] Al-Hertani, H., Al-Khalili, D. and Rozon, C. Accurate total static leakage current estimation in transistor stacks. *In Proc. Int. Conf. on Computer Systems and Applications*, 2006, pp. 262-65.

[3] Thorp, T. J., Yee, G. S. and Sechen, C. M. Design and synthesis of dynamic circuits. *IEEE Trans. on VLSI Systems*, v. 11, n. 1, p. 141-149, Feb. 2003.

[4] Reis, A. I. and Anderson, O. C. *Library Sizing*. US Patent number: 8015517, Filing date: Jun 5, 2009, Issue date: Sep 6, 2011, Application number: 12/479,603.

[5] Roy, R., Bhattacharya, D. and Boppana, V. Transistor-level optimization of digital designs with flex cells. *IEEE Trans. on Computers* , vol.38, no.2, pp. 53- 61, Feb. 2005.

[6] Golumbic, M. C., Mintz, A. and Rotics, U. An improvement on the complexity of factoring read-once Boolean functions. *Discrete Appl. Math*, 2008, Vol. 156, n. 10, p. 1633-1636.

[7] Sentovich, E., et al. *SIS: A system for sequential circuit synthesis*. Technical Report No. UCB/ERL M92/41, EECS Department, University of California, Berkeley, 1992.

[8] Martins, M. G. A., Da Rosa Junior, L. S., Rasmussen, A., Ribas, R. P. and Reis, A. I. Boolean Factoring with Multi-Objective Goals. *In: IEEE Int. Conf. on Computer Design (ICCD 2010)*, p. 229-234.

[9] Zhu, J. and Abd-El-Barr, M. On the optimization of MOS circuits. *IEEE Trans. on Circuits and Systems: Fundamental Theory and Applications, Theory Appl.*, vol. 40, no. 6, pp. 412–422, 1993.

[10] Da Rosa Junior, L. S., Marques, F. S., Schneider, F., Ribas, R. P. and Reis, A. I. A Comparative Study of CMOS Gates with Minimum Transistor Stacks. *In: 20th Symp. on Integrated Circuits and Systems Design (SBCCI 2007)*, p. 93-98.

[11] Possani, V. N., Souza, R. S., Domingues Junior, J. S., Agostini, L. V., Marques, F. S. and Da Rosa Junior, L. S. Optimizing Transistor Networks Using a Graph-Based Technique. *Journal of Analog Integrated Circuits and Signal Processing (ALOG)*, May 2012, v. 73, p. 841–850, 2012.

[12] Kagaris, D. and Haniotakis, T. A Methodology for Transistor-Efficient Supergate Design. *IEEE Trans. on Very Large Scale Integration (VLSI) Systems*, p. 488-492, 2007.

[13] *Catalog of 53 Handmade Optimal Switch Networks*, 2012. Retrieved October 15, 2012, from Logic Circuit Synthesis Labs, Federal University of Rio Grande do Sul: http://www.inf.ufrgs.br/logics/docman/53_NSP_Catalog.pdf

[14] Martins, M. G. A., Callegaro, V., Ribas, R. P. and Reis, A. I. Efficient method to compute minimum decision chains of Boolean functions. *In: 21st Great Lakes Symposium on VLSI (GLSVLSI 2011)*, p. 419-422.

[15] Possani, V. N., Callegaro, V., Reis, A. I., Ribas, R. P., Marques, F. S. and Da Rosa Junior, L. S. NSP Kernel Finder - A Methodology to Find and to Build Non-Series-Parallel Transistor Arrangements. *In: 25th Symp. on Integrated Circuits and Systems Design (SBCCI 2012)*, p. 1-6.

[16] Da Rosa Junior, L. S., Schneider, F., Ribas, R. P. and Reis, A. I. Switch Level Optimization of Digital CMOS Gate Networks. *In: 10th IEEE Int. Symp. on Quality Electronic Design (ISQED 2009)*, p. 324-329.

[17] Sasao, T. and Butler, J. T. Worst and Best Irredundant Sum-of-Products Expressions. *IEEE Trans. on Computers*, vol. 50, no. 9, 2001.

Electronics for a Greener Planet

Christian Piguet
CSEM
Neuchâtel
Switzerland
+41 32 720 5227
Christian.piguet-at-csem.ch

ABSTRACT

Our goal must be the usage of ICT technologies to reduce significantly the worldwide energy consumption and carbon emission. Predictions about energy consumption in 2050 show that nearly 100% renewable energy is possible. On one hand, ICT devices themselves can be much more energy efficient in their usage but also in their fabrication. However, a more significant role for these systems will be to reduce the pressure over our planet in domains like transportation, agriculture, lighting, home control and health monitoring. Nevertheless, people behavioral changes are also definitely required for achieving sustainability. It is also obvious that engineers are the only persons capable of improving planet health by proposing intelligent ICT systems, but today they are not credited for this; it is therefore time for changing the image of engineers.

Categories and Subject Descriptors

K.6.1 [**Management of Computing and Information Systems** Project and People Management

General Terms

Management, Performance, Design, Security, Human Factors.

Keywords

Energy consumption, Oil peak, Autonomous Devices, Renewabl Energy, Grey energy, PC reuse, datacenters, Cloud Computin; Mobile, Chip production, Photovoltaic, Image of engineer Wireless sensor networks, Sustainable agriculture.

1. INTRODUCTION

We have to ask ourselves what we can do in ICT (Information & Communication Technologies) to reduce the worldwide energy consumption and carbon emission for saving our planet and obviously if we have a chance to be successful.

For ICT, one has to focus on electricity consumption which is today 15 to 18% of global energy worldwide, even 25% in advanced countries and predicted to be 50 % in the future. One can say that ICT consumes about 2% of total energy, but is predicted to be 10% in the future [1]. This is clearly not sustainable, and it turns out that one has to move to renewable energy. Therefore, two actions have to be taken urgently: a dramatic reduction of electricity consumption and the use of renewable energy by using only autonomous devices. It is a key point, as it is a big threat for societies to depend too much on imported petrol and gas or energy dependence in general. More you consume energy, more you are dependent and weak. So everything autonomous is clearly a way to go.

WWF and EcoFys [2], in a 2011 report, predict that 95% renewable energy is possible in 2050 (Fig. 1) while having 15% lower energy demand compared to 2005. It is sure a very big change, as we would have no more nuclear, oil, gas and coal energies! One has to move to only bio energy, hydropower, geothermic, solar, wind and waves! Even economists predict that oil will be gone in 50 years.

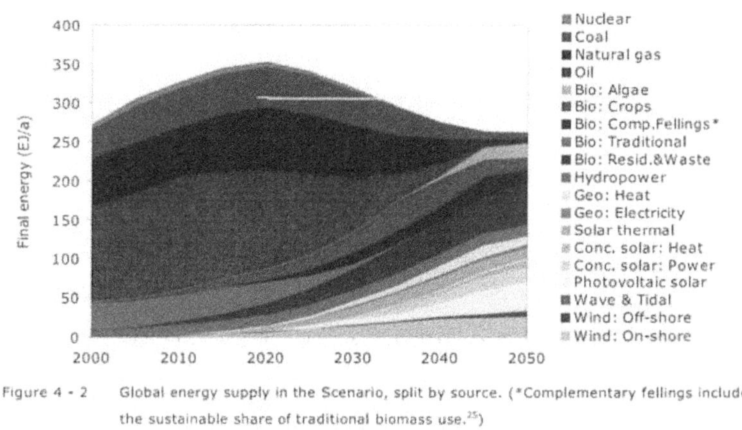

Figure 1: 95% renewable energy 2050 (WWF, EcoFys)

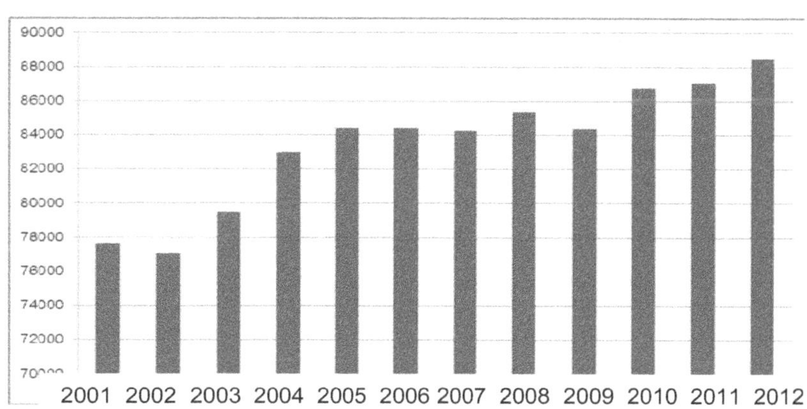

Figure 2: Oil Peak (Thousands of barrels per day)

Oil peak was reached apparently between 2005 and 2009 with 84 millions of barrels per day, but one can see that we have a little bit more (88 millions of barrels per day) in 2010-2012 (Fig. 2). However, we can be convinced that we are close to the oil peak.

Figure 3 shows the increase of electricity production until 2050 (it is not decreased!!), but nuclear, oil, gas and coal have disappeared! Only bio, geo, hydropower, solar, wind and waves would produce enough electricity.

It is therefore highly recommended to change our behaviors, like:

- Buy and produce only clean energy
- Buy only energy efficient equipment and devices
- Invest only in renewable energy, energy-efficient devices and buildings
- Stop food waste. Choose food which is produces locally. Eat less meat.
- Reduce, re-use and recycle all kind of materials. Avoid what is not needed.
- Greater use of public transport and to reduce mobility
- Action plans to promote research and development in renewable energy
- Buy things with little or no packaging
- Buy local and seasonal products

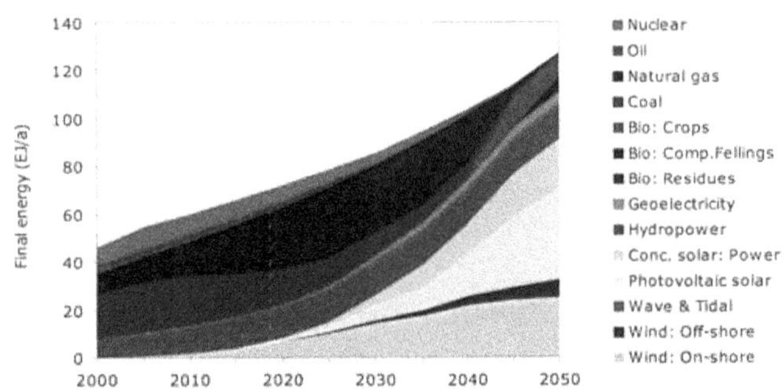

Figure 4 - 7 Global power supply in the Energy Scenario.

Figure 3: Electricity Production

Another very important point is the amount of grey energy (or fabrication energy) that is consumed during the fabrication of devices. Does everybody know which grey energy is used for fabricating a car or a house? Sure not. For instance, a car requires 400'000 liters of water and 30 tons of raw materials. A house requires 30 to 50 years of its annual energy consumption. It turns out that fabrication energy is more and more important, as we try to reduce the usage energy consumption. We will see some examples in the following: PC fabrication, tablets, integrated circuits, photovoltaic panels. In products bought every day for living, it is assumed that 63% of the energy is fabrication energy. To this, we have to add the energy for eliminating the product!! So it is not sure that it is good to buy a Prius versus a tiny car!

One objection to a reduction of the number of bought new devices is that the market will slow down and that the impact of the growth will be detrimental. Sure that some sectors like nuclear and oil will significantly be decreased in 2050, but new high profitable sectors like renewable energies will grow significantly. Various scenarios [13] show that the number of jobs in the energy sector could grow from a base scenario at 8.7 millions of jobs to 10.6 or even 11.9 millions of jobs for scenarios with more and more renewable energy and less fossil and nuclear energies.

2. PC, DATACENTERS, GAMES, MOBILE

2.1 PC reuse

Today PC consume worldwide a lot of power, approximately 150 billions of kWh/year. However, grey energy is not negligible. To fabricate one PC requires 240 kg of fossil energy and 1500 liters of water. The grey energy is 6 to 7 Gjoules. The usage energy is much smaller! It is about 0.8 to 1.5 Gjoules for your PC at home and 1.4 to 2.7 Gjoules for your office PC. One can think to recycle a PC of three years, and the energy cost is about 1.7 Gjoules. Consequently, for energy savings, it is much better to recycle your PC after 3 years than buying a new one [8]. Perhaps you would change the processor, but the keyboard, the display, the memory chips, etc... could be recycled without any problem. The good news is that today about 40% of PC are reused and recycled by some specialized companies. But there are 150 millions obsolete PCs currently sitting idle in warehouses and storerooms.

2.2 Data Centers

Data centers including big servers represent today 1.2% of worldwide electrical energy and are expected to grow in a non-sustainable way at 15%/year. A Google search [3] is estimated to consume 0.3 watt*hour (Google) to 2 watts*hours (even 8 watts*hours). The number of Google searches is estimated to 0.5 to 3 billion per day. So we could have about 1 to 6 millions of kWh per day (or 1.5 to 8 GJoules per year) just for Google search.

For data centers, the cost is more and more in energy cost and cooling becomes very expensive (water cooling for instance). One can think to recycle the dissipated heat for houses (A 10 MW datacenter would be sufficient for 700 houses).

A clean energy index measuring how much renewable energy is used for various data centers has been published by Greenpeace [9]. Yahoo is first with 56.4% renewable energy, DELL with 56.3%. Google 39.4%, facebook 36.4%, twitter 21.3 %, Apple 15.3%, Microsoft 13.9%, Amazon 13.5 % and IBM 12.1%. But many of these companies, to improve their image, are constructing new datacenters with 85 to 100% renewable energy (Amazon, Google, Yahoo, Microsoft). Looking at states and countries, a same analysis can be achieved: Washington state is first with 77% renewable energy, followed by Oregon 65%, Sweden 54%, California 44%. But France is only 14%, Germany too, UK 5%, North Carolina 3.6% and Illinois 1%. North Carolina has 61% coal and 31% nuclear, the kWh cost is low, and many datacenters (Apple, Google) are in North Carolina!! Many data centers start to have solar panels (Fig. 4), but it is a small contribution to their consumed energy (for instance, 5.6% for a 80 MW datacenter in Phoenix).

Cloud Computing is often associated to green computing. But is this really true? To use computing resources delivered over a network called the cloud seems to be reasonable. Many users are sharing servers and applications that could work much more efficiently (80% of the time) compared to local datacenters that have many idle machines, as they have been designed for a peak computation. This point is important, as even with a 5% workload, servers consume 50% of their peak power. In addition, local datacenters are using most of the time quite old machines. Also cooling is more efficient for big datacenters. Economies of scale are obtained with cloud computing.

However, looking carefully to datacenters for cloud computing, one can see that a lot of power is wasted. Servers are working even if they are nothing to do, up to 50% of the consumed power is for cooling, and finally, cloud computing encourages more compute consumption and therefore more energy consumption. One can say that if you use 80% of the energy per unit, but you have 1000% more capacity, at the end of the day, you are consuming much more energy. To have your files, photographs, music and movie son the hard disk of your PC is surely less energy consuming than to have them in the cloud. An on-site server room that is run with energy efficiency best practices may be a greener alternative to a 'brown cloud. Finally, although cloud computing is often assumed to be a form of "green computing", there is no published study to substantiate this assumption.

Figure 4: Solar Panels on a Datacenter

2.3 Games, Apps, Portable Devices

A lot of energy (5 to 25%) is wasted in stand-by modes (TV, PC, games, printers) and it is estimated that 35 TWh/year could be saved. If other wastes of energy are considered, such useless usage of ICT devices, one can say that about 80% of energy is wasted. It turns out that an "avatar" in Second Life consumes as much as a Brazilian person. In portable phones, free Apps are known to exhaust the battery very rapidly for providing useless GPS-based publicity (70 to 90% of energy is used for publicity).

The energy consumed by portable devices (for instance 6 billion of portable phones) is very small compared to datacenters, but another threat is the waste of batteries. 160 millions of tons are produced per year. From this, only 35% are recycled. It is therefore mandatory to design autonomous devices like it is planned in many projects. In the next 10 years, it is planned to increase the scavenged energy from 100 uW/cm^2 to 10 mW/cm^2, while reducing computation and communication energy by a factor of 1000. Receiving the news on a tablet wirelessly compared to reading a newspaper on paper reduces the carbon footprint from 32 to 140 times [12]. If you communicate by teleconferences versus travelling by planes will reduce 200X the carbon footprint!

3. SUSTAINABLE CHIP PRODUCTION

3.1 Chip Production

Regarding high performance microprocessors, most of the consumed energy (3.8 GJoules for 2 years) is the usage energy. However, for Wireless Sensor Network (WSN) chips, one has a very different situation: the fabrication or grey energy is about 100 KJoules over a total of 180 KJoules.

Assuming a fabrication yield of 80%, the energy for producing a 1 cm^2 chip is 4.2 kWh, which is equivalent to 1.7 liters of petrol. For this chip (1 cm^2), 18 to 27 liters of water are required. The estimated number of fabricated chips per year worldwide is estimated to an equivalent of 14 billion of 1 cm^2 chips produced per year. It is based on the fact that we are producing per year about one billion of transistors per person on Earth, so with 7 billion people, one ends up with $7*10^{18}$ transistors produced per year. With a density of 0.5 billion of transistors per 1 cm^2 chip, we are producing an equivalent of 14 billion of 1 cm^2 chips produced per year. So it turns out that $24*10^9$ liters of petrol are required for chip industry (it is $1200*10^9$ for car industry). For water, it is 400 billion of liters required per year (20'000 billion for car industry). So the requirements for chip industry are 2% in petrol and 2% in water compared to car industry.

3.2 Solar Panels

Some people think that photovoltaic panels do not pay back the energy that was used for their fabrication. In fact, the energy payback is 1-3 years, depending on solar panel type and location. So during 30 years lifetime, PV will produce 10 to 30 times the energy that they originally used for their fabrication.

4. ICT FOR PLANET HEALTH

ICT technologies could be very useful to reduce the energy usage as well as the waste of water and other precious materials. In various domains, such as energy production and distribution, agriculture, health environment and transportation, engineers have to contribute to develop much better techniques. However, it seems that engineers are not seen as the guys that could change our world into a much better world.

4.1 Bad Image of Engineers

Today, our image of engineers is not good [10]. Consequently, there are less people in engineering classes, and except in India and China, still a very few girls. What a big threat! Engineers are the people that design new devices that all people worldwide are using today. In addition, they are the guys that will invent the new techniques to reduce energy and water waste [11]. To have only lawyers, managers and finance boys in our society is not sustainable: we definitely need engineers to create something to produce and to sell!

In U.S., if 750 engineers/year for 1 million people are finishing today their engineering studies, it will be less in the next years. In China and India, it is respectively 500 and 200, but growing. The number of women is 20% in USA, 17% in France, 10% in Japan, less than 10% in Switzerland. What is the problem? We have a big problem of image; people believe that engineering studies are difficult, quite far for femininity image and engineers not sensitive to social and environmental issues.

Some years ago, a questionnaire was addressed to 40'000 people with the following question: do engineers contribute to:

- Protection of environment : 83% NO
- Quality of life: 78 % NO
- Sensitive to social problems: 72% NO
- Contributions for saving lives: 86% NO

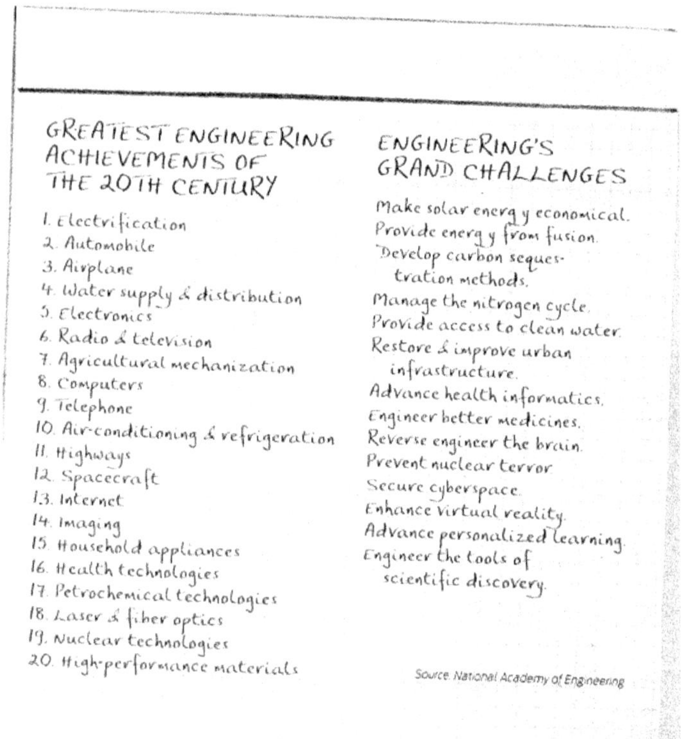

GREATEST ENGINEERING ACHIEVEMENTS OF THE 20TH CENTURY

1. Electrification
2. Automobile
3. Airplane
4. Water supply & distribution
5. Electronics
6. Radio & television
7. Agricultural mechanization
8. Computers
9. Telephone
10. Air-conditioning & refrigeration
11. Highways
12. Spacecraft
13. Internet
14. Imaging
15. Household appliances
16. Health technologies
17. Petrochemical technologies
18. Laser & fiber optics
19. Nuclear technologies
20. High-performance materials

ENGINEERING'S GRAND CHALLENGES

Make solar energy economical.
Provide energy from fusion.
Develop carbon sequestration methods.
Manage the nitrogen cycle.
Provide access to clean water.
Restore & improve urban infrastructure.
Advance health informatics.
Engineer better medicines.
Reverse engineer the brain.
Prevent nuclear terror.
Secure cyberspace.
Enhance virtual reality.
Advance personalized learning.
Engineer the tools of scientific discovery.

Source: National Academy of Engineering

Fig. 5: Big Achievements of Engineers

But it is completely wrong! Engineers are the people inventing many devices and medicines to save lives as well as many techniques for reducing energy and water consumption and carbon emission (Fig. 5). But we are not recognized for what we are doing! Sure there is also an old image of engineers inventing nuclear energy, Fukushima, and this does not help!

We are pretty sure that this bad image of engineers explains why less and less people want to start engineering studies (except in China and India) and why we have so few women. So please, engineers, you have to propagate the fact that engineers are working to combat climate change and to promote renewable energy. If engineers would be considered as capable of providing solutions to these very big challenges, sure that their image would be much better!

4.2 ICT for Saving Energy

ICT technologies and Wireless Sensor Networks can be used in lighting (Fig. 5), in agriculture and in many other domains such as health surveillance and buildings or plants monitoring. The lighting is estimated to 230 billion $ per year worldwide. With intelligent lighting, for which the light is provided only if there is somebody (but not a dog or a cat) moving along the street, 60% savings are achievable. Vision sensors in the street lights with recognition algorithms can do that already today. It means that we could save 38 TWh of electricity in Europe.

4.3 ICT for Agriculture

The largest thread today is the production of food. World food production is growing and can catch up with much of population growth. However, most of the productivity yield is based on availability of oil.... as agriculture consumes too much water, pesticides, fertilizers, all based on petrol. Today, without petrol, no more food!! And we are at the peak of oil production today (Fig. 2).

ICT technologies and more specifically Wireless Sensor Networks (WSN) could be used for monitoring agriculture, basically the water distribution with pumps working with petrol [5, 6], not to put too much inorganic fertilizers that fabricated with petrol [7] and not to waste too much pesticides that are also based on petrol. And we are at the peak of oil production today.

Only 2.5% of the water on earth is fresh water, the rest is salt water, and 2/3 of fresh water is contained in glaciers and snow cover. From the remaining 1/3 of fresh water, about 70% is consumed by agriculture. That is a quite huge amount of water, estimated at 5×10^{15} liters per year (5'000 km^3).

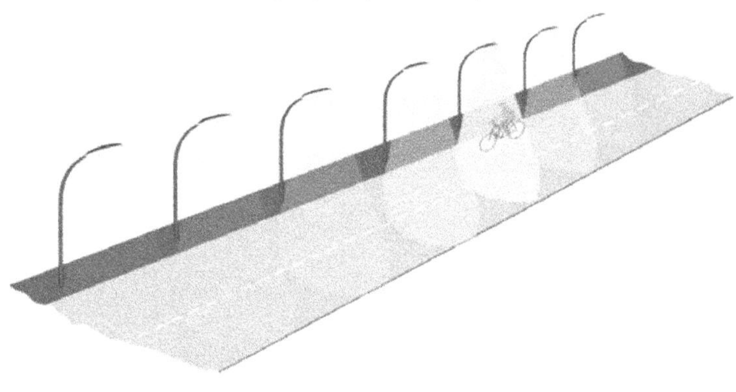

Figure 6: Intelligent Lighting

So by using ICT and WSN, it was shown that for instance water savings of up to 30-60% could be achieved. Such experience was conducted for a 1500 ha area in Spain. If this ratio could be saved everywhere, 30% to 60% of 5×10^{15} liters per year would represent a huge reduction in water consumption.

By the way, fresh water savings will be of the greatest importance in the next decades, due to more and more polluted water.

Regarding inorganic fertilizers, the estimation today is a use of 140×10^6 tons per year. Savings up to 45% are possible while using WSN to monitor the quantity of fertilizers which is just required. In U.S., inorganic fertilization consumes about a third of total energy input for crop production. According to another reference, fertilizer contributes up to 45% of the commercial energy used worldwide in agricultural production.

Regarding pesticides, 1.1 billion of tons was consumed in 1994. It is today 400'000 tons in Europe. These pesticides are very toxic and the number of deaths is estimated to 20'000 deaths per year (this number is probably 10 to 20X too low). 70% of fruits and vegetables contain traces to pesticides. Consequently, we have to use WSN to control the water, fertilizers and pesticides usage by putting them only when and where they are required. Fig. 7 shows a node of WSN used for test purpose in a vineyard in Switzerland.

It is used to measure temperature, insulation, humidity, ventilation, soil moisture and transmit these data to a base station.

The agriculture of today is not sustainable regarding its usage of water and energy, which is a large waste of precious resources. ICT technologies will be capable of offering solutions for achieving a large reduction of water and energy consumption.

ICT solutions are capable of implementing precise agriculture that will not waste fertilizer, pesticide and water when it is not necessary and will be capable of delivering the right amount of water, fertilizer, etc… at the right moment.

Today, biofuel is still produced in many countries and has been considered as a good move in the direction of a smarter planet. However, it is not at all sustainable to see people using biofuel for transportation while other people in other countries suffer from famine. So in a very next future, one has to decide if you prefer to use cars or to eat something.

Figure 7: WSN developed in CSEM

5. CONCLUSION

ICT is obviously only a part of the solution. Behavioral changes are required too. As 1400 liters of water are required for one kilo of corn and one kilo of beef requires 16'000 liters of water, why to eat meat every day? Meat production was 44 millions of tons in 1950 and it is 220 tons today. In USA, 70% of cereals are for animals, in India only 2%. And big scandal: one billion of tons of food are wasted every year!! Another scandal is the amount of food which is thrown away in stores, restaurants and even at home: about 30%!

6. REFERENCES

[1] A. Lingamneni, « Implementing Energy parsimonious Circuits through inexact Designs », Master thesis, Rice Univ. USA, 2011

[2] WWF, « The Energy Report: 100% Renewable Energy by 2050 », 2011

[3] http://en.wikipedia.org/wiki/Google_search

[4] David Bol et al », « Application-Aware LCA of Semiconductors: Life-cycle Energy of Microprocessors from High-performance 32 nm CPU to Ultra-Low-Power 130 nm MCU », IEEE International Symposium on Sustainable Systems and Technologies, 6 p., 2011.

[5] "ICT as an enabler for smart water management", ITU-T Technology Watch Report, October 2010. http://www.scribd.com/doc/40602248/ICT-as-an-Enabler-for-Smart-Water-Management

[6] Wang N. et al., "Wireless sensors in agriculture and food industry—Recent development and future perspective", Computers and Electronics in Agriculture, vol. 50, 2006, pp. 1-14.

[7] Gellings C.W., "Energy efficiency in fertilizer production and use", EFFICIENT USE AND CONSERVATION OF ENERGY - Energy Efficiency in Fertilizer Production and Use - Clark W. Gellings, Kelly E. Parmenter. http://www.eolss.net/ebooks/Sample%20Chapters/C08/E3-18-04-03.pdf

[8] Vasily Moshnyaga, "Untraditional Approach to Computer Energy Reduction", PATMOS 2008, Lisbon.

[9] Greenpeace, "How Clean is Your Cloud?", April 2012

[10] John Cohn, "Kids today, Engineers tomorrow?", ISSCC 2009, Plenary 1.4, Feb. 9, 2009, San-Francisco, pp. 29-35.

[11] C. Piguet, "Ingénieur, un métier "has been", LE TEMPS, Jeudi 28 janvier 2010, page 13.

[12] Toffel, M., Horvath, A., "Environmental Implications of Wireless Technologies: News Delivery and Business Meetings." Env.Sci.Techn., 38(11), 2004, pp. 2961-2970

[13] Report "Energy [r]evolution", Greenpeace International, European Renewable Energy Council (EREC), June 2010, page 70.

[14] Massoud Pedram, "Energy Efficient Datacenters", *IEEE Trans. on Computer Aided Design*, Vol. 31, No. 10, Oct. 2012.

[15] James Urquhart, "Cloud computing's green paradox", CNET News, January 7, 2010. http://news.cnet.com/8301-19413_3-10428065-240.html

Assignment of Adjustable Delay Buffers for Clock Skew Minimization in Multi-Voltage Mode Designs

Jin-Tai Yan
Department of Computer Science and Information Engineering, Chung-Hua University, Hsinchu, Taiwan, R. O. C.

Zhi-Wei Chen
College of Engineering, Chung-Hua University, Hsinchu, Taiwan, R. O. C.

ABSTRACT

It is well known that clock skew minimization becomes critical in high-performance VLSI designs. In this paper, the assignment of adjustable delay buffers(ADBs) is applied to minimize the clock skew in a buffered clock tree in a multi-voltage mode design. Given a buffered clock tree, based on the assignment flexibility of the delay value on an ADB, bottom-up ADB assignment is firstly proposed to insert ADBs to minimize the clock skew by assigning the delay values of the inserted ADBs for each power mode. Furthermore, bottom-up ADB elimination is proposed to eliminate the redundant ADBs to minimize the number of the inserted ADBs in a multi-voltage mode design while maintaining the minimized clock skew. Compared with Su's heuristic algorithm and Lim's optimal algorithm, the experimental results show that our proposed algorithm uses less CPU time to reduces 9.3% of the used ADBs and 1.3%~1.6% of the average latency on the average, respectively.

Categories and Subject Descriptors

B.7.2 [**Integrated Circuits**]: Design Aids – *Placement and routing*

Keywords: Clock tree, Adjustable delay buffer, Skew minimization

1. INTRODUCTION

In general, it is necessary for synchronous circuits to design a skew-minimized clock tree. The clock skew in a clock tree is the difference between the minimum and the maximum arrival times of the clock signal. It is known that clock skew may cause time violation on sequential elements in a synchronous circuit. Hence, clock skew minimization becomes critical in high-performance VLSI designs. To reduce the clock skew in a clock tree, a set of buffers can be inserted on the branch nodes to construct a buffered clock tree.

Due to process or thermal variation, several tuning approaches[1-7] are proposed to insert adjustable delay buffers(ADBs) to improve timing yield. For Intel's processors[1-3], the clock distribution networks compensate the process-induced clock skews by inserting the adjustable delay buffers(ADBs). Additionally, Takahashi et. al[4] propose a genetic-based algorithm to adjust the clock timing by inserting the feasible delay elements. To systematically insert the ADBs into a buffered clock tree, some statistical approaches[5-6] are proposed to analyze the tunable range of the inserted ADBs for the improvement of the timing yield. Chakraborty et. al[7] propose an efficient approach to compensate the dynamic thermal clock skew by using the tunable delay buffers.

In a multi-voltage mode design, some circuit modules may be allowed to operate in different voltages for low-power requirements. Generally speaking, the design of a clock tree in a multi-voltage mode design can be simply obtained by satisfying the worst-case power mode and considering the extra effort for the other modes. However, the simple used approach cannot efficiently deal with some specific requirements. Recently, several efficient algorithms[8-12] are proposed to design a buffered clock tree using ADBs for clock skew minimization in a multi-voltage mode design. Su et. al[8-9] firstly design a clock tree using ADBs to compensate the delay variation in different power modes. In this design, an efficient algorithm is proposed to heuristically find the feasible locations for ADBs and optimally determine the delay values of the inserted ADBs on each power mode for clock skew minimization. Furthermore, Lin et. al[10-11] further design a tunable clock tree structure by adopting ADBs. In this design, an efficient algorithm is proposed to insert ADBs for the minimization of clock skew and buffer area in a multi-voltage mode design. However, the heuristic algorithms[8-11] cannot minimize the number of the inserted ADBs for clock skew minimization in a multi-voltage mode design. In the ADB assignment for clock skew minimization, Lim et. al[12] propose an optimal algorithm to minimize the number of the inserted ADBs and allocate the inserted ADBs in a multi-voltage mode design. However, the proposed optimal algorithm does not minimize the assigned delay values of the inserted ADBs. Because the delay range on an ADB decides the implemented area of the ADB, it is known that the delay range on an ADBs is discrete and limited. Due to the larger delay values on ADBs, the number of the inserted ADBs may increase and the latencies on most of the sequential elements become longer in a multi-voltage mode design.

In this paper, the assignment of adjustable delay buffers(ADBs) is applied to minimize the clock skew in a buffered clock tree in a multi-voltage mode design. Given a buffered clock tree, based on the assignment flexibility of the delay value on an ADB, bottom-up ADB assignment is firstly proposed to minimize the clock skew by assigning the delay values of the inserted ADBs for each power mode. Furthermore, bottom-up ADB elimination is proposed to eliminate the redundant ADBs to minimize the number of the inserted ADBs in a multi-voltage mode design while maintaining the minimized clock skew. Compared with Su's heuristic algorithm[9] and Lim's optimal algorithm[12], the experimental results show that our proposed algorithm uses less CPU time to reduces 9.3% of the used ADBs and 1.3%~1.6% of the average latency on the average, respectively.

2. PROBLEM FORMULATION

It is well known that the ADB assignment in a buffered clock tree can be applied to minimize the clock skew for each power mode in a multi-voltage mode design. Generally speaking, the

GLSVLSI'13, May 2–3, 2013, Paris, France.

larger the delay value on an ADB in a buffered clock tree is, the larger the implemented area of the corresponding ADB in the final layout is. Clearly, the ADB with the larger delay value must use more layout area and may not be implemented inside limited available space in a complicated design. Hence, it is necessary for a buffered clock tree to reduce the layout area of the inserted ADBs in a complicated design. Since the layout area of an inserted ADB depends on the delay value on an ADB in a buffered clock tree, the total delay value on all the inserted ADBs must be further minimized for the ADB assignment in a buffered clock tree.

Given a buffered clock tree, T_B, with an initial set of m buffers, a set of n connecting flip-flops and c sets of initial latencies on the flip-flops for c power modes in a multi-voltage mode design, the problem of minimizing the clock skews for c power modes is to replace a feasible set of buffers in T_B with a set of ADBs and assign the delay values onto the inserted ADBs for c power modes such that the number of the inserted ADBs is minimized and the total delay value on the inserted ADBs is minimized while the clock skew is minimized for each power mode.

As illustrated in Figure 1(a), given a buffered clock tree, T_B, with an initial set of 10 buffers, $\{B_1, B_2, B_3, B_4, B_5, B_6, B_7, B_8, B_9, B_{10}\}$, a set of 12 connecting flip-flops, {A, B, C, D, E, F, G, H, I, J, K, L}, and 2 set of initial latencies, {2, 5, 4, 6, 6, 9, 10, 14, 9, 10, 11, 12} and {5, 7, 8, 9, 4, 10, 10, 15, 5, 9, 4, 6}, on the 12 flip-flops for 2 power modes in a multi-voltage mode design, it is clear that the initial clock skews are 12 and 11 for 2 power modes, respectively. Based on the assignment flexibility of the delay value on an ADB, the number of the inserted ADBs in the given buffered clock tree can be optimally obtained as 3 and the original buffers, B_2, B_4 and B_7, can be replaced with 3 inserted ADBs, ADB_1, ADB_2 and ADB_3, to minimize the clock skews, respectively. As illustrated in Figure 1(b), the delay values of the 3 inserted ADBs can be optimally assigned as (+5, +5), (+2, 0) and (+3, +5) for 2 power modes in a multi-voltage mode design, respectively. It is clear that the final clock skews can be minimized as 5 and 6 for 2 power modes, respectively.

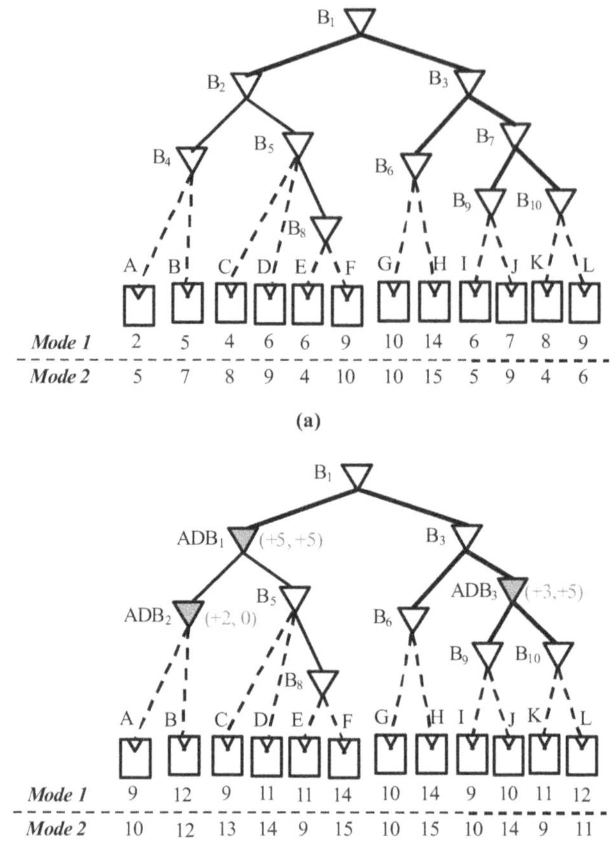

Figure 1 Assignment of delay values on ADBs for clock skew minimization in a multi-voltage mode design

3. ASSIGNMENT OF ADBS FOR CLOCK SKEW MINIMIZATION

In a given buffered clock tree, a feasible set of buffers can be replaced with a set of ADBs and the delay values of the inserted ADBs can be optimally assigned to minimize the clock skew in a multi-voltage mode design. In the ADB assignment in a given buffered clock tree for clock skew minimization, the clock skew can be minimized by using *bottom-up ADB assignment*. Under the minimized clock skew, the number of the inserted ADBs can be minimized and the delay values on the inserted ADBs can be assigned by using *bottom-up ADB elimination*.

3.1 Bottom-Up ADB Assignment for Clock Skew Minimization

In a given buffered clock tree, it is known that a set of buffers are assigned onto the branch nodes in the clock tree. For a given buffer, B_i, the *lower(upper) latency*, $l_i(u_i)$, at the buffer, B_i, can be defined as the minimum(maximum) latency inside a clock sub-tree rooted at the buffer, B_i. Furthermore, the *latency range*, $[l_i, u_i]$, at the buffer, B_i, can be defined as an interval from a lower latency, l_i, to an upper latency, u_i, inside a clock sub-tree rooted at the buffer, B_i. Hence, the *latency skew* at the buffer, B_i, can be defined as the difference of the minimum and maximum latencies and obtained as $u_i - l_i$. If the given buffer, B_i, is a root buffer in a buffered clock tree, the latency skew at the buffer, B_i, will be the clock skew in the buffered clock tree.

As illustrated in Figure 2(a), if a buffer, B_i, only connects to a set of p flip-flops with their latencies, d_1, d_2,.., and d_p, no buffer inside the clock sub-tree can be replaced with any ADB to minimize the latency skew at the buffer, B_i, and the *latency range*, $[l_i, u_i]$, at the buffer, B_i, can be intuitively obtained as $[Min\{d_1,..,d_p\}, Max\{d_1,..,d_p\}]$. On the other hand, if a buffer, B_i, connects to a set of p flip-flops with their latencies, d_1, d_2,.., and d_p, and a set of q buffers, B_1, B_2,.., and B_q, with their latency ranges, $[l_1, u_1]$, $[l_2, u_2]$,.., $[l_q, u_q]$, as illustrated in Figure 2(b), a feasible set of buffers inside the clock sub-tree can be replaced with a set of ADBs and the delay values of the inserted ADBs can be assigned to minimize the latency skew at the buffer, B_i.

For the minimization of the latency skew at the buffer, B_i, initially, a set of p flip-flops with their latencies, d_1, d_2,.., and d_p, can be treated as a dummy buffer, B_0, with its latency range, $[l_0, u_0]$, where $l_0 = Min\{d_1,..,d_p\}$ and $u_0 = Max\{d_1,..,d_p\}$. Furthermore, all the latency ranges, $[l_1, u_1]$, $[l_2, u_2]$,.., $[l_q, u_q]$, at the buffers, B_1, B_2,.., and B_q, can be sorted into an interval list according to all the lower latencies in an increasing order. For the construction of the latency range at the buffer, B_i, an iterative process can be done as follows: Given two sorted latency ranges, $[l_j, u_j]$ and $[l_k, u_k]$, $l_j \leq l_k$, at the two buffers, B_j and B_k, if the interval, $[l_k, u_k]$, is not fully included inside the interval, $[l_j, u_j]$, the interval, $[l_j, u_j]$, can be slid by a positive distance, $\text{Min}\{l_k - l_j, u_k - u_j\}$, to maximize the intersection of two corresponding intervals in an interval-merging operation. To minimize the latency skew at the buffer, B_i, the buffer, B_j, must be replaced with an ADB and the delay value of the inserted ADB can be assigned as $Min\{l_k-l_j,u_k-u_j\}$. Furthermore, the two latency ranges, $[l_j, u_j]$ and $[l_k, u_k]$, must be deleted from the sorted interval list and the merging latency range can be obtained as $[Min\{l_j+Min\{l_k-l_j, u_k-u_j\}, l_k\}, Max\{u_j+Min\{l_k-l_j, u_k-u_j\}, u_k\}]$ and added into the sorted interval list. After at most $(q-1)$ merging operations, the merging latency range of the q latency ranges, $[l_1, u_1]$, $[l_2, u_2]$,.., $[l_q, u_q]$, can be obtained by replacing a set of buffers with ADBs and assigning the delay values of the inserted ADBs. Finally, without sliding the latency range, $[l_0, u_0]$, at the dummy buffer, B_0, the latency range, $[l_i, u_i]$, at the buffer, B_i, can be obtained by using a merging operation.

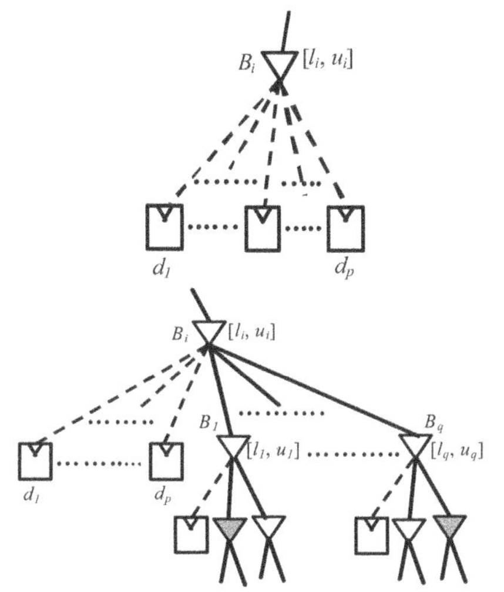

(a) (b)

Figure 2 Latency range and ADB assignment for latency skew minimization

For a given buffered clock tree, T_B, with a set of m buffers and a set of n connecting flip-flops, a buffer-connection graph, $G_B(V_B, E_B)$, can be constructed as follow: each vertex, v_i, in V_B represents a buffer in the buffered clock tree and each directed edge, e_i, in E_B represents the connection relation between two connecting buffers in the buffered clock tree. For the assignment of the inserted ADBs in a given buffered clock tree, initially, a sorted list of all the buffers in the buffered tree can be obtained by using a topological sorting in the corresponding buffer-connection graph. By finding the latency range at a given buffer for latency skew minimization, the construction of the latency ranges with ADB assignment on all the buffers can be done in bottom-up ADB assignment. Finally, a feasible set of buffers in the buffered clock tree can be replaced with a set of ADBs and the delay values of the inserted ADBs can be assigned. As a result, the clock skew in the buffered clock tree can be optimally obtained. Hence, the inserted ADBs in a given buffered clock tree for clock skew minimization can be assigned by running the algorithm, *ADB_Assignment,* as follows:

Algorithm ADB_Assignment;
Input: a buffered clock tree, T_B, with a set of m buffers and a set of n connecting flip-flops and the latencies of all the flip-flops;
{ Construct a buffer-connection graph, G_B, for the buffered clock tree, T_B;
Sort all the buffers in the buffered clock tree, T_B, by using a topological sorting in G_B;
for (all the buffers in a bottom-up manner)
{ Select a sorted buffer in a bottom-up manner;
Find the latency range at the selected buffer for latency skew minimization;
};
Output a buffered clock tree with ADB assignment and Obtain the optimal clock skew in the buffered clock tree;
}

Theorem 1: Given a buffered clock tree with a set of m buffers and a set of n connecting flip-flops, the algorithm, *ADB_Assignment,* can assign the inserted ADBs to minimize the clock skew in $O(m+n)$ time.

Refer to the buffered clock tree in Figure 1(a), for bottom-up ADB assignment in the first power mode as illustrated in Figure 3(a), initially, a sorted list of buffers, B_1, B_2, B_3, B_5, B_7, B_4, B_8, B_6, B_9 and B_{10}, can be obtained. Firstly, the latency ranges of the buffers, B_4, B_8, B_6, B_9 and B_{10}, can be intuitively obtained as [2, 5], [6, 9], [10, 14], [6, 7] and [8, 9] according to the latencies of the 10 flip-flops. Furthermore, the latency ranges of the buffers, B_5 and B_7, can be obtained as [4, 9] and [8, 9] according to the latencies of the 2 flip-flops and the latency ranges of the buffers, B_8, B_9 and B_{10}. After assigning two latency ranges of the buffers, B_5 and B_7, the buffer, B_9, must be replaced with an ADB, ADB_4, to minimize the latency skew and the delay value of the inserted ADB, ADB_4, can be assigned as +2. Next, the latency ranges of the buffers, B_2 and B_3, can be obtained as [4, 9] and [10, 14] according to the latency ranges of the buffers, B_4, B_5, B_6 and B_7. After assigning two latency ranges of the buffers, B_2 and B_3, the buffers, B_4 and B_7, must be replaced with two ADBs, ADB_2 and ADB_3, to minimize the latency skews and the delay value of the inserted ADBs, ADB_2 and ADB_3, can be assigned as +2 and +2. Finally, the latency range of the buffer, B_1, can be obtained as [9, 14] according to the latency ranges of the buffers, B_2 and B_3. After assigning the latency range of the buffer, B_1, the buffer, B_2, must be replaced with an ADB, ADB_1, to minimize the latency skew and the delay value of the inserted ADB, ADB_1, can be assigned as +5. As a result, the clock skew in the first power mode can be obtained as 5. Similarly, for bottom-up ADB assignment in the second power mode as illustrated in Figure 3(b), firstly, the latency ranges of the buffers, B_4, B_8, B_6, B_9 and B_{10}, can be obtained as [5, 7], [4, 10], [10, 15], [5, 9] and [4, 6]. Furthermore, the latency ranges of the buffers, B_5 and B_7, can be obtained as [4, 10] and [5, 9]. After assigning two latency ranges of the buffers, B_5 and B_7, the buffer, B_{10}, must be replaced with an ADB, ADB_3, to minimize the latency skew and the delay value of the inserted ADB, ADB_3, can be assigned as +1. Next, the latency ranges of the buffers, B_2 and B_3, can be obtained as [4, 10] and [10, 15]. After assigning two latency ranges of the buffers, B_2 and B_3, the buffer, B_7, must be replaced with an ADB, ADB_2, to minimize the latency skew and the delay value of the inserted ADB, ADB_2, can be assigned as +5. Finally, the latency range of the buffer, B_1, can be obtained as [9, 15]. After assigning the latency range of the buffer, B_1, the buffer, B_2, must be replaced with an ADB, ADB_1, to minimize the latency skew and the delay value of the inserted ADB, ADB_1, can be assigned as +5. As a result, the clock skew in the second power mode can be obtained as 6.

3.2 Bottom-Up ADB Elimination under Minimized Clock Skew

In bottom-up ADB assignment for clock skew minimization, it is known that the necessary ADBs are inserted in a bottom-up manner. Clearly, the ADBs can be locally inserted for clock skew minimization and a sequence of the local ADB assignments may not lead to a minimized result in ADB count. Hence, some inserted ADBs for clock skew minimization may be redundant and the redundant ADBs can be further eliminated under the minimized clock skew.

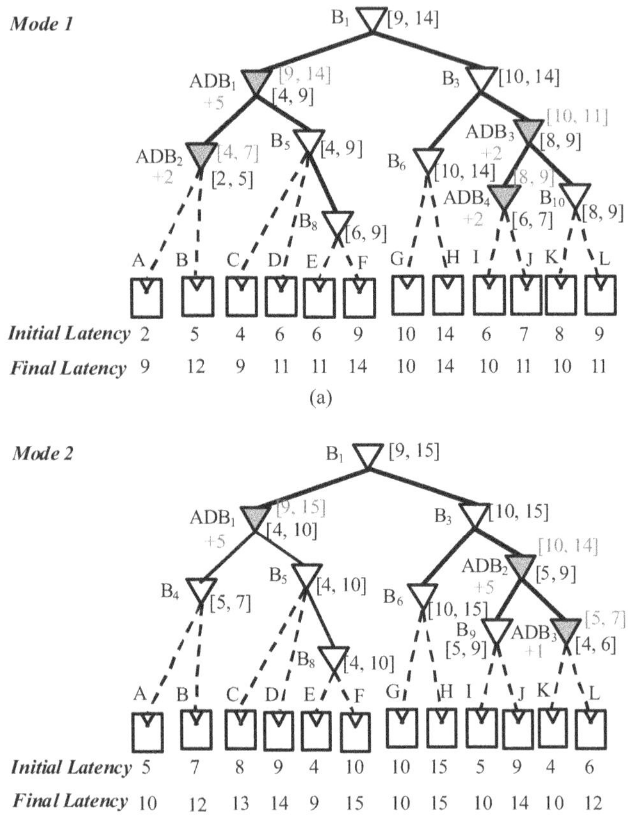

(b)

Figure 3 Bottom-up ADB assignment for clock skew minimization

Under the minimized clock skew, the delay value on any inserted ADB can be distributed into the buffer nodes in the next level in a buffered clock tree by using an *ADB distribution operation.* It is assumed that an inserted ADB, ADB_i, with the delay value, v_i, in a buffered clock tree connects to a set of r flip-flops, $\{FF_1,\ldots, FF_r\}$, a set of s ADBs, $\{ADB_1,.., ADB_s\}$, with the assigned delay values, v_1, v_2,.., v_s, and a set of t buffers, $\{B_1,\ldots, B_t\}$. In an ADB distribution operation, the delay value, v_i, on the inserted ADB, ADB_i, can be distributed into the buffers and ADBs in the next level by replacing the ADB, ADB_i, with a buffer, B_i, introducing a dummy ADB, ADB_0, to connect r flip-flops, FF_1,…and FF_r, replacing the t buffers with t ADBs, ADB_{s+1},.. and ADB_{s+t}, and adding its delay value, v_i, into the delay values on the s original ADBs and the $(t+1)$ introduced ADBs as illustrated in Figure 4(a). In contrast, if a buffer only connects a set of ADBs in the next level, some connected ADBs can be further eliminated by using an *ADB elimination operation* under the minimized clock skew. In an ADB elimination operation, given a buffer, B_i, connecting to a set of s ADBs, $\{ADB_1,.., ADB_s\}$, with the assigned delay values, v_1,.., v_s, some ADBs with the minimal delay value, v_{min}, can be eliminated by replacing the buffer, B_i, with an inserted ADB, ADB_i, assigning the delay vale, v_{min}, onto the inserted ADB, ADB_i, replacing the eliminated ADBs with buffers and decreasing the minimal delay value, v_{min}, from the delay vales on the ADBs, ADB_1,…and ADB_t, as illustrated in Figure 4(b).

For a buffered clock tree, T_{ADB}, with ADB assignment, given the latency range, $[l_{opt}, u_{opt}]$, at the root buffer, the minimized clock skew can be obtained as $u_{opt} - l_{opt}$. For any ADB, ADB_i, with the latency range, $[l_i, u_i]$, and the assigned delay value, v_i, in the buffered clock tree, the *slack range*, $[sl_i, su_i]$, at the ADB, ADB_i, can be defined as an interval representing the delay slack on the clock sub-tree rooted at ADB_i from the *lower slack*, sl_i, to the *upper slack*, u_i, and obtained as $[l_{opt} - l_i, u_{opt} - u_i]$. Furthermore, the *delay range*, $[dl_i, du_i]$, at the ADB, ADB_i, can be defined as an interval representing the possible delay value on ADB_i from the *lower delay*, dl_i, to the *upper delay*, du_i, and obtained as $[sl_i + v_i, su_i + v_i]$. As there exists a non-empty delay range, $[dl_i, du_i]$, at ADB_i, some connecting ADBs in the next level may be further eliminated by using an ADB elimination process for the given delay range, $[dl_i, du_i]$, at ADB_i.

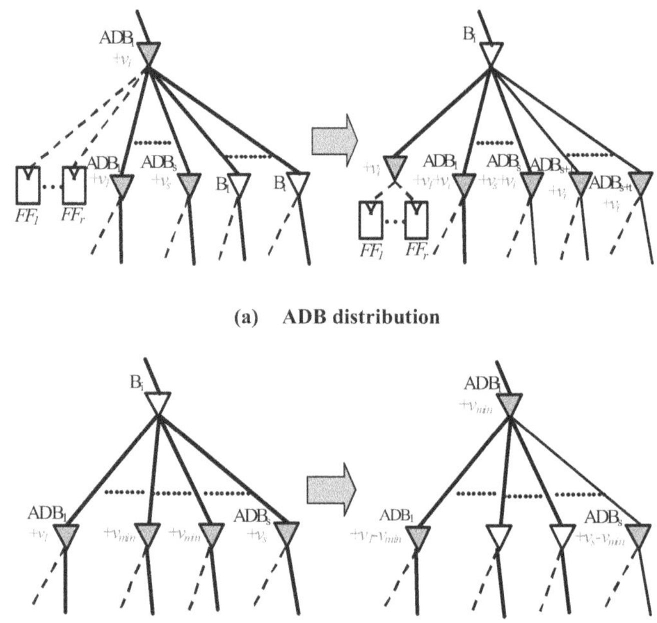

(a) ADB distribution

(b) ADB elimination

Figure 4 ADB distribution and elimination under minimized clock skew

It is assumed that an inserted ADB, ADB_i, with the delay range, $[dl_i, du_i]$, in a buffered clock tree connects to a set of r flip-flops, $\{FF_1,\ldots, FF_r\}$, a set of s ADBs, $\{ADB_1,.., ADB_s\}$, with the assigned delay values, v_1, v_2,.., v_s, and a set of t buffers, $\{B_1,\ldots, B_t\}$. In an ADB elimination process as illustrated in Figure 5, firstly, by using an ADB distribution operation for the ADB, ADB_i, the ADB, ADB_i, can be replaced with a buffer, B_i, the delay ranges of the original ADBs, ADB_1,.., and ADB_s, can be obtained as $[dl_i + v_1, du_i + v_1]$,... and $[dl_i + v_s, du_i + v_s]$, respectively, and the delay range of any introduced ADB, ADB_0, ADB_{s+1},.., or ADB_{s+t}, can be obtained as $[dl_i, du_i]$. Furthermore, to eliminate some original ADBs and all the introduced ADBs, the delay value on any original ADB, ADB_j, $1 \le j \le s$, can be further assigned as $dl_i + v_c$ if $v_j \le du_i - dl_i$ or $dl_i + v_j$, if $v_j > du_i - dl_i$, where $v_c = Max\{v_j \mid v_j \le du_i - dl_i, 1 \le j \le s\}$. Besides that, the delay value on any introduced ADB, ADB_j, j=0 or $s+1 \le j \le s+t$, can be further assigned as $dl_i + v_c$. Finally, by using an ADB elimination operation, all the introduced ADBs can be eliminated, the buffer, B_i, can be replaced with an ADB, ADB_i, and the delay value on the inserted ADB, ADB_i, can be assigned as $+(dl_i + v_c)$. If the delay value, v_j, on the ADB, ADB_j, $1 \le j \le s$, is greater than v_c, the delay value, v_j, on the ADB, ADB_j, can be reassigned as $+(v_j - v_c)$. On the other hand, if the delay value, v_j, on the ADB, ADB_j, $1 \le j \le s$, is equal to v_c, the ADB, ADB_j, can be eliminated and replaced with a buffer.

For a given buffered clock tree, T_{ADB}, with a set of m buffers and ADBs and a set of n connecting flip-flops, a ADB-connection graph, $G_{ADB}(V_{ADB}, E_{ADB})$, can be constructed as follow: each vertex, v_i, in V_{ADB} represents an ADB in the buffered clock tree and each directed edge, e_i, in E_{ADB} represents the connection relation from an ADB to its connecting ADB in the buffered clock tree. For the elimination of the redundant ADBs in the buffered clock tree, initially, a sorted list of all the ADBs can be obtained by using a topological sorting in the corresponding ADB-connection graph. By eliminating some redundant ADBs at an ADB-ADB connection with a non-empty delay range in an elimination process, a sequence of ADB elimination processes can be sequentially done in bottom-up ADB elimination. Finally, a set of redundant ADBs in the buffered clock tree can be eliminated and the delay values of the inserted ADBs can be reassigned. Therefore, the redundant ADBs in a buffered clock tree with ADB assignment can be eliminated by running the algorithm, *ADB_Elimination,* as follows:

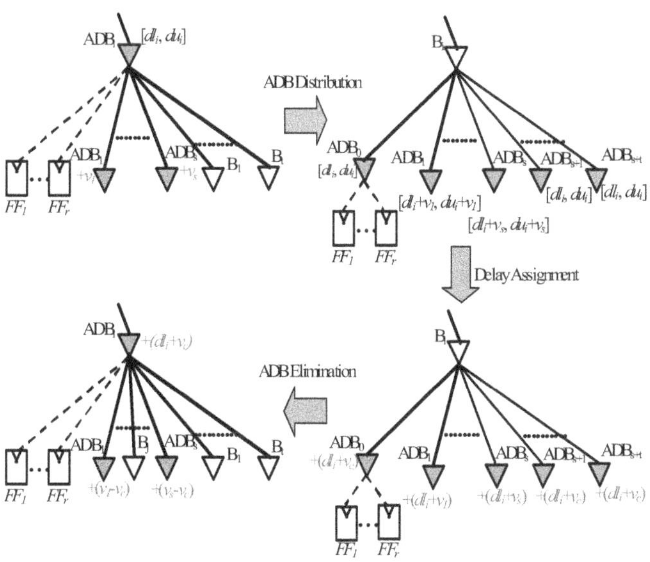

Figure 5 ADB elimination process under minimized clock skew

Algorithm ADB_Elimination;
Input: a buffered clock tree, T_{ADB}, with a set of m buffers and ADBs and a set of n connecting flip-flops;
{ Construct an ADB-connection graph, G_{ADB}, for the buffered clock tree, T_{ADB};
Sort all the ADBs in the buffered clock tree, T_{ADB}, by using a topological sorting in G_{ADB};
for (all the ADB-ADB connections in a bottom-up manner)
{ Select an ADB-ADB connection with a non-empty delay range in a bottom-up manner;
Run ADB elimination process at the selected ADB-ADB connection;
};
Output the buffered clock tree with ADB assignment and Reassign the delay values on the inserted ADBs;
}

Theorem 2: Given a buffered clock tree a set of m buffers and ADBs and a set of n connecting flip-flops, the algorithm, *ADB_Elimination,* can eliminate the redundant ADBs in the buffered clock tree in $O(m+n)$ time.

Refer to the buffered clock trees with ADB assignment in Figure 3, for bottom-up ADB elimination in the first power mode as illustrated in Figure 6(a), initially, a sorted list of ADBs, ADB_1, ADB_2, ADB_3 and ADB_4, can be obtained. Firstly, the ADB, ADB_3, in the buffered clock tree can be considered to eliminate the connected ADB, ADB_4, in an elimination process. In the elimination operation, the delay range at the ADB, ADB_3, can be obtained as [1, 5]. By distributing the delay range, [1, 5] at the ADB, ADB_3, into the ADB, ADB_4, and the buffer, B_{10}, in an ADB distribution operation, the delay ranges of the ADB, ADB_4, and the introduced ADB, ADB_5, can be obtained as [3, 7] and [1, 5], respectively. Furthermore, the delay values of the ADBs, ADB_4, and ADB_5, can be both assigned as +3. By using an ADB elimination operation for the ADBs, ADB_4, and ADB_5, in, the delay value of the ADB, ADB_3, can be reassigned as +3 and the ADB, ADB_4, can be eliminated in the first power mode. Similarly, for bottom-up ADB elimination in the second power mode as illustrated in Figure 6(b), initially, a sorted list of ADBs, ADB_1, ADB_2 and ADB_3, can be obtained. Firstly, the ADB, ADB_2, in the buffered clock tree can be considered to eliminate the connected ADB, ADB_3, in an elimination process. In the elimination operation, the delay range at the ADB, ADB_2, can be obtained as [4, 6]. By distributing the delay range, [4, 6] of the ADB, ADB_3, into the ADB, ADB_3, and the buffer, B_9, in an ADB distribution operation, the delay ranges of the ADB, ADB_3, and the introduced ADB, ADB_4, can be obtained as [5, 7] and [4, 6], respectively. Furthermore, the delay values of the ADBs, ADB_3, and ADB_4, can be both assigned as +5. By using an ADB elimination operation, the delay value of the ADB, ADB_3, can be reassigned as +5 and the ADB, ADB_3, can be eliminated in the second power mode.

4. EXPERIMENTAL RESULTS

For the problem of assigning a feasible set of ADBs into a buffered clock tree to minimize the clock skew in a multi-voltage mode design, Su's heuristic algorithm[9], Lim's optimal algorithm[12] and our proposed algorithm have been implemented by using standard C++ language and run on a Intel Core2 Quad Q9450 2.66GHz machine with 2GB memory. In the experiments, the clock sinks are from five industrial circuits, *circuit01*, *circuit02*, *circuit03*, *circuit04* and *circuit05*, and the tested buffered clock trees are generated using Synopsys IC compiler. For each tested benchmark, it is assumed that each circuit block can operate in two different voltage levels, 0.96V and 1.20V, and the tested circuit can work in four different power modes. Additionally, it is assumed that each ADB can be adjusted with a granularity of 10ps. In the computation of the latencies in flip-flops, it is assumed that the capacitance of each sink in the clock tree is set as 35fF and the wire capacitance in the clock tree is set as 0.2 fF/um.

For five tested clock trees, Table I shows the experimental results on the number of the inserted ADBs and the average latency for the inserted ADBs with the assignment of the delay values in the final buffered clock trees. Compared with Su's heuristic algorithm[9], the experimental results show that our proposed algorithm uses less CPU time to reduces 9.3% of the inserted ADBs and 1.3% of the average latency on the average, respectively. Compared with Lim's optimal algorithm[12], the experimental results show that our proposed algorithm uses less CPU time to minimize the number of the inserted ADBs and reduces 1.6% of the average latency on the average.

5. CONCLUSIONS

Based on the assignment flexibility of the delay value on an ADB, bottom-up ADB assignment is firstly proposed to insert ADBs to minimize the clock skew for each power mode in a multi-voltage mode design. Furthermore, bottom-up ADB elimination is proposed to eliminate the redundant ADBs to minimize the number of the inserted ADBs in a multi-voltage mode design while maintaining the minimized clock skew.

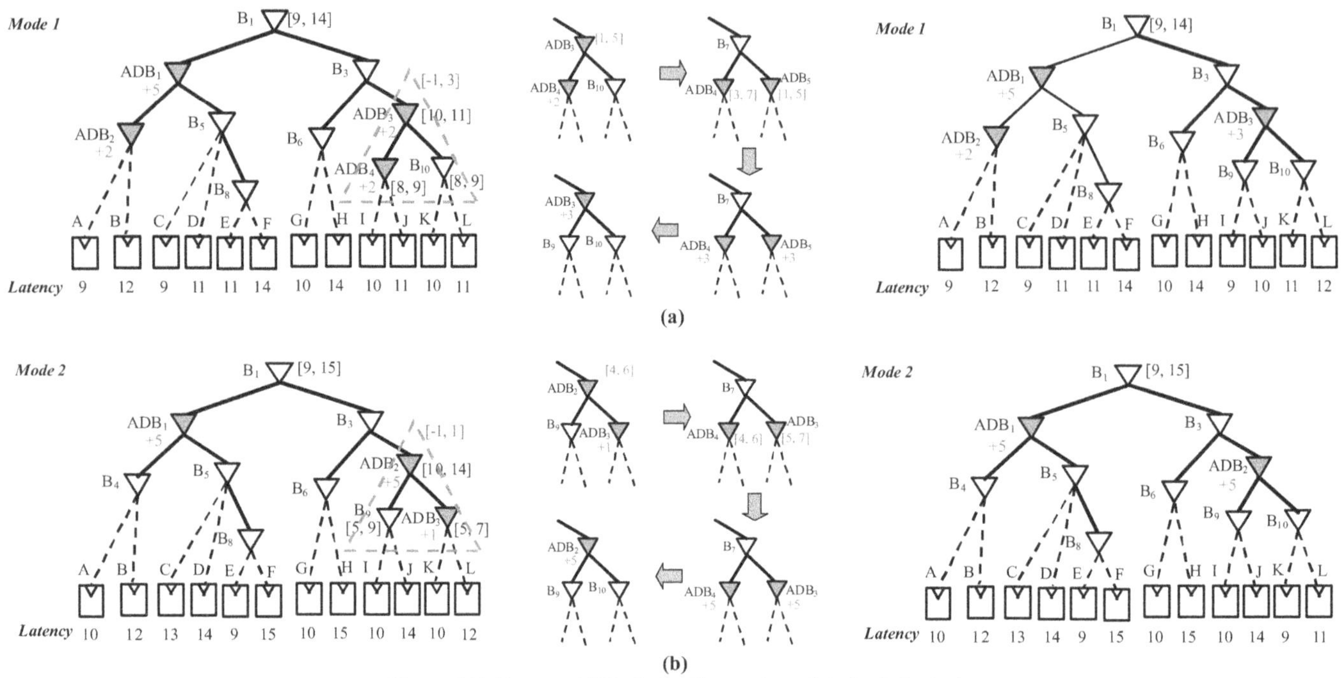

Figure 6 Bottom-up ADB elimination under minimized clock skew

6. REFERENCES

[1] S. Tam, S. Rusu, U. N. Desai, R. Kim, J. Zhang, and I. Young, "Clock generation and distribution for the first IA-64 microprocessor," *IEEE Journal of Solid-State Circuits*, Vol. 35, No. 11, pp.1545–1552, 2000.

[2] N. A. Kurd, J. S. Barkatullah, R. O. Dizon, T. D. Fletcher, and P. D. Madland, "A multigigahertz clocking scheme for Pentium 4 microprocessor," *IEEE Journal of Solid-State Circuits*, Vol. 36, No. 11, pp.1647–1653, 2001.

[3] P. Mahoney, E. Fetzer, B. Doyle, and S. Naffziger, "Clock distribution on a dual-core multi-threaded Itanium-family processor," *International Solid-State Circuits Conference*, pp.292–293, 2005.

[4] E. Takahashi, Y. Kasai, M. Murakawa, and T. Higuchi, "A post-silicon clock timing adjustment using genetic algorithms," *International Symposium on VLSI Circuits*, pp.13–16, 2003.

[5] J. L. Tsai, D. Baik, C. P. Chen, and K. K. Saluja, "A yield improvement methodology using pre and post-silicon statistical clock scheduling," *International Conference on Computer-Aided Design*, pp.611–618, 2004.

[6] J. L. Tsai, L. Zhang, and C. Chen, "Statistical timing analysis driven post-silicon-tunable clock-tree synthesis," *International Conference on Computer-Aided Design*, pp.575–581, 2005.

[7] A. Chakraborty, K. Duraisami, A. Sathanur, P. Sithambaram, L. Benini, A. Macii, E. Macii and M. Poncino, "Dynamic thermal clock skew compensation using tunable delay buffers," *International Symposium on Low Power Electronics and Design*, pp.162-167, 2006.

[8] Y. S. Su, W. K. Hon, C. C. Yang, S. C. Chang and Y. J. Chang, "Value assignment of adjustable delay buffers for clock skew minimization in multi-voltage mode designs," *International Conference on Computer-Aided Design*, pp.535-538, 2009.

[9] Y. S. Su, W. K. Hon, C. C. Yang, S. C. Chang and Y. J. Chang, "Clock skew minimization in multi-voltage mode designs using adjustable delay buffers," *IEEE Transaction on Computer-Aided Design*, Vol. 29, No. 12, pp. 1921–1930, 2010.

[10] K. Y. Lin, H. T. Lin and T. Y. Ho, "An efficient algorithm of adjustable delay buffer insertion for clock skew minimization in multiple dynamic supply voltage designs," *Asia-South Pacific Design Automation Conference*, pp.825-830, 2011.

[11] K. Y. Lin, H. T. Lin, T. Y. Ho and C. C. Tsai, "Load-balanced clock tree synthesis with adjustable delay buffer insertion for clock skew reduction in multiple dynamic supply voltage designs," *ACM Trans. on Design Automation of Electronic Systems, Vol. 17, No. 3, Article 34*, 2012.

[12] K. H. Lim and T. Kim, "An optimal algorithm for allocation, placement and delay assignment of adjustable delay buffers for clock skew minimization in multi-voltage mode designs," *Asia-South Pacific Design Automation Conference*, pp.503-508, 2011.

TABLE I EXPERIMENTAL RESULTS OF ASSIGNMENT OF ADBS FOR CLOCK SKEW MINIMIZATION IN A MULTI-VOLTAGE MODE DESIGN

Circuits (#FFs/#Bufs)	Original Clock Tree		Su's heuristic algorithm[9]				Lim's optimal algorithm[12]				Our optimal algorithm			
	Worst skew (ps)	Average. latency (ps)	Worst skew (ps)	#ADBs	Average. latency (ps)	CPU Time (s)	Worst skew (ps)	#ADBs	Average. latency (ps)	CPU Time (s)	Worst skew (ps)	#ADBs	Average. latency (ps)	CPU Time (s)
circuit01 (1042/117)	278.16	816.21 (100%)	37.23	37(100%)	833.52 (102.1%)	2.67	36.43	34(91.9%)	824.16 (101.0%)	1.62	34.67	34(91.9%)	820.27 (100.5%)	1.27
circuit02 (1276/134)	361.45	847.62 (100%)	31.67	46(100%)	861.78 (101.7%)	3.35	32.48	41(89.1%)	882.34 (104.1%)	2.74	30.79	41(89.1%)	856.55 (101.0%)	1.98
circuit03 (1587/139)	327.38	906.43 (100%)	28.52	41(100%)	923.69 (101.9%)	2.97	27.67	37(90.2%)	916.37 (101.1%)	2.53	26.32	37(90.2%)	912.21 (100.6%)	1.83
circuit04 (1862/152)	435.68	1063.51 (100%)	32.74	65(100%)	1084.12 (101.9%)	4.26	31.99	59(90.8%)	1101.28 (103.6%)	3.72	30.42	59(90.8%)	1072.61 (100.9%)	2.65
circuit05 (2151/185)	343.54	1207.23 (100%)	29.38	82(100%)	1231.28 (102.0%)	4.83	30.06	75(91.5%)	1225.19 (101.5%)	4.36	27.94	75(91.5%)	1210.62 (100.3%)	3.18

Multi-Corner Multi-Voltage Domain Clock Mesh Design

Can Sitik
Electrical and Computer Engineering
Drexel University
Philadelphia, PA, USA
as3577@drexel.edu

Baris Taskin
Electrical and Computer Engineering
Drexel University
Philadelphia, PA, USA
taskin@coe.drexel.edu

ABSTRACT

This paper introduces a novel multi-voltage domain clock mesh design methodology that is effective under multiple process corners. In multi-voltage designs, a single clock mesh that spans multiple voltage domains is infeasible due to the incompatibility of voltage levels of the clock drivers on the electrically-shorted mesh–each voltage domain requires a separate mesh. The skew among these isolated meshes need to be matched and a novel premesh tree synthesis is required to tolerate the impact of PVT variations exacerbated due to the separation of clock meshes for multiple voltage levels. The experiments performed with the largest three ISCAS'89 benchmark circuits operating at 500 MHz, 90 nm technology and 3 process corners show that: 1) The multi-voltage domain clock mesh can achieve up to 42% lower power on average with 39.04 ps skew, as low as ≈1.95% of the clock period, on average over a typical single voltage domain clock mesh, and 2) multi-corner optimized multi-voltage domain clock mesh can decrease the global skew of all corners by 190.42 ps on average over a multi-voltage domain clock mesh optimized for a single corner with a 15% degradation in power consumption necessary for variation-tolerance.

Categories and Subject Descriptors

B.7.1 [**Hardware**]: INTEGRATED CIRCUITS—*Types and Design Styles*

General Terms

Design

Keywords

VLSI, multi-corner clocking, multi-voltage design, low-power

1. INTRODUCTION

In high performance IC clock design, the tradeoff between the performance and the power consumption is well-studied [1–9]. In ASIC design, clock tree topologies are preferred due to their low power profile with less wire demand and flexibility to be combined with low power techniques such as multi-voltage domain design [10]. On the other hand, in high-end microprocessor design, clock topologies with redundancies are preferred due to their tolerances against process variations with the cost of extra wire capacitance. In the clock tree topologies with spines and cross-links [1, 9], selected branches are shorted to decrease the skew mismatch in the presence of variations, whereas in clock mesh topologies, every branch is shorted in the design to provide a global improvement of the skew [2–4, 6–8]. Literature on clock meshes focuses on design methodologies for mesh reduction and optimization, and ignores multi-corner analysis, as the ubiquitous shorting of branches provides remarkable robustness under PVT variations. However, this assumption only holds true for a traditional, single-Vdd clock mesh. When a system-on-chip style multi-Vdd single-clock domain design is considered, or for a design employing the low power technique of multi-voltage design, a clock network built with the mesh-topology becomes uncharacteristically susceptible to PVT variations. This susceptibility emerges as a multi-voltage clock mesh is constructed with disjoint premesh trees and meshes, each synchronizing an individual voltage domain.

In this work, a novel multi-corner multi-voltage clock mesh design methodology is presented. Multi-voltage clock mesh design is not a straight-forward process: First, a single clock mesh is not feasible as its electrically-shorted mesh wires cannot drive the voltage sinks operating at different voltage levels. Separate meshes are needed for each domain. Second, the skew among these domains must be balanced, which is a challenge that arises due to the isolation of the premesh trees among the voltage domains. Third, the skew introduced by the variation must be analysed and tuned considering multiple process corners, as isolated premesh trees in different voltage domains have different tolerances against the PVT variations. A previous work in [11] addressed the first two objectives but fails to consider the third objective of variation-tolerance. Variation-awareness of the multi-voltage clock mesh topology necessitates a novel design scheme, which is proposed in this paper to alleviate the impacts on timing variation, which are further exacerbated due to the presence of multiple voltage domains. Hence, a novel methodology that addresses all three of these objectives concurrently is necessary to synthesize a variation-tolerant, power-efficient multi-voltage domain clock network.

The experiments performed on the largest circuits of ISCAS'89 with 90 nm models at 500 MHz highlights two important benefits of multi-corner multi-voltage domain clock mesh design: 1) Proposed multi-corner multi-voltage domain clock mesh can achieve up to 42% lower power at three process corners with a 39.04 ps skew on average, that resides in the 2% budget of the clock period, 2) proposed method can achieve 190.42 ps lower skew in average and a much tighter slew compared to optimizing multi-voltage domain clock mesh design at only one process corner, with a 15% degrada-

GLSVLSI'13, May 2–3, 2013, Paris, France.

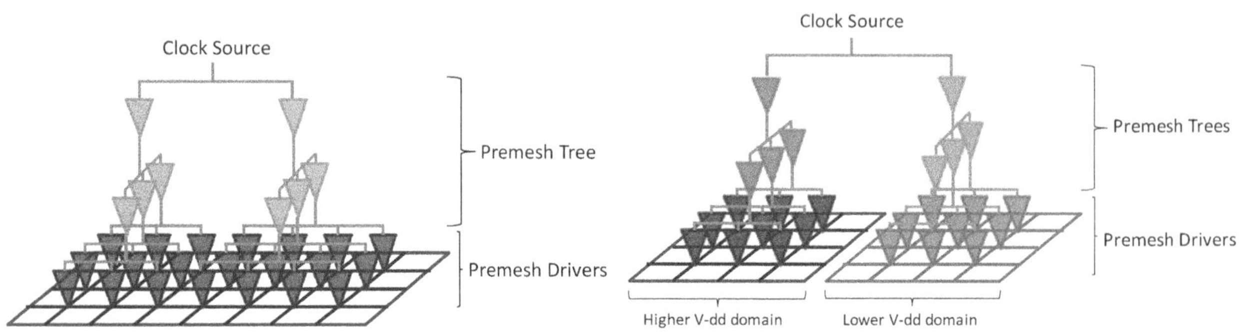

(a) Typical Single-Voltage Domain Clock Mesh Topology. (b) Proposed Multi-Voltage Domain Clock Mesh Topology.

Figure 1: Clock Mesh Topologies

tion in the power consumption at that corner (but functional as the skew is corrected not to be larger than an allowable skew budget).

The rest of the paper is organized as follows. In Section 2, the preliminaries of the clock mesh design and the challenges of multi-voltage clocking are briefed. In Section 3, the proposed multi-corner multi-voltage domain clock mesh design methodology is explained in detail. In Section 4, experimental results are presented. The paper is finalized with concluding remarks in Section 5.

2. PRELIMINARIES AND BACKGROUND

The clock mesh topologies and their challenges are described in Section 2.1. The effect of multiple voltage-domains on the clock buffers is discussed in Section 2.2.

2.1 Clock Mesh Design

There are two major design specifications of the clock mesh design shown in Figure 1. First is the power consumption, introduced by the RC effect of the redundant mesh wires, and the second is the clock skew, introduced by the variations and geometric mismatches.

2.1.1 Power Dissipation in Clock Mesh

The power consumption of a clock mesh can be approximated as follows:

$$P_{total} \approx \alpha c_{total} f V_{dd}^2 \tag{1}$$

where α is the switching factor, f is the operating frequency and V_{dd} is the supply voltage. The switching capacitance c_{total} is:

$$c_{total} \approx c_{mesh} + c_{stub} + c_{pmt} + \sum_{\forall i} c_i \tag{2}$$

where the last term stands for the total capacitance of the sink registers, and c_{mesh}, c_{stub} and c_{pmt} are the total capacitance of the mesh wires, stub wires and the premesh tree, respectively.

Under the same switching activity α and the same frequency f, there are two ways to decrease the power dissipation of the synthesized clock mesh network. First is to decrease the total capacitance, for which [2, 6] propose mesh wire reduction, [3] proposes stub wire reduction and [4, 7] propose both. Second is the supply voltage reduction, which can be realized using multi-voltage domains as described in this paper. Although multi-voltage mesh has unique challenges, it is more promising than capacitance reduction because 1) most of the switching capacitance is at the clock sinks, and thus, mesh or stub wire length reduction does not reduce the dominant switching capacitance component, and 2) supply voltage reduction scales power quadratically, rather than linearly. Besides, voltage domains are not proposed to exclusively reduce the clock power but as a necessity due to the voltage domains being created for the logic (e.g. an SoC or multi-Vdd design) and not for the clock itself. It is an added incentive that these multiple voltage domains can help reduce the power dissipation of a clock mesh. The culprit is the undesired increase in clock skew between the meshes on each individual voltage domain, which is addressed in this paper.

2.1.2 Skew in Clock Mesh

The global clock skew in the mesh is estimated as:

$$t_{skew} = t_{skew}^{pmt} + t_{skew}^{mesh} + t_{skew}^{stub} \tag{3}$$

where t_{skew}^{pmt}, t_{skew}^{mesh} and t_{skew}^{stub} are the skews introduced by the difference between the maximum and the minimum delays on the premesh tree of the mesh, the difference between the maximum and the minimum delays on the mesh from a premesh driver to a stub wire connection point on the mesh and the difference between the maximum and the minimum delays from a connecting point of a stub wire to a sink register, respectively. The skew introduced by t_{skew}^{mesh} and t_{skew}^{stub} can be reduced by using dense meshes in each domain, however, decreasing the skew introduced by t_{skew}^{pmt} is a challenge for multi-voltage meshes as the premesh trees depicted in Figure 1(b) are isolated.

2.2 Insertion Delay and Its Variation in Multi-Voltage Designs

Any multi-voltage clocking algorithm has the following challenges. First, the gate delay changes with the switching of the supply voltage between voltage domains. Second, the variation in the gate delay is not the same for the identical gates with different supply voltages. Clock skew is defined as the difference between the maximum and the minimum insertion delays, which may fall on different branches within separate voltage domains. To that end, it is proposed in this paper that the insertion delay profile of a clock mesh network be carefully investigated to optimize the skew, as this investigation reveals how to design these premesh tree branches within separate voltage domains in the presence of PVT variations of multiple process corners. In order to analyse how the insertion delay varies depending on the applied voltage, a motivational example is presented with 2 voltage domains, each with a simple 2-level clock tree with 16 sinks, as depicted in Figure 2.

In this motivational example, one domain is supplied with 1.2V and the other is with 0.8V, and SPICE models of *SAED90nm EDK library* of *Synopsys* are used [12]. This simple example is synthesized by *IC Compiler* of *Synopsys* using the same size of clock buffers. This circuit is simulated in *CustomSim XA* simulator of

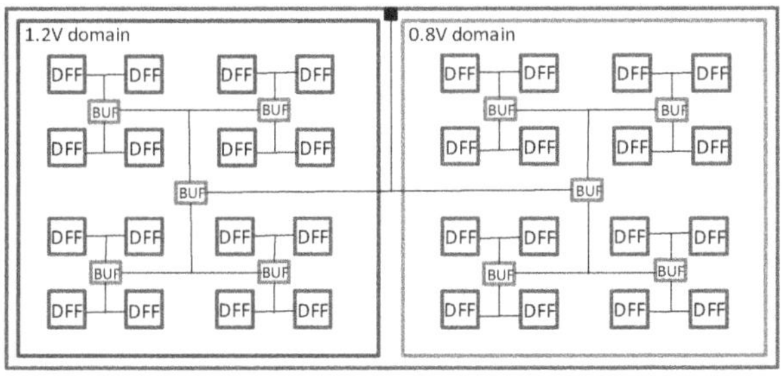

Figure 2: Simple 2-level Clock Tree with 16 Sinks

Table 1: Insertion Delay Profile of the Motivational Clock Tree Example at 2 Domains in Figure 2, the Max and the Min Insertion Delays that Define the Global Skew are Marked with Bold

	Best (ps)	Nom (ps)	Worst (ps)
Min ins delay in 0.8V D.	104.73	157.77	301.71
Max ins delay in 0.8V D.	**106.64**	**161.50**	**309.57**
Inc. compared to Best	-	≈ 55	≈ 203
Min ins delay in 1.2V D.	**101.36**	**130.71**	**223.46**
Max ins delay in 1.2V D.	101.65	131.29	224.65
Inc. compared to Best	-	≈ 30	≈ 122
Global Skew	5.28	30.79	86.11

Synopsys at 500 MHz, and the insertion delay is measured in 3 different process corners: The best, nominal and the worst cases of the process. The insertion delay profile monitored is shown in Table 1.

From Table 1, it is observed that the insertion delay variation is higher in the 0.8V domain which also increases the skew switching from the best case to the worst case: Maximum insertion delay increases by $\approx$203 to 309.57 ps, whereas the minimum insertion delay only increases by $\approx$122 to 223.46 ps. In this undesirable case, the maximum insertion delay increases the largest, as it is affected by PVT variations the most going from the best to worst case. The minimum insertion delay, on the other hand, increases the least, which is also undesirable. It is desirable, instead, that the maximum insertion delay path be delayed the least and the minimum insertion delay be delayed the most, to contain the skew under PVT variations going from the best to the worst case. The basis for the novel premesh tree methodology in this paper is that the maximum insertion delay path should be placed in the higher voltage domain at the best case *on purpose*, as paths in the higher voltage domain get affected less by PVT variations, and the skew budget is lower at all corners. To demonstrate this postulate, the insertion delay of the 1.2V domain is increased gradually by upsizing the first level of the two-level motivational clock tree first, and then both levels together to observe these two desirable delay profiles. It is expected that the insertion delay in the 1.2V domain increases by a delay offset in all PVT corners, which may increase the skew in the best case corner, but decrease the skew in the worst case, which means a decrease in the global skew.

From Table 2, it is seen that the global skew is decreased when the insertion delay is gradually increased in the 1.2V domain by upsizing the clock buffers. It is clear that, a novel clock network synthesis methodology should control the insertion delay of the different domains by either adding/removing buffers or upsizing/downsizing over a naive, single-corner optimized tree. Notice how the minimum and the maximum insertion delays (marked in bold) are consistently in the same voltage domain in all three corners in the undesired case, shown in Table 1. In the desired case, shown in Table 2, the delays vary between paths in either voltage domains for PVT corners, the skew being affected less from PVT variations. Although the skew is increased in the best corner (from 5.28 ps to 25.4 ps), global skew is defined as the worst case of all corners: Thus, global skew is improved from 86.11 ps to 39.47 ps, when the insertion delay is shifted by an upsizing. This example concludes that optimizing the skew at all three corners simultaneously requires an early arrival of the clock signal in the lower voltage domain at the best case so as to have a balanced skew at the worst case, and for the skew budget to be lower. Multi-corner optimization is available only in single-voltage mesh synthesis in the standard EDA tools. Therefore, the multi-corner optimized premesh tree synthesis methodology for the multi-voltage clock mesh designs developed in this paper fills this gap.

3. METHODOLOGY

A new methodology to synthesize a multi-corner mesh-based clock distribution network in a multi-voltage design flow is proposed. By considering multiple process corners at the same time, variation-tolerance is targeted. The steps of the proposed methodology are as follows: Given an initial placement and the mesh size,

1. Premesh drivers are placed at each intersection, and sized considering the slew constraints (Algorithm 1).
2. Premesh tree is synthesized iteratively considering all process corners to minimize the global skew (Algorithm 2).

For the multi-voltage clock mesh, local meshes are placed for each domain with their own premesh trees. The roots of the premesh trees of each domain can be connected to a master root through level shifters, however, these level shifters are not strictly necessary. The virtual root of each domain can function as a level shifter, instead, if the source voltage swing is higher than that of the supply voltage at the root. In order to eliminate the extra power cost, the level shifters are not used in this work, as shown in Figure 1(b). The highest voltage in the design is assumed to be the voltage swing of the clock source connection of the multi-voltage clock mesh design without loss of generality of the presented work. Buffer driver selection and sizing, and the premesh tree synthesis procedures are presented in Section 3.1 and Section 3.2, respectively. Section 3.1 is a simple algorithm that does not define the novelty of this work but complements the design flow. Note that, the proposed methodology is independent of the premesh driver sizing algorithm; any alternative, more (or less) sophisticated method for premesh driver sizing can be seamlessly integrated, which demonstrates the practicality of the proposed method for automation purposes. Furthermore, the algorithms presented in this work are designed for any number of voltage domains, however it is important to note that the number of voltage domains is 2 in the selected technology library, which is also the same library and setup in [11], against which the results are compared.

3.1 Premesh Driver Selection

In this step, a simple heuristic is used to select the premesh driver buffers. First, maximum size buffers in the library are placed at each intersection point of the clock mesh, then, the buffer sizes are decreased iteratively until the maximum slew exceeds the slew constraint at any of the corners. Note that, this algorithm is very similar to the premesh driver selection algorithm in [11], with one

Table 2: Improved Skew Numbers with a Delay Offset in the Insertion Delay of the 1.2V Domain

	Only first level upsized by one			Both two levels upsized by one		
	best (ps)	nom (ps)	worst (ps)	best (ps)	nom (ps)	worst (ps)
Min insertion delay in 0.8V domain	**104.64**	157.52	301.12	**104.64**	**157.51**	301.16
Max insertion delay in 0.8V domain	106.53	**161.29**	**308.93**	106.56	161.28	**308.98**
Min insertion delay in 1.2V domain	115.23	**145.07**	**239.61**	129.89	164.77	**269.51**
Max insertion delay in 1.2V domain	**115.53**	145.62	240.77	**130.04**	**165.12**	270.21
Global Skew	10.89	16.22	68.16	25.4	7.61	39.47

Algorithm 1 Premesh Driver Selection

Input: Mesh size for each voltage domain i, buffer library, slew constraint $slew_{const}$
Output: The sizes of each premesh driver buffers
  for Each voltage domain i **do**
    Place maximum size buffer at each intersection
    Calculate $slew^i_{max}$
    while $slew^i_{max} < slew_{const}$ **do**
      Replace all buffers with the next smaller buffer cell in the library
      Update $slew^i_{max}$
    end while
  end for
  Return the selected driver for each domain i

definite difference: [11] starts with minimum size buffers and performs upsizing whereas in this work, maximum size buffers are selected at the beginning and downsizing is performed iteratively. This approach decreases the number of iterations as the slew constraints are much tighter in this work compared to [11] and larger size buffers are needed. Furthermore, this similarity makes fair comparisons to [11], and is necessary to highlight the elegance of multi-corner optimized premesh trees (Section 3.2) over single-corner optimized premesh trees. The slew constraint $slew_{const}$ is set to its typical value of 5% of the clock period, although it can be set to another value depending on the performance requirements. Note that this constraint may be degraded by the synthesized premesh tree in the following procedure, yet the slew target needs to be selected in this stage as it is used as a guide that drives the premesh tree synthesis stage. In the premesh driver selection procedure, all the buffers are sized down at the same time with an identical scale for the sake of simplicity. This procedure is shown in Algorithm 1.

Algorithm 2 Multi-Corner Multi-Voltage Domain Premesh Tree Synthesis

Input: Premesh drivers for each domain, assigned voltage for each domain, buffer library for each domain i and each corner c, slew constraint $slew_{const}$
Output: Premesh tree for each domain i
  Set N = 1
  while $slew_{max} > slew_{const} + slew_{margin}$ **do**
    Place $N \times N$ max size buffers at the (n-1)st level of each domain i
    $N = N + 1$
  end while
  if *max insertion delay is in the same domain for all cases* **then**
    while *max insertion delay is still at the same domain j* **do**
      Add a max size buffer of that voltage domains other than j
    end while
  end if
  $curr_{skew} = \max_{\forall c}(max_{\forall i}(delay^i_{max})$ - $(\min_{\forall i}(delay^i_{min}))$
  $prev_{skew} = T$
  while $k_j < num - of - levels - in - j$ **do**
    Downsize the k-th level of the max insertion delay domain j
    $prev_{skew} = curr_{skew}$
    $curr_{skew} = \max_{\forall c}(max_{\forall i}(delay^i_{max})$ - $(\min_{\forall i}(delay^i_{min}))$
    if $curr_{skew} > prev_{skew}$ **then**
      Undo that step, mark k-th level as done
      $k_j = k_j + 1$
    end if
  end while
  while $curr_{skew} > prev_{skew}$ **do**
    Add parallel buffers at the max insertion delay domain at the first level
    $prev_{skew} = curr_{skew}$
    $curr_{skew} = \max_{\forall c}(max_{\forall i}(delay^i_{max})$ - $(\min_{\forall i}(delay^i_{min}))$
  end while

3.2 Multi-Corner Premesh Tree Synthesis

Premesh tree synthesis is the novel step of this proposed work that synthesizes a premesh tree to optimize the global skew considering all process corners simultaneously. This stage also considers the slew at the sinks, not to degrade the slew more than an allowed margin within the slew budget, as the maximum slew (Section 3.1) of the design is not completely independent from the premesh tree.

It is important to note the significance of the proposed skew-and-slew budgeted multi-corner, multi-voltage clock mesh design methodology within the current state of the art clock mesh design automation: In the multi-voltage design flow, standard EDA tools cannot match the insertion delays of the premesh trees driving clock meshes that belong to different voltage domains even in a single corner [11]. In this work, an iterative algorithm is proposed to accomplish this task at all process corners. In the algorithm, the premesh tree is synthesized iteratively, until the lowest overall (e.g. global) skew at all corners is achieved. This procedure is shown in Algorithm 2.

In this algorithm, first the (n-1)st level of the premesh tree is synthesized with the maximum size buffers in an N×N topology. The iterations starts (generically) with N=1, and it is increased by 1 (also generically) at each iteration until the slew constraint with the degradation margin is met at all corners. Then, the insertion delays are measured to see if the maximum insertion delay path is in the same domain for all cases. If so, a maximum size buffer is added in the next level of each domain other than the maximum insertion delay domain because it is an undesired case, as explained in Section 2.2. The purpose of selecting maximum size buffers in this stage is the fact that having less number of levels in the premesh tree with larger size buffers is more variation-tolerant than having more number of levels with smaller buffers. Buffer adding

Table 3: Created Benchmark Circuits

	0.8V Domain	1.2V Domain
Benchmark 1	s35932	s38417
Benchmark 2	s38417	s35932
Benchmark 3	s35932	s38584
Benchmark 4	s38584	s35932

stage ends when the maximum insertion delay paths are no longer encountered in the same domain for different corners. The buffer addition stage is succeeded by the buffer sizing stage. In the buffer sizing stage, the buffers are downsized starting from the first level of the domain whose maximum insertion delay path sets the global skew. If downsizing improves the skew, that step is performed. Otherwise, that step is undone and that level is marked as "done". Downsizing continues with the next maximum insertion delay domain and finalizes when all levels of all domains are marked as "done". After that, pruning stage starts. In this last stage, the number of levels or the sizes of the buffers are unchanged, but parallel buffers of the same sizes are added at the first level of the maximum insertion delay domain iteratively as long as the global skew is improved. The algorithm is an heuristic and does not guarantee an optimal solution. However, the algorithm is scalable with a complexity of O(n) for n voltage domains and m process corners. The complexity is linear because the algorithm visits every domain constant number of times, at each stage of the algorithm and simulates for the best and the worst cases only. Certain simplifications are done, such as choosing identical size buffers, which degrade the quality of the solution, but are sufficient to demonstrate the novelty of this work.

4. EXPERIMENTAL RESULTS

The proposed methodology is implemented with Tcl in order to inter-operate with standard EDA tools and tested on the benchmark circuits created with two voltage domains. The selected *SAED 90nm EDK Library* [12] has two voltage levels, a high voltage at 1.2V and a low voltage at 0.8V. For testing purposes, 4 benchmark circuits are created by placing different combinations of the three largest circuits of ISCAS'89 benchmarks, s35932, s38417 and s38584 within two voltage domains. These combinations are shown in Table 3.

The RTL level designs are synthesized using *Design Compiler* of *Synopsys* and the placement of the circuits are performed using *IC Compiler* of *Synopsys*. The skew and the power analysis are performed using *CustomSim XA* Simulator of *Synopsys* at the SPICE accuracy level with the SPICE models. Due to the tight slack constraints at the domain that operates at 0.8V, 500 MHz is selected as the operating frequency. In order to verify the variation-tolerance, all the tests are performed at 3 different process corners:

1. Best Corner (BC): V=V_{dd}+10%, T=-40°C, fast transistors
2. Nominal Corner (NC): V=V_{dd}, T=25°C, typical transistors
3. Worst Corner (WC): V=V_{dd}-10%, T=125°C, slow transistors

In order to compare the quality of results with the standard single-voltage mesh and [11], the following procedure is performed: First, the same circuits are synthesized with a single-voltage domain mesh whose premesh drivers are sized using the proposed method given in Section 3.1. For consistency, the premesh trees are synthesized using *IC Compiler* in the multi-corner mode. Then, a custom implementation of the multi-voltage domain clock mesh design algorithm in [11] is applied on the same circuits to compare to the

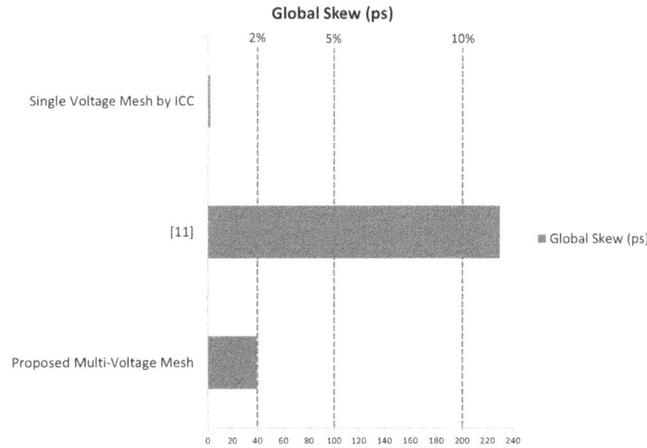

Figure 3: Skew Comparison vs. Typical Skew Budgets

proposed work. The maximum slew $slew_{const}$ is selected as 100 ps, with a degradation margin allowance in premesh tree synthesis of 20% totalling to a slew budget of 120 ps, as explained in Section 3.1 and Section 3.2. The mesh size is selected as 10×10, similar to [11], to have a fair comparison.

The proposed methodologies are applied on the 4 benchmarks in Table 3. The premesh buffer sizing (Section 1) takes 2 iterations, and the premesh tree synthesis stage (Section 3.2) takes 9 iterations at most to converge. The experimental results are shown in Table 4 and Table 5. Recall that the contemporary practical objective in clock distribution network design is to optimize power within a given skew/slew bound. It is visible from Table 4, that a single voltage mesh is the best in terms of clock skew, as the single voltage mesh (famously) has a very high variation-tolerance. The important contribution of this paper, is that, when multi-voltage designs are implemented, the clock mesh can preserve its variation-tolerance while reducing the power dissipation through multi-Vdd design. The proposed methodology can achieve up to 42% less power consumption in the clock network compared to a single voltage mesh. The skew increases from 1.73 ps to 39.04 ps on average, shown in Table 5, yet still as low as ≈1.95% of the clock period so well within the skew budget. In comparison, note that the previous single-corner work in [11] does not satisfy the 5% (or 10%) skew budget at 229.46 ps, ≈11.5% of the clock period, on average, highlighted in Figure 3. It is concluded that [11] is useful in demonstrating the feasibility of a multi-voltage mesh but the process was completed at a single corner. That methodology in [11] does not generate functional meshes at the multi-corner analysis. Similar to multi-corner multi-mode (MCMM) versions of design automation routines, the multi-corner multi-voltage mesh design methodology proposed in this paper employs novel and different algorithms to achieve demonstrated results.

It is reported in Table 4 that the power dissipation in the multi-voltage clock network is much lower compared to a single voltage clock mesh. The total power savings on the multi-voltage circuit will be even higher considering the additional savings on the logic, which are not reported here (as total power savings depend on the particular voltage domain partitioning technique and may sway the results). There is a 15% degradation in power consumption in the nominal case over the single-corner design in [11], as optimization at multiple corners is considered. The 15% power degradation, thus, is not a penalty but a necessity to achieve a slew and a skew constrained design. The slew constraint in [11] is 120 ps in

Table 4: Power Comparison over Single-Voltage Domain Mesh and [11] at 3 corners, BC, NC and WC

Circuits	SV Mesh Synthesized by *ICC*			[11]			Proposed Work		
	BC(mW)	NC(mW)	WC(mW)	BC(mW)	NC(mW)	WC(mW)	BC(mW)	NC(mW)	WC(mW)
Benchmark 1	467.64	112.69	201.89	272.56	66.52	137.90	304.19	82.97	132.72
Benchmark 2	502.04	128.33	220.41	219.09	80.28	148.54	254.88	95.96	140.25
Benchmark 3	385.19	93.17	178.90	197.55	48.34	104.18	231.41	64.51	102.70
Benchmark 4	434.58	106.73	184.59	212.60	72.29	145.62	239.84	88.26	132.65
	Avg. % Improvement			49%	40%	32%	42%	25%	35%

Table 5: Skew Comparison over Single-Voltage Domain Mesh and [11] at 3 corners, BC, NC and WC, Overall Skew is Bold

Circuits	SV Mesh Synthesized by *ICC*			[11]			Proposed Work		
	BC (ps)	NC (ps)	WC (ps)	BC (ps)	NC (ps)	WC (ps)	BC (ps)	NC (ps)	WC (ps)
Benchmark 1	**1.30**	0.97	0.63	**228.03**	21.06	177.31	**40.17**	20.16	28.80
Benchmark 2	1.54	2.65	**4.03**	**227.41**	21.64	178.73	**38.88**	18.54	30.94
Benchmark 3	**2.53**	1.90	1.19	**229.58**	20.60	179.16	30.26	9.99	**39.11**
Benchmark 4	**1.56**	1.25	0.76	**232.82**	16.42	170.09	30.26	10.14	**38.00**
Avg. Skew			1.73			229.46			39.04
	Degradation					-227.73			-37.31

the nominal case, but it is observed to be 240 ps in the worst case, which requires an extravagant 12% slew budget. In comparison, the worst case slew budget in this paper is 120 ps, which is satisfied at 108 ps, 5.4% of the period, in the worst case. For variation-awareness, and the saved timing slack, the power consumption is degraded by 15% in the nominal case. This degradation is 7% in the best case corner, and there is a 3% improvement in the worst case corner, which highlights the multi-corner optimization objective of the proposed work.

5. CONCLUSIONS

A new methodology is proposed to synthesize variation-tolerant, multi-voltage, single clock domain clock meshes for high performance ICs. The proposed method enables a multi-corner clock mesh synthesis on designs with multiple voltage domains. This is a substantial novelty in demonstrating the problem with multi-voltage designs and their synchronization with clock meshes and developing design automation routines for the methodology whose unavailability in industrial tools also demonstrated in experiments. Slew-and-skew budgeted premesh clock tree design at all process-voltage-temperature (PVT) corners is proposed for high practicality. In this work, a simple algorithm is used to size the premesh drivers to highlight the elegance of multi-corner optimized multi-voltage domain clock meshes. Therefore, this methodology can easily be combined with existing buffer sizing algorithms for improved results and a wide practical applicability. If more number of voltage levels is available, the proposed methodology can achieve a wide range in the skew vs. power curve by exploiting different voltage levels for different performance requirements.

6. REFERENCES

[1] A. Rajaram and D. Z. Pan, "Variation tolerant buffered clock network synthesis with cross links," in *Proceedings of the International Symposium on Physical Design (ISPD)*, 2006, pp. 157–164.

[2] A. Abdelhadi, R. Ginosar, A. Kolodny, and E. G. Friedman, "Timing-driven variation-aware nonuniform clock mesh synthesis," in *Proceedings of the ACM Great Lakes Symposium on VLSI (GLSVLSI)*, May 2010, pp. 15–20.

[3] M. R. Guthaus, G. Wilke, and R. Reis, "Non-uniform clock mesh optimization with linear programming buffer insertion," in *Proceedings of the ACM/IEEE Design Automation Conference (DAC)*, June 2010, pp. 74–79.

[4] Y. Teng and B. Taskin, "Clock mesh synthesis method using the Earth Mover's Distance under transformations," in *Proceedings of the IEEE International Conference on Computer Design (ICCD)*, Oct. 2012, pp. 121–126.

[5] J. Lu, Y. Teng, and B. Taskin, "A reconfigurable clock polarity assignment flow for clock gated designs," *IEEE Transactions on Very Large Scale Integration (TVLSI) Systems*, vol. 2, no. 6, pp. 1002–1011, June 2012.

[6] G. Venkataraman, Z. Feng, J. Hu, and P. Li, "Combinatorial algorithms for fast clock mesh optimization," *IEEE Transactions on Very Large Scale Integration Systems (TVLSI)*, vol. 18, no. 1, pp. 131–141, January 2010.

[7] J. Lu, X. Mao, and B. Taskin, "Integrated clock mesh synthesis with incremental register placement," *IEEE Transactions on Computer-Aided Design (TCAD)*, vol. 31, no. 2, pp. 217–227, Feb. 2012.

[8] N. Kurd, J. Barkarullah, R. Dizon, T. Fletcher, and P. Madland, "A multigigahertz clocking scheme for the Pentium 4 microprocessor," *IEEE Journal of Solid-State Circuits (JSSC)*, vol. 36, no. 11, pp. 1647–1653, Nov. 2001.

[9] G. Shamanna, N. Kurd, J. Douglas, and M. Morrise, "Scalable, sub-1w, sub-10ps clock skew, global clock distribution architecture for Intel Core i7/i5/i3 microprocessors," in *Proceedings of the IEEE Symposium on VLSI Circuits (VLSIC)*, June 2010, pp. 83–84.

[10] C.-C. Tsai, T.-H. Lin, S.-H. Tsai, and H.-M. Chen, "Clock planning for multi-voltage and multi-mode designs," in *Proceedings of the International Symposium on Quality Electronic Design (ISQED)*, March 2011, pp. 1–5.

[11] C. Sitik and B. Taskin, "Multi-voltage domain clock mesh design," in *Proceedings of the IEEE International Conference on Computer Design (ICCD)*, Oct. 2012, pp. 201–206.

[12] *Synopsys 90nm Generic Library*, Synopsys Inc., 2009.

Combating NBTI-induced Aging in Data Caches

Shuai Wang, Guangshan Duan, Chuanlei Zheng, and Tao Jin
State Key Laboratory of Novel Software Technology
Department of Computer Science and Technology
Nanjing University
swang@nju.edu.cn, {guangshan_duan, zhengchl, taojin}@smail.nju.edu.cn

ABSTRACT

The negative bias temperature instability (NBTI) in CMOS devices is one of most prominent sources of aging mechanisms, which can induce severe threats to the reliability of modern processors at deep submicron semiconductor technologies. Due to the unbalanced duty cycle ratio of the SRAM cells, the data cache suffers a heavy NBTI stress and this will further exacerbate the aging effect in the data cache. In this paper, an aging-aware design is proposed to combat the NBTI-induced aging in the data cache. First, the detailed lifetime behaviors of the cachelines in the data cache are studied. Then, different schemes are proposed to mitigate the negative aging effects by balancing the duty cycle ratio of the SRAM cells in the cachelines according to their different lifetime phases. By applying our proposed idle-time-based cacheline invalidation, early write-back, and bit-flipping schemes, the duty cycle ratio of the data cache can be well balanced. By adopting the drowsy scheme for invalidated cachelines, our design can also reduce the power consumption significantly, which will further optimize the thermal behavior and aging effect of data caches.

Categories and Subject Descriptors

B.8.1 [**Hardware**]: Performance and Reliability—*Reliability, Testing, and Fault-Tolerance*

General Terms

Reliability

Keywords

Data caches; negative bias temperature instability; low power; duty cycle balancing

1. INTRODUCTION

In deep submicron semiconductor technologies, the aging effect in CMOS devices has become one of major challenges in new microprocessor designs [4]. Recent research has shown that the lifetime reliability of CMOS devices can be degraded by aging mechanisms such as bias temperature instability (BTI), hot-carrier injection, and gate-oxide wearout [23, 17, 15]. The negative bias temperature instability (NBTI) has proven to be one of the most critical failure mechanisms affecting future technologies.

The NBTI affects the pMOS device when negative voltage is applied at the gate (logic '0') and the NBTI-induced aging is proportional to the stress time and the switching activity of the device. If the SRAM cells hold the same value for a long time, which means a highly biased duty cycle ratio, it will cause some devices under heavy NBTI stress and exacerbate the aging effect. Since data caches in modern processors are implemented with SRAM cells and hold the data for program execution, protecting the data cache against NBTI-induced aging is very important for reliable processor design. Due to the uneven use of the cachelines and the presence of the narrow-width values [9], the data cache suffers a highly biased duty cycle ratio thus a heavy NBTI stress. However, the NBTI-induced degradation of device reliability cannot be mitigated simply by adopting some traditional techniques, such as guardbanding, which may incur significant reduction in circuit speed. Therefore, the focus of this work is the microarchitectural solution to balance the duty cycle ratio and mitigate the aging stress.

In this work, an aging-aware data cache (AADC) design is proposed to combat the lifetime degradation in the performance and reliability of the SRAM cells in the data cache by duty cycle balancing. We first conduct the detailed analysis on the lifetime behaviors of the data cache and divide the cachelines in the data caches into three groups, clean cachelines, dirty cachelines, and invalid cachelines. For the clean cachelines in the data cache, we propose to do the idle-time-based cacheline invalidation first and then bit-flip these invalidated cachelines periodically. For the dirty cachelines, we propose to do the idle-time-based early write-back first, and then do the invalidation and apply the bit-flipping. For the invalid cachelines, we can just do the bit-flipping periodically. By carefully choosing the idle-time and bit-flipping time intervals, the average duty cycle ratio of the data cache can be well balanced with negligible performance and energy overheads, and thus the NBTI degradation of the data cache will be significantly mitigated. Further, to reduce the power consumption of the data cache, we adopt the drowsy scheme for the invalidated cachelines in our design, and the low-energy states of the SRAM cells will further alleviate the aging effect in the data cache.

GLSVLSI'13, May 2–3, 2013, Paris, France.

The rest of the paper is organized as follows. In the next section, related work on aging-aware/NBTI-aware designs is discussed. In Section 3, we provide detailed designs of our aging-aware data cache. The experimental results and discussion are presented in Section 4. Section 5 draws the conclusion.

2. RELATED WORK

The NBTI-induced instability in the SRAM cells was studied and a data flipping scheme was proposed in [14]. In their scheme, the data in inverted mode need to be flipped back during the read operation. Therefore, extra XOR gates were added to do the inverting, which will increase the cycle time and the power consumption of the SRAM cells. In our AADC design, no bit-inverting is needed during the data cache access, thus it has no impact on the performance and power efficiency of the data cache. In [1], a design of NBTI-aware processor named Penelope was proposed and evaluated. Penelope protects the memory structures, such as registers and cache blocks, in the processor by utilizing the idle time of these resources. Therefore, it has the limitation in balancing the duty cycle ratios of the heavily in-use memory structures like the data cachelines. However, our AADC provides the solutions for both idle and in-use cachelines. In [21], duty cycle balancing designs were proposed for register files by exploiting the narrow-width data in the processor. Shin et al. proposed a redundancy scheme to improve the lifetime of the SRAM caches against the NBTI-induced wearout [19]. Duty cycle balancing scheme proposed in [12] targeted at the aging-effect reduction in instruction caches by exploring the lifetime behaviors, while our work is a microarchitecture solution to data caches. Their schemes cannot be directly applied to data caches, since the lifetime behaviors are much more complicated in data cache compared to those in instruction caches. Moreover, we also consider the low power designs and their effects in aging reduction. In [11], a holistic approach named Colt was proposed to balance duty cycle ratios of devices in modern processors by applying the complement mode to data path, control path, and storage hierarchy. Although Colt does not need to do the bit inverting when the data are fetched, extra XOR gates are still required to do the bit-flipping. In our AADC, the bit-flipping is done by writing all zeros or ones to these cachelines, thus no extra XOR gates will be involved.

3. AGING-AWARE DATA CACHE (AADC)

3.1 Motivation

Previous work has studied that caches are often occupied by more '0' than '1' [14, 1]. Our experimental results also demonstrate that the duty cycle ratio of data caches is not balanced to the best case (i.e., 50%) in our simulated microprocessor. Therefore, the pMOS devices in data caches will be affected by the negative bias at most of the time and suffer a high NBTI-stress. The traditional guardbanding technique requires a large guardband in SRAM V_{DDMIN}, which is too expensive and may limit some low power designs, such as supply voltage scaling. Customized NBTI-resilient SRAM cells were proposed in [2, 20]. Recent work [16, 7, 8] also exploited low-energy states of the SRAM cells for mitigating the aging effect. However, all the previous schemes target at alleviating the NBTI-induced aging effects in general SRAM cell structures. There is no scheme targeting at aging-aware design specifically for the L1 data cache by utilizing the access pattern and lifetime behavior of the data cache. Therefore, we propose a microarchitecture solution to balance the duty cycle ratios and combat the NBTI-induced aging of the SRAM cells in the data cache.

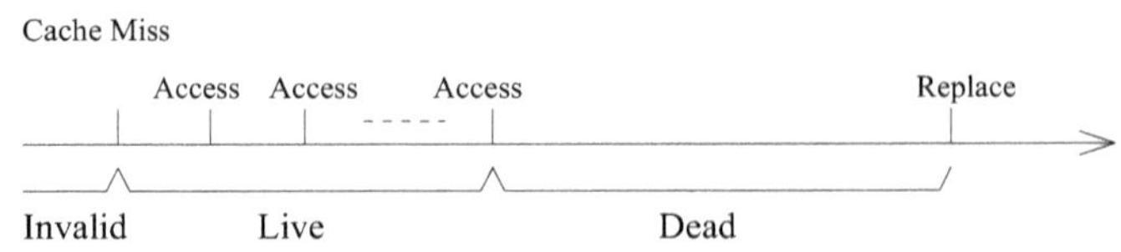

Figure 1: The lifetime of a data item in the data cache.

3.2 Lifetime Behavior of the Data Cache

In [3, 24, 22], the lifetime behaviors of L1 caches have been broadly studied, especially for improving their reliability against soft errors. Due to the variety of access patterns in the L1 data cache, such as read, write, replace, and write-back, the lifetime model of the L1 data cache in their studies is quite complicated, which makes the data cache difficult to be analyzed and optimized. Therefore, we simplify the lifetime model of cachelines in the data cache and divide their lifetime into the following three phases, `Live`, `Dead`, and `Invalid`, similar to the analysis in [13].

- `Live`: lifetime phase between first access and last access of a data item,
- `Dead`: lifetime phase between the last access and the replacement of a data item,
- `Invalid`: lifetime phase when the data item is in the invalid state.

Figure 1 shows the correlation among three lifetime phases for typical data cache activities, and the access (A) can be a cache read (R) or a cache write (W). Notice that the data item in the data cache can be a cacheline, a word, a byte, or a single bit. Although [22] claims that a byte-level analysis is accurate for the lifetime characterization for the data cache, we choose the a cacheline-level model in our AADC study for two reasons: a) the cost of controlling a byte-level bit-flipping is too high, so we choose bit-flipping for each cacheline, and b) the target of our work is to mitigate the NBTI-induced aging in the data cache, a coarse-grained lifetime models like cacheline-based model should be enough for this study.

3.3 NBTI-Aware Designs for Different Lifetime Phases

Based on the lifetime categorization of the data cache, we adopt different strategies to different lifetime phases in order to mitigate the NBTI stress of the SRAM cells with minimum performance and energy overheads. For the cachelines in the invalid states, we propose to simply bit-flip these cachelines periodically. Since the data in these invalid cachelines will not be needed for program execution, they do not need to be flipped back, even if they are in the complement mode when the cachelines will become valid after a cacheline replacement from the L2 cache. Our bit-flipping is just writing all zeros or ones to these cachelines, thus we do not need extra XOR gates to do the flipping, which makes our design more power and area efficient compared to the previous schemes in [14, 11].

For the cachelines in the valid states, we cannot simply apply the similar bit-flipping because the data in these cachelines may be needed during future cache accesses. For the cachelines in the `Live` phase, if we do our flipping scheme (writing zeros or ones) to the cachelines, the original data will be lost. If we use the inverting scheme proposed in [14, 11], extra XOR gates for inverting are required. For the clean cachelines in the `Dead` phase, they are actually not needed in the future. Because the data in the clean cacheline is not updated by the CPU, the data will be just discarded at the replacement. This provides possibilities to do our flipping scheme for these cachelines in the `Dead` phase. However, the problem is that we cannot determine which read operation to the clean cacheline is the last read during the program execution. Therefore we cannot know when to start applying our bit flipping scheme.

In [22], a clean cacheline invalidation (CCI) scheme was proposed to reduce the vulnerability factor of the clean cachelines in the data cache by invalidating the cachelines after being idle for some predefined intervals. The CCI scheme is based on the observation that most read-read (RR) instances have small intervals (less than 1K cycles) and these RR instances with small intervals only contribute a small percent of the overall RR time. Different from their scheme, we adopt the cacheline invalidation (CI) scheme in order to do our bit flipping and balance the duty cycle ratios of the cachelines. For the clean cacheline in the data cache, if it remains idle for certain predefined interval, we propose to invalidate it and then do the our bit flipping schemes similar to these invalid cachelines. By applying our cacheline invalidation and flipping (CIF) scheme, most of `Dead` phase in the clean cacheline will be converted into the invalid phase, so the NBTI stress can be mitigated by applying our bit flipping scheme. Moreover, the `Live` phase in the clean cacheline will be reduced if a small invalidation interval is used. Therefore, part of the `Live` phase will be converted into the invalid phase and its aging effects can also be alleviated. The duty cycle ratio of the remaining `Live` phase is not optimized by our CIF scheme. However, the remaining `Live` phase only contributes a small percentage (less than 10% in our study) to the cacheline lifetime, so the overall duty cycle ratio of the SRAM cells in the clean cachelines can be well balanced.

For the dirty cachelines in a write-back data cache, the data in these cachelines are still needed and will be written back into the L2 cache at the replacement. Therefore, we cannot apply our CIF scheme to the dirty cachelines directly. Instead, we propose to do the idle-time-based early write-back (EWR) [22] first, and then do the invalidation and bit-flipping. Similar to the clean cachelines, due to the small percentage of the `Live` phase after the early write-back and invalidation, the overall duty cycle ratio of the SRAM cells in the dirty cachelines will be also well balanced. For a write-through data cache, the situation is much simple. Since all cachelines are clean in the write-through data cache, we can just apply our CIF scheme to balance the duty cycle ratio.

3.4 Microarchitecture of the AADC

The key issues in the AADC design are how to do the early write-back (EWR) for the dirty cachelines, the cacheline invalidation (CI) for the clean cachelines, and the bit-flipping for the invalid cachelines. Figure 2 shows the block diagram of our AADC design. We use the valid bit (V) in the tag array to control whether the EWR/CI or the bit-flipping scheme should be applied to each cacheline. For the valid cacheline (V = 1), an N-bit global counter (IT for idle-time based) ticked by the clock signal and a two-bit local counter ticked by the global counter every 2^N cycles are introduced for each cacheline. The local counter is reset to zeros once the cacheline is accessed. If the local counter saturates, we use the dirty bit (D) to control either EWR+CI is performed for the dirty cacheline (D = 1), or only CI is performed is performed for the clean cacheline (D = 0). After that, the valid bit V is set to zero, and the local counter is also reset to zero. For the invalid cacheline (V = 0), a global counter (BF) is used for the bit-flipping. The BF counter and the cacheline state zero bit (Z) work together to determine whether all zeros or ones should be updated into the entire cacheline. If the BF counter saturates and the Z bit is equal to one, indicating that currently the data in the cacheline are all zeros, all ones will be written into the cacheline in order to balance the duty cycle ratios of the SRAM cells in the cacheline, and the Z bit will be set to zero. If the BF counter saturates and the Z bit is equal to zero, all zeros should be written into the cacheline and the Z bit will be set to one. Note that in order to minimize the area overhead of our AADC design, we choose the same idle time interval for EWR and CI. Therefore, only one two-bit local counter is needed for each cacheline and it can be shared by using D bit for dirty and clean cachelines.

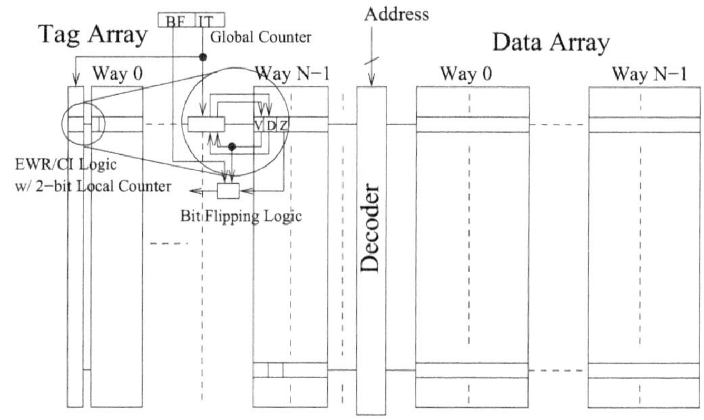

Figure 2: Microarchitectural schematic of the proposed AADC.

3.5 Power Optimization

Previous aging reduction solutions have studied the aging benefits provided by the low-energy states of the SRAM cells [16, 7, 8]. If energy saving or leakage control schemes [10, 13] are adopted in data caches, the aging effect will be further mitigated. In our AADC design, the cachelines will not be needed after the invalidation, which makes them very suitable for applying the energy saving schemes, such as the drowsy scheme. Therefore, we propose to adopt the drowsy scheme to further reduce the aging effect of the data cache, i.e., applying the drowsy scheme to these invalidated cachelines. Moreover, the leakage control scheme will also result in the temperature reduction in the data cache, which can further alleviate the aging.

3.6 Area, Performance, and Power Overheads of the AADC

For the area overhead, since no extra XOR gates or inverting operation are needed in our AADC design, the overhead is mainly from one extra Z bit showing the current state (all zeros or ones) for the invalid cachelines, and the two-bit local counter to implement the EWR+CI or CI for each cacheline. The space overheads of the global counter BF and IT are negligible because they are shared by the entire data cache. The space overheads of the Z bit and the two-bit local counter for each cacheline is also not too costly. For instance, for a data cache with a 64-byte cacheline, the space overhead of our AADC compared to the data array is only 3 bits out of 64 bytes ($3/(64 \times 8) = 0.6\%$).

For the performance overhead, since the data in the invalid cacheline have no impact on processor execution and our bit-flipping operation is not in the critical path, the performance will not be degraded. However, the early write-back and cacheline invalidation schemes do have the impact on the performance, because the invalidation operations may cause additional cache misses, if the invalidated cachelines need to be accessed by the CPU in the near future. Therefore, we need to carefully choose the idle interval for EWR and CI in order to maximize the lifetime aging mitigation and minimize the performance degradation. Note that the drowsy scheme in our AADC has no performance impact, since all the cachelines in drowsy modes are invalid and will not need to be waken up during accesses.

The bit-flipping operation in our AADC scheme is the major contribution of the power overhead. In general, if a large time interval for bit-flipping is used, the power overhead will be reduced. However, if the time interval is too large, duty cycle balancing for the data cache of our AADC scheme will become less effective. Therefore, detailed experiments need to be conducted in order to choose a proper bit-flipping interval for power efficiency.

4. EXPERIMENTAL EVALUATION

4.1 Experimental Setup

We derive our simulators from SimpleScalar V3.0 [6] to model a high-performance microprocessor similar to Alpha 21364. Table 1 gives the detailed configuration of the simulated microprocessor. To evaluate the power efficiency of our AADC design, a modified version of the Wattch power model [5] is used for power profiling (at 32nm technology) during the simulation. For experimental evaluation, we use the SPEC CPU2000 benchmark suite compiled for the Alpha Instruction Set Architecture. Ten benchmarks are randomly selected for our experimental evaluation. We use the reference input sets for this study. Each benchmark is first fast-forwarded to its early single simulation point specified by SimPoint [18]. We use the last 100 million instructions during the fast-forwarding phase to warm-up if the number of skipped instructions is more than 100 million. Then, we simulate the next 100 million instructions in detail.

4.2 Experimental Results and Analysis

In order to apply our AADC design, we first need to conduct the lifetime behavior analysis on the data cache and this characterization is performed at the cacheline level. Our experimental results show that most of the cachelines in the data cache are valid (in-use) during the execution. As shown in Figure 3, 99.5% of the cachelines in the data cache are valid on the average. The `Live` and `Dead` phase are the lifetime phases when the cachelines are in the valid state. Figure 3 shows that the `Live` phase accounts for about 24.6% of a cacheline's lifetime and the `Dead` phase contributes about 74.9% on the average. Therefore, in order to apply different effective aging mitigation schemes according to the different lifetime behaviors of the cachelines, we first divide the cachelines into two groups in our AADC study: valid and invalid cachelines.

For the invalid cachelines, we propose to simply bit-flip these cachelines periodically. However, as we discussed in Section 3, we need to choose a proper bit-flipping time interval in order to achieve high-balanced duty cycle ratio of the invalid cachelines as well as minimize the overheads. If

Table 1: Parameters of the simulated processor.

Processor Core	
Datapath Width	4 inst. per cycle
Int Issue Queue	20 entries
FP Issue Queue	15 entries
Load/Store Queue	64 entries
Active list (ACL)	80 entries
Int Register File	80 registers
FP Register File	72 registers
Function Units	4 IALU, 2 IMULT/IDIV 2 FALU, 1 FMULT/FDIV/FSQRT 2 MemPorts
Branch Predictor	
Branch Predictor	Alpha 21264 tournament predictor 32-entry RAS
BTB	2048-entry 2-way
Memory Hierarchy	
L1 I/DCache	64KB, 2 ways, 64B blocks, 2 cycles
L2 UCache	4MB, 8 ways, 128B blocks, 12 cycles
Memory	225 cycles first chunk, 12 cycles rest
TLB	Fully-assoc., 128 entries

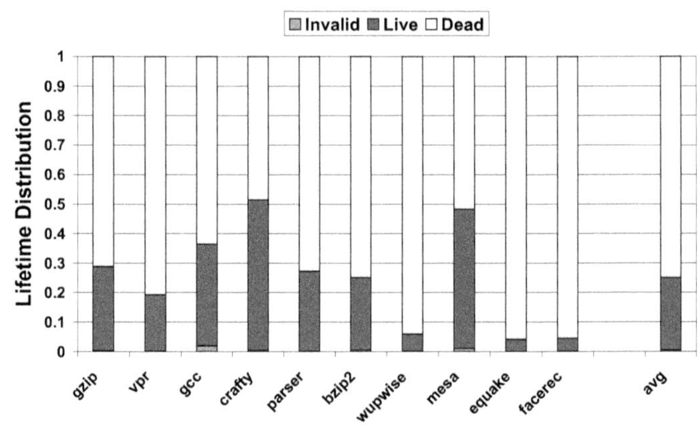

Figure 3: The lifetime distribution of the cachelines in the data cache.

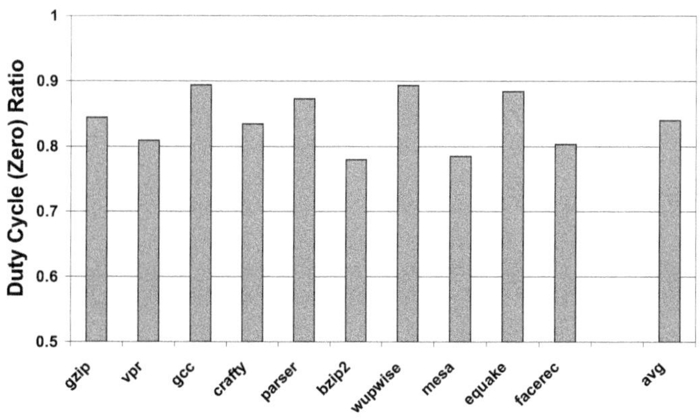

Figure 4: The average stress duty cycle (zero) ratio for valid cachelines.

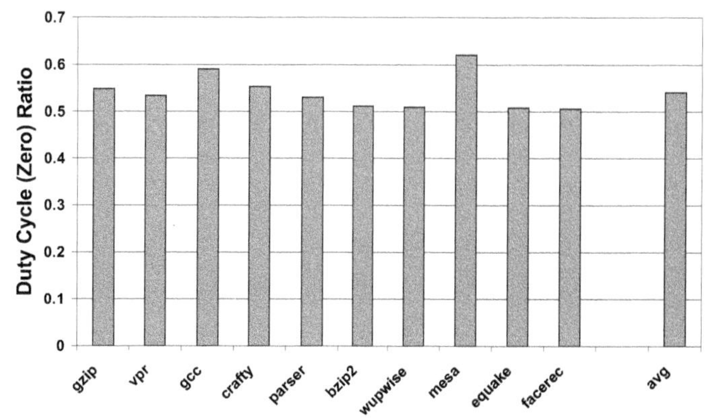

Figure 5: The average stress duty cycle (zero) ratio for all cachelines after applying the proposed AADC scheme.

we choose a small interval, the power overhead will be high, but the duty cycle ratio will be more perfectly balanced. If a large interval is used, the power overhead will be reduced, but the duty cycle ratio of the cachelines will not be well balanced. Based on our experimental results, a 40K-cycle interval for bit-flipping has negligible power and performance overheads with nearly perfect duty cycle balancing capability. Therefore, we choose the 40K-cycle bit-flipping interval for our AADC design.

For the valid cachelines, our experimental results in Figure 4 show that the average stress duty cycle (zero) ratio is 84.0%, which needs to be further reduced. For clean cachelines, based on the fact that most of the read-read (RR) instances have small intervals, we proposed to use an idle-time-based cacheline invalidation (CI) scheme to invalidate the valid cachelines after being idle for some predefined intervals in Section 3. By applying the CI scheme, most of the duty cycles of clean cachelines will be converted into the duty cycles of invalid cachelines. Then, we can further balance them by applying our bit-flipping scheme. However, similar to the bit-flipping scheme, the problem is how to choose the proper invalidation interval that can reduce the `RR` phase significantly with negligible performance degradation. Our experimental results show that if a small 500-cycle interval is chosen, the `RR` phase can be significantly reduced to 0.5% from the original 13.7%, but the performance loss is high, 5.4% on the average. This high performance loss is mainly caused by the high pipeline stall penalty due to the increased data cache misses incurred by the CI scheme, which is not affordable in modern high-performance processor. However, if a large 64K-cycle interval is adopted, the performance loss is less than 0.3 %, while the `RR` phase increases to 6.3%. Based on our experimental results, 4K-cycle is a good choice for the clean cacheline invalidation. The performance loss is under 0.7% and the `RR` phase is reduced from 13.7% to 2.4%.

For dirty cachelines, we proposed to adopt the idle-time-based early write-back (EWR) scheme [22] first, and then apply the invalidation and bit-flipping. Similar to the idle-time chosen for clean cacheline invalidation, we conduct a study based on different idle times and the experimental results show that 4K-cycle is also a good choice for the early write-back, which can effectively reduce the `Live` phase in dirty cachelines with negligible performance overheads.

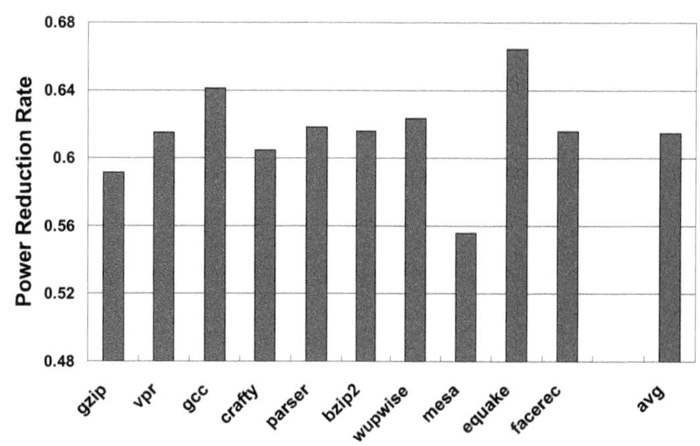

Figure 6: The power reduction rate by applying the drowsy scheme.

Therefore, as we discussed in Section 3, we choose a 4K-cycle interval for both idle-time-based cacheline invalidation and early write-back to minimize the area overhead of our AADC design. After the cacheline invalidation, we use the same 40K-cycle interval for bit-flipping in order to achieve duty cycle balancing. Our experimental results in Figure 5 show that our AADC design can reduce the average stress duty cycle ratio to 54.1% for all cachelines in the data cache with the performance loss under 0.8%. Previous study has shown that the gate-oxide failure probability is proportional to the device stress time [15]. Therefore, we can expect a similar MTTF (mean time to failure) improvement for the data cache, which is 48% in our study.

For further power saving, we propose to adopt the drowsy scheme to these invalidated cachelines. We scale the power numbers provided in [10] for this study. Since the data in invalidated cachelines of our AADC design will not be needed during the drowsy mode, the performance overhead due to the wake-up operations for drowsy scheme can be ignored or overlapped with the cache miss penalty. Figure 6 shows that our AADC design can achieve a 61.5% power reduction for the data cache, which can further mitigate the aging effect.

5. CONCLUSION

The NBTI-induced aging effect is becoming a critical threat to the performance and reliability for future CMOS devices. The SRAM data caches in modern processors suffer high aging stresses due to the unbalanced duty cycle ratio of the devices. In this paper, we propose an aging-aware data cache (AADC) design to combat the NBTI-induced aging by duty cycle balancing. Based on our detailed study on the lifetime behaviors of the cachelines in the data cache, different aging reduction schemes are proposed for different lifetime phases. By applying our proposed idle-time-based invalidation for clean cachelines, early write-back and invalidation for dirty cachelines, and bit-flipping scheme for invalid cachelines, the duty cycle ratio of the entire data cache can be well balanced to ~50% with minimized overheads. For further power saving, we adopt the drowsy scheme to the invalidated cachelines in our AADC design and it can achieve a 61.5% power reduction in the data cache. Therefore, the NBTI degradation of the data cache can be significantly alleviated.

6. ACKNOWLEDGMENTS

This work was supported in part by a grant from Chinese NSF Award 61100035.

7. REFERENCES

[1] J. Abella, X. Vera, and A. Gonzalez. Penelope: The nbti-aware processor. In *Proceedings of IEEE/ACM International Symposium on Microarchitecture*, pages 85–96, 2007.

[2] J. Abella, X. Vera, O. Unsal, and A. Gonzalez. Nbti-resilient memory cells with nand gates for highly-ported structures. In *Workshop on Dependable and Secure Nanocomputing*, June 2007.

[3] A. Biswas et al. Computing architectural vulnerability factors for address-based structures. In *Proc. of the IEEE International Symposium on Computer Architecture*, June 2005.

[4] S. Borkar. Designing reliable systems from unreliable components: the challenges of transistor variability and degradation. *IEEE Micro*, 25(6):10–16, Nov. 2005.

[5] D. Brooks, V. Tiwari, and M. Martonosi. Wattch: a framework for architectural-level power analysis and optimizations. In *Proc. International Symposium on Computer Architecture*, 2000.

[6] D. Burger and T. M. Austin. The simplescalar tool set, version 2.0. Technical Report 1342, Computer Sciences Department, University of Wisconsin, 1997.

[7] A. Calimera, M. Loghi, E. Macii, and M. Poncino. Dynamic indexing: Concurrent leakage and aging optimization for caches. In *Proceedings of the ACM/IEEE International Symposium on Low-Power Electronics and Design*, pages 343–348, August 2010.

[8] A. Calimera, M. Loghi, E. Macii, and M. Poncino. Partitioned cache architectures for reduced nbti-induced aging. In *Proceedings of the Design, Automation and Test in Europe*, pages 938–943, March 2011.

[9] O. Ergin et al. Exploiting narrow values for soft error tolerance. *IEEE Computer Architecture Letters*, 5(2), July-Dec. 2006.

[10] K. Flautner, N. Kim, S. Martin, D. Blaauw, and T. Mudge. Drowsy caches: Simple techniques for reducing leakage power. In *Proc. the 29th International Symposium on Computer Architecture*, pages 148–157, Anchorage, AK, May 2002.

[11] E. Gunadi, A. Sinkar, N. Kim, and M. Lipasti. Combating aging with the colt duty cycle equalizer. In *Proceedings of the IEEE/ACM Int. Symp. on Microarchitecture*, pages 103–114, 2010.

[12] T. Jin and S. Wang. Aging-aware instruction cache design by duty cycle balancing. In *Proceedings of IEEE Computer Society Annual Symposium on VLSI*, pages 195–200, 2012.

[13] S. Kaxiras, Z. Hu, and M. Martonosi. Cache decay: Exploiting generational behavior to reduce cache leakage power. In *Proc. the Int'l Symposium on Computer Architecture*, pages 240–251, 2001.

[14] S. V. Kumar, C. H. Kim, and S. S. Sapatnekar. Impact of nbti on sram read stability and design for reliability. In *Proceedings of Int. Sym. on Quality Electronic Design*, 2006.

[15] E. Minami et al. Circuit-level simulation of tddb failure in digital cmos circuit. *IEEE Trans. on Semiconductor Manufacturing*, 8(3), Aug 1995.

[16] A. Ricketts, J. Singh, K. Ramakrishnan, N. Vijaykrishnan, and D. K. Pradhan. Investigating the impact of nbti on different power saving cache strategies. In *Proceedings of the Design, Automation and Test in Europe*, pages 592–597, March 2010.

[17] E. Rosenbaum et al. Effect of hot-carrier injection on n- and pmosfet gate oxide integrity. *IEEE Electron Device Letters*, 12(11), Nov 1991.

[18] T. Sherwood et al. Automatically characterizing large scale program behavior. In *Proc. of ASPLOS X*, October 2002.

[19] J. Shin et al. A proactivewearout recovery approach for exploiting microarchitectural redundancy to extend cache sram lifetime. In *Proceedings of International Symposium on Computer Architecture*, pages 353–362, 2008.

[20] T. Siddiqua and S. Gurumurthi. Recovery boosting: A technique to enhance nbti recovery in sram arrays. In *Proceedings of the IEEE Computer Society Annual Symposium on VLSI*, July 2010.

[21] S. Wang et al. Low power aging-aware register file design by duty cycle balancing. In *Proceedings of Conference on Design, Automation and Test in Europe*, pages 546–549, 2012.

[22] S. Wang, J. Hu, and S. G. Ziavras. On the characterization and optimization of on-chip cache reliability against soft errors. *IEEE Transactions on Computers*, 58(9):1171–1184, September 2009.

[23] W. Wang et al. The impact of nbti on the performance of combinational and sequential circuits. In *Proceedings of the Design Automation Conf.*, 2007.

[24] W. Zhang. Computing cache vulnerability to transient errors and its implication. In *Proc. of the 20th IEEE International Symposium on Defect and Fault Tolerance in VLSI Systems*, Oct. 2005.

Performance/Reliability Trade-off in Superscalar Processors for Aggressive NBTI Restoration of Functional Units

Simone Corbetta, William Fornaciari
Politecnico di Milano – Dipartimento di Elettronica e Informazione
Via Ponzio 34/5, 20133 Milano, Italy
{scorbetta, fornacia}@elet.polimi.it

ABSTRACT

Negative-Bias Temperature Instability has become a serious reliability concern in modern processors design, and in the last decade many research effort has been spent in developing circuit-level and architecture-level strategies to mitigate the induced delay variation of nanoscale circuits. At the architecture level, work has been proposed to alleviate this by appropriate dynamic instruction scheduling techniques. However, their benefit is bounded to the available redundancy, limiting their attractiveness for a cost-effective VLSI solution. This paper presents an in-depth analysis of a performance reliability trade-off FSM design, that is able to attain the desired level of reliability improvement according to the ILP performance constraints. Power-gating is used for aggressive NBTI restoration. Extensive experimental results show several performance reliability trade-off examples in a broad range of application scenarios.

Categories and Subject Descriptors

B.8 [**Performance and Reliability**]: Reliability, Testing, and Fault-Tolerance

General Terms

Design, Reliability

Keywords

Power-gating, reliability performance trade-off, NBTI

1. INTRODUCTION

As microelectronics scales down to 45 nm: and beyond, reliability is perceived as a major design concern, since the integration capability of VLSI circuits becomes strictly bound to the reliability of the circuit itself. Thanks to continuous improvements in manufacturing technology and design methodologies, scaled circuits exploit very high performance profiles at the cost of an increased power consumption density, mining their lifetime. Several failure mechanisms have been studied in nanoscale circuits, while others are still under in-depth investigation: interconnect wear-out [1] and device wear-out [2] are considered first-class design concerns in reliable designs. One of the most serious failure mechanisms is Negative-Bias Temperature Instability (NBTI). Shrinking geometries and increasing electric field across the oxide makes aging of PMOS devices a challenging source of unreliability at long-term [2]. The NBTI-induced aging manifests itself as a time-dependent increase in the absolute threshold voltage V_{th} of the transistor. This effect reduces performance modifying timing in logic circuits, and reducing operation margins in SRAM cells [23].

The solution to NBTI analytical models have guided researchers to find mitigation strategies at the circuit and architecture level: the usage duty-cycle of silicon devices has been shown to have direct influence on their degradation [22], and this dependence having a much stronger relationship at scaled supply voltages. For this reason, in the last decade NBTI-aware duty-cycle optimization gained great attention from the community (see Section 4), and several works have been proposed so far in the context of superscalar processors, where the available redundancy is a starting point for uniform stress optimization.

1.1 Key observations

Key observations are reported in this section, driving the research work proposed in this paper. In the following, we are going to consider a generic superscalar processor with an out-of-order pipeline model and an issue width of N instructions.

1.1.1 ILP performance of applications

For any given N-wide issue superscalar processor, at most N instructions can be put in execution at each cycle, bounded by the available bandwidth that is in turn determined by the microarchitecture configuration of the processor. The amount of instructions in execution depends on the available instruction-level parallelism in the application, that is generally constrained by the memory accesses and data hazards. There exist different classes of applications that are generally known to exploit an high ILP. However, published results show that in reality only a little fraction of the available bandwidth is used, even considering high-performance applications [15, 14, 10]. In these cases, on the average, only $n < N$ units are busy during run-time at each clock cycle. However, even the unused ones experience an NBTI stress, since there will be an input bitmask at specific internal nodes stressing a portion of the PMOS in the circuits.

1.1.2 Physical redundancy

Considering $n < N$ active units, the best attainable duty-cycle from an NBTI view-point is given by n/N, assuming that each of the available N units is uniformly stressed in round-robin fashion.

GLSVLSI'13, May 2–3, 2013, Paris, France.

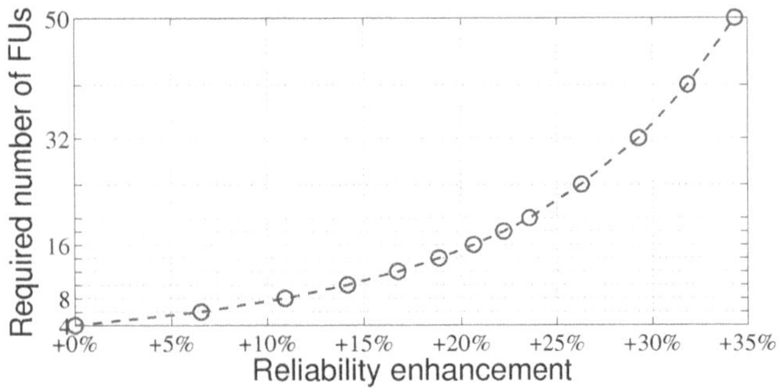

Figure 1: Area reliability relationship: the area is the one required to reduce duty-cycle for NBTI-induced reliability.

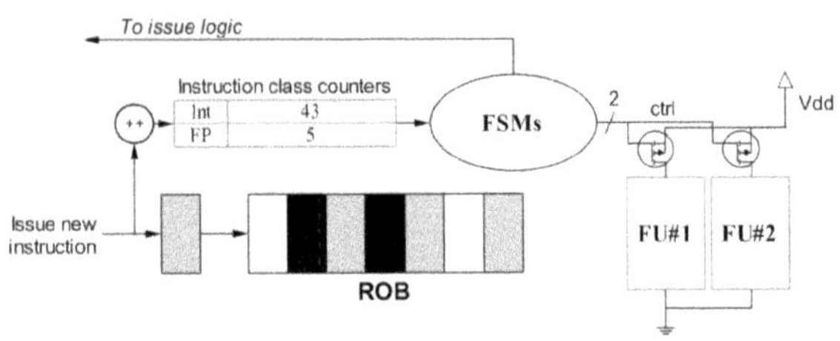

Figure 2: Architecture overview of the proposed FSM-based control strategy, highlighting the interfaces to microarchitecture blocks in an out-of-order processor.

Lower duty-cycle, indeed, is one of the control knobs to reduce the NBTI-induced threshold degradation. From a design view point, to reduced the duty-cycle usage one option is to increase the number N of physically available units. However, due to dependence of NBTI to duty-cycle, a huge amount of physical units is required to achieve a sensitive improvement in reliability. To quantitatively show this observation, preliminary results are reported in Figure 1. The vertical axis reports the number N of functional units required such that the target reliability improvement is the one desired on the horizontal axis. This relationship is based on the long-term NBTI degradation model [22]. It is clear that sensible reliability improvement is not a cost-effective solution in any real ASIC design.

1.2 Contributions and paper structure

Starting from key observations in the previous section, this paper proposes an FSM-based power-gating control mechanism for out-of-order superscalar processors to achieve a performance reliability trade-off: since the number of allocated units is lower than the number of physically available ones, those unused can be power-gated, in order to reduced the supply voltage and get aggressive NBTI restoration. Meanwhile, this strategy is cost-effective since it requires low area and power overhead, if compared to a base processor design. The remainder of this paper is organized as follows. Section 2 reports and discusses the proposed FSM-based design for dynamic performance reliability control. Extensive experimental results are given in Section 3. Related work on NBTI-aware policies is given in Section 4. Last, conclusions and future works are given in Section 5.

2. PROPOSED STRATEGY

The proposed FSM-based strategy is based on a performance reliability controller that selects a subset of the available functional units to be switched off according to the performance requirements of the processor. The logical architecture of the proposed strategy is reported in Figure 2. The performance reliability trade-off is determined by proper configuration of the states of a centralized FSM. An FSM can be attached to each class of functional unit, according to reference scenario constraints. The purpose of the FSM is to control which of the physically available units will be switched off, to tune reliability against performance requirements. In particular, there exists one FSM for each class of functional unit that is desired to be controlled, such that for instance the reference performance driving states transition is tied to the integer instructions for an integer FSM. From here on, we will consider the integer FSM without loss of generality.

2.1 Performance tracking

The class-related performance of the application/processor pair is tracked through the enumeration of the instructions belonging to that particular and that have been already inserted in the reorder buffer (ROB) for later retirement. Each time a new instruction is dispatched, its class (i.e., the functional unit type that accommodates its execution) is used to update local counters: a counter is incremented each time an integer instruction is added to the ROB, while it is decremented upon later retirement of that instruction. The percentage of integer instructions that are currently stored in the ROB determines the number of integer units that will be switched-off. Since the ROB can host up to a finite number of entries (depending on the microarchitecture configuration), its depth is logically divided into ranges: one range is defined for each state in the FSM. In general, the higher the number of ranges, the finer the granularity we might achieve in the control process. Ranges are designed to partition the whole depth of the ROB, and never intersect. These intervals are used in conjunction with state-specific saturation counters to determine when a state transition can occur. At each clock cycle, the number of (integer) instructions in the ROB is checked against the current state range: if this number falls within the upper (lower) state range, then the saturation counter is increased (decreased), otherwise it remains unchanged. Upon reaching the saturation counter value (or 0), a state transition is scheduled upward (downward). The effective state transition will then take place if and only if the number of clock cycles in the current state is higher than the wake-up latency (called the wake-up latency constraint) *and* the number of units to be gated is available (i.e., not currently busy). This might take an unspecified number of clock cycles according to the current status of the units and the state statistics. In general, the number of states in the FSM equals the number N of physically available units, where each state $j = \{1, , 2, ..., N\}$ is responsible of switching off exactly $N - j$ units. Thus we also assume that there will be at least one active unit in the processor.

2.2 Performance/reliability control parameters

In an FSM, the performance tracking support is coupled with reliability control circuitry, determining the number of physical units that should be switched off. The FMS can be customized at design-time through appropriate configuration of its parameters. There exist different control knobs that can be proportionally customized, and the intersection of these build up the desired performance reliability trade-off (see Section 3 for details on this).

The configurable parameters are limited to a simple set of configuration registers for easy of customization. Range partitioning in each state can be used to weight the probability for a state to remain active with the reliability control that the state exploits. Furthermore, the number of units to be switched off in each state (the power-gating factor), can be tuned as well with the same objective. The designer would in general jointly tune both of them in a joint. Finally, the saturation counters for upward state transition can be still used to weight application phases. Aside these design-time parameters, the wake-up latency constraint is much more related to the reference architecture and technology scenario; for this reason, it should be customized according to the ASIC flow results. In the experimental results, however, we will use these parameters to explore the design space of the power-gating control.

2.3 Microarchitecture integration

The FSM is logically interposed between the dispatch execution stages: during dispatch, instructions are added to the ROB, and the per-class counters are updated. The FSM is then reacting upon every new instruction at each clock cycle. The update of its internal status register might determine a change in the combinatorial network driving header sleep transistors on the controlled units: signals on the common control bus are used to generate sleep or wake-up signals. To guarantee uniform usage of the units at long stress times, priority rotation logic is embedded in the FSM, thus no other modification is required at the unit RTL design. The design of header PMOS transistors for power gating has already been addressed in literature [5], as well as the influence of aging on sleep transistor has been already studied in [25]. This is indeed out of the scope of this paper.

3. RESULTS

This sections reports and discusses a set of quantitative analysis of the proposed FSM-based power-gating control. In particular, the following results will be shown:

- Section 3.2 will show an in-depth analysis of the performance reliability trade-off for a particular reference application, considering different FSM configurations and the attainable IPC versus NBTI-induced delay degradation;
- Section 3.3 demonstrates that the benefits of a particular FSM configuration have a strong application dependence, considering a broad range of test cases from different benchmark suites;
- Section 3.4 will present the impact that the wake-up latency has on the number of clock cycles the FSM stays in lower states;
- Section 3.5 reports preliminary results on the digital design of the proposed FSM approach, considering standard-cell flow at 45 nm technology.

3.1 Experimental setup

Experimental results have been obtained using a modified version of the GEM5 simulator (available at www.m5sim.org): support for power-gating of selected functional units has been integrated in the release version, and the FSM model has been added. We use X86 emulation in system-call mode, thus simulating bare-metal execution of single-thread workloads in single-core environment. The common microarchitecture configuration used during our experiments is given in Table 1: parameter settings have been selected comparing real processors configuration and published configurations [17]. We then adopted the NBTI model presented in [3], developed as a stand-alone Matlab model to be used as a post-processing unit after detailed simulation has been performed. Technology parameters used to compute the final V_{th} value are selected according to ITRS predictions [13] for 32 nm process: supply voltage V_{DD} is set to 1.2 V, oxide thickness t_{ox} 1.2 nm, initial threshold voltage 160 mV and operating temperature 383 K, as previously done in literature for NBTI analysis [4, 12].

Table 1: Processor microarchitecture and memory hierarchy configuration common to experimental results.

Parameter	Value
Core	Out-of-order pipeline
Fetch/Decode/Dispatch width	4
Issue width	6
Commit width	8
ROB size	96
Branch predictor	Global w/ history
Integer/FP register file	128/80
# of integer ALU/MultDiv	2/1
# of floating-point ALU/MultDiv	1/1
# of SIMD unit	4
# of L1 ports	4
Cache line size	64B
L1 I/D cache	64kB/128kB
L2 cache	2MB

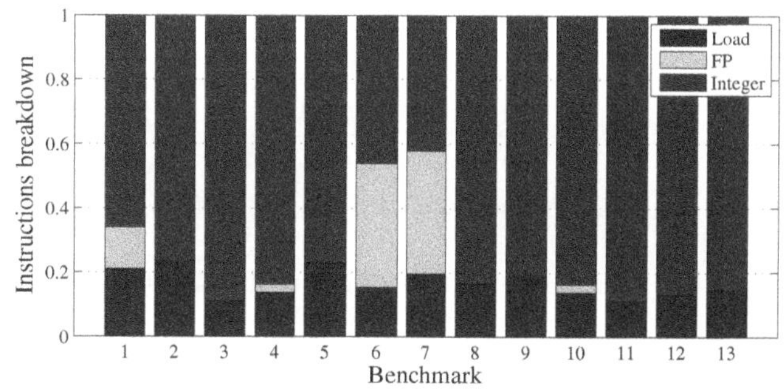

Figure 3: Instructions breakdown for different applications of selected benchmark suites.

The performance reliability trade-off curve induced by the configuration of the power-gating controller is highly dependent on the class of application that the processor is running. An exhaustive analysis of the entire design space is out of the scope of this paper. For this reason, without losing in generality, the impact of the FSM configuration is analyzed in details for an integer-intensive matrix multiplication problem. Then, several different applications from WCET [7], SPLASH-2 [21] and MiBench [8] benchmark suites will be compared against a given configuration, to show the dependence of the benefits on the application class, and to cover a broad range of use cases. Instructions breakdown are given in Figure 3 for future reference.

3.2 Performance/reliability trade-off

The main objective of the FSM circuitry is to provide a mean to control the number of gated units that should be switched-off according to the performance requirements of the application/processor pair in the reference sampling period. Sampling is done continuously at each clock-cycle, and the configuration of the saturation counters, lower and upper bounds in each state (see Section 2) determine the final state transitions pattern. In general, considering two homogeneous functional units, the higher the number of cycles the FSM remains in the lower state (representative of lower performance), the lower will be the final performance (IPC) of

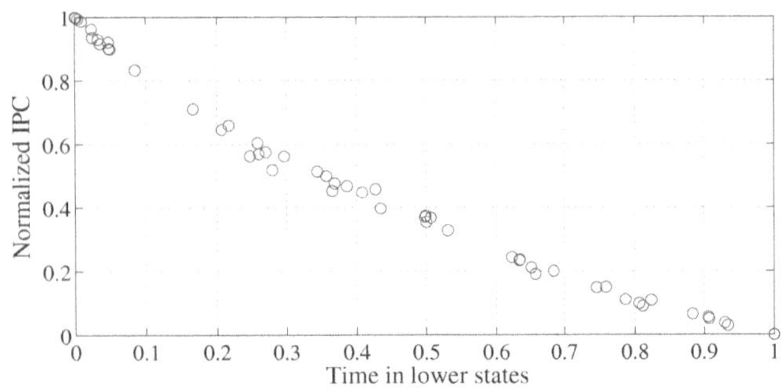

(a) Performance in lower stats

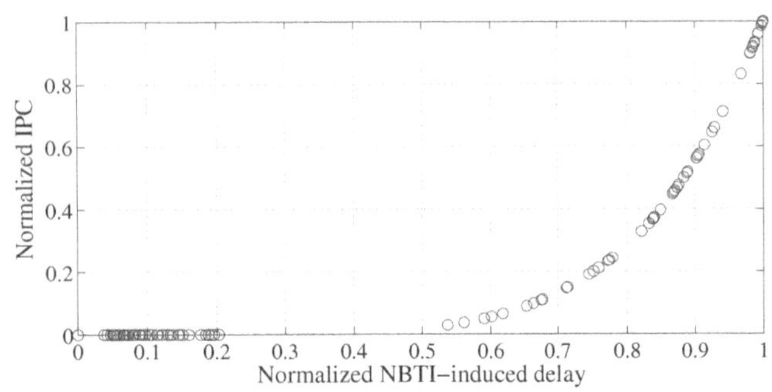

(b) Performance/reliability trade-off

Figure 4: Performance and reliability of the target processor under different FSM configurations and parameters.

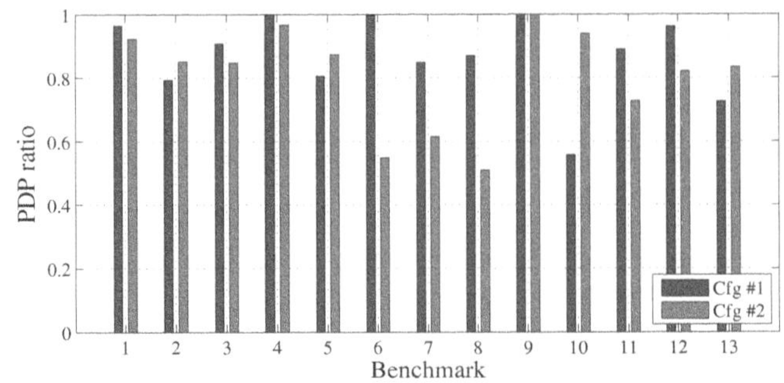

Figure 5: Performance and degradation sensitivity to running application, considering a common FSM configuration.

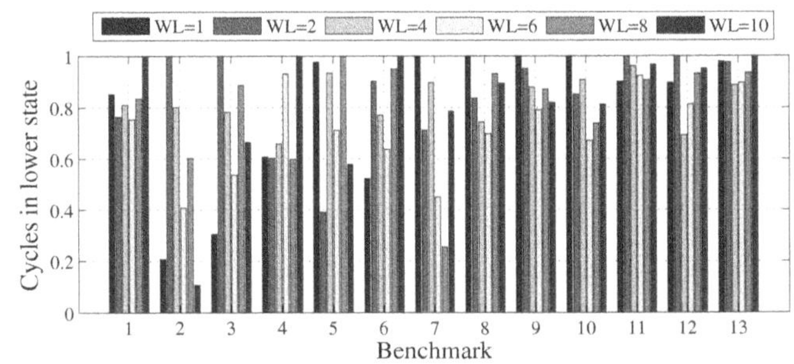

Figure 6: Wake-up latency impact on the number of cycles FSM remains in lower states, for different benchmarks.

the processor. This is indeed shown in Figure 4(a), considering a 250x250 matrix multiplication example and a total of 192 different FSM configurations. The IPC is lower when the time spent in lower states is higher, since the FSM has been designed in such a way that the power-gating factor decreases at higher states.

By tuning the number of functional units to be gated in each state, and the FSM parameters (i.e., saturation value, ranges of acceptance in each state and the like), the synthesized IPC differs and the NBTI-induced degradation will be affected: longer execution times (i.e., lower IPC) will generate higher degradation in principle, but since the functional units are not filled at their maximum pace, we can switch off unused units and get lower NBTI degradation. This is shown in Figure 4(b). Normalized degradation data is computed considering the effects of threshold voltage degradation on gate delay, following the α power-law model for short-channel effects [16].

3.3 Application dependence

Purpose of this section is to demonstrate that performance reliability trade-off induced by the proposed FSM power-gating control is highly dependent on the running application class. We performed an in-depth analysis of different benchmark suites to cover a broad range of applications for high-performance, real-time and embedded scenarios. Each application is simulated with the same common FSM configuration, and results are compared to test the application dependence. For each application, we first compute the NBTI-induced delay as done in Section 3.2, and compute the *Performance-Delay Product* (PDP) metric as a ground for quantitative comparison. The PDP metric is defined as the product of commit IPC (performance factor) and NBTI-induced delay (degradation factor), and it is proportional to the classical *Energy-Delay Product* (EDP) by means of some coefficient relating performance and power consumption. Figure 5 reports the obtained results. The vertical axis reports the ratio of the PDP metric when power-gating is applied to the PDP metric when no power-gating control is considered (but still degradation is experienced): the closer the ratio to unity, the lower the overall benefits of the selected FSM configuration on the application behavior and processor degradation.

From Figure 5 it is clear how the application has great impact on the benefits of a particular FSM configuration. However, an exhaustive exploration of the design-space for optimal configuration assignment is neither feasible nor valuable, since the configuration space is too wide, and running applications change at runtime. This gives then free room for dynamic adaptation of the FSM configuration to running applications pattern.

3.4 Impact of wake-up latency

The wake-up latency of a power-gating design is a function of several design and technology considerations [18]. Furthermore, the actual wake-up latency is known only after the back-end flow of a classical ASIC design has been performed. Nevertheless, there has been experimental evidence in reference designs reporting a minimum wake-up latency of 10 clock cycles [9], avoiding power and performance overhead. We then use this as a starting value for our evaluation, and estimate during simulation the impact that an increasing number of break-even cycles has on the performance of the power-gated design. In particular, we force the FSM to remain in a state (see Section 2) for an increasing number of cycles, ranging from 10 to 100, and then evaluate the number of clock cycles the FSM remains in lower states. Results are given in Figure 6. For each benchmark, 6 different wake-up latency configurations are given (considering 1X and 10X multiplication factor with respect to 10 clock cycles base case); data on the vertical axis is given normalized to the maximum number of clock cycles in the lowest state

Table 2: Synthesis results against PTM-based 45 nm standard-cell library.

Hierarchy	Area	Power
Top	$2.206\,mm^2$ (+1%)	91.51 W (+5%)
FSM	$23\,\mu m^2$ (19%)	3.66 mW (19%)
Rotation logic	$64\,\mu m^2$ (25%)	3.8 mW (19%)
Power-gating net	$23\,\mu m^2$ (8%)	0.6 mW (3%)
Other	$106\,\mu m^2$ (48%)	9 mW (59%)

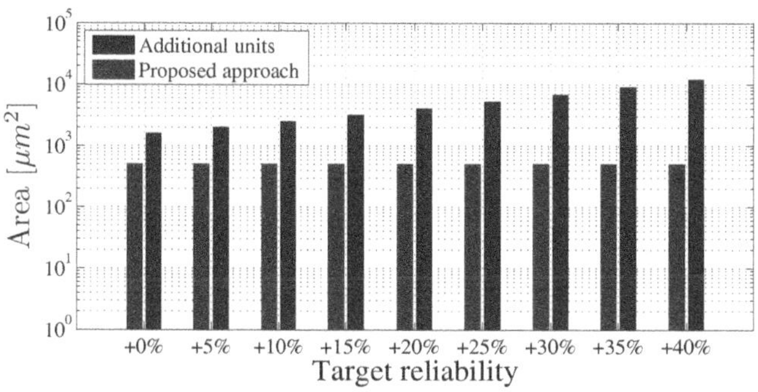

Figure 7: Area requirements comparison between the proposed approach and the use of additional units.

for that particular benchmark in that wake-up latency configuration. Two conclusions are clear from the data: at first, as before, the same configuration has a different influence on the application behavior, according to different applications. Furthermore, it is clear that within the same application there is not a strictly monotonic relationship between the number of cycles in lower states and the number of cycles to wake an unit up. This is mainly due to the different phases an application undergoes: the probability to wake-up a unit depends on the wake-up latency constraint (since we force the FSM to remain in a state for at least the same amount of cycles), *and* on the chance that the ROB gets to be filled with integer instructions (in case on integer units). In this way, the higher the number of cycles to be in a state (i.e., the wake-up latency), the higher the probability to incur in an application phase in which the saturation counter decreases *or* increases.

3.5 Digital design

The proposed FSM controller has been deigned and integrated in a representative out-of-order superscalar processor, from the FabScalar project toolset [6]. The configuration of the processor follows the simulation configuration given in Table 1. The design has then been synthesized against a 45 nm standard-cell library based on the PTM model [11], using Cadence RTL Compiler flow. Preliminary power, timing and area results are obtained to assess the validity of the approach to be integrated in a real full processor design. Table 2 show the synthesis results at different hierarchy levels: area and power values of the top design are given as absolute and percentage values with respect to the original unmodified design. Values of power and timing refer to typical PVT settings at 45 nm (1.0 V reference supply voltage and operating temperature 25°C). From this preliminary analysis, the FSM accounts for less than 1% area overhead, but still a 5% higher power consumption. Notice that, however, the critical path does not change, and thus maximum operating frequency is not affected.

Notice that the area overhead introduced by the proposed design is much lower than the introduction of additional spare units, at the same reliability levels. Indeed, Figure 7 shows the area overhead of the proposed reliability strategy compared to the area overhead required by an introduction of an additional number of functional units to accommodate the same target reliability improvement (see Section 1.1).

4. RELATED WORKS

The long-term model [22] highlights a true dependence of NBTI aging on the stress/recovery phases alternation: the duty-cycle parameter then expresses the probability of having a stress signal (i.e., a negative voltage or a logical 0 in CMOS inverters) at the input node of a circuit, determining the amount of threshold voltage that gets degraded over time.

Solutions to this issue have been proposed in literature at different levels of abstraction. At transistor level, designers have a few amount of design parameters to deal with NBTI issues [1], but performance requirements have to be generally traded off for reliability aspects. At the circuit level, the duty-cycle can be controlled through the Input Vector Control (IVC) methodology, by which inputs are periodically fed with appropriate bit-masks to wear circuit nodes uniformly [24]. The authors show that benefits are achieved, and both reliability is increased and leakage power is reduced. In memory structures, e.g. SRAM cells, the adoption of IVC approach ensures higher stability. The IVC technique is a general methodology that can be applied in different environments; however it requires precise knowledge of the stress history of each node within the circuit (or within critical paths of the circuit), making it less attractive for VLSI systems. The stress/phase alternation is also addressed at higher levels. The authors in [4], in particular, present different techniques to mitigate aging in microprocessors focusing on dynamic instruction scheduling strategies. Experimental results on representative out-of-order models show that round-robin allocation produces the lowest degradation due to optimal duty-cycle, since functional units experience a comparable amount of degradation as opposed to classical fixed priority scheduling. Time-dependent strategies are also analyzed: by *logically* switching off units after they are used (i.e., they are considered unavailable for a fixed number of clock cycles) they achieve reliability benefits with low performance overhead. However, they do not consider switching them off effectively, e.g. through power gating mechanisms, and as long as supply voltage is present benefits are limited. Thanks to the advance of multi-core architectures, authors in [20] perform NBTI-driven application scheduling at system-level, based on the concept of *workload capacity*: the degradation is assumed to be computed at run-time, and cores are ranked according to a delay-driven performance metrics. Similar balancing schemes are adopted to schedule and alternate cores between busy and idle phases [19].

5. CONCLUSIONS

NBTI mitigation is an important topic for nanoscale circuits and aggressive scaled VLSI designs. Circuit-level and architecture-level techniques are based on proper optimization of the duty-cycle of stress/recovery phases. However, main limitations of NBTI-directed policies do not make them affordable for cost-effective aggressive mitigation. This paper presents an FSM-based power-gating control scheme that achieves the desired performance reliability trade-off in a superscalar out-of-order processor. Experimental results have shown the benefits of power-gating to trade reliability improvement against performance requirements. However, the strong application dependence requires adaptation of the FSM configuration parameters to dynamically changing application scenarios.

6. ACKNOWLEDGMENTS

This research work is partially supported by European Community Seventh Framework Programme (FP7/2007-2013), under agreements no. 248716 (`2PARMA` project `www.2parma.eu`) and Artemis no. 100230 (`SMECY` project `www.smecy.eu`).

7. REFERENCES

[1] M. Alam, K. Kang, B. Paul, and K. Roy. Reliability- and process-variation aware design of vlsi circuits. In *14th International Symposium on the Physical and Failure Analysis of Integrated Circuits.*, pages 17 –25, july 2007.

[2] M. A. Alam and S. Mahapatra. A comprehensive model of PMOS NBTI degradation. *Microelectronics Reliability*, 45(1):71–81, 2005.

[3] S. Bhardwaj, W. Wang, R. Vattikonda, Y. Cao, and S. Vrudhula. Predictive modeling of the nbti effect for reliable design. In *IEEE Custom Integrated Circuits Conference.*, pages 189 –192, sept. 2006.

[4] T.-B. Chan, J. Sartori, P. Gupta, and R. Kumar. On the efficacy of nbti mitigation techniques. In *Design, Automation Test in Europe Conference Exhibition (DATE), 2011*, pages 1 –6, march 2011.

[5] D.-S. Chiou, Y.-T. Chen, D.-C. Juan, and S.-C. Chang. Sleep transistor sizing in power gating designs. In *ASIC, 2007. ASICON '07. 7th International Conference on*, pages 1326 –1331, oct. 2007.

[6] N. K. Choudhary, S. V. Wadhavkar, T. A. Shah, H. Mayukh, J. Gandhi, B. H. Dwiel, S. Navada, H. H. Najaf-abadi, and E. Rotenberg. Fabscalar: composing synthesizable rtl designs of arbitrary cores within a canonical superscalar template. In *Proceedings of the 38th annual international symposium on Computer architecture*, ISCA '11, pages 11–22, New York, NY, USA, 2011. ACM.

[7] J. Gustafsson, A. Betts, A. Ermedahl, and B. Lisper. The Mälardalen WCET Benchmarks - Past, Present and Future. In *Proceedings of the 10th International Workshop on Worst-Case Execution Time Analysis*, July 2010.

[8] M. R. Guthaus, J. S. Ringenberg, D. Ernst, T. M. Austin, T. Mudge, and R. B. Brown. Mibench: A free, commercially representative embedded benchmark suite. In *Proceedings of the IEEE International Workshop on Workload Characterization.*, pages 3–14, Washington, DC, USA, 2001.

[9] Z. Hu, A. Buyuktosunoglu, V. Srinivasan, V. Zyuban, H. Jacobson, and P. Bose. Microarchitectural techniques for power gating of execution units. In *Proceedings of the 2004 international symposium on Low power electronics and design*, ISLPED '04, pages 32–37, New York, NY, USA, 2004. ACM.

[10] A. Kejariwal, A. Veidenbaum, A. Nicolau, X. Tian, M. Girkar, H. Saito, and U. Banerjee. Comparative architectural characterization of spec cpu2000 and cpu2006 benchmarks on the intel core 2 duo processor. In *Embedded Computer Systems: Architectures, Modeling, and Simulation, 2008. SAMOS 2008. International Conference on*, pages 132 –141, july 2008.

[11] W. Zhao and Y. Cao. Predictive technology model for nano-cmos design exploration. *J. Emerg. Technol. Comput. Syst.*, 3(1), Apr. 2007.

[12] S. Li, J. H. Ahn, R. D. Strong, J. B. Brockman, D. M. Tullsen, and N. P. Jouppi. Mcpat: an integrated power, area, and timing modeling framework for multicore and manycore architectures. In *Proceedings of the 42nd Annual IEEE/ACM International Symposium on Microarchitecture*, MICRO 42, pages 469–480, New York, NY, USA, 2009. ACM.

[13] MASTAR - Model for Assessment of CMOS Technology and Roadmaps `http://www.itrs.net/models.html`.

[14] A. Nair and L. John. Simulation points for spec cpu 2006. In *Computer Design, 2008. ICCD 2008. IEEE International Conference on*, pages 397 –403, oct. 2008.

[15] T. K. Prakash and L. Peng. Performance characterization of spec cpu 2006 benchmarks on intel core 2 duo processor. *Computers and Software Engineering, ISTAT Transactions on*, 2(1):36–41, 2008.

[16] K. Saluja, S. Vijayakumar, W. Sootkaneung, and X. Yang. Nbti degradation: A problem or a scare? In *21st International Conference on VLSI Design*, pages 137 –142, jan. 2008.

[17] R. Sasanka, S. V. Adve, Y.-K. Chen, and E. Debes. The energy efficiency of cmp vs. smt for multimedia workloads. In *Proceedings of the 18th annual international conference on Supercomputing*, ICS '04, pages 196–206, New York, NY, USA, 2004. ACM.

[18] Y. Shin, J. Seomun, K.-M. Choi, and T. Sakurai. Power gating: Circuits, design methodologies, and best practice for standard-cell vlsi designs. *ACM Trans. Des. Autom. Electron. Syst.*, 15(4):28:1–28:37, Oct. 2010.

[19] J. Sun, A. Kodi, A. Louri, and J. Wang. NBTI aware workload balancing in multi-core systems. In *International Symposium on Quality of Electronic Design*, pages 833 –838, march 2009.

[20] J. Sun, R. Lysecky, K. Shankar, A. Kodi, A. Louri, and J. Wang. Workload capacity considering nbti degradation in multi-core systems. In *15th Asia and South Pacific Design Automation Conference*, pages 450 –455, jan. 2010.

[21] The modified SPLASH-2 benchmark suite `http://www.capsl.udel.edu/splash/`.

[22] W. Wang, V. Reddy, A. Krishnan, R. Vattikonda, S. Krishnan, and Y. Cao. Compact modeling and simulation of circuit reliability for 65-nm cmos technology. *Device and Materials Reliability, IEEE Transactions on*, 7(4):509 –517, dec. 2007.

[23] W. Wang, S. Yang, S. Bhardwaj, R. Vattikonda, S. Vrudhula, F. Liu, and Y. Cao. The impact of nbti on the performance of combinational and sequential circuits. In *Proceedings of the 44th annual Design Automation Conference*, DAC '07, pages 364–369, New York, NY, USA, 2007. ACM.

[24] Y. Wang, X. Chen, W. Wang, V. Balakrishnan, Y. Cao, Y. Xie, and H. Yang. On the efficacy of input vector control to mitigate nbti effects and leakage power. In *International Symposium on Quality of Electronic Design*, pages 19 –26, march 2009.

[25] M.-C. Lee, Y.-G. Chen, D.-K. Huang, and S.-C. Chang. Nbti-aware power gating design. In *Design Automation Conference (ASP-DAC), 2011 16th Asia and South Pacific*, pages 609 –614, jan. 2011.

Delay Model for Reconfigurable Logic Gates Based on Graphene PN-junctions

Sandeep Miryala, Andrea Calimera,
Enrico Macii, Massimo Poncino
Politecnico di Torino, 10129, Torino, ITALY

ABSTRACT

In this paper we address the problem of modeling the timing behavior of a new class of reconfigurable logic gates based on electrostatically controlled graphene pn-junctions. These gates naturally behave as a 2-to-1 multiplexer in which the polarity of the input select line can be dynamically reconfigured. Interconnection of multiple gates and proper assignments of the inputs signals allow to implement all the basic Boolean logic functions, and, at a larger scale, any digital circuit.

Exploiting the symmetric structure of the graphene device, we first identify the main in-to-out timing arcs of the gate, then we derive the parametric models that accurately describe the propagation delay across them as function of the input signal transition slope and the output capacitive load. Delay equations are fit to SPICE-level simulation data by means of bilinear and linear interpolation methods; the obtained regression coefficients are stored and used as substitutes of the huge lookup-tables typically adopted in CMOS timing libraries.

Experimental results show that the proposed model allows accurate timing analysis with average and peak errors below 0.6% and 3.1% respectively.

Categories and Subject Descriptors

B.6.3 [**Design Aids**]: Simulation

Keywords

Graphene, PN junction, reconfigurable gate, timing model

1. INTRODUCTION

Transparent and stretchable, with high thermal-mechanical resistance and unique electrical qualities [1], graphene is emerging as one of the viable candidate to replace silicon in the post-CMOS electronic era. First isolated in 2004 [9, 10, 8], graphene is available in nature in the form of stacked sheets, i.e., graphite. It shows a two-dimensional (2D) structure in which carbon atoms are packed in an honeycomb lattice with a bond length a_{C-C} of 1.42Å, Figure 1 (a). The resulting energy band structure E is analytically reported in the Equation 1 and plotted in Figure 1 (b)[1].

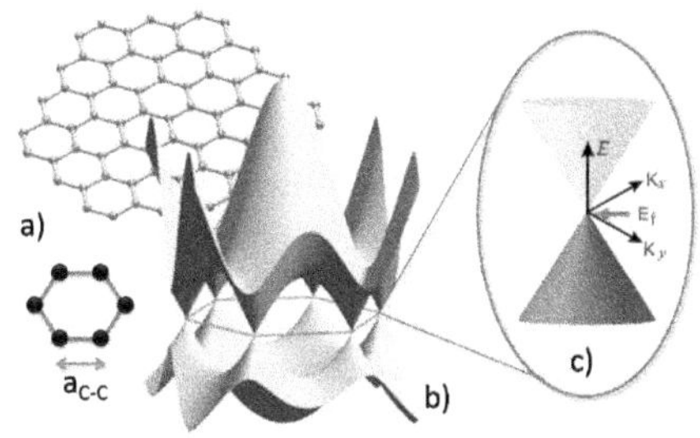

Figure 1: The graphene honeycomb lattice (a), its energy band structure (b) and the linear energy dispersion around the Fermi Energy E_F (c)

$$E^{\pm}(\mathbf{k}) = \pm\gamma\sqrt{1 + 4\cos\left(\frac{\sqrt{3}a}{2}k_x\right)\cos\left(\frac{a}{2}k_y\right) + 4\cos^2\left(\frac{a}{2}k_y\right)} \quad (1)$$

The positive and negative energy branches are the conduction band (E^+) and the valence band (E^-) respectively; the vector $\mathbf{k} = \{k_x, k_y\}$ represents the 2D wavevector; γ is a fitting parameter whose value can range from 2.7eV to 3.3eV; $a = \sqrt{3}a_{C-C}$ is the side length of the parallelogram representing the primitive cell in the Bravis lattice [12]. From the plot one can notice that conduction and valence curves touch each other near the edges of the Brillouin Zone, i.e., at zero energy, where the Fermi energy E_F passes (Figure 1 (c)).

The gapless energy spectrum gives graphene semi-metallic properties that do not allow the material to implement the *off* state. Nevertheless, recent works have jointly exploited (i) the massless propagation of carriers in graphene, and (ii) the existence of a tunneling property called *Klein's tunneling*, to devise an electrostatically controlled pn-junction [6, 2] that allows the control of the current flow. Such a pn-junction serves as basic element for a more complex reconfigurable gate, the target of this work, that conceptually implements a reconfigurable 2-to-1 multiplexer [13] (RG-MUX hereafter). The RG-MUX consists of a graphene sheet with co-planar back-gates through which it is possible to alter the doping profile of the graphene sheet; this allows to select which of the two input signals connected at the front metal

[1]Interested readers can refer to [11] and [12] for a formal derivation, which is out of the scope of this work.

contacts of the device is forwarded to the output. The analysis reported in [13], [3] and [7] indicates that the RG-MUX can outperform CMOS technologies below the 22nm node in terms of both delay and power consumptions.

The main contribution of this work is to provide a compact timing model for the RG-MUX. It is worth emphasizing that the proposed model has been conceived with the aim of supporting future logic synthesis tools for graphene ICs, and not as a substitute of analytical equations for device-level simulations. The methodology we followed recalls the NLDM-based approach (NLDM stands for Non-Linear Delay Model) widely employed in CMOS technologies for the creation of standard cells timing libraries. The symmetric structure of the device allowed us to group the timing arcs of the RG-MUX in two main classes, one including *back-to-output* timing arcs, i.e., those from back-gates to the output, the other including *front-to-out* timing arcs, i.e., those that traverse the device from the front metal contacts to the output. SPICE-level characterization data obtained simulating a verilog-A model of the device have been used for model fitting. We demonstrate that the regression coefficients obtained using bilinear and linear interpolation (for back-to-out and front-to-out arcs respectively) are constant over the entire characterization space, thereby enabling LUT-free static timing analysis.

Experimental results show that the proposed model allows accurate timing analysis with an average and peak error below 0.6% and 3.1% respectively w.r.t. SPICE simulations.

2. BACKGROUND

2.1 PN-Junction Graphene Device

Figure 2 shows 3D, back and front views of the RG-MUX as introduced in [13]. The device consists of a graphene sheet, three back-gates, $\bar{U}$, S and U, that are isolated from the graphene by a thick layer of oxide, and three upper metal-to-graphene contacts, A, B, (inputs) and Z (output).

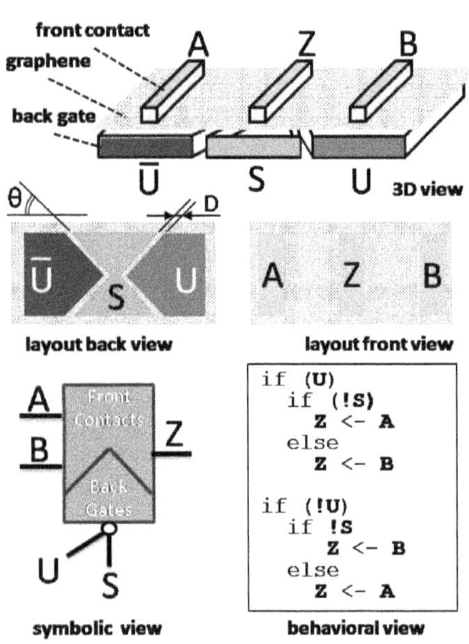

Figure 2: RG-MUX design views.

The voltages applied at the back-gates $\bar{U}$, S and U, control the doping profile of the two resulting back-faced pn-junctions, S-$\bar{U}$ on the left and S-U on the right; they perform an electrostatic doping [1] through which the Fermi energy E_F can be shift down in the valence band, or up in the conduction band, leading to p-type or n-type graphene respectively.

Pins $\bar{U}$ and U are the *configuration pins* and are always controlled by a complemented logic voltage; the pin S works as selective pin of the MUX. When U='1' ($\bar{U}$='0'), the graphene region above the U gate results as n-doped (the Fermi energy E_F moves up in the conduction band due to positive control voltage), while that above the $\bar{U}$ gate as p-doped (the Fermi energy E_F moves down in valence band due to a negative control voltage). Notice that the '1'-logic value is $+V_{dd}/2$ and the '0'-logic value is $-V_{dd}/2$. Under such configuration, when S='0', the central graphene becomes p-type forming a pp-junction on the left and a pn-junction on the right; the pp-junction shows a low resistance (R_{pp}), whereas the pn-junction a high resistance (R_{pn}). Therefore, the output Z follows the input signal associated with the smallest resistive path, i.e., A. On the contrary, when S ='1', the central graphene becomes n-type forming a low resistive nn-junction on the right (R_{nn}); this forces the output Z to follow B. A dual behavior is observed when U='0' ($\bar{U}$='1'). Figure 2 gives the symbolic and behavioral view of the gate.

2.2 Electrical Model

Borrowing the work of [13], we implemented a Verilog-A compact device model whose equivalent circuit is drawn in Figure 3.

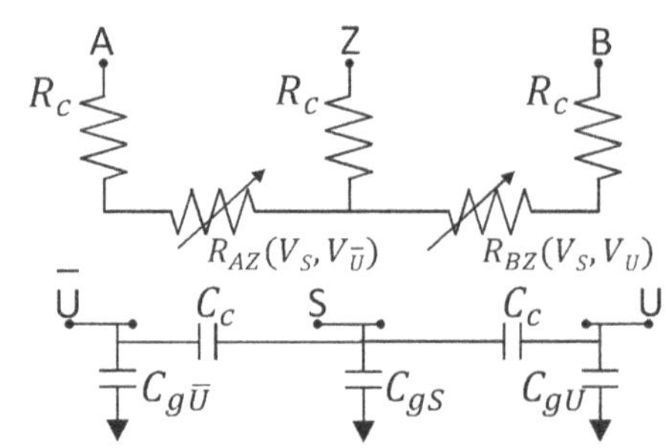

Figure 3: RG-MUX electrical model.

The two resistors R_{AZ} and R_{BZ} model the resistive graphene path from the input pins A and B to the output Z respectively. Their value ranges from $R_{nn} = R_{pp}$ (that is the on state) when the back-gates are polarized with concordant voltages, to R_{pn} (that is the off state) when the back-gates are polarized with discordant voltages.

The value of R_{nn} is given by:

$$R_{nn} = R_0/N_{ch} \tag{2}$$

where $R_0 = \frac{h}{4q^2}$ is the quantum resistance per mode and N_{ch} is the number of excited modes in the graphene sheet. The following table reports the equation of N_{ch} for R_{AZ} and R_{BZ}:

R_{AZ}	R_{BZ}
$N_{ch} = \frac{W}{\pi} min(K_{F,S}, K_{F,\bar{U}})$	$N_{ch} = \frac{W}{\pi} min(K_{F,S}, K_{F,U})$
$K_{F,\bar{U}} = \frac{2\pi E_{F,\bar{U}}}{h\nu_F}$	$K_{F,U} = \frac{2\pi E_{F,U}}{h\nu_F}$
$K_{F,S} = \frac{2\pi E_{F,S}}{h\nu_F}$	

where W is the device width, while $K_{F,\bar{U}}$, $K_{F,S}$, $K_{F,U}$ are the Fermi wave vectors on the graphene regions above the back-gates $\bar{U}$, S and U respectively; all them depend on the Fermi energy E_F whose level is function of the voltage V_g applied at the corresponding back gate as:

$$E_F = \frac{1}{\gamma t_{ox}}\left(\sqrt{\epsilon^2 + 2\gamma\epsilon q|V_g|t_{ox}} - \epsilon\right) \quad (3)$$

In the above equation t_{ox} is the oxide thickness, ϵ is the permittivity of the oxide, the constant $\gamma = 4\pi q^2/h^2\nu_F^2$, h is the Planck's constant, ν_F is the Fermi velocity which is of the order of 10^6m/s.

The R_{pn} can be estimated including the transmission probability T_θ of the carriers across the pn-junction:

$$R_{pn} = \frac{R_o}{N_{ch}T(\theta)} \quad (4)$$

$T(\theta)$ depends on (i) the angle θ between the electron's wave vector (K_F) and the normal of the junction and (ii) the distance D of the transition regions (please refer to back view in Figure 2). Equation 5 gives the expression for $T(\theta)$.

$$T(\theta) = \begin{cases} \cos^2(\theta)e^{-\pi D min(K_{F,S},K_{F,\bar{U}})\sin^2\theta} & \text{for } R_{AZ} \\ \cos^2(\theta)e^{-\pi D min(K_{F,S},K_{F,U})\sin^2\theta} & \text{for } R_{BZ} \end{cases} \quad (5)$$

$K_{F,\bar{U}}$, $K_{F,S}$, $K_{F,U}$ are the same used for N_{ch}, while θ is 45° by construction.

The electrical model also includes parasitics of the metal contacts. R_c (assumed to be 10Ω) represents the resistance of a metal-to-graphene contact [13] and it has been applied at the three front metal contacts, A, B and Z.

The lumped capacitance C_g at the back-gates (i.e., $C_{g\bar{U}}$ at $\bar{U}$, C_{gS} at S, C_{gU} at U) consists of the series of the oxide capacitance C_{ox} and the quantum capacitance of the graphene sheet C_q, i.e., $C_g = 1/(C_{ox}^{-1} + C_q^{-1})$. C_{ox} is the conventional parallel plate capacitance and its value per unit of area is given by $C_{ox} = \frac{\epsilon}{t_{ox}}$, where ϵ and t_{ox} are the dielectric constant and the dielectric thickness respectively. The quantum capacitance C_q is a function of the density of carriers at the Fermi energy E_F (Equation 3) and its value per unit of area is given by $C_q = \gamma E_F$.

3. DELAY MODEL

Figure 4 shows the four different timing arcs in the RG-MUX gate: from the select pin S to the output Z (*S-to-Z*); from the reconfiguration pin U to the output pin Z (*U-to-Z*), from input data pins A or B to output pin Z (*A-to-Z* or *B-to-Z*).

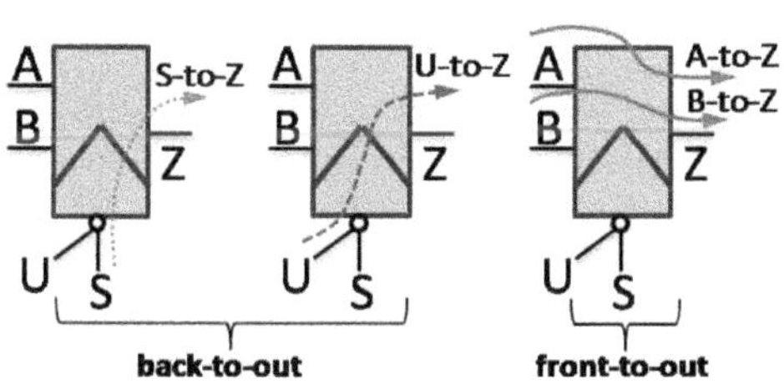

Figure 4: In-to-out RG-MUX timing arcs.

Thanks to the symmetric structure of the device (i) all the timing arcs show same low-to-high (LH) and high-to-low (HL) propagation delays and (ii) the arcs *A-to-Z* and *B-to-Z* are electrical equivalent, hence, they have same propagation delay. For these properties, we can simplify the modeling procedure by focusing on LH delays only and ignoring the equivalent path *B-to-Z* during characterization.

Another important property is that the timing behavior along the arcs is primarily determined by the type of terminals involved. This allows us to identify two main categories: *back-to-out* arcs, i.e., those involving the back-gates (*S-to-Z* and *U-to-Z*), which show a non-linear relationship w.r.t. input transition times t_{in} and capacitive load C_l; *front-to-out* arcs, i.e., those involving the front metal contacts (*A-to-Z* and *B-to-Z*), which show a linear behavior w.r.t. C_l only. For the former category we use a non-linear delay model (NLDM), whereas a linear delay model (LDM) is successfully employed for the latter category.

3.1 NLDM for back-to-out timing arcs

Table 1 resumes LH back-to-out timing arcs.

Table 1: Back-to-out timing arcs.

	A	B	S	U	Z
S-to-Z	0	1	0→1	1	0→1
U-to-Z	0	1	0	0→1	0→1

As introduced in Section 2, the device shows an RC electrical behavior. As per the Elmore delay model [5] the propagation delay Dp is proportional to $R_{device} \cdot C_l$, with R_{device} the in-to-out device resistance and C_l the output load capacitance at pin Z. R_{device} is not a constant, but it changes dynamically with the voltages applied at the back gates S and U, like the channel resistance of MOS transistors changes with the gate voltage. The input signals are modeled by a linear ramp with slope that varies linearly with the transition time t_{in}. This results into $Dp \propto R_{device}(t_{in}) \cdot C_l$, which is a non linear equation that well fits with the non-linear delay model (NLDM) Synopsys employed for CMOS gates characterization [4]:

$$Dp(t_{in}, C_l) = A \cdot t_{in} + B \cdot C_l + C \cdot t_{in} \cdot C_l + D \quad (6)$$

Each timing arc has four coefficients associated with it (A, B, C, D in Equation 6) that are calculated using the bilinear interpolation method pictorially described in Figure 5. A 2D lookup table (LUT) stores the delay measures obtained from SPICE. Each entry in the table represents the delay of the timing arc at predefined points in the $\{t_{in}, C_l\}$ characterization space. When determining the response of the device for a given coordinate (t_{in-x}, C_{l-x}), the four points in the table that bound the desired coordinate are determined: (t_{in1}, C_{l1}), (t_{in1}, C_{l2}), (t_{in2}, C_{l1}), and (t_{in2}, C_{l2}) in Figure 5. The four coordinates determine four entries in the characterization table, i.e., f_{11}=LUT(t_{in1}, C_{l1}), f_{12}=LUT(t_{in1}, C_{l2}), f_{21}=LUT(t_{in2}, C_{l1}), f_{22}=LUT(t_{in2}, C_{l2}) in Figure 5.

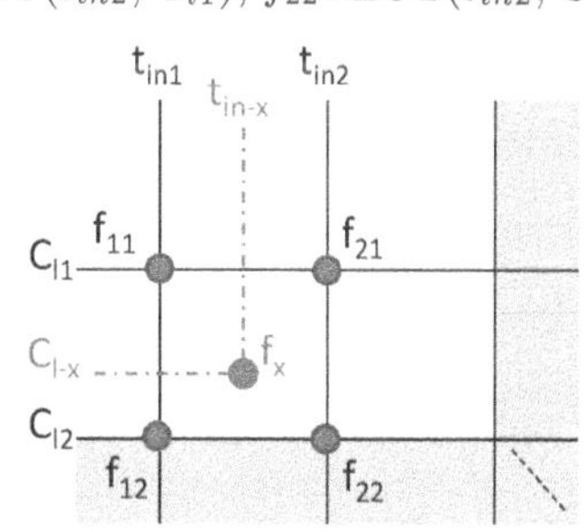

Figure 5: Bilinear interpolation.

The coefficients A, B, C and D of the characteristic equation that describes the gate's delay in the region enclosed in the square (f_{11}, f_{12}, f_{21}, f_{22}) are finally calculated by solving the system of equations reported below:

$$\begin{bmatrix} f_{11} \\ f_{12} \\ f_{21} \\ f_{22} \end{bmatrix} = \begin{bmatrix} t_{in1} & C_{l1} & t_{in1} \cdot C_{l1} & 1 \\ t_{in1} & C_{l2} & t_{in1} \cdot C_{l2} & 1 \\ t_{in2} & C_{l1} & t_{in2} \cdot C_{l1} & 1 \\ t_{in2} & C_{l2} & t_{in2} \cdot C_{l2} & 1 \end{bmatrix} \times \begin{bmatrix} A \\ B \\ C \\ D \end{bmatrix} \tag{7}$$

which results in the following regression formulas:

$$\begin{cases} A = \frac{C_{l2}(f_{21}-f_{11})+C_{l1}(f_{12}-f_{22})}{(t_{in2}-t_{in1})(C_{l2}-C_{l1})} \\ B = \frac{t_{in2}(f_{12}-f_{11})+t_{in1}(f_{21}-f_{22})}{(t_{in2}-t_{in1})(C_{l2}-C_{l1})} \\ C = \frac{(f_{11}-f_{12})+(f22-f21)}{(t_{in2}-t_{in1})(C_{l2}-C_{l1})} \\ D = \frac{(t_{in2}C_{l2}f_{11}-t_{in1}C_{l2}f_{21}-t_{in2}C_{l1}f_{12}+t_{in1}C_{1}f_{22})}{(t_{in2}-t_{in1})(C_{l2}-C_{l1})} \end{cases} \tag{8}$$

For CMOS technologies the selection of the data stored in the 2D characterization LUT plays an important role in determining the accuracy of the model. This is due to the fact that the error on the NLDM coefficients may increase significantly when the four coordinates used as interpolation points (f_{11}, f_{12}, f_{21}, f_{22}) are too far from the coordinate of interest (t_{in-x}, C_{l-x}). Obviously, for a given characterization range, i.e., $t_{in} \in [t_{in-min} : t_{in-max}]$; $C_l \in [C_{l-min} : C_{l-max}]$, the larger the number of entries in the LUT, the more accurate the delay estimation [4]. However, as we show later in the experimental section, this issue is immaterial for the RG-MUX. We prove that using the four coefficients calculated at the most outer corners of the characterization space, i.e., (t_{in-min}, C_{l-min}), (t_{in-min}, C_{l-max}), (t_{in-max}, C_{l-min}), (t_{in-max}, C_{l-max}), we get negligible errors across the entire range, whatever the size of the LUT. This represents a key feature that might enable LUT-free static timing analysis in future RG-MUX based designs flows.

3.2 LDM for front-to-out timing arc

Table 2 resumes the two equivalent LH front-to-out timing arcs.

Table 2: Front-to-out timing arcs.

	A	B	S	U	Z
A-to-Z	$0 \to 1$	0	0	1	$0 \to 1$
B-to-Z	0	$0 \to 1$	1	1	$0 \to 1$

As one can observe from the table, both S and U are kept at fixed values. Hence, differently from the back-to-out arcs, the resulting device resistance R_{device} is a constant, that is, it does not change with the input signals and their transition time t_{in}. This results into $Dp \propto R_{device} \cdot C_l$, which is a linear equation that well fits with a linear delay model (LDM):

$$Dp(C_l) = E \cdot C_l + F \tag{9}$$

The bilinear interpolation described for back-to-out arcs reduces to a simple one-dimension linear interpolation. A 1D LUT stores the delay measures obtained from SPICE simulations. Each entry represents the delay at predefined C_l points in the characterization space. When determining the response of the device for a given C_{lx}, the two points in the table that bound the desired coordinate are determined, i.e., C_{l1} and C_{l2}. The corresponding delay values at these points, i.e., $f_1 = LUT(C_{l1})$ and $f_2 = LUT(C_{l2})$, are then used to calculate the coefficients E and F using the following formulas:

$$\begin{cases} E = \frac{(f_2-f_1)}{(C_{l2}-C_{l1})} \\ F = f_1 - E \cdot C_{l1} \end{cases} \tag{10}$$

As for *back-to-out* arcs, we show that the optimal selection of the C_l points used for the construction of the characterization LUT does not represent an issue.

4. MODEL VALIDATION

4.1 Methodology flow

Figure 6 shows the flow we implemented for testing and validating the proposed timing models.

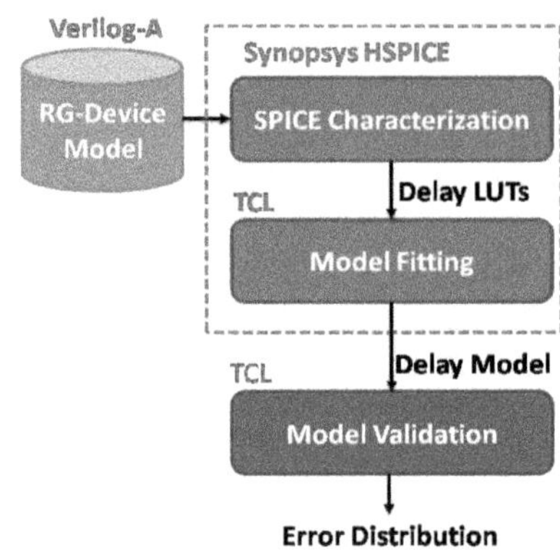

Figure 6: Methodology flow and tools.

For the simulation we used a standard SPICE netlist in which the RG-MUX, instantiated with a macro containing the Verilog-A model presented in Section 2, is driven by piece-wise linear voltage sources that emulate input signals. The characterization spans in the range $t_{in} \in \{10ps : 25ps\}$, $C_l \in \{1e^{-20}F : 1e^{-16}F\}$. A total of 50 equally spaced points in these ranges are considered for characterization. The output generated by Synopsys HSPICE are then collected in dedicated 50x50 LUTs; each timing arc has a LUT associated with it.

The interpolation methods presented in Section 3 have been integrated within procedures written in TCL, the de-facto standard language for modern EDA tools. The model fitting tool takes as input the characterization LUT of the timing arc under analysis and returns the model coefficients for a specific t_{in-x}, C_{l-x} point of interest.

The model validation stage consists of dedicated algorithms that calculate the error distribution across the characterization LUT when using a fixed set of coefficients. The main goal is to understand if it is possible to find a fixed set of coefficients that work over the entire characterization space with a reasonable error. In other words, we want to check if is possible to implement LUT-free timing analysis.

Algorithms 1 and 2 report the pseudo-code used in the validation stage for *back-to-out* and *front-to-out* timing arcs respectively. Let us first consider the former case, i.e., Algorithm 1. It takes as input the characterization LUT, and two arrays Cl and tin that contains the indices of the LUT. At each iteration of the **for** loop i, the model coefficients are calculated using as entries the four corners ($f11$, $f12$, $f21$, $f22$ in the code) of a submatrix of length i having the top-left corner stacked in the origin ($f11 = LUT(tin(1), Cl(1))$). The size of the submatrix increases with i from 2x2 to 50x50. The obtained coefficients (line 8) are then used in the two inner **for** loops j and k that iterate over all the points in the LUT for calculating the average and peak interpolation errors. The error is defined as the the difference between

Algorithm 1 NLDM error analysis

```
1: INPUT: LUT, tin, Cl, LUT_Size;;
2: for i = 2 to LUT_Size do
3:   f11=LUT[tin(1),cl(1)];
4:   f12=LUT[tin(1),cl(i)];
5:   f21=LUT[tin(i),cl(1)];
6:   f22=LUT[tin(i),cl(i)];
7:   calculate A, B, C, D;
8:   err_total=0;
9:   err_max=0;
10:   for j=1 to LUT_Size do
11:     for k=1 to LUT_Size do
12:       Dp= A*tin(k)+ B*Cl(j)+ C*tin(k)*Cl(j)+ D;
13:       err=abs((Dp - LUT(j,k));
14:       err_tot = err_tot + err;
15:       if err>err_max
16:           err_max=err;
17:       end if
18:     end for
19:   end for
20:   Avg_Err[i]=err_tot/(LUT_Size*LUT_Size);
21:   Peak_Err[i]=err_max;
22: end for
```

the delay calculated using NLDM (line 13) and that measured with HSPICE ($LUT(j,k)$). *Avg_Err* and *Peak_Err* and then stored in two dedicated arrays (line 21, 22).
Algorithm 2 refers to *front-to-out* timing arcs for which the propagation delay depends only on the load Cl. The code works as for the NLDM case, but submatrix now reduces to a 1D array (tin is now fixed). At each iteration of the **for** loop i, the two model coefficients are calculated using as entries the first and last items ($f1$ and $f2$) of a subarray of length i having the first item stacked at 1 ($f1 = LUT(Cl(1))$). The size of the array increases with i from 2 to 50. The inner **for** loop j is then used to calculate average and peak errors across the 1D LUT. *Avg_Err* and *Peak_Err* and then stored in two dedicated arrays.

Algorithm 2 LDM error analysis

```
INPUT: LUT, Cl, LUT_Size;
for i = 2 to LUT_Size do
  f1=LUT(Cl(1));
  f2=LUT(Cl(i));
  calculate E, F;
  err_tot=0;
  err_max=0;
  for j=1 to LUT_Size do
    Dp= E*cl(j)+ F;
    err=abs((Dp - LUT(j,1));
    err_tot=err_tot+err;
    if err>err_max
        err_max=err;
    end if
  end for
  Avg_Err[i]=err_tot/(LUT_Size);
  Peak_Err[i]=err_max;
end for
```

4.2 Results

As intuitively described in Section 3, due to the symmetric structure of the device (i) timing arcs have same LH and HL delays and (ii) the arcs *A-to-Z* and *B-to-Z* are equivalent. To sustain this claim, Table 3 shows the delays of the *front-to-out* arcs measured with HSPICE for min/max input transition times (columns t_{in-min} and t_{in-max}) and min/max load capacitance (columns C_{l-min} and C_{l-max}) of the characterization space. It is worth emphasizing that the obtained data are compliant with the numbers reported in [3]. It can be inferred that (i) LH and HL delays are the same, and (ii) there is no delay difference between the two *front-to-out* timing arcs, therefore, the delay model and the model coefficients that hold for *A-to-Z* are also valid for *B-to-Z*. As additional comment, one can observe that delays are independent from t_{in}, therefore, we can safely apply the LDM.

Table 3: LH and HL delays for front-to-out arcs.

		t_{in-min}		t_{in-max}	
		A-to-Z	B-to-Z	A-to-Z	B-to-Z
C_{l-min}	LH	1.09e-14	1.09e-14	1.06e-14	1.06e-14
	HL	1.09e-14	1.09e-14	1.06e-14	1.06e-14
C_{l-max}	LH	13.8e-14	13.8e-14	13.8e-14	13.8e-14
	HL	13.8e-14	13.8e-14	13.8e-14	13.8e-14

A more interesting analysis is graphically proposed in the Figures 7, 8, 9, which show the average (top) and peak (bottom) error distribution as returned from Algorithms1 and 2, i.e., the arrays $Avg\_Err[i]$ and $Peak\_Err[i]$.
For all the timing arcs (*S-to-Z* in Figure7, *U-to-Z* in Figure8, and *A-to-Z* in Figure 8) the percentage of error decreases for larger i, i.e., as the square length of the submatrix (subarray for LDM) get larger. Minimum average and peak errors, obtained for i=50, are always below 0.6% and 3.1% respectively, that is, accurate enough to integrate these models along with their coefficients in future LUT-free STA tools for graphene digital ICs.
Table 4 shows the model coefficients that guarantee the minimum error, namely, those extracted using as interpolation points the most outer corner of the characterization LUT (i.e., i=50). Notice that the arcs *S-to-Z* and *U-to-Z* use NLDM, hence have coefficients A, B, C and D, whereas *A-to-Z* and *B-to-Z* uses LDM which has two coefficients E and F. Since the RG-MUX is symmetric w.r.t. inputs A and B, the coefficients for the two *front-to-out* timing arcs are same.

Table 4: Model coefficients for Low-to-High and High-to-Low in-to-out propagation delays.

Back-to-out	A	B	C	D
S-to-Z	0.36	$6.30 \cdot 10^{3}$	$-4.53 \cdot 10^{13}$	$-2.00 \cdot 10^{-14}$
U-to-Z	0.36	$4.38 \cdot 10^{3}$	$2.13 \cdot 10^{13}$	$6.22 \cdot 10^{-16}$
Front-to-out	E	F	-	-
A-to-Z	1380	$-1.7 \cdot 10^{-16}$	-	-
B-to-Z	1380	$-1.7 \cdot 10^{-16}$	-	-

As further proof of model accuracy, Figure 10 finally plots model (dotted lines) vs. HSPICE (discrete points) as function of C_l; model coefficients are those reported in Table4. It can be seen that both the NLDM and the LDM well match with the SPICE responses of the device.

5. CONCLUSIONS

In this paper we applied conventional LUT-based timing modeling approaches to identify the delay model of a new class of pn-junction based reconfigurable graphene devices.

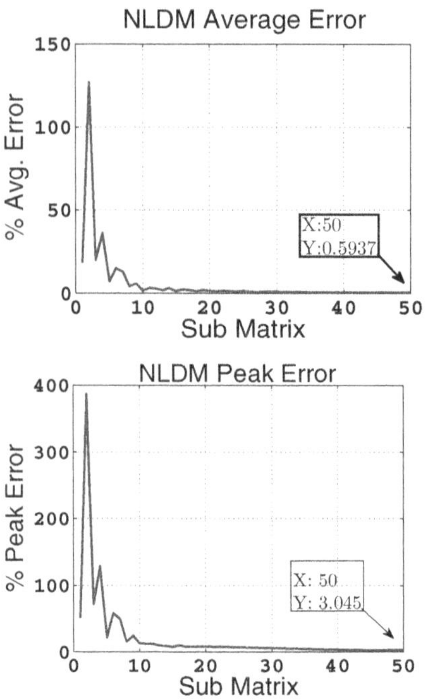

Figure 7: S-to-Z: average (top) and peak (bottom) error

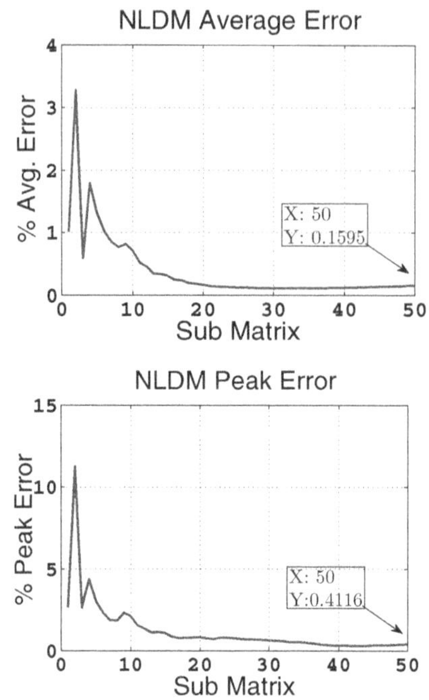

Figure 8: U-to-Z: average (top) and peak (bottom) error

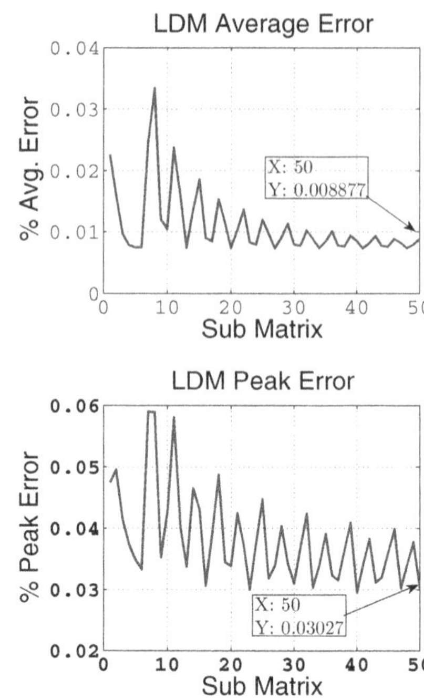

Figure 9: A-to-Z: average (top) and peak (bottom) error

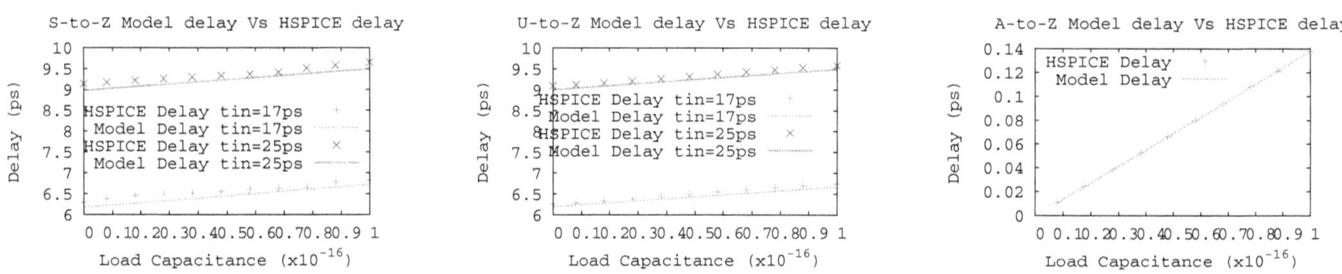

Figure 10: Model vs. HSPICE: S-to-Z(left), U-to-Z(center), A-to-Z(right)

We implemented a dedicated strategy that allowed us to extract the regression coefficients of the models and validate them for a wide range of input signal transition times and load capacitances.

The analysis has shown that the proposed model allows accurate delay estimation for all the in-to-out timing arcs of the device (average and peak estimation errors below 0.6% and 3.1%) with no need of storing large characterization LUT as typically done in todays' CMOS timing libraries.

6. REFERENCES

[1] A. H. Castro Neto, F. Guinea, N. M. R. Peres, K. S. Novoselov, and A. K. Geim. The electronic properties of graphene. *Reviews of Modern Physics*, 81:109–162, Jan. 2009.

[2] V. V. Cheianov, V. Fal'ko, and B. L. Altshuler. The focusing of electron flow and a veselago lens in graphene p-n junctions. *Science*, 315(5816):1252–1255, 2007.

[3] P. Chenyun and A. Naeemi. Device- and system-level performance modeling for graphene p-n junction logic. In *ISQED'12: IEEE International Symposium on Quality Electronic Design*, pages 262–269, Mar. 2012.

[4] J. F. Croix and D. F. Wong. Fast and accurate technique to optimize characterization tables for logic synthesis. In *DAC'97: ACM/IEEE Design Automation Conference*, pages 337–340, Jun. 1997.

[5] W. C. Elmore. The transient response of damped linear networks with particular regard to wideband amplifiers. *Journal of Applied Physics*, 19(1):55–63, Jan. 1948.

[6] M. I. Katsnelson, K. S. Novoselov, and A. K. Geim. Chiral tunneling and the klein paradox in graphene. *Nature Physics*, 2(9):620–625, Sep. 2006.

[7] S. Miryala, M. Montazeri, A. Calimera, E. Macii, and M. Poncino. A verilog-a model for reconfigurable logic gates based on graphene pn-junctions. In *DATE'13: ACM/IEEE Design, Automation and Test in Europe*, pages 1–4, Mar. 2013.

[8] K. S. Novoselov, A. K. Geim, S. V. Morozov, D. Jiang, M. I. Katsnelson, I. V. Grigorieva, S. V. Dubonos, and A. A. Firsov. Two-dimensional gas of massless dirac fermions in graphene. *Nature*, 438(7065):197–200, 2005.

[9] K. S. Novoselov, A. K. Geim, S. V. Morozov, D. Jiang, Y. Zhang, S. V. Dubonos, I. V. Grigorieva, and A. A. Firsov. Electric field effect in atomically thin carbon films. *Science*, 306(5696):666–669, 2004.

[10] K. S. Novoselov, D. Jiang, F. Schedin, T. Booth, V. Khotkevich, S. Morozov, and A. K. Geim. Two-dimensional atomic crystals. In *Proceedings of the National Academy of Sciences of The United States of America*, pages 10451–10453, 2005.

[11] H.-S. Philip Wong and D. Akinwande. *Carbon nanotubes and graphene device physics*. Cambridge University Press, 2010.

[12] R. Saito, G. Dresselhaus, and M. Dresselhaus. *Physical properties of carbon nanotubes*. World Scientific, 1998.

[13] S. Tanachutiwat, J. U. Lee, W. Wang, and C. Y. Sung. Reconfigurable multi-function logic based on graphene p-n junctions. In *DAC'10: ACM/IEEE Design Automation Conference*, pages 883–888, Jun. 2010.

Analysis of the Area-Delay Performance of Hybrid Nanoelectronic Memory Cores Used in Field Programmable Gate Arrays

Qin Wang, Arne Heittmann and Tobias G. Noll
Chair of Electrical Engineering and Computer Systems
RWTH Aachen University
Schinkelstr. 2, D-52062 Aachen, Germany

ABSTRACT
In this paper, an area-delay metric (AT) of hybrid memristive/CMOS memory architectures is discussed. The proposed memory circuit can be used as a lookup table in field programmable gate arrays (FPGAs) and is modeled by a passive nanoelectronic crossbar comprising resistive switches (RS) as memory elements. In particular, resistive switches which are based on the electrochemical metallization effect (ECM) are assumed. At the periphery, CMOS circuits are included with provide appropriate voltage levels for robust read and write operations. The optimization of the CMOS periphery was done for a 40-nm CMOS technology, and especially in regard to the physical properties of ECM cells which allows for the derivation of a realistic AT metric for the memory core circuit. For different architectural choices, the evaluation of the AT metric shows, that the area overhead which is caused by the peripheral CMOS circuits significantly determines the total circuit area. Under almost all conditions the required silicon area per bit is far beyond $4F^2$ (F: lithographic resolution), but for particular conditions the effective area per bit is smaller than the area per bit which is required for a corresponding SRAM core circuit.

Categories and Subject Descriptors
B.7.1 [Integrated Circuits]: Gate Arrays;

General Terms
Theory, Performance

Keywords
nanoelectronics, resistive switches, hybrid circuit, electrochemical metallization, scaling

1. INTRODUCTION
Resistive switches (RS) have gained considerable attention since these devices combine non-volatility with multilevel storage capabilities in an area efficient way. RS are commonly seen as an area-efficient successor to well-established storage devices.

GLSVLSI'13, May 2–3, 2013, Paris, France.

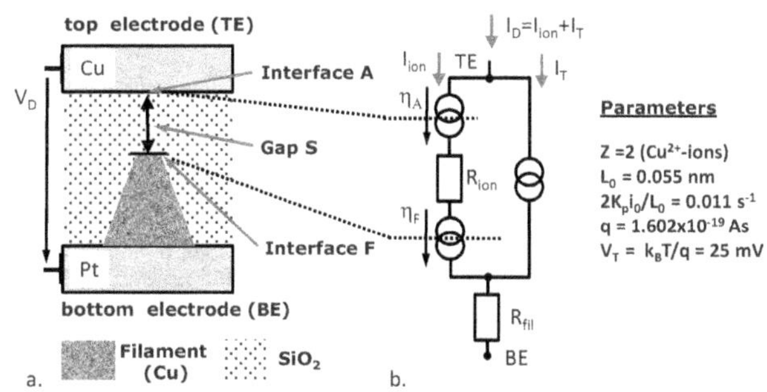

Figure 1: a. Cross section of an ECM device based on $Cu/SiO_2/Pt$. b. equivalent circuit which models ionic and tunnel current over a gap of size S.

Their comparatively simple setup may help to overcome the technological as well as economical limits anticipated for scaling CMOS beyond the 10nm node [1]. Potentially, an RS can be implemented with an area occupation of $4F^2$ (F: lithographic resolution), and crossbars comprising an array of RSs provide the highest possible storage density. Since RS are passive devices their utilization requires additional circuits. If RS are used in memory applications, circuits for addressing, signal conversion, and signal amplification are required which cause a considerable area overhead. Although a whole bunch of potential applications of RSs were reported in literature, such as logic which is based on look-up tables (LUTs) and programmable wiring junctions [2], and specialized applications such as filters [3] and neuromorphic circuits [4], less is known about a realistic performance metric (including area, power consumption and delay) of hybrid RS/CMOS circuits. Since such a performance metric is required in order to elaborate a successful roadmap of requirements as well as applications for hybrid circuits, in this work a first step towards a realistic performance metric is provided. The proposed metric includes a simplified area and delay model of small-capacity hybrid RS/CMOS-memory arrays comprising up to 64 bit storage capacity, which could be used as lookup tables (LUTs) in reconfigurable field programmable gate arrays (FPGAs). Based on a physical model for a particular class of RSs and a specification in regard to voltage margins for robust read and write operations, the required CMOS periphery was optimized towards a minimal area-delay product. The results constitute the set of potential area-delay tradeoffs and allows for the comparison with a state-of-the-art SRAM-based LUT.

2. THE ECM DEVICE MODEL
In this work, passive crossbars (pCBs) comprising resistive switches, which are based on the electrochemical metallization

effect (ECM) [5-8], were analyzed. Fig. 1a shows a sketch of the cross section of an ECM cell, which consists of three layers: two different metallic electrodes are on top and bottom, and the electronically insulating material in the middle acting as the ion conducting layer. Typically, the active electrode is copper (Cu), the inert electrode is platinum (Pt), the switching layer is silicon dioxide (SiO_2). If a positive voltage (V_D) is applied between the active and the inert electrode, copper ions (Cu^{2+}) move into the ion conducting layer and drift towards the counter electrode. The deposition of ions on the counter electrode results in the continuous growth of a copper filament. The distance between the top electrode (interface A) and the counter electrode (interface F) is indicated by a gap of size S. The current density for the charge transfer across the electrolyte-electrode interface F during the cathodic reduction is described by the Butler-Volmer-equation [6]

$$I_{ion} = A_{fil} \cdot i_0 \cdot \left(e^{\alpha \cdot z \cdot \eta_F / V_T} - e^{(\alpha-1) \cdot z \cdot \eta_F / V_T} \right). \quad (1)$$

In (1) A_{fil} denotes the area of the filament surface, i_0 is the reaction rate at zero voltage, V_T is the temperature voltage, z the charge count of the involved ions, η_F the potential difference across the interface F (outer Helmholtz layer) and α a symmetry factor [9]. A similar equation holds for the interface A. If the areas at the interfaces A and F are considered to be equal, the potential differences η_F as well as η_A are equal to $V_D/2$. Since the ionic charge transport is associated with material transport towards the filament (Farady's law), the filament growth is proportional to the ion current density and results in an effective shrink of the gap size S by

$$\frac{dS}{dt} = -K_p \cdot \frac{I_{ion}}{A_{fil}}. \quad (2)$$

If S becomes sufficiently small the electronic current is conducted via tunneling through this gap [6]. The tunnel current over this barrier can be approximated to

$$I_T \approx \frac{q^2}{4 \cdot \pi \cdot h} \cdot e^{-S/L_0} \cdot \frac{A_{fil}}{S \cdot L_0} \cdot V_D. \quad (3)$$

In (3) the definition

$$L_0 = \frac{1}{\sqrt{2 m_e \cdot q \cdot \phi}} \cdot \frac{h}{4 \cdot \pi} \quad (4)$$

was used. I_T, V_D, ϕ, and S denote the tunnel current, the applied device voltage, the barrier height, and the gap size. In (4) A_{fil}, m_e, q, and h denote the active filament surface, the effective electron mass, elementary charge, and Plack's constant. The I-V characteristics (3) is linear and can be described by

$$I_D = G \cdot V_D. \quad (5)$$

If the RS is used as a memory element at least two states have to be distinguished. A logical '0' is represented by a

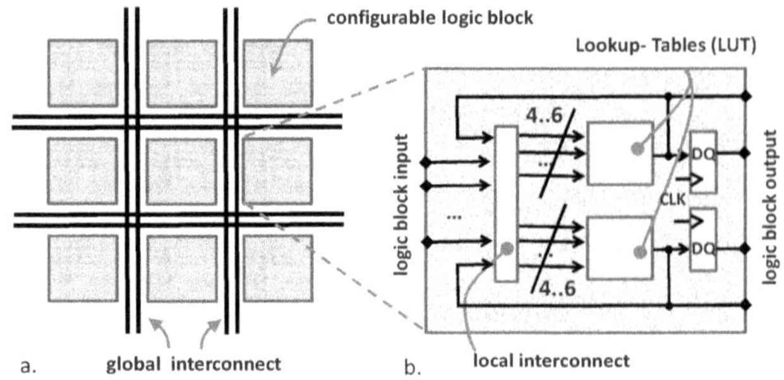

Figure 2: a. FPGA architecture comprising configurable logic blocks and global interconnect; b. Logic block structure with Look-Up-Table (LUT) and local interconnect.

high resistive state (HRS, conductance G_{OFF}), and a logical '1' is represented by a low resistive state (LRS, conductance G_{ON}). By variation of S between $1\text{nm} > S > 0.1\text{nm}$ the conductance G of the ECM device can be varied between more than 6 orders of magnitude. This has two consequences. First, the HRS as well as LSR can be separated by a high G_{ON}/G_{OFF} ratio which has advantages in sensing of stored states in small capacity memory arrays. Second, the magnitude of G_{ON} can be chosen almost freely which is helpful in order to match the electrical properties of the peripheral active circuits.

In order to reliably detect the state of an RS by a sensing circuit during a read operation some requirements which regard to the precision of the programmed conductance have to be fulfilled [10]. Due to variability at the nanoscale the programming of an ECM device is done using an active feedback loop by application of a sequence of short voltage pulses [10]. Since the programming of a state is done in an iterative and controlled way it is not necessary to provide an exact programming voltage. However, the effective write time of a write cycle is significantly affected as soon as the programming voltage deviates from the nominal programming voltage in the range of the temperature voltage V_T.

3. LUT ARCHITECTURE

Small capacity memories are typically found in field programmable gate arrays (FPGAs). The functionality of an FPGAs can be (re-)programmed after silicon fabrication is completed. A typical layout of the FPGA is given by an array of configurable logic blocks that implement the desired functionality of the circuit, and a hierarchy of an user-programmable routing network that provides the interconnect architecture between the logic blocks, see Fig. 2a. Most state-of-the-art FPGAs use logic blocks comprising look-up-tables (LUTs) for the implementation of Boolean functions with limited number of input bits, see Fig. 2b. Typically, each LUT maps a 4 bit (or 6 bit) input onto a single output bit [11].

The architecture of an RS-based LUT is shown in Fig.3. The LUT consists of a passive crossbar (pCB) with $m+1$ word lines and n bit lines comprising $n \times (m+1)$ memory cells in total.

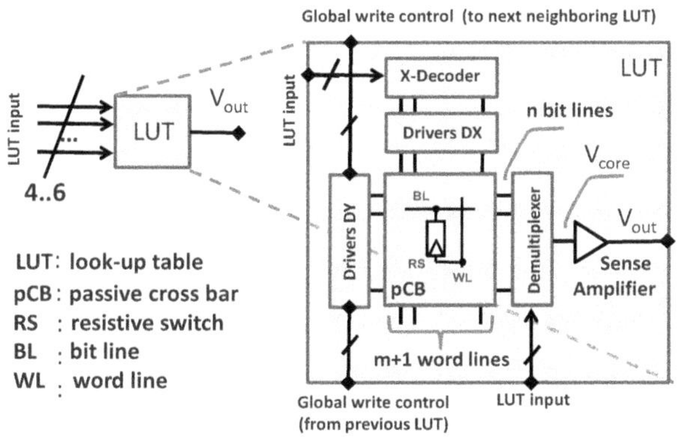

Figure 3: Architecture of a LUT which is based on a passive crossbar comprising resistive switches.

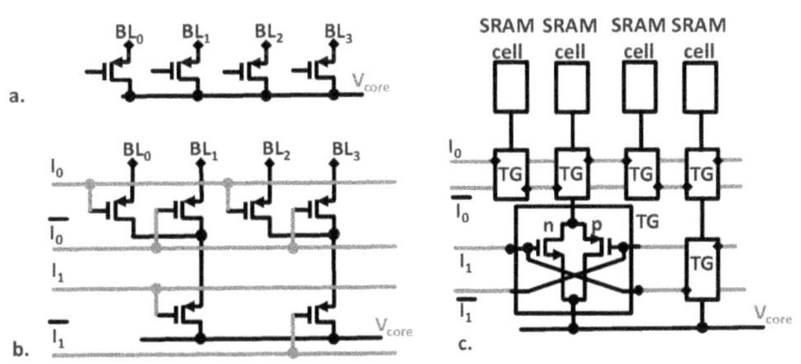

Figure 4: Demultiplexer circuits. a. flat demultiplexer, used for n =4. b. tree demultiplexer, used for n = 8,16, and 32. c. tree demultiplexer with transfer gates, for comparison of SRAM-based LUT architecture.

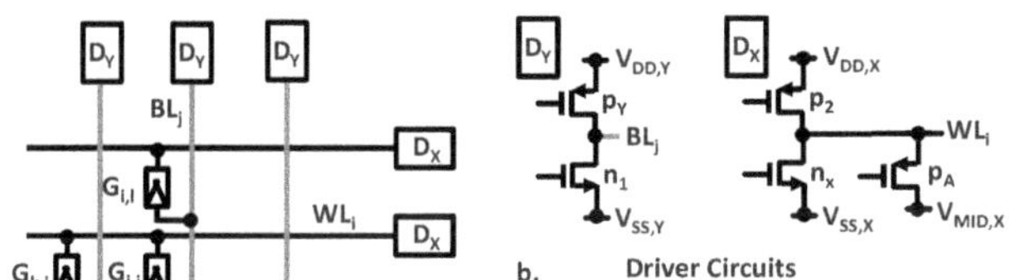

	$V_{DD,Y}$	$V_{SS,Y}$	$V_{DD,X}$	$V_{SS,X}$	$V_{MID,X}$
write 1	$2/3V_P$	0	V_P	$1/3V_P$	V_P-V_R
write 0	V_P	$1/3V_P$	$2/3V_P$	0	V_P-V_R
read	V_P	0	V_P	$1/3V_P$	V_P-V_R

c.

Figure 5: a. pCB with driver circuits DX, DY, and demultiplexer. b. circuit implementation of DX and DY. c. voltage levels of the supply lines for read and write operation.

Here, nxm memory cells are used to store the desired mapping from the input to the output. The remaining word line comprising n RSs is necessary in order to preserve one reference conductance per bit line which is required to be set to the LRS if all cells connected to the respective bit line are in the HRS. Otherwise, the reference conductance is set to the HRS. This scheme guarantees that the effective internal driving conductance of a bit line is at least G_{ON}. During a read operation, i.e. the circuit maps the LUT input (4 or 6 bit) onto the output bit, some input bits are used to select appropriate voltage levels for the word lines via the X-decoder. The remaining input bits of the LUT are used to select a particular bit line via the demultiplexer, cf. Fig.4a-b. After a short delay, the voltage V_{core} is equal to the voltage level obtained at the selected bit line.

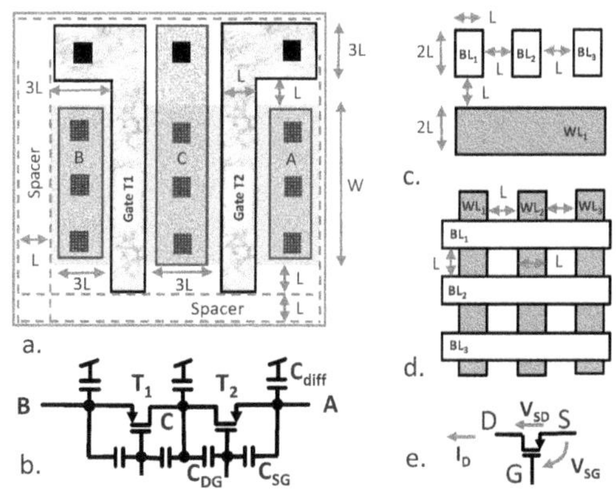

Figure 6: a. minimum layout (top view) of two transistors with common drain electrode C, b. circuit diagram of (a) including parasitic diffusion capacitance C_{diff}, source-gate capacitance C_{SG}, and drain-gate capacitance C_{DG}, c. interconnect model, two layers (cross section) d. interconnect model (top view) e. pmos transistor symbol, voltages and drain current I_D.

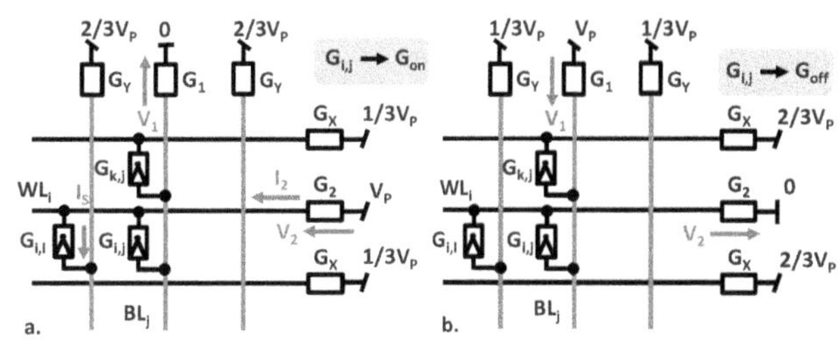

Figure 7: a. equivalent circuit for setting G_{ij} to G_{ON}. b. equivalent circuit for setting G_{ij} to G_{OFF}

The voltage V_{core} represents the associated logic state and is transformed via the sense amplifier to a full-swing output signal which can be used for driving the following logic stages. Note, that the internal voltage levels of the ECM-based LUT are typically much smaller (i.e. in the range of 200mV for the read operation in order to keep the cell's state stable) than the supply voltage of the CMOS circuits (i.e. nominal 0.9V for 40-nm CMOS, [1]). Therefore, additional circuits are necessary which transform the CMOS levels into appropriate internal voltage levels that comply with the electrical requirements of the ECM cells.

In order to change the state of an ECM cell (i.e. during the configuration of a LUT) a programming voltage is applied to selected cells. A positive voltage sets the conductance of a selected cell to the ON state (G_{ON}), and a negative voltage sets the conductance to the OFF state (G_{OFF}). From the analysis made in [10] it turns out that programming voltages in the range of 1V are sufficient to change the cell's state within microseconds, i.e. the CMOS supply voltage is sufficient to be used for writing. For the subsequent discussion the programming voltage is set to V_P=V_{DD}=0.9V and the read voltage is set to V_R=0.2V.

3.1 Area Model of ECM-based LUT

Fig.5 shows a detailed view of the driver circuits DX as well as DY. Dependent on the operation (read or write) the circuits DX as well as DY provide the required voltage levels for word lines and bit lines. DY is only active during the configuration phase and is turned off during a read operation. The control of DY is realized via global write control signals which are shared between neighboring LUTs. DX, in turn, is active during read and write and is controlled via the X-decoder.

In order to derive the area metric for the proposed LUT architecture the sizes of the circuits DY, DX, the X-decoder and the demultiplexer, have to be determined. For the following analysis the sense amplifier is excluded.

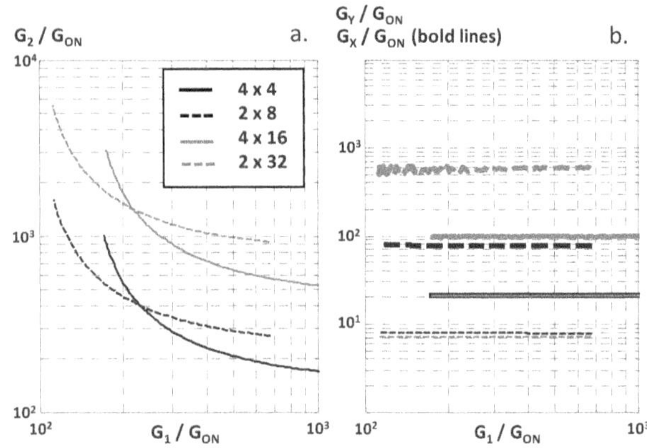

Figure 8: Driving conductances in relation to the ON-conductance G_{ON}. a. G_2 versus G_1 for different pCB configurations mxn. b. G_y, G_x (bold) versus G_1 for different configurations.

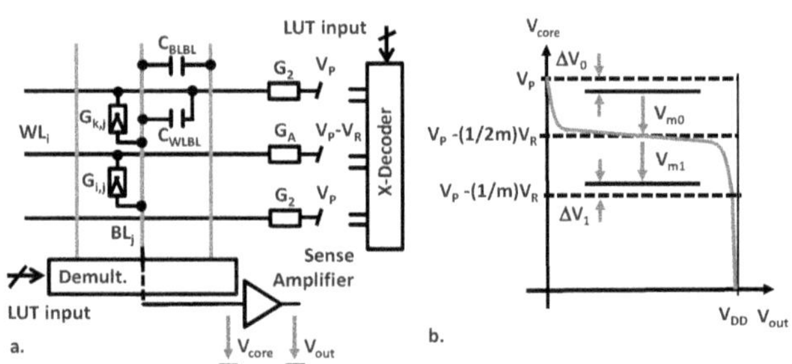

Figure 9: a. word line configuration for accessing G_{ij}. b. sense amplifier transfer characteristics (the red curve), worst case core voltage and signal margins V_{m1} (reading a one) and V_{m0} (reading a zero).

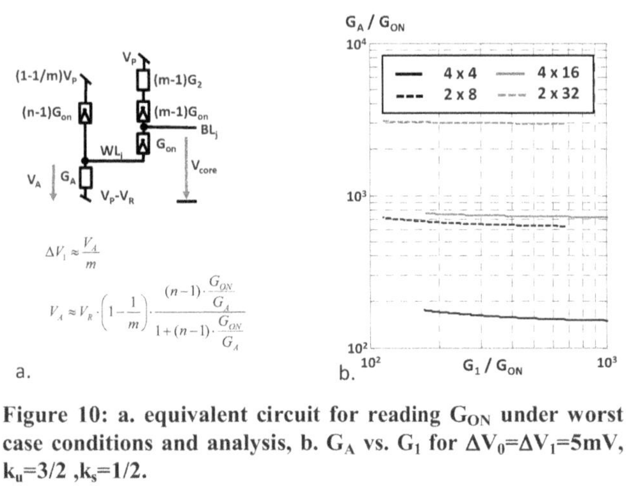

Figure 10: a. equivalent circuit for reading G_{ON} under worst case conditions and analysis, b. G_A vs. G_1 for $\Delta V_0 = \Delta V_1 = 5mV$, $k_u = 3/2$, $k_s = 1/2$.

Based on a specification with regard to voltage margins for both the read and write operation the transistor dimensions (i.e. the channel width W assuming a fixed minimum channel length L_p) are derived. Since it is necessary to electrically connect the transistors the transistor area model includes the minimum required area overhead which is necessary to connect drain, gate, and source area, as well as the area which is necessary to electrically separate different transistors by spacers, cf. Fig. 6a. Then, the effective area of a transistor comprising a width W is (note: L represents the lithographic resolution which is different from the effective physical length L_p):

$$A_{Tran} = 6 \cdot L \cdot (W + 6 \cdot L). \tag{6}$$

All transistors act as switches, i.e. for a closed switch (which is conducting) the gate node is connected to 0V (for a PMOS transistor) or to the supply voltage V_{DD} (for an NMOS transistor). For open (non-conducting) switches the gate node is connected to the opposite supply level. The (almost linear) conductance of a switch is modeled by (PMOS-switch)

$$G_p \approx \frac{W}{L_p} \cdot \mu_{eff} \cdot C_{ox} \cdot (V_{SG} - V_{TH}) \approx \frac{\partial I_D(V_{SG}, V_{SD})}{\partial V_{SD}}. \tag{7}$$

In (7) $W, L_P, \mu_{eff}, C_{ox}, V_{SG}$, and V_{TH} represent the transistor width, the physical gate length, the effective carrier mobility, the gate capacitance per unit area, the source gate voltage (cf. Fig.6e), and the effective transistor threshold voltage [12]. An equation similar to (7) can be derived for an NMOS transistor.

3.1.1 The Write Operation

The write operation is illustrated in Fig.7a (write LRS). Here, the aim is to set the conductance G_{ij}, which is located at the intersection between word line i and bit line j, to the LRS. For the associated driver circuits DX and DY the transistors n_1 as well as p_2 are conducting while for all other word lines and bit lines the associated transistors p_y as well as n_x are conducting. The transistors p_A are switched off. Ideally, the voltage across G_{ij} equals the programming voltage V_P, while the effective device voltage for all other cells in the pCB is approximately equal to $1/3V_P$. Hence, the conductive state of unselected cells is left almost unchanged while an adaptation of G_{ij} towards G_{ON} is initiated. However, the conductances of the driving transistors are still finite. Due to other cells connected to word line WL_i as well as to bit line BL_j which also draw a current $\sim 1/3V_PG$, the overall current obtained at the interface of the selected bit line BL_j and word line WL_i, respectively, potentially leads to a significant voltage drop across the driving transistors (modeled by G_1 and G_2). Then, the effective programming voltage across G_{ij} is reduced. Consequently, the required time to set G_{ij} to G_{ON} is enlarged. In view of (2), the relation of an effective programming voltage V, the required time T to set G_{ij}, the maximum programming voltage V_P, and the minimal setting time T_{min} is given by

$$e^{\alpha z V / 2V_T} \cdot T = e^{\alpha z V_P / 2V_T} \cdot T_{\min}, \tag{8}$$

which results in (cf. Fig.7a)

$$(V_1 + V_2) = V_P - V = \frac{2}{\alpha \cdot z} \cdot V_T \cdot \ln\left(\frac{T}{T_{\min}}\right) =: k_s V_T. \tag{9}$$

In order to limit the required (worst case) programming time, the overall voltage drop across G_1 and G_2 has to be bounded to an upper limit, which has to be fulfilled in any case. The maximum current through G_1 and G_2 is obtained if all unselected conductances are equal to G_{ON}. Then, by fulfilling the upper limit for V_1+V_2, G_1 as well as G_2 are related by the following approximation:

$$\frac{\frac{n+2}{3}-k_S\frac{V_T}{V_P}}{\frac{G_2}{G_{ON}}+n-1}+\frac{\frac{m+2}{3}-k_S\frac{V_T}{V_P}}{\frac{G_1}{G_{ON}}+m-1}=k_S\frac{V_T}{V_P} \cdot \quad (10)$$

Under identical conditions for write, G_2 as well as G_1 can be traded against each other which allows for the area optimization in view of the proposed area model. However, as can be concluded from Fig. 7, also the conductance G_x as well as G_y should have an impact on the current distribution on the word lines and bit lines.

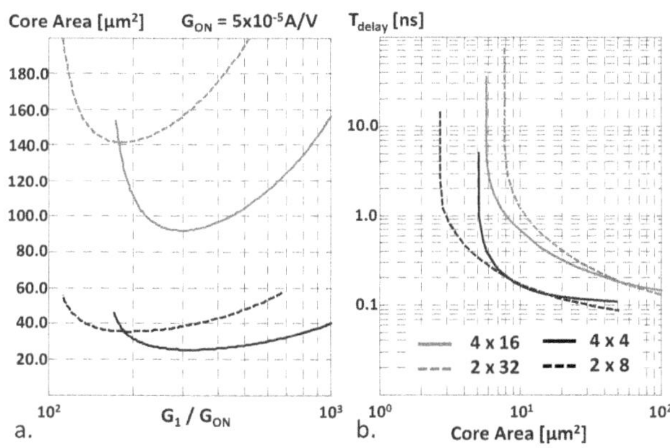

Figure 11: a. total circuit area including DX, DY, X-decoder, demultiplexer, and pCB for specific G_{ON} and $\Delta V_0=\Delta V_1=5mV$, $k_u=3/2$,$k_s=1/2$. B. area vs. delay for minimal area-delay product for different array configurations assuming $\Delta V_0=\Delta V_1=5mV$, $k_u=3/2$, $k_s=1/2$.

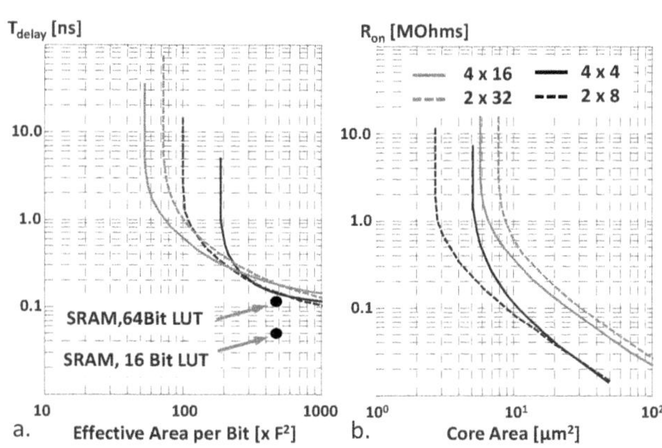

Figure 12: a. Effective area per bit in relation to F^2 (F: lithographic resolution), comparison with SRAM-based implementation @V_{DD}=0.9V. b. Minimum On-resistance R_{ON}

At least, G_x as well as G_y are required to ensure a limited voltage drop across unselected cells, which should be close to $1/3V_P$; also for unselected cells a voltage margin has to be specified. Here, G_x as well as G_y have to be sized in such a way that under all conditions the voltage drop across unselected cells is less than

$$|V(G_{kl})| \leq \frac{1}{3}V_P + k_u \cdot V_T \cdot \quad (11)$$

For the following discussion k_s is set to k_s=1/2 and k_u is set to k_u=3/2. Then, by this exemplary choice, the worst case write cycle is enlarged by ~30%. Note, that the particular specification of k_u is less significant for the circuit performance if k_u is larger than k_s, (which is a reasonable condition since the unselected cells do not change their conductance too much at $1/3V_P+k_uV_T$). A specification of k_u is required for the numerical optimization which was executed to find the appropriate relations between all driving conductances. Fig. 8 shows the results. For the given parameters, G_X as well as G_Y are at least one order of magnitude smaller than G_1 as well as G_2.

If the transistor n_x is sized such that the conductance of n_x equals G_2, and if transistor n_y is sized such that the conductance of n_y equals G_1, the reset operation can be realized in a simple way, which can be explained as follows. By reversing all voltage levels using the mapping $V \leftarrow (V_{DD}-V)$ (cf. Fig.7b) G_{ij} adapts towards G_{OFF}. With respect to Fig.5b transistor p_y has to be activated instead of n_1, transistor n_x has to be activated instead of p_2, and so on, but the effective conductances have to be maintained which is achieved if the proposed symmetry holds.

3.1.2 The Read Operation

During a read operation the DY circuits are turned off. Only the DX circuits provide voltage levels to the word lines while a demultiplexer (cf. Fig.4 and Fig.9a) connects the selected bit line and the output node. For fast demultiplexing either a flat architecture (cf. Fig.4a, for n=4) or a tree architecture (cf. Fig. 4b, for n>4) is used (also note, that m>4 is not useful since the obtained voltage levels on the bit lines are too small). The voltage of the word line which is connected to the addressed conductance G_{ij} is set to V_P-V_R while all other word lines voltages are set to V_P. The main advantage of this scheme is that the transistor p_2 can be reused for the read operation and ensures a stable voltage level on the word lines. Additionally, transistor p_A is now needed (modeled by G_A) which finally has to be sized. For infinite conductances G_A and G_2, the output voltage V_{core} depends not only on the conductance to be read but on all conductances G_{kj} connected to bit line j:

$$V_{core} = V_p - V_R G_{ij} / \sum_{k=0}^{m-1} G_{kj} \cdot \quad (12)$$

If G_{ij} is in the HRS and at least one other conductance in the LRS (note: this can be accomplished by the cells connected to the additional word line) the output voltage V_{core} is close to V_P, cf. Fig. 9b. If G_{ij} is in the LRS the voltage V_{core} varies between

$$V_p - V_R \cdot \frac{1}{m} \geq V_{core} \geq V_p - V_R \cdot \quad (13)$$

The upper limit is the critical configuration for reading a "one" since the output voltage is very close to the threshold of the sense amplifier which results in the highest sensitivity to noise. The sizing of G_A has to ensure that under all possible conditions the voltage V_{core} does not exceed the upper limit in (13) by ΔV_1, which is the specification of the margin for reading a "one".

The worst case situation (which maximizes V_{core}) is given if the current through G_A is maximized which is given if all conductances on the remaining bit lines are in the LRS, too, cf. Fig. 10a. For given G_{ON}, G_1, G_2 and a defined margin ΔV_1 (e.g. ΔV_1=5mV) G_A can be derived by analyzing the equivalent circuit shown in Fig. 10a. In Fig.10b, G_A is depicted in dependence of G_1 which shows only a weak dependence on G_1.

By setting G_{ON} and by variation of G_I the total circuit area can be derived for different options. In Fig.11a, the expected circuit area is exemplarily shown for $G_{ON}=5x10^{-5}$ A/V. The circuit area always shows a specific minimum.

4. AREA-DELAY PERFORMANCE

By specifying G_{ON}, ΔV_0, ΔV_1, k_u, and k_s and by variation of G_I different transistor dimensions can be derived by using (7). From the transistor sizing a model can be derived which predicts the capacitive load of particular nodes along the critical path. The model which was developed so far includes diffusion capacitances of source and drain areas as well as the source-gate and drain-gate capacitances of transistors located in DX, DY, and the demultiplexer circuit as well as wiring capacitances associated inside of the pCB, cf. Fig.6c,d. For L=40nm the bit line-bit line capacitance per cell (assuming SiO_2 as dielectrics) was calculated to $C_{BL,BL}$= 7.5aF, while the bit line-word line coupling capacitance per cell was calculated to $C_{WL,WL}$=4.3aF, cf. Fig. 9a.The diffusion capacitance models is based on the equation [12]

$$C_{diff} = C_j \cdot W \cdot 3L + C_{jsw} \cdot (W + 6L) + C_{jG} \cdot W \quad (14)$$

for a transistor with channel width W (cf. Fig.6a) and typical parameters C_j=1.6x10^{-3} F/m^2, C_{jsw}=1.0x10^{-10} F/m, and C_{jG}=2.0x10^{-10} F/m. An NMOS transistor with unit W/L-ratio was assumed to provide a conductance of 0.9x10^{-4} A/V at V_{DD}=0.9V, while the conductance of the PMOS pendant was calculated to be 1.3 times less than that of the NMOS transistor [1]. Based on these parameters the signal path was modeled by an RC chain, and the transition from accessing a logic zero to accessing a logic one (same bit line) was simulated assuming one conductance along the bit line in the LRS while all other conductances were in the HRS. In order to determine the delay, the output signal V_{core} had to reach 90% of the final level. The obtained delay was multiplied by the associated area. The minimum of the obtained product (for constant G_{ON} and by variation of G_I) was taken as the optimal solution for a given G_{ON}. In Fig.11b the results are shown. Compared with standard SRAM implementations of LUTs the RS-based implementations have clear advantages in regard to the area requirements, but the delay performance is by 1-2 orders of magnitude worse. The explanation of this observation is given through Fig.12b. For reasonable core areas, the ON-resistance $R_{ON}=1/G_{ON}$ is significantly larger than 10kOhms which at least one order of magnitude larger than the driving capability of a transistor with unit-W/L ratio. Smaller ON-resistances require larger driving transistors (cf. Fig.8) which directly results in a larger core areas, and - by (14) - in larger capacitive loads on the bit lines. The main advantage of pCBs can be expected for the area performance at the penalty of a significant reduction of speed. The effective cell are per bit is far beyond 4F^2, but for particular conditions the effective area per bit is smaller than the area per bit which is required for a corresponding SRAM-based LUT.

5. ACKNOWLEDGMENTS

The authors like to thank Rainer Waser, Stephan Menzel and Eike Linn from RWTH IWE II for providing the fundamental ECM device model as well as representative parameters for simulation

6. REFERENCES

[1] ITRS roadmap, http://www.itrs.net

[2] S. Tanachutiwat, M.Liu, W. Wang, „FPGA based on Integration of CMOS and RRAM", IEEE Trans. On Very large Scale Integration (VLSI) Systems, Vol.19, No.11, pp.2023-2032, Nov. 2011

[3] T.Driscaoll, J.Quinn, S.Klein, H.T.Kim, B.J.Kim, Y.V.Pershin, M.DiVentra, D.N.Basov, „Memristive Adaptive Filters", Applied Physics Letters 97, 093502, 2010

[4] S.Yu, Y.Wu, R.Jeyasingh, D.Kuzum, H.-S. P. Wong, „An Electronic Synapse Device Based on Metal Oxide Resistive Switching Memory for Neuromorphic Computation", IEEE Transactions on Electron Devices, Vol. 58, No.8, pp 2729-2737, Aug. 2011

[5] C.Schindler, M.Meier, R.Waser, M.N. Kozicki, „Resistive Switching in Ag-Ge-Se with extremely Low Write Current", NVMTS,pp.82-85, 2007

[6] S.Menzel, U.Böttger, and R.Waser, „Simulation of Multilevel Switching in Electrochemical Metallization Memory Cells", Journal of Applied Physics Vol. 111, pp.014501-014501-5, Jan. 2012

[7] R.Waser, "Electrochemical and Thermochemical Memories", IEDM, pp 1-4, 2008.

[8] Waser, R., Dittmann, R., Staikov, G. and Szot, K., „Redox-Based Resistive Switching Memories – Nanoionic Mechanisms, Prospects, and Challenges", Adv. Mater., 21:2632–2663, 2009

[9] Bockris, John O'M., Reddy, Amulya K.N., Gamboa-Aldeco, Maria E., „Modern Electrochemitry 2A", 2nd ed., Springer, 2000

[10] A.Heittmann, T.G.Noll, „A Monte Carlo Analysis of a Write Method used in Passive Nanoelectronic Crossbars", Procs. of ACM/IEEE Nanoarch, 2012, Amsterdam, The Netherlands, 2012

[11] Ahmed, E.; Rose, J.; , „The effect of LUT and cluster size on deep-submicron FPGA performance and density", *Very Large Scale Integration (VLSI) Systems, IEEE Transactions on* , vol.12, no.3, pp.288-298, Mar. 2004

[12] S.C Liu, J.Kramer, G.Indiveri, T.Delbrück, R.Douglas, „Analog VLSI:Circuits and Principles", Cambridge, MA: MIT Press, 2002

DRMA: Dynamically Reconfigurable MPSoC Architecture

Lawrance Zhang
Jude Angelo Ambrose
Jorgen Peddersen
Sri Parameswaran
School of Computer Science and Engineering, UNSW
Sydney, Australia
zhangx@cse.unsw.edu.au

Roshan Ragel
Swarnalatha Radhakrishnan
Department of Computer Engineering, University of Peradeniya
Peradeniya, Sri Lanka
roshanr@cse.unsw.edu.au

Kewal K. Saluja
Dept. of Elect. & Computer Engr. University of Wisconsin-Madison
Madison, WI, US
saluja@engr.wisc.edu

ABSTRACT

Embedded systems are ubiquitous and are deployed in a large range of applications. Designing and fabricating Integrated Circuits (ICs) targeting such different range of applications is expensive. Designers seek flexible processors which efficiently execute a multitude of applications. FPGAs are considered affordable, but design cost, high reconfiguration delay and power consumption are all prohibitive. In this paper, we propose a novel ASIC based flexible MPSoC architecture, which can execute separate tasks in parallel, and it can be configured to execute single task with wide data widths or execute multiple tasks with varying data widths. The architecture presented, called Dynamically Reconfigurable MPSoC Architecture (DRMA), can be rapidly reconfigured through instructions. We present applications as case studies to showcase the flexibility and efficacy of DRMA. Results show for an additional area overhead of about 5%, the system is capable of working as four 32-bit processors, a single 128 bit processor or as a pipelined processing system.

Categories and Subject Descriptors

C.1 [PROCESSOR ARCHITECTURES]: Parallel processors, Pipeline processors, Adaptable architectures

Keywords

MPSoC, Dynamical Reconfiguration, Embedded System, Multiprocessor, Parallelism, Word Size Variation

1. INTRODUCTION

The cost of designing and testing of the modern multiprocessor system is increasing rapidly [5]. The number of chips deployed per design has to be in the millions before the cost can be amortized. Since a number of applications serve a much smaller demographic than the aforementioned millions, a single design which is reusable (and reconfigurable) is necessary for future systems. FPGAs meet this goal by allowing reusability and reconfigurability. However, each design still has to be completed at the RTL level and hardware must be verified. This design effort and time required is costly. Additionally, the power consumption costs of FPGAs are prohibitive.

In this paper, we propose an alternative reconfigurable MPSoC design which is reusable. Our data-width reconfigurable MPSoC architecture, referred to as Dynamic Reconfigurable Multiprocessor Architecture (DRMA), allows dynamical reconfiguration, provides instruction level parallelism (ILP) and thread level parallelism (TLP), and enables pipelining of processors. Compared to state-of-the-art reconfigurable MPSoC architectures, such as Voltron [32], TFlex [14], CoreFusion [13] and Bahurupi [21], which aggregate processors to perform multi-issue executions (for example: to mix two 2-way processor to one 4-way processor), DRMA integrates multiple processors at runtime to support varying word sizes (e.g.it integrates 2 processors to createa a 64 bit word size processor). DRMA allows reconfiguration at real time. The system necessitated additional hardware, though the overhead is small.

Note that the primary objective of this exercise is the flexibility. We are able to provide many 32 bit cores which can be used to execute separate applications. A pair of 32 bit processors can be combined together to create a 64 bit processor (not all instructions are 64 bits wide- but a fair number are 64 bits). Four processors can be combined together to make a 128 bit processor. We could also pipeline processors in a novel way. Thus our main goal is to allow flexibility. There is a cost in terms of time and area, but these can be amortized for particular applications. In application families, such as encryption, this feature can be exploited for large bit width operands and DRMA can provide speed improvements.

In this paper, we demonstrate the DRMA architecture with four 32-bit processors using the PISA instruction set. A detailed comparison to other architectures is also performed. The kernels of few small programs are scheduled and executed on this architecture and the results are compared against single 128bit word size and single 32bit word size processors. We show the advantage of flexibility by varying the bitwidth for differing applications.

The remainder of the paper is organized as follows: In Section 2, we give a motivational example for MPSoC runtime reconfigurability. In Section 3, related work in reconfiguring MPSoC system is discussed. In Section 4, we present the DRMA architecture followed by design methodology and implementation in Section 5. Then the experimental results for three case studies are presented in Section 6 and finally the paper is concluded in Section 7.

2. MOTIVATIONAL EXAMPLE

2.1 Motivation of Varying Bit

With the developments in 3D graphics technology and physical simulation systems, applications with large data size (such as 64, 128 and 256 bits) will become needed. Additionally, there is a plethora of 32 bit applications which will continue to prevail. Modern block-wise encryption applications for security involve large amounts of instructions that can be naturally operated as fewer instructions with large word size processors [22] [31]. It is these widely varying bit width operations, and the cost of making new masks that motivate this work.

2.2 Example

For simplicity, let us assume that we have three code segments as in Figure 1(a), (b) and (c) each with a set of four independent operations. Word sizes in Figure 1(a) and (b) are fixed to 32 bits and 128 bits respectively, whereas in (c) it varies between operations.

Let us schedule the given code segments on a 32bit x 4 DRMA architecture that could be dynamically configured to behave like 4 x

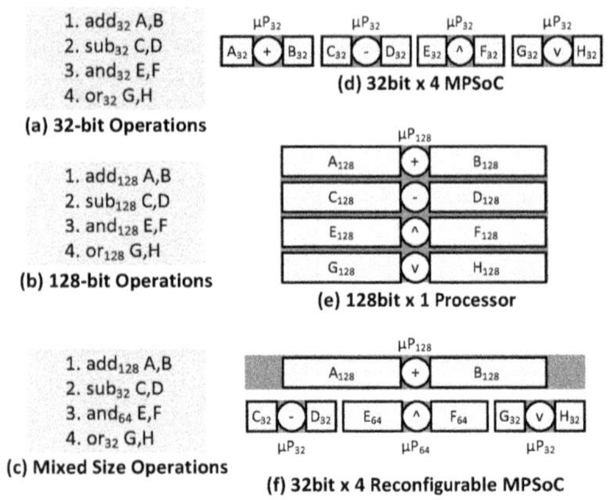

Figure 1: Motivational Example

32bit independent processors or a 128bit processor or or a combination thereof in-between.

While the code segment in Figure 1(a) can be efficiently scheduled on the 4 x 32bit independent processors (DRMA configured in separate mode) as shown in Figure 1(d), the segment in Figure 1(b) can be scheduled on the 1 x 128bit processor (DRMA configured in complete combined mode) as shown in Figure 1(e).

The code segment in Figure 1(c) has four operations with word sizes 128, 32, 64 and 32 bits respectively and they can be efficiently scheduled on processors with the same word sizes as shown in Figure 1(f) where operations 2, 3 and 4 are scheduled together. It is worth noting that the given processor behaves both as a 128bit processor and later as 2 x 32bit and 1 x 64bit processors, and it is only possible due to the *dynamic* reconfigurability in DRMA.

The details of reconfigurability, data sharing and synchronization of DRMA are explained in later sections.

3. RELATED WORK

3.1 Pre-MPSoCs Era

In the pre-MPSoCs era, many methods were developed to speed up applications through parallelism within one processor. Such parallel mechanisms can be classified into methods that use Instruction Level Parallelism (ILP) and those that use Data Level Parallelism (DLP). In case of ILP, the independent instructions available in the program are scheduled/executed in parallel (such as in Very Large Instruction Word (VLIW) architecture [9] or SuperScalar [27] architectures) and in case of DLP the operations are executed in parallel on independent data [23] (such as Intel's MMX Single Instruction Multiple Data (SIMD) architecture [20]). Given that the VLIW implementations on ASICs are not flexible [6], Anjam et al. presented an FPGA based softcore VLIW processor. Although the design in [1] is flexible and extendable through static and dynamic partial reconfigurations, such features come with a relatively large amount of reconfiguration time of around ten thousands of cycles.

Despite the fact that larger data widths can be handled by SIMD extensions such as MMX [20], AltiVec [7] or Synchroscalar [19], instructions that support packing and unpacking data for such extensions incur 75-85% of the dynamic instruction stream of the processor [29]. In [11] Huang et al. proposed a method to reduce this overhead by adding new custom instructions in the ISA of the processor. This technique adds to the large number of hardware accelerators proposed in the past [15, 24, 30] that makes processors flexible and customizable at the same time. However, such techniques will not completely eliminate the overhead mentioned earlier (75-85%). With DRMA, since it can work like a SIMD extension through reconfiguration and each processor element (PE) in DRMA architecture processes one certain part of data with its own memory element, there will be no overhead of packing and unpacking data.

3.2 FPGA Based Reconfiguration

MPSoCs on Field Programmable Gate Arrays (FPGAs) is a popular reconfigurable hardware design technique [8]. They are flexible (due to reconfigurability), have low time to market and have relatively low cost cost. FPGAs can accommodate 80-100 soft-core processors in a single chip, and NoCs are often used as a solution to manage such large number of cores [18]. However, they have lower performance compared to their ASIC counterparts.

Singh et al. [26] argued that there are different combinations of simultaneously active applications in different situations. They require specific configuration for the FPGA to meet the performance requirement and there might be large number of situations which need to be supported by the system, requiring their specific configurations. The appropriate configuration needs to be loaded into the reconfigurable platform at run-time, as and when required. The authors demonstrate the designing of reconfigurable NoC-based MPSoCs and programming it for real-life applications. Their experiments showed that the NoC reconfiguration kept overheads to a minimum, while utilizing 85% of the total resources of a Virtex-5 FPGA.

FPGA's functionality and flexibility come from the reconfigurability at gate level, however, building large and advanced circuits based on gate arrays may result in undesired performance penalties and power consumption [2].

3.3 Reconfigurable PEs or Modules

In [28], Sun et al. proposed a multiprocessor system built using extendable processors. Extendable processors are configurable base processors where custom instructions can be implemented to extend their instruction sets. By reconfiguring the extendable part, the processor can be optimized and evolved further. Liang et al. [16] improved the extendable cores into MPSoCs by exploiting the imbalance in execution time and custom instructions requirement among processors. It proposes an architecture with a shared reconfigurable fabric to extend instruction set. However, the bottleneck of multiprocessor systems with extendable processors is the design complexity. Custom-instruction selection has complex interdependencies with task assignment and scheduling, improper selection may result in a sharp decrease in the efficiency of the whole system. The reconfigurability available in Sun et al.'s proposal is at the instruction set level of individual processors and not at the system level.

The *PipeRench* architecture is described in [10]. The architecture consists of a series of simple PEs, which only include ALUs and register files. Different functions can be implemented by reconfiguring the PEs. This architecture can be used to achieve static parallelism with the overhead of reconfiguration time. The *ADRES* architecture presented in [17] and [4] are similar to that of *PipeRench*. However, *ADRES* employs a VLIW processor to fulfil ILP and reconfigurable array to fulfil DLP, so that the VLIW processor can gain benefits from both parts. Nevertheless, manual intervention is needed to achieve best performance and the reconfigurable parallelism is limited to the sizes of reconfigurable arrays. [2] presents an expression-grained reconfigurable array (EGRA) architecture. The array can be configured to form a better equalization of critical paths over different elements and to a flexible design space. To some extent, it is similar to ADRES. However, it focuses on the performance of multi stage computations, while ADRES explores strong instruction-level parallelism.

In [21], the authors propose a multiprocessor based reconfigurable architecture known as *Bahurupi* that focuses on instruction-level and thread-level parallelism (TLP). *Bahurupi* merges processors together to achieve ILP and divides them back for TLP. However, *Bahurupi* cannot perform large word size operations like DRMA. Moreover, it has fixed clusters while clusters in DRMA can be formed by reconfiguration.

The first related work ([28]) presented in this subsection supports reconfigurability in individual PEs at the instruction set level. The other two projects ([10, 17]) have simple PEs (ALU and register file) which are reconfigurable at the system level but cannot work as individual fully-fledged processors. Compared to [21], DRMA focuses on different features.

The focus of DRMA is dynamic reconfigurability at the system level with complete processors as PEs. We intend to achieve reconfigurability at run time with ease by introducing communication ports to PEs that can be controlled via instructions, making DRMA both flexible and simple. More details about the architecture are presented in Section 4.

3.4 Contributions

- The architecture proposed in DRMA is a novel architecture for MPSoC in which processors can be combined together for word-size variation.
- DRMA is a dynamically reconfigurable architecture that allows the configuration to be done cycle by cycle at beginning of an instruction execution with little overhead.
- DRMA architecture can take advantage of both instruction level parallelism and data level parallelism.

4. ARCHITECTURE

Both the processor and the instruction set architectures of DRMA are presented in this section followed by a description of how the DRMA would function and then how the PEs are synchronized when they work on larger word sizes.

4.1 Processor Architecture

Each processor (or PE) in the DRMA architecture, is a pipelined RISC type processor. Each PE contains separate data and instruction memories and can be treated as a standardized module. Additionally, they are driven by a central clock.

The features of DRMA claimed earlier (such as operating like a 4 x 32bit processors or a single 128bit processor as in Figure 1) are achieved by programming the PEs into one of their four different modes: (1) Separate Mode (*SMode*); (2) Initial Mode (*IMode*); (3) Terminal Mode (*TMode*); and (4) Combined Mode (*CMode*).

In the PISA instruction set, the instructions are 64 bits wide. The first two bits are used to indicate the modes which are used. Bits *"00"* for *SMode*, bits *"01"* for *IMode*, bits *"10"* for *TMode* and bits *"11"* for *CMode* are used. PEs work in *SMode* so that task level parallism can be achieved. *IMode* enables ILP. Together *IMode*, *CMode* and *TMode* enables word-size integration.

Figure 2 depicts the peripherals that could support the four modes of operations. The peripherals (and therefore the corresponding ports) shown are enabled and disabled by the modes set through instructions. The I/O ports named *DataIO* and the *CtrlO* ports are implemented to communicate computation-relevant data and the control signal separately in word-size varying mode. In both word-size varying mode and ILP mode, the load/store instruction will be synchronised for multi-clock action. The I/O block named *SyncIO* is deployed to handle synchronisation issues. The synchronisation protocol will be discussed in Section 4.4.

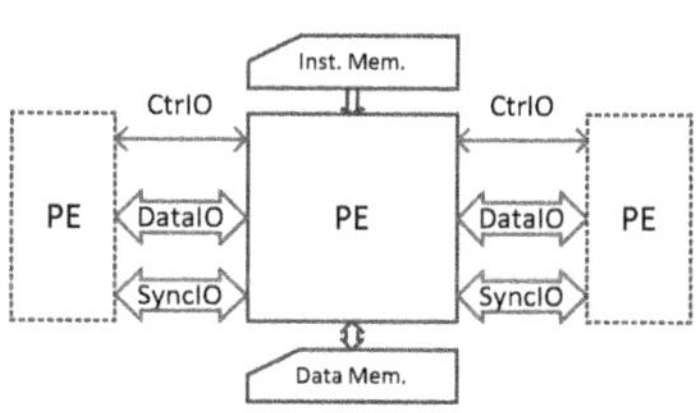

Figure 2: The Peripherals of the Processing Element

4.2 Instruction Set Architecture

Since the dynamic reconfigurability of the DRMA architecture is achieved by instructions, the ISA of each PE plays a vital role in the realization of word-size variation. The functioning of the PEs could be categorized into four, based on the type of instructions they execute. In this section, four most common instruction types are analysed as typical examples: **[TYPE$_1$]** instructions that involve either load/store or bitwise operations (such as $or_{128}\ A, B$); **[TYPE$_2$]** instructions that involve carry in/out (such as $add_{128}\ A, B$); **[TYPE$_3$]** instructions that involve multiple bit transfers between PEs (such as $srl_{128}\ A, B$); and **[TYPE$_4$]** instructions that need communication among PEs for decision making and sharing (such as branch/jump $beq_{128}\ A, B$).

4.3 Functioning of DRMA

Let us assume that the MPSoC has four PEs and we connect all of them together to form a larger data width (four 32bit processors are combined together to form a single 128bit processor). Then, each small PE will fetch identical instructions from their respective memories, fetch data from the same address from their respective data memories or register files separately. Depending upon the instruction type some data will be exchanged (for example: shift, carry taken, etc.) during the execution and the final result in each processor will be modified as they process parts of one piece of large-word-size data. From an external standpoint, they work together as if it is a larger word processor.

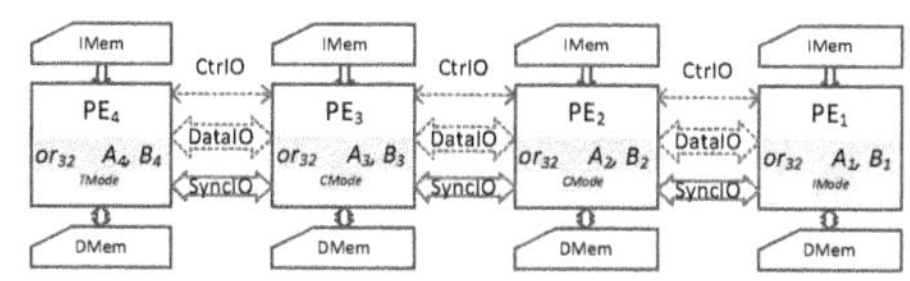

Figure 3: DRMA for TYPE$_1$ Instructions (example: $or_{128}\ A, B$)

Figure 3 depicts how a TYPE$_1$ instruction (such as $or_{128}\ A, B$) would behave. The data that is to be loaded to registers A and B (that are of larger word size) will be divided into four smaller word sizes each: A into (A_1, A_2, A_3, A_4) and B into (B_1, B_2, B_3, B_4). Given that there is NO need of data communication in TYPE$_1$ instruction for performing the computation, the *DataIO* and *CtrlO* are disabled. *SyncIO* will only be enabled for multi cycle instructions (such as $load$) and not utilized for single cycle instructions (such as or). In addition, as shown in Figure 3 while the far left PE works in *TMode* and the far right PE works in *IMode* to define the boundary of this combination group, the other two PEs will work in *CMode*.

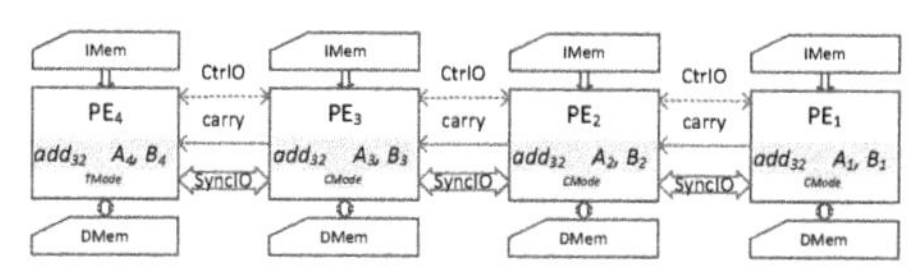

Figure 4: DRMA for TYPE$_2$ Instructions (example: $add_{128}\ A, B$)

Figure 4 depicts how a TYPE$_2$ instruction (such as $add_{128}\ A, B$) would behave. The only difference in TYPE$_2$ instructions compared to TYPE$_1$ is the carry bit each smaller word size instruction produces. Since, such carries have to be integrated in the computations (they have to be sent by the right PE and received by the left PE), a single bit wire from the *DataIO* port is enabled as shown in Figure 4. The rest of the functioning of TYPE$_2$ is similar to that of TYPE$_1$ instructions. The transfer of carry signal increases only a small amount of clock width.

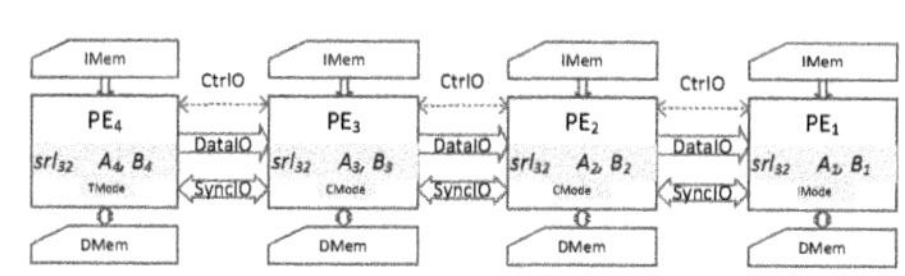

Figure 5: DRMA for TYPE$_3$ Instructions (example: $srl_{128}\ A, B$)

Figure 5 depicts how a TYPE$_3$ instruction (such as $srl_{128}\ A, B$) would behave. Here, since the instruction is a logical right shift, the shifted data has to be send across PEs and this is achieved by enabling the *DataIO* port in the right direction. The right most PE here works in *IMode* and the left most works in *TMode*. The rest of the functioning is similar to the previously explained modes.

Figure 6 depicts how a TYPE$_4$ instruction (such as $beq_{128}\ A, B$) would function. Similar to TYPE$_3$ instruction, the final decision will be made based on the computation of all the PEs. Partial conditional

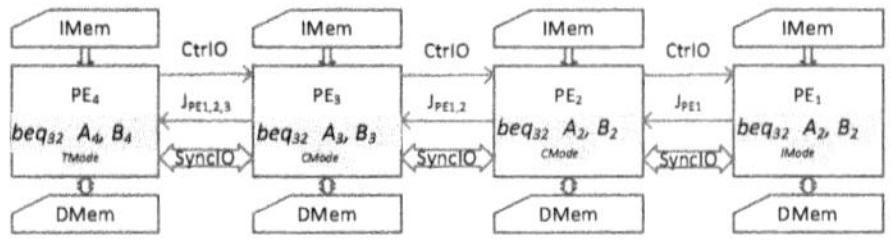

Figure 6: DRMA for TYPE$_4$ Instructions (example: $beq_{128}\ A, B$)

results from the right PEs will be sent to the left PEs and the leftmost PE (the PE in *TMode*, PE$_4$ in Figure 6) will make the decision for the whole instruction. Now the decision is sent back from PE$_4$ to the rest of the PEs through *CrtIO* port. The rest of the functioning of TYPE$_4$ instructions are similar to that of the previous types.

4.4 PE Synchronisation

For multiple cycle instructions, such as Load/Store, the system has to be synchronized. This is particularly important in instructions such as memory instructions because it may take varying cycles that the designer may not be aware of. For example, when the instruction involves memory loading, the processor will stall until it receives the signal 'ACK' (acknowledgement) from the data memory. A PE combined together with others has to wait for its own 'ACK' signal and other 'ACK' signals attached to the other PEs, because the processor system may lose synchronization if any one finishes its loading operation and fetches the next operation while others still wait for the right data from its own memory. In DRMA, the synchronization block will store the 'ACK' signal once it is received, and then transfer the information from the right PE to the left. The far left PE of this group will send a control signal to finish the stall of all PEs in one group, only when every PE has received its own 'ACK' signal.

5. DESIGN METHODOLOGY AND IMPLEMENTATION

The design flow of the DRMA architecture is shown in Figure 7. First of all, a base ISA is selected for our PEs (note that our PEs are fully fledged processors) and an ASIP design tool is used to build the processors that are capable of behaving as processing elements for DRMA. We used ASIPmeister [12], an ASIP design tool for building each PE.

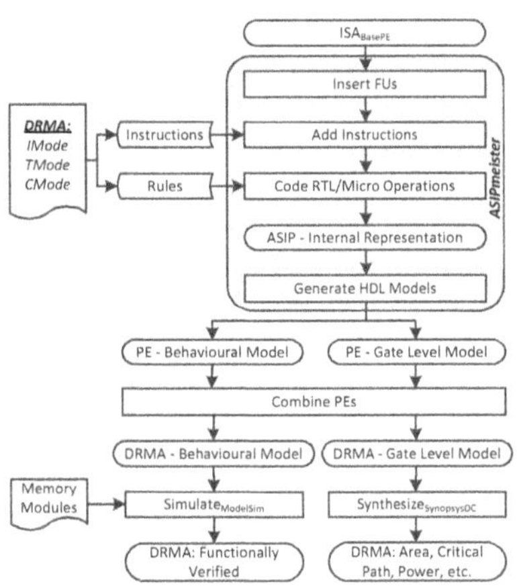

Figure 7: The Design Flow of DRMA architecture

As shown in the design flow, the instructions that support DRMA will be copied four times; one for each mode and the instructions' micro-operations will be altered. The micro-operations will not only provide the operations of the instruction (as they typically do) but also indicate the connections among processors. For example, the *core* of the micro-operations for 'addCMode' instruction (add instruction in *CMode*) is shown in Figure 8(a) and the corresponding logic is shown in Figure 8(b). For the *CMode* 'add' instruction, the *carry in* will be received from the right PE and the *carry out* should be passed on to the left PE as shown in Figure 8(b).

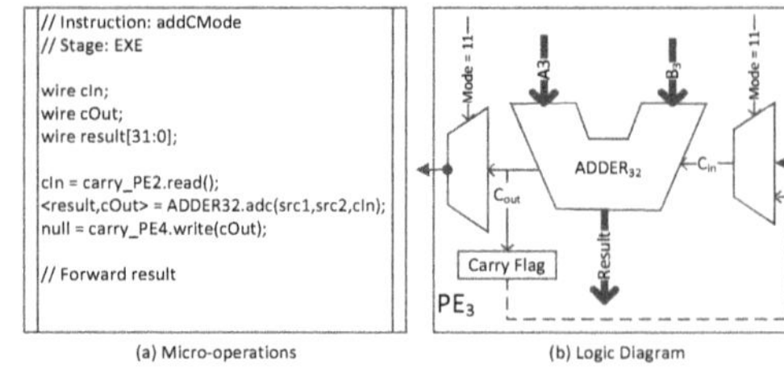

Figure 8: The 'add' instruction in the *CMode*

The output from the ASIPmeister tool is the VHDL files for synthesis and simulation. Then the integration of PEs to build DRMA is done manually which is both synthesized and simulated using commercial chip design tools as shown in Figure 7.

We used the design flow explained to implement DRMA with PISA instruction set [3] (only integer instructions) and we had four PEs in our system. PISA is a 32bit architecture and therefore the largest data width we could achieve with DRMA is 128 bits. Each PE in the architecture had six pipeline stages and their own data and instruction memory, as explained earlier in Section 4. The DRMA architecture is clocked at the fastest possible frequency to allow the longest operation to be finished within this period. To enable the change in clock periods, we need to have a multiplexor which can change differing clock signals.

Register Type	lb, lbu, lh, lhu, lw, dlw, sb, sh, sw, dsw dsz, jr, jalr, add, addu, sub, subu, mfhi, mthi, mflo mtlo, and, or, xor, nor, sllv, srlv, srav, slt, sltu
Immediate Type	lb, lbu, lh, lhu, lw, dlw, lwl, lwr, sb, sh sw, dsw, dsz, swl, swr, beq, bne, bleg, bgtz, bltz bgez, j, jal, sll, srl, sra

Table 1: Instructions Supported by DRMA

Table 1 lists the instructions that are available for DRMA type of operations (the instructions that works in all four modes). Not all instructions used by the PEs are available in the DRMA architecture. We have not supported instructions such as multiply and divide ('mult','multu','div', and 'divu') due to their complex nature when they are needed to be combined. In addition, we do not handle instructions with immediate operands ('addi', 'addiu', 'andi', 'ori', 'xori', 'slti', 'sltiu', and 'lui'). We leave both these categories as our future work. At present long word instructions such as 128 bit multiply is performed on one single PE, and a software function is instantiated.

6. EXPERIMENTS AND RESULTS

6.1 Synthesis

We evaluate the hardware parameters of DRMA (with reconfigurable four processors) and compare it against non-configurable processors (μP-32bit and μP-128bit). We synthesized the RTL and analyzed it using Design Compiler from Synopsys. μP-32bit is the original PISA processor (32-bit), and μP-128bit is a processor which supports 128-bit word size with PISA instructions, with ALU and memory interfaces modified to support 128-bit word sizes. The multiplier in the 128-bit implementation is very large with a large delay (not pipelined). In order to make a fair evaluation we removed the multiplier to calculate the critical path. The results are summarized in Table 2. Column 1 lists the properties analyzed and column 2 presents the numbers for DRMA. Column 3 and column 4 present details for 128-bit and 32-bit single processors respectively. As shown in the table the cost of DRMA platform with four PEs is roughly four times the cost of the original 32-bit processor in area and power consumption, which indicates that the additional cost to create DRMA is low (no more than 5%). Even though the cost of counterpart μP-128bit in area and power consumption is smaller than DRMA, DRMA provides greater flexibility.

	DRMA	μP 128bit	μP 32bit
Combinational area (# of cells)	157863	123818	36315
Noncombinational area (# of cells)	136000	102531	33627
Total (# of cells)	293863	226349	69942
Ratio	4.2×	3.2×	1×
Global Operation Voltage = 1 v			
Average dynamic power(mW)	113.4	86.7	28.6
Average leakage power(mW)	1.3	1.0	0.3
Average Total Power(mW)	114.7	87.7	28.9
Ratio	3.97×	3.04×	1×
Requested Arrival Time = 1 ns			
Clock Width (ns)	1.69	1.47	1.29
Ratio	1.31×	1.14×	1×

Table 2: Hardware Parameters from Synthesis

6.2 Case Studies

We took three different case studies to show the efficacy of DRMA: i) three applications(AES, CRC32 and IPv6) are chosen to test the word variation capability of DRMA and is compared against μP-128bit and μP-32bit; ii) all three applications are executed on the DRAM architecture in differing modes to see the efficacy of DRMA; and three, a pipelined JPEG application is demonstrated to show a pipeline architecture being supported by DRMA.

6.2.1 *Case 1*

We examined three applictions in Case 1. The characteristics of the applications are: i) a highly parallel encryption 'kernel' of AES (AES 128 bit encryption after round keys generation); ii) CRC32 (32 bit Cyclic Redundancy Check) application; iii) an IPv6 packet filter program that matches 128 bit IPv6 addresses against a series of rules. The applications are experimented on the three different architectures. We simulated the RTL models of the three architectures (as explained in Section 5) using ModelSim

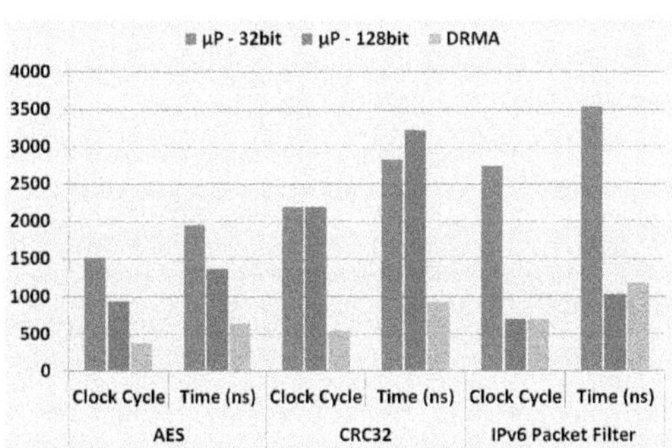

Figure 9: The Comparison in Performance

Figure 9 shows the number of clock cycles incurred and the corresponding execution times for the three architectures we considered. For AES, the DRMA architecture working in word size variation mode (set PEs to *SMode*, *CMode*, and *TMode* respectively) outperforms both of its counterparts by 3.99× and 2.46× in clock cycles. The speed up is because DRMA can process 128 bit data directly while, μP-32bit spends more than four times the number of instructions to process all 32 bit parts of 128 bit data and then encrypt them. DRMA can also process 32 bit data in parallel while μP-128bit can only process one instruction regardless of the word size. In CRC32, since it is a pure 32 bit applications, the architectures with large bit width (such as μP-128bit) will not gain any improvement in performance. DRMA is distinct in its small area and energy consumption as μP-32bit, because we can use one processor in *SMode* and then turn off redundant PEs or dedicate them to other tasks. For IPv6 packet filter, which is a 128 bit application, the DRMA architecture is as good as μP-128bit and outperforms μP-32bit by 3.92× in the number of clock cycles.

6.2.2 *Case 2*

We use the above three applications to evaluate the total efficiency of DRMA in executing a series of applications.It is not uncommon to see such three applications executing together in a standard system. Where data is transported, encrypted and checked for errors. In this section, we mix the three applications into a single application: i) IPv6 filter, if the rule is valid then fetch 128 bit data and store; ii) AES encryption: the data is encrypted as 128-bit blocks (we hand modified the AES code to take advantage of 128 bit encryption); iii) CRC32: create checksum for the 128-bit data and four, store and record — the system will store the encoded data and CRC32 value for back-up or for transmission, as well as record the order information. We can execute these in three different ways. First, we ignore the dependency between applications and allow each core to work in *parallel* (set them to *SMode*) (see Figure 10(a). AES was occupying roughly three forth of total instructions, hence becoming the bottleneck in common MPSoC. This architecture achieves a speed up of 1.52× compared to executing the application in a single PE. In the second sub-case, DRMA is configured into a fixed *pipeline* architecture, named by implementing FIFO to connect adjacent PEs as shown in Figure 10(b), where the output of one processor is fed to the next one. A 1.20× speed up is achieved, once again constrained by the large delay of AES (other processors were idling). In the third case shown in Figure 10(c), DRMA can be reconfigured dynamically through its instructions into different configurations: working in word size variation mode for a 128 bit application in IPv6 filter and AES encryption; working in a pipelined fashion in CRC and working separately in data store, the speed up compared to μP-32bit is 4.7×, mainly because DRMA was controlled to dynamically optimize its architecture for different applications. Figure11 depicts the clock cycles taken for the three configurations compared to μP-32bit as baseline.

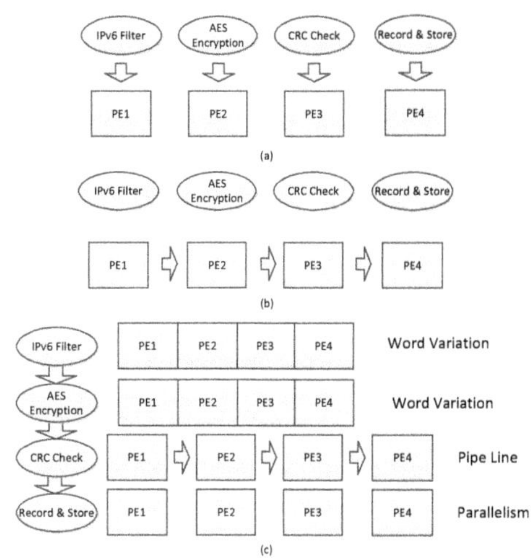

Figure 10: The Comparison of Different Configurations

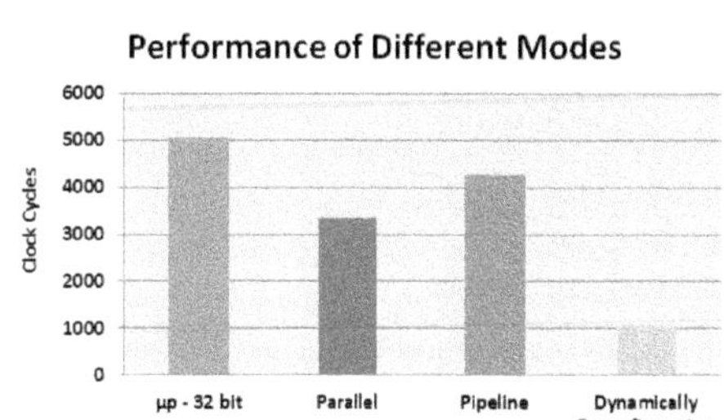

Figure 11: The Results for Case 2

6.2.3 *Case 3*

Pipeline architectures are suited for data streaming applications in MPSoCs [25]. We implemented the JPEG encoding application in the DRMA architecture like Figure 12 with each core executing one stage. Compared to μP-32bit, the speed up is 4.11×. The result is shown in Table3

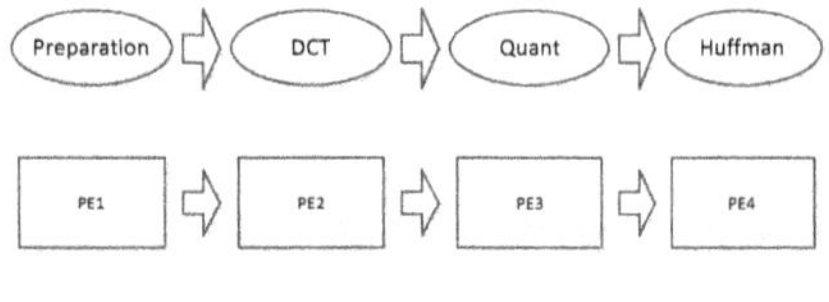

Figure 12: The Configuration in Case 3

	μP-32bit	DRMA
Clock Cycles	47.1M	11.47M
Speed Up	1×	4.11×

Table 3: The Results of JPEG Encoding

6.3 Discussion

The *Flexibility* of DRMA is shown through three case studies: case 1 shows benefits of word width variation and ILP; case 2 shows the ability of the DRMA to be configured in differing ways; and case 3 shows an application where a configuration of pipeline is suited..

The flexibility afforded by DRMA results in a certain level of efficiency. While the examples were small (due to the lack of a suitable compiler), the careful selection of configuration for a particular application(s) can be quite efficient. It is currently not possible to simulate DRMA in a modified system level simulator such as the multi-core version of SimpleScalar simulator [3], therefore all simulations were carried out at the RTL level. Therefore, DRMA cannot be directly compared to other architectures. Since DRMA can be configured to be a processor from a 32 bit to a 128 bit processor and can exploit both ILP and TLP, to build a compiler for this architecture would involve multiple compiler technologies for each configurations and optimization. We are beginning to attempt this, and do not underestimate the challenge.

The clock cycle should be used a comparative guide rather than as a rule. There are many optimizations possible, and those have not been implemented. ASIPmeister only offers a single cycle multiplier and divider. This means that large bitwidth multiply instruction or divide instruction would increase the clock period. Currently the applications used do not use 128 bit multiplication, thus does not affect our test. Since multiply instruction is uncommon, we could theoretically use a 32bit one and run it multiple times to get a 128 bit result. In the future we plan to integrate a pipelined 128 bit multiplier.

7. CONCLUSIONS

In this paper, we presented a reconfigurable architecture, named DRMA, for MPSoCs targeting flexibility and reconfiguration. The main features of DRMA are: i) the ability to vary word size; ii) configurations are controlled by the executing instruction (not a separate instruction) thus eliminating configuration time; iii) a simple and modular architecture with only 5% overhead in area. Due to the reconfigurability which is based on instructions, DRMA is efficient and flexible. We used three case studies with different word sizes to demonstrate the efficacy of DRMA.

Acknowledgements

We would like to thank Tuo Li from the School of Computer Science and Engineering for his advice in the ASIP processors.

8. REFERENCES

[1] F. Anjam, M. Nadeem, and S. Wong. A VLIW softcore processor with dynamically adjustable issue-slots. In *2010 International Conference on Field-Programmable Technology (FPT)*, pages 393 -398, Dec. 2010.

[2] G. Ansaloni, P. Bonzini, and L. Pozzi. EGRA: A coarse grained reconfigurable architectural template. *IEEE Transactions on Very Large Scale Integration (VLSI) Systems*, 19(6):1062 -1074, 2011.

[3] T. Austin, E. Larson, and D. Ernst. Simplescalar: an infrastructure for computer system modeling. *IEEE Computer*, 35(2):59 -67, Feb 2002.

[4] F. Bouwens, M. Berekovic, B. De Sutter, and G. Gaydadjiev. Architecture enhancements for the adres coarse-grained reconfigurable array. In *Proceedings of the 3rd international conference on High performance embedded architectures and compilers*, HiPEAC'08, pages 66–81. Springer-Verlag, 2008.

[5] M. Campbell-Kelly. Chips and change: how crisis reshapes the semiconductor industry. *Business History*, 52(6):1017–1019, 2010.

[6] L. H. Chen and O. T.-C. Chen. Design methodology of a hardware-efficiency VLIW architecture with highly adaptable data path. In *48th Midwest Symposium on Circuits and Systems, 2005*, pages 1223 -1226 Vol. 2, Aug. 2005.

[7] K. Diefendorff, P. Dubey, R. Hochsprung, and H. Scale. Altivec extension to powerpc accelerates media processing. *IEEE Micro*, 20(2):85 -95, Mar/Apr 2000.

[8] T. Dorta, J. Jimenez, J. Martin, U. Bidarte, and A. Astarloa. Overview of fpga-based multiprocessor systems. In *International Conference on Reconfigurable Computing and FPGAs*, pages 273 -278, Dec. 2009.

[9] J. A. Fisher. Very long instruction word architectures and the ELI-512. *SIGARCH Comput. Archit. News*, 11(3):140–150, June 1983.

[10] S. C. Goldstein, H. Schmit, M. Budiu, S. Cadambi, M. Moe, and R. R. Taylor. Piperench: A reconfigurable architecture and compiler. *IEEE Computer*, 33(4):70–77, 2000.

[11] L. Huang, L. Shen, Z. Wang, W. Shi, N. Xiao, and S. Ma. Sif: Overcoming the limitations of simd devices via implicit permutation. In *2010 IEEE 16th International Symposium on High Performance Computer Architecture (HPCA)*, pages 1 -12, Jan. 2010.

[12] M. Imai. ASIPMeister: A configurable processor core development system. In *Proc. ITI of 3rd International Conference on Information & Communications Technology (ICICT 2005), Cairo, Egypt*, 2005.

[13] E. Ipek, M. Kirman, N. Kirman, and J. Martinez. Accomodating workload diversity in chip multiprocessors via adaptive core fusion. In *Workshop on Complexity-Effective Design*, pages 10–21, 2006.

[14] C. Kim, S. Sethumadhavan, D. Gulati, D. Burger, M. Govindan, N. Ranganathan, and S. Keckler. Composable lightweight processors. In *40th Annual IEEE/ACM International Symposium on Microarchitecture, 2007.*, pages 381 -394, Dec. 2007.

[15] J. S. Kim and M. Sunwoo. Three low power asip processor designs for communications, video, and audio applications. In *International Conference on Design Technology of Integrated Systems in Nanoscale Era, 2007. DTIS.*, pages 241 -244, 2007.

[16] C. Liang and T. Mitra. Shared reconfigurable fabric for multi-core customization. In *Design Automation Conference (DAC), 2011 48th ACM/EDAC/IEEE*, pages 830 -835, 2011.

[17] B. Mei, S. Vernalde, D. Verkest, and R. Lauwereins. Design methodology for a tightly coupled vliw/reconfigurable matrix architecture: a case study. In *Proceedings Design, Automation and Test in Europe Conference and Exhibition, 2004*, volume 2, pages 1224 – 1229 Vol.2, Feb. 2004.

[18] G. Mplemenos and I. Papaefstathiou. Mplem: An 80-processor fpga based multiprocessor system. In *16th International Symposium on Field-Programmable Custom Computing Machines, 2008*, pages 273 -274, 2008.

[19] J. Oliver, R. Rao, D. Franklin, F. T. Chong, and V. Akella. Synchroscalar: Evaluation of an embedded, multi-core architecture for media applications. *J. Embedded Comput.*, 2(2):157–166, Apr. 2006.

[20] A. Peleg and U. Weiser. MMX technology extension to the intel architecture. *IEEE Micro*, 16(4):42 -50, Aug. 1996.

[21] M. Pricopi and T. Mitra. Bahurupi: A polymorphic heterogeneous multi-core architecture. *ACM Trans. Archit. Code Optim.*, 8(4):22:1–22:21, Jan. 2012.

[22] A. Sagheer, S. Al-Rawi, and O. Dawood. Proposing of developed advance encryption standard. In *Developments in E-systems Engineering (DeSE), 2011*, pages 197 -202, Dec. 2011.

[23] E. Salami and M. Valero. A vector-mu;SIMD-VLIW architecture for multimedia applications. In *International Conference on Parallel Processing, 2005*, pages 69 -77, 2005.

[24] G. Schewior, H. Flatt, C. Dolar, C. Banz, and H. Blume. A hardware accelerated configurable ASIP architecture for embedded real-time video-based driver assistance applications. In *International Conference on Embedded Computer Systems (SAMOS), 2011*, pages 209 -216, 2011.

[25] S. L. Shee and S. Parameswaran. Design methodology for pipelined heterogeneous multiprocessor system. In *44th ACM/IEEE Design Automation Conference, 2007.*, pages 811 -816, 2007.

[26] A. Singh, A. Kumar, T. Srikanthan, and Y. Ha. Mapping real-life applications on run-time reconfigurable NoC-based MPSoC on FPGA. In *2010 International Conference on Field-Programmable Technology (FPT)*, pages 365 -368, Dec. 2010.

[27] J. Smith and G. Sohi. The microarchitecture of superscalar processors. *Proceedings of the IEEE*, 83(12):1609 -1624, Dec. 1995.

[28] F. Sun, S. Ravi, A. Raghunathan, and N. Jha. Application-specific heterogeneous multiprocessor synthesis using extensible processors. *IEEE Transactions on Computer-Aided Design of Integrated Circuits and Systems*, 25(9):1589 -1602, 2006.

[29] D. Talla, L. John, and D. Burger. Bottlenecks in multimedia processing with SIMD style extensions and architectural enhancements. *IEEE Transactions on Computers*, 52(8):1015 – 1031, Aug. 2003.

[30] X. Xie, J. Williams, and N. Bergmann. Asymmetric multi-processor architecture for reconfigurable system-on-chip and operating system abstractions. In *International Conference on Field-Programmable Technology, 2007. ICFPT 2007.*, pages 41 -48, Dec. 2007.

[31] M. Zaharuddin, R. Rahman, and M. Kassim. Technical comparison analysis of encryption algorithm on site-to-site ipsec vpn. In *International Conference on Computer Applications and Industrial Electronics (ICCAIE), 2010*, pages 641 -645, Dec. 2010.

[32] H. Zhong, S. Lieberman, and S. Mahlke. Extending multicore architectures to exploit hybrid parallelism in single-thread applications. In *IEEE 13th International Symposium on High Performance Computer Architecture, 2007*, pages 25 -36, Feb. 2007.

A Compact FPGA-based Montgomery Multiplier over Prime Fields*

Miguel Morales-Sandoval
Laboratorio de Tecnologías de la Información
CINVESTAV-Tamaulipas
Ciudad Victoria, México 87130
mmorales@tamps.cinvestav.mx

Arturo Díaz-Pérez
Laboratorio de Tecnologías de la Información
CINVESTAV-Tamaulipas
Ciudad Victoria, México 87130
adiaz@tamps.cinvestav.mx

ABSTRACT

This work describes a compact FPGA hardware architecture for computing modular multiplications over GF(p) using the Montgomery method. The multiplier is parameterizable, allowing to evaluate the hardware design for different prime fields using different radix of the form $\beta = 2^k$. The design uses only three $k \times k$ multipliers and three $2k$-bit adders. The hardware organization of the multiplier maximizes the use of the multipliers processing iteratively the multiplicand, multiplier and modulus. The parametric design allows to study area-performance trade offs, in order to meet system requirements such as available resources, throughput, and efficiency. The proposed multiplier achieves a 1024-bit modular multiplication in 15.63 μs using $k = 32$. Compared to the most compact FPGA implementation previously reported, our proposed design uses 79% less FPGA resources with better efficiency expressed as Mbps/Slice.

Categories and Subject Descriptors

B.2 [**Arithmetic and logic structures**]: High speed arithmetic—*cost/performance*

General Terms

Security

Keywords

Finite field arithmetic, Prime fields, Montgomery multiplication, FPGA, coprocessor

1. INTRODUCTION

The Montgomery method [11] is considered the fastest one to perform modular multiplication. It has been used in public key cryptosystems, particularly to perform exponentiation of the form $g^e \bmod M = (g \times g \times \cdots \times g) \bmod M$ over finite fields. Exponentiation is a common operation used in the Diffie-Hellman key exchange protocol [3], for encryption and digital signatures in RSA [13] and digital signature schemes like DSA [1]. In [9] several algorithms were proposed for practical implementation (in software) of the Montgomery method. Also, several hardware implementations of Montgomery algorithms are found in the literature. The implementations in FPGAs are mainly motivated since these devices are reconfigurable, flexible and physically secure. FPGAs allow the implementation of this kind of algorithms in a broad range of applications with different area-performance requirements.

The main approach in the literature has been the implementation of Montgomery algorithms as fast as posible, which usually requires great amount of FPGA resources, making those solutions unsuitable for embedded or mobile applications, where small circuits are desired. In [10], Mondal et al. proposed a 256-bit Montgomery multiplier. They used three different configurations for the multiplier, exploring full parallel, semi-parallel and pipelined designs. They implemented a 256 $\times$ 256 multiplier that executes the multiplications at steps 1 and 2 of Montgomery algorithm, and used multiplexers for routing signals from and to internal registers. Their multiplier was implemented using built in multipliers in the FPGA used, requiring from 4 to 16 DSP blocks and using more than 1500 slices. A similar approach is followed by Gong et al. in [4], where a 256-bit Montgomery modular multiplier is designed using the 18×18-bit multipliers embedded in the Altera Cyclone3 FPGA. Their design uses 81 embedded multipliers, operates at 30.38 MHz and uses more than 60% of the available resources in the device. One of the well studied approaches for implementing Mongomery multipliers in hardware is the systolic approach, which main objective is to achieve regular structures, scalabe designs and high clock frequencies. This approach is used in [6], where the smallest multiplier design for a 1024-operand size uses 5357 LUTs, operates at 106.4 MHz and requires 1088-cycles to achieve a Montgomery multiplication.

In this work, we explore the design of compact Montgomery multipliers in FPGAs. Starting form the formal definition of the Montgomery algorithm and using a radix $\beta = 2^k$ we derive a formal equation to compute a Montgomery multiplication iteratively, processing k-bit digits from the multiplier, multiplicand and modulus, similar to the Finely Integrated Operand Scanning (FIOS) method presented in [9]. Few works have considered the compact design of Mont-

*Miguel Morales-Sandoval is funded by the Council for Science and Technology of Mexico (CONACyT), project FORDECYT 2011-01-174509.

gomery multipliers. To the authors' knowledge, the most recent work of a compact Montgomery multiplier was reported in [12], where the Coarsely Integrated Operand Scanning (CIOS) [9] method is implemented in a Spartan3E FPGA. Different to the algorithm used in [12], we take advantage of parallelism on data and propose a new iterative algorithm for Montgomery multiplication that requires only two loops instead of three as in the CIOS method. For a nk-bit modulus, a main loop parses the k-bit digits (Y_i) from the multiplicand Y. Inside this loop, another loop computes in parallel the operations ($z = X \times Y_i$) and ($z + q \times M$), processing each k-bit digit X_j, M_j from the multiplier x and modulus M respectively. While in [12] the multiplier is based on a systolic architecture, in our design the multiplier consist of a single processing unit. For a 1024-bit operand size, using radix $\beta = 2^{32}$, our design requires only 327 slices from the FPGA xc3s500e, achieving a Montgomery multiplication in 15.63μs. Compared to [12], our design is 4.75 times smaller, twice slower but with better efficiency, doubling the bits processed in a second per slice used.

The rest of this paper is organized as follows. Next section describes the Montgomery method for modular multiplication. Section 3 describes the proposed Montgomery multiplication architecture. Section 4 presents the implementation results and comparison against related works. Finally, section 5 concludes this work and points out future work.

2. MONTGOMERY MULTIPLICATION

Given two numbers X, Y, the Montgomery method computes $A = X \times Y \times 2^{-n} \bmod M$, being n some fixed integer. By doing this, trial division is avoided and replaced by less costly operations, for example, by a shift operation. The original Montgomery algorithm [11] consists of the following steps:

1. $z \leftarrow X \times Y$
2. $q \leftarrow z \times M' \bmod R$
3. $A \leftarrow (z + U \times M)/R$.
4. $A \leftarrow A - M$ if $A > M$.

At step 1, multiplication is performed without modular reduction. This operation results in a number greater than M that can be computed using any multiplication algorithm. In the second step, a second multiplication is required to compute a number q such that at step 3, $(z + q \times M)$ results in a number divisible by R (the n less significant bits are zeros), performing this division by a simple shift to the right operation. At step 2, M' satisfies $R \times R^{-1} + M \times M' = 1$. M' can be computed as $-M^{-1} \bmod \beta$.

Lets assume that the numbers X, Y, M are expressed as $(n+1)$ symbols using $\beta = 2^k$ as radix, and represented as $M = (M_n \cdots M_0)_\beta = \sum_{i=0}^{n} \beta^i M_i$, $X = (X_n \cdots X_0)_\beta = \sum_{i=0}^{n} \beta^i X_i$ and $Y = (Y_n \cdots Y_0)_\beta = \sum_{i=0}^{n} \beta^i Y_i$. In [15], C. Walter presented an improved time-constant algorithm for computing a Montgomery multiplication in an iterative fashion (see algorithm 1). An additional iteration is performed to avoid the final subtraction in the original Montgomery method. The requirement is that $0 \leq X, Y < 2 * M$. In algorithm 1, the main loop at line 2 process each digit Y_i from Y, and computes $X \times Y \bmod \beta^{-(n+1)} \bmod M$ in $n+1$ iterations. At step 4, division by β is implemented as a shift operation since the LSB of the sum is equal to zero. At each iteration i, both q_i and A_{i+1} are computed. From a sequential approach, these two operations could be performed by the set of operations described in equations 1 - 7.

The next section describes the derivation of a compact hardware architecture for computing equations 1 - 7, exploiting the parallelism, reusing operations and removing unnecessary memory resources.

Algorithm 1 Iterative Montgomery algorithm

Require: integers X and Y, with $0 < X, Y < 2 * M$, $R = \beta^{n+1}$ with $\gcd(M, \beta) = 1$, and $M' = -M^{-1} \bmod \beta$
Ensure: $A = X \times Y \times R^{-1} \bmod M = \sum_{i=0}^{n} a_i \beta^i$

```
1: A_0 ← 0
2: for i ← 0 to n do
3:    q_i ← (a_0 + X_0 × Y_i) × M' mod β
4:    A_{i+1} ← ([A_i + X × Y_i] + q_i × M)/β
5: end for
6: return A_{n+1}
```

$$t_1 = X \times Y_i \quad (1)$$
$$t_2 = a_0 + X_0 \times Y_i \quad (2)$$
$$q_i = t_2 \times M' \bmod \beta \quad (3)$$
$$t_4 = M \times q_i = (r_n r_{n-1} \cdots r_1 r_0) \quad (4)$$
$$t_5 = A_i + t_1 = (s_n s_{n-1} \cdots s_1 s_0) \quad (5)$$
$$t_6 = t_4 + t_5 \quad (6)$$
$$A_{i+1} = t_6/\beta \quad (7)$$

3. NOVEL COMPACT MONTGOMERY MULTIPLIER

The variables t_1, q_i and t_4 are multiplications between a k-bit number and a $[(n+1) * k]$-bit number. Note that t_2 is a $2k$-bit number. Since q_i is the result of a reduction modulo β, just the k least significant bits (LSB) from t_2 are considered for the computation of q_i. Indeed, M' is a k-bit value, constant during all the multiplication that can be computed in advance. Multiplications for t_1, q_i and t_4 can be implemented for example, using the schoolbook method (shift and add), processing k-bits at a time from X and M.

Using the polynomial representation of X in radix β, the operation $X \times Y_i$ can be expressed as:

$$X \times Y_i = \left(\sum_{j=0}^{n} \beta^j X_j\right) \times Y_i = \sum_{j=0}^{n} \beta^j (X_j \times Y_i)$$
$$= \beta^n (X_n \times Y_i) + \cdots + \beta (X_1 \times Y_i) + (X_0 \times Y_i) \quad (8)$$

Each term $(X_j \times Y_i)$ is of size $2k$-bits. The accumulative sum of terms in equation 8 is shown graphically in figure 1 (a). Despite the fact that in a multiplication the smaller operand is considered as the multiplier for reducing the number of terms in the sum, in this work we consider the contrary, we considered X and M as the multiplier. This allows to use a smaller multiplier, that in this case is a $k \times k$-multiplier. This comes with a penalization in the latency.

The circuit that implements the computation of equation 8 using the schoolbook method is shown in figure 1 (b). It is required a shift right register ZX of size $[2k + (n+1)k]$-bits, which can be viewed as two joined registers, the high part of $2k$-bits (register Z) and the low part of size $(n+1)k$-bits (register X). The register X stores the $(n+1)$ k-bit digits $X_n \cdots X_1 X_0$ of variable X and Z stores the two k-bit digits $Z_1 Z_0$ of the product $X_j \times Y_i$. In a shift right operation on ZX, X_0 is discarded and $X_n = Z_0$. The value Z_1 is

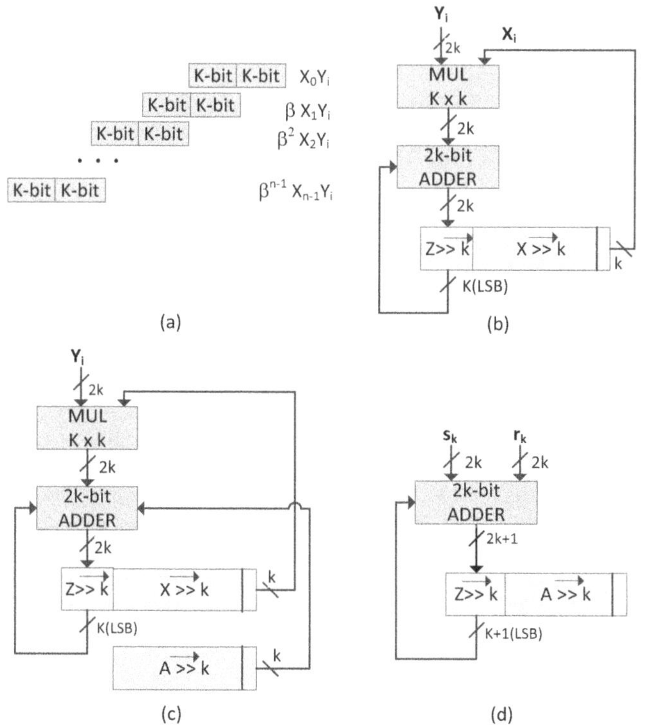

Figure 1: (a) Partial products of $X \times Y_i$. (b) Hardware realization of $X \times Y_i$. (c) Parallel computation of t_1, t_2 and t_5. (d) Hardware computation of t_6.

used in the accumulative sum with the next partial product $X_{j+1} \times Y_i$ (see figure 1 (a)). In figure 1 (b), the same register X is used to store the multiplier and the product, which saves memory resources.

Instead of waiting that all the terms of t_1 be computed for computing t_5, both computations can be performed in parallel. The operation $t_5 = A_i + X \times Y_i$ can be expressed as in equations 9. In order to compute t_5 and t_1 at the same time, it is required a shift right register for A_i and process a digit A_j at a time. According to equation 9, at each iteration j the digit a_j must be added to the product $(X_j \times Y_i)$ together with Z_1. This is achieved replacing the 2-input adder in figure 1 (b) by a 3-input adder. The new hardware module to compute t_5 and t_1 concurrently is shown in figure 1 (c), obtaining at each iteration a digit s_j. After $(n+1)$ iterations, equation 9 will be fully completed and the result stored in register X.

$$A_i + X \times Y_i = \sum_{j=0}^{n} \beta^j (a_j + (X_j \times Y_i)) \qquad (9)$$

In the circuit shown in figure 1 (c), at iteration $j = 0$, the value $t_2 = s_0$ is available from the 3-in adder. This value can be stored in a k-bit register and used to compute $q_i = s_0 \times M'$. By using a multiplier as the one shown in figure 1 (b), t_4 can be computed considering M_j as the multiplier and q_i as the multiplicand, computing one digit r_j at each iteration j and storing the results in a register ZM. Since t_4 and t_5 are independent operations, they can be computed in parallel.

After $(n+1)$ iterations, t_4 and t_5 are finally computed, so

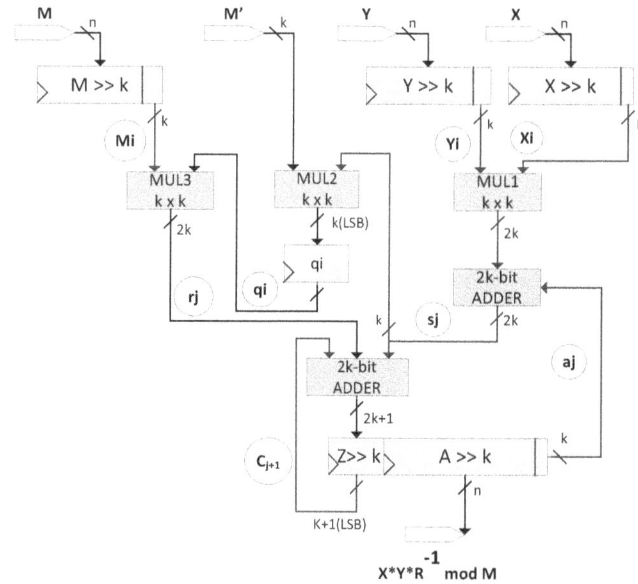

Figure 2: Compact hardware architecture to compute iteratively a Montgomery multiplication.

$t_6 = t_4 + t_5$ could be computed. The final result is divided by $\beta = 2^k$, what means discarding the k less significant bits from t_6 to obtain A_{i+1}. However, it is not necessary to wait $(n+1)$ cycles for computing t_6. Each value r_j and s_j from t_4 and t_5 respectively allows to compute t_6 iteratively in parallel. s_j and r_j are $2k$-bit numbers that can be represented in radix β as $s_1 s_0$ and $r_1 r_0$ respectively. Indeed, both of them are partial products.

The sum $(s_j + r_j)$ is a $(2k+1)$-bit number that can be expressed as $(c_{j+1} << k) + t_{6j}$, where c_{j+1} is a $(k+1)$-bit number and t_{6j} is the k-bit digit of t_6 computed at iteration j. c_{j+1} can be seen as a $(k+1)$-bit carry value to be used in the accumulative sum with the next term $(s_{j+1} + r_{j+1})$. A shift right register can be used to store both the value A_i and the next value A_{i+1}. As a new digit a_j from A is used to compute $(s_j + r_j)$, a shift right operation on register A will provide the next digit a_{j+1} and t_{6j} can be used to fill A, that is, $a_n = t_{6j}$. After $(n+1)$ iterations, all digits from A_i will be processed and the final result in register A will be t_6. The circuit that implements all this operations for computing t_6 iteratively is shown in figure 1 (d). It is worth to mention that all digits from A_i are accessed through the most to the right k-bit digit of register A, that is, a_0.

With this modifications, t_6 is computed after $(n+1)$ iterations, sharing memory resources with variable A. Then, the next is to compute t_6/β, which can be achieved by means of an additional shift right operation over register A. For this last shift operation on A the fill value is c_{j+1} instead of t_{6j}, that is $a_n = c_{n+1}$. At each iteration i, the modulus and multiplier are loaded again into registers X and M respectively. After iteration $i = n$, the operation $X \times Y \times R^{-1} \bmod M$ is finally computed.

Taking into consideration all previous optimizations, we formulate a new algorithm to compute modular Montgomery multiplications processing k-bit digits form X, Y, and M at a time. It is listed in algorithm 2. The corresponding circuit that executes algorithm 2 is shown in figure 2.

Algorithm 2 Novel iterative Montgomery algorithm

Require: integers $X = \sum_{i=0}^{n} X_i\beta^i$, $Y = \sum_{i=0}^{n} Y_i\beta^i$ and $M = \sum_{i=0}^{n} M_i\beta^i$, with $0 < X, Y < 2 * M$, $R = \beta^{n+1}$ with $\gcd(M, \beta) = 1$, and $M' = -M^{-1} \bmod \beta$

Ensure: $A = \sum_{i=0}^{n} a_i\beta^i = X \times Y \times R^{-1} \bmod M$

```
1:  A ← 0
2:  for i ← 0 to n do
3:     c_0 ← 0
4:     for j ← 0 to n do
5:        s_j ← [a_0 + X_j × Y_i] /*a_0 takes each a_j value */
6:        if j = 0 then
7:           q_i ← (s_j × M') mod β
8:        end if
9:        r_j ← q_i × M_j
10:       {c_{j+1}, t_{6j}} ← s_j + r_j + c_j
11:       A ← SHR(A)
12:       a_n ← t_{6j}
13:    end for
14:    A ← SHR(A) /*A ← A/β */
       a_n ← c_{n+1}
15: end for
16: return A
```

The value $M' = -M^{-1} \bmod \beta$, being a k-bit value, can be pre-computed since in many cryptographic applications the modulus is fixed and known in advance.

The performance of circuit in figure 2 can be improved by adding register to the output of multipliers `MUL1` and `MUL3`, at the cost of one extra clock cycle in the latency of loop at line 4 in algorithm 2. All the operations involved inside the body of that loop require only combinatorial logic, so instructions 5, 7, 9, 10, 11 and 12 occur during a single clock cycle. At each clock raising, the register A is updated with a shift to the right operation, delivering the next digit a_j as a_0 and filling with t_{6j}. At line 7, one register is needed to store the value of q_i, for which an enable signal is set to '1' just for the first iteration $j = 0$. After q_i is computed in the first clock cycle, we start processing each digit from X, A and M in $(n+1)$ iterations. As `MUL1` and `MUL3` are pipelined, the entire loop at line 4 of algorithm 2 is executed in $(n+1)+1+1$ clock cycles. One more cycle is required to perform the division by β at line 14. After processing all the $(n+1)$ digits from Y, the whole computation of a Montgomery multiplication takes $(n+1)*((n+1)+1+1+1) = (n+1)*((n+1)+3) = (n+1)^2 + 3(n+1)$, being $(n+1)$ the numbers of digits in the operands. For example, for 1024-bit numbers and using a radix $\beta = 2^{32}$, we have that $(n+1) = 32$ and the latency to compute a 1024-bit Montgomery multiplication is $32^2 + 3 * 32 = 1120$ clock cycles. All the dataflow in Algorithm 2 is controlled by a finite state machine with three states: one for load data into registers and initializations, used at each iteration i; another state for reading the digits X_j, a_j, M_j (basically routing data during the loop in line 4); and a third one to perform the final division at line 14 and evaluate the end of the outer loop.

4. RESULTS

We designed a VHDL code generator for the circuit shown in figure 2 to describe different Montgomery multipliers. Different configurations *size-radix* were tested to show the trade-offs among clock rate, minimum latency, and area. We considered $size \in \{1024, 512, 256\}$ and *radix* of the form $\beta = 2^k$ with $k \in \{2,4,8,16,32\}$, which are common sizes and radixes used in literature for testing and for practical applications. The VHDL designs were synthesized using the Xilinx ISE design suite 14.2 for the XC3S500e FPGA. All the VHDL designs were verified using a reference software implementation in Java, using the `BigInteger` class to perform all the modular arithmetic.

Figure 3 shows implementation results for our proposed compact Montgomery Multiplier. Although we are interested in reduced area implementations, the results presented in figure 3 were taken using the speed optimization criteria since the savings in slices using the area optimization criteria (0.2%) was not significant compared with the gaining in clock frequency (5%).

We addressed three implementation approaches:

Architecture 1 (CSR) - The memory blocks for X, Y and M, were implemented as circular shift registers of k-bits using FPGA's standard logic, while using the 18x18 embedded multipliers into the FPGA for implementing the three $k \times k$-bit multipliers. A was implemented as a shift to the left register using standard logic. The shifting capacity of all registers is enabled/disabled using a single control signal. The use of circular shift registers avoid loading registers X and M at each iteration i.

Architecture 2 (BRam) - The memory requirements for X, Y and M were mapped to memory blocks (BRAM) available in the FPGA. However, register A keeps implemented as in the *CSR* design. More standard logic from the FPGA device was required for accessing the memory blocks without adding new states to the finite state machine for control.

Architecture 3 (2Muls) - In the third approach, we modified the datapath in Architecture 2 to use only two multipliers. Since `MUL2` is used only once at the beginning of an iteration i (line 8 in algorithm 2), we decided to use a single multiplier to compute r_j and q_i multiplexing the input signals for `MUL2` and `MUL3` (see figure 2). It was done with the aiming of reducing the area resources in the multiplier design.

Implementation results are shown in figure 3. It is observed from figure 3 (a) that most of the slices occupied by architecture *CSR* are used to implement the three circular shift registers. As the operand size grows, the slices used also grow linearly. The amount of area slices is considerable reduced for architecture *BRam*. In this case, minor area usage is reached for radix $\beta = 2^8$. In the case of architecture *2Muls*, instead of reducing the slices respect to architecture *BRam*, these were increased, around 0.05%-12.5%. That is why the slices used by architecture *2Muls* are not shown in figure 3. In the case of the frequency achieved, the clock rate is lower for architecture *2Muls* compared to architecture *BRam* as it can be observed from figure 3 (b). The clock rate results for architecture *CSR* are not shown in figure 3 (b) due to the fact that these are closed to the ones obtained for architecture *BRam*. From these results, we conclude that the best implementation approach considered both low area requirements and high clock frequencies is architecture *BRam*.

For architecture *BRam*, a significant increasing in area resources only occurs when the radix increases. This penalty in area could be justified by the reduction in the latency implied by the use of a bigger radix. On the contrary, the area resources keep very similar even though the operand size doubles, as it can be observed in figure 3 (a). This can be verified from the curves for the operand sizes 256, 512 and 1024 which are very closed each other particularly for k equals to 4, 8, 16 and 32.

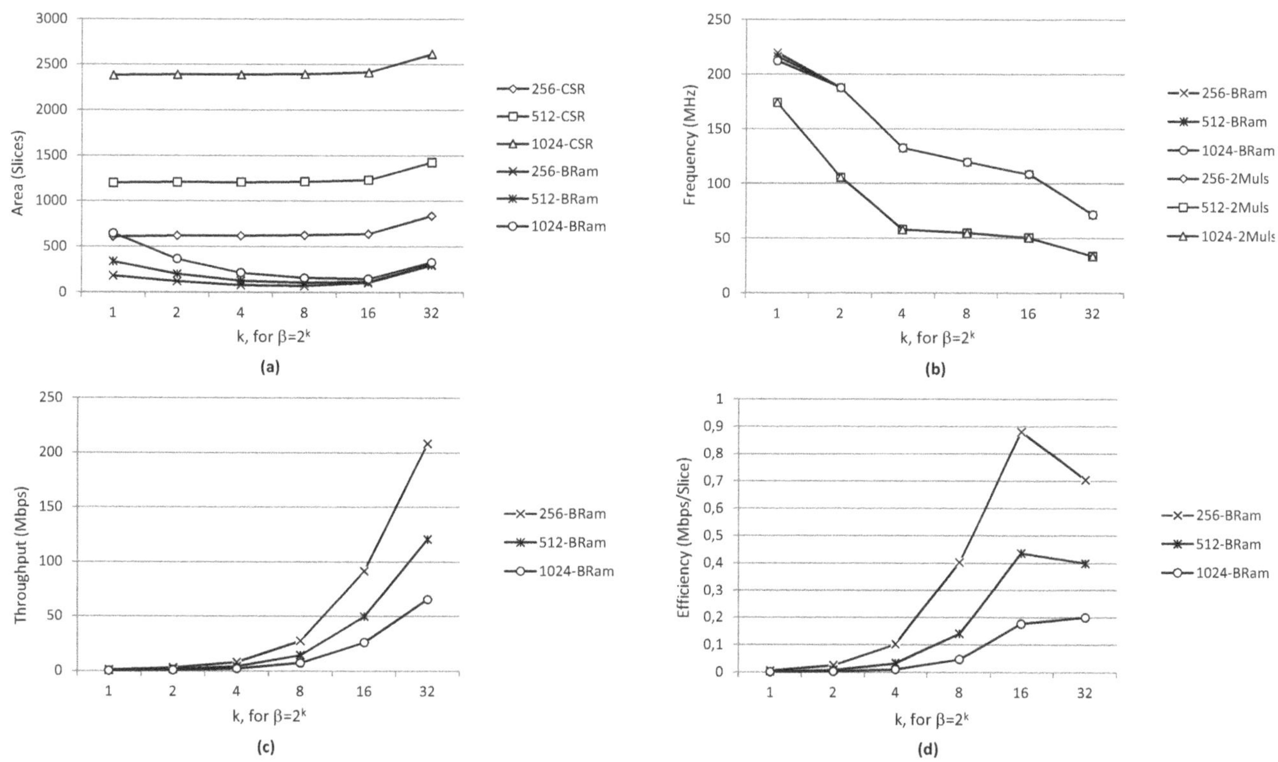

Figure 3: Implementation results. (a) Slices used by the *CRS* and *BRam* architectures. (b) Frequency achieved by the *BRam* and *2Muls* architectures. (c) Throughput and (d) efficiency exhibited by the *BRam* architecture.

Figure 3 (c) shows the throughput achieved by architecture *BRam*, expressed as the number of megabits per seconds processed (Mbps). Higher throughput is reached when the radix is bigger, for example, the multiplier can reach a throughput above 50Mbps for β equals 2^{16} or 2^{32}. Although better throughput is achieved as the radix gets bigger, a greater radix leads to more area usage and a penalty in the clock frequency. A metric to evaluate the relation between area and performance achieved is the efficiency, expressed as Mbps/Slices. For architecture *BRam*, its efficiency is shown in figure 3 (d), where it is observed that better efficiency is obtained for a radix 2^{16} justifying its corresponding cost in area.

In the literature, few works have considered the design of compact hardware architectures of Montgomery multipliers for prime fields. The most similar work to the one presented here is [12]. architecture *BRam* can be compared against the results achieved in [12] since the conditions are very similar. This comparison is shown in table 1. For an operand size of 1024 bits, our proposed multiplier uses around 4 times less area resources than [12], requiring twice the time to compute a Montgomery Multiplication. However, our design makes better use of area resources expressed by the efficiency achieved, which is above twice that the one achieved in [12].

In table 2 we provide a comparison of our proposed compact Montgomery multiplier (figure 2, architecture *BRam*) against previous Montgomery implementations, considering both software and FPGA implementations. Although a comparison against those works is not fair due to the different implementation platforms used, we provide that comparison just as a reference. In the case of software implementations

Table 1: Comparison of our proposed design against the most compact Montgomery multiplier on the xc3s500e FPGA.

	[12]	This($k = 16$)	This($k = 32$)
Size	1020	1024	1024
Mults.	10	3	11
BRams	4	3	3
Slices	1553	147	327
Time(μs)	7.62	39.41	15.63
Mbps	133.8	25.98	65.48
Mbps/Slices	0.086	0.176	0.20

where the processors used are typically found in embedded or mobile systems [14, 2, 8], the comparison allows to justify the motivation to perform Montgomery multipliers as dedicated hardware modules. In the case of FPGA implementations [6, 10, 5, 4], the comparison shows the enormous saving in area resources by our proposed multiplier, which makes it suitable for mobile and embedded applications. In table 2 we also show implementation results for the FPGA xc5vlx50, which has been also considered in [5]. It is worth to mention that Virtex5 and Spartan3E slices are different, four 6-in LUTs and four flip-flops in Virtex5 and 4-in LUTs and two flip-flops in Spartan3E. architecture *BRam* was better synthesized on the xc5vlx50 using radix $\beta = 2^{16}$, requiring both fewer slice flip-flops and slice-LUTs, and achieving a clock frequency of 102.42MHz.

5. CONCLUSIONS

A typical hardware implementation approach for the Montgomery algorithm to compute $X \times Y \times 2^{-n} \bmod M$ is to

Table 2: Comparison of our proposed design against software and FPGA implementations.

Work	Oper. size	Platform	FPGA resources	Freq. (MHz)	Time μs
[14]	1024	ARM	-	80	570
[14]	256	ARM	-	80	42.3
[2]	256	PentiumII	-	400	1.57
[8]	256	DSP	-	200	2.68
This work $k = 32$	256	xc3S500e	296 slices 3 BRams	71.61	1.22
This work $k = 32$	512	xc3S500e	302 slices 3 BRams	71.61	4.24
This work $k = 32$	1024	xc3S500e	327 slices 3 BRams	71.61	15.63
This work $k = 16$	256	xc5vlx50	34 slices 3 BRams	153.109	1.98
This work $k = 16$	512	xc5vlx50	37 slices 3 BRams	153.109	7.31
This work $k = 16$	1024	xc5vlx50	94 slices 3 BRams	102.42	10.93
[5]	521	xc5vlx50	957 slices	54	9.64
[6] $k = 2$	1024	Virtex-II	5,356 LUTs	106.4	0.56
[10]	256	xchvhx250T	1572 slices 64DSP48E	69.44	0.014
[4]	256	Cyclone3	23405 LEs 81Mults	30.38	0.1
[7]	1024	Virtex-II	4178slices 65PEs	100	10.88

decompose Y in n/k digits, $\beta = 2^k$ being the radix used, while X and M are used in their full widths. Systolic architectures result in high performance but also in high consumption of computational resources in the FPGA. In this work we considered the partition of the multiplier, multiplicand and modulus into digits of size n/k and processing a digit at a time, like in a software approach but exploiting the parallelism in the Montgomery algorithm. The operands partition allowed to use only three $k \times k$-bit multipliers, reducing the amount of area at expenses of performance, that can be tolerated in most mobile or embedded applications.

After implementing three different versions for an improved digit-serial Montgomery algorithm, the best implementation approach was to use memory blocks (BRAM) available in most of the available FPGAs to store X, Y and M. The area required by our implementation is considerably smaller than the reference FPGA implementations, saving around 79% FPGA resources. Further, the performance achieved by our proposed compact Montgomery multiplier outperforms software implementation over processors typically found in embedded or mobile applications.

In our experiments, we found that the most compact Montgomery multiplier is obtained using a radix 2^8, but a radix 2^{32} allowed to obtain the best performed multiplier. The best trade-off between area and performance is achieved by a multiplier using 2^{16} as radix, as it exhibits the best efficiency. As future work, we planned to use this compact Montgomery multiplier in the construction of a compact FPGA accelerator for the RSA digital signature scheme.

6. REFERENCES

[1] *Digital signature standard (DSS)*. National Institute of Standards and Technology, Washington, 2000.

[2] M. Brown, D. Hankerson, J. López, and A. Menezes. Software implementation of the NIST elliptic curves over prime fields. In *Proceedings of the 2001 Conference on Topics in Cryptology: The Cryptographer's Track at RSA*, CT-RSA 2001, pages 250–265, London, UK, UK, 2001. Springer-Verlag.

[3] W. Diffie and M. Hellman. New directions in cryptography. *Information Theory, IEEE Transactions on*, 22(6):644 – 654, nov 1976.

[4] Y. Gong and S. Li. High-throughput FPGA implementation of 256-bit Montgomery modular multiplier. In *IEEE on Second International Workshop on Education Technology and Computer Science*, pages 173 –177, jan. 2010.

[5] M. Hamilton, W. Marnane, and A. Tisserand. A comparison on FPGA of modular multipliers suitable for elliptic curve cryptography over GF(p) for specific p values. In *2011 International Conference on Field Programmable Logic and Applications*, pages 273 –276, sept. 2011.

[6] M. Huang, K. Gaj, and T. A. El-Ghazawi. New hardware architectures for Montgomery modular multiplication algorithm. *IEEE Trans. Computers*, 60(7):923–936, 2011.

[7] M. Huang, K. Gaj, S. Kwon, and T. El-Ghazawi. An optimized hardware architecture for the Montgomery multiplication algorithm. In *Proceedings of the 11th International Conference on Public Key cryptography*, PKC'08, pages 214–228, Berlin, Heidelberg, 2008. Springer-Verlag.

[8] K. Itoh, M. Takenaka, N. Torii, S. Temma, and Y. Kurihara. Fast implementation of public-key cryptography on a DSP TMS320C6201. In *Proceedings of the First International Workshop on Cryptographic Hardware and Embedded Systems*, CHES '99, pages 61–72, London, UK, UK, 1999. Springer-Verlag.

[9] C. K. Koç, T. Acar, and B. S. Kaliski, Jr. Analyzing and comparing Montgomery multiplication algorithms. *IEEE Micro*, 16(3):26–33, June 1996.

[10] A. Mondal, S. Ghosh, A. Das, and D. R. Chowdhury. Efficient FPGA implementation of Montgomery multiplier using DSP blocks. In *Proceedings of the 16th International Conference on Progress in VLSI Design and Test*, VDAT'12, pages 370–372, Berlin, Heidelberg, 2012. Springer-Verlag.

[11] P. L. Montgomery. Modular multiplication without trial division. *Math. Computation*, 44:519–521, 1985.

[12] E. Oksuzoglu and E. Savas. Parametric, secure and compact implementation of RSA on FPGA. In *International Conference on Reconfigurable Computing and FPGAs*, pages 391 –396, dec. 2008.

[13] R. L. Rivest, A. Shamir, and L. Adleman. A method for obtaining digital signatures and public-key cryptosystems. *Commun. ACM*, 21(2):120–126, Feb. 1978.

[14] A. F. Tenca and C. K. Koç. A scalable architecture for modular multiplication based on Montgomery's algorithm. *IEEE Trans. Computers*, 52(9):1215–1221, 2003.

[15] C. D. Walter. Montgomery's multiplication technique: How to make it smaller and faster. In *CHES'99*, pages 80–93, 1999.

On the Design of Modulo 2^n-1 Cubing Units

Evangelos Vassalos
Electronics Laboratory, Department of Physics
University of Patras
Patras, Greece
P.O. Box 26500
vassalos@upatras.gr

Dimitris Bakalis
Electronics Laboratory, Department of Physics
University of Patras
Patras, Greece
P.O. Box 26500
bakalis@physics.upatras.gr

ABSTRACT

A hardware intensive operation that is frequently employed in several arithmetic processing applications is the computation of the third power (cube) of an operand. Residue Number System can be used in such applications. In this paper we present two methods for designing modulo 2^n-1 cubing units that are based on reducing the size of the partial product matrix. The first method leads to faster circuits while the second one leads to smaller circuits for medium and large values of n. We also present minimized logic functions for efficiently designing modulo 2^n-1 cubing units for small wordlengths (n<6).

Categories and Subject Descriptors

B.2.4 [**Arithmetic and Logic Structures**]: High-Speed Arithmetic – *cost/performance.*

Keywords

Residue Number System, Modulo Arithmetic, Cubing Units

1. INTRODUCTION

Computing the cube of an operand is often required in Digital Signal Processing (DSP) applications, in graphics applications, in telecommunications as well as in cryptography. Typical cases can be found in multiplicative-divide functional units for elementary function approximation as well as in the computation of higher power functions that are used for encryption/decryption in cryptography [4] [6] [10].

Residue Number System (RNS) has been proposed as an alternative for the design of DSP and telecommunication applications enhancing their performance and reducing their power dissipation [1] [2] [9] [12]. An RNS-based system parallelizes the arithmetic processing load by splitting the inputs in smaller numbers according to a chosen set of moduli values and performing the arithmetic processing separately for each modulo value. A value of the 2^n-1 form is present in almost every moduli set of an RNS-based system. To this end, the design of units for modulo 2^n-1 arithmetic is strongly required.

Contemporary systems can benefit from the use of a dedicated cubing unit in order to achieve faster speed and lower power dissipation [3]. A cubing unit can be designed following several approaches. For small wordlengths, either direct realizations of minimized logic functions or look-up tables can be used. However, these approaches are inefficient for medium and large wordlengths. Another solution is to utilize either a multiplier or a multiplier along with a squarer unit. The resulting cubing units however are unnecessarily complex and slow. Another approach is to derive the partial product bits required for the cubing operation, exploit the redundancy that exists in them to reduce their number and utilize a tree adder-based approach to sum the reduced partial products. Several design methods have been presented in the past to increase the performance of binary cubing units based on the last approach [7] [8] [11].

In this paper we present, for the first time in the open literature, two methods for designing modulo 2^n-1 cubing units. The first one leads to faster circuits while the second one leads to smaller circuits for medium and large values of n. We also present minimized logic functions for the design of modulo 2^n-1 cubing units for small wordlengths.

The remaining of the paper is organized as follows. Section 2 presents some preliminaries on binary cubing while Section 3 presents novel design methods for modulo 2^n-1 cubing units. Section 4 evaluates the proposed methods and presents experimental results. Section 5 concludes the paper.

2. PRELIMINARIES

In binary, the cubing operation of an unsigned n-bit wide operand, $A=a_{n-1}\ldots a_0$, is defined as:

$$A^3=\left(\sum_{i=0}^{n-1}a_i2^i\right)^3=\sum_{i=0}^{n-1}\sum_{j=0}^{n-1}\sum_{k=0}^{n-1}a_ia_ja_k2^{i+j+k} \tag{1}$$

The partial product matrix (PPM) produced by the above equation consists of n^3 terms. However, as shown in [7] and [11] its size can be reduced by grouping terms with identical bits since $a_ia_ia_i=a_i$ and $a_ia_ja_j=a_ia_j$. Equation (1) is then rewritten as:

$$\begin{aligned}A^3=&\sum_{i=0}^{n-1}a_i2^{3i}+3\sum_{i=0}^{n-2}\sum_{j=i+1}^{n-1}a_ia_j(2^{2i+j}+2^{i+2j})\\&+3\sum_{i=0}^{n-2}\sum_{j=i+1}^{n-2}\sum_{k=j+1}^{n-1}a_ia_ja_k2^{i+j+k+1}\end{aligned} \tag{2}$$

where the first term corresponds to single-bit terms, the second term corresponds to double-bit terms (pairs) and the third one to three-bit terms (triplets) of the PPM.

Two methods have been proposed to implement equation (2) [7] [11]. The method presented in [7] at first compresses the pairs and triplets (denoted hereafter as 3× terms) to a sum and carry vector

GLSVLSI'13, May 2–3, 2013, Paris, France.

using an adder tree. Then this carry-save result along with its two times multiple and the single-bit terms (denoted hereafter as 1× terms), that is, a total of 5 vectors are reduced to two vectors by a second adder tree and a fast two-input adder is used to derive the final result. This method produces a PPM with: (i) n bits due to the single-bit terms, (ii) $n(n-1)$ bits due to pairs and (iii) $\frac{1}{6}n(n-1)(n-2)$ bits due to triplets. Hence the PPM consists of a total number of bits ($NB_{[7]}$) equal to:

$$NB_{[7]} = NB_{1\times} + NB_{3\times} = \frac{1}{6}n^3 + \frac{1}{2}n^2 + \frac{1}{3}n$$

where $NB_{1\times}$ and $NB_{3\times}$ are the number of bits of the 1× and 3× terms respectively.

The second method introduced in [11], instead of initially adding the 3× terms and then adding their sum with the 1× terms, it combines the pairs, the triplets and the single-bit terms all at once. Since $3=1\times 2^1+1\times 2^0$, each bit that corresponds to a pair or triplet is repeated to the next to the left column. All these bits along with the single-bit terms are then reduced to two vectors by an adder tree and a final adder is used to derive the result. The resulting PPM has a total number of bits ($NB_{[11]}$) equal to:

$$NB_{[11]} = NB_{1\times} + 2NB_{3\times} = \frac{1}{3}n^3 + n^2 - \frac{1}{3}n \, .$$

This method has the advantage of using less levels in the adder tree, compared to the total number of levels required by the two adder trees of [7], leading to faster circuits. This is achieved at the expense of a larger PPM.

A third method has been explored in [8] for binary cubing. Several simplifications and transformations are proposed that can be sequentially applied to the PPM in order to reduce its size.

3. MODULO 2^n-1 CUBING

In this section we present two methods for designing modulo 2^n-1 cubing units starting from the two above-mentioned techniques for binary cubing [7] [11]. The third subsection deals with modulo 2^n-1 cubing units for small values of n.

Assume that $X=x_{n-1}\ldots x_0$ denotes a modulo 2^n-1 operand with $X\in[0,2^n-1)$. The cube of X taken modulo 2^n-1 is equal to:

$$\left|X^3\right|_{2^n-1} = \left| \left|\sum_{i=0}^{n-1} x_i 2^{3i}\right|_{2^n-1} + \left|3\sum_{i=0}^{n-2}\sum_{j=i+1}^{n-1} x_i x_j (2^{2i+j} + 2^{i+2j})\right|_{2^n-1} + \left|3\sum_{i=0}^{n-2}\sum_{j=i+1}^{n-2}\sum_{k=j+1}^{n-1} x_i x_j x_k 2^{i+j+k+1}\right|_{2^n-1} \right|_{2^n-1} \quad (3)$$

3.1 Modulo 2^n-1 Cubing Units – Method 1

Using the method of [11], equation (3) can be written as:

$$\left|X^3\right|_{2^n-1} = \left| \left|\sum_{i=0}^{n-1} x_i 2^{3i}\right|_{2^n-1} + \left|\sum_{i=0}^{n-2}\sum_{j=i+1}^{n-1} x_i x_j (2^{2i+j} + 2^{i+2j} + 2^{2i+j+1} + 2^{i+2j+1})\right|_{2^n-1} + \left|\sum_{i=0}^{n-2}\sum_{j=i+1}^{n-2}\sum_{k=j+1}^{n-1} x_i x_j x_k (2^{i+j+k+1} + 2^{i+j+k+2})\right|_{2^n-1} \right|_{2^n-1} \quad (4)$$

Since for a bit w and an unsigned integer p it holds that $\left|w2^p\right|_{2^n-1} = w2^{|p|_n}$, all bits having weights greater that 2^{n-1} can be placed to the n least significant bit positions. Therefore, the modulo 2^n-1 cubing operation described in equation (4) produces an initial PPM with only n columns and a total number of bits (NB_1) equal to $NB_{[11]}$. Concerning the maximum height of the PPM, H_1, we consider two cases:

Case i: if $|n|_3 \neq 0$, that is n is not a multiple of 3, then the PPM is totally balanced, that is, all columns have the same height and H_1 is equal to $H_1^i = \frac{1}{3}(n^2-1)+n$.

Case ii: when $|n|_3 = 0$, the PPM is not balanced and its maximum height is defined by the number of pairs and triplets produced in columns with weight equal to 2^{3h+2}, h=0,1,…,(n/3)-1. In this case H_1 is equal to $H_1^{ii} = 2\left\lfloor \frac{(n-2)(n-1)}{6} \right\rfloor + 2n+1$, where $\lfloor X \rfloor$ denotes the floor value of X, that is, the largest integer not greater than X.

We can exploit the inherent folding property of the modulo 2^n-1 arithmetic $\left|w2^p\right|_{2^n-1} = w2^{|p|_n}$ and the patterns of common bits that appear in each column due to this folding in order to further reduce the size of the PPM. Table 1 presents some *Reduction Rules* (*RR*) that can be used for this purpose. Subscripts i, j, k and m refer to integers with values in range [0, n), $\bar{x}$ corresponds to the one's complement of a bit x and $\vee$ denotes a logic OR operation respectively. Due to lack of space, their proofs are given in [14] and are based on [7], [8] and [11].

We consider the two cases separately.

Case i: $|n|_3 \neq 0$. When n is not a multiple of 3, the PPM is totally balanced. Its size can be reduced by first applying *RR*1 to the n $x_i x_j$ pairs with $j=|i+1|_n$, i=0, ..., n-1. Then *RR2* can be applied once in every column. These two rules lead to a PPM height reduction equal to 4. *RR3* is then repeatedly applied, again to all columns, for a total of (n-4) times reducing the height further by 2(n-4). The final PPM still remains totally balanced. It consists of $NB^i_{1,final} = n \times H^i_{1,final} = \frac{1}{3}n^3 - n^2 + \frac{11}{3}n$ bits and its height is equal to $H^i_{1,final} = \frac{1}{3}(n^2-1) - n + 4$.

Let us present an explanatory example so as to clarify the proposed PPM reduction steps. For simplicity we assume n=5. The initial PPM produced according to Equation (4) is shown in Figure 1(a). It consists of 65 bits and its height is equal to 13. The shaded boxes indicate the terms that *RR*1 is to be applied to. The resulting PPM with a reduced by 2 height is shown in Figure 1(b). The shaded boxes now indicate the terms that are subjected to *RR2*. *RR2* also reduces the height by two (Figure 1(c)). Finally, *RR3* is applied (n-4) times, in this example only once, in every

Table 1. Reduction Rules

$x_i x_j + x_i x_j + 2x_i x_j = 4x_i x_j$	*RR*1
$x_i x_j + x_i x_j x_k + x_i x_k + 2x_i x_j x_k = 4x_i x_j x_k + x_i(x_j \vee x_k)$	*RR2*
$x_i x_j + x_i x_j x_k + x_i x_m x_k + 2x_i x_j x_k = 4x_i x_j x_k + x_i(x_j \bar{x}_k \vee x_m x_k)$	*RR3*
$x_j x_k + x_i x_j x_k + x_i + 2x_i x_j x_k = 4x_i x_j x_k + x_i \vee x_j x_k$	*RR4*

2^4	2^3	2^2	2^1	2^0
x_3	x_1	x_4	x_2	x_0
x_4x_0	x_3x_2	x_1x_0	x_4x_3	x_2x_1
x_4x_0	x_3x_2	x_1x_0	x_4x_3	x_2x_1
x_3x_2	x_1x_0	x_4x_3	x_2x_1	x_4x_0
x_2x_1	x_3x_0	x_3x_1	x_1x_0	x_4x_3
x_2x_0	x_4x_0	x_3x_2	x_4x_1	x_4x_2
x_4x_1	x_4x_2	x_2x_0	x_3x_0	x_3x_1
x_4x_2	x_2x_0	x_4x_1	x_3x_1	x_2x_0
x_3x_0	x_3x_1	x_3x_0	x_4x_2	x_4x_1
$x_2x_1x_0$	$x_4x_3x_0$	$x_4x_2x_0$	$x_4x_1x_0$	$x_4x_3x_2$
$x_4x_3x_1$	$x_4x_2x_1$	$x_3x_2x_1$	$x_3x_2x_0$	$x_4x_3x_1$
$x_4x_3x_0$	$x_4x_2x_0$	$x_4x_1x_0$	$x_3x_1x_0$	$x_3x_1x_0$
$x_4x_2x_1$	$x_3x_2x_1$	$x_3x_2x_0$	$x_4x_3x_2$	$x_2x_1x_0$

(a) Initial PPM for $\left|X^3\right|_{2^5-1}$

2^4	2^3	2^2	2^1	2^0
x_3	x_1	x_4	x_2	x_0
x_1x_0	x_4x_3	x_2x_1	x_4x_0	x_3x_2
x_2x_1	x_3x_0	x_3x_1	x_1x_0	x_4x_3
x_2x_0	x_4x_0	x_3x_2	x_4x_1	x_4x_2
x_4x_1	x_4x_2	x_2x_0	x_3x_0	x_3x_1
x_4x_2	x_2x_0	x_4x_1	x_3x_1	x_2x_0
x_3x_0	x_3x_1	x_3x_0	x_4x_2	x_4x_1
$x_2x_1x_0$	$x_4x_3x_0$	$x_3x_2x_1$	$x_4x_1x_0$	$x_4x_3x_2$
$x_4x_3x_0$	$x_3x_2x_1$	$x_4x_1x_0$	$x_4x_3x_2$	$x_2x_1x_0$
$x_4x_3x_1$	$x_4x_2x_1$	$x_4x_2x_0$	$x_3x_2x_0$	$x_3x_1x_0$
$x_4x_2x_1$	$x_4x_2x_0$	$x_3x_2x_0$	$x_3x_1x_0$	$x_4x_3x_1$

(b) PPM after *RR*1 is applied

2^4	2^3	2^2	2^1	2^0
x_3	x_1	x_4	x_2	x_0
x_1x_0	x_4x_3	x_2x_1	x_4x_0	x_3x_2
x_4x_1	x_4x_2	x_2x_0	x_3x_0	x_3x_1
x_4x_2	x_2x_0	x_4x_1	x_3x_1	x_2x_0
x_3x_0	x_3x_1	x_3x_0	x_4x_2	x_4x_1
$x_2(x_1 \vee x_0)$	$x_0(x_4 \vee x_3)$	$x_3(x_2 \vee x_1)$	$x_1(x_0 \vee x_4)$	$x_4(x_3 \vee x_2)$
$x_3x_2x_1$	$x_4x_2x_1$	$x_4x_2x_0$	$x_3x_2x_0$	$x_3x_1x_0$
$x_4x_3x_1$	$x_4x_1x_0$	$x_4x_3x_2$	$x_2x_1x_0$	$x_4x_3x_0$
$x_4x_2x_1$	$x_4x_2x_0$	$x_3x_2x_0$	$x_3x_1x_0$	$x_4x_3x_1$

(c) PPM after *RR*2 is applied

2^4	2^3	2^2	2^1	2^0
x_3	x_1	x_4	x_2	x_0
x_1x_0	x_4x_3	x_2x_1	x_4x_0	x_3x_2
x_4x_2	x_2x_0	x_4x_1	x_3x_1	x_2x_0
x_3x_0	x_3x_1	x_3x_0	x_4x_2	x_4x_1
$x_2(x_1 \vee x_0)$	$x_0(x_4 \vee x_3)$	$x_3(x_2 \vee x_1)$	$x_1(x_0 \vee x_4)$	$x_4(x_3 \vee x_2)$
$x_1(x_4\overline{x}_3 \vee x_3x_2)$	$x_4(x_2\overline{x}_1 \vee x_1x_0)$	$x_2(x_0\overline{x}_4 \vee x_4x_3)$	$x_0(x_3\overline{x}_2 \vee x_2x_1)$	$x_3(x_1\overline{x}_0 \vee x_4x_0)$
$x_4x_2x_0$	$x_3x_2x_0$	$x_3x_1x_0$	$x_4x_3x_1$	$x_4x_2x_1$

(d) Final PPM for $\left|X^3\right|_{2^5-1}$

Figure 1. Proposed PPM reduction steps for $\left|X^3\right|_{2^5-1}$.

column to the shadowed terms of Figure 1(c). Figure 1(d) presents the final PPM for the $\left|X^3\right|_{2^5-1}$. It consists of only 35 bits and its maximum height has been reduced from 13 to 7 while it remains balanced. Examples for other values of n can be found on [14].

Case ii: $|n|_3 = 0$. The PPM in this case is partially balanced. It consists of $(n/3)$ repetitions of a group of three consecutive columns of different heights. Specifically, the columns with weights 2^{3h}, 2^{3h+1}, 2^{3h+2}, h=0,1,...,$(n/3)$-1, have heights equal to $H_1^{ii}-1$, $H_1^{ii}-3$ and H_1^{ii} respectively (see for example the PPM for n=6 in Figure 2(a) where H_1^{ii}=19). The maximum height of the PPM can be also reduced. This can be achieved by first applying *RR*1 as in *Case i* where $|n|_3 \neq 0$ (Figures 2(a)&(b)). Then, *RR*2 can be applied three times to every column with weight 2^{3h+1} (Figures 2(b)&(c)), *RR*4 can be applied three times to columns with weight 2^{3h} (Figures 2(c)&(d)) and *RR*3 can be applied once to columns with weight 2^{3h+2}. After these reductions the PPM has a height equal to $H_1^{ii}-8$ and is totally balanced as shown in Figure 3 for n=6. For n>6, *RR*3 can then be reapplied (n-6) times to all columns so as to further reduce the height by 2(n-6). Finally, the height of the PPM in this case is reduced to $H_{1,final}^{ii} = 5 + 2\left\lfloor \frac{(n-2)(n-1)}{6} \right\rfloor$ and since PPM is balanced, it consists of $NB_{1,final}^{ii} = n \times H_{1,final}^{ii}$ bits.

The implementation of the proposed modulo 2^n-1 cubing circuits (see Figure 4(a)) requires at first some simple logic gates to derive the bits of the reduced PPM. Then a Dadda adder tree is utilized in order to compress the PPM's bits to two n-bit wide vectors (sum and carry). The adder tree consists of $D(H_{1final})$ end-around carry carry save adder (EAC CSA) levels, where the $D(t)$ function denotes the number of levels that are required in a Dadda adder tree for reducing t vectors to two [5]. In *Case i*, the Dadda adder tree requires $\frac{1}{3}n^3 - n^2 + \frac{5}{3}n$ Full-Adders (FAs), while in *Case ii* $3n + 2n\left\lfloor \frac{(n-2)(n-1)}{6} \right\rfloor$ FAs are required. Finally, a fast two-operand modulo 2^n-1 adder is utilized to add the sum and carry vectors and produce the result.

3.2 Modulo 2^n-1 Cubing Units – Method 2

If the method proposed in [7] is employed for implementing equation (3), then the PPM for the modulo 2^n-1 cubing after applying the folding property $\left|w2^p\right|_{2^n-1} = w2^{|p|_n}$ is n columns wide and consists of $NB_{[7]}$ bits. Similarly to the previous method, when n is not a multiple of 3 the PPM is totally balanced, while when n is a multiple of 3 it is not. The maximum height among all columns of the PPM for both cases of n is equal to $H_2 = 1 + \left\lfloor \frac{1}{6}(n^2 + 3n) \right\rfloor$. Note that the PPM actually consists of two separate parts: the part of the 1× terms and the part of the 3× terms. Since each 3× term appears only once inside the PPM and no patterns of similar bits appear in it (see for example the PPMs for n=5 and 6 that are given in Figures 5 and 6 or those given in [14]), no height reductions are possible in a systematic manner.

In case where $|n|_3 \neq 0$ the reduction of the 3× terms to an n-bit wide sum and an n-bit wide carry vector is achieved with an adder tree of $D(H_2$-1), EAC CSA levels consisting of ($NB_{3\times}$-2n) FAs, or equivalently $\frac{1}{6}(n^3 + 3n^2 - 16n)$ FAs. Then, the sum and carry vectors along with their two times multiples and the 1× terms are

2^5	2^4	2^3	2^2	2^1	2^0
x_2x_1	x_5x_0	x_1	x_1x_0	x_5x_1	x_0
x_2x_1	x_2x_1	x_3	x_1x_0	x_3x_2	x_2
x_4x_3	x_2x_0	x_5	x_3x_2	x_3x_1	x_4
x_4x_3	x_4x_3	x_1x_0	x_3x_2	x_5x_4	x_2x_1
x_5x_0	x_4x_2	x_3x_2	x_5x_4	x_5x_3	x_4x_3
x_5x_0	x_4x_0	x_5x_4	x_5x_4	x_1x_0	x_5x_0
x_3x_1	x_3x_0	x_3x_0	x_4x_2	x_4x_1	x_4x_1
x_4x_0	x_5x_2	x_5x_2	x_2x_0	x_3x_0	x_3x_0
x_2x_0	x_4x_1	x_4x_1	x_4x_0	x_5x_2	x_3x_1
x_5x_3	$x_5x_4x_0$	x_2x_0	x_5x_1	$x_5x_1x_0$	x_5x_2
x_5x_1	$x_2x_1x_0$	x_4x_2	x_3x_1	$x_3x_2x_1$	x_5x_3
x_4x_2	$x_4x_3x_2$	x_4x_0	x_5x_3	$x_5x_4x_3$	x_5x_1
$x_5x_4x_0$	$x_5x_3x_0$	$x_5x_3x_0$	$x_5x_1x_0$	$x_3x_2x_0$	$x_3x_2x_0$
$x_2x_1x_0$	$x_5x_2x_1$	$x_5x_2x_1$	$x_3x_2x_1$	$x_4x_1x_0$	$x_4x_1x_0$
$x_4x_3x_2$	$x_4x_3x_1$	$x_4x_3x_1$	$x_5x_4x_3$	$x_5x_4x_2$	$x_5x_4x_2$
$x_3x_1x_0$	$x_5x_3x_1$	$x_4x_2x_1$	$x_4x_2x_1$	$x_4x_2x_0$	$x_3x_1x_0$
$x_5x_4x_1$		$x_4x_3x_0$	$x_5x_2x_0$		$x_5x_3x_2$
$x_5x_3x_1$		$x_5x_2x_0$	$x_4x_2x_0$		$x_5x_4x_1$
$x_5x_3x_2$			$x_4x_3x_0$		

(a) Initial PPM for $\left|X^3\right|_{2^6-1}$

2^5	2^4	2^3	2^2	2^1	2^0
x_3x_1	x_1x_0	x_1	x_4x_2	x_2x_1	x_0
x_4x_0	x_3x_2	x_3	x_2x_0	x_4x_3	x_2
x_2x_0	x_5x_4	x_5	x_4x_0	x_5x_0	x_4
x_5x_3	x_5x_0	x_3x_0	x_5x_1	x_5x_1	x_4x_1
x_5x_1	x_2x_1	x_5x_2	x_3x_1	x_3x_2	x_3x_0
x_4x_2	x_2x_0	x_4x_1	x_5x_3	x_3x_1	x_5x_2
$x_5x_4x_0$	x_4x_3	x_2x_0	$x_5x_1x_0$	x_5x_4	x_3x_1
$x_2x_1x_0$	x_4x_2	x_4x_2	$x_3x_2x_1$	x_5x_3	x_5x_3
$x_4x_3x_2$	x_4x_0	x_4x_0	$x_5x_4x_3$	x_1x_0	x_5x_1
$x_3x_1x_0$	x_3x_0	$x_5x_3x_0$	$x_4x_2x_1$	x_4x_1	$x_3x_2x_0$
$x_5x_4x_1$	x_5x_2	$x_5x_2x_1$	$x_5x_2x_0$	x_3x_0	$x_4x_1x_0$
$x_5x_3x_1$	x_4x_1	$x_4x_3x_1$	$x_4x_2x_0$	x_5x_2	$x_5x_4x_2$
$x_5x_3x_2$	$x_5x_4x_0$	$x_4x_2x_1$	$x_4x_3x_0$	$x_5x_1x_0$	$x_3x_1x_0$
	$x_2x_1x_0$	$x_4x_3x_0$		$x_3x_2x_1$	$x_5x_3x_2$
	$x_4x_3x_2$	$x_5x_2x_0$		$x_5x_4x_3$	$x_5x_4x_1$
	$x_5x_3x_0$			$x_3x_2x_0$	
	$x_5x_2x_1$			$x_4x_1x_0$	
	$x_4x_3x_1$			$x_5x_4x_2$	
	$x_5x_3x_1$			$x_4x_2x_0$	

(b) PPM after *RR*1 is applied

2^5	2^4	2^3	2^2	2^1	2^0
x_3x_1	x_1x_0	x_1	x_4x_2	x_2x_1	x_0
x_4x_0	x_3x_2	x_3	x_2x_0	x_4x_3	x_2
x_2x_0	x_4x_0	x_5	x_4x_0	x_1x_0	x_4
x_5x_3	x_3x_0	x_3x_0	x_5x_1	x_4x_1	x_4x_1
x_5x_1	x_5x_2	x_5x_2	x_3x_1	x_3x_0	x_3x_0
x_4x_2	x_4x_1	x_4x_1	x_5x_3	x_5x_2	x_5x_2
$x_3x_1x_0$	$x_5(x_4\vee x_0)$	x_2x_0	$x_4x_2x_1$	$x_5(x_1\vee x_0)$	x_3x_1
$x_5x_4x_1$	$x_2(x_1\vee x_0)$	x_4x_2	$x_5x_2x_0$	$x_3(x_2\vee x_1)$	x_5x_3
$x_5x_3x_1$	$x_4(x_3\vee x_2)$	x_4x_0	$x_4x_2x_0$	$x_5(x_4\vee x_3)$	x_5x_1
$x_5x_3x_2$	$x_5x_3x_0$	$x_5x_3x_0$	$x_4x_3x_0$	$x_3x_2x_0$	$x_3x_2x_0$
	$x_5x_2x_1$	$x_5x_2x_1$		$x_4x_1x_0$	$x_4x_1x_0$
	$x_4x_3x_1$	$x_4x_3x_1$		$x_5x_4x_2$	$x_5x_4x_2$
	$x_5x_3x_1$	$x_4x_2x_1$		$x_4x_2x_0$	$x_3x_1x_0$
		$x_4x_3x_0$			$x_5x_3x_2$
		$x_5x_2x_0$			$x_5x_4x_1$
		$x_5x_1x_0$			$x_5x_4x_0$
		$x_3x_2x_1$			$x_2x_1x_0$
		$x_5x_4x_3$			$x_4x_3x_2$

(c) PPM after *RR*2 is applied

2^5	2^4	2^3	2^2	2^1	2^0
x_3x_1	x_1x_0	x_2x_0	x_4x_2	x_2x_1	x_3x_1
x_4x_0	x_3x_2	x_4x_2	x_2x_0	x_4x_3	x_5x_3
x_2x_0	x_4x_0	x_4x_0	x_4x_0	x_1x_0	x_5x_1
x_5x_3	x_3x_0	$x_1\vee(x_5x_2)$	x_5x_1	x_4x_1	$x_0\vee(x_4x_1)$
x_5x_1	x_5x_2	$x_3\vee(x_4x_1)$	x_3x_1	x_3x_0	$x_2\vee(x_3x_0)$
x_4x_2	x_4x_1	$x_5\vee(x_3x_0)$	x_5x_3	x_5x_2	$x_4\vee(x_5x_2)$
$x_3x_1x_0$	$x_5(x_4\vee x_0)$	$x_4x_2x_1$	$x_4x_2x_1$	$x_3(x_2\vee x_1)$	$x_3x_1x_0$
$x_5x_3x_0$	$x_2(x_1\vee x_0)$	$x_4x_3x_0$	$x_4x_1x_0$	$x_5(x_4\vee x_3)$	$x_5x_3x_2$
$x_5x_4x_1$	$x_4(x_3\vee x_2)$	$x_5x_2x_0$	$x_5x_2x_0$	$x_5(x_1\vee x_0)$	$x_5x_4x_1$
$x_5x_3x_1$	$x_5x_3x_1$	$x_5x_1x_0$	$x_4x_2x_0$	$x_4x_2x_0$	$x_5x_4x_0$
$x_5x_3x_2$		$x_3x_2x_1$	$x_4x_3x_0$		$x_2x_1x_0$
$x_5x_2x_1$		$x_5x_4x_3$	$x_3x_2x_0$		$x_4x_3x_2$
$x_4x_3x_1$			$x_5x_4x_2$		

(d) PPM after *RR*4 is applied

Figure 2. Proposed PPM reduction steps for $\left|X^3\right|_{2^6-1}$.

2^5	2^4	2^3	2^2	2^1	2^0
x_4x_0	x_1x_0	x_2x_0	x_2x_0	x_2x_1	x_3x_1
x_2x_0	x_3x_2	x_4x_2	x_4x_0	x_4x_3	x_5x_3
x_5x_3	x_4x_0	x_4x_0	x_5x_1	x_1x_0	x_5x_1
x_5x_1	x_3x_0	$x_1\vee(x_5x_2)$	x_3x_1	x_4x_1	$x_0\vee(x_4x_1)$
x_4x_2	x_5x_2	$x_3\vee(x_4x_1)$	x_5x_3	x_3x_0	$x_2\vee(x_3x_0)$
$x_3(x_1\overline{x}_0\vee x_5x_0)$	x_4x_1	$x_5\vee(x_3x_0)$	$x_4(x_2\overline{x}_1\vee x_1x_0)$	x_5x_2	$x_4\vee(x_5x_2)$
$x_5x_4x_1$	$x_5(x_4\vee x_0)$	$x_5x_2x_0$	$x_5x_2x_0$	$x_3(x_2\vee x_1)$	$x_5x_3x_2$
$x_5x_3x_1$	$x_2(x_1\vee x_0)$	$x_4x_3x_0$	$x_4x_2x_0$	$x_5(x_4\vee x_3)$	$x_5x_4x_1$
$x_5x_3x_2$	$x_4(x_3\vee x_2)$	$x_5x_1x_0$	$x_4x_3x_0$	$x_5(x_1\vee x_0)$	$x_5x_4x_0$
$x_5x_2x_1$	$x_5x_3x_1$	$x_3x_2x_1$	$x_3x_2x_0$	$x_4x_2x_0$	$x_2x_1x_0$
$x_4x_3x_1$	$x_4x_2x_1$	$x_5x_4x_3$	$x_5x_4x_2$	$x_3x_1x_0$	$x_4x_3x_2$

Figure 3. Final PPM for $\left|X^3\right|_{2^6-1}$

reduced to two final summands by a second adder tree with $D(5)$=3 levels comprised of extra $(4n+n)-2n = 3n$ FAs. It must be noted that the formation of the two times multiple of the sum and carry vectors requires zero area and delay since it can be requires hardwired. Finally, the overall adder tree of the cubing unit requires $(D(H_2\text{-}1)+3)$ levels and a total of $\frac{1}{6}(n^3+3n^2+2n)$ FAs.

For the values of n that are multiples of 3, the PPM is not balanced (see for example Figure 6) and its maximum height is equal to the height of the columns with weights 2^{3h+1}, h=0,1…,(n/3)-1. The 1× terms form three vectors with n/3 non-zero bits each with weights $2^{|3i|_n}$, i=0,…,n-1. These vectors can be reduced to only one n-bit wide vector with n/3 FAs. The 3× terms are reduced to a sum and carry vector by an adder tree with a total of $\frac{1}{6}(n^3+3n^2-16n)$ FAs and $D(H_2)$ levels. Similarly to the previous case, these two vectors, their two times multiples and the vector produced by the reduction of the 1× terms are driven to a second adder tree to be further reduced to two final summands (see Figure 4(b)). The reduction of the 1× terms does not contribute to the overall delay of the adder tree since it can be done in parallel with the reduction of the 3× terms. The overall adder tree of the cubing unit in this case consists of $D(H_2)$+3 EAC CSA levels requiring a total of $\frac{1}{6}(n^3+3n^2+4n)$ FAs.

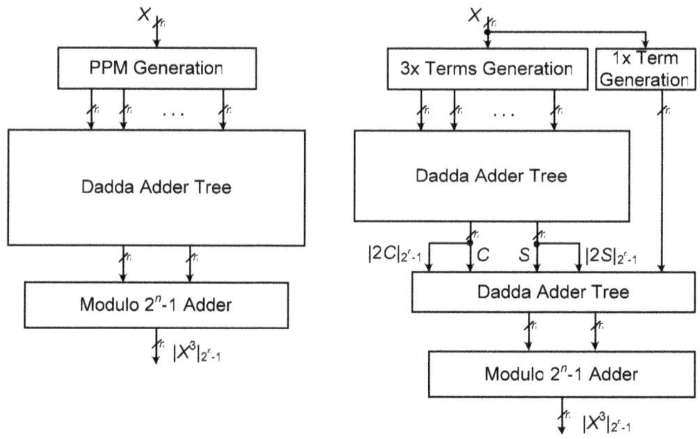

(a) (b)

Figure 4. Implementation of (a) Method 1 and (b) Method 2.

2^4	2^3	2^2	2^1	2^0	
x_3	x_1	x_4	x_2	x_0	1×
x_4x_0	x_3x_0	x_2x_0	x_1x_0	x_3x_1	3×
x_2x_1	x_4x_0	x_1x_0	x_4x_1	x_2x_1	
x_2x_0	x_4x_2	x_3x_1	x_3x_0	x_4x_2	
x_4x_1	x_3x_2	x_3x_2	x_4x_3	x_4x_3	
$x_2x_1x_0$	$x_4x_2x_1$	$x_3x_2x_1$	$x_3x_2x_0$	$x_3x_1x_0$	
$x_4x_3x_1$	$x_4x_3x_0$	$x_4x_2x_0$	$x_4x_1x_0$	$x_4x_3x_2$	

Figure 5. PPM for $\left|X^3\right|_{2^5-1}$ using Method 2.

2^5	2^4	2^3	2^2	2^1	2^0	
		x_1			x_0	1×
		x_3			x_2	
		x_5			x_4	
x_5x_0	x_4x_0	x_3x_0	x_2x_0	x_1x_0	x_4x_1	3×
x_3x_1	x_2x_1	x_5x_2	x_1x_0	x_5x_1	x_3x_0	
x_2x_1	x_2x_0	x_4x_1	x_4x_2	x_3x_2	x_5x_2	
x_5x_3	x_5x_0	$x_5x_3x_0$	x_4x_0	x_3x_1	$x_4x_1x_0$	
x_5x_1	x_4x_3	$x_5x_2x_1$	x_3x_2	x_5x_4	$x_3x_2x_0$	
x_4x_3	x_4x_2	$x_4x_3x_1$	x_5x_4	x_5x_3	$x_5x_4x_2$	
$x_3x_1x_0$	$x_2x_1x_0$		$x_5x_2x_0$	$x_5x_1x_0$		
$x_5x_4x_1$	$x_5x_4x_0$		$x_4x_3x_0$	$x_4x_2x_0$		
$x_5x_3x_2$	$x_5x_3x_1$		$x_4x_2x_1$	$x_3x_2x_1$		
	$x_4x_3x_2$			$x_5x_4x_3$		

Figure 6. PPM for $\left|X^3\right|_{2^6-1}$ using Method 2.

3.3 Modulo 2^n-1 Cubing Units for Small n

Both the proposed methods that have been previously described are suitable for medium and large values of n ($n\geq 6$). For small wordlengths ($n<6$), more efficient cubing circuits can be designed by utilizing minimized logic equations of the outputs. These can be derived by a Boolean function minimizing technique such as the Quine-McCluskey technique, since the modulo 2^n-1 cubing operation is a single- operand operation with an n-bit wide result. To this end, we present in Table 2 minimized logic functions for the outputs of modulo 2^n-1 cubing circuits for values of n equal to 2, 3, 4 and 5. We assume in Table 2 a double zero representation, that is both 0…0 and 1…1 are valid modulo 2^n-1 representations of zero. Symbol ⊕ denotes a logic XOR operation.

4. EVALUATION

At first we comment on the area and delay requirements of the cubing units designed according to the proposed methods. For theoretical evaluations, we neglect the logic gates for deriving the partial product bits of the PPM as well as the final 2-input modulo 2^n-1 adder and consider only the adder tree for the reduction of the PPM which consists the main area and delay overhead of the circuits.

Table 2. Minimized logic functions for the outputs of modulo 2^n-1 cubing circuits for small wordlengths

n	$C=\left\|X^3\right\|_{2^n-1}=c_{n-1}\ldots c_0,\quad X=x_{n-1}\ldots x_0$
2	$c_0=x_0$
	$c_1=x_1$
3	$c_0=x_2\oplus x_1\oplus x_0$
	$c_1=x_2x_1\vee x_2x_0\vee x_1x_0$
	$c_2=c_1$
4	$c_0=\overline{x}_1\,(x_0\vee x_3x_2)\vee(x_3\vee x_2)x_0$
	$c_1=\overline{x}_0\,(x_3\vee x_2x_1)\vee(x_2\vee x_1)x_3$
	$c_2=\overline{x}_3\,(x_2\vee x_1x_0)\vee(x_1\vee x_0)x_2$
	$c_3=\overline{x}_2\,(x_1\vee x_0x_3)\vee(x_0\vee x_3)x_1$
5	$c_0=x_0\,\overline{x}_1\,\overline{x}_3\vee x_2x_4(x_0\vee x_1\,\overline{x}_3)\vee x_1x_0(x_3\vee\overline{x}_4\,\overline{x}_2)\vee x_3\,\overline{x}_1\,(x_2\oplus x_4)$
	$c_1=x_2\,\overline{x}_3\,\overline{x}_0\vee x_4x_1(x_2\vee x_3\,\overline{x}_0)\vee x_3x_2(x_0\vee\overline{x}_1\,\overline{x}_4)\vee x_0\,\overline{x}_3\,(x_4\oplus x_1)$
	$c_2=x_4\,\overline{x}_0\,\overline{x}_2\vee x_1x_3(x_4\vee x_0\,\overline{x}_2)\vee x_0x_4(x_2\vee\overline{x}_3\,\overline{x}_1)\vee x_2\,\overline{x}_0\,(x_1\oplus x_3)$
	$c_3=x_1\,\overline{x}_2\,\overline{x}_4\vee x_3x_0(x_1\vee x_2\,\overline{x}_4)\vee x_2x_1(x_4\vee\overline{x}_0\,\overline{x}_3)\vee x_4\,\overline{x}_2\,(x_3\oplus x_0)$
	$c_4=x_3\,\overline{x}_4\,\overline{x}_1\vee x_0x_2(x_3\vee x_4\,\overline{x}_1)\vee x_4x_3(x_1\vee\overline{x}_2\,\overline{x}_0)\vee x_1\,\overline{x}_4\,(x_0\oplus x_2)$

Table 3 summarizes the maximum height of the PPM (H) and the overall number of FAs (NFAs) required by the adder trees of the modulo 2^n-1 cubing units according to the design methods 1 and 2 of Sections 3.1 and 3.2, respectively. The Dadda adder tree of the modulo 2^n-1 cubing circuits of Section 3.1 consists of $D(H_1)$ EAC CSA levels in order to reduce the PPM's bits to two n-bit wide vectors. As far as the cubing units of Method 2 are concerned, when $|n|_3\neq 0$ the overall adder tree requires $D(H_2\text{-}1)+3$ EAC CSA levels while when $|n|_3=0$ it consists of $D(H_2)+3$ EAC CSA levels. Figure 7 presents the area and delay requirements of the adder trees in terms of (a) the number of EAC CSA levels and (b) the number of FAs required by the adder trees for various wordlengths ($n\geq 6$). Evidently, the cubing units that are based on Method 1 are faster than those of Method 2 for all cases of n, due to the exploitation of the common patterns that arise between the bits of the PPM. However they are expected to be larger for medium and large wordlengths, since their PPM consists of more bits.

Table 3. (a) Maximum PPM height and (b) total number of FAs required by the adder trees of the proposed cubing units.

	Modulo 2^n-1 Cubing Units Method 1 (Section 3.1)	Modulo 2^n-1 Cubing Units Method 2 (Section 3.2)
H	$\frac{1}{3}(n^2-1)-n+4$, $\|n\|_3\neq 0$	$1+\left\lfloor\frac{1}{6}(n^2+3n)\right\rfloor$
	$5+2\left\lfloor\frac{(n-2)(n-1)}{6}\right\rfloor$, $\|n\|_3=0$	
$NFAs$	$\frac{1}{3}n^3-n^2+\frac{5}{3}n$, $\|n\|_3\neq 0$	$\frac{1}{6}n^3+\frac{1}{2}n^2+\frac{1}{2}n$, $\|n\|_3\neq 0$
	$3n+2n\left\lfloor\frac{(n-2)(n-1)}{6}\right\rfloor$, $\|n\|_3=0$	$\frac{1}{6}n^3+\frac{1}{2}n^2+\frac{2}{3}n$, $\|n\|_3=0$

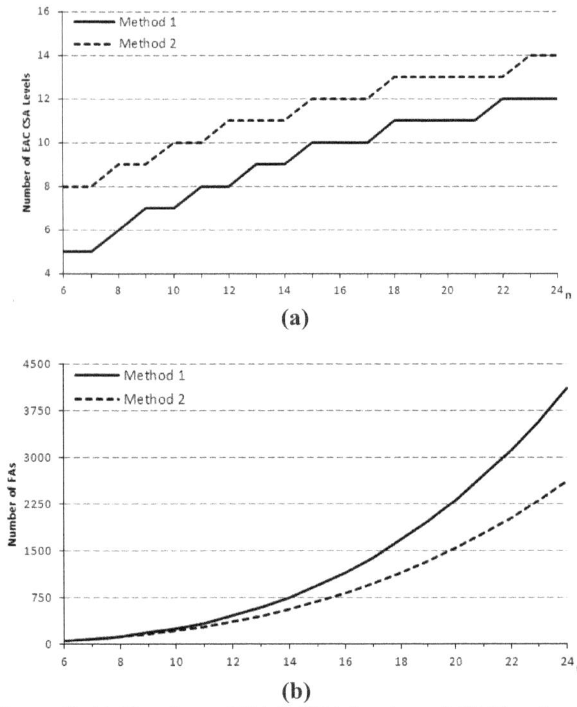

Figure 7. (a) Number of EAC CSA levels and (b) Number of FAs, required by the adder trees of the proposed cubing units.

The above evaluation can only be considered as indicative. To this end, modulo 2^n-1 cubing units for several wordlengths ($5<n<10$) were described in HDL. Specifically, both methods of Sections 3.1 and 3.2 were examined. Each circuit was synthesized and mapped to a 90 nm CMOS implementation technology [13] and was recursively optimized for speed using a bottom-up approach. An area recovery step was finally applied. Table 4 presents the attained area, delay and average power dissipation (assuming the same operating frequency of 400 MHz for all circuits) results for the proposed cubing circuits.

The attained experimental results are consistent with the theoretical estimations, that is, they confirm that the modulo 2^n-1 cubing units of Method 1 are always faster, while for $n\geq 8$ they require more area, compared to the respective units of Method 2. Evidently, the simplified logic functions for the smaller wordlengths ($n<6$) given in Section 3.3 are very efficient.

5. CONCLUSIONS

For the first time in the literature, methods for designing modulo 2^n-1 cubing units have been presented. They are based on corresponding methods for binary cubing and the application of modulo 2^n-1 arithmetic properties and reduction rules. The two proposed methods can be utilized for medium and large wordlengths for deriving fast or small circuits respectively. For completeness, simplified logic functions for cubing units with small wordlengths have been also provided.

6. ACKNOWLEDGMENTS

This work was supported by the Caratheodory Programme of the University of Patras (D.178).

Table 4. CMOS VLSI results of proposed cubing units

n	Modulo 2^n-1 Cubing Units Method 1 (Section 3.1)			Modulo 2^n-1 Cubing Units Method 2 (Section 3.2)		
	Area (μm^2)	*Delay* (*ns*)	*Power* (μW)	*Area* (μm^2)	*Delay* (*ns*)	*Power* (μW)
6	5108	1.73	600	5441	2.31	670
7	7187	1.77	930	7273	2.34	990
8	10121	1.96	1450	9983	2.56	1440
9	15032	2.34	2187	14536	2.62	2180

7. REFERENCES

[1] J.-C. Bajard, and L. Imbert. A full RNS implementation of RSA. *IEEE Trans. Comput.*, 53(6): 769-774, June 2004.

[2] G.C. Cardarilli, A. Nannarelli, and M. Re. Residue Number System for low-power DSP applications. In *Proc. Asilomar Conf. on Signals, Syst. & Comput.*, pages 1412-1416, November 2007.

[3] A. Deshpande, and J. Draper. Comparing squaring and cubing units with multipliers. In *Proc. IEEE Midwest Symp. Circuits and Syst.*, pages 466-469, August 2012.

[4] M.D. Ercegovac, T. Lang, J.-M. Muller, and A. Tisserand. A. Reciprocation, square root, inverse square root, and some elementary functions using small multipliers. *IEEE Trans. Comput.* 49(7):628-637, July 2000.

[5] K. Hwang. *Computer Arithmetic: Principle, Architecture and Design*. Wiley, New York, NY, 1979.

[6] T.-J. Kwon, J. Sondeen, and J. Draper. Floating-point division and square root implementation using a Taylor-series expansion algorithm. In *Proc. IEEE Conf. Electronics, Circuits and Syst.*, pages 702-705, 2008.

[7] A. Liddicoat, and M. Flynn. Parallel square and cube computations. In *Proc. Asilomar Conf. Signals, Syst. and Comput.*, volume 2, pages 1325-1329, 2000.

[8] R. Michard, A.Tisserand, and N. Veyrat-Charvillon. New identities and transformations for hardware power operators. In *Proc. SPIE 6313 Advanced Signal Process. Algorithms, Archit. and Implementations XVI*, pages 1-10, August 2006.

[9] R. Modugu, Y. -B. Kim, and M. Choi. Design and performance measurement of efficient IDEA (International Data Encryption Standard) crypto-hardware using novel modular arithmetic components. In *Proc. IEEE Instrument. Measur. Tech. Conf.*, pages 1222-1227, May 2010.

[10] R.L. Rivest, A. Shamir, and L. Adleman. A method for obtaining digital signatures and public-key cryptosystems. *Commun. ACM.* 21(2):120-126, February 1978.

[11] J.E. Stine, and J.M. Blank. Partial product reduction for parallel cubing. In *Proc. IEEE Computer Society Annual Symp. VLSI*, pages 337-342, May 2007.

[12] T. Stouraitis, and V. Paliouras. Considering the alternatives in low-power design. *IEEE Circ. & Dev. Mag.* 17(4): 22-29, July 2001.

[13] Synopsys Inc., SAED 90nm EDK, 2011, https://www.synopsys.com/apps/protected/university/members.html.

[14] E. Vassalos, and D. Bakalis. Examples on the design of modulo 2^n-1 cubing units. 2012. http://www.ellab.physics.upatras.gr/downs/TR.pdf

An Error Tolerant CAM with NAND Match-Line Organization

Aristides Efthymiou
Dept. of Computer Science
University of Ioannina
45110, Ioannina
Greece
efthym@cs.uoi.gr

ABSTRACT

Content Addressable Memories (CAMs) are valuable components in applications where fast, highly-associative searches are required. While common SRAMs can be protected from soft-errors using error-correcting codes (ECC), CAMs are much harder to protect against such effects. The main problem is that the comparison of a stored memory word to a search word, when both are encoded by ECCs, is far too complicated and expensive to be performed directly by circuits at each row of the CAM. The proposed design uses a modified NAND-type, dual match-line organization which keeps track of up to 1 bit mismatch per row. As CAM words are encoded using an ECC of Hamming distance of 3, a single bit error can be "corrected" during the search operation, by ignoring any single-bit mismatch. The design is analyzed in detail with regard to charge-sharing noise, speed and power and the cost of error-tolerance is investigated against a state-of-the art NAND-type CAM. Compared against the best-known, error-tolerant CAM, which uses a NOR match-line organization, the proposed circuit is faster and more energy efficient.

Categories and Subject Descriptors

B.3.2 [**Memory Structures**]: Design Styles—*Associative Memories*; B.8.1 [**Performance and Reliability**]: Reliability, Testing, and Fault-Tolerance

Keywords

CAM

1. INTRODUCTION

Content Addressable Memories (CAMs) are used in multiple applications which require fast searches on large sets of data, such as cache memories and network routing tables. As is the case with most memory blocks, CAMs are susceptible to soft-errors which can modify the stored values and, thus, affect the integrity of the device and possibly the system in which they are used.

To protect against such problems, memories can store data encoded by a special error-correcting code (ECC). The most commonly used ECCs can correct 1 bit and detect 2-bit errors. When a word is read from a memory, its integrity can be checked and if it cannot be corrected, appropriate measures can be taken, for example raise a processor exception for system software to deal with the problem.

In contrast to common memories, the most frequent operation in CAMs is associative search. During a search any words with correctable bit errors should be corrected first and then compared to the search input. However, although comparing two unencoded words is straightforward (a bitwise comparison followed by a logic AND), error correction is too complicated to be done in parallel for every word.

There have been few attempts at addressing this problem by changing the CAM circuit [9],[6], all of which are based on a NOR-type match-line organization. The problem with NOR organization is that it has a high activity factor: the CAM words, which do not match, consume significant energy per search. Typically most words of a CAM do not match, so this leads to poor energy-efficiency.

This paper presents a novel, error-tolerant CAM with a NAND-type match-line organization, capable of ignoring single bit mismatches in a word. Therefore if the data stored in the CAM is encoded by an ECC with a minimum Hamming distance of 3, searches can still be successful even at the presence of single bit errors. To the best of the author's knowledge, this is the first NAND-type error-tolerant CAM ever proposed. Detailed simulations confirm that it compares favorably to a NOR-type error-tolerant CAM with regard to both speed and energy efficiency.

Section 2 provides background on CAM design, describes the two CAM match-line organizations and discusses the effect of soft-errors in CAMs. Related work is discussed in Section 3. Section 4 introduces the proposed error-tolerant, NAND-type CAM, analyzes a noise issue due to charge-sharing and describes the critical paths of the circuit. Section 5 compares the proposed design to a well-known NOR-type error-tolerant CAM and evaluates the cost of adding error tolerance to a state-of-the-art NAND-type CAM. Finally, Section 6 concludes the paper.

2. BACKGROUND

CAMs [10] are memory arrays similar to SRAMs, which, in addition to reads and writes, can perform word searches. For this purpose a CAM cell contains an SRAM cell and

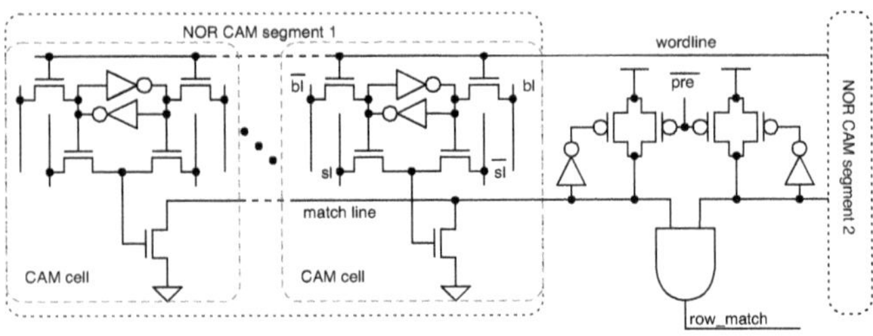

Figure 1: NOR-type CAM organization.

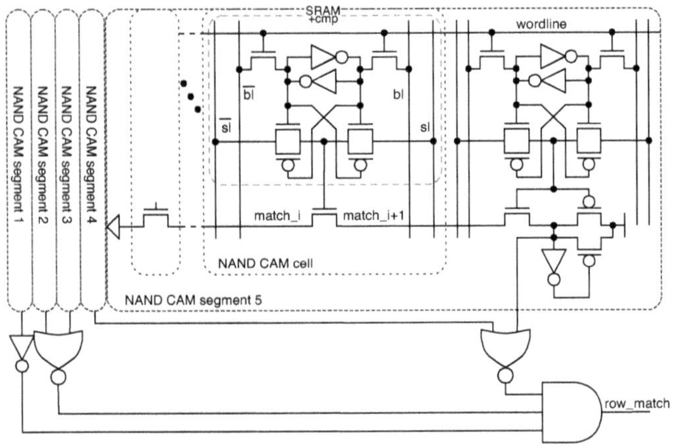

Figure 2: NAND-type CAM organization [4].

a comparator. The input search word is provided on complementary search-lines, which span the height of the array, in a manner similar to the SRAM bitlines. In some CAM organizations a single pair of lines per word bit is used both for bit-lines and search-lines, however, low-power CAMs use separate search-lines to avoid the load of the SRAM cells' access transistors for the most frequent search operation. The comparator of each cell of a row (word) drives the row's match-line, which indicates whether the whole row matches the input word or not.

There are two main organizations of the match-line. In the NOR organization, the match-line is first precharged and then any cell comparator which does not match, pulls down the match-line indicating a whole row mismatch. If the match-line stays charged, all bits in the row can be presumed to match. The operation is thus identical to that of a dynamic NOR gate. Figure 1 shows the NOR organization with the common optimization where the match-line is split into two segments which are evaluated in parallel and ANDed together to produce the final match output.

In the NAND organization the match line is split at every CAM cell by a switch controlled by the cell's comparator. Usually the switch is implemented as a NMOS pass transistor and a logic 0 is used to indicate a match. If any cell does not match, the logic 0 does not reach the output of the match-line, which is usually precharged to logic 1, indicating a miss. A true NAND organization would be too slow. Therefore the NAND match line is split into segments which are evaluated in parallel. The final row match signal is produced by combining the segment outputs using a high fan-in NOR gate, since match in a NAND segment is indicated by 0. Figure 2 shows a NAND-type CAM [4] with its match-line split into five segments. A cell comparator built with transmission gates is not a requirement of the NAND organization and it could also be used for the NOR organization. A good discussion of different NAND and NOR CAM cell implementations can be found in [4].

In both organizations the cell comparisons start after one of the complementary search-lines (sl, $\overline{sl}$) is driven high. During the match-line precharge phase, all search-lines are low, so that the cell comparators are forced to indicate a mismatch so that they do not discharge the match lines.

The advantage of the NOR organization is high-speed. Its major disadvantage is the very high activity factor: all match-lines are precharged in each cycle and most of them will be discharged after a search, as in most applications at most one entry can match. Moreover, the large number of parallel transistors on the NOR match-line cause increased leakage power consumption. In order to counter the leakage, the staticiser/keeper must be relatively large. However, if only 1 bit does not match, a single NMOS must be able to flip the NOR output. These conflicting design constraints limit the maximum fan-in of the dynamic NOR, so the match line would need to be split into a number of segments and they will be combined to produce the final match signal.

The NAND organization has a much lower activity factor and leakage power consumption. Furthermore, because a match is indicated by a change in signal value, the circuits following the search, e.g. driving the word-line of an adjacent SRAM, are simpler and more energy efficient [4].

Hybrid CAMs [5][3] combine a small NAND match-line segment with a large NOR segment and a mechanism which does not evaluate the NOR part unless the NAND part indicates a match. This significantly reduces the activity factor of the CAM but they still suffer to some extend from the leakage and transistor sizing problems of the NOR organization.

2.1 Soft-error effects on CAMs

A soft-error, which flips a bit in a CAM entry, can cause one of the following [6] [11]:

False hit – A search matches the wrong CAM entry because the soft-error changed a bit in the CAM entry. If, as is very common, the search result drives an SRAM, parity bits of the CAM entry can be stored in the SRAM, in addition to the data stored there [12]. A false hit can then be detected by checking these bits and, assuming there is enough redundancy, the CAM entry can be corrected by overwriting it at a later cycle.

False miss – A search operation misses in the CAM because the soft-error changed the value of the CAM entry that would normally match. This case is impossible to distinguish from a true miss; one would have to read and check all entries for errors. If, for example, the CAM is used in a write-back cache, a false miss could cause serious problems: the access will be directed down the memory hierarchy which could contain stale data.

Multiple hits – the bit-flip changes a CAM entry to be identical to another entry. This can be detected with extra circuits in the CAM. For example, CAMs in OpenSparc caches detect multiple hits and can raise exceptions so that supervisor software can address the problem. Correction is possible as in the false hit case, although it would be more complicated since all matching entries would have to be checked to determine which ones to correct.

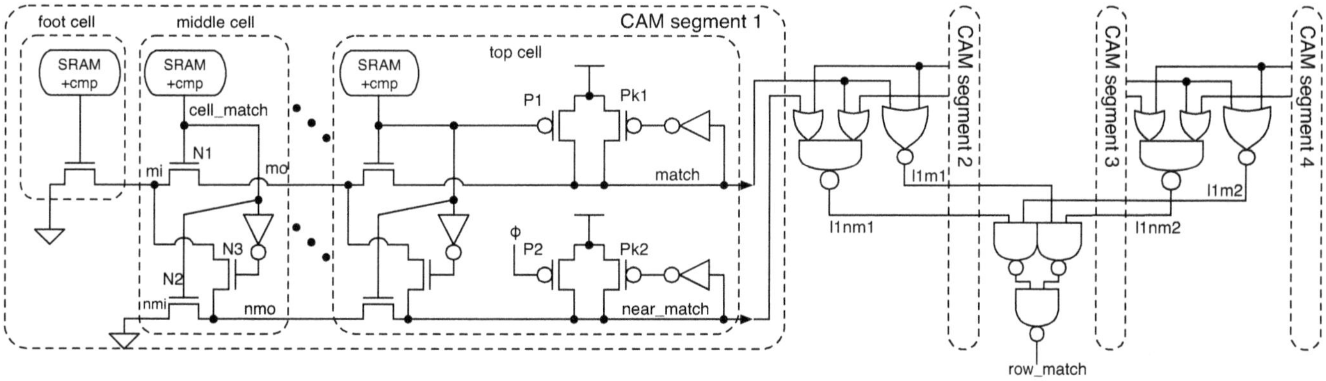

Figure 3: A 4-segment row of the proposed CAM.

If the CAM allows multiple hits and there is a priority mechanism to choose amongst them, a wrong hit [11] may happen: for example if the soft-error changes the highest priority word so that it does not match, the CAM will incorrectly output the next entry in the priority list.

3. RELATED WORK

The hardest problem for building a reliable CAM is false misses. Much work has focused on avoiding it. One such solution is ensuring that a false miss may cause extra delay, but it does not affect correctness; for example, write-though caches can speed up memory access on a hit, but are not essential for correctness. Another solution is to periodically scrub the CAM [8] to reduce the probability of errors occurring. Although such methods are useful, they are not capable of eliminating the problem. They have significant cost in energy consumption and may delay access to the cache when a scrubbing operation is underway. Moreover write-through caches can be less efficient than write-back in a number of situations.

Another work [11] leaves the CAM unchanged and checks that it is error free using redundant structures such as Bloom filters [1]. The advantage of this method is that low-level redesign of CAM blocks is not required. However, there is significant area overhead due to the Bloom filter and the glue logic and there is also a time penalty.

Another line of research is to modify the CAM itself so that it can tolerate soft errors [9] [6]. Pagiamtzis *et al.* [9] presented a soft-error tolerant NOR-type CAM. In addition to the original key, the CAM also stores enough parity bits so that stored codewords have a Hamming distance of 4, i.e. keys which differ by 1 bit will be stored as codewords which differ by 4 bits. When a search for key K is performed, if there has been a single bit-flip to a matching entry, the codewords of the search key and the stored entry will only differ by 1 bit. Since all other entries will have at least 4 different bits compared to the CAM entry holding K, any single bit flip at another, non-matching, entry would cause a minimum of 3 bit difference. In [9] a current-race, match-line sense amplifier is presented which examines the time it takes to charge the match-line capacitance in order to determine the difference between a 1 bit (at most) mismatch and a 3 (or more) bit mismatch. The former is reported as a hit, therefore solving the false miss problem and the latter is reported as a miss, thus solving the false hit problem.

The method proposed in this paper uses redundant encoding (with distance 3) but the match-line is NAND-type, so as to be more energy efficient. Moreover, the current-race match-line sense amplifier can be sensitive to noise and device variation, while the proposed method is based on full 2-level logic. A detailed comparison is given in Section 5.

Finally, there has been work on fault tolerant ternary CAMs (TCAM) [2], [7], [13]: these are similar to (binary) CAMs but can also store (or search for) a "don't care" value which matches any binary value. They use 2 SRAM cells per bit and a special encoding for the 3 possible values. Because there is some redundancy already present in each bit, these papers propose a number of encodings used to make TCAMs tolerant to soft-errors. Although one could use a reliable TCAM in place of a CAM and ignore the "don't care" value, a TCAM is approximately twice as large as a CAM, even without the addition of any parity bits. Thus, a solution suited to a (binary) CAM is advantageous.

4. PROPOSED ERROR-TOLERANT CAM

The proposed error-tolerant CAM requires that stored data and search inputs are encoded with extra parity bits so that valid codewords have a Hamming distance of 3. In addition, the common NAND match-line is augmented with a *near-match* line which indicates that, at most, 1 bit in the chain does not match. As explained earlier, the redundant encoding and the ability to differentiate between *up to* 1 bit mismatch and *more than* 1 bit mismatch can solve both false hit and false miss problems, under the condition that only a single bit of the CAM row can be affected by errors.

The proposed match-line is shown in Figure 3. The SRAM cell and comparator are identical to the standard NAND CAM cell (Fig. 2). The extra circuits for the near-match line are: a second pass transistor (N2) for the near-match line, a pass transistor (N3) connecting the incoming (full) match line, *mi*, to the outgoing near-match line, *nmo* and an inverter to drive N3 when the cell does not match. The comparator is implemented with complementary pass gates to provide full gate overdrive for the match-line pass transistors and avoid leakage in the added inverter.

When a cell matches, it simply connects the incoming match-line *mi* segment to the outgoing *mo* and similarly for the near-match segments *nmi, nmo.* Therefore it just propagates the results of the cells to its left, without any changes.

When a cell does not match, it breaks the connection of both match lines and connects the incoming match-line to the outgoing near-match line. Therefore, if all the cells to the left match, only the near-match output line will become asserted. If the cells to the left indicate a near-match, transistor N2 will break the connection to the near-match output *nmo*. There will be a connection between *mi* and *nmo* at this cell, but there is no path from *mi* to the ground because the previous cells do not match. Therefore the near-match output will not be asserted if there are more than one mismatches in the segment.

The near-match chain is shorter by one cell, at the foot of the NMOS chain, on the left. The foot cell is treated as don't care since two mismatches are required to block the near-match line.

The output node of a segment is indirectly precharged by a PMOS driven by the comparator output of the top cell. As all search-lines are driven low in the first-phase of the CAM search cycle, the comparator will output a logic 0 regardless of the value stored in the cell, thus cutting off the rest of the chain and charging the top node. Charge-sharing is the major problem here: the charge of the top node may spread to the previous node when the top cell matches and potentially further down the NMOS stack as more cells match, even if the segment, as a whole, does not match. This potentially produces a false result, as the voltage of the top node, which is the output of the segment, drops proportionally to the loss of charge. As in [4] a weak keeper PMOS, driven by an inverter, is added to each node (Pk1, Pk2 in Fig. 3). Although, this makes it slightly harder for the NMOS stack to pull down the output, it substantially improves the charge-sharing noise problem. Despite the charge-sharing problem, this can be both the most energy efficient and fastest precharge option for NAND CAMs [4]. A detailed analysis of charge sharing in the match-lines of the proposed CAM is given in Section 4.1.

The (full) match-line of the proposed circuit uses the above precharge approach. The near-match line requires a modification: the comparator of the top cell cannot drive the precharge PMOS because if the top cell is the only one that does not match, it would prevent the near-match output from being asserted. Therefore, the precharge PMOS for the near-match output (P2 in Fig. 3) is driven by a clock signal. The clock signal need only connect to the top-of-the-chain cells, so the energy expended by clock power is not as significant as in other precharge options.

The last CAM component is the circuit which combines the segment match signals into the final match signal. A full match is easy to determine: all segments must match, so a large NOR, decomposed into simpler gates, is used. Combining the individual near-match signals into a full-row near-match output is more complicated; the only case for a row near-match to occur is when only one segment indicates near-match and all other segments indicate a full-match. For 2 segments the combined near-match is $\overline{m_1} \cdot \overline{nm_2} + \overline{nm_1} \cdot \overline{m_2}$, where m_i, nm_i are the match and near-match signals of segment i respectively. (Note that all match signals are active low.) This can be implemented efficiently with an or-and-invert (OAI) complex gate for up to 3 segments.

For wider CAMs, more levels of logic are needed. Near-match (and match) signals of every group of 2 or 3 segments are generated using OAI gates (NOR for match). The near-match signals of the groups are active high, so they can be combined by AND-OR logic (using 2 levels of NANDs) to produce the row near-match signals. Figure 3 shows these circuits for a CAM row composed of 4 segments. Only the near-match output is generated as this would be the corrected signal used by subsequent circuits as the row's match indicator. The figure implies a possible layout arrangement with logic placed between segments of the CAM.

4.1 Charge-sharing noise analysis

There are two charge-sharing paths, one for each of the match lines. For the full match-line the worst case is when only the foot cell of the segment does not match; the charge stored at the output node during the precharge phase will be shared by all intermediate nodes in the NMOS chain, dropping the voltage of the output node. If the drop is substantial, the circuit will not function correctly, as the subsequent stages will interpret the output as a logic 0. The final output voltage depends on the capacitance of each node and their initial voltages/charges. If the output node has a much higher capacitance than the other nodes, the drop is relatively small. To produce the worst possible case, the charge sharing cycle is preceded by a full-match cycle, so that all intermediate nodes are fully discharged. Note that because of the transistor (N3) connecting the near-match line to the match-line, the second node of the match-line is precharged (to $V_{dd} - V_{tn}$) by the near-match top-cell precharger. This makes the charge-sharing problem of this path somewhat less important.

The second charge-sharing path is for the near-match output. The situation is similar to the match path, but when there is an off transistor (N2) on the near-match stack there is an on transistor (N3) leading to the full-match line. So the worst case charge-sharing happens when there is a path from the near-match output to the last node of the full match-line through a single intermediate node which is not matching. The worst case path corresponds to the case where two cells do not match: the one at the foot of the stack and any middle cell. If the top cell mismatches, the 2nd node of the match line is always precharged to $V_{dd} - V_t$ so the charge sharing problem is slightly better.

Charge-sharing noise forces an upper limit on the length of the NMOS stack in a segment and, thus, on the number of cells in a segment. Spice level simulations indicate that the near-match output of a 6-deep segment falls below $V_{dd}/2$ in the worst case process corner of a 90nm technology, while the match-output drops below $V_{dd}/2$ at a stack length of 10. The worst case process corner is fast NMOS, slow PMOS, because this allows the charge to move quickly away from the top node while the PMOS keeper is weak. In the typical corner, the maximum stack length is 8.

4.2 NAND match-line speed

The worst evaluation path for a NAND match-line occurs when all the cells match and the output is pulled down through the relatively long chain of NMOS transistors. If the intermediate nodes of the match-line are at a higher voltage due to prior charge-sharing, the NMOS chain will take even longer to discharge. Therefore the worst case delay happens when a match cycle immediately follows a mismatch cycle which has left the intermediate nodes to as high voltage as possible due to charge-sharing. This happens when the 2 leftmost cells do not match, ensuring the near-match line does not get discharged at all. Otherwise, during the

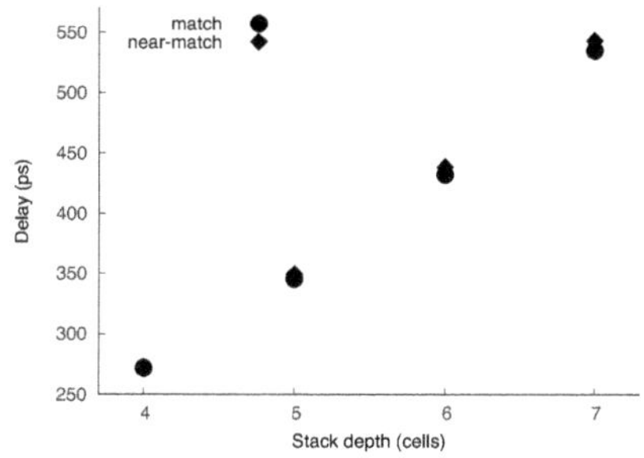

Figure 4: Delays of match lines by stack size.

precharge phase, when all search-lines are at logic 0 and the cells are forced to indicate a mismatch, the charge of each intermediate node of the match-line drops because it is shared with the corresponding near-match intermediate node, through transistor N3. Figure 4 shows the worst-case delay for match and near-match lines by number of cells per segment.

5. EVALUATION

In order to evaluate the proposed error-tolerant CAM, a 32x31b CAM was designed and simulated. As required by a code of Hamming distance 3, out of the 31 bits of a CAM row, 26 are the actual data bits and the remaining 5 are parity bits. Each CAM row is organized as 8 segments: 7 of length 4 and one of length 3. The local outputs of the segments are combined to produce the final (near) match output, by using the logic described earlier. Gates of fan-in of 3 were used to reduce the combining logic to 2 levels.

A full row of the CAM was designed at transistor level and parasitic capacitances were estimated for the match-line nodes. Search-line drivers were loaded with the full 32 cells of each column and estimated parasitic capacitance. Spice-level simulation reports a delay of 487ps from the clock signal controlling the search-line drivers to the final near-match output of the row.

A detailed energy evaluation requires full parasitic extraction from layout and real search data. The latter is much more important for NAND-type CAMs as the segments rarely become fully discharged, so much of their power consumption comes from partial discharges, which are very data dependent. Previous work has showed large differences in energy consumption between random and real inputs as well as strong energy dependence on the distribution of address bits to segment cells [4].

Table 1 summarizes the match-line energy expended for a number of different scenarios. 'Hit' and 'NearHit' are self-explanatory. 'Miss: all-cells' corresponds to the case where all cells of the row mismatch; this is the lowest energy consuming case. 'Miss: charge-sharing' means that the row does not match and worst case charge-sharing is occurring on all segments. 'Miss: worst-case' is when 2 segments 'near match' while all other segments match.

In the absence of high-level simulation data, the following scenario will be used for energy estimations: For each search, only a single CAM entry matches, while all the other entries miss; 1/5 of these misses are 'worst-case' and 4/5 are 'charge-share' misses.

In order to evalute the overhead of the error tolerance feature of the proposed CAM, an equivalent NAND-type CAM, based on [4], was also designed and simulated. This baseline CAM only needs the actual 26 data bits per row, which are organized in 5 segments of 4 cells each and 2 segments of 3 cells. Removing the parity bits makes the baseline at least 20% smaller. It also requires simpler CAM cells and simpler logic to combine partial outputs, so the area overhead of the proposed CAM would be significantly higher.

The worst case delay (from clock signal enabling the search-line to final row match signal output) of the baseline CAM is 374ps, so error tolerance adds a delay penalty of 30%.

Table 1 shows that the match-line energy of the proposed CAM is about 2.7 times higher for a matching row and between 1.9 and 2.6 for a non-matching row. As most rows do not match and the worst-case mismatch is infrequent, the average energy overhead following the scenario described earlier is 2.15. Given that match-line energy is approximately 50% of the total CAM energy, for the whole CAM the energy overhead for adding error tolerance drops to approximately 1.6. Overall, the proposed CAM is quite expensive compared to an energy efficient unprotected CAM. However, if error protection is required, the cost is clearly lower than using multiple unprotected units.

The best known error-tolerant CAM is [9], shown in Fig. 5. It uses a current-race match-line sensing scheme, which discharges the match line at the first phase of the clock and, in the evaluation phase, a current source tries to charge the line, while mismatching transistors try to keep it discharged. A carefully designed dynamic latch senses when less than 2 cells discharge the match-line and reports a match.

As this method requires a Hamming distance of 4, it requires at least one more cell per row than the proposed method. However, the CAM cells themselves are simpler and there is no need for extra logic to combine segment outputs. So overall, the silicon area required for the NOR-type error tolerant CAM will be lower than the proposed CAM.

Nevertheless, this method has a number of disadvantages. The delay for a near-match, i.e. single cell mismatch in a row, is significantly worse than that of a full-match so that near-matches get enough time to charge the match-line at an appropriate voltage level for the dynamic latch to sense as a match. Therefore a relatively infrequent case will dominate the cycle time of the circuit. Given that only a single bit mismatch in a row is allowed, the match-line cannot be split into sections in order to speed up the operation, as in standard NOR-type CAMs.

Current-race match-lines consume significant power when there are mismatches: the most common case in a CAM. Furthermore, the current-race lasts for the whole evaluation phase, which has to be quite long to accommodate the slow

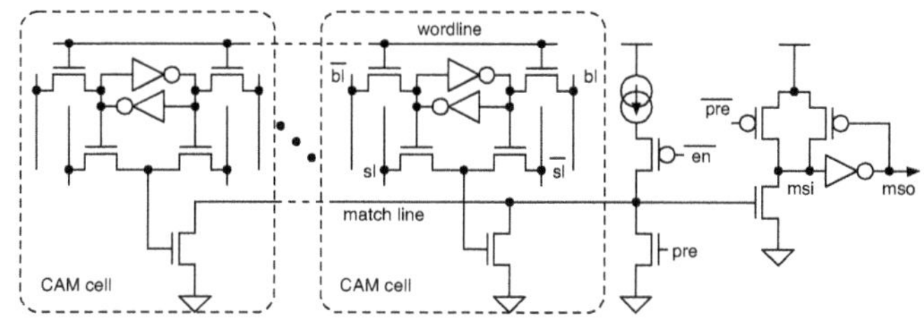

Figure 5: NOR-type match-line with current-race sensing (based on [9]).

Table 1: Match-line energy and delay

	Proposed (cells: 31)	baseline (cells: 26)	NOR (cells: 32)
Energy (fJ) / row			
Miss: all cells	2.5	1.1	113
Miss: charge-share	69.7	36.6	
Miss: worst-case	183	69.6	
Hit:	224	83	76
Near-Hit	202	NA	108
Total (32 rows)	3087	1422	3579
Match delay (ps)	487	374	339
Near-match delay		NA	692

near-match case. Finally, current race makes the circuit sensitive to device variation and noise.

A 32x32b CAM based on [9] was designed and simulated using a spice-level simulator at the same technology as the proposed CAM. Instead of limiting the voltage swing of the search-lines, a 9-transistor CAM cell was used with a single match-line pull down transistor, which has the same effect: increases the on-resistance of the match-line pull-down transistor. The delay from the enable signal ($\overline{en}$, Fig. 5) to the time a full-match is reported at the output (mlso) is 339ps, while for a near-match is 692ps. Thus, while in the usual, error-free case, the NOR-type CAM is 43% faster than the proposed, its cycle time has to follow the worst case, which is 1.42 times slower than the proposed CAM.

The match-line energy (third column of Table 1) varies between 76fJ and 113fJ, depending on how many cells match. Although matching rows consume less energy compared to the proposed CAM, the vast majority will not match and such rows expend up to 45 times more energy ('Miss all-cells'). Using the values in Table 1 and the scenario described earlier, one can calculate that when the number of rows exceeds 8, the proposed CAM is more energy efficient. For example, comparing the energy for 64 rows of the two CAM implementations, the proposed is 20% more efficient.

6. CONCLUSIONS

Soft-errors are a significant threat to the integrity of a system and memory structures in particular. Common memories can be protected by using ECCs to encode the data and checking the integrity of a memory word whenever it is read. However, Content Addressable Memories (CAMs) are harder to protect especially for their most common operation, associative search.

This paper proposes an error-tolerant CAM which can ignore single bit errors in its words during search operations. This is achieved by using a dual match-line organization which keeps track of up to 1 bit mismatch per word. The match lines follow the NAND-type organization which is more energy efficient compared to the better known NOR match-line organization.

Although the cost of adding error-tolerance is high compared to an unprotected NAND-type CAM, the proposed CAM is 40% faster and more energy efficient compared to the state of the art error-tolerant NOR-type CAM [9].

7. ACKNOWLEDGEMENTS

This research has been co-funded by the European Union (European Social Fund) and Greek national resources under the framework of the "Thales" project of the "Education & Lifelong Learning" Operational Program.

8. REFERENCES

[1] B. H. Bloom. Space/time trade-offs in hash coding with allowable errors. *Commun. ACM*, 13(7):422–426, July 1970.

[2] A. Bremler-Barr, D. Hay, D. Hendler, and R. M. Roth. PEDS: a parallel error detection scheme for TCAM devices. *IEEE/ACM Transactions on Networking*, 18(5):1665–1675, Oct. 2010.

[3] Y.-J. Chang and Y.-H. Liao. Hybrid-type CAM design for both power and performance efficiency. *Very Large Scale Integration (VLSI) Systems, IEEE Transactions on*, 16(8):965 –974, Aug. 2008.

[4] V. Chaudhary and L. T. Clark. Low-power high-performance NAND match line content addressable memories. *IEEE Transactions on Very Large Scale Integration (VLSI) Systems*, 14(8):895–905, Aug. 2006.

[5] A. Efthymiou and J. D. Garside. An adaptive serial-parallel CAM architecture for low-power cache blocks. In *Proceedings of the International Symposium on Low Power Electronics and Design*, pages 136–141, Aug. 2002.

[6] L. Hung, M. Goshima, and S. Sakai. Mitigating soft errors in highly associative cache with CAM-based tag. In *IEEE International Conference on Computer Design*, pages 342 – 347, Oct. 2005.

[7] S. Krishnan, R. Panigrahy, and S. Parthasarathy. Error-correcting codes for ternary content addressable memories. *IEEE Transactions on Computers*, 58(2):275 –279, Feb. 2009.

[8] S. S. Mukherjee, J. Emer, T. Fossum, and S. K. Reinhardt. Cache scrubbing in microprocessors: Myth or necessity? In *10th IEEE Pacific Rim International Symposium on Dependable Computing*, pages 37–42, 2004.

[9] K. Pagiamtzis, N. Azizi, and F. Najm. A soft-error tolerant content-addressable memory (CAM) using an error-correcting-match scheme. In *Custom Integrated Circuits Conference*, pages 301 –304, Sept. 2006.

[10] K. Pagiamtzis and A. Sheikholeslami. Content-addressable memory (CAM) circuits and architectures: A tutorial and survey. *IEEE Journal of Solid-State Circuits*, 41(3):712–727, Mar. 2006.

[11] S. Pontarelli and M. Ottavi. Error detection and correction in content addressable memories by using bloom filters. *IEEE Transactions on Computers*, PP(99):1. DOI:10.1109/TC.2012.56, 2012.

[12] N. Saxena, C.-W. Chang, K. Dawallu, J. Kohli, and P. Helland. Fault-tolerant features in the HaL memory management unit. *IEEE Transactions on Computers*, 44(2):170 –180, Feb. 1995.

[13] I. Sideris and K. Pekmestzi. Cost effective protection techniques for TCAM memory arrays. *IEEE Transactions on Computers*, 61(12):1778–1788, 2012.

Efficient Modulo 2^n+1 Multiplication for the IDEA Block Cipher

Kiamal Pekmestzi
National Technical University of Athens
15780 Athens, Greece
pekmes@microlab.ntua.gr

Constantinos Efstathiou
Technological Institute of Athens
12210 Athens, Greece
cefsta@teiath.gr

Nikos Moschopoulos
Kostas Tsoumanis
National Technical University of Athens
15780 Athens, Greece
Nikos@microlab.ntua.gr

ABSTRACT

International Data Encryption Algorithm (IDEA) is a popular and secure cryptography algorithm, suitable for hardware implementation. IDEA comprises of modulo 2^{16} additions, bitwise exclusive-OR operations and modulo 2^{16}+1 multiplications of 16-bit words. Among them, modulo 2^{16}+1 multiplication is the most time, space and power consuming operation. In this work, we propose an efficient modulo 2^n+1 modified Booth multiplication algorithm which is adapted to operands used in the IDEA. The IDEA multiplier based on the proposed modulo 2^n+1 multiplication algorithm yields area and power advantages of up to 12% and 14% respectively, compared to the already proposed modulo 2^n+1 multiplier designs. The implementation of a single round of the IDEA block cipher based on the proposed multiplier verifies the area and power advantages over the implementations based on existing modulo 2^n+1 multipliers.

Categories and Subject Descriptors

B.6.1 [Logic Design]: Design Styles—*Combinational logic;*
B.7.1 [Integrated Circuits]: Types and Design Styles—*Algorithms implemented in hardware, VLSI (very large scale integration)*

Keywords

Parallel multiplier, IDEA, encryption, modulo arithmetic

1. INTRODUCTION

The IDEA block cipher [1] is a symmetric-key algorithm which encrypts 64-bit plain to 64-bit cipher blocks using a 128-bit key. The security of IDEA relies on combining operations from three different groups of 2^{16} elements, namely integer addition modulo 2^{16}, bitwise exclusive-OR of 16-bit operands and modulo 2^{16}+1 multiplication. The encryption and decryption process of IDEA comprises eight rounds with the same structure and a final output transformation. Both encryption and decryption incorporate four modulo 2^{16}+1 multipliers, three of which are on the critical path, as shown in Fig. 1. Therefore, improving the modulo 2^{16}+1 multiplier significantly improves the entire IDEA implementation in hardware. The number representation adopted in the IDEA block cipher instructs that zero values are not defined while vector $\underbrace{000...0}_{16}$ represents the number 2^{16}. Many studies [2]-[10] have been proposed to improve the performance of the modulo 2^n+1 multiplier used in the IDEA block cipher. The design proposed in [2] is based on a conventional $(n+1)\times(n+1)$ multiplication and has the drawback of using a wide $(2n+1)$-bit adder and two n-bit parallel adders at the final stage of the multiplier to correct the result. An improvement of [2] is presented in [3] where (4:2) compressors are used for the implementation of the carry save addition tree and only one n-bit parallel adder is needed for the correction. Since zero value operands are not used, the IDEA modulo 2^n+1 multiplication can be performed using the modulo 2^n+1 multiplication algorithms for operands in the diminished-1 representations proposed in [4]-[6].

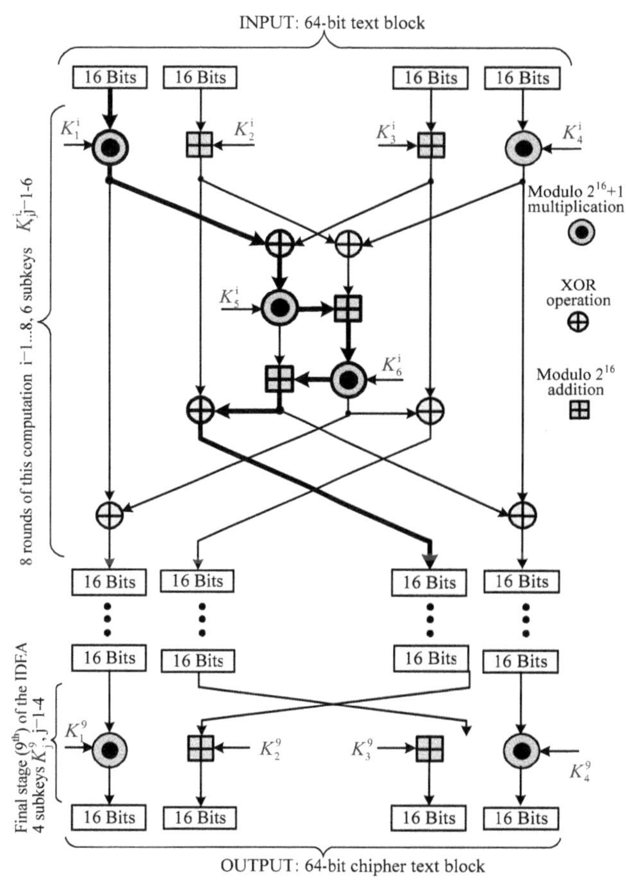

Fig. 1 The block diagram of the IDEA block cipher.

This approach has the disadvantage that it requires area, power and time consuming translators between the operand representation adopted in the IDEA block cipher and the diminished-1 representation. Dedicated modulo 2^n+1 multipliers for operands represented in the format adopted in the IDEA algorithm are given in [7], [8]. The methodology proposed in [7] has the drawback that uses two rows of n-bit 2-to-1 multiplexers which are located on the critical path increasing the area and time complexity. In [8], the partial products of the multiplier are simplified, but two extra rows of n logic gates, a "special" adder (incrementer by 2) implementing the zero handling circuit and a row of n two-input OR gates are required to produce the final result.

An efficient way to implement the IDEA multiplication algorithm is converting the operands which are in the form $\underbrace{000...0}_{16}$ into their weighted representation $\underbrace{1000...0}_{17}$ and use the modulo 2^n+1 multiplier for weighted operands as proposed in [9], [10], for n=16. The design in [9] is the most efficient architecture without encoding, while the design in [10] is the most efficient one using modified Booth encoding of the multiplier. This conversion is achieved using zero value detecting circuits (NOR gates) which compute bits with weight 2^{16}.

In this work, a new modulo 2^n+1 modified Booth multiplication algorithm is proposed. Our design compares favorably for n=16 to the IDEA multipliers based on the modulo 2^n+1 multiplication algorithms given in [9], [10].

The paper is organized as follows. In section 2 the proposed modulo 2^n+1 modified Booth multiplication algorithm is presented. In section 3, implementation details are given. In sections 4,5 the comparisons to the state-of-the-art designs are presented while section 6 concludes our work.

2. DESIGN METHODOLOGY

In the following text we denote as X_W an integer in the range $(0, 2^n]$ and as $X = x_{n-1}x_{n-2}\dots x_1x_0$ its representation adopted in the IDEA algorithm. For $0 \le X_W < 2^n$, X is the weighted binary representation of X_W, whereas for $X_W = 2^n$, $X = \underbrace{000\dots0}_{n}$. Thus, the following relations hold:

$X = |X_W|_{2^n}$ and $X_W = x_n 2^n + X$, where

$x_n = \overline{x_{n-1} \vee x_{n-2} \vee \dots \vee x_0}$ is a zero value indication of X.

Considering that $A_W, B_W \in (0, 2^n]$, $Q_W = |A_W \times B_W|_{2^n+1}$ is their product modulo 2^n+1, and A, B, Q are the representations adopted in the IDEA. Then,

$Q_W = |A_W \times B_W|_{2^n+1} = |(a_n 2^n + A) \times (b_n 2^n + B)|_{2^n+1}$ or

$$Q_W = \left|A \times B + a_n 2^n B + b_n 2^n A + a_n b_n 2^{2n}\right|_{2^n+1}. \tag{1}$$

Using $|2^n|_{2^n+1} = -1$, and $|2^{2n}|_{2^n+1} = 1$, we get that

$$Q_W = \left||A \times B|_{2^n+1} - a_n B - b_n A + a_n b_n\right|_{2^n+1}. \tag{2}$$

Since $|A \times B|_{2^n+1} = \left|(A+1) \times |B|_{2^n+1} - B\right|_{2^n+1}$ the next relation holds:

$$Q_W = \left|\left|(A+1) \times |B|_{2^n+1}\right|_{2^n+1} - B - a_n B - b_n A + a_n b_n\right|_{2^n+1}. \tag{3}$$

The partial products of $\left|(A+1) \times |B|_{2^n+1}\right|_{2^n+1}$ are reduced by using Modified Booth (MB) encoding of $|B|_{2^n+1} = \left|\sum_{i=0}^{n-1} b_i 2^i\right|_{2^n+1}$. For even values of n, we have that

$$\begin{aligned}
|B|_{2^n+1} &= \left|(b_0 - 2b_1) + \sum_{i=1}^{n/2-1} (b_{2i-1} + b_{2i} - 2b_{2i+1}) 2^{2i} + b_{n-1} 2^n\right|_{2^n+1} = \\
&= \left|(b_0 - 2b_1 - b_{n-1}) + \sum_{i=1}^{n/2-1} (b_{2i-1} + b_{2i} - 2b_{2i+1}) 2^{2i}\right|_{2^n+1} = \\
&= \left|b_0 - 2b_1 - b_{n-1} + 4b_1 + (b_2 - 2b_3)4 + \sum_{i=2}^{n/2-1} (b_{2i-1} + b_{2i} - 2b_{2i+1}) 2^{2i}\right|_{2^n+1} = \\
&= \left|b_0 + b_{n-1} - 2(b_{n-1} - b_1) + (b_2 - 2b_3)4 + \sum_{i=2}^{n/2-1} (b_{2i-1} + b_{2i} - 2b_{2i+1}) 2^{2i}\right|_{2^n+1}.
\end{aligned}$$

Using the relation $b_{n-1} - b_1 = -2\overline{b}_{n-1} \cdot b_1 + b_{n-1} \oplus b_1$ we get that

$$\begin{aligned}
|B|_{2^n+1} &= \left|[b_0 + b_{n-1} - 2 \cdot (b_{n-1} \oplus b_1)] + 4\overline{b}_{n-1} \cdot b_1 + (b_2 - 2b_3) \cdot 2^2 + \sum_{i=2}^{n/2-1} \boldsymbol{b}_i^{MB} \cdot 2^{2i}\right|_{2^n+1} = \\
&= \left|[b_0 + b_{n-1} - 2 \cdot (b_{n-1} \oplus b_1)] + (\overline{b}_{n-1} \cdot b_1 + b_2 - 2b_3) \cdot 2^2 + \sum_{i=2}^{n/2-1} \boldsymbol{b}_i^{MB} \cdot 2^{2i}\right|_{2^n+1} = \\
&= \left|\boldsymbol{b}_0^{MB} \cdot 2^0 + \boldsymbol{b}_1^{MB} \cdot 2^2 + \sum_{i=2}^{n/2-1} \boldsymbol{b}_i^{MB} \cdot 2^{2i}\right|_{2^n+1} = \left|\sum_{i=0}^{n/2-1} \boldsymbol{b}_i^{MB} \cdot 2^{2i}\right|_{2^n+1}
\end{aligned}$$

where $\boldsymbol{b}_i^{MB} \in \{-2, -1, 0, +1, +2\}$ are given by the next relations:

$$\begin{aligned}
&\boldsymbol{b}_0^{MB} = b_0 + b_{n-1} - 2 \cdot (b_{n-1} \oplus b_1), \\
&\boldsymbol{b}_1^{MB} = \overline{b}_{n-1} \cdot b_1 + b_2 - 2b_3 \text{ and} \\
&\boldsymbol{b}_i^{MB} = b_{2i-1} + b_{2i} - 2b_{2i+1} \text{ for } i \ge 2.
\end{aligned} \tag{4}$$

For odd values of n, we have that

$$|B|_{2^n+1} = \left|\sum_{i=1}^{\frac{n-1}{2}} (b_{2i-1} + b_{2i} - 2b_{2i+1}) 2^{2i} + (b_0 - 2b_1)\right|_{2^n+1} = \left|\sum_{i=0}^{\frac{n-1}{2}} \boldsymbol{b}_i^{MB} \cdot 2^{2i}\right|_{2^n+1}$$

where

$$\boldsymbol{b}_i^{MB} = b_{2i-1} + b_{2i} - 2b_{2i+1} \text{ for } 0 \le i \le \frac{n-1}{2} \text{ and } b_n = b_{-1} = 0. \tag{5}$$

Concluding, operand $|B|_{2^n+1}$ is modified Booth encoded as

$$|B|_{2^n+1} = \left|\sum_{i=0}^{\lceil n/2 \rceil - 1} \boldsymbol{b}_i^{MB} 2^{2i}\right|_{2^n+1} \tag{6}$$

where $\boldsymbol{b}_i^{MB}$ for n even and odd are given by (4) and (5) respectively.

From (3) and (6) we obtain that

$$\left|(A+1) \times |B|_{2^n+1}\right|_{2^n+1} = \left|\sum_{i=0}^{\lceil n/2 \rceil - 1} \left|(A+1)\boldsymbol{b}_i^{MB} \cdot 2^{2i}\right|_{2^n+1}\right|_{2^n+1}. \tag{7}$$

The computation of the terms $\left|(A+1)\boldsymbol{b}_i^{MB} \cdot 2^{2i}\right|_{2^n+1}$ is based on the following Lemma:

Lemma 1: Let X be an integer in the range $[0, 2^n-1]$ and $x_{n-1}x_{n-2}\ldots x_1x_0$ its binary representation. Then,

$$\left|s(X+1)2^j\right|_{2^n+1}=\begin{cases} x_{n-j-1}x_{n-j-2}\ldots x_0\overline{x}_{n-1}\ldots\overline{x}_{n-j-1}\overline{x}_{n-j}+1, \text{ if } s=+1 \\ \overline{x}_{n-j-1}\overline{x}_{n-j-2}\ldots\overline{x}_0 x_{n-1}\ldots x_{n-j-1}x_{n-j}+1, \text{ if } s=-1 \\ \underbrace{111\cdots1}_{n-j}\underbrace{00\cdots0}_{j}+2^j+1, \text{ if } s=0 \end{cases}$$

Lemma 1 can be proved by following the methodology of a similar Lemma given in [10]. According to Lemma 1 the partial products $\left|(A+1)\boldsymbol{b}_i^{MB}2^{2i}\right|_{2^n+1}$ in relation (7) can be replaced by their corresponding PP_i operands in Table 1, each one requiring a correction equal to 1, except those with a zero value requiring a correction of $2^{2i}+1$. That is,

$$\left|(A+1)\boldsymbol{b}_i^{BM}\cdot 2^{2i}\right|_{2^n+1}=\left|PP_i+1+z^{2i}\right|_{2^n+1} \tag{8}$$

where each term z_{2i} is equal to 1 when its corresponding $\boldsymbol{b}_i^{MB}$ term is equal to zero. Terms z_{2i} are derived by slightly modifying the Booth Encoding blocks. Therefore,

$$\begin{aligned} Q_W&=\left|\sum_{i=0}^{\lceil n/2\rceil-1}PP_i+\lceil n/2\rceil+Z_n+\left|a_nb_n-B-a_nB-b_nA\right|_{2^n+1}\right|_{2^n+1}= \\ &=\left|\sum_{i=0}^{\lceil n/2\rceil-1}PP_i+\lceil n/2\rceil+Z_n+R\right|_{2^n+1} \end{aligned} \tag{9}$$

where $Z_n=\sum_{i=0}^{\lceil n/2\rceil-1}z_{2i}2^{2i}$ and $R=\left|-B-a_nB-b_nA+a_nb_n\right|_{2^n+1}$.

Operand Z_n is of the form $0z_{n-2}0\ldots0z_20z_0$ for n even and of the form $z_{n-1}0\ldots0z_20z_0$ for n odd.

Table 1. Formation of the partial products

$\boldsymbol{b}_i^{MB}$	***Meaning***	$\boldsymbol{PP_i}$	$\boldsymbol{cor_i}$
0	0	$\underbrace{11\ldots1}_{n-2i}\underbrace{00\ldots0}_{2i}$	$1+2^{2i}$
+1	$\left\|(A_{n-1:0}+1)2^{2i}\right\|_{2^n+1}$	$\underbrace{a_{n-1-2i}a_{n-2-2i}..a_0}_{n-2i}\underbrace{\overline{a}_{n-1}..\overline{a}_{n+1-2i}\overline{a}_{n-2i}}_{2i}$	+1
-1	$\left\|-(A_{n-1:0}+1)2^{2i}\right\|_{2^n+1}$	$\underbrace{\overline{a}_{n-1-2i}\overline{a}_{n-2-2i}..\overline{a}_0}_{n-2i}\underbrace{a_{n-1}..a_{n+1-2i}a_{n-2i}}_{2i}$	+1
+2	$\left\|(A_{n-1:0}+1)2^{2i+1}\right\|_{2^n+1}$	$\underbrace{a_{n-2-2i}..a_1a_0}_{n-2i-1}\underbrace{\overline{a}_{n-1}..\overline{a}_{n-2i}\overline{a}_{n-1-2i}}_{2i+1}$	+1
-2	$\left\|-(A_{n-1:0}+1)2^{2i+1}\right\|_{2^n+1}$	$\underbrace{\overline{a}_{n-2-2i}..\overline{a}_1\overline{a}_0}_{n-2i-1}\underbrace{a_{n-1}..a_{n-2i}a_{n-2i-1}}_{2i+1}$	+1

For the sum $R=\left|-B-a_nB-b_nA+a_nb_n\right|_{2^n+1}$, we have:

For $(a_n, b_n)=(0, 0)$,

$$\begin{aligned} R&=\left|-B\right|_{2^n+1}=\left|\sum_{i=0}^{n-1}(-b_i)2^i\right|_{2^n+1}=\left|\sum_{i=0}^{n-1}(\overline{b}_i-1)2^i\right|_{2^n+1} \\ &=\left|\sum_{i=0}^{n-1}\overline{b}_i2^i-\sum_{i=0}^{n-1}2^i\right|_{2^n+1}=\left|\overline{B}-2^n+1\right|_{2^n+1}=\left|\overline{B}+2\right|_{2^n+1}. \end{aligned} \tag{10}$$

For $(a_n, b_n)=(0, 1)$, $R=\left|-A\right|_{2^n+1}=\left|\overline{A}+2\right|_{2^n+1}$. (11)

For $(a_n, b_n)=(1, 0)$,

$$\begin{aligned} R&=\left|-2B\right|_{2^n+1}=\left|-(b_{n-1}2^n+b_{n-2}2^{n-1}+\ldots+b_12^2+b_02^2)\right|_{2^n+1}= \\ &=\left|-(b_{n-2}2^{n-1}+\ldots+b_12^2+b_02^2-b_{n-1})\right|_{2^n+1}= \\ &=\left|-(b_{n-2}2^{n-1}+\ldots+b_12^2+b_02^2+\overline{b}_{n-1})+1\right|_{2^n+1}= \\ &=\left|-B_L+1\right|_{2^n+1}=\left|\overline{B}_L+3\right|_{2^n+1} \end{aligned} \tag{12}$$

where $B_L=b_{n-2}\,b_{n-3}\ldots b_0\overline{b}_{n-1}$.

For $(a_n, b_n)=(1, 1)$ which means that $A=B=0$, we obtain $R=1$.

According to the above, $R=R_1+2a_n\overline{b}_n+\overline{a}_n+1$, where

$$R_1=\overline{a}_n(\overline{b}_n\overline{B}\vee b_n\overline{A})\vee a_n\overline{b}_n\overline{B}_L. \tag{13}$$

Finally,

$$\begin{aligned} Q_W&=\left|\sum_{i=0}^{\lceil n/2\rceil-1}PP_i+Z_n+R_1+2a_n\overline{b}_n+\overline{a}_n+\lceil n/2\rceil+1\right|_{2^n+1}= \\ &=\left|\sum_{i=0}^{\lceil n/2\rceil-1}PP_i+Z_n^*+R_1+2a_n\overline{b}_n+\lceil n/2\rceil+1\right|_{2^n+1} \end{aligned} \tag{14}$$

where $Z_n^*=Z_n+\overline{a}_n$. If n is even, then for $a_n=0$, $Z_n^*=Z_n+1$ or since $x+1=2x+\overline{x}$

$$Z_n^*=0z_{n-2}0\ldots0z_2z_0\overline{z}_0, \tag{15}$$

while for $a_n=1$, $Z_n^*=0z_{n-2}0\ldots0z_20z_0$. (16)

Relations (15), (16) are unified as:

$$Z_n^*=0z_{n-2}0\ldots0z_2(\overline{a}_nz_0)(\overline{a_n\oplus z_0}).$$

For $a_n\overline{b}_n=1$, it holds that $\overline{a}_nz_0=0$. Then, these terms can be ORed instead of added. Concluding, for n even, $Z_n^*=0z_{n-2}0\ldots0z_2(\overline{a}_nz_0\vee a_n\overline{b}_n)(\overline{a_n\oplus z_0})$, while for n odd,

$Z_n^*=z_{n-1}0\ldots0z_2(\overline{a}_nz_0\vee a_n\overline{b}_n)(\overline{a_n\oplus z_0})$. Therefore,

$$Q_W=\left|\sum_{i=0}^{\lceil n/2\rceil-1}PP_i+Z_n^*+R_1+\lceil n/2\rceil+1\right|_{2^n+1}. \tag{17}$$

The derived $\lceil n/2\rceil$ partial products PP_i and the operands R_1, Z_n^* are added by an Inverted End Around Carry (IEAC) Carry Save Adder (CSA) tree, which produces a Carry, Sum vector pair (C, S). The IEAC CSA stages is composed of n full adder (FA), the most significant output carry of which is inverted and repositioned to the least significant bit position. According to the relation $\overline{c}_{n,i}=\left|-c_{n,i}+1\right|_{2^n+1}=\left|c_{n,i}2^n+1\right|_{2^n+1}$ each output carry inversion and reposition adds to the result a one. The reduction of the derived $\lceil n/2\rceil+2$ partial products is achieved by $\lceil n/2\rceil$ CSA stages. Therefore, $\lceil n/2\rceil$ carries are inverted and repositioned. Then, the output $\left|C+S\right|_{2^n+1}$ of the IEAC CSA tree is equal to the modulo 2^n+1 sum of these operands augmented by $\lceil n/2 \rceil$. That is,

$|C+S|_{2^n+1} = \left| \sum_{i=0}^{\lceil n/2 \rceil -1} PP_i + Z_n^* + R_1 + \lceil n/2 \rceil \right|_{2^n+1}$. Consequently,

$$Q_W = |C+S+1|_{2^n+1}. \quad (18)$$

The output vectors (C, S) of the CSA tree are added by an inverted end around carry adder [7] which computes the sum $\left||C+S+1|_{2^n+1}\right|_{2^n} = |Q_W|_{2^n}$. Since $Q = |Q_W|_{2^n}$, this adder computes the representation of Q_W adopted in the IDEA.

3. IMPLEMENTATION DETAILS

The block diagram of the derived modulo 2^n+1 modified Booth multiplier for operands in the representation adopted in the IDEA block cipher is shown in Fig. 2.

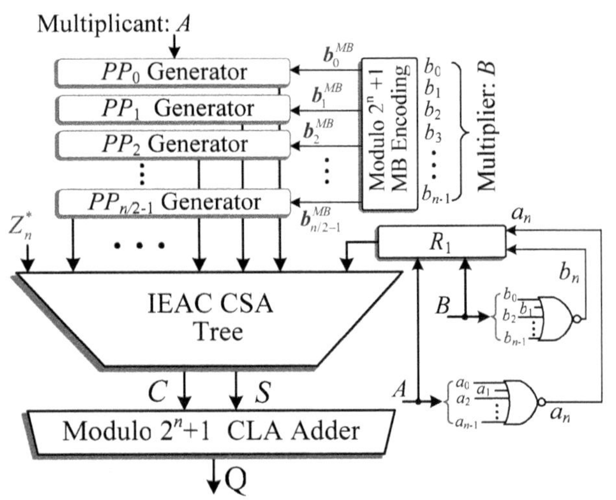

Fig. 2. Block diagram of the proposed modified Booth modulo 2^n+1 IDEA multipliers.

The MB digits b_i^{MB}=-2w+x+y are encoded according to the truth table given in Table 2. In the same table, the values of bits z_{2i} are also given. A possible gate level implementation of the modified Booth Encoding (BE) block is drawn in Fig. 3. The z_{2i} signals required by our algorithm are generated by simply NORing the *one* and *two* signals of the BE block. The BE blocks used in our designs are simpler and operate faster than the ones used in [5], [10].

Table 2. Modified Booth Encoding

$b_i^{MB}=-2w+y+x$			Encoding				Digit
w	x	y	s	two	one	z_{2i}	
0	0	0	0	0	0	1	+0
0	0	1	0	0	1	0	+1
0	1	0	0	0	1	0	+1
0	1	1	0	1	0	0	+2
1	0	0	1	1	0	0	-2
1	0	1	1	0	1	0	-1
1	1	0	1	0	1	0	-1
1	1	1	1	0	0	1	-0

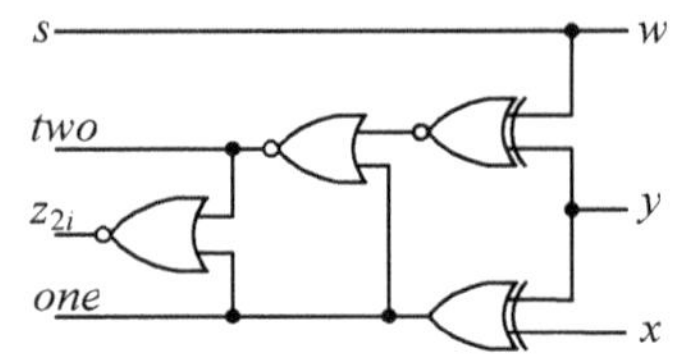

Fig. 3. An implementation of the BE block.

According to (8), for n even, the bits w, x and y of the digit b_i^{MB} for i=0, i=1 and i≥2 are given in Table 3.

Table 3. MB digit terms in weighted modulo 2^n+1 numbers

b_i^{MB} =-2w+y+x	w	x	y
b_0^{MB} (i=0)	$b_{n-1} \oplus b_1$	b_0	b_{n-1}
b_1^{MB} (i=1)	b_3	b_2	$\overline{b}_{n-1} \cdot b_1$
b_i^{MB} for i≥2	b_{2i+1}	b_{2i}	b_{2i-1}

Let $pp_{i,n-1}pp_{i,n-2}\cdots pp_{i,1}pp_{i,0}$ be the binary representation of the partial products PP_i. A possible implementation of the MB Selector (BS) block which computes bit $pp_{i,j}$ is shown in Fig. 4a and the corresponding truth table is given in Table 4. Bit $a_{i,j}$ is obtained by rotating a_j 2i places to the left, that is $a_{i,j} = a_{|j-2i|_{\bmod n}}$. According to Table 1, the output bits $pp_{i,0}$ through $pp_{i,2i-1}$ of the BS blocks must be inverted. Also the output $\overline{s_i \oplus a_{i,2i-1}} = \overline{s_i \oplus a_{n-1}}$ of the BS block that generates the term $pp_{i,2i-1}$, required in the case that $b_i^{MB}=+2$ or -2, is driven to the next BS block, inverted ($s_i \oplus a_{n-1}$).

Table 4. Truth table of the BS block

s_i	two_i	one_i	$pp_{i,j}$
0	0	0	1
0	0	1	$a_{i,j}$
0	1	0	$a_{i,j-1}$
1	1	0	$\overline{a}_{i,j-1}$
1	0	1	$\overline{a}_{i,j}$
1	0	0	1

The implementation of the $r_{1,j}$ bit of operand R_1 is shown in Fig. 4b according to (13).

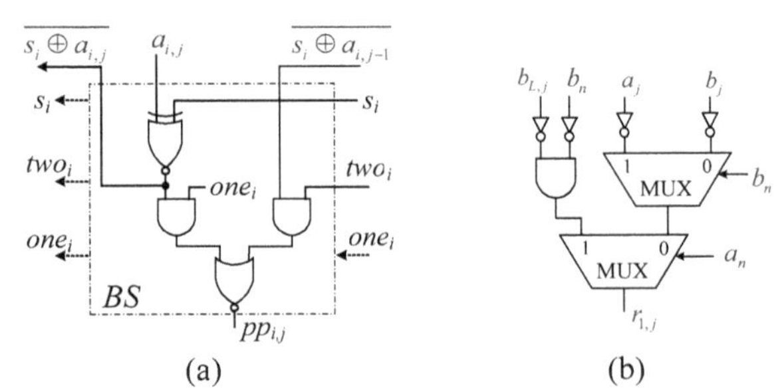

Fig. 4. Implementation of (a) the block BS and (b) the $r_{1,j}$ bit of operand R$_1$.

Example. As a schematic of the modulo $2^{16}+1$ multiplier is difficult to be drawn, our methodology is exemplified by the design of a modulo 2^6+1 (=65) modified Booth multiplier for operands in the IDEA representation, shown in Fig. 5. The boxes labeled as $r1$ in Fig. 5 compute the $r_{1,j}$ bits of operand R_1. The small circles at the outputs of the *BS* blocks and the full adder modules mean that they are used inverted in the next stage.

4. IDEA MULTIPLIER COMPARISONS

In this section, we compute the area, time and power complexity of the proposed 2^n+1 multipliers for n=16. We also compare them against the modulo $2^{16}+1$ multipliers designed according to the methodologies in [9], [10] properly modified to handle IDEA operands.

The comparison data are generated as follows. The proposed multipliers and the designs based on the multiplies in [9], [10] have been described in structural Verilog HDL with registered inputs and outputs, verified for their correctness and synthesized with the Synopsys Design Compiler tool using a TSMC 90nm standard cell library in typical conditions. The tool is instructed to synthesize the considered designs for the best possible frequency. Power measurements have been extracted by driving the synthesized netlists to Synopsys PrimeTime-PX and applying a large sample of random input operands in the range of $[0, 2^{16}-1]$.

Table 5. IDEA multiplier Area (um²) synthesis results

Period (ns)	Proposed	[9]	[10]	Gains compared to [9]	Gains compared to [10]
1.78	8253				
1.83	7358	8202		10%	
1.87	7305	7966	8307	8%	12%
1.9	7075	7923	7653	11%	8%
2.0	6618	7498	7106	12%	7%
2.1	6314	7246	6757	13%	7%
2.2	6110	6466	6378	6%	4%

The area and power dissipation of registers and clock network are not included in the measurements. The derived results are summarized in Tables 5 and 6 for area and power respectively. Our experiments indicate that the proposed multipliers yield at high performance an area advantage of up to 10% and 12% compared to previous designs in [9] and [10] respectively for the same clock period. The corresponding average power gains are 3% and 14% respectively. Furthermore, the proposed design

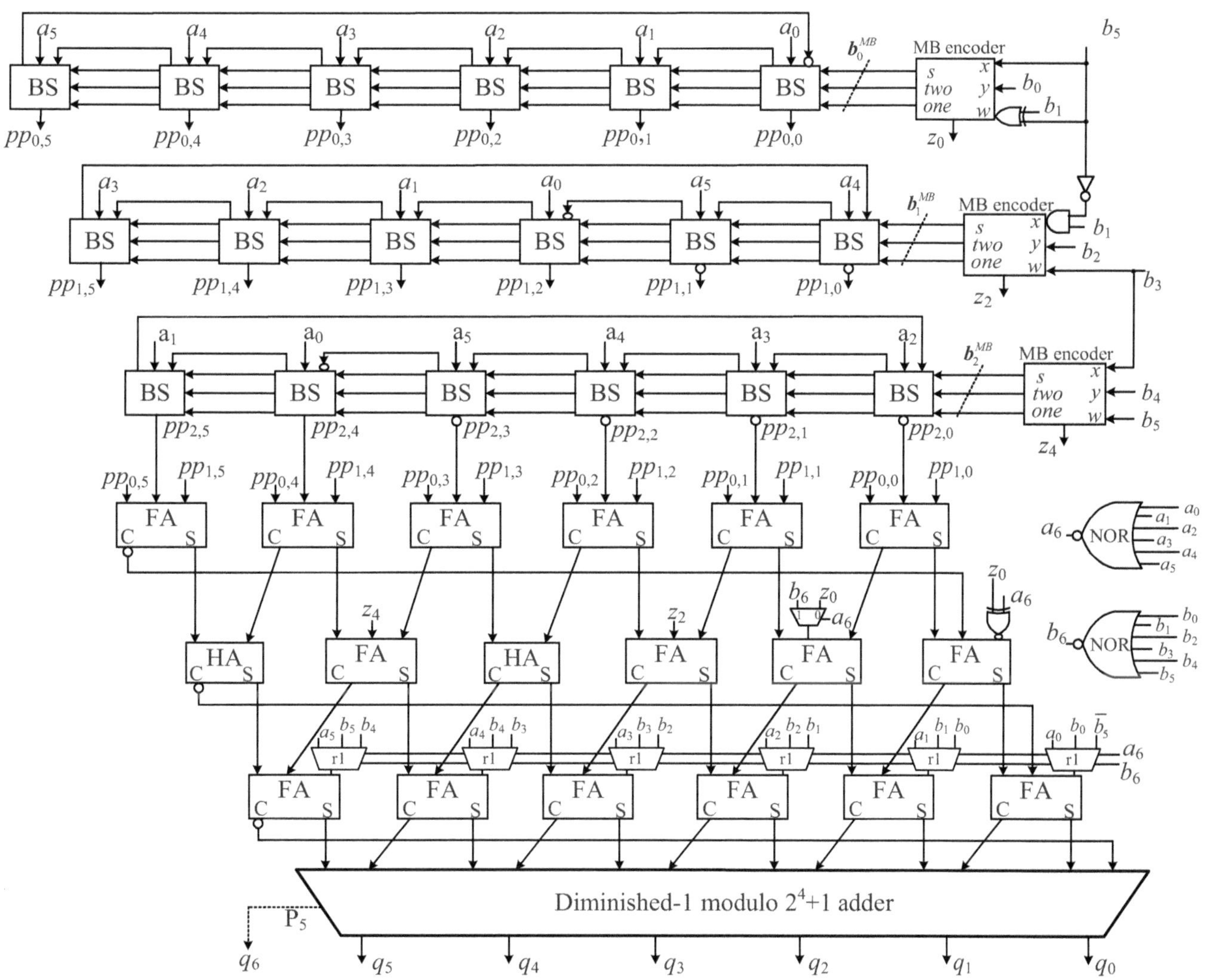

Fig. 5. Design of the modulo 65 modified Booth multiplier demonstrating the modulo 2^n+1 IDEA multiplier.

reaches the best timing than both competitors (1.78 ns). Considering that the critical path of a combinatorial implementation of the single round of the IDEA algorithm comprises 3 modulo $2^{16}+1$ multipliers, the 90 ps per module results to some hundreds of KHz faster circuit.

Table 6. IDEA multiplier Power (mW) synthesis results

Period (ns)	Proposed	[9]	[10]	Gains compared to	
				[9]	[10]
1.78	6.97				
1.83	5.85	6.06		3%	
1.87	5.69	5.68	6.59	0%	14%
1.9	5.42	5.57	6.03	3%	10%
2.0	4.73	4.92	5.55	4%	15%
2.1	4.24	4.36	5.06	3%	16%
2.2	3.91	4.14	4.32	6%	9%

For our comparison we use conventional implementations of the CSA trees using FA (3:2 compressor) stages for the carry save addition of the derived partial products. We also use inverted end around parallel-prefix adder with fast carry increment [7] as a final stage adder (FSA). The adder's propagate signals are defined as $p_i = a_i \oplus b_i$. The prefix carry computation unit is designed according to the Sklansky algorithm [11]. In practice, more efficient implementations can be derived if the CSA trees are designed using (4:2), (5:2) and (7:2) compressors in a way similar to that proposed in [12]. More efficient final stage adders can also be used as those proposed in [13].

5. IDEA IMPLEMENTATION AND COMPARISONS

In order to confirm the efficiency of the proposed multipliers for the implementation of the IDEA algorithm, one complete IDEA round, as depicted in Fig. 1, is implemented using all considered multipliers. Inputs and outputs were registered and the circuits have been synthesized at various speeds. Table 7 presents the experimental results at the fastest speed achieved by all designs.

Table 7. Experimental for the implementation of IDEA

Multiplier	Delay (ns)	Area (um^2)	Power (mW)
[9]	5.98	30738	23.01
[10]	5.98	32250	24.46
Proposed	5.98	27902	21.08

The results in Table 7 indicate that the circuit of one round of IDEA using the proposed multiplier is of 13% and 9% smaller than the designs using the multipliers in [9] and [10], while providing a 6% and 11% saving in power dissipation respectively. It is also important to note that, the implementation based on the proposed multiplier achieves the lowest delay (5.65ns) while the one based on [9] starts at 5.73ns.

6. CONCLUSIONS

In this work a new modulo 2^n+1 multiplication algorithm adapted to the IDEA block cipher is proposed. The proposed modulo 2^n+1 multiplier which uses modified Booth encoding, for n=16, yields an area and power advantage of up to 12% and 14% respectively compared to the already proposed modulo 2^n+1 multiplier designs. Also, the implementation of the one round IDEA block cipher based on the proposed multiplier achieves an area and power advantage of up to 13% and 11% respectively.

7. ACKNOWLEDGMENTS

This research has been co-funded by the European Union (European Social Fund) and Greek national resources under the framework of the "Archimedes III: Funding of Research Groups in TEI of Athens" project of the "Education & Lifelong Learning" Operational Programme.

8. REFERENCES

[1] X. Lai, "On the design and security of block ciphers", ETH Series in Information Processing, Hartung-Gorre Verlang Konstantz, 1992.

[2] A. V. Curiger, H. Bonnenberg, and H. Kaeslin, "Regular VLSI architecture for multiplication modulo (2^n+1)", *IEEE Journal of Solid-State Circuits*, vol. 26, no. 7, pp. 990-994, July 1991.

[3] M. Bahrami and B. Sadeghiyan, "Efficient modulo 2^n+1 multiplication schemes for IDEA", in Proc. of *IEEE International Symposiums on Circuits and Systems* (ISCAS), pp. 653-656, 2000.

[4] C. Efstathiou, I. Voyiatzis, N. Sklavos, "On the modulo 2^n+1 multiplication for diminished-1 operands", in Proc. of the *2nd IEEE Int. Conference on Signals Circuits and Systems* (SCS), 2008.

[5] J. W. Chen, R. H. Yao, "Efficient modulo 2^n+1 multipliers for diminished-1 representation", *IET Circuits, Devices, Systems*, vol. 4, no. 4, pp. 291-300, July 2010.

[6] S. Mukherjee, B. Sahoo, "A novel modulo (2^n+1) multiplication approach for IDEA cipher", *International Journal of Programmable Devices, Circuits and Systems*, vol. 2, no. 11, Nov. 2010.

[7] R. Zimmermann, "Efficient VLSI implementation of modulo (2^n+1) addition and multiplication", *Proc. of IEEE Symp. on Computer Arithmetic*, pp. 158–167, April 1999.

[8] Y.-J. Chen, D.-R. Duh, Y. S. Han, "Improved modulo (2^n+1) Multiplier for IDEA", *Journal of Information Science and Engineering*, vol. 23, no. 3, pp. 907-919, 2007.

[9] H. T. Vergos, C. Efstathiou, "Design of efficient modulo 2^n+1 multipliers", *IET Computers & Digital Techniques*, vol. 1, no. 1, pp. 49-57, Jan. 2007.

[10] [C. Efstathiou, N. Axelos, K. Pekmestzi, "On the design of modulo 2^n+1 multipliers", in Proc. of *14th Euromicro Conference on Digital System Design* (DSD), pp. 453-459, 2011.

[11] J. Sklansky, "Conditional sum addition logic", *IRE Trans. Electronic Computers*, vol. 9, no. 6, pp. 226-231, June 1960.

[12] W. Hong, R. Modugu, and M. Choi, "Efficient Online Self-Checking Modulo 2^n+1 Multiplier Design", *IEEE Trans. on Computers*, vol. 60, no. 9, pp. 1354-1365.

[13] H. T. Vergos, C. Efstathiou, "Efficient Modulo 2^n+1 Adder Architectures", *Integration, the VLSI Journal*, vol. 42, no. 2, pp. 149–157, Feb. 2009.

An Asymmetric Adaptive-Precision Energy-Efficient 3DIC Multiplier

Gopi Neela
Information Sciences Institute
University of Southern California
Los Angeles, USA
gneela@isi.edu

Jeffrey Draper
Information Sciences Institute
University of Southern California
Los Angeles, USA
draper@isi.edu

ABSTRACT

Decades of research in optimizing multipliers for power, speed, and area efficiency, and the continuing push for further enhancing multipliers reflects their importance. Energy-efficient computing requires optimization of every single logic component and hence, an energy-efficient 64-bit multiplier is proposed. This design exploits the highly frequent occurrence of low-precision operands, by dynamically adapting to asymmetric precision to save energy. Depending on input operands, it functions in three modes: 32×32, 64×64, or asymmetric 32×64/64×32. Results show that the asymmetric multiplier uses up to 42.4% less switching energy, and is overall up to 33.4% more energy-efficient than the baseline design. A 3-dimensional integrated circuit (3DIC) version of the asymmetric multiplier is also introduced. It is designed using a "design for 3D" methodology to mitigate some of the challenges of 3DIC. The average interconnect length and chip footprint of the 3DIC are 17.3% and 46.5% smaller than the traditional IC implementation.

Categories and Subject Descriptors

B.2.m [**ARITHMETIC AND LOGIC STRUCTURES**]: Miscellaneous; B.7.1 [**INTEGRATED CIRCUITS**]: Types and Design Styles—*Advanced technologies, VLSI*

Keywords

Multiplier; 3DIC; Chip Stacking; Energy Efficient; Precision

1. INTRODUCTION

Energy efficiency is among the most important goals of current semiconductor research. The key reasons being (i) recent explosion of mobile computing, and (ii) the push for Exascale computing. Both these technologies demand higher energy efficiency, the former for longer battery life, and the latter for low operational cost. Low energy consuming logic blocks such as multipliers are an essential building block for such systems. Multipliers, signed or unsigned, are very important functional units for both integer and floating-point computational units. These units are critical components of any high-performance microprocessor (CPU, GPU, etc.). A tremendous amount of research has been dedicated to optimizing multipliers for low power, high speed, and area efficiency [1, 2, 3]. However, continued effort is being made for further improving multipliers. The research presented in this paper concentrates on building an input data dependent 64-bit multiplier, that asymmetrically adapts to operand precisions to save energy. To the best knowledge of the authors, the proposed asymmetric 32×64 mode of operation is a new feature. Also the proposed design is implemented using 45nm IBM SOI standard cell libraries and the results presented in this paper are based on simulation traces of the post-route back-annotated simulation.

The proposed multiplier architecture is also inspired by 3-dimensional integrated circuit (3DIC) technology. It is designed for 3DIC implementation to further enhance the performance of the multiplier. 3DIC technology is a very promising alternative to sustain the trend of integration and energy-efficiency, while remaining cost-effective, in the era of slower transistor scaling [4, 5]. In a 3DIC, multiple active layers are stacked vertically to form a single chip. The vertical interconnections are made using through-silicon-vias (TSVs). These 3DICs with TSVs have tremendous potential to be energy-efficient [4]. Furthermore, a 3DIC offers other advantages, like the ability to stack heterogeneous technologies into a single chip and enhanced chip security to protect intellectual property against reverse engineering [6]. Also, a logic-on-logic stacked 3DIC can provide a cost-effective solution [7] for growing yield concerns of devices at smaller technology nodes. However, 3DIC technology suffers from some challenges including thermal issues and lack of CAD tools [8]. Hence, the presented 3DIC multiplier is designed for efficient 3DIC implementation following a "design for 3D" approach, to offset some of the obstacles of 3DIC technology.

The subsequent section presents the background, followed by the proposed architecture. Section-4 presents the asymmetric 3DIC multiplier design, and the "design for 3D" approach. The implementation details and experiment setup are given in sections 5 and 6, respectively. Finally, the results are analyzed in section-7.

2. BACKGROUND

Most modern microprocessors performing 64-bit operations have a native 64-bit multiplier unit for multiplication instructions. However, they still support 32-bit or even half word (16-bit) multiplication instructions. These lower operand width multiplications are often executed using the 64-bit multiplier, or in some cases using an additional multiplier unit. The processor typically relies on software to indicate whether the multiplication is a 64-bit operation or not, depending on the instruction type. Over the years, researchers have identified that most input operands for 64-bit multiplication instructions have small inherent precision. In [1] the authors analyzed the operand values of multiplication instructions across the SPEC benchmark suite and concluded that nearly 58% of the 64-bit multiplication instructions have both operand values with precision less than or equal to 32-bit. Moreover, the authors in [2, 3] did a similar analysis and showed that most of the times the input operands are small as well as positive. Although a N-bit multiplier produces a 2N-bit product, many architectures only return the lower N bits of the product and an overflow flag [9], because most of the time the least significant N bits are sufficient for representing the product. For N=64, this implies that when one of the input operand precisions is greater than 32-bit, it is highly probable that the other is less than 32-bit, otherwise there will be many overflows. Therefore, most 64-bit multiplications require 32×32 or 32×64 multiplication, and rarely 64×64. These findings inspired the design of the proposed asymmetric adaptive-precision multiplier.

Like any emerging technology, 3DIC also faces several challenges, including thermal issues, no manufacturing standards, absence of standardized design rules, and lack of 3DIC-specific CAD tools [8]. However, careful design partitioning techniques can enable the use of existing tools and standard cell libraries to build useful 3DICs [5]. So, as part of earlier research, an experimental 3DIC implementation of an existing floating-point unit (FPU) design was conducted to identify the challenges in building a 3DIC using current tools [10]. This unit was originally designed for a traditional IC implementation. During this exercise it was observed that, for the most part current tools are adequate, but a 3DIC-specific design strategy would have benefited the implementation. Furthermore, a "design for 3D" strategy can also be used to overcome other obstacles of 3DIC technology like thermal issues. Hence, such techniques are used to build the proposed 3DIC design to combat some of the challenges currently present in 3DIC technology.

3. ARCHITECTURE

The motivation for the proposed multiplier is twofold: one is to take advantage of the narrow-width operands to save energy, and the other is to design for 3DIC implementation. The architectural details are given in this section.

The main idea is to dynamically detect the precision of each operand and perform one of the following operations: (i) 32×32 multiplication: If both operand precisions are less than or equal to 32 bits; (ii) 32×64 or 64×32 asymmetric multiplication: Only if one of the operand precisions is less than 32 bits; and (iii) 64×64 multiplication: Otherwise. This functionality is achieved by building a 64-bit multiplier using four 32-bit multipliers. Also, such a hierarchical design is well suited for 3DIC implementation, which is discussed in section-4. The proposed design functions in two halves, one half which detects the operand precisions and selectively updates the inputs of only the required 32-bit multipliers. As a result, no dynamic energy is consumed by the other 32-bit multipliers whose inputs remain unchanged. The second half of the design combines the partial products generated by the four 32-bit multipliers, based on the type of operation to be performed: 32×32, 32×64, 64×32, or 64×64.

Additionally, to achieve high throughput, the multiplier is pipelined into two stages. Attributing to the structure of the proposed design, it is a natural choice to designate the first half of the design as pipeline stage-1 and the other half as stage-2. Nevertheless, if required, the design can be further pipelined to match the specifications of a target system into which it may be integrated. Before getting into further details of the architecture, the following sections focus on unsigned multiplication because, most of the times multiplications are of positive numbers [2, 3], and also FPUs use a sign-magnitude multiplier, which is essentially an unsigned multiplier. However, sections 3.1 and 7, present the details of extending the proposed design to a signed multiplier, and its results, respectively. Henceforth, Ah and Bh represent the higher half, i.e., 32 most significant bits (MSB) of inputs A and B, respectively. Similarly, Al and Bl represent the lower half, i.e., 32 least significant bits (LSB) of A and B, respectively. Now that a top-level idea of the design is presented, the following paragraphs discuss each stage in detail.

Stage-1: Figure-1(a) shows stage-1 of the pipelined design. The preprocessing logic includes two 32-bit zero detectors with outputs Za and Zb, respectively. These two signals control the multiplier's mode of operation as follows:

Za	Zb	:	*Mode of Operation*
0	0	:	64×64
0	1	:	64×32 with $M1 = AhBl$
1	0	:	32×64 with $M1 = AlBh$
1	1	:	32×32

It should be noted that both 32×64 and 64×32 operations require only two 32-bit multipliers to compute the output. In these cases, signal Za controls the input selection for the 32-bit multiplier that computes $M1$. This allows the logic to remain the same in pipeline stages for both cases of asymmetric multiplication, i.e., 32×64 and 64×32. Henceforth, 32×64 is used to represent both the cases. $E1$ and $E2$ signals enable the stage-1 input pipeline registers only when a new computation is required, avoiding unwanted activity to save energy. Additionally, these 32-bit multipliers' outputs that drive the input pipeline registers of stage-2 also remain unchanged, further increasing the energy savings.

Stage-2: This half is responsible for combing the four partial products: $M0$, $M1$, $M2$, and $M3$, generated by the 32-bit multipliers to produce the required *Product*. Based on various modes of operation, stage-2 functions as follows:

1. *32×32 Multiplication*: $Product = \{64'd0, M0\}$.

2. *Asymmetric 32×64 Multiplication*: Figure-2 illustrates the partial products combination to produce the output.

$Product = \{32'd0,\ \ M1 + \{32'd0, M0[63:32]\},\ \ M0[31:0]\}$

3. *64×64 Multiplication*: Figure-1(b) illustrates the multiplication operation, where an array of 64 3:2 compressors reduce $M1$, $M2$, and $\{M3[31:0], M0[63:32]\}$ to generate a *Sum* and *Carry* streams. Then, a 64-bit adder is used to further reduce partial products as shown in figure-1(b), to generate A_{sum} and C_{out}, the sum and carry outputs, respectively. The *Product* is obtained as follows:

$Product = \{M3[30:0] + C_{out},\ A_{Sum},\ Sum[0],\ M0[31:0]\}$

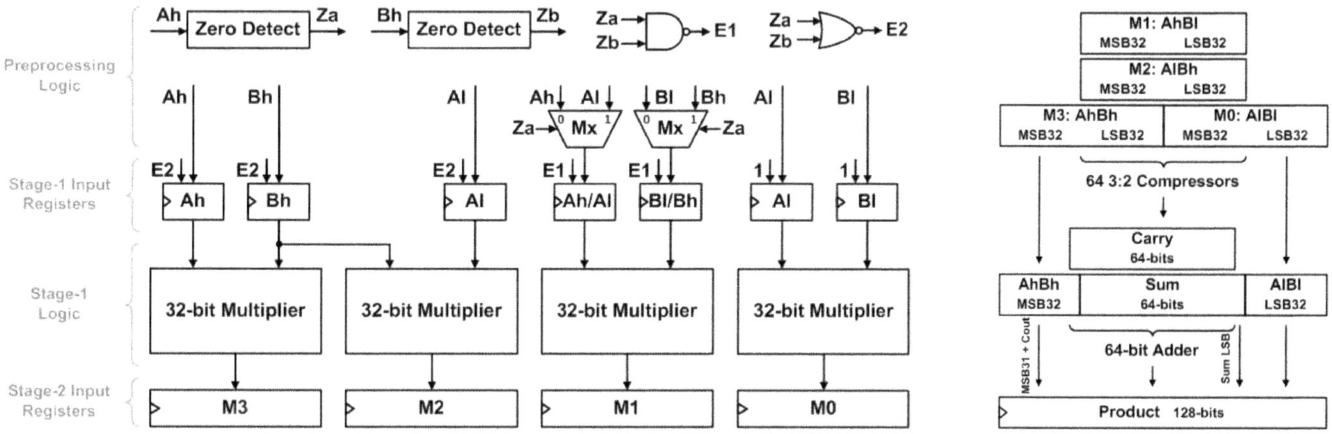

(a) Stage-1 of the proposed architecture

(b) Stage-2: 64×64 operation

Figure 1:

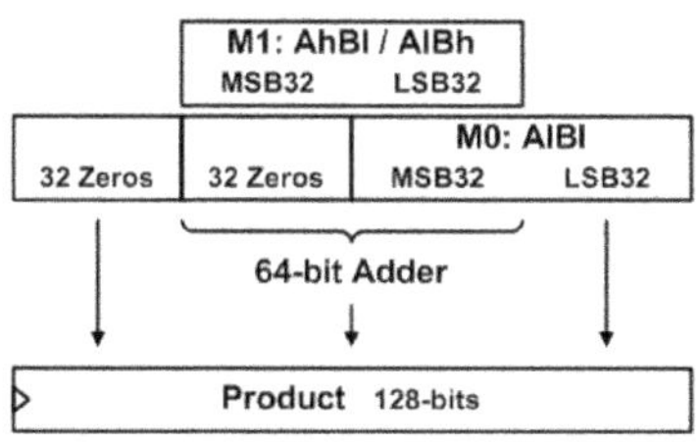

Figure 2: Stage-2: 32×64 operation

3.1 Analysis

The ability to dynamically adapt multiplier precision comes with an area overhead. The asymmetric multiplier requires two 32-bit zero detectors, multiplexers to control inputs, and additional stage-1 registers. Most of this overhead occurs in the preprocessing logic. As indicated in [1], a good portion of this overhead (zero detectors and multiplexers in preprocessing logic) can potentially be already present in the microprocessor that can be modified to assist the proposed multiplier in the processor architecture. However the results and analysis presented throughout this paper are based on a design that includes this overhead. So, when this multiplier is integrated into a typical microprocessor, the area penalty can be minimized and thus further improve the energy.

As mentioned earlier, the multiplier design can be extended to handle 2's Complement based signed multiplication operations. This requires a few additional logic blocks in the preprocessing logic that convert the 2's complement number to an unsigned number and keep track of the sign. Given the probability of negative operands is low [2, 3], the number of times these computations are required is less, hence the dynamic energy penalty could be low. Additionally, another logic block is required that converts the product generated in stage-2 to a 2's complement notation. This is required only when one of the inputs is negative. Again, the chances of this additional computation are low.

Furthermore, the proposed architecture can be used for building a FPU. A 53-bit unsigned multiplier, required for a double precision FPU, can be designed using four 27-bit multipliers in a manner similar to the proposed 64-bit multiplier. All single-precision operations, which require a 24-bit multiplier, will be computed using 27×27 mode of operation. The double precision multiplication operations will be executed in one of the three modes, depending on values of mantissas. For further enhancing the 53-bit multiplier, it can be built using one 24×24, two 24×29, and one 29×29 multipliers. This architecture will be able to dynamically switch between the following modes: 24×24, 24×29, 29×29, 24×53, 29×53, and 53×53, depending on the input operands and the FPU precision, thus resulting in a highly energy-efficient, asymmetric adaptive-precision multiplier.

Clock-Gating: All the enable signals in the proposed architecture can be used as control signals for clock-gating. The enable signals help in blocking unwanted data to appear at the flip-flops, but if these signals are used for clock-gating then additional dynamic energy savings in flip-flop cells would result in higher energy-efficient design.

4. 3DIC MULTIPLIER

The architecture of the multiplier presented in the previous section is inspired by 3DIC technology. The hierarchical architecture is designed not only to take advantage of narrow width operands, but also to enable an efficient 3DIC design that can exploit the benefits of extending to the third dimension. In the following sections, the 3DIC design methodology is explained, along with the details on how the proposed design addresses challenges of 3DIC.

The fundamental step in building a 3DIC is to determine the number of tiers. Given the size of the multiplier design is relatively small when compared to designs like multi-core processors, a two-tier 3DIC is adequate. Also, this multiplier design may not have extremely long interconnects that could benefit from having a higher number of tiers, i.e., three or more. Moreover, it is also very challenging to implement a 3DIC with more than two tiers using current tools. Hence, the multiplier is partitioned into two tiers to form a 3DIC.

Design partitioning is the most crucial step in designing for a 3DIC. It affects the design flow and performance of the target 3DIC. There are many ways in which a design can be partitioned, ranging from coarse to fine grained, depending on the granularity of the blocks that are to be stacked: core-level, functional-unit blocks (FUBs), logic gates, and transistors [5]. Given the multiplier itself is a functional-unit block, it has to be split across the layers at a logic gate level. But, such a partitioning granularity demands new

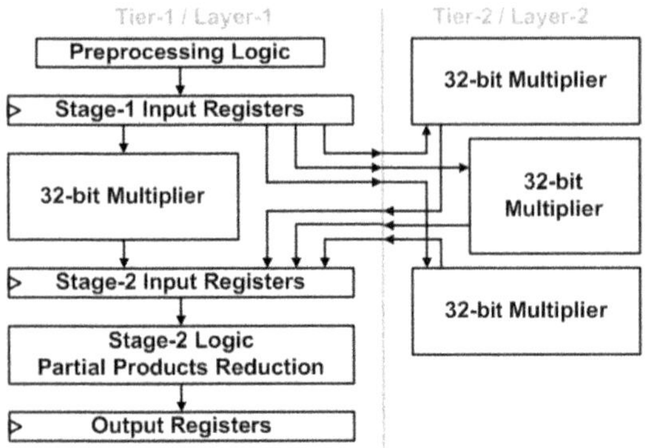

Figure 3: 3DIC multiplier block diagram

design methodologies and layout tools [5]. Thanks to the hierarchical architecture of the proposed 64-bit multiplier, it contains four smaller FUBs (32-bit multipliers). Hence the design is split as follows (see figure-3): *Tier-2*, comprising the three 32-bit multipliers that compute $M1$, $M2$, and $M3$; *Tier-1*, containing the rest of the logic. Therefore, the above mentioned challenges of splitting the multiplier across the tiers are avoided. Also, this choice of design partitioning is consistent with earlier research [11], where it was proposed that partitioning the design around a sequential element is advantageous over partitioning between combinational logic. It is better to have each interconnect between the tiers either sourced or sunk by a sequential unit in one of the tiers, rather than having both ends as part of a combinational logic. In this 3DIC design, the inter-layer signals going from tier-1 to tier-2 are the outputs of stage-1 pipeline registers (sequential element), and the signals going from tier-2 to tier-1 are inputs to the stage-2 pipeline registers (sequential element). The reason for manually partitioning the design is due to the absence of automatic design partitioning tools. Further details are presented in the following sub-section.

4.1 Design for 3D

As stated in the previous section, the structure of the 64-bit multiplier simplified the crucial design partitioning task for building a 3DIC. Similarly, there are several other factors that influenced the architecture of the proposed design, with the aim of making it amenable to 3DIC implementation.

One of the major road blocks in 3DIC technology is the thermal issue. It is very hard to manage the heat dissipated in the tiers that are far from the heat sink. One way to avoid thermal hot spots is to minimize the logic activity in the layers that are far from the heat sink. By the virtue of this 3DIC design, the activity in layer-2 (farther from the heat sink) is kept to a minimum. Layer-2 is fully active only in 64×64 mode of operation, which is rare. Most of the time the multiplier operates in 32×32 mode, and the 32-bit multiplier that computes in this mode is located in layer-1, the one closest to the heat sink. Therefore, by-design, the probability of thermal hot spots is greatly reduced.

Clock tree synthesis across multiple tiers is one of the challenging tasks for tools to accomplish. Given the 3DIC multiplier has purely combinational logic in tier-2, a clock signal is not required in tier-2. Therefore, the clock tree synthesis problem is reduced to a single layer (tier-1), which current tools can handle seamlessly. Moreover, the clock tree distribution spans across a smaller area, when compared to a

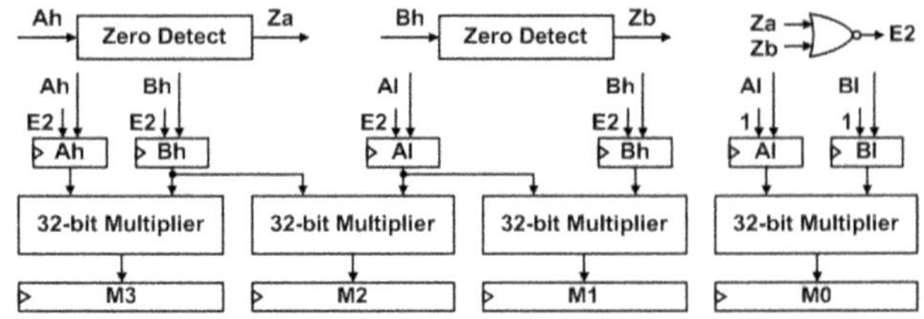

Figure 4: Stage-1 of the symmetric 2DIC design

corresponding 2DIC implementation. As a result the number of clock-buffers required to distribute and balance the clock across the chip is less, thereby further enhancing the energy-efficiency.

5. IMPLEMENTATION

To make a fair comparison and show the superiority of the proposed design the following multiplier alternatives were implemented for experimental analysis.

(1) *Baseline Design*: A two-stage pipelined, 64-bit unsigned Synopsys DesignWare multiplier component that uses an efficient tree-based architecture.

(2) *Asymmetric 2DIC Design* (A2DIC): Traditional 2DIC implementation of the proposed asymmetric multiplier. For an unbiased comparison, the 32-bit multipliers are implemented using Synopsys DesignWare single-stage unsigned multiplier components that have the same architecture as that of the baseline. So, any improvements on the baseline design can also be applied to these 32-bit multipliers.

(3) *Symmetric 2DIC Design* (S2DIC): To demonstrate the advantages of supporting an asymmetric operation, a symmetric variant of the proposed design is implemented. This version does not support the 32×64 mode of operation. Its stage-1 architecture is shown in figure-4. The preprocessing logic now updates either all 32-bit multipliers inputs (64×64 mode) or just the lower 32-bit multiplier (32×32 mode). In the 64×64 mode of operation, the second stage executes in the same fashion as that of the asymmetric version to generate the *Product* output, as shown in figure-1(b).

(4) *3DIC Design*: 3DIC version of the asymmetric multiplier; its implementation details are given in section-5.1.

All 2DIC designs are implemented using a standard VLSI design flow with the help of Cadence and Synopsys suites of CAD tools. The designs are synthesized to 45nm IBM SOI technology standard cell libraries, and placed-and-routed with a target clock period of 1ns.

5.1 3DIC Implementation

The 3DIC design is targeted to the same 45nm technology standard cell libraries that are used for building the 2DIC counterparts. First, the 2DIC asymmetric multiplier design RTL code is manually partitioned into two layers, as shown in figure-3. Then, each layer is synthesized independently, and verified together for any timing violations. Finally, a layout is obtained using Cadence First Encounter, with both layers bonding face-to-face.

In, face-to-face bonding, the interconnects between the layers are made using top-level metal layer micro bumps. When one layer is flipped and bonded to the other layer, these bumps should align properly to form a contact. This requires the bumps locations in one layer to be a mirrored version of the other. But, current tools are incapable of automatically finding an optimal signal-bump assignment for inter-layer interconnects. They either follow a random assignment or at best a greedy approach (assigning an available bump nearest to the point where the signal is generated) for signal-bump assignment in one layer, and force the other layer to use the mirrored version. A better solution is used in building the proposed 3DIC design where each signal is assigned a bump manually. Also, there are no standard design rules to guide tools, to layout these bumps at 45nm technology. Hence, the required rules (related to the size and pitch of bumps) are borrowed from Tezzaron's 130nm 3DIC technology. Once the bump grid is determined, each inter-layer signal is assigned a bump manually. Then the layers are independently placed and routed to build layouts required for 3DIC. Like 2DIC design, 3DIC design is also built for a target clock period of 1ns. The post-route netlists are then analyzed together for timing verification. As a final verification step, Cadence Conformal is used for a logical equivalence check of the 3DIC design against the original 2DIC design RTL source code.

After all the designs are implemented, a post-route simulation environment is developed as explained in the subsequent section. Then, each design is thoroughly tested to generate various results that are analyzed in section-7.

6. SIMULATION ENVIRONMENT

A SDF back-annotated simulation of post-route netlists is performed using the Cadence NCsim simulator. The simulation trace is captured in a value-change-dump (VCD) file. Then, power consumption is analyzed using the Synopsys PrimeTime-PX tool that takes the SPEF RC parasitics file (generated by Cadence First Encounter layout tool), and the VCD file, as inputs. The former is used for back-annotating the post-route netlist, while the latter provides the actual signal transitions. Using the signal activity, accurate average power consumption is measured, which includes net switching power, cell internal power, and leakage power. Then total energy consumption is obtained by multiplying the average power consumption with total simulation time.

Once such an accurate simulation environment is established, realistic test cases are required to generate meaningful results. Each test case is composed of 100 thousand (100,000) test vectors, with each vector containing two 64-bit inputs. To emulate a practical workload, these test cases are generated using a random number generator that has a distribution similar to the one presented in [1]. About, 58% of the test vectors contain both inputs that have operands of precision 32-bit or less. However, this test case does not capture the distribution of test vectors that have one input with operand precision greater than 32-bit while the other operand precision is less than 32-bit. This is very important to demonstrate the usefulness of the asymmetric mode of operation. So, the following criterion is used: The remaining 42% of the test vectors have either one or both input operands with precision greater than 32-bit. Let's define P as follows, P is the fraction of test vectors of the above defined 42% of test vectors that have one input with operand precision less than 32-bit. Hence, $P \times 42\%$ of total test vectors in a test case execute in the 32×64 mode of operation. Using this criterion, P is swept from 1/10 to 9/10, in steps of 1/10, to generate 9 different test cases each containing 100,000 test vectors. Each design, including the baseline is then simulated with these 9 test cases, and the results obtained are presented in the following section. While it is difficult to specify a P that applies to all workloads, for most workloads the P value is quite high because most programs running on 64-bit machines rarely encounter overflow exceptions [9], implying that when one of the input operand precisions is greater than 32-bit for a multiply operation, the other must be less than 32-bit most of the time. These instructions will be executed in asymmetric mode.

7. RESULTS

Table-1 shows the average switching and total power consumption measured using the VCD files obtained by simulating the designs at a clock frequency of 1GHz. Clearly, all the proposed variants outperform the baseline design, in terms of power. Using these values total energy consumption is calculated. The A2DIC, 3DIC, and S2DIC designs consume up to 33.4%, 31.7%, and 24.6% less energy than the baseline design, respectively. Figure-5 shows the plot of percentage of switching (dynamic) energy saved when compared to the baseline design versus the P value. These results prove that the asymmetric mode of operation is very fruitful, as A2DIC provides nearly 9% additional energy savings than S2DIC.

Table 1: Net switching (NS) and total power (TP)

P	Baseline		A2DIC		S2DIC		3DIC	
	NS	*TP*	*NS*	*TP*	*NS*	*TP*	*NS*	*TP*
1/10	17.4	39.1	12.3	29.9	12.2	29.5	12.0	31.0
2/10	17.1	38.6	11.9	29.1	12.1	29.3	11.6	30.1
3/10	16.8	38.0	11.4	28.2	12.0	29.1	11.1	29.2
4/10	16.5	37.4	11.0	27.4	11.8	28.8	10.7	28.3
5/10	16.0	36.5	10.3	26.1	11.6	28.3	10.0	26.9
6/10	15.7	35.9	9.9	25.3	11.4	27.9	9.6	26.0
7/10	15.4	35.2	9.4	24.4	11.1	27.5	9.1	25.1
8/10	15.0	34.4	8.9	23.4	10.9	26.9	8.6	24.0
9/10	14.7	33.8	8.5	22.5	10.7	26.5	8.2	23.1

P = (Asymmetric operations / Operations that are not 32×32).
Power units: *mW*

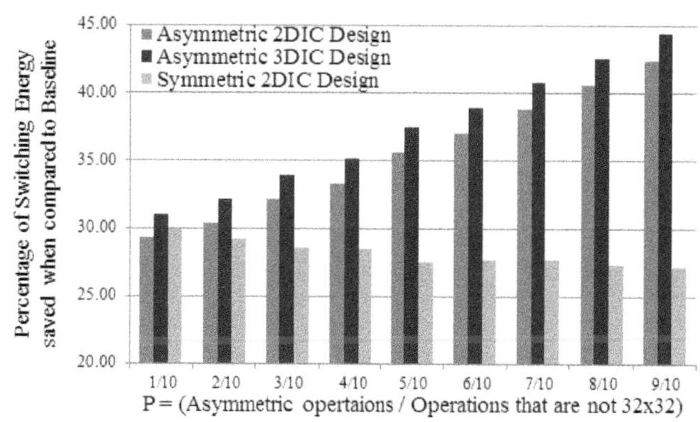

Figure 5: Switching energy savings

Table-2 presents the comparisons between various designs, with respect to footprint, cell density, cell area, average connection length (ACL), and leakage power. Predictably, the leakage power increased with cell area, but the difference is marginal. As expected, the 3DIC demonstrates its ability to have a high device density achieved through a lower footprint. The 3DIC is 36.7%, 46.5%, and 44.8%, smaller in terms of chip size than baseline, A2DIC, and S2DIC designs, respectively.

Table 2: Comparisons between various designs

Design	Footprint $\mu m \times \mu m$	Density %	Cell Area μm^2	ACL μm	Leakage mW
Baseline	174×168	76	22216	6.613	2.96
A2DIC	188×184	74	25598	5.223	3.28
S2DIC	185×181	75	25114	5.571	3.27
3DIC	136×136	71	26264	4.318	3.57

As stated above the 3DIC chip is almost half the size of the A2DIC chip, but it has about 2.6% more cell area. The ACL of 3DIC is 17.3% smaller than that of A2DIC. When compared to A2DIC, 3DIC consumes 2.9% less switching energy, but uses 3.1% more overall energy, due to larger cell internal power dissipation. Energy efficiency can be heavily boosted by a more mature and standardized 3DIC technology. The size and pitch used for the micro bumps are that of Tezzaron's 130nm technology. As shown in figure-6 the bump density is quite low, and the available bumps for layout are small in number. The inter-layer interconnects are widespread across the chip (blue bumps in figure-6), causing signals to travel long distances. For 3DIC layout, the place-and-route tool inserted buffers or resized cells (made larger) that drive the bumps, to satisfy timing constraints, there by reducing ACL as well as wire load capacitance, and hence reducing switching energy. But the cell area required for 3DIC layout has increased, causing cell internal energy consumption to increase. This is evident from the fact that the cell area of the post-synthesis 3DIC is slightly less (<1%), while the post-route is 2.6% larger, when compared to A2DIC. The penalty for increased cell area dominated the savings achieved by reduced ACL causing the overall energy consumption to increase. However, as the technology evolves the bumps size and pitch will scale down [12], enhancing the 3DIC performance.

Additionally, signed variants of the baseline design and the asymmetric multiplier are implemented. Results show that the signed asymmetric 2DIC design consumes about 25.0% and 30.4% less energy than the signed baseline design for $P = 5/10$ and $9/10$, respectively. If only energy consumed in switching is considered the savings increase to 33.1% and 40.3%, respectively. However, the signed A2DIC has 26.9% more area than the baseline design.

8. CONCLUSIONS

A 1GHz 64-bit asymmetric adaptive-precision multiplier is proposed, which saves energy by exploiting the frequent presence of low-precision operands. Depending on the precision of operands the multiplier dynamically switches between three modes of operation: 32×32, 64×64, or asymmetric 32×64, to reduce energy consumption. Results show up to 33.4% energy-efficiency and up to 42.4% less switching energy when compared to the baseline design. Additionally, a 3DIC asymmetric multiplier design is proposed. A "design for 3D" approach is used to offset some of the challenges of 3DIC technology. The 3DIC is built using 45nm standard cell libraries. In spite of following a conservative approach for 3DIC implementation, results show reduced switching energy consumption, 17.3% smaller average interconnect length, and an impressive 46.5% smaller chip footprint, when compared to the corresponding 2DIC implementation. A 3DIC-specific process technology and tools, along with clock-gating promise further improvements in area and energy-efficiency.

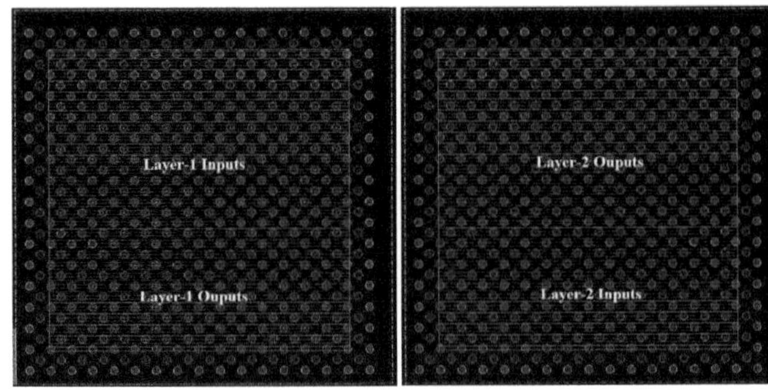

Figure 6: Face-to-Face bonding bumps: blue-signals, white-unused, red-VDD, yellow-Ground

9. REFERENCES

[1] D. Brooks and M. Martonosi. Dynamically exploiting narrow width operands to improve processor power and performance. In *High-Performance Computer Architecture, 1999. Proceedings. Fifth International Symposium On*, pages 13 –22, Jan 1999.

[2] Yijun Liu and S. Furber. The design of a low power asynchronous multiplier. In *Low Power Electronics and Design, 2004. ISLPED '04. Proceedings of the 2004 International Symposium on*, pages 301 –306, Aug. 2004.

[3] D.R. Kelly, B.J. Phillips, and S. Al-Sarawi. Approximate multiplication and division for arithmetic data value speculation in a RISC processor. *Algorithm-Architecture Matching for Signal and Image Processing*, pages 95–116, 2011.

[4] A. W. Topol et al. Three-dimensional integrated circuits. *IBM Journal of Research and Development*, 50(4.5):491 –506, July 2006.

[5] Gabriel H. Loh, Yuan Xie, and Bryan Black. Processor design in 3D die-stacking technologies. *Micro, IEEE*, 27(3):31 –48, May-June 2007.

[6] 3D-ICs and Integrated Circuit Security http://www.tezzaron.com/about/papers/3d-ics and integrated circuit security.pdf.

[7] R. Patti. Homogeneous 3D integration. *Three Dimensional System Integration: IC Stacking Process and Design*, page 51, 2010.

[8] 3D ICs with TSVs - Design Challenges and Requirements. *Cadence White Paper.*

[9] J.L. Hennessy and D.A. Patterson. *Computer architecture: a quantitative approach.* Morgan Kaufmann, 2011.

[10] G. Neela and J. Draper. Challenges in 3DIC implementation of a design using current CAD tools. In *Circuits and Systems (MWSCAS), 2012 IEEE 55th International Midwest Symposium on*, pages 478 –481, Aug. 2012.

[11] Gopi Neela and Jeffrey Draper. Logic-on-logic partitioning techniques for 3-dimensional integrated circuits. In *The IEEE International Symposium on Circuits and Systems (ISCAS)*, May 2013.

[12] S. Gupta, M. Hilbert, S. Hong, and R. Patti. Techniques for producing 3D ICs with high-density interconnect. In *Proceedings of the 21st International VLSI Multilevel Interconnection Conference*, 2004.

A Geometric Approach to Chip-Scale TSV Shield Placement for the Reduction of TSV Coupling in 3D-ICs

Caleb Serafy, Bing Shi and Ankur Srivastava
University of Maryland, College Park, MD, USA
{cserafy1, bingshi, ankurs}@umd.edu

ABSTRACT

In 3D ICs, interlayer communication is achieved using through-silicon-vias (TSVs), which can suffer from cross coupling if placed naïvely. In this paper, cross coupling between TSVs is modeled, and a chip-scale TSV coupling mitigation scheme is presented using TSV shielding. A geometric coupling model is developed which quickly estimates the pairwise coupling between TSVs, unlike circuit models of coupling that have been proposed in previous works. Our geometric model's ability to make fast accurate estimations of chip-scale cross coupling make it a good model to use for shield placement optimization.

A shield placement algorithm is presented which reduces TSV coupling by formulating a min cost flow (MCF) problem based on the proposed model. Our algorithm is compared to another shield placement algorithm presented in [9] which is based on a circuit model of coupling. Experimental results show that the algorithm proposed here is able to reduce the total cross coupling in a layout an average of 4.59x more than the other algorithm while using the same number of shields. Alternatively, our algorithm uses an average of 88% less shields to shield a layout to the same degree as the shielding schemes produced by the other algorithm.

Categories and Subject Descriptors

B.7.2 [**Integrated Circuits**]: Design Aids

Keywords

TSV coupling; TSV shielding; 3D-IC; Vertical Integration

1. INTRODUCTION

Recently vertical stacking of ICs has become a viable option for reducing the length of global interconnects in ICs. Vertical integration can be used to integrate different technologies into a single IC, for example stacking DRAM on top of processor cores [11]. Even stacking layers of the same technology can be advantageous such as in the case of stacking processor cores vertically to reduce inter-core communication latency [8].

Vertically stacking ICs requires vertical interconnects, and the most common type is the through-silicon-via (TSV). A TSV is a long piece of copper surrounded by an insulation layer which extends vertically through the silicon substrate. TSVs can cause coupling between one another, leading to signal integrity issues that could cause a chip to fail. A common practice for reducing coupling in a 3D-IC layout is to insert "shield" TSVs which help to electrically isolate the signal-carrying TSVs in the system. Circuit models of TSV coupling have been presented in previous works [5, 9], but these models are too complex to estimate the magnitude of coupling between *all* pairs of TSVs in a layout quickly and thus are not good models for chip-scale shield placement optimization.

GLSVLSI'13, May 2–3, 2013, Paris, France.

In this paper we develop a geometric model of TSV coupling which is simple enough to use at chip-scale while still being accurate enough to yield good results. What is meant by "geometric" is that **the model only considers the position of TSVs**, and does not consider electrical parameters. Unlike a circuit model of coupling, our geometric model does not involve solving systems of equations, which allows the quality of shield placement schemes to be quickly estimated. Using our geometric coupling model, we design a chip-scale shield placement algorithm. The contributions of this work are:

1. To the best of our knowledge this is the first work to present a coupling model fast enough to estimate chip-scale coupling between *all* pairs of TSVs in a layout. A chip-scale estimate of TSV coupling can lead to much more optimized shield placements than those presented in previous works.
2. A chip-scale shield placement algorithm is presented which uses the proposed geometric coupling model to formulate a minimum cost flow (MCF) problem. Solving the MCF formulation yields a shield placement scheme that can satisfy the shielding needs of each TSV pair, while minimizing the total number of shields added to the design.
3. Our proposed algorithm produces better results than the algorithm presented in [9] because it is able to model the chip-scale effects of placing a shield at a specific point. Our proposed algorithm generates shield layouts which can provide 4.59x more shielding than the layouts produced by the other algorithm. Alternatively, our proposed algorithm can match the degree of shielding produced by the other algorithm while using 88% less shields.

The remainder of this paper is laid out as follows: The background and motivation for this work are presented in Section 2. Section 3 presents the developed geometric coupling model and Section 4 presents the full-chip shield placement algorithm proposed in this paper. Finally, experimental results are presented in Section 5 and the paper concludes in Section 6.

2. BACKGROUND AND MOTIVATION

TSV coupling can cause a circuit to fail or degrade its performance in a few different ways. Coupling between signal carrying TSVs can cause the logic value at the input of a transistor gate to change, which could possibly result in incorrect logic or short circuit current. Coupling between a signal TSV and a power TSV could cause variation in VDD or GND rails, which could increase leakage current. These observations motivate us to design methods of reducing TSV coupling in 3D-ICs.

The degree of cross coupling between a pair of TSVs can be reduced by inserting shields between the TSVs. A shield can be any piece of metal which obstructs EM wave propagation. A shield need not be driven by a voltage source, so adding shields to a design does not increase the complexity of wire routing. Guard rings have been used as shields in 2D-ICs to prevent transistor coupling, however Cho et al. showed in [4, 5] that guard rings are not applicable to TSV coupling due to the vertical nature of the TSV structure. For these reasons, in this work we consider using undriven TSVs as shields.

Khan, et al. have suggested TSV shielding using coaxial TSVs [7], Tungsten ground plugs and thicker TSV liners [6], but they do not discuss the manufacturing details of creating such structures. Liu et al. consider reducing cross coupling by increasing the drive strength of TSV drivers in [9]. This approach can sufficiently reduce cross coupling, but at the expense of higher power consumption. Liu et al. also suggest a coupling reduction scheme in [10], which ensures that no TSVs are placed close enough together to cause significant coupling by using a force driven placement refinement algorithm.

2.1 Liu Algorithm

Most previous work concerning TSV coupling has presented circuit models of coupling [5, 9]. Circuit models are very accurate, but have a long solve time because they require solving systems of differential equations. For this reason, an electrical coupling model is not practical for shield placement optimization, where the coupling model must be solved repeatedly. In this paper we propose a geometric model of TSV cross coupling which is simple enough to be used to develop chip-scale coupling-aware shield placement strategies.

To the best of our knowledge the only prior work to present a chip-scale shield placement algorithm is [9], but it uses a circuit model of coupling. In this work we refer to the algorithm presented in [9] as the Liu algorithm, named after its author. The Liu algorithm estimates the degree of coupling at each signal TSV in a layout, and then inserts shields around those TSVs found to be suffering from too much coupling. The Liu algorithm constrains shield TSVs to only be placed in groups of eight surrounding a signal TSV.

The Liu algorithm estimates the coupling intensity at each TSV by solving a circuit model of coupling between pairs of TSVs one at a time. The algorithm examines each TSV in decreasing order of coupling intensity, and places eight shields around it if it has the following three properties: (1) coupling intensity of the signal TSV is above some threshold, (2) the TSV is not already shielded and (3) no other signal TSVs within a certain distance from the current TSV have already been shielded. Due to the complexity of the electrical coupling model, the Liu algorithm does not recalculate the coupling profile after shielding each signal TSV, so property (3) is the Liu algorithm's attempt to consider a shield's effect on neighboring signal. After all TSVs have been considered and the TSVs to be shielded have been identified, the algorithm inserts eight shield TSVs around each identified signal TSV.

One problem with the Liu algorithm is that it often overshields. When a TSV is identified as having high coupling intensity, the algorithm does not consider which TSV is causing the cross coupling, and the algorithm simply places shields all around the identified TSV. This can lead to extreme overshielding of signal TSVs, which is very wasteful of area. Furthermore, the Liu algorithm restricts shields to be placed only in the area directly surrounding signal TSVs, which severely limits the algorithm's ability to simultaneously shield more than one coupling pair with a single shield. Finally, the Liu algorithm considers TSV coupling one pair at a time, and so it is unable to capture the interaction between multiple signal TSVs. A signal TSV can act as a shield when it is not changing voltage level, but the Liu algorithm fails to consider this phenomenon. In the geometric coupling model proposed in this work, the shielding contribution of signal TSVs **is** considered.

3. GEOMETRIC MODELING

Throughout this work, the metric used to quantify TSV coupling is the insertion loss scattering parameter (S parameter S_{12}). The insertion loss between two TSVs is defined as the amount of energy leaving one TSV due to an injection of energy into the other. When the insertion loss is large, a change in voltage on one TSV will cause a significant change in voltage on the other TSV, so insertion loss is a good metric for representing coupling intensity.

Our geometric model is created by fitting simulated insertion loss parameters of example structures to analytical functions. Simulations are carried out using Ansoft HFSS [1], an FEM solver for Maxwell's equations. An HFSS structure is modeled as a multi-port network, and the insertion losses

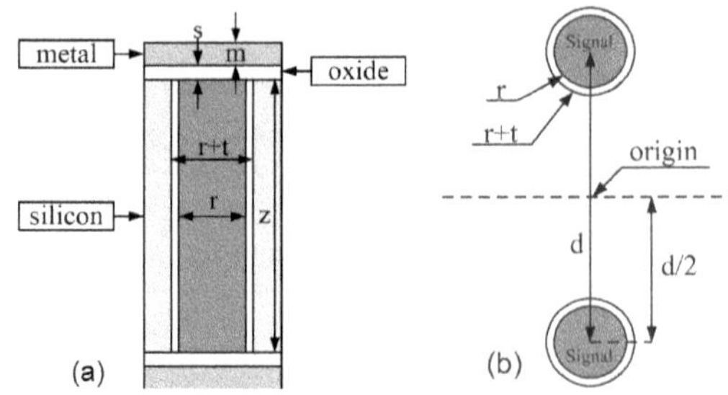

Figure 1: Unshielded HFSS structure (a) side (b) top views

between each pair of ports are calculated. The insertion loss is a frequency dependent value, so it is worth noting that all simulations were carried out at 10 GHz.

We define the shielding coefficient of the ith TSV pair, $SC(i)$, as the **additive inverse** of the insertion loss parameter between the top ports of those TSVs. We consider the additive inverse of the insertion loss parameter so that the shielding problem can be formulated as a maximization problem. Models for calculating insertion loss are presented in the following subsections. The authors would like to stress that the model of insertion loss for an arbitrary number of shields is **not a simple superposition model**. It does more than just consider each shield one at a time and sum the results. The model **considers all shields simultaneously** and considers the interaction between shields. Moreover as each pair of TSVs is considered, all other TSVs in the design (**including other signal TSVs**) are considered to be shields, and their shielding contributions are considered.

Name	Value	Description
r	$1\mu m$	TSV Radius
t	$0.2\mu m$	TSV Liner Thickness
z	$50\mu m$	Substrate Thickness
s	$3\mu m$	Oxide Thickness
m	$0.75\mu m$	Metal Thickness

Table 1: HFSS Simulation Parameters

In Sections 3.1 through 3.4 we describe how this work models insertion loss. We develop a model for the shielding coefficient of a pair of signal TSVs with zero, one or two shields by fitting HFSS data. Then we extend the two shield model to an arbitrary number of shields and test the validity of our extension. Figure 1(a) shows a cross section of a single TSV in our HFSS setup, and defines the parameters r, t, z, s and m. The specific values used in our simulations are shown in Table 1. It is important to note that the models developed in this paper are only accurate for the specific TSV parameters listed in Table 1, but the method described here for obtaining these equations would still be valid, and could be repeated if the TSV parameters changed.

3.1 Unshielded Coupling Model

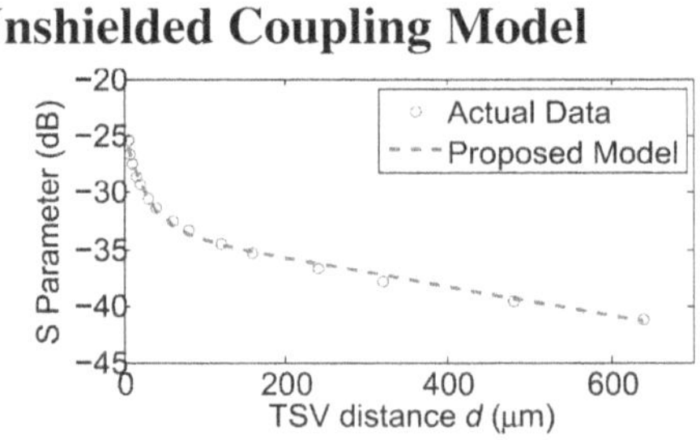

Figure 2: Insertion loss of an unshielded TSV pair

Characterization of an unshielded pair of TSVs was achieved by setting up a simple HFSS structure consisting of two signal TSVs and sweeping the distance d between them. Figure 1(b) shows the unshielded HFSS setup and defines the TSV distance d. The data from this sweep seemed to indicate a negative exponential relationship between insertion loss and d when d is small and a linear relationship when d is larger (see Figure 2). Thus the data were fit to a function of the form $\alpha \times \beta^{-d} + \delta d + \kappa$, where α, β, δ, and κ are fitting parameters. All fitting in this paper is done by minimizing the mean square error (MSE) between the model and the data. The best fit function is presented in Equation (1),

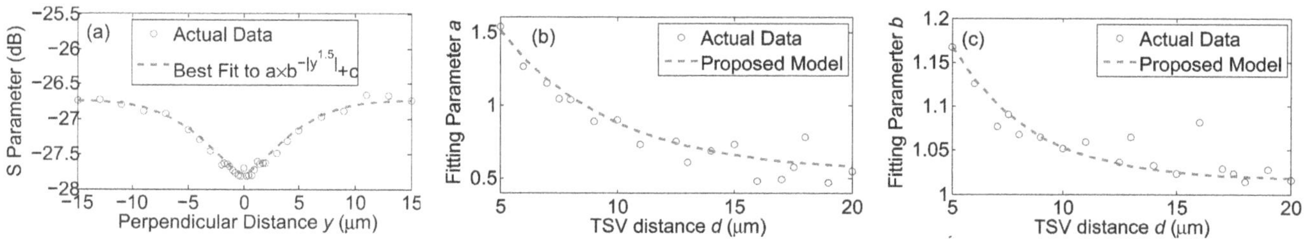

Figure 3: (a) Insertion loss parameter of a TSV pair with one shield at perpendicular distance y. ($d = 7.5\mu m$) (b) Fitting parameter a vs. d. (c) Fitting parameter b vs. d.

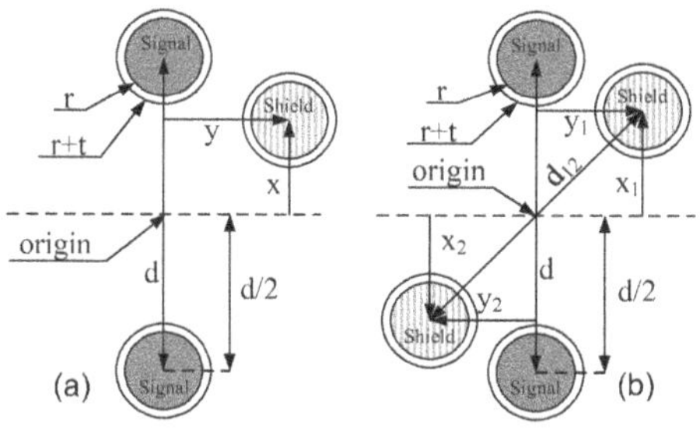

Figure 4: (a) Single and (b) Double shield HFSS structures

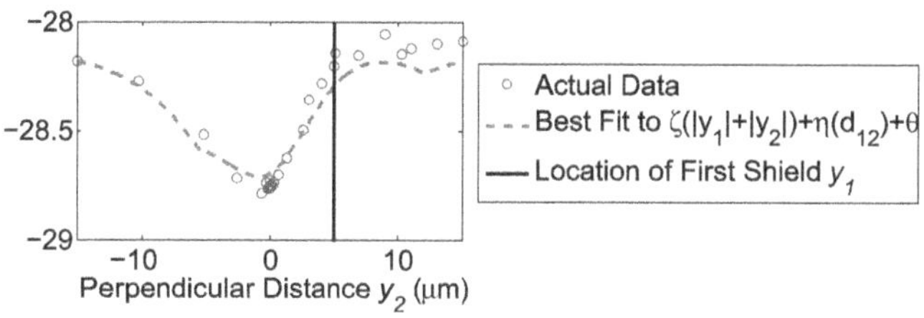

Figure 5: Insertion loss S parameter of a TSV pair with two shields: one at perpendicular distance $5\mu m$ and the other at distance y_2. ($d = 10\mu m$)

where $S_0(d)$ is the insertion loss parameter of an unshielded pair of TSVs separated by distance d. Figure 2 shows a close fit between the proposed model and the HFSS data.

$$S_0(d) = 8.79 \times 1.04^{-d} - 0.0126(d) - 33.2 \quad (1)$$

3.2 Single Shield Coupling Model

Characterization of a TSV pair with a single shield was achieved by adding a third TSV to the unshielded HFSS setup. Figure 4(a) shows the HFSS setup with one shield and defines the variables d, y and x. The HFSS results produced by sweeping x indicated that the shield's effect on insertion loss is significant *only* when $|x| < \frac{d}{2}$. Furthermore sweeping x inside of this range did not significantly effect the insertion loss parameter This implies that as long as $|x| < \frac{d}{2}$, the shielding coefficient is roughly independent of x. Thus the effect of a shield on a coupling pair need not be considered unless x is within that range, and the single shield shielding coefficient is a function of y and d, not x. However, we will show in the following sections that **when more than one shield is effecting a coupling pair, the value of x must be considered**.

The HFSS results produced by sweeping y indicated a negative exponential relationship between the insertion loss and y (see Figure 3(a)). For different values of d, a set of (y, S) pairs was fit to a negative exponential function of the form $-a \times b^{-|y^{1.5}|} + S_0(d)$, where a and b are fitting parameters. A plot of one such fitting is shown in Figure 3(a). This fitting procedure was repeated for many values of d and resulted in a set of $(d, \{a, b\})$ pairs which are the specific values of (a, b) that give a good fit when the distance between signal TSVs is d. The values a and b seemed to be negative exponential functions of d, and they are fit to Equation (2) and (3). Figure 3(b) and Figure 3(c) show the actual values of a and b vs d and the proposed model of $a(d)$ and $b(d)$. Equation (4) defines $S_s(d, y)$, the change in the insertion loss parameter of a pair of TSVs separated by distance d when a shield is added at perpendicular distance y from the signal TSVs. Equation (5) defines $S_1(d, y)$, the overall insertion loss parameter for a pair of TSVs with one shield.

$$a(d) = 2.81 \times 1.24^{-d} + 0.545 \quad (2)$$

$$b(d) = 0.616 \times 1.32^{-d} + 1.02 \quad (3)$$

$$S_s(d, y) = -a(d) \times b(d)^{-|y^{1.5}|} \quad (4)$$

$$S_1(d, y) = S_0(d) + S_s(d, y) \quad (5)$$

3.3 Double Shield Coupling Model

Characterization of a TSV pair with two shields was achieved by adding a second shield to the single shield HFSS setup. Figure 4(b) shows the setup and defines the variables d, y_1, y_2, x_1 and x_2. The HFSS results indicated that two shields did not contribute to insertion loss independently (i.e. **superposition is not an accurate model**), and there is some overlap between their shielding abilities. In other words $S_2 = S_0(d) + S_s(y_1, d) + S_s(y_2, d) - M$ where M is the overlap. A plot of insertion loss vs shield location when $y_1 = 5\mu m$ and y_2 is swept is presented in Figure 5.

It can be observed that when $y_2 > y_1$, the insertion loss is independent of y_2, but once $y_2 < y_1$ it becomes a strong function of y_2. When y_1 and y_2 have the same sign, the farther shield does not significantly affect shielding because it is unable to offer any shielding not already offered by the closer shield. Thus y_2 does not affect the insertion loss until shield two is the closer shield. As y_2 continues to decrease, the second shield provides shielding to more and more areas that the first shield does not (i.e. overlap decreases), and y_2 becomes a significant factor contributing to insertion loss. Alternatively, when y_1 and y_2 have opposite signs, they do not have significant overlap and both shields contribute significantly to the insertion loss S parameter.

This experiment was repeated with different values of x_1 and x_2. These variables were found to significantly effect the results. Although the actual values of x_1 and x_2 do not effect S_s and S_0, the overlap M seems to be a function of the distances y_1 and y_2 as well as the distance between the two shields d_{12}, which depends on x_1 and x_2. We attempt to characterize this overlap term by fitting a function of the form $\zeta(|y_1| + |y_2|) + \eta(d_{12}) + \theta$ to the HFSS results, where ζ, η and θ are fitting parameters. Due to the extremely long simulation times of HFSS, the two shield sweep was carried out only at $d = 10\mu m$. The equation for overlap $M(y_1, y_2, d_{12})$ is presented in Equation (6).

$$M(y_1, y_2, d_{12}) = \max[0, -0.02(|y_1| + |y_2|) - 0.04(d_{12}) + 0.65] \quad (6)$$

The max operator in Equation (6) is used to ensure that the overlap between two shields is always positive. Negative overlap implies that two shields could offer more shielding together than the sum of either one alone, which was not observed. The final model for the insertion loss of a TSV pair with two shields is presented in Equation (7). The max operator in Equation (7) is used to ensure that adding a shield will not decrease the shielding coefficient of a pair of TSVs, because this phenomenon was not observed.

$$S_2(y_1, y_2, d_{12}, d) = S_0(d) + S_s(y_1) + \max[0, S_s(y_2) - M(y_1, y_2, d_{12})] \quad (7)$$

3.4 Multiple Shield Coupling Model

Based on the overlap concept introduced in Section 3.3, a natural extension of Equation (7) would be $S_3(y_1, y_2, y_3, d_{12}, d_{13}, d_{23}, d) = S_2(y_1, y_2, d_{12}, d) + \max[0, S_s(y_3) - M(y_1,$

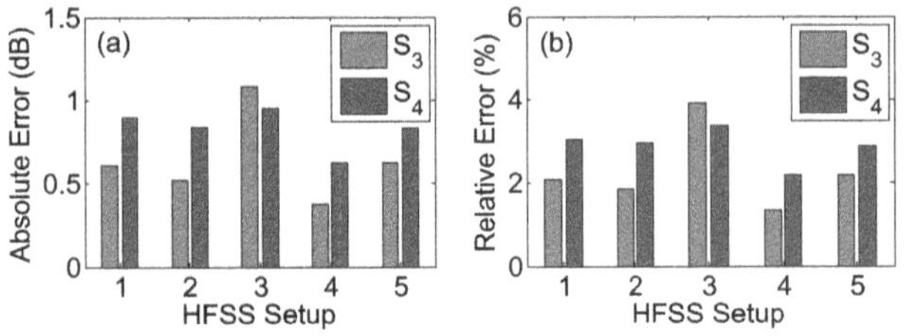

Figure 6: Model error of S_3 and S_4 (a) absolute (b) relative

$y_3, d_{13}) - M(y_2, y_3, d_{23}) + M^*]$, where M^* is the third order overlap between all three shields. The third order overlap is added (not subtracted) for the same reason that the third order intersection is added when calculating the magnitude of the union of three sets: $|A \cup B \cup C| = |A| + |B| + |C| - |A \cap B| - |A \cap C| - |B \cap C| + |A \cap B \cap C|$. When subtracting the second order overlap terms, the third order overlap gets subtracted three times, so it must be added back.

$$S_3(y_1, y_2, y_3, d_{12}, d_{13}, d_{23}, d) = S_0(d) + S_s(y_1) + \max[0, S_s(y_2) - M(y_1, y_2, d_{12})] + \max[0, S_s(y_3) - M(y_1, y_3, d_{13}) - M(y_2, y_3, d_{23})] \quad (8)$$

We assume M^* is negligibly small, which results in Equation (8). This assumption will lead our model to slightly **underestimate** the shielding ability of more than two shields because we neglect to add back the third order overlap term, but we will show that the accuracy is still reasonable. Underestimation of shielding coefficients is better than overestimation because it leads to using more shields than necessary, which wastes some area, as opposed to using less shields than necessary, which might result in an unusable chip. If we assume that all overlap terms of order three or greater are negligibly small, then Equation (8) would naturally extend to the case with n shields as in Equation (9).

$$S_n(d, y_1, \ldots, y_n, d_{12}, \ldots, d_{1n}, \ldots, d_{(n-1)n}) = S_0(d) + \sum_{i=1}^{n} \max\left[0, S_s(d, y_i) - \sum_{j=1}^{i-1} M(y_i, y_j, d_{ij})\right] \quad (9)$$

To validate this model, we generate 10 HFSS setups, five with three shields and five with four shields, and compare the actual insertion loss parameter with the model prediction. Figures 6(a) and 6(b) show the absolute and relative errors of the models for different HFSS setups. Although there is roughly 1dB of absolute error between the model and the actual data, the absolute error is roughly constant across configurations which means it is easy to correct for. As predicted, we see that the second order truncation in our models is **overestimating the insertion loss**, which means that **shielding coefficients are being underestimated**.

3.5 Shielding Coefficient

Now that we have formally defined the geometric model for the insertion loss of a pair of TSVs with an arbitrary number of shields, we can formally define the shielding coefficient which will be used in our proposed algorithm introduced in Section 4. The definition of $SC(i)$ is given in Equation (10), where n is the number of TSVs shielding coupling pair i.

$$SC(i) = -S_n(d, y_1, \ldots, y_n, d_{12}, \ldots, d_{1n}, \ldots, d_{(n-1)n}) \quad (10)$$

4. PROPOSED SHIELDING ALGORITHM

In Section 2.1, the Liu algorithm was introduced. In that section some problems with the Liu algorithm were identified, which we reiterate here:

1. It identifies TSVs which suffer from coupling, but doesn't identify the coupling source, and thus overshields the TSV by surrounding it with shields instead of only placing shields in the direction of the coupling source.
2. It only places shields close to signal TSVs, which limits their ability to shield multiple coupling pairs simultaneously.
3. It does not consider a signal TSV's ability to act as shield.

In this section we describe the TSV shield placement algorithm we developed using Equation (10). The algorithm identifies pairs that require shielding, and shields are added to the design to maximize the sum of all shielding coefficients, TS.

$$TS = \sum_{\forall i} \min[T, SC(i)] \quad (11)$$

The general flow of our proposed algorithm is presented in Figure 7(a) and enumerated here as follows:

1. Each pair of TSVs requiring shielding define a unique region. A region is a set of points on the layout grid such that if a shield were placed at this point it would significantly shield the TSV pair represented by the region.
2. A min cost flow (MCF) problem is formulated which chooses a point from each region such that placing a shield at that point would offer the greatest improvement to TS.
3. The MCF problem is solved iteratively by perturbing the costs of the MCF formulation between iterations so as to reduce the number of shields. The details concerning the MCF formulation and iteration are explained in Section 4.1.
4. The result from the MCF iteration is further improved by low temperature simulated annealing.
5. If the change in TS is below some threshold then the algorithm terminates, if not it repeats from step 1.

Our proposed algorithm starts by generating the initial shielding coefficients between each pair of TSVs using Equation (10). The algorithm defines a region for each pair of signal TSVs whose shielding coefficient is less than coupling threshold T. More precisely, the region corresponding to a TSV pair separated by distance d would consist of all points with perpendicular distance $y_{max}(d) + \rho$ or less, as shown in Figure 7(b). The value $y_{max}(d)$ is the maximum value of y such that placing a shield at distance y could potentially shield a TSV pair of length d above threshold T. The algorithm parameter ρ can be tuned to control the size of regions. The parameter ρ must be large enough to offer a reasonable region size when y_{max} is small, but not large enough to include points that offer poor shielding. The mapping between points and regions is used to set up an iterated MCF formulation which is described in Section 4.1.

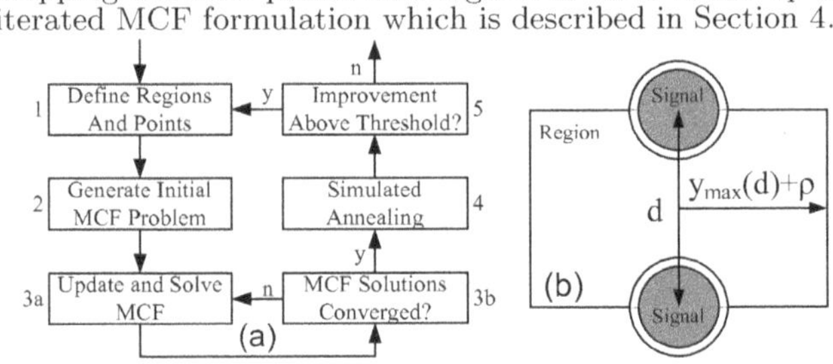

Figure 7: (a) Flowchart of proposed algorithm (b) Visualization of points and regions

4.1 MCF Formulation

We define a flow network which has a node for each region (region node), a node for each point (point node), a source and sink node. A zero-cost, unit-capacity arc exists between the source and each region node. A zero-cost arc is inserted between each region node and all the point nodes that make up that region, and a finite-cost arc connects each point node to the sink. In this formulation, m units of flow are inserted at the source, where m is the number of regions. Each unit of flow is routed from the source to a unique region node and then to the sink through some intermediate point node. Each unit of flow represents a TSV pair's need for shielding. TSV pair i's need for shielding is satisfied by placing a shield at the point corresponding to the point node through which flow passes from region node i to the sink. The described network topology is illustrated in Figure 8, where u represents capacity and c represents cost.

The cost of each arc between a point node and the sink is set such that the point nodes chosen to carry flow correspond to the points in the layout where shields should be placed to sufficiently shield the design while minimizing the total number of shields. Initially, the cost on the arc between each point node and the sink is inversely proportional to the

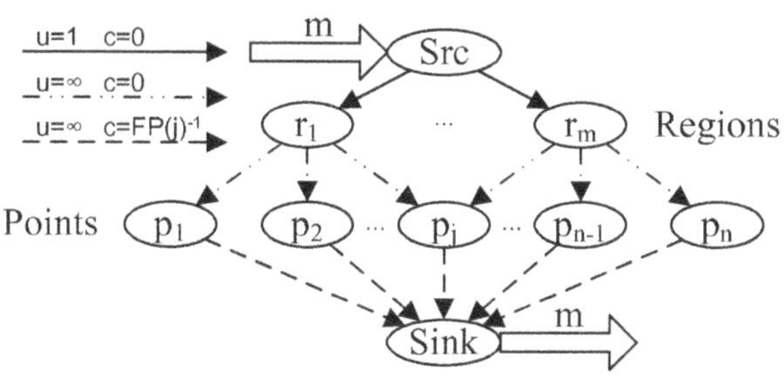

Figure 8: Example MCF formulation

degree of coupling improvement provided by placing a shield at that point (formally introduced as *SP* in the following paragraphs). In this way the initial solution will choose to place shields where they offer the most benefit. However, it is possible that significantly less shields could be used while only slightly decreasing the quality of the solution. On subsequent iterations, the cost on arcs between point nodes and the sink are **inversely proportional to the amount of flow on that arc during the last iteration**. In this way we motivate multiple region nodes to send flow to the same point node by lowering a point node's cost when it has already been chosen in the previous iteration. This iterative approach reduces the number of shields added to the system while maintaining a sufficient level of shielding. Each MCF formulation is solved using `MCF` [3].

Each point within a region will effect the coupling of the associated TSV pair to a different degree, with points closer to the center line of the region providing better shielding. We quantify the degree of improvement offered to region i by point j in Equation (12). The value $w(i,j)$ is the shielding potential of point j to region i. It is defined as the ratio between (1) the increase in $SC(i)$ (while ignoring overlap) due to placing a shield at point j and (2) the total amount of shielding that region i needs to meet the coupling tolerance T. The value y_{ij} is the perpendicular distance between the point j and the line passing through both signal TSVs represented by region i. The value d_i is the distance between the two signal TSVs. The cost of adding a shield to a design should be proportional to the amount of shielding it would provide to the *entire* design. Thus we define the total shielding potential $SP(j)$ of point j as the sum of $w(i,j)$ across all regions. This is shown in in Equation (13). During the first iteration of the MCF formulation, the cost on the arc between point node j and the sink is the multiplicative inverse of $SP(j)$. By assigning arc costs in this way, points with large values of total shielding potential $SP(j)$ will be chosen for shield placement during the first iteration. Then on subsequent iterations the costs are perturbed such that region nodes are incentivized to send flow through point nodes corresponding to points that can shield multiple regions simultaneously in order to reduce the total number of shields added to a design.

$$w(i,j) = \frac{S_s(y_{ij}, d_i)}{T - SC(i)} \quad (12) \qquad SP(j) = \sum_{\forall i} w(i,j) \quad (13)$$

The cost of adding a shield to a layout is independent of how many TSV pairs that shield effects because a shield consumes the same amount of area regardless of its shielding potential. To reflect this fact in the MCF formulation, the cost of sending flow through a point node (i.e. placing a shield at that point) should be independent of the amount of flow passing through it. However, in an MCF formulation, the cost of sending flow across an edge is by definition proportional to the amount of flow across it. To be able to use an MCF formulation, but still accurately model cost, we propose iteratively solving and perturbing the MCF problem until the solutions converge.

The idea behind perturbing is the following: if the flow through a point node increases from one iteration to the next, the respective cost should decrease such that the cost per shield (flow through point node i multiplied by the cost of point node i) remains constant. This is accomplished by defining a flow potential for each point and using this value instead of shielding potential to calculate the edge costs in each iteration beyond the first. The flow potential $FP(j)$ of point j is the sum of shielding potential across all regions that sent flow through point j in the previous iteration. To formally state this we define I_j as the set of all regions that sent flow to point j during the previous iteration and then present Equation (14). Thus when the amount of flow through point node j increases, $FP(j)$ also increases, which means the cost on the arc between point node j and the sink decreases. The max operator in Equation (14) prevents the costs on an edge from becoming infinite when it does not carry flow. We initialize I_j as the set of all regions, so on the first iteration $FP(j) = SP(j)$.

$$FP(j) = \max\left[1, \sum_{i \in I_j} w(i,j)\right] \quad (14)$$

4.2 Comparison to Liu Algorithm

Our proposed algorithm has overcome the flaws in the Liu algorithm, which were enumerated at the beginning of Section 4. The geometric model proposed in this work defines a shielding coefficient between each *pair* of TSVs, whereas the Liu algorithm defines a coupling intensity for each *individual* TSV. Thus our proposed algorithm is aware of the source of coupling and can place shields with this in mind, whereas the Liu algorithm simply places shields all around a TSV suffering from high coupling intensity.

In light of our geometric model's ability to identify coupling sources, our proposed algorithm lifts the restriction that shields must be placed close to a signal TSV. This allows the algorithm to place a shield such that it simultaneously shields multiple TSV pairs. Thus the algorithm can consider the trade-off between using one TSV that can shield multiple pairs, but only to a moderate degree on each pair, or using multiple TSVs that each shield a single pair to a high degree. This will prevent our proposed algorithm from overshielding when a shielding coefficient is only slightly under the coupling threshold.

Finally, when the proposed algorithm calculates the initial shielding coefficient of each pair of TSVs, it considers all the other signal TSVs to be shields, which is to say it assumes that the other signal TSVs are not changing voltage at the same time as the TSVs under consideration. This assumption is justified because even if the other signal TSVs are changing at the same time there is an equal probability that they switch in either direction, so on average these coupling sources cancel each other out.

5. EXPERIMENTS

In this work we execute our proposed algorithm and the Liu algorithm on TSV layouts which are generated from the IBM-Place 2.0 Benchmark Suite [2]. Each benchmark in the suite consists of a list of cells, and from this list we can calculate the points in the layout which are not covered by any cells (i.e. the white space). Even though IBM-Place benchmarks are 2D in nature, we use their whitespace distribution to generate TSV locations in a single layer of a 3D IC, which are used as inputs to the algorithms. For each benchmark, a set of TSVs are randomly allocated from the white space distribution.The number of signal TSVs in each benchmark layout is presented in the two leftmost columns of Table 2.

5.1 Metrics

In order to compare the two algorithms, it is necessary to define the metrics we will use for comparison. Total shielding TS was defined in Equation (11) as the sum of all shielding coefficients across all TSV pairs. The change in TS after applying the shielding scheme generated by each algorithm will be the primary metric used to compare the two algorithms. Since TS increases as more shields are added, we compare the total shielding of both algorithms when they are constrained to insert the same number of shields, in order to give a fair comparison. It is worth noting that overshielding a TSV pair cannot increase TS due the the max operator in Equation (11). Thus the ideal value of TS would be $T \times N^2$, where N is the number of signal TSVs in the design. For each benchmark, we calculate the

		Change in Total Shielding				Number Below Threshold			Total Shields Added		
Layout	# TSVs	Liu	Ours	Ideal	Improve	Liu	Ours	Reduce	Liu	Ours	Less
ibm01	83	19.5	140.6	205.9	619%	84	60	29%	155	10	94%
ibm02	97	31.3	193.7	294.4	518%	124	92	26%	233	13	94%
ibm07	144	51.9	179.4	248.0	246%	102	69	32%	185	22	88%
ibm08	184	36.3	177.8	212.8	389%	94	50	47%	192	19	90%
ibm09	186	69.8	262.1	355.7	276%	141	91	35%	271	33	88%
ibm10	244	79.5	260.3	329.1	227%	145	77	47%	273	40	85%
ibm11	177	127.6	307.7	477.0	141%	175	134	23%	303	73	76%
ibm12	219	43.8	245.1	337.1	459%	179	91	33%	302	25	92%
Avg.					359%			34%			88%

Table 2: Experimental Results

maximum possible increase in TS and compare this value to the increase in TS offered by each shielding algorithm. Some alternative ways to compare the algorithms would be to consider the difference in shield count when both algorithms produce the same value for TS, or the number of shielding coefficients still below threshold after adding a set number of shields to each layout.

5.2 Results and Discussion

Figures 9(a) and 9(b) show zoomed in views of the same spot on the `ibm01` layout after applying the shielding schemes generated by the Liu algorithm and our proposed algorithm. In order to provide a fair comparison, these two schemes both use 155 shield TSVs. The figures illustrate that the Liu algorithm is severely overshielding some TSVs in the layout while leaving other pairs completely unshielded. Our proposed algorithm is able to effectively shield many more coupling pairs without much overshielding because it considers the position of *each* shield in the design whereas the Liu algorithm places shields in groups of eight and quickly reaches its shield quota before sufficiently shielding the layout. Furthermore, our proposed algorithm is able to achieve coupling reduction to a degree not possible with the Liu algorithm by placing many shields along the line running through a pair of signal TSVs. This effect can be seen in the top right corner of Figure 9(b).

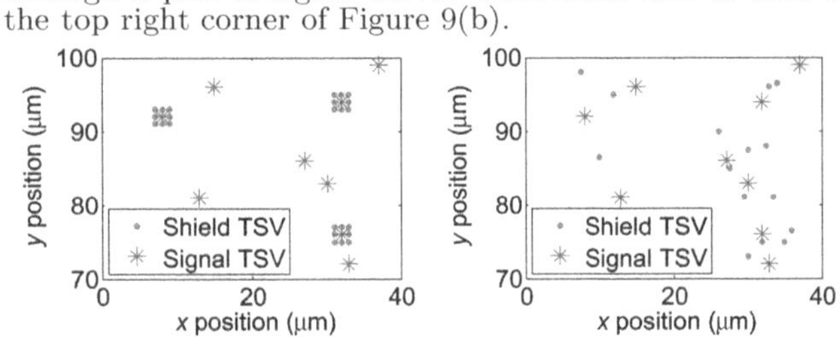

Figure 9: Shield placement schemes produced for `ibm01` (a) Liu algorithm (b) proposed algorithm

A comparison of the total shielding improvement due to the shield placement generated by each algorithm is presented in columns 3-7 of Table 2. Our proposed algorithm improves TS on average 4.59x more than the Liu algorithm. The maximum possible improvement to TS is also listed in the table, in order to put the results from the two algorithms in context. A comparison of the number of TSV pairs with shielding coefficients less than threshold is presented in columns 8-10 of Table 2. On average our proposed algorithm results in only 66% as many signal TSV pairs with coupling parameters below threshold.

Finally a comparison of total shields added is presented in columns 11-13 of Table 2. In this case a fair comparison would be the number of shields needed by either algorithm to produce the same improvement to TS. So the reported number of TSVs used in our proposed algorithm is really the number of TSVs required to match the results of the Liu algorithm. On average our proposed algorithm can achieve the same degree of shielding using 88% less shields.

Figure 10(a) shows a plot of total shielding improvement vs number of shields added in `ibm01`, and Figure 10(b) shows the cumulative distribution of shielding coefficients after adding 155 shields. We can see that the Liu algorithm gets much worse improvement per shield and has more shielding coefficients below threshold as compared to our proposed algorithm.

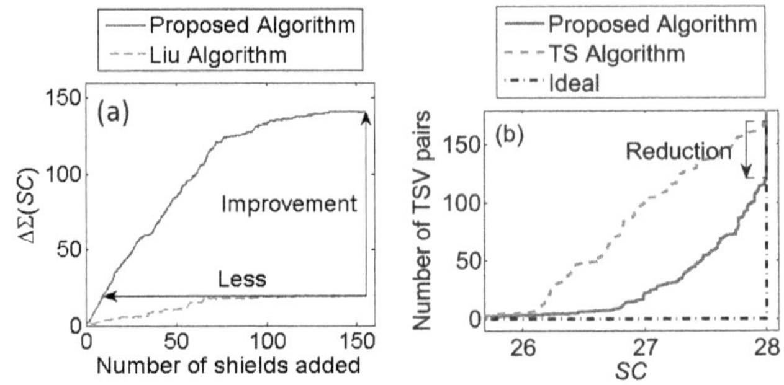

Figure 10: Results from `ibm01` when $T = 28\ dB$ (a) Total shielding improvement (b) Cumulative distribution of shielding coefficients when 155 shields are used.

6. CONCLUSIONS

In this work we propose a geometric model of TSV coupling, and consider TSV shielding. The geometric model can be quickly evaluated, making it suitable for chip-scale shield placement optimization. An algorithm for placing shields in a design that considers the a shields effect on *all* coupling pairs is presented. Our algorithm is able to improve TSV coupling an average of 4.59x more than the algorithm presented in [9] when using the same number of shields, or it can match their results while using an average of 88% less shields.

7. REFERENCES

[1] Ansys HFSS. `http://www.ansys.com/Products/Simulation+Technology/Electromagnetics`.

[2] IBM-Place 2.0 Benchmarks. `http://er.cs.ucla.edu/benchmarks/ibm-place2/`.

[3] MCF - A network simplex implementation. `http://typo.zib.de/opt-long_projects/Software/Mcf/`.

[4] J. Cho, J. Kim, T. Song, et al. Through silicon via (TSV) shielding structures. In *Electrical Performance of Electronic Packaging and Systems (EPEPS), 2010 IEEE 19th Conf. on.*

[5] J. Cho, E. Song, K. Yoon, et al. Modeling and Analysis of Through-Silicon Via (TSV) Noise Coupling and Suppression Using a Guard Ring. *Components, Packaging and Manufacturing Technology, IEEE Trans. on.*

[6] N. Khan, S. Alam, and S. Hassoun. Mitigating TSV-induced substrate noise in 3-D ICs using GND plugs. In *Quality Electronic Design (ISQED), 2011 12th Int. Symposium on.*

[7] N. Khan, S. Alam, and S. Hassoun. Through-Silicon Via (TSV)-induced noise characterization and noise mitigation using coaxial TSVs. In *3D System Integration, 2009. 3DIC 2009. IEEE Int. Conf. on.*

[8] Y.-J. Lee and S. K. Lim. Timing analysis and optimization for 3D stacked multi-core microprocessors. In *3D Systems Integration Conf. (3DIC), 2010 IEEE Int.*

[9] C. Liu, T. Song, J. Cho, et al. Full-chip TSV-to-TSV coupling analysis and optimization in 3D IC. In *Proc. of the 2011 48th Design Automation Conf.* ACM.

[10] C. Liu, T. Song, and S. K. Lim. Signal integrity analysis and optimization for 3D ICs. In *Quality Electronic Design (ISQED), 2011 12th Int. Symposium on.*

[11] G. Loh. 3D-Stacked Memory Architectures for Multi-core Processors. In *Computer Architecture, 2008. ISCA '08. 35th Int. Symposium on.*

Thermal Stress Aware 3D-IC Statistical Static Timing Analysis

Bing Shi and Ankur Srivastava
University of Maryland, College Park, MD, USA
{bingshi, ankurs}@umd.edu

ABSTRACT

It is widely known that fabrication and thermal variations influence circuit delay. In three dimensional circuits (3D-ICs), due to the incorporation of through-silicon-vias (TSVs), thermal stress also becomes an increasing contributor to gate delay. As a result, thermal variations cause not only direct impact to the circuit parameters, they also cause 3D-IC stress variations resulting in an additional source of variability which needs to be accounted for in statistical static timing analysis (SSTA). In this paper, we study the impact (both direct and indirect – through thermal stress) of thermal variations on gate delay, and propose Monte Carlo (MC) simulation based fabrication, temperature and thermal stress variations aware SSTA methodology that accounts for this complex effect. We show that thermal stress variations cause extra $-4 \sim 10\%$ delay changes compared with the SSTA that only accounts for fabrication and direct thermal impacts, indicating that thermal stress should be considered together with direct temperature effect. We then develop a more efficient canonical thermal stress aware SSTA method that can achieve good accuracy and 843x speedup compared with MC based method.

Categories and Subject Descriptors

B.7.2 [**Integrated Circuits**]: Design Aids

Keywords

3D-IC, TSV, thermal stress, SSTA

1 Introduction

In the 3D-IC, TSVs are incorporated to enable interlayer communications and delivery of power/ground. Copper, due to its low resistivity, is a commonly used material for TSV fill. Since the chips are usually annealed at the temperature level much higher than their operating temperature, when cooling down from the annealing temperature, thermal stress occurs due to the coefficient of thermal expansion (CTE) mismatch between the TSV fill material (e.g. copper) and silicon. Thermal stress influences carrier mobilities, hence affecting gate delays. The impact of stress on gate/circuit delay is complex, depending on the intensity of stress, the location of gates and TSVs and the type of transistor (NMOS or PMOS). [1] studied the impact of thermal stress on mobilities, and performed static timing analysis considering stress effect. On the other hand, temperature also affects the gate delay by changing the mobility and threshold voltage [2]. From these studies, we can see that temperature causes both *direct* and *indirect* impacts on gate delay. The direct impact is that temperature directly influences gate delay through changing the mobility and threshold voltage. The indirect impact is that temperature influences mobility (and hence gate delay) indirectly through thermal stress.

It is widely known that the fabrication and thermal variations have significant impact on circuit delay. As a result, statistical approaches that account for fabrication and thermal variations have been developed for timing analysis (SSTA) [3][4]. However the existing works on thermal aware SSTA only consider the direct impact of thermal variations on gate delay, while ignoring the indirect impact of temperature – thermal stress effect. While in conventional 2D-ICs the impact of thermal stress is minimal, in 3D-ICs, hundreds of TSVs are distributed across the chip [5], and TSV induced thermal stress becomes an increasing contributor to gate delay. TSV induced stress also varies due to thermal variations. Such phenomenon adds further variations to gate delay and complicates the timing analysis in 3D-ICs.

In this work, we investigate fabrication variability, temperature and thermal stress aware SSTA that accounts for these complex influences. Specifically, the contributions of our work are as follows:

1. The widely known thermal impact on circuit delay is monotonically increasing – higher temperature leads to increased delay. However, the impact of stress on gate/circuit delay is more complex, depending on the intensity of stress, location of gates and TSVs, as well as the type of transistor. By studying the thermal stress effect on gate delay, we found that, due to the existence of thermal stress, the impact of temperature on *longest path delay* is more complex and no longer monotonically increasing necessarily (as Fig. 1 shows).

2. To account for the impact of temperature and thermal stress variations, we develop Monte Carlo (MC) simulation based thermal and stress aware SSTA methodology. Through our thermal and stress aware SSTA, we show that thermal stress variations cause extra $-4 \sim 10\%$ delay changes compared with the SSTA that only accounts for fabrication and direct thermal impacts, indicating that thermal stress has significant impact on the path delay to the extent that it should be considered together with direct temperature effect.

3. In order to develop a computationally efficient approach for fabrication, thermal and stress aware SSTA, we develop a strategy for using the linear canonical SSTA. To apply canonical SSTA, we develop a linear gate delay model that can account for both direct and indirect temperature effects. Using the linear gate delay model, the gate delay distribution is represented using canonical form. We then develop canonical thermal and stress aware SSTA approach. Comparing the delay distributions estimated by canonical SSTA and MC method, the average error of mean and standard deviation are 0.18% and 4.87% respectively. Our canonical SSTA also achieves 843x speedup.

The organization of this paper is as follows. In section 2, we introduce thermal stress and the impact of fabrication

GLSVLSI'13, May 2–3, 2013, Paris, France.

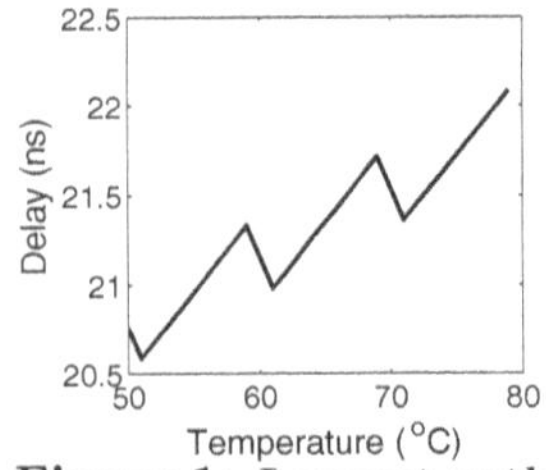

Figure 1: Longest path delay for different thermal conditions

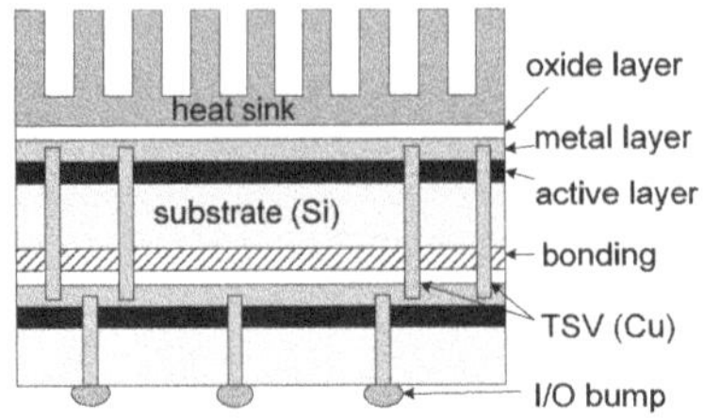

Figure 2: Basic structure of 3D-IC with TSVs

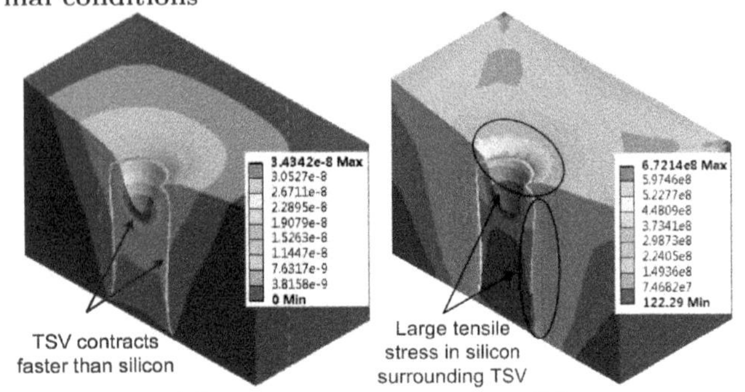

(a) (b)

Figure 3: (a) Deformation and (b) Tensile stress inside and surrounding TSVs

and thermal variations on gate delay. We study the thermal stress effect on gate and path delay in section 3. Section 4 gives the motivation of our work. We explore MC based and canonical thermal stress aware SSTA in sections 5 and 6. The experimental results are given in section 7.

2 Background

2.1 TSV induced thermal stress

Fig. 2 shows an example of a two-tier stacked 3D-IC. Each active layer contains functional units such as cores, caches. TSVs are inserted in 3D-IC to enable signal/power/ ground interconnections among different layers.

The TSV fill material such as copper (Cu) and silicon substrate have different CTE. After cooling down from the annealing temperature, both TSV and silicon substrate contract/deform. However, copper contracts much faster than silicon (due to CTE mismatch) and pulls the silicon surrounding TSV/silicon interface. This causes tensile stress in silicon surrounding TSVs. Similarly, the CTE mismatch also causes tensile stress inside TSVs. To illustrate the phenomenon, we simulated a basic TSV structure surrounded by silicon at 50^oC using ANSYS (assuming the annealing temperature is 275^oC) [6]. The resulting deformation and stress field are given in Fig. 3, which shows that large tensile stress occurs in the silicon near the TSV/silicon interface.

A few recent works have tried to characterize or model the TSV induced thermal stress [7][8]. For instance, [8] develops closed form expression to calculate thermal stress and [7] proposes a finite element method (FEM) combined with superposition based approach for full chip stress estimation.

2.2 Effect of variations on gate delay

Fabrication variations: The delay of a gate is a function of many circuit parameters such as oxide thickness, effective channel-length, threshold voltage and carrier mobility. These parameters vary due to the fabrication process, causing variations in gate delay. Hence statistical timing analysis approaches are used to account for these variations [3].

Some of the circuit parameters, such as the threshold voltage and mobility are sensitive to the temperature of transistor and the stress applied.

Impact of thermal variation on gate delay: The chip-level runtime thermal profile exhibits both temporal and spatial variations. This because some modules dissipate more

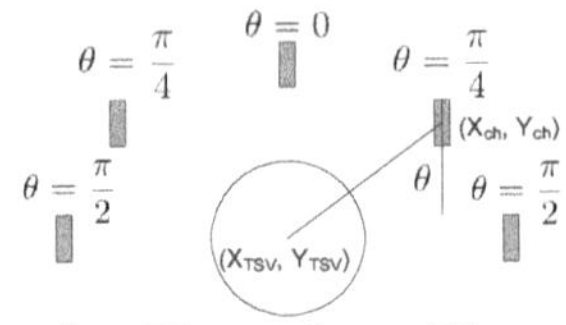

β	NMOS	PMOS
$\theta = 0$	0.5	0.6
$\theta = \pi/4$	0.75	-0.1
$\theta = \pi/2$	1	-1

Figure 4: Channel position with respect to TSV (θ), and values of orientation coefficient $\beta(\theta)$

power than others, which results in spatial thermal variation. Moreover different applications and data result in different switching activities, leading to temporal thermal variations.

Temperature influences circuit parameters such as threshold voltage v_{th} and mobility μ, and therefore chip-level thermal variations will result in further gate delay variations in addition to fabrication variations. Generally increase in temperature results in reduced threshold voltage and mobility. The dependence of mobility μ and threshold voltage v_{th} on temperature is modeled as [2]:

$$\mu(T) = \mu(T_0)(\frac{T_0}{T})^{\epsilon}, \qquad v_{th}(T) = v_{th}(T_0) - \delta(T - T_0) \qquad (1)$$

where T_0 is the reference temperature, ϵ and δ are constants.

Due to the impact of thermal variations on circuit delay, thermal aware timing analysis approaches have been investigated [9][4]. For instance, [9] assumes the voltage and thermal variations are deterministic, and uses the thermal profile to adjust gate delay and improve timing analysis. [4] proposes thermal aware statistical timing analysis algorithm that accounts for both fabrication and thermal variations.

3 Thermal stress effect on gate delay: a detailed study

As shown in [10][1], the temperature induced stress (thermal stress) also influences carrier mobilities significantly, hence changing the gate delay. The mobility change ($\Delta\mu$) due to thermal stress depends on the intensity and orientation of applied stress, and the type of transistor [10][1]:

$$\frac{\Delta\mu_j}{\mu_j} = \Pi \sum_{i \in TSV(j)} S_{rr,ij} \times \beta(\theta_{ij}) \qquad (2)$$

Here Π is a coefficient indicating the sensitivity of mobility change to applied stress, $TSV(j)$ represents the set of TSVs that influence gate j, $S_{rr,ij}$ is the radial stress applied on gate j caused by i-th TSV, θ_{ij} is defined as the angle between the line connecting the center of TSV i and transistor j, and the channel direction of transistor j (as Eq. 3 and Fig. 4 show).

$$\theta_{ij} = \arctan\frac{|X_{TSV,i} - X_{ch,j}|}{|Y_{TSV,i} - Y_{ch,j}|} \qquad (3)$$

where $(X_{TSV,i}, Y_{TSV,i})$ and $(X_{ch,j}, Y_{ch,j})$ indicate the coordinates of TSV center and transistor channel.

In Eq. 2, β is an orientation coefficient whose value depends on θ and also the type of transistor. Fig. 4 gives the value of β for different types of transistors and θ values.

These studies show that temperature causes two effects on gate delay. The first effect is that temperature *directly* influences the gate delay through changing the mobility and threshold voltage. The second effect is that temperature influences mobility (and hence gate delay) *indirectly* through thermal stress.

To study the *direct* and *indirect* impacts of temperature on gate delay, we tested the worst case gate delays (the worst case between rising and falling delays) under different temperature levels and at different positions with respect to a single TSV using Alg. 1 (we used a basic inverter):

This detailed simulation study enables us to evaluate the impact of temperature on gate delay at different angles and distances. Fig. 6(a) and 6(b) show the gate delay increase versus θ when the gate is located at the distance of $1.5\mu m$ and $30\mu m$ from the TSV/silicon interface at different temperature levels. The figures show that, for the same temperature level, the gates at different angles experience different delay increases. θ plays an important role in deciding gate delay increase. For instance, when $T = 22^oC$ at

Algorithm 1 Characterizing impact of temperature and TSV position on gate delay

1. Instantiate a testing structure as Fig. 5 shows in ANSYS;
2. Set uniform temperature environment $T = 22^oC$ and annealing temperature $T_a = 275^oC$;
3. Calculate the stress field $\vec{S}$ at temperature T;
4. For $\theta = 0$ to 2π:
4.1. Calculate mobility and threshold voltage for this $(T, \vec{S}, \theta)$ combination using Eq. 1 and 2 (assuming $90nm$ technology);
4.2. Calculate gate delay increase w.r.t. gate delay calculated at reference temperature $T_0 = 22^oC$ and ignoring stress effect;
5. If $T < 100^oC$, $T = T + \Delta T$, repeat steps 3-5.

distance $= 1.5\mu m$, gate delay increases by 80% at $\theta = \pi/2$ and $3\pi/2$, while it *reduces* by 20% at $\theta = 0$, π and 2π. Moreover, the gate delay variation is large near the TSV, and reduces as the distance to TSV increases. For example, the gate delay increase changes from -20% to 80% for different θ when *distance* $= 1.5\mu m$, while it changes from -5% to 20% when *distance* $= 30\mu m$. This is because the effect of thermal stress decays with distance.

Fig. 6(c) and 6(d) show the increase in gate delay versus temperature at different θ. Interestingly, when the gate is close to TSV, increase in temperature either increases or reduces the gate delay depending on θ. That means, θ also plays a significant role in deciding the impact of temperature on gate delays. For example, at $\theta = 0$, higher temperature results in gate delay increase just as the conventional *direct* thermal impact. However, at $\theta = \pi/2$, increase in temperature actually reduces gate delay, which is opposite to the conventional trend. This is because, at $\theta = \pi/2$, stress results in degradation in PMOS hole mobility. When temperature increases (hence thermal stress reduces), such degradation effect on hole mobility is mitigated, which indicates reduced gate delay. When the gate is close to TSV, this stress effect dominates, and hence increase in temperature leads to gate delay reduction in this case. As the distance increases, thermal stress decays and the conventional *direct* thermal impact begins to take hold, which increases delay with higher temperature. This finding will have significant impact on how timing analysis is performed, since now we have gates that are against the conventional trend.

Summary of conclusions:

Based on these observations, we arrive at the following conclusions about the impact of temperature on delay.

1. Conventionally, if we ignore the thermal stress, the impact of temperature on gate and path delay is a monotonically increasing function – higher temperature leads to increased delay. However, due to the existence of TSV induced thermal stress in 3D-ICs, the impact of temperature on gate and path delay becomes more complex.
2. Based on their locations, the gates can be categorized into three types:
   (a) For those gates far away from TSV, thermal stress is small and the *direct* impact of temperature dominates. Hence high temperature usually leads to increased delay for such gates.
   (b) For those gates that are far but not very far (e.g. $20 - 50\mu m$) from TSV, the thermal impact is a combination of *direct* and *indirect* impacts.
   (c) For the gates near TSVs, both temperature and thermal stress influence gate delay. The influence not only depends on temperature level, but also strongly depends on the position of gates and TSVs. For those gates around $\theta = \pi/2$ and $3\pi/2$, as Fig. 6(c) shows, gate delays are actually reduced due to increased temperature, which is opposite to the conventional trend.
3. Because of such complex influence, the impact of temperature on path delay is no longer monotonic. When temperature increases, some gates along the path may become slower, while some other gates may become faster. Hence the path delay depends on the location of gates along the path, and is no longer monotonic.

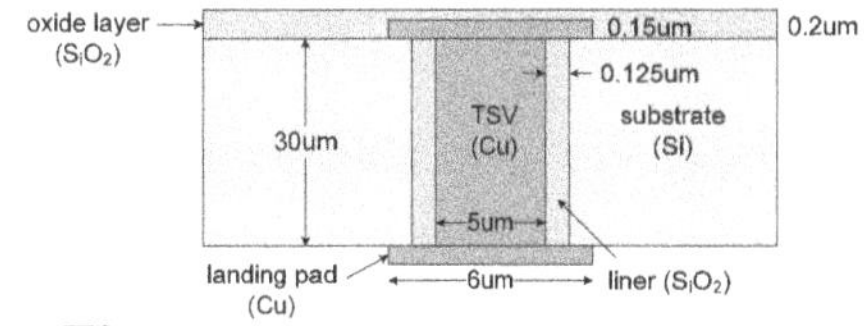

Figure 5: Testing TSV structure

4 Motivation

The existing works on thermal aware SSTA only focus on the *direct* impact of thermal variation on delays, while ignoring the *indirect* impact of temperature - thermal stress effect. The existence of TSVs in 3D-ICs causes thermal stress on the chip, and the stress field varies due to thermal variations. Due to the thermal stress, the impact of temperature on circuit delay is more complex (depending on stress intensities and angles) and no longer monotonically increasing. Such phenomenon changes the timing analysis in 3D-ICs.

Furthermore, since the thermal stress is a function of temperature, the variations in temperature due to the fluctuations in application workloads cause variations in stress field. In our experiment, we found that 80^oC thermal variation (from 20^oC to 100^oC) results in about 30% thermal stress variations near the TSV. This adds further variations to gate delays. Especially for those gates close to TSVs, the thermal stress induced gate delay variation is even more severe than direct temperature impact. As the number of TSVs incorporated in 3D-IC increases, more gates are affected by this new phenomenon, and stress becomes an increasing contributor to gate delays, thereby further complicating the 3D-IC timing analysis. In Yang's work [1], they investigated thermal stress aware timing analysis. However they focused on the static case, and failed to consider the stress variation caused by thermal variation and also ignored the direct impact of temperature on gate delay.

In our work, we investigate SSTA methodologies that account for fabrication, temperature and thermal stress variations. We first develop Monte Carlo simulation based SSTA methodology that accounts for the effects of temperature, thermal stress and fabrication randomness in section 5. In section 6, we approximate the gate delay as a linear function of temperature and develop canonical thermal and stress aware SSTA algorithm.

5 Monte Carlo simulation based methodology for thermal and stress aware SSTA

5.1 Simulation setup:

In this section, we investigate Monte Carlo simulation based methodology for thermal and stress aware SSTA. Assuming we are given the 3D-IC structure where the gates are already placed and TSVs are allocated. In the Monte Carlo simulation, each sample corresponds to one 3D-IC thermal profile sample together with fabrication variable samples. The approaches to generate fabrication variable and thermal profile samples will be explained in the following two subsections. The Monte Carlo simulation calculates the longest path delay for each sample using the framework in Fig. 7.

1) Modeling fabrication randomness: The gate delay depends on circuit parameters such as oxide thickness t_{ox}, effective channel-length L_{eff}, threshold voltage v_{th} and carrier mobility μ, etc, which are impacted by fabrication randomness. These fabrication randomness can be incorporated by expressing gate delay using the following form:

$$delay(\vec{F}, R) = d_0 + \sum_{\forall i \in PC(\vec{F})} a_{F,i} F_i + a_R R \quad (4)$$

where d_0 is the nominal gate delay when there is no fabrication variations. It is a function of the aforementioned circuit parameters $d_0 = f(\mu, v_{th}, t_{ox}, L_{eff})$. $F_{i,\forall i}$ are principle components (PC) of global fabrication random variables, and $a_{F,i}$ indicates the sensitivity of gate delay to the variation of F_i. R represents the local randomness associated with this gate. This model is similar to other statistical gate delay models [11].

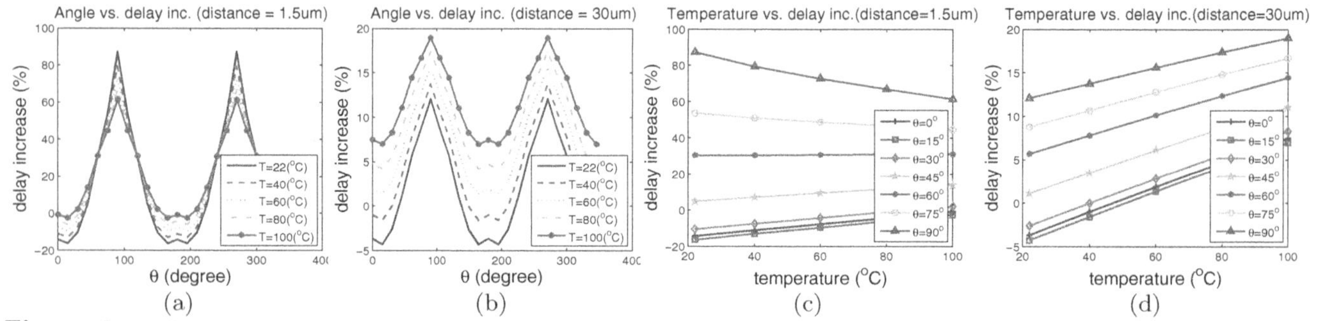

Figure 6: Effect of temperature and position on gate delay, (a)(b) gate delay versus θ at $distance = 1.5\mu m$ and $distance = 30\mu m$, (c)(d) gate delay versus $temperature$ at $distance = 1.5\mu m$ and $distance = 30\mu m$

2) Modeling thermal variation and its impact on gate delay: In direct thermal effect, temperature influences gate delay through mobility and threshold voltage. Assuming the 3D-IC is discretized into $I \times J \times K$ grids. The thermal profile is basically a vector containing the temperature at all grids. We assume the probability density functions (PDF) of thermal variables at all grids are known. Various simulation and analytical techniques can be carried out to extract the random characteristics of the temperature variables on each grid including the mean, variance, covariance etc [4]. Temperature distribution can also be computed from power profile if the random characteristics of power profile is known. In order to do that, we could use power data combined with 3D-IC thermal resistive model to estimate the temperature profile statistics [12].

Given the thermal profile sample, the mobility and threshold voltage changes of each gate is calculated using Eq. 1.

3) Thermal stress profiling and its impact on gate delay: In order to estimate the impact of thermal variation on stress variation, and hence estimating the associated gate delay variation, for each temperature sample, which corresponds to a thermal profile vector discussed in the previous subsection, we need to calculate the impact of stress induced by each TSV on each gate. Eq. 2 models the impact of stress on mobility (it is rewritten here):

$$\frac{\Delta\mu_j}{\mu_j} = \Pi \sum_{i \in TSV(j)} S_{rr,ij} \times \beta(\theta_{ij})$$

To calculate the mobility change of each gate due to stress, for each TSV (in a certain region of vicinity from this gate), we calculate its impact on stress and mobility of this gate independently, and then add them up as Eq. 2 shows. Hence for each TSV, we should compute its associated stress field for the given temperature sample separately.

Once we obtained the stress field associated with each TSV, we calculate the stress induced mobility change. The mobility change due to thermal stress depends on the intensity of applied stress as well as the locations of transistors and TSVs as Eq. 2 and Fig. 6 shows. To calculate the mobility change of each gate j, we first find the radial stress ($S_{rr,ij}$) applied on this gate caused by each neighboring TSV i based on the stress field calculated for TSV i. We also calculate the angle (θ_{ij}) between gate j and TSV i based on their locations. Given the intensity of radial stress $S_{rr,ij}$ (whose calculation will be discussed subsequently) and angle θ_{ij}, we can calculate the stress induced mobility change of gate j using Eq. 2.

Given the mobility and threshold voltage changes due to temperature and stress obtained from this and previous subsections, the new mobility μ' and threshold voltage v'_{th} can be easily calculated. The new gate delay model considering fabrication, thermal and stress variations is expressed as:

$$delay(\vec{F}, \vec{T}, \vec{S}, R) = d'_0(\mu'(\vec{T}, \vec{S}), v'_{th}(\vec{T})) + \sum_{\forall i \in PC(\vec{F})} a_{F,i}F_i + a_R R \quad (5)$$

Here $\vec{S} = (\vec{S}_1, ..., \vec{S}_{TSV})$, where $\vec{S}_i$ represents the stress field vector for TSV i. $\vec{T}$ represents the thermal profile vector.

Fundamentally, we need to compute the stress field for each TSV separately.

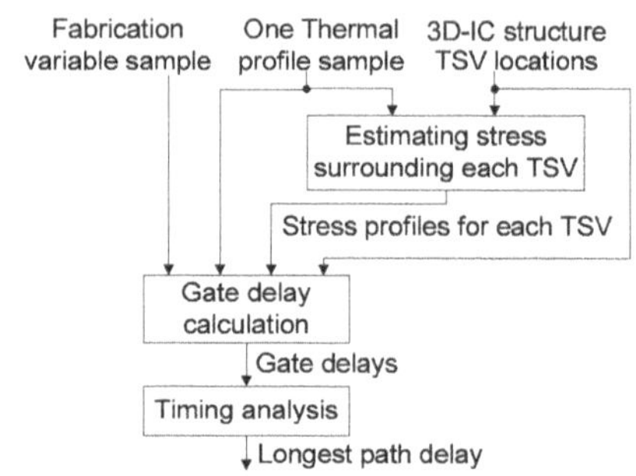

Figure 7: Monte Carlo based thermal and stress aware SSTA for one sample

5.2 Modeling and estimation of thermal stress

While some works are interested in getting the full chip stress profile, in this problem, Eq. 2 forces us to look at the stress field associated with each TSV separately, since the mobility change not only depends on the intensity of stress, but also the angle between TSV and gate. Therefore, in this stress estimation step, we calculate the stress field caused by each individual TSV based on the given thermal profile sample.

Thermal stress is basically caused by deformations. The interdependency between temperature, deformation and thermal stress are governed by a set of complex equations. The details of these governing equations are given by [13]. In order to solve the stress field for a given TSV and thermal profile, we divide the 3D chip into grids assuming the temperature and stress within each grid are uniform. In this case, the thermal stress field $\vec{S}_i$ for each TSV i can be approximated as a linear function of temperature profile vector $\vec{T}$. This relationship is obtained by discretizing the models of temperature, deformation and thermal stress given in [13]. The details have been omitted for brevity.

$$\vec{S}_i = \mathbf{L_{i,1}} \cdot \vec{T} + \mathbf{L_{i,2}} \cdot T_0 \quad (6)$$

$\mathbf{L_{i,1}}, \mathbf{L_{i,2}}$ are matrices decided by the material properties and TSV location. Using this model, we can estimate the thermal stress field for each TSV by solving a set of linear equations. The stress field obtained by Eq. 6 is expressed in Cartesian coordinate system. To estimate stress-induced mobility change using Eq. 2, we should calculate the radial stress S_{rr} by converting the Cartesian system expression to Cylindrical system whose center is the TSV center. The conversion is given in [7].

This process is repeated for each TSV. Then for each gate we compute the mobility change and hence gate delay which is a function of the stress field and thermal profile sample.

Note that other approaches for thermal stress profiling such as FEM based approach proposed in [7] are also applicable to our SSTA framework. However, [7] fails to consider thermal variations surrounding each TSV.

5.3 Longest path delay estimation

Given the delay of each individual gate, we calculate the longest path delay of the circuit. This is done by firstly sorting the gates in topological order, and then propagate the signal arrival time from the primary inputs to primary

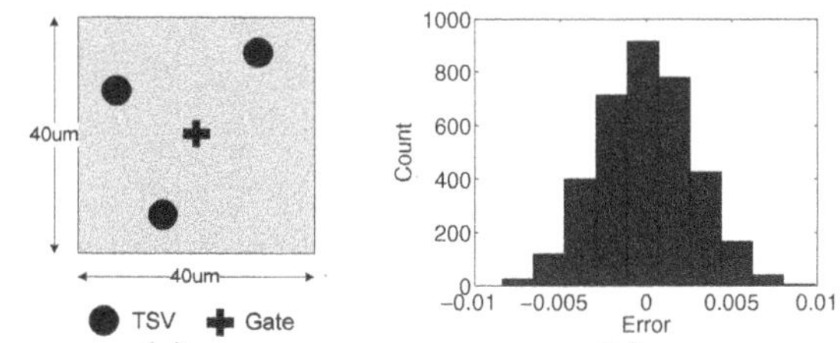

(a) (b)

Figure 8: (a) Testing structure for gate delay modeling, (b) Error of linear gate delay model

outputs based on the circuit topology and gate delays. The latest signal arrival time at primary outputs becomes the longest path delay.

6 Canonical SSTA approach

The MC based approach, though gives a good estimation of the circuit delay distribution, is very time consuming. Therefore, in this section, we investigate a canonical SSTA approach that is much faster than MC simulation and also gives very good approximation to the MC simulation result.

6.1 Modeling gate delay as linear function of temperature

To apply canonical SSTA approach [11], we attempt to approximate the gate delay as a linear function of temperature profile vector $\vec{T}$.

Firstly, to demonstrate the feasibility of linear gate delay model, we tested gate delay under random thermal profiles. In the test, we simulated with a $40 \times 40 \mu m^2$ area. A gate is located in the center and three TSVs are allocated as Fig. 8(a) shows. We generate 2500 random thermal profile samples and calculate the gate delay for each of the samples using the approach in section 5. Then we perform linear regression to compute the coefficients of the linear gate delay model. In order to test the accuracy of achieved linear model, we generate **another** 3600 thermal profile samples, compute the gate delay associated with each sample using both the approach in section 5 and the linear gate delay model, and then calculate the error between them. The distribution of error for the 3600 samples is plotted in Fig. 8(b). It shows that the error is concentrated near 0 and the maximum error is less than 1%, indicating that the gate delay can be approximated as a linear function of the temperature with good accuracy.

Note that our model incorporates both direct and indirect (stress induced) effects of temperature. Therefore, the gate delay is a function of not just it's own temperature, but also the temperature at its neighboring regions, since the temperature at neighboring TSVs affects the stress applied on this gate. This is why we model the gate delay as a linear function of the entire thermal profile vector $\vec{T}$.

6.2 Gate delay in canonical form

By modeling the gate delay as a linear function of temperature across the neighboring region, the gate delay distribution is expressed as:

$$delay(\vec{F}, \vec{T}, R) = \sum_{\forall j} a_{T,j} T_j + \sum_{\forall i \in PC(\vec{F})} a_{F,i} F_i + a_R R \quad (7)$$

Similar as Eq. 4, $F_{i,\forall i}$ represent principle components of global fabrication variables, R represents the local fabrication variable, $a_{F,i}$ and a_R are their sensitivity coefficients. $T_{j,\forall j}$ represent the temperature variables at each grid. Parameters $a_{T,j}$ are the coefficients in linear gate delay model. They indicate the impact of the temperature at each grid on gate delay.

Basically based on the location of each gate, the delay model would have to be generated separately (since delay depends on stress which is a location dependent quantity).

1) Canonical SSTA for Gaussian thermal profiles: If the distribution of thermal profile vector $\vec{T}$ is jointly Gaussian, we can represent the gate delay using canonical form. Since the canonical representation of gate delay needs uncorrelated principle components, we can perform a principle component analysis (PCA) on temperature and fabrication variations together to compute the PCA components and represent the gate delay with uncorrelated principle components.

By extracting the mean values of the principle components and scaling the sensitive coefficients, the canonical gate delay can be converted to standard form:

$$delay(\vec{Z}, R) = a_0 + \sum_{\forall i \in PC(\vec{F}, \vec{T})} a_i Z_i + a_R R \quad (8)$$

where $Z_{i,\forall i}$ denote the normalized principle components (for both temperature and fabrication variables) that follow standard Gaussian distribution $\mathcal{N}(0, 1)$. Now R represents the *normalized* local fabrication variable that follows standard Gaussian distribution as well.

Once we have obtained the gate delay in standard form, we can use the algorithm proposed in [11] to estimate the two basic operations in SSTA (SUM and MAX). Details are omitted here for brevity.

2) Canonical SSTA for non-Gaussian thermal profiles: As illustrated in [4], due to the impact of leakage, temperature profile follows log-normal distribution. Under such consideration, the method proposed in [4] can be used for computing the SUM and MAX operations.

7 Experimental results

In our experiment, we assume a two-layered 3D-IC. We use IBM-PLACE 2.0 circuits with placement information as the benchmark circuit [14]. For each test, we choose two circuits from $ibm01 - ibm12$ circuits, each circuit corresponds to one 3D-IC layer. Based on the placement information, we extract the whitespace distribution within layout, which is basically the potential TSV locations. The size of keep-out-zone is $2\mu m$, which indicates that gates cannot be allocated within a distance of $2\mu m$ from TSV edge. From this information, we generate the TSV distribution. The number of TSVs varies from $100 - 1000$. To generate the thermal profile, we divide the chip into $50 \times 50 \times 3$ grids, the temperature distribution in each grid is assumed Gaussian in the range $20 - 100^o C$. The standard deviation of fabrication variables is 3%. The annealing temperature is $275^o C$. Details of the benchmarks are omitted for brevity.

7.1 Effect of thermal and stress variations on SSTA

Firstly, we test the impact of thermal and stress variations on the longest path delay by comparing four SSTA approaches: a) SSTA considering only **F**abrication variations (**SSTA-F**); b) SSTA considering only **T**hermal and **F**abrication variations (**SSTA-TF**); c) SSTA considering only thermal **S**tress and **F**abrication variations (**SSTA-SF**); d) SSTA considering **T**hermal, thermal **S**tress and **F**abrication variations (**SSTA-TSF**).

For each of the four approaches, we use Monte Carlo method illustrated in section 5 to estimate the longest path delay. The distributions of longest path delay for each benchmark are plotted in Fig. 9. Fig. 9 shows that there is great difference between the delay distributions estimated by the four approaches, and hence both temperature and thermal stress have significant impact on the path delay to the extent that they should be considered together. In many benchmarks (e.g., benchmarks 1-3 in Fig. 9(a)-(c)), the longest path delay computed by SSTA-TSF is larger than that obtained by SSTA-TF, which indicates that thermal stress increases longest path delay. However in some benchmarks (e.g., benchmark 4 in Fig. 9(d)), the longest path delay computed by SSTA-TSF is even smaller than that obtained by SSTA-TF, indicating that thermal stress actually reduces the longest path delay. This is because the delay is a strong function of TSV distribution, if the TSVs are distributed in such a way that the mobility of many gates is enhanced, then thermal stress can actually result in reduced circuit delay.

Table 1 compares the mean and worst case values of longest path delay computed by these four methods. In the table,

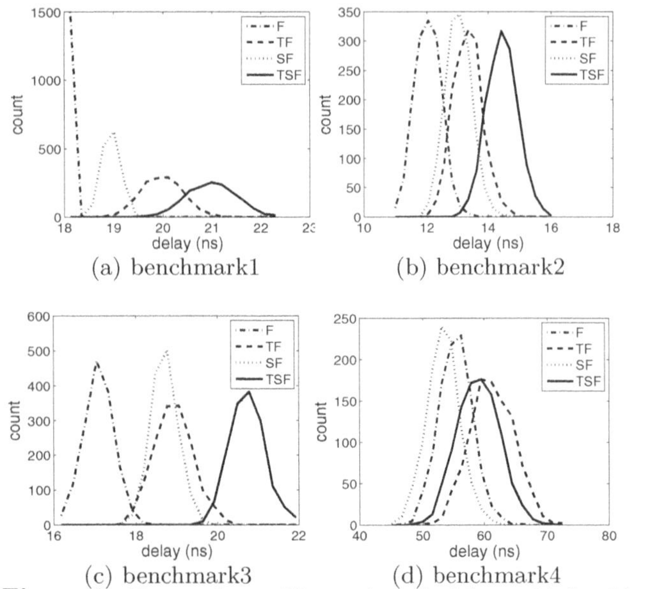

(a) benchmark1 (b) benchmark2

(c) benchmark3 (d) benchmark4

Figure 9: Comparison of longest path delays obtained by four SSTA approaches

Table 1: Comparison of mean and worst circuit delay estimated by four SSTA approaches (delay: ns)

Bench mark	SSTA-F		%Inc SSTA-TF		%Inc SSTA-SF		%Inc SSTA-TSF	
	mean	worst	mean	worst	mean	worst	mean	worst
1	18.04	18.07	10.70	17.42	4.99	16.30	16.35	24.00
2	12.05	13.62	10.69	8.95	8.28	7.39	19.86	18.52
3	17.09	18.41	10.71	9.92	9.42	9.06	21.29	19.92
4	55.33	65.21	10.62	12.35	-3.61	-3.09	6.63	8.73

"%Inc" indicates the mean (worst) delay increase compared with the based case SSTA-F. Comparing the mean delay, the SSTA-TF method results in 10% delay increase, while SSTA-SF method changes the delay by $-4 \sim 9\%$. SSTA-TSF method results in $6 \sim 21\%$ delay increase, which indicates that ignoring either of them can lead to inaccuracy in the calculated longest path delay.

7.2 Performance of our canonical SSTA

We then evaluate the accuracy and speedup of our canonical SSTA proposed in section 6 by comparing the distributions of longest path delay calculated by our SSTA approach and MC method (shown in Fig. 10), as well as the runtime used by these two methods. Fig. 10 shows that the longest path delay distributions calculated by these two methods are very similar. Table 2 compares mean, standard deviation (std) of the delay distributions and the runtime for the two methods. The percentage of error "% Error" is calculated by: $\%Error = (Canonical - MC)/MC \times 100\%$. The average error of mean and std are 0.18% and 4.87% respectively, indicating that our canonical SSTA approach is quite accurate. Comparing the runtime, our canonical SSTA achieves 843x speedup.

8 Conclusion

In this paper, we studied the impact of thermal and stress variations on gate delay, and proposed MC simulation based thermal and stress aware SSTA methodology. We showed that both thermal and stress variations have significant impact on circuit delay to the extent that they should be considered together. We then developed an efficient canonical thermal and stress aware SSTA method that achieves 843x speedup compared with MC based SSTA.

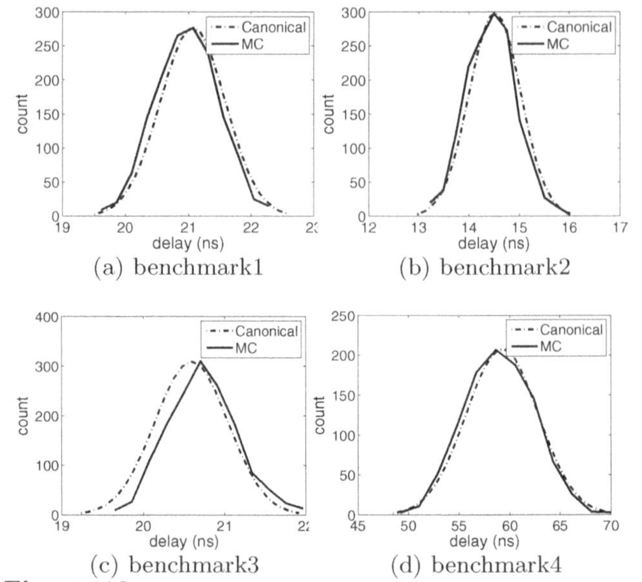

(a) benchmark1 (b) benchmark2

(c) benchmark3 (d) benchmark4

Figure 10: Delay distribution achieved by canonical SSTA and MC simulation

Table 2: Comparison of canonical SSTA with MC simulation (delay: ns, runtime: s)

Bench mark	MC simulation			Canonical SSTA			% Error		speed up
	mean	std	runtime	mean	std	runtime	mean	std	
1	20.99	0.50	5488	21.07	0.52	8.09	0.39	3.61	678x
2	14.44	0.49	11258	14.52	0.51	12.93	0.52	4.92	871x
3	20.73	0.43	25200	20.59	0.45	41.69	-0.63	5.26	604x
4	59.00	3.41	119000	59.26	3.60	97.65	0.45	5.71	1219x
Average							0.18	4.87	843x

9 References

[1] J.-S. Yang, K. Athikulwongse, Y.-J. Lee, S. K. Lim, and D. Z. Pan, "Tsv stress aware timing analysis with applications to 3d-ic layout optimization," in *Proceedings of the 47th Design Automation Conf. (DAC'10)*, 2010.

[2] N. Weste and D. Harris, "Cmos vlsi design: A circuits and systems perspective," *Addison Wesley*, 2010.

[3] H. Chang, V. Zolotov, S. Narayan, and C. Visweswariah, "Parameterized block-based statistical timing analysis with non-gaussian parameters, nonlinear delay functions," in *Design Automation Conf. (DAC'05)*, pp. 71–76, 2005.

[4] A. Rogachev, L. Wan, and D. Chen, "Temperature aware statistical static timing analysis," in *Int. Conf. on Computer-Aided Design (ICCAD'11)*, pp. 103–110, 2011.

[5] M. Pathak, Y.-J. Lee, T. Moon, and S. K. Lim, "Through-silicon-via management during 3d physical design: When to add and how many?," in *Int. Conf. on Computer-Aided Design (ICCAD'10)*, pp. 387–394, 2010.

[6] "Ansys fem tool," in *http://www.ansys.com/*.

[7] M. Jung, J. Mitra, D. Z. Pan, and S. K. Lim, "Tsv stress-aware full-chip mechanical reliability analysis and optimization for 3d-ic," in *Design Automation Conf. (DAC'11)*, pp. 188–193, 2011.

[8] S. Ryu, K. Lu, X. Zhang, J. Im, P. Ho, and R. Huang, "Impact of near-surface thermal stresses on interfacial reliability of through-silicon-vias for 3-d interconnects," *IEEE Trans. on Device and Materials Reliability*, pp. 35–43, 2011.

[9] B. Lasbouygues, R. Wilson, N. Azemard, and P. Maurine, "Temperature-and voltage-aware timing analysis," *IEEE Trans. on Computer-Aided Design of Integrated Circuits and Systems,*, vol. 26, pp. 801–815, 2007.

[10] H. Irie, K. Kita, K. Kyuno, and A. Toriumi, "In-plane mobility anisotropy and universality under uni-axial strains in n- and p-mos inversion layers on (100), (110), and (111) si," in *IEEE Int. Electron Devices Meeting*, pp. 225–228, 2004.

[11] C. Visweswariah, K. Ravindran, K. Kalafala, S. G. Walker, and S. Narayan, "First-order incremental block-based statistical timing analysis," in *Design Automation Conf. (DAC'04)*, pp. 331–336, 2004.

[12] P. Wilkerson, M. Furmanczyk, and M. Turowski, "Compact thermal modeling analysis for 3d integrated circuits," in *Int. Conf. Mixed Design of Integrated Circuits and Syst.*, 2004.

[13] J. Lau, C. Wong, J. Prince, and W. Nakayama, *Electronic packaging: design, materials, process, and reliability*. McGraw-Hill, 1998.

[14] "Ibm-place 2.0 benchmark," in *http://er.cs.ucla.edu/benchmarks/ibm-place2/*.

Effect of TSV Fabrication Technology on Power Distribution in 3D ICs

Suhas M. Satheesh
High Speed Fabrics Team
NVIDIA
Santa Clara, California 95050
ssatheesh@nvidia.com

Emre Salman
Department of Electrical and Computer Eng.
Stony Brook University
Stony Brook, New York 11794
emre@ece.sunysb.edu

ABSTRACT

The design implications of two distinct through silicon via (TSV) fabrication methods (via-first and via-last) have been investigated for power delivery in a 3D system. Different geometry, connectivity, and filling materials have been considered to develop equivalent electrical models for both via-first and via-last based power distribution networks. Based on these models, a valid design space has been developed where power supply noise is satisfied and physical area overhead is minimized. Under constant power supply noise, a via-last based power network occupies 7.5% less area. However, in addition to causing routing blockages, a via-last based power network exhibits high sensitivity to design parameters due to a high quality factor. Alternatively, a via-first based power network requires a large number of TSVs, but exhibits relatively more predictable behavior due to a lower quality factor (higher damping).

Categories and Subject Descriptors

B.7.m [**Integrated Circuits**]:

General Terms

Design; 3D IC; TSV; Power noise

1. INTRODUCTION

Three-dimensional (3D) integration is an emerging paradigm for developing multi-functional systems at a lower cost. Utilizing the vertical dimension not only increases the integration density, but also reduces the length and number of global interconnects. This reduction enhances system performance (due to reduced interconnect delay) while lowering the power consumption (due to reduced switching capacitance) [1–3]. Furthermore, through silicon via (TSV) based 3D technologies enable heterogeneous integration that expands the application domain from high performance computing to relatively lower power systems-on-chip consisting of, for example, sensors, analog/RF circuits, memory, and digital processing blocks, as depicted in Fig. 1.

A significant circuit- and physical-level challenge in 3D ICs is reliable power delivery to each plane. Maintaining the power supply voltage variations within 5-10% of the global V_{DD} is highly challenging due to three reasons [4]: 1) higher supply current, 2) lower supply voltage around or below one volt, and 3) relatively high TSV impedances that vary depending upon the particular TSV fabrication method.

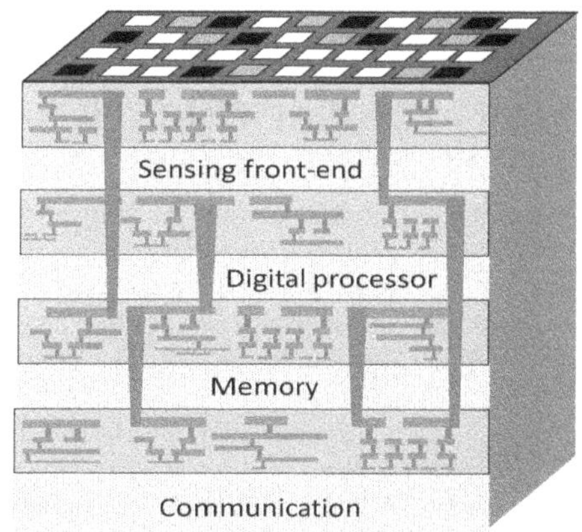

Figure 1: Three-dimensional integration of diverse planes using through silicon via technology [1].

Previous works on 3D power delivery have focused on different power distribution topologies, effect of TSV geometry, and 3D decoupling capacitor placement techniques [5–7]. A primary limitation of the existing works is the absence of a strong link between proposed design guidelines and distinct TSV fabrication methods. Each TSV technology exhibits unique characteristics, thereby requiring different design constraints. Pavlidis and De Micheli have demonstrated the advantage of alternative low impedance current paths in via-first TSVs [8]. These additional current paths have been exploited, resulting in a 22% reduction in the number of intraplane vias or alternatively, a 25% decrease in the required decoupling capacitance. Different filling materials for via-first and via-last TSVs, however, have not been considered. Furthermore, a simplified model has been assumed for the power distribution network within each plane.

The primary contributions of this paper are three-fold: (1) electrical characteristics and connectivity differences of via-first and via-last TSVs are considered during the design process of a robust 3D power distribution network. (2) Design space that satisfies power supply noise while minimizing the physical area is determined for both via-first and via-last TSVs. Note that area overhead is an important concern in TSV based 3D ICs, particularly for low power, low cost applications. (3) Related tradeoffs are identified both at the fabrication and circuit levels, providing design guidelines in developing 3D power networks.

The rest of the paper is organized as follows. The characteristics

Table 1: Characteristics of via-first and via-last TSVs [9–11].

Parameter	Via-first	Via-last
Filling material	Doped polysilicon	Copper
Processing temperature	High	Low
Sensitivity to contamination	Low	High
Manufacturability	Difficult	Established
Formation	Before FEOL	After BEOL
Electrical characteristics	Highly resistive	Inductive
Landing metal	M_1	M_{Top}
Take-off metal	M_{Top}	M_{Top}

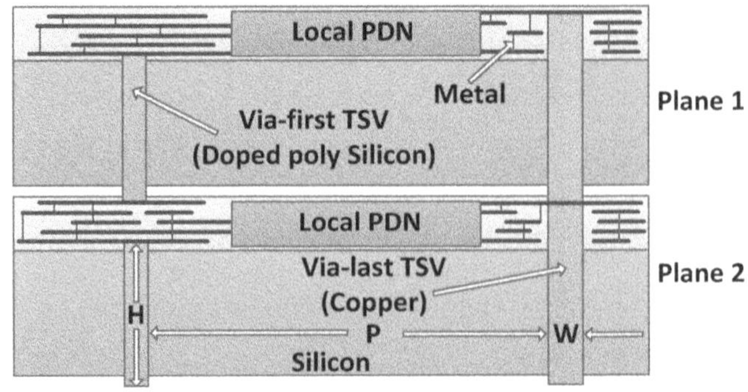

Figure 2: Illustration of via-first and via-last TSV fabrication technologies.

of the via-first and via-last technologies are reviewed in Section 2. Electrical models for 3D power distribution for both technologies are described in Section 3. The results of the proposed analysis are provided in Section 4, and the paper is concluded in Section 5.

2. PRIMARY CHARACTERISTICS OF TSV TECHNOLOGIES

Each TSV technology exhibits unique challenges in designing a 3D power distribution network. These differences arise due to distinct fabrication techniques. The fabrication characteristics and relative physical dimensions of via-first and via-last TSVs are summarized, respectively, in Sections 2.1 and 2.2. These properties are also listed in Table 1.

2.1 Via-First TSV

In the via-first method, TSVs are fabricated before the transistors are patterned in silicon, *i.e.*, prior to front-end-of-line (FEOL) [9, 10]. Thus, TSVs fabricated with the via-first technique do not pass through the metalization layers, as depicted in Fig. 2. This characteristic is an important design advantage since via-first TSVs do not affect the routability of the signals. In via-first technology, the TSV of a plane is connected between the first metal layer of the current plane and the top most metal layer of the previous plane. Polysilicon is typically used as the filling material due to its ability to withstand high temperatures [9, 10]. Via-first TSVs are less sensitive to contamination since both the filling and substrate material are the same [10]. The physical dimensions of via-first TSVs are smaller than via-last TSVs [11]. Via-first TSVs, however, are highly resistive due to the use of polysilicon as the filling material.

2.2 Via-Last TSV

In the via-last approach, TSV formation occurs after the metalization layers are fabricated, *i.e.*, after BEOL [9–11]. Thus, as opposed to via-first TSVs, via-last TSVs pass through the metal layers, causing metal routing blockages, as depicted in Fig. 2 [8, 11]. A lower resistivity filling material such as copper is used since high temperature FEOL and BEOL processes are performed before the via formation [9, 10]. The use of copper as a filling material makes the process sensitive to both temperature (must be maintained less than 230°C) and contamination [10]. Despite exhibiting relatively low resistance, the inductive characteristics of via-last TSVs are relatively more significant than via-first TSVs due to greater physical dimensions [11]. The physical connection between the TSV and metal layers is typically achieved at the top most metal layer.

3. ELECTRICAL MODELS FOR 3D POWER DISTRIBUTION

The models that are used to analyze power supply noise for each TSV technology are described in this section. System level model, including the orientation of the planes and the dimensions of the dies are discussed in Section 3.1. Electrical models used for TSVs, substrate, and the power distribution network within a plane are provided, respectively, in Sections 3.2, 3.3, and 3.4. Finally, the model for the load circuit is described in Section 3.5.

3.1 System Level Model

A 3D system in 32 nm CMOS technology consisting of eight memory planes and one plane for the processor core is considered. The power supply voltage V_{DD} is equal to 1 volt. Devices in each plane occupy an area of 120 mm^2. The system has 1 (gigabyte) GB of DRAM spread uniformly across eight memory planes. Each memory plane has 1 (gigabit) Gb DRAM. Every plane is divided into 32 modules of equal size. Each of these modules consumes an area of 1500×2500 μm^2. This topology is depicted in Fig. 3. For via-first technology, the TSVs are placed beneath the active circuit, as illustrated in Fig. 3(a) whereas in via-last technology, the power and ground TSVs are distributed on both sides of each module, as depicted in Fig. 3(b). Note that although the TSVs are placed beneath the active circuit in via-first technology, these TSVs consume additional device area. Thus, in Fig. 3(a), the area represented by each DRAM module includes both the 1500×2500 μm^2 DRAM area and the TSV area. A nine metal layer process is considered where the metal thickness and aspect ratio are determined according to an industrial 32 nm CMOS technology [12].

The processor plane is closest to the heat sink and farthest from the package due to its high switching activity. The orientation of the memory planes such as face-to-face or face-to-back exhibits a design tradeoff. The first memory plane can face the package where the metal layers are directly connected without the TSVs to the package pads or it can face the adjacent DRAM plane. In the second option, as considered in this work, TSVs are required to connect the first memory plane with the package pads. First option eliminates these TSVs, but the additional current paths available in via-first technology [8] cannot be exploited in distributing power. These additional current paths exist only when the current flow is from the lowest to the highest metal layer in a plane. Another consideration is the orientation between the processor plane and the adjacent memory plane. A face-to-back approach maintains the symmetry of the 3D system, but the communication bandwidth between the two planes is limited. Alternatively, in a face-to-face approach, processor and memory can communicate with the metal layers without requiring TSVs, thereby enhancing the communication bandwidth. This scheme, as considered in this work, also reduces the overall number of TSVs, compensating the additional TSVs between the first memory plane and package. Thus the processor plane does not house any TSVs. The 3D processor-memory

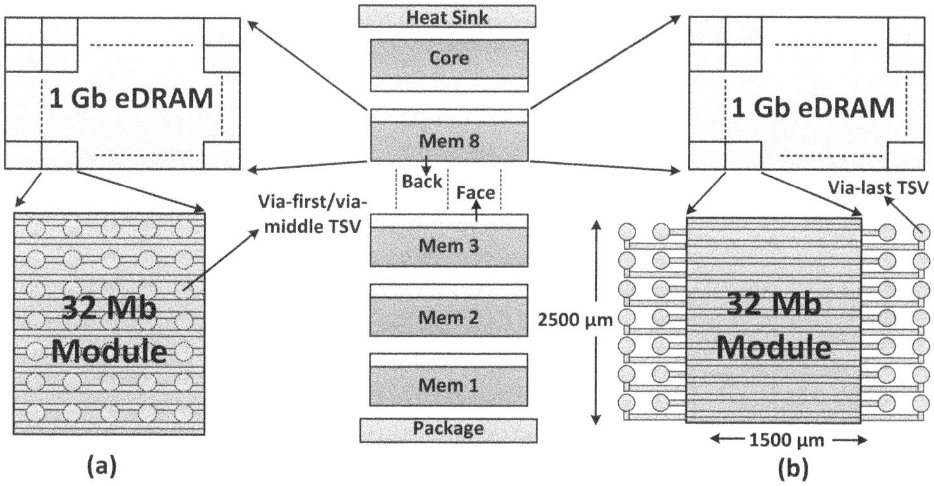

Figure 3: Illustration of the 3D processor-memory stack: (a) via-first TSVs and (b) via-last TSVs.

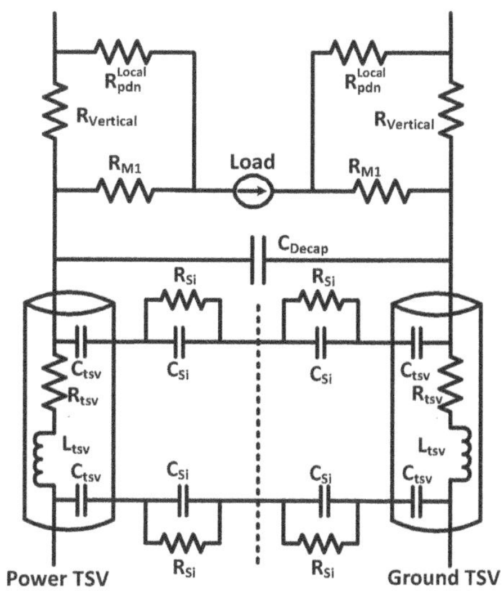

Figure 4: Equivalent power distribution network.

stack is illustrated in Fig. 3. The parasitic package resistance and inductance are, respectively, 3 mΩ and 100 pH at both the power and ground supplies.

In the rest of this paper, the analysis is performed for an area of 1500×2500 μm^2, which corresponds to one of the 32 modules. The procedure is similar for the remaining 31 modules. All of the references to the amount of decoupling capacitance represent the capacitance in each module. Similarly, a reference to the number of TSVs represents the number of power TSVs in each module. An equivalent electrical model corresponding to the power distribution network of this portion of the system is illustrated in Fig. 4. This model consists of the TSVs, substrate, power distribution network within a plane, switching load circuit, and decoupling capacitance, as described in the following sections.

3.2 TSV Model

A TSV is typically represented as a cylinder with a diameter W and depth H. Aspect ratio of a TSV is given by H/W. The minimum distance between the two TSVs is determined by the pitch P, which is typically twice the TSV diameter. The TSV model consists of a resistance R_{tsv} and inductance L_{tsv} due to the filling material, and a capacitance to the substrate C_{tsv} due to the thin dielectric layer [13]. R_{tsv} is determined by [14],

$$R_{tsv} = \sqrt{(R_{AC}^{tsv})^2 + (R_{DC}^{tsv})^2}, \tag{1}$$

where the DC resistance R_{DC}^{tsv} and AC resistance R_{AC}^{tsv} are, respectively,

$$R_{DC}^{tsv} = \frac{\rho_f H}{\pi (W/2)^2}, \tag{2}$$

$$R_{AC}^{tsv} = \frac{\rho_f H}{2\pi (W/2)\delta_{tsv}}. \tag{3}$$

ρ_f is the conductivity of the filling material and the skin depth δ_{tsv} is [14]

$$\delta_{tsv} = \frac{1}{\sqrt{\pi f \mu_f \rho_f}}, \tag{4}$$

where f is the frequency and μ_f is the permeability of the filling material. The TSV inductance L_{tsv} is [15]

$$L_{tsv} = \frac{\mu_o}{4\pi}[2H \ln(\frac{2H + \sqrt{(W/2)^2 + (2H)^2}}{W/2}) + (W/2 - \sqrt{(W/2)^2 + (2H)^2})], \tag{5}$$

where μ_o is vacuum permeability. The TSV capacitance C_{tsv} is determined from the cylindrical capacitor formula as [16]

$$C_{tsv} = \frac{2\pi\varepsilon_{ox} H}{\ln(\frac{W/2 + t_{ox}}{W/2})}, \tag{6}$$

where ε_{ox} is the oxide permittivity. Note that the skin effect is neglected for via-first TSVs due to its sufficiently small TSV diameter. The primary characteristics of the two TSV technologies are also listed in Table 2.

3.3 Substrate Model

The substrate is modeled as an RC impedance, where substrate capacitance, C_{si}, and substrate resistance, R_{si} are, respectively [16],

$$C_{si} = \frac{\pi\varepsilon_{si} H}{\ln\left(\frac{2P}{W/2} + \sqrt{(\frac{2P}{W/2})^2 - 1}\right)}, \tag{7}$$

Table 2: Model parameters for TSVs [9,15].

Parameter	Via-first	Via-last
Diameter W	4 μm	10 μm
Height H	10 μm	60 μm
Pitch P	8 μm	20 μm
Oxide thickness t_{ox}	0.2 μm	0.2 μm
TSV resistance R_{tsv}	5.7 Ω	20 mΩ
TSV inductance L_{tsv}	4.18 pH	34.94 pH
TSV capacitance C_{tsv}	23 fF	283 fF
Aspect ratio AR	2.5:1	6:1
Material resistivity ρ_f	7.2 $\mu\Omega$m	16.8 nΩm
Substrate resistance R_{si}	8.81 kΩ	1.76 kΩ
Substrate capacitance C_{si}	1.19 fF	6 fF
Co-efficient of thermal expansion CTE	Low	17 ppm/K

Table 3: Power distribution network resistances within a plane.

Parameter	Via-first	Via-last
Network resistance R_{pdn}^{Local}	1.97 mΩ	8.51 mΩ
Vertical resistance $R_{Vertical}$	1.97 mΩ	NA
M1 Path Resistance R_{M1}	73 $\mu\Omega$	∞

$$R_{si} = \frac{\varepsilon_{si}\rho_{si}}{C_{si}}. \tag{8}$$

where $\varepsilon_{si} = 105 \times 10^{-12}$ F/m and $\rho_{si} = 10$ Ωcm are, respectively, silicon permittivity and substrate resistivity. These numbers are listed in Table 2.

3.4 Plane-Level Power Distribution Network

As shown in Fig. 2, in the via-first method, local metal layers are utilized to connect the two TSVs located in adjacent planes whereas via-last based TSVs pass through the metal layers. As depicted in Fig. 4, the local power network is modeled with three resistances, which are determined based on 32 nm technology parameters [12]. An interdigitated grid topology is adopted where a stack of vias, located at each intersection of M_9 and M_8, transmits the power supply voltage down to the first metal layer, and eventually to the switching transistors. A single stack of vias from M_1 to M_9 has 16 Ω resistance [17]. Note however that there is a large number of parallel paths. The value of resistances in Fig. 4 resulting from simulation results are listed in Table 3.

3.5 Switching Circuit

The switching circuit for both memory and processor is modeled as a piecewise linear current source, as shown in Fig. 5. The DRAM consumes 3 W of power uniformly distributed across the eight stacks [5]. Alternatively, the processor consumes 90 W of power. 30% of the overall power is due to static power dissipation whereas the remaining portion is due to dynamic power consumption [5]. As illustrated in Fig. 5, a triangular current waveform is assumed. The DC current I_{dc} and peak current I_{peak} are determined based on, respectively, the static and dynamic power consumption. The power and switching current characteristics for the DRAM and the processor are listed in Table 4. Note that the power characteristics in this table are for the entire system whereas the current characteristics represent the current in each of the 32 modules.

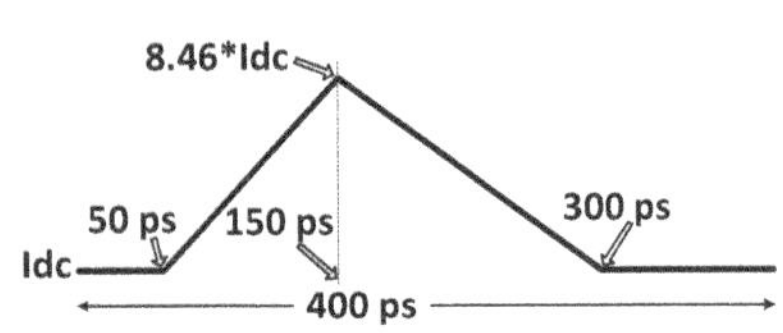

Figure 5: Piecewise linear current source model to mimic switching circuit.

Table 4: Power parameters used to estimate the characteristics of switching current [5].

Parameter	DRAM	Processor
Total power	3 W	90 W
Static Power	0.9 W	27 W
Dynamic Power	2.1 W	63 W
Static Current	28.125 mA	843.75 mA
Peak Current	238.125 mA	7.14 A
Operating Frquency	2.5 GHz	2.5 GHz

4. SIMULATION RESULTS

The dependence of peak noise on decoupling capacitance and number of TSVs, design space that satisfies the target power supply noise, and the area overhead due to TSVs and decoupling capacitance are provided in this section. The power supply noise is observed across the current source located at the processor plane which is the farthest node from the power supply pads. In transient analysis, the tolerable power supply noise is assumed to be 100 mV at the power and ground planes. Decoupling capacitance is implemented using MOS capacitors in an industrial 32 nm technology node with a capacitance density of 39.35 fF/μm^2 as determined from the equivalent oxide thickness (EOT) of the technology [12]. Results for via-first and via-last technologies are discussed, respectively, in Sections 4.1 and 4.2.

4.1 Via-First TSV

Use of doped polysilicon as the filling material in via-first technology causes the TSVs to be highly resistive. Shorter TSV height makes via-first TSVs relatively less inductive. Due to these two reasons, a via-first based power network is typically over-damped.

4.1.1 Valid Design Space to Satisfy Target Power Supply Noise

The peak noise surface as a function of decoupling capacitance and number of TSVs is plotted in Fig. 6(a) for via-first TSVs. Since the power distribution network is over-damped, the response is monotonic. Thus, peak noise decreases as the number of TSVs and/or decoupling capacitance increase. To determine the valid design space, a contour at 100 mV peak noise is extracted from the noise surface, as depicted in Fig. 6(b). Any point above the curve satisfies the noise constraint whereas the shaded region should be avoided.

4.1.2 Optimum Number of TSVs and Decoupling Capacitance

Since multiple valid pairs of number of TSVs and decoupling capacitance exist, a pair that minimizes the overall area overhead can be chosen. The area overhead is depicted in Fig. 7 for each point on the contour of Fig. 6(b). As demonstrated in this figure, a specific design point exists that minimizes the area overhead. This design point corresponds to 2750 TSVs and 2.7 nF of decoupling capacitance, producing an area overhead of approximately 9%. Note that if an arbitrary pair is chosen from the contour in Fig. 6(b), the area overhead can be as high as 16%.

An important design requirement is to have a significantly high number of via-first TSVs to reduce current per TSV, which increases the required physical area. Note however that 6.25 times more via-first TSVs than via-last TSVs can be placed within a given area due to smaller via-first TSV dimensions. Also note that a large

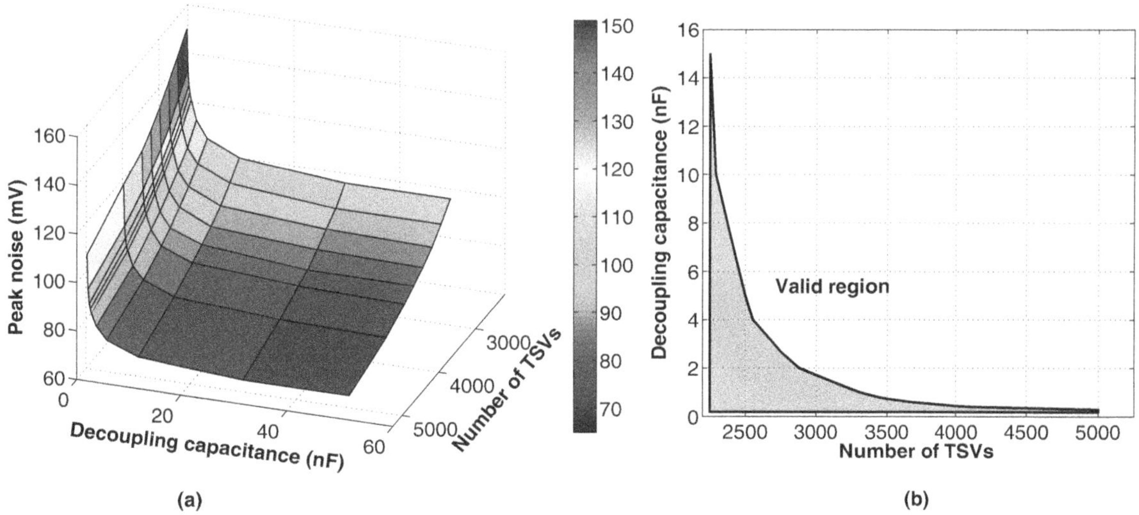

Figure 6: Peak noise as a function of number of via-first TSVs and decoupling capacitance: (a) surface plot and (b) contour plot at 100 mV.

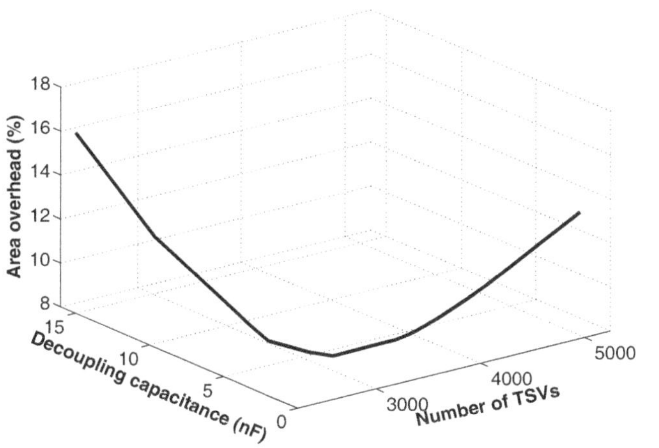

Figure 7: Area overhead in via-first technology.

amount of decoupling capacitance is required to reduce transient *IR* noise. An advantage of a via-first based power distribution network is the high damping factor due to high TSV resistance and low TSV inductance. Thus, decoupling capacitance effectively suppresses the transient noise, producing a low peak-to-peak noise, as listed in Table 5.

4.2 Via-Last TSV

Via-last TSV technology exhibits significantly different noise characteristics as compared to via-first TSV technology due to differences in filling material and physical dimensions. Due to significantly less resistance and a relatively large inductance, a power distribution network with via-last TSVs is typically under-damped.

4.2.1 *Valid Design Space to Satisfy Target Power Supply Noise*

The peak noise surface as a function of decoupling capacitance and number of TSVs is plotted in Fig. 8(a) for via-last TSVs. As opposed to via-first TSVs, the noise surface is non-monotonic with multiple peaks where the noise exceeds the design objective. The noise contour at 100 mV is illustrated in Fig. 8(b) to determine the valid design space where the peak noise constraint is satisfied. The unshaded region represents a valid combination of number of TSVs

Table 5: Valid design space that satisfies the power supply noise while minimizing physical area.

TSV type	Area penalty	Number of TSVs	Decoupling capacitance (nF)	Peak-to-peak noise (mV)
Via-first	9.06%	2750	2.7	33.42
Via-last	1.54%	76	0.385	134.72

and decoupling capacitance. As illustrated in this figure, at a specific decoupling capacitance, an increase in the number of TSVs can violate the noise constraint due to a lower damping factor. In via-last technology, the power supply noise is highly sensitive to number of TSVs and decoupling capacitance, particularly at relatively low decoupling capacitances. Furthermore, peak-to-peak noise is also significantly higher, as listed in Table 5.

4.2.2 *Optimum Number of TSVs and Decoupling Capacitance*

The area overhead, however, is significantly lower in via-last TSVs since the number of required TSVs is less. The dependence of area overhead on number of TSVs and decoupling capacitance for the valid design points are represented in Fig. 9. Note that this relationship is highly complicated due to the non-monotonic behavior of the power supply noise. This behavior suggests that it is highly critical to appropriately choose the number of TSVs and amount of decoupling capacitance in a via-last based power network. The design points that correspond to the minimum area overhead while satisfying the power supply noise are listed in Table 5.

5. CONCLUSIONS

Distinct design requirements of via-first and via-last TSVs have been identified for robust power delivery in 3D ICs. Electrical models have been developed considering different connectivity characteristics, physical geometry, and filling materials. A valid design space that satisfies target power supply noise while minimizing physical area has been identified for both TSV technologies assuming a 32 nm processor-memory stack. Despite the low parasitic resistance of via-last TSVs, a power network based on via-last TSVs produces signal routing blockages and exhibits high inductive behavior, thereby producing a non-monotonic design space. Thus, the power network exhibits high sensitivity to design parameters. Alternatively, a power network based on via-first TSVs is typically

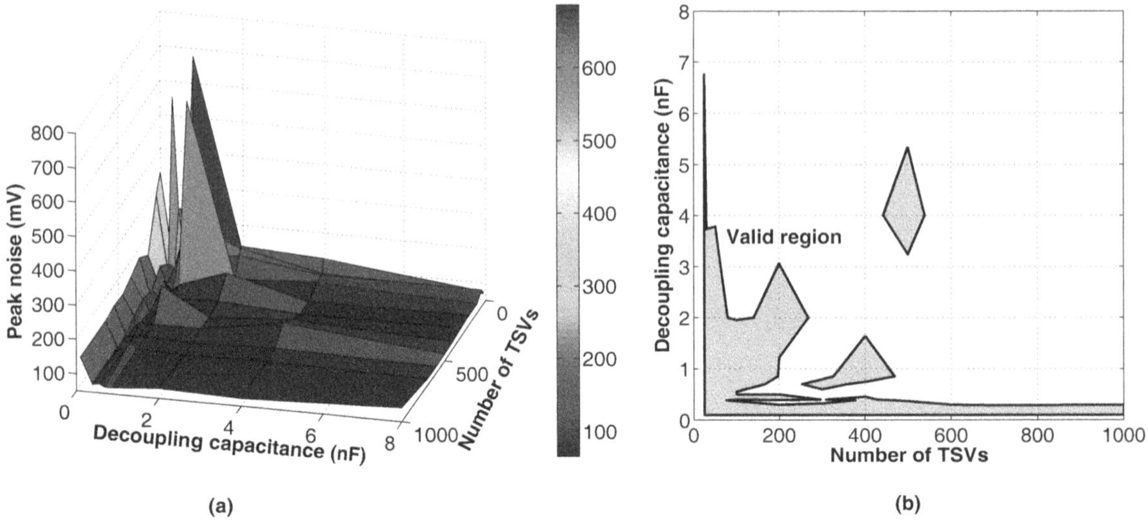

Figure 8: Peak noise as a function of number of via-last TSVs and decoupling capacitance: (a) surface plot and (b) contour plot at 100 mV.

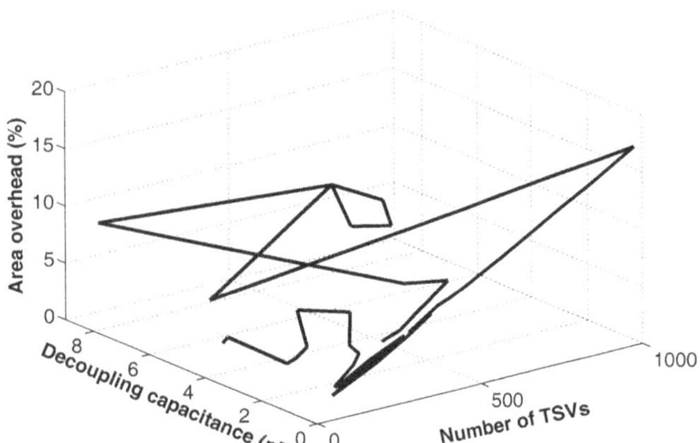

Figure 9: Area overhead in via-last technology.

over-damped, producing a monotonic and more predictable design space. Due to high TSV resistance, a via-first based power network can satisfy the power supply noise at the expense of 7.52% additional area overhead.

6. REFERENCES

[1] V. F. Pavlidis and E. G. Friedman, *Three-Dimensional Integrated Circuit Design*, Morgan Kaufmann, 2009.

[2] A. W. Topol *et al.*, "Three-dimensional Integrated Circuits," *IBM Journal of Research and Development*, Vol. 50, No. 4/5, pp. 491–506, July/September 2006.

[3] J. A. Burns *et al.*, "A wafer-scale 3-D circuit integration technology," *IEEE Transactions on Electron Devices*, Vol. 53, No. 10, pp. 2507–2516, October 2006.

[4] E. Salman and E. G. Friedman, *High Performance Integrated Circuit Design*, McGraw-Hill, 2012.

[5] Q. Wu and T. Zhang, "Design Techniques to Facilitate Processor Power Delivery in 3-D Processor-DRAM Integrated Systems," *IEEE Transactions on Very Large Scale Integration (VLSI) Systems*, Vol. 19, No. 9, pp. 1655–1666, September 2011.

[6] N. H. Khan, S.M. Alam, and S. Hassoun, "Power Delivery Design for 3-D ICs Using Different Through-Silicon Via (TSV) Technologies," *IEEE Transactions on Very Large Scale Integration (VLSI) Systems*, Vol. 19, No. 4, pp. 647–658, April 2011.

[7] G. Huang, M. Bakir, A. Naeemi, H. Chen, and J. D. Meindl, "Power Delivery for 3D Chip Stacks: Physical Modeling and Design Implication," *Proceedings of the IEEE Electrical Performance of Electronic Packaging*, pp. 205–208, October 2007.

[8] V. F. Pavlidis and G. De Micheli, "Power Distribution Paths in 3-D ICs," *Proceedings of the ACM Great Lakes Symp. on VLSI*, pp. 263–268, 2009.

[9] A. Agarwal, R. B. Murthy, V. Lee, and G. Viswanadam, "Polysilicon Interconnections (FEOL): Fabrication and Characterization," *Proceedings of the IEEE Electronics Packaging Technology Conference*, pp. 317–320, Dec. 2009.

[10] M. Kawano *et al.*, "Three-Dimensional Packaging Technology for Stacked DRAM With 3-Gb/s Data Transfer," *IEEE Transactions on Eletron Devices*, Vol. 55, No. 7, pp. 1614–1620, July 2008.

[11] D. H. Kim, S. Mukhopadhyay, and S. K. Lim, "Through-silicon-via Aware Interconnect Prediction and Optimization for 3D Stacked ICs," *Proceedings of the International Workshop on System Level Interconnect Prediction*, pp. 85–92, September 2009.

[12] S. Natarajan *et al.*, "A 32nm Logic Technology Featuring 2nd-Generation High-k + Metal-Gate Transistors, Enhanced Channel Strain and 0.171 μm^2 SRAM Cell Size in a 291Mb Array," *Proceedings of the IEEE International Electron Devices Meeting*, pp. 1–3, December 2008.

[13] I. Savidis and E. G. Friedman, "Closed-Form Expressions of 3-D Via Resistance, Inductance, and Capacitance," *IEEE Transactions on Electron Devices*, Vol. 56, No. 9, pp. 1873–1881, September 2009.

[14] J. Kim, J. Cho, and J. Kim, "TSV Modeling and Noise Coupling in 3D IC," *Proceedings of the IEEE Electronic System-Integration Technology Conference*, pp. 1–6, September 2006.

[15] G. Katti, M. Stucchi, K. De Meyer, and W. Dehaene, "Electrical Modeling and Characterization of Through Silicon via for Three-Dimensional ICs," *IEEE Transactions on Electron Devices*, Vol. 57, No. 1, pp. 256–262, January 2010.

[16] J. S. Pak *et al.*, "PDN Impedance Modeling and Analysis of 3D TSV IC by Using Proposed P/G TSV Array Model Based on Separated P/G TSV and Chip-PDN Models," *IEEE Transactions on Comp., Packaging and Manufacturing Technology*, Vol. 1, No. 2, pp. 208–219, February 2011.

[17] "NCSU 45 nm process design kit," [Online]. Available: `http://www.eda.ncsu.edu/wiki/FreePDK45:Metal_Layers`.

Multi-Port FinFET SRAM Design

Yirong Zhao
Electrical and Computer Engg.
University of Pittsburgh
Pittsburgh, PA
yiz52@pitt.edu

Jiayin Li
Electrical and Computer Engg.
University of Pittsburgh
Pittsburgh, PA
jil135@pitt.edu

Kartik Mohanram
Electrical and Computer Engg.
University of Pittsburgh
Pittsburgh, PA
kmram@pitt.edu

Abstract

Multi-port SRAMs are essential for caches and shared data structures, especially in modern multi-core SoCs. The FinFET device, which offers high threshold voltage and high on/off current ratio, is a promising candidate for multi-port SRAMs for fast read/write speed, high cell density, and low power consumption. In this paper, we perform the first study of multi-port FinFET SRAMs, including the double-ended multi-port FinFET SRAM and three single-ended multi-port FinFET SRAMs with isolated read ports. We evaluate the static noise margin, leakage current, and read/write performance of these structures with the predictive technology model for FinFETs. Our results show that single-ended multi-port FinFET SRAMs with isolated read ports offer improved read stability and flexibility over the double-ended structure at the expense of write performance. By increasing the size of the access transistors, we show that the single-ended structures can achieve equivalent write performance to the double-ended structure for 9% area overhead.

Categories and Subject Descriptors

B.3.1 [**MEMORY STRUCTURES**]: Semiconductor memories—*Static memory (SRAM)*

Keywords

SRAM; FinFETs; Multi-port; Performance; Power

1. Introduction

High-speed scalable multi-port embedded memories that support simultaneous read and write accesses per cycle are integral to modern microprocessors and SoCs [1]. However, scaling of SRAM with planar CMOS faces many challenges, such as short channel effects (SCE), gate induced drain leakage (GIDL), and band-to-band tunneling (BTBT). These result in high leakage current and low threshold voltage that impact cell stability. In multi-port SRAMs, increasing the number of ports exacerbates these problems, due to the increase in leakage current, parasitic capacitance, and number of bit lines. Although there are techniques to deal with these problems in multi-port SRAM design [2–5], they remain insufficient in practice to overcome the limitations imposed by CMOS scaling.

This research was supported by awards CCF-1211099 and CCF-1208933 from the National Science Foundation.

GLSVLSI'13, May 2–3, 2013, Paris, France.

FinFET SRAM designs have been proposed and studied extensively to deal with scalability problems [6], primarily by leveraging three features. (1) IG (Independent Gate) mode [6], which allows modifying the back-gate voltage to control the threshold voltage and improve read/write performance [6]. (2) The high on/off current ratio of FinFET, due to its better control over short channel effects, which provides opportunities to reduce standby power [7]. (3) The small parasitic capacitance, due to the fact that FinFET is highly depleted, decreases the bit line load capacitance significantly, which reduces read and write access latency [6]. Thus, FinFET is an ideal alternative to planar CMOS in SRAM designs.

Improving the read/write latency and cell stability remain the major challenges when multi-porting FinFET SRAMs. Fundamentally, multi-port SRAM structures can be classified into two types: double-ended structures and single-ended structures. Double-ended structures have two bit lines on each port and single-ended structures have only one. For read operations, as the number of access transistors increases, the read time increases when serving several reads simultaneously, and the internal nodes suffer from the destructive read. To address these issues, single-ended SRAM with isolated read ports has been proposed, where the destructive read is eliminated and read time is reduced [8]. For write operation, one bit line can be eliminated, thus reducing the area and wires in single-ended multi-port SRAM designs. However, single-ended structures find it difficult to break the feedback loop between the two inverters, which increases the write latency. To address this issue, write assist structures have been proposed in the literature [9].

This paper makes the following contributions. First, we perform the first study of multi-port FinFET SRAMs: one double-ended multi-port FinFET SRAM and three single-ended multi-port FinFET SRAMs with isolated read ports. To the best of our knowledge, this is the first work evaluating read/write acceleration through multi-porting in FinFET SRAMs. Second, we evaluate these structures with the predictive technology model for FinFETs [10] using metrics such as static noise margin, leakage current, and read/write time. We measure the impact of different sizes of transistors on the performance of FinFET SRAMs. Based on simulation results, we can conclude that single-ended multi-port FinFET SRAM with isolated read ports has the advantage over the double-ended multi-port FinFET SRAM in the read operation, such as robust read access and high flexibility in read port configuration. However, its drawback is that its write performance is worse than the double-ended structure, due to its weakness in breaking the feedback loop in write operations. Finally, we show that the write performance of single-ended structures with one read port can be improved to the same level of the double-ended structure for 9% area overhead by sizing the access transistors appropriately.

2. Multi-port SRAM background

Multi-port SRAMs and multi-port register files, are widely implemented in current multi-core processors. They support simultaneous read and write operations in one cycle, which significantly increases read/write access throughput from cores to the memory. A quad-port shared memory based on FPGAs was proposed [1], which increases the speed of write and read by almost 66.7%, compared to a single-port memory. Multi-port register files were implemented in the AMD K7 processor [11] and the Itanium Microprocessor [12].

There are two types of cell structure in multi-port SRAMs: the double-ended structure and the single-ended structure. The double-ended structure can easily break the feedback loop in the write operation using its two bit lines. Meanwhile, the single-ended structure has a short wire delay and a small cell area, due to its single-ended bit line and word line [9]. As SRAM cells are scaled down, single-ended multi-port SRAMs and register files with isolated read ports become more attractive. For instance, a single-ended multi-port register file was built in the Itanium Microprocessor [12], and a single-ended 34 bit $\times$ 64 bit 10R/6W register file was proposed [13]. However, to avoid the longer write latency in the single-ended structure [9], the double-ended structure is preferred in multi-port SRAM design to provide faster access, e.g., the register file in the AMD K7 processor [11] and the 8T double-ended dual-port SRAM cell [14]. In this paper, single-ended cells with isolated read ports are referred to as single-ended structures. Note that the double-ended structure with isolated read ports in not considered due to its large area overhead.

To accelerate read operations in a double-ended structure SRAM cell, we can duplicate access transistors to compose more ports. Thus, we can read the value in one cell from multiple ports simultaneously when they are enabled. The total access time of two simultaneous reads in the same row but not in the same cell can be reduced largely when two word lines are enabled at the same time. However, the challenge is that the total access time of two simultaneous reads in the double-ended structure is longer than one single read in the single-port cell. This is because adding one read port brings another bit line to the cell, making it more difficult to pull down the pre-charged bit lines, due to the larger capacitance. As the number of ports increases, the read access time for simultaneous read grows. Single-ended structures, however, do not suffer from such a problem. Due to isolated read ports, the cell provides gate voltage to read port transistors while read port transistors do not affect the cell. The bit line is only driven by the read port transistors instead of the cell. Since read ports are isolated with each other, they only need to drive their own bit line, providing equivalent driving ability for simultaneous multiple reads to one single read. As a result, increasing ports does not increase the simultaneous read time. Furthermore, the destructive read problem gets worse in simultaneous read operations in double-ended structures. Adding access transistors in double-ended structures makes the internal node affected seriously by the supply voltage of every bit line. Therefore, the single-ended structure is better than the double-ended structure in terms of read acceleration.

For write acceleration in the double-ended structure, duplicating access transistors is the preferred approach. However, the time of one write slightly increases in multi-port cells because the extra access transistor adds extra capacitance to the internal node. The simultaneous write can occur in the same row, but not in the same cell, because it is impossible to write multiple values into one cell in one cycle. Thus, the total write time for two simultaneous writes in the same row but different cells is close to the time needed for one single write in the single-port structure, while the total write time for two writes in the same cell still remains the same as the total time for two writes in the single-port structure. For the single-ended structure, the critical issue in write acceleration is the need of write-assist structures to reduce the write access time, because the feedback loop in the single-ended structure is strong.

A common challenge in both read and write multi-porting is the half-select problem: for a cell in retention, the word line voltage may be high, when other cells in the same row are being read or written to, which enables the bit line supply voltage to affect the internal node [15]. As the number of ports increases, the cell is more vulnerable to the half-selected problem. To address the issue, a divided word line configuration was introduced to completely eliminate the condition of disturbed access in the unselected cell using hierarchical blocks of word line selection logics [16], to mitigate power and area overhead.

In summary, as the number of ports increases, FinFETs are a promising technology to replace CMOS for two reasons: 1) the read and write delay does not increase significantly, because there is less parasitic capacitance added to the internal node capacitance in write acceleration and the bit line capacitance in read acceleration when using FinFET and (2) due to its high threshold voltage, multi-port FinFET SRAM has a high cell stability against the destructive read under both simultaneous read and half-selected conditions.

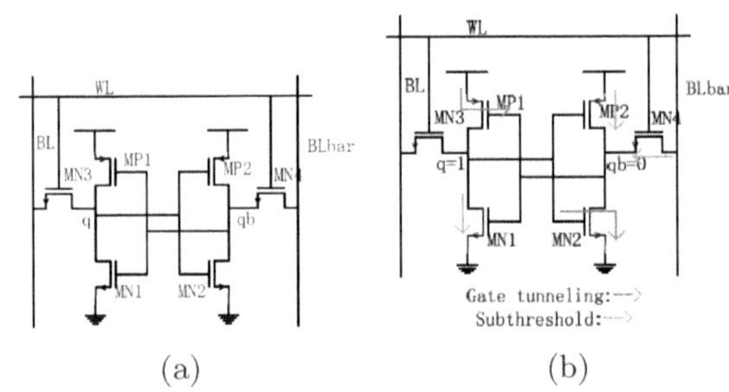

Figure 1: (a) 6T classical SRAM cell [9]. (b) leakage current model for 6T SRAM cell [9].

3. Classical 6T FinFET SRAM cell

In this section, we evaluate the classical single-port 6T FinFET SRAM. For the FinFET technology, there are two popular models: UFDG model [17] and PTM [10]. We use the double-gate PTM 22nm technology model and HSPICE as the platform in our FinFET SRAM simulations. In our simulations, based on the equation $W_{eff} = (2 \times h_{fin} + W_{fin}) \times N_{fin}$, where the fin height h_{fin} and fin width W_{fin} is fixed, we can change the number of fins N_{fin} to get a reasonable effective channel width W.

Figure 1(a) illustrates the classical FinFET SRAM cell. Generally, cell stability, power, and read/write performance are common metrics in evaluating an SRAM cell. Cell stability is evaluated in terms of static noise margin (SNM) [9], measuring the DC disturbance in the unselected condition and the read condition. It is obtained by measuring the two

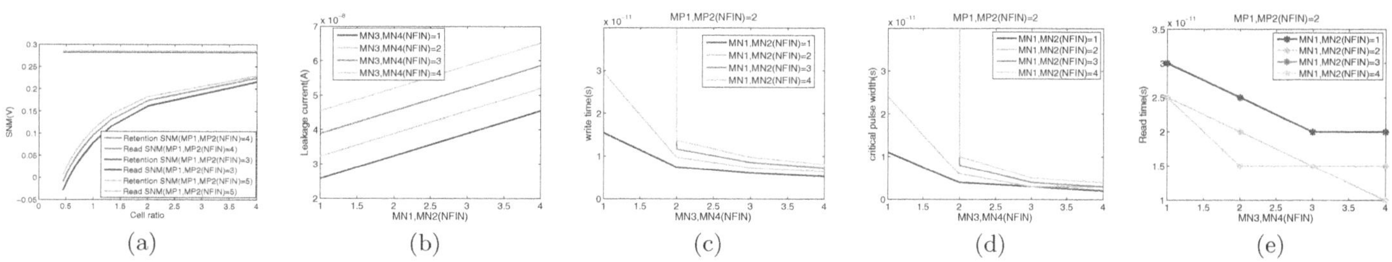

Figure 2: (a) Retention and read SNM of the 6T SRAM cell. (b) Leakage current of the 6T SRAM cell.(c) Write time of the 6T SRAM cell. (d) Critical pulse width of the 6T SRAM cell. (e) Read time of the 6T SRAM cell. (The red curve and the blue curve fully overlap).

squares in butterfly curves [9]. There are mainly two types of SNM that should be considered in SRAM designs: retention SNM and read SNM. The SNM in half selected condition (half-selected SNM) is the same as read SNM in classical FinFET SRAM. The static power, in terms of leakage current, is mainly sub-threshold current and gate tunneling current, as shown in Figure 1(b). In FinFET, the gate tunneling current is negligible, compared to the sub-threshold one. Read performance is evaluated in terms of read time, which is the time between the rise of word line and the time the sense amplifier can detect the voltage difference of the two bit lines [18]. We evaluate the write performance using two metrics: write time and critical pulse width. Write time is the time between half the rise of word line and half the rise of node 'qb' when writing a '0' [6]. Critical pulse width [19] is the smallest word line pulse width to successfully write a '0' or a '1' to the cell.

Cell stability: Figure 2(a) shows the result of retention SNM and read SNM of the classical FinFET SRAM with different cell ratios, which is the ratio of the size of the pull-down transistor to the size of the access transistor [9]. As the cell ratio grows, the read SNM also increases. Figure 2(a) also shows the pull-up transistors have limited impact on read SNM. For retention SNM, it is almost independent to the cell ratio.

Power: Figure 2(b) shows the sum of two types of leakage current in this FinFET SRAM cell, in which the sub-threshold is dominant. The sub-threshold current is proportional to its conductance, which is proportional to the transistor width. Thus, the leakage current is proportional to the width of the transistors.

Read time: In the simulation setup, the bit line capacitance is set based on the estimation of a standard 128×128 SRAM array. Figure 2(c) shows the read time when we change the size of access transistors. We observe that read time decreases when we increase the width of pull-down transistors (MN1 and MN2), and it increases when we decrease the width of access transistors.

Write performance: As shown in Figure 2(c) and Figure 2(d), when we increase both the widths of pull-down transistors, the write time and the critical write pulse becomes wider, which is contrary to the case of read time. However, when we increase the size of access transistors, both read and write time decreases. When the cell ratio is too small, such as $N_{fin} \geq 3$ for MN1/2 and $N_{fin} \leq 2$ for MN3/4, the values of write time and the critical pulse width are almost infinite, which means the write operation fails. Note that when evaluating the read/write performance, we just change the size of pull-down transistors, instead of changing cell ratio, which is not directly related to the read/write performance. Changing the size of pull-up transistors (MP1 and MP2) has similar impact to sizing the pull-down transistors and is not reported here.

4. Multi-port FinFET SRAM Cells

This section compares the double-ended and single-ended FinFET multi-port SRAM structures to the classical 6T SRAM.

4.1 Double-ended multi-port FinFET SRAM

In a double-ended multi-port SRAM, two extra access transistors are added to the classical FinFET SRAM cell to compose an extra port in the double-ended multi-port FinFET SRAM cell. In this sub-section, we present simulation results comparing this structure to the 6T cell.

Cell stability: Figure 3(a) shows the result of the retention SNM, the one-read SNM and the two-read SNM when the width of the pull-up transistors is fixed. One-read SNM of a cell is the SNM when there is only one read access is served in the column. It may cause the half selected condition if the read is served in another cell. Two-read SNM of a cell is the SNM when there are a read access is served in this cell while there is another read/write access is served in another port of the same cell or in another cell of the same column. In this case, if there is a read and a write served in the same column, it is a severer half selected condition, because the cell is disturbed simultaneously by the read word line and the write word line. In Figure 3(a), the two-read SNM is much worse than the one-read SNM. Therefore, when we continuously increase the number of access transistors without any additional structures to avoid the half-selected condition, the cell becomes fragile.

Leakage current: Figure 3(b) shows that increasing the width of pull-down transistors or access transistors leads to the growth of the leakage current. Also, the extra two access transistors in the 8T cell compared the 6T cell contribute in higher leakage current.

Write performance and read time: In Figure 3(c) and (d), write time and CPW is slightly larger than that of the 6T FinFET SRAM, because the two extra access transistors brings extra capacitance to the cell, which increases the charging time. The read time of one read in either of the ports is not included in the graph, because it is the same to the read time in the classical 6T FinFET SRAM. However, we include the results of the read time in simultaneous reads scenario in Figure 3(e), compared with the read time in the 6T FinFET SRAM, which shows that the read time gets worse in the double-ended multi-port FinFET SRAM when we simultaneously read in multiple ports.

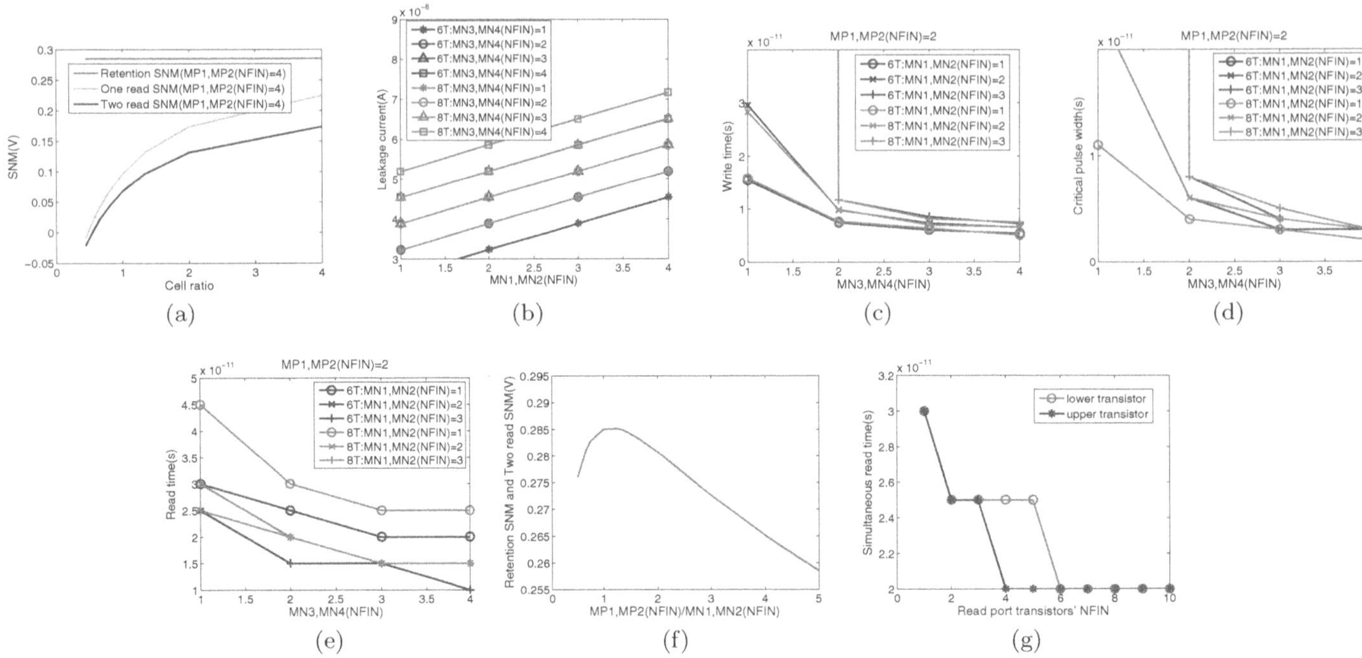

(a) (b) (c) (d) (e) (f) (g)

Figure 3: (a) SNM, (b) leakage current, (c) write time, (d) CPW, and (e) simultaneous read time of the double-ended FinFET SRAM cell with one and two sets of access transistors; (f) retention and two read SNM and (g) simultaneous read time of three single-ended structures.

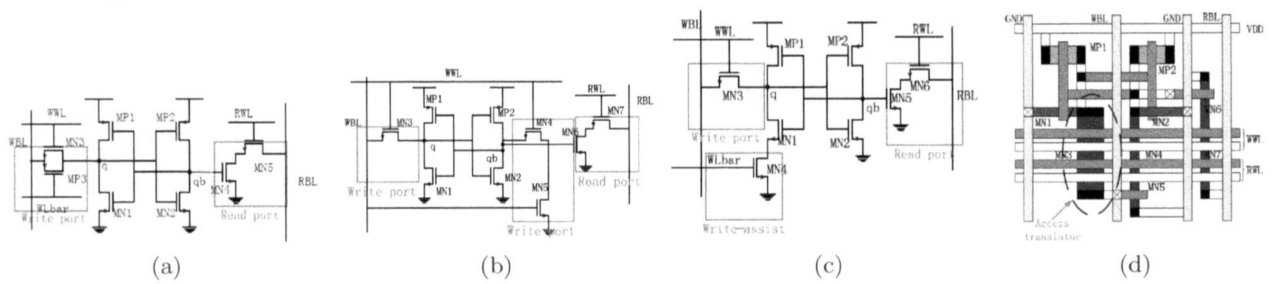
(a) (b) (c) (d)

Figure 4: (a) Single-ended multi-port FinFET SRAM structure 1 [9]. (b) Single-ended multi-port SRAM structure 2 [8,9]. (c) Single-ended multi-port SRAM structure 3 [9]. (d) Layout of structure 2 based on [20].

4.2 Single-ended multi-port FinFET SRAM

In this sub-section, we introduce three different single-ended multi-port SRAM structures, as shown in Figure 4. Write-assist techniques are implemented in all these three structures to enhance the write performance. There are four combinations for at most two write ports and two read ports: 1R/1W, 1R/2W, 2R/1W, and 2R/2W. We compare the simulation results of 2R/1W and 2R/2W structures, since the performance of structures with one or two read ports is almost the same, except for leakage current, which linearly increases as read ports increase in our simulation.

In Figure 4((a), a transmission gate is used to replace the classical nMOS access transistor, which can write a strong '0' and '1' to the cell. In Figure 4(b), we use MN4 and MN5, two nMOS transistors, to create a pseudo-double-ended structure when writing a '1'. In Figure 4(c), an extra write assist transistor MN4 is added as a switch, which turns off the feedback loop when writing '1'. The extra transistor is shared through a whole row, reducing the area [8].

Cell stability: Since read ports are isolated, the retention SNM and the two-read SNM are the same, regardless of the change in N_{fin} of the other transistors other than the read port transistors. The only parameter that influences the retention SNM and the two-read SNM is the ratio of the size of pull-up transistors to the pull-down transistors. And for these SNMs, three structures have the same performance because they have the identical internal structure. Figure 3(f) shows that the SNM has a peak value, which locates at the ratio range between 1 and 1.5. This is the point that the resistance of the pull-down transistor and the pull-up transistor is almost the same. Half-selected condition can happen in single-ended structures when the write word line is enabled in a whole row, affecting unselected cells. When the voltage of 'qb' is changing from '0' to '1', node 'q' cannot be pulled down to ground due to the impact of the high voltage of the bit line. However, this condition rarely happens, because of no bit line on the side of 'qb' causing the DC disturbance. Therefore, we do not include the half-selected SNM of single-ended structures in this paper.

Read time: Based on simulation results, the only parameter that significantly affect the simultaneous read time is the size of read port transistors. However, the lower transistor and the upper transistor in the read port also have minor effects as reported in Figure 3(g).

Leakage current: It is measured in the scenarios when '0' or '1' is stored, since the cell is asymmetric. We only evaluate the result of changing the size of read port transistors

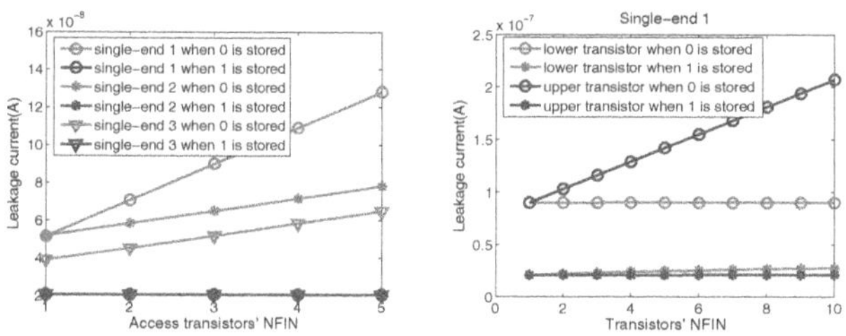

(a) (b)

Figure 5: (a) Leakage current versus the access transistors of the three single-ended structures. (b) Leakage current versus the read port transistors of single-ended structure 1.

and access transistors in 2R/1W structures, because increasing the write port transistors linearly increases the leakage current. In Figure 5(a) and Figure 5(b), the result shows that when '0' is stored, increasing either the size of the upper read port transistor or the access transistors causes high leakage current. The impact of two internal inverters on leakage current in single-ended structures is similar to that of the double-ended structure. When we add a write port, the leakage current increases by a certain amount when '0' is stored. In the case when '1' is stored, extra write ports do not impact the leakage current, since write port transistors, which have no voltage difference between their two terminals, do not contribute to leakage current.

Critical pulse width (CPW) and write time: They are also evaluated in the scenarios of writing '0' or '1'. The feedback inverter and the forward inverter have different influences in the write performance. For each inverter, we set the size of pMOS transistor to twice the size of the nMOS transistor. In Figure 6(a) and (e), feedback transistors seriously affect the write time and the CPW of structure 2 and 3 when writing '0'. They rapidly increase to infinity as the size of inverter increase. That is because when the feedback inverter is too strong, the access transistor cannot pass '0' to the internal node. Comparatively, the effect when writing '1' is not obvious for structure 2 and 3, because the feedback inverter is off in structure 3 and pseudo double-ended structure is enabled in structure 2 when writing '1'. In structure 1, the transmission gate can pass either strong '0' or strong '1' at the cost of leakage current. Meanwhile, for the forward inverter, when we change the size of transistors, the change in write performance is less significant than that of the feedback inverter, because there is no conflicts between the effect of the access transistors and the forward inverter. However, as shown in Figure 6(b) and (f), the effect of changing the size of access transistors is opposite to that of feedback inverters on CPW and write time. Finally, in Figure 6(c) and (g), increasing the size of the lower transistor in the read port increases the write time and the CPW, because the gate capacitance is increased. Compared to forward inverters, it has a larger impact on write performance. In addition, if we add an extra write port, the write performance is different between writing '0' and writing '1'. It is better when writing '1' and worse when writing '0'. Figure 6(d) and (h) shows the write time and the CPW of the first two single-ended structures.

4.3 Multi-port vs. single-port FinFET SRAMs

In this sub-section, based on comprehensive comparisons, we show that the single-ended multi-port FinFET SRAM provides the best tradeoffs between read acceleration and the read cell stability. Further, for less than 10% area overhead in sizing the access transistors, it can deliver equivalent write time to all other cells.

Leakage current: As double-ended and single-ended multi-port structures have more transistors than the 6T FinFET SRAM cell, they generate more leakage current with the same transistor size. Single-ended structure is a little worse than the double-ended one. However, there is a method for single-ended structure to reduce the leakage current, while maintaining the read time and write time at a reasonable level. That is to decrease the size of feedback inverters to its minimum value. In contrast, for the double-ended FinFET SRAM structure or the classical 6T FinFET SRAM, this approach is not feasible, because the read time and the write time change in opposite directions when we change the internal inverters size, and reducing the size of access transistors increases both the read time and the write time.

Read time: The read time of FinFET SRAM cells is in the order of 10^{-11} s. Single-ended multi-port FinFET SRAM is the best candidate in multi-port FinFET SRAM structures for getting the similar read time of classical 6T FinFET SRAM cell. There are two advantages of single-ended structures regarding to read operations. The first one is that, regardless of the number of read ports added, the simultaneous read time is identical to that of the single-ended structure with only one read port. The other advantage is that we only need to select proper sizes of transistors for read ports to ensure the reasonable read time in single-ended structure while maintaining leakage current and cell stability. In double-ended structures, the read time is related to internal transistors and access transistors. In order to make the simultaneous read time close to the read time in classical 6T FinFET SRAM cell in the double-ended structure, we should increase the size of access transistor, which seriously increases power and decreases cell stability.

Cell stability: For retention SNM, the result of the multi-port structures is similar to that of 6T FinFET SRAM. The difference is in read SNM. In the single-ended multi-port structures, we can easily achieve the same read SNM as retention SNM, because of the isolated read ports. In the double-ended structure, the simultaneous read SNM is worse, compared to the single-ended structure, due to the isolated read port in the single-ended structure. Meanwhile, in the single-ended structures, the half selected condition decreases its cell stability and make it worse than that of the double-ended structure, because there is almost no DC disturbance on 'qb' node while in the double-ended structure. The half selected SNM is much worse as the number of ports increases in single-ended structures.

Write performance: Both CPW and write time is in the order of 10^{-11} to 10^{-12} s in all the structures simulated with the FinFET models. We observe that in Figure 3(c) and (d), increasing the ports of write and read does not have a significant impact on the write time in both double-ended and single-ended multi-port structures. And we should set the size of access transistors in single-ended structures to 2 to 3 times that of double-ended structures, or make the feedback inverter of single-ended structures even smaller, in order to achieve the same write performance of the double-ended structure. For example, consider the layout of structure 2 with one isolated read based on the classical 6T cell [20] in Figure 4(d), where sizing the access transistors three times adds 9% area overhead.

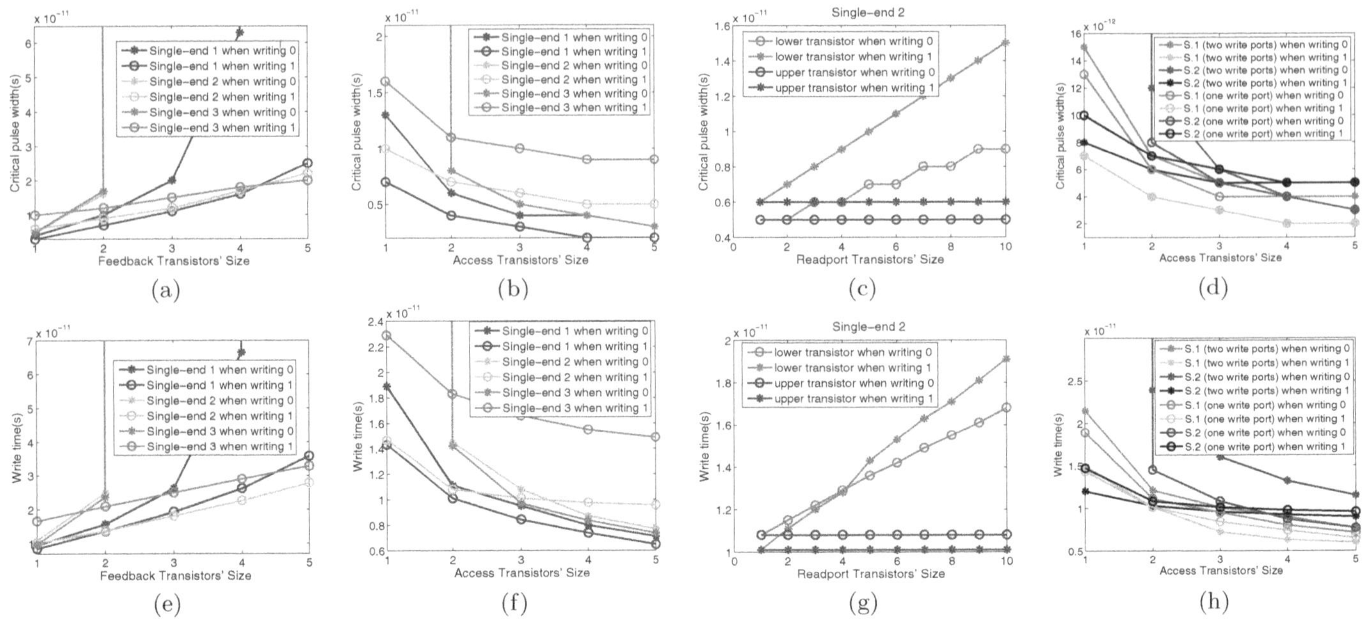

Figure 6: (a) CPW of single-ended structures for feedback inverter sizing; (b) CPW of single-ended structures for access transistor sizing; (c) CPW of structure 2 for read port transistor sizing; (d) CPW of structures 1 and 2 (1–2 write ports) for access transistor sizing; (e) Write time of single-ended structures for feedback inverter sizing; (f) Write time of single-ended structures for access transistor sizing; (g) Write time of structure 2 for read port transistor sizing; (h) Write time of structures 1 and 2 (1–2 write ports) for access transistor sizing.

5. Conclusions

Our evaluations of read and write acceleration in different multi-port FinFET SRAM structures illustrate the impact on read/write performance, leakage current, and cell stability. Based on simulation results with the PTM FinFET model, single-ended multi-port FinFET SRAM with isolated read ports is a good choice for multi-port design, since for similar leakage current, write time, and 9% area overhead, it performs better in read operation, offers higher flexibility in the configuration of read acceleration, and provides better cell stability than double-ended multi-port FinFET SRAM.

References

[1] P. Manoilov and P. Krivoshieva, "Shared memory design for multicore systems," in *International Scientific Conference Computer Science*, pp. 302–307, 2008.

[2] H. Yang, M. Chang, S. Lai, H. Wang, and W. Hwang, "A low-power low-swing single-ended multi-port SRAM," in *Intl. Symposium on VLSI design, automation and test*, pp. 1–4, 2007.

[3] F. Lai, Y. Chuang, and S. Chen, "A new design methodology for multiport SRAM cell," *IEEE Trans. Circuits and Systems*, vol. 41, pp. 677–685, November 1994.

[4] F. Lai, Y. L. Chuang, and S. J. Chen, "A new design methodology for multiport SRAM cell," *IEEE Trans. Circuits and Systems*, vol. 41, pp. 677–685, November 1984.

[5] C. C. Chao and B. A. Wooley, "A 1.3-ns 32-word x 32-bit three-port BiCMOS register file," *IEEE Journal of Solid-state Circuits*, vol. 31, pp. 758–766, June 1996.

[6] R. Joshi, K. Kim, and R. Kanj, "FinFET SRAM design," in *International Conference on VLSI Design*, pp. 440–445, 2010.

[7] K. Kim, C.-T. Chuang, J. Kuang, H. Ngo, and K. Nowka, "Low-power high-performance asymmetrical double-gate circuits using back-gate-controlled wide-tunable-range diode voltage," *IEEE Trans. Electron Devices*, vol. 54, pp. 2263–2268, September 2007.

[8] J. Singh, D. S. Aswar, S. P. Mohanty, and D. K. Pradhan, "A 2-port 6T SRAM bitcell design with multi-port capabilities at reduced area overhead," in *Proc. Intl. Symposium on Quality Electronic Design*, pp. 131–138, 2010.

[9] N. Tzartzanis, "Static memory design," in *High-Performance Energy-Efficient Microprocessor Design* (V. G. Oklobdzija and R. K. Krishnamurthy, eds.), ch. 4, Springer, 2002.

[10] "PTM-MG multi-gate model," 2012. Please visit the URL http://ptm.asu.edu/ for further details.

[11] M. Golden and H. Partovi, "A 500 mhz, write-bypassed, 88-entry, 90-bit register file," in *Proc. Symposium on VLSI circuits*, pp. 105–108, 1999.

[12] E. S. Fetzer and J. T. Orton, "A fully-bypassed 6-issue integer datapath and register file on an Itanium microprocessor," in *Proc. Intl. Solid-state Circuits Conference*, pp. 421–478, 2002.

[13] N. Tzartzanis, W. W. Walker, H. Nguyen, and A. Inoue, "A 34word x 64b 10R/6W write-through self-timed dual-supply-voltage register file," in *Proc. Intl. Solid-state Circuits Conference*, pp. 338–357, 2002.

[14] H. Bajwa and X. Chen, *Area-Efficient Dual-Port Memory Architecture for Multi-Core Processors*. PhD thesis, City University of New York, 2007.

[15] T. Suzuki, H. Yamauchi, Y. Yamagami, K. Satomi, and H. Akamatsu, "A stable 2-port SRAM cell design against simultaneously read/write-disturbed accesses," *IEEE Journal of Solid-state Circuits*, vol. 43, pp. 2109–2119, September 2008.

[16] M. Yoshimoto, K. Anami, H. Shinohara, T. Yoshihara, H. Takagi, S.Nagao, S. Kayano, and T. Nakano, "A divided word-line structure in the static RAM and its application to a 64 K full CMOS RAM," *IEEE Journal of Solid-state Circuits*, vol. 18, pp. 479–485, October 1983.

[17] SOI Group of University of Florida, *UFDG MOSFET Model User Guide (Linux Version 3.71)*, 2003. Please visit the URL http://www.soi.tec.ufl.edu for further details.

[18] S. Yaldiz, U. Arslan, X. Li, and L. Pileggi, "Efficient statistical analysis of read timing failures in SRAM circuits," in *Proc. Intl. Symposium on Quality Electronic Design*, pp. 617–621, 2009.

[19] J. Wang, S. Nalam, and B. H. Calhoun, "Analyzing static and dynamic write margin for nanometer SRAMs," in *Proc. Intl. Symposium on Low Power Electronics and Design*, pp. 129–134, 2008.

[20] N.-H. E. Weste and D. Harris, "Array subsystems," in *CMOS VLSI design: a circuits and systems perspective (4th edition)* (M. Goldstein and M. Suarez-Rivas, eds.), ch. 11, Pearson and Addison-Wesley, 2012.

Modeling Symmetrical Independent Gate FinFET using Predictive Technology Model

Mohammad Yousef Zarei*, Reza Asadpour*
Siamak Mohammadi*, Ali Afzali-Kusha*, Razi Seyyedi**
*School of ECE, University of Tehran, Tehran, Iran **Institute of CS, University of Bonn, Germany
{my.zarei, r.asadpour}@ece.ut.ac.ir, {smohamadi, afzali}@ut.ac.ir, seyyedi@informatik.uni-bonn.de

ABSTRACT

Predicting MOSFET models plays a pivotal role in circuit design and its optimization. Independent Gate FinFETs (IGFinFET) are interesting for designers as they are more flexible than Common Multi-Gate FinFETs (CMGFinFET) in digital circuit design. In this work, we implement a model for symmetrical IGFinFET using CMGFinFET model based on Multi-Gate Predictive Technology Model (PTM-MG). This model has been developed from TCAD IGFinFET, based on previously published experimental results of CMG-FinFET. Different basic gates in SG (shorted gate), LP (low power), IG (low area), and IG/LP modes have been designed using the implemented model. For LP, IG, and IG/LP NAND gates, the leakage power is reduced by 89%, 26%, and 67%, respectively in comparison to SG. To show that our model does not have any convergence problem for large circuits, we used ISCAS'85 benchmark suite. The results show that for independent gate in high performance PTM-MG library, on average we can save up to 24% in the number of transistors and lower the total power by 42%

Categories and Subject Descriptors

B.7.1 [**Hardware**]: Integrated Circuits— Types and Design Styles, Advanced Technologies

Keywords

FinFET, Independent Gate, Vertical Gate, Multi-Gate, 3D transistors, Predictive Models, SPICE

1. INTRODUCTION

As CMOS scaling continues beyond 25nm node, challenging issues such as gate leakage [1] Short Channel Effects (SCE) and process variation arise. Among alternative structures, FinFETs are promising due to their better control of gate over channel, and hence less SCE and immunity to process variation because of lightly doped body [2]. Therefore, it is indispensible to develop a model for FinFETs as a candidate to replace conventional CMOS technology in the future.

Among different types of FinFET structures, IGFinFET can be used in various kinds of applications such as Multiple Threshold CMOS (MTCMOS) [3], utilizing dual V_{th} for low power schemes [4], high speed domino logics [5], and merging a couple of transistors in series which leads to reduction of capacitance and/or area footprint of logic gates [4]. The feature that one gate can be adjusted to determine the V_{th} allows the usage of intrinsic body, eliminates the V_{th} adjustment implantation and prohibits random dopant fluctuation [6]. Moreover, as in [7], [8] the process flow of IGFinFET fabrication can be co-integrated with FinFET technology. The most important part of IGFinFET process flow is the separation of gates which can be done by a Chemical Mechanical Polishing (CMP) and a self stopping mechanism to make sure the fin is undamaged [8].

According to planar CMOS predictive modeling, Berkeley Predictive Technology Modeling (BPTM) [9] and Arizona State University (ASU) PTM [10] have been developed for up to 20nm technology node. Also University of Florida Double-Gate (UFDG) SPICE model is available. It is physics-based and shows a perfect agreement with experimental measurements of FinFETs [11]. To this day, the compact model of BSIM IGFinFET has not been publicly released. UFDG model needs some tuning to be used in different technology nodes. Also this model is a double-gate MOSFET model and is unable to model FinFET properties and suffers from convergence problem for 22nm node [12]; however, our model is based on FinFET structure. Another available tool for simulating circuits is TCAD. It is highly accurate, but the simulation time for large circuits is too high and there is a constraint on the number of allowed nodes.

We use a method to transform the CMGFinFET to IGFinFET. Our goal in this work is to develop a Predictive Technology Model for symmetrical IGFinFET based on PTM-MG [13]. The presented model uses SPICE for simulation; therefore, the simulation time is lower than the other models for 25nm node technology and beyond. As a consequence, all implemented ideas using IGFinFET TCAD models for beyond 25nm, can be used in simulation of large circuits saving time of simulation, without losing the accuracy.

The rest of the paper is organized as follows. Section 2 is a background about IGFinFETs an analysis of V_{th}, and describes the model parameters similar to BSIM-CMG (Berkeley Short-channel IGFET Common Multi-Gate)[14]. Section 3 presents the methodology of our developed model and validating it using TCAD. In Section 4, we explain libraries implemented by the model and evaluation of our model us-

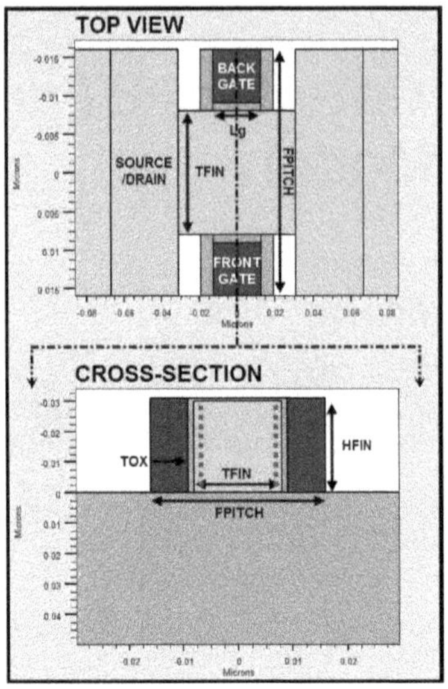

Figure 1: Top and cross-sectional view of IGFinFET

ing ISCAS'85 suite benchmarks. Finally, in Section 5 we conclude this work and present some future guidelines.

2. BACKGROUND

2.1 Symmetrical Independent Gate Physical Structure and Parameters

Etching the top part of the gate on fin channel changes conventional symmetrical CMG-FinFETs to symmetrical IGFinFETs (i.e. the gate oxide for both sides has the same thickness). Figure 1 shows top and cross sectional view of SOI IGFinFET, in which are depicted the primary parameters. Primary parameters variation has a significant impact on behavior, while there are secondary parameters, for fine tuning [13]. By applying gate voltage in IGFinFET, the sides of the fin (doted parts in cross-sectional view) are affected by the electrical field and the channel (inversion layer) is created. Thus, the effective width is W_{eff} = 2*HFin. For multi-fin transistors the W_{eff} is equal to 2*n*HFin, where n is the number of fins.

2.2 V_{th} Analysis of FinFETs

By solving Poisson equation in channel region for 3 dimensions and assuming depletion approximation, the potential in channel direction can be written as [14]:

$$\varphi(x) = \varphi_0 exp(\pm\frac{x}{\lambda_l}) \quad (1)$$

where φ_0 is a constant. λ_l is called "Natural length" and can be calculated using (2)[14]:

$$\lambda = \sqrt{\frac{\varepsilon_{Si}}{n.\varepsilon_{OX}} t_{OX} t_{Si}} \quad (2)$$

where ε_{Si} and ε_{OX} are silicon and oxide permittivity, respectively and t_{OX} and t_{Si} are oxide and silicon thickness, respectively. n represents number of gates around channel. The less the λ, the more the attenuation of drain potential inside channel and therefore, SCE can be mitigated by decreasing t_{OX}, t_{Si}, or using high-k material.

Analysis of V_{th} in [4] shows that for achieving low current when one gate is disabled and high current when both gates are enabled concurrently, t_{Si} and t_{OX} should be reduced and increased, respectively. However, the impact of increasing t_{OX} will be attenuated as the control of gate over the channel is diminished for undoped devices [15].

V_{th} can be obtained using (3)[4]:

$$V_{th} = -\phi_{ms} + \frac{Q_D}{C_{OX}} + V_{inv} + V^{QM} - V^{SCE} \quad (3)$$

Where ϕ_{ms} is the difference between electrode and silicon work-function, V_{inv} is a constant that shows extra surface potential needed to bring enough charge into channel for the strong inversion for transistor to reach threshold, Q_D is the depletion charge, C_{OX} is the gate capacitance, V^{QM} represents quantum mechanical effects, and V^{SCE} is the short channel effect. Using some simplifications, the only terms that have significant role in V_{th} are ϕ_{ms} and V^{SCE}. Decreasing t_{Si} would reduce SCE and raise V_{th}. In summary using simplified subthreshold slope S formula in different two modes i.e. disabling one gate and activating both gates [16], shows that increasing t_{OX} leads to higher V_{th}. Although, it has a reverse point for total leakage current and due to Drain Induced Barrier Lowering (DIBL) the leakage current increases. Hence, there is an optimum value for t_{OX}.

In IGFinFET structures the gates are not tied together, however, the electrostatic correlation between gates is strong because of narrow fin. In other words, if one gate is activated, developed channel under this gate would make the other gate easier to be turned on. Also, if one gate is disabled, the input capacitance is approximately half of that of the activated transistor. Therefore the circuit speed may rise due to reduction of load on the driver circuit [4].

3. NEW PREDICTIVE METHODOLOGY

There are no dual independent gate transistors beyond 25nm technology that can be utilized for large circuits. In this section we explain how to transform CMGFinFET to IGFinFET. We used published results the same as of PTM-MG [13] and TCAD modeling, Hence the accuracy is acceptable.

A technique has been used in [17] to show effects of back gate in common gate transistors. The technique is similar to body biasing in classical bulk technologies [18] [19]. Basically it is a Voltage Controlled Voltage Source (VCVS) on gate line of CMG transistors. This approach is not accurate enough for circuit design. Therefore, to implement IGFinFET model from CMGFinFET model using this technique, we should generalize this method. Figure 2 shows a flowchart describing prediction model development. Details of implementing IGFinFET model will be explained in next subsection.

3.1 Development of Predictive Model

Due to the lack of silicon data for IGFinFET from foundries; we used previously published 25nm CMGFinFET technology node data [20]. We assumed that if effective widths of CMG and IG were the same, I-V characteristics of them would be identical. Hence, based on published results of CMG transistors, TCAD IG transistors can be created. The first step of our model development is to fit PTM-MG model using I-V characteristics of TCAD.

In IGFinFET for a constant V_{ds} we read I_{ds} for an ar-

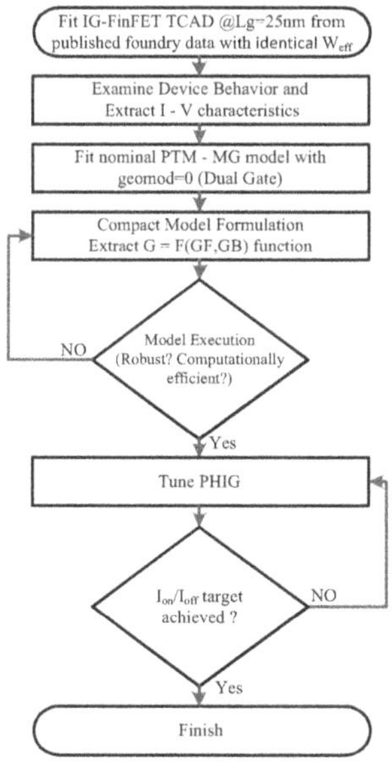

Figure 2: Flowchart of predictive technology model development [13]

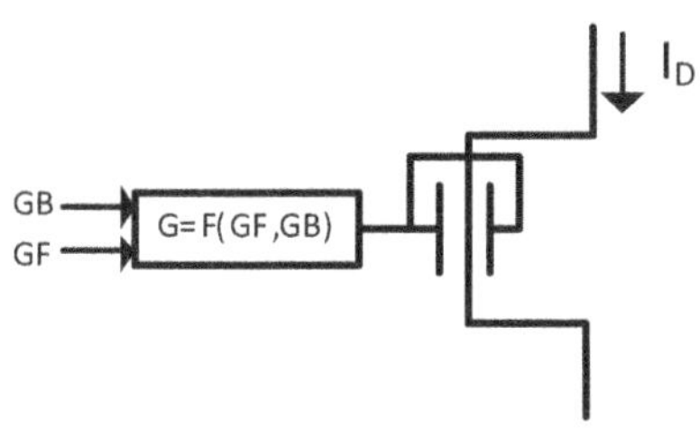

Figure 3: Transistor structure and Verilog-A module for mapping GB,GF to G

bitrary front and back gate voltage. If we can find a gate voltage in CMGFinFET for the same I_{ds} and V_{ds}, then it is the corresponding value of IGFinFET front and back gate voltage. Hence, we can define a mapping from GF, GB space to G space in the form of G=F(GF, GB) for constant V_{ds} using I-V_{gs} characteristics, in which GF and GB are the potentials of IGFinFET and G is the corresponding gate potential of CMGFinFET. Utilizing this, we will map the current of IGFinFET to CMGFinFET for the same V_{ds}. If gate voltages are constant (GF, GB, G) the I-V_{ds} characteristic would be identical. Thus, the mapping from GF, GB space to G space is independent of source-drain voltage. Because of symmetrical structure of IGFinFET developed in TCAD the mapping has symmetry over GF, GB parameters. To be practically used in circuit, the mapping can be implemented as a voltage controlled voltage source at the gate or as a module in Verilog-A as shown in Figure 3.

For delay consideration, input capacitance should be added. Each input gate node is connected to a CMGFinFET with half W_{eff}. To make sure these added devices have a negligible impact on noise and power consumption, they would be supplied by a voltage source other than main supply voltage of the circuit.

To prevent SCE and especially DIBL and to achieve subthreshold slope less than 90mV/dec, L_g/T_{Fin} ratio is approximately 1.5 [21]. The leakage current depends on various parameters especially threshold voltage. For simplicity and controlling the leakage the same as in PTM-MG [13], only PHIG parameter which is gate work function has been changed to acquire the desired I_{on} and I_{off} ratio.

Table 1: PTM-MG Verification

Data source	P-TYPE FINFET	N-TYPE FINFET
L_g(nm)	25	25
V_{DD}(V)	1	1
EOT (nm)	1.1	1.1
TFIN (nm)	15	15
HFIN (nm)	30	30
I_{on} (model)($nA/\mu m$)	944	1300
I_{on} (TCAD)($nA/\mu m$)	923	1262
I_{on} (TSMC)($nA/\mu m$)	925	1296
I_{off} (model)($nA/\mu m$)	70.4	41
I_{off} (TCAD)($nA/\mu m$)	69.2	43
I_{off} (TSMC)($nA/\mu m$)	69	41
Worst Case Error	2.2%	4.87%

3.2 Validation of Predicted Model

Validation of I-V characteristics have been done using TCAD model based on experimental published results. To reduce the simulation time as in [22], we used 2D model instead of 3D model. TCAD model is developed from set of parameters of [13] and experimental data of [20], assuming that if W_{eff} are the same then I-V characteristic would be identical. FinFET parameters have been tuned to have the best I_{on} and I_{off} fitting in comparison to experimental results. Table 1 shows a validation of TCAD model vs. the results in [20]. Figure 4 shows I-V characteristics of an IGFinFET with L_g=25nm PTM model and TCAD model.

We have fitted the PTM-MG on TCAD results and used the fitted PTM model to find the parameters needed for the Verilog-A module. Because both PTM-CMG and our TCAD model are based on same experimental results [20], the developed module can be used along with two original PTM-MG libraries i.e. Low STandby Power (LSTP) and High Performance (HP) in circuit simulations.

4. PTM EVALUATION

IGFinFET transistor flexibilities lead to wide usage in digital circuits. Due to independent gate voltage, IGFinFET can be used for different schemes such as low-power, low-area, and high-performance. As a result, circuit area and parasitic capacitance would decrease in static and dynamic circuits. Using back-gate voltage, the front-gate threshold voltage can be controlled, accordingly leakage current reduces [23] and speed increases in interconnection [24]. In this section different schemes of circuit design using IGFinFETs and different NAND gates are explained. As a consequence, a standard cell based on IGFinFET model for various digital circuits has been developed in LSTP and HP libraries. Furthermore, for benchmarking ISCAS'85 suite is used and the results have been evaluated.

4.1 Exploring IGFinFET Features

DC transfer characteristic i.e. I_{ds} versus V_{gfs} for different back-gate voltages (V_{gbs}) of the model is shown in Figure 5 for V_{dd} =1V for both LSTP and HP libraries. V_{gfs} is the potential difference between front gate and source terminal. Source and drain nodes are connected to ground and voltage

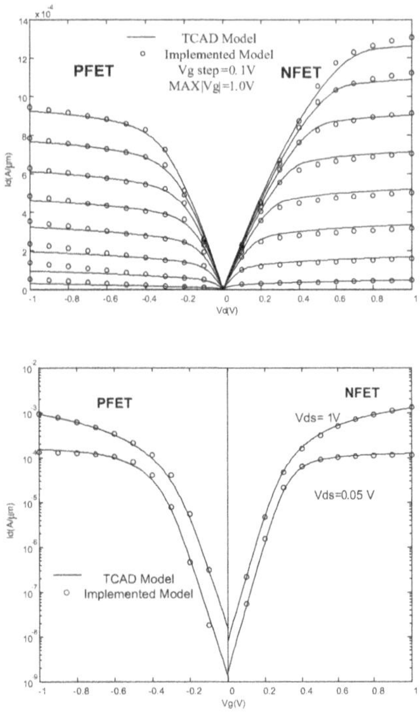

Figure 4: Implemented PTM fitted on TCAD simulation results

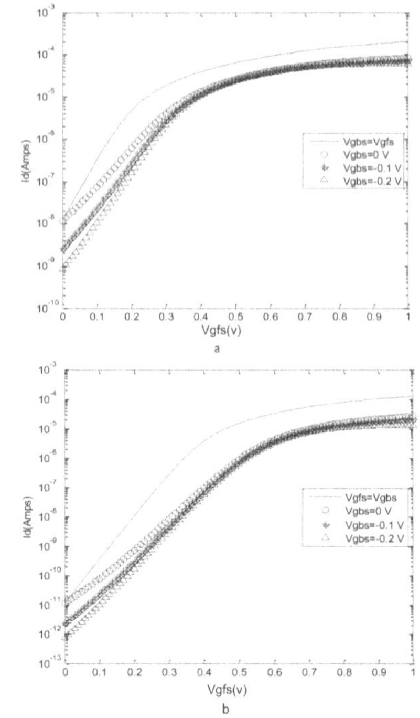

Figure 5: I-V_{gfs} of 25nm node NFinFET for different back gate voltages (a) HP library (b) LSTP library

supply, respectively. Because of dependence of temperature to switching activity and self-heating effect in FinFET structures [25], all the simulations have been done in 70 °C. As it can be seen in Figure 5.a the change in *off* and *on* current is considerable for different functional modes. The highest *on* current is when two gates are tied together, which is 70% more than when V_{gbs} is negative or zero biased. In contrast to leakage current which reduces considerably as V_{gbs} decreases, the *on* current is approximately the same for V_{gbs}=0V or V_{gbs}= -0.2V. The reason is exponential dependence of leakage current on threshold voltage that descends by lowering back-gate voltage. The same results from p-type FinFETs have been extracted.

As it is shown in Figure 5.b the overall behavior of LSTP library is the same as HP, but when the back gate voltage is reverse biased the *off* and *on* current would reduce drastically. Consequently, the delay will rise and the total power will decrease dramatically. In the rest of the paper we will use only HP PTM-MG for evaluation and show there is no convergence problem in our model.

4.2 Design of Logic Gates

FinFET can be used in three general modes in which gates can be manipulated to improve performance, power, or area. 1- SG (Shorted Gate): gates are connected together and the circuit topology is the same as CMOS circuits. 2- IG (low area): gates have independent potentials. If the threshold voltage is low the transistor may work as two parallel transistors. If threshold voltage is high it can conduct as two series transistors [4]. Using this feature, we can decrease the number of transistors in a digital circuit without changing the functionality. Hence, in this work we call this mode low area. 3- LP: reverse biasing of back gate (0-V_{th}=V_{low} and V_{dd}+V_{th}=V_{high} for NFET and PFET respectively) would decrease *off* and *on* current dramatically which leads to power reduction. This mode also increases the delay and can be used in non-critical paths of circuit. 4- IG/LP: it combines both LP and IG modes [25] and has both their features.

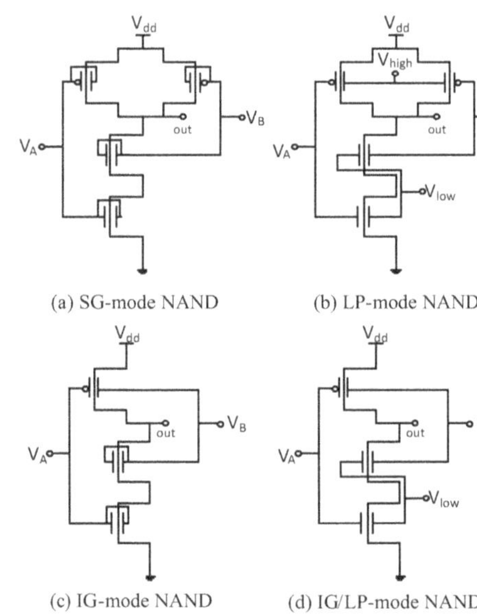

Figure 6: Different NAND gate configurations

In this subsection, we investigate NAND gate structures implemented by IGFinFETs. Four NAND gate configurations are shown in Figure 6. Figure 6.a shows the conventional NAND gate implemented by using IGFinFET in the place of CMOS transistors. Figure 6.b is the LP NAND

Table 2: Characteristics of NAND gates for HP library

Design	I_{OFF}(nA)				FO0(ps)			FO4(ps)			FO20(ps)		
	11	00	10	01	T_P	T_{Rise}	T_{Fall}	T_P	T_{Rise}	T_{Fall}	T_P	T_{Rise}	T_{Fall}
SG	133.6	8.36	52.5	46.48	2.54	4.09	3.82	22.98	19.2	29.72	41.91	81.4	132.3
LP	13.73	0.76	5.3	2.55	4.4	5.25	5.2	26.09	34.91	61.48	112.55	158.6	284.7
IG	66.85	8.36	52.49	46.48	2.8	4.1	4.02	12.32	27.02	28.96	54.19	120.8	131.6
IG/LP	66.84	0.76	5.3	2.55	3.09	3.6	3.92	22.62	25.65	60.13	101.91	117.7	283

Table 3: Power, Delay and PDP of ISCAS'85 suite using different modes of HP library with L_g=25nm

Design	SG			LP			IG			IG/LP		
	Power (μW)	Delay (ps)	PDP (10^{-15})	Power (μW)	Delay (ps)	PDP (10^{-15})	Power (μW)	Delay (ps)	PDP (10^{-15})	Power (μW)	Delay (ps)	PDP (10^{-15})
C432	210.0	116.2	24.4	137.1	199.7	27.4	182.4	125.3	22.9	131.7	165.3	21.8
C499	220.2	69.5	15.3	154.5	120.1	18.6	210.5	71.5	15.0	153.9	109.6	16.9
C880	299.4	101.2	30.3	197.5	179.8	35.5	251.2	105.9	26.6	194.4	145.1	28.2
C1908	889.6	106.3	94.6	804.8	179.7	144.6	788.1	111.5	87.9	765.0	159.7	122.1
C2670	1299.7	76.7	99.7	912.5	129.8	118.4	1058.5	77.6	82.1	914.7	114.7	104.9
C3540	1897.0	141.0	267.5	1291.5	235.3	303.9	1634.6	148.2	242.2	1310.7	210.7	276.1
C5315	2848.8	82.4	234.9	1882.0	146.4	275.5	2345.4	88.4	207.3	1872.0	112.8	211.1
C6288	9738.3	368.6	3589.0	4242.4	614.0	2604.6	7307.9	383.3	2800.7	4122.4	505.4	2083.4
C7552	4620.8	103.3	477.2	3483.9	167.1	582.0	3870.6	104.5	404.3	3358.9	141.2	474.3
Average	2447.1	129.5	537.0	1456.3	219.1	456.7	1961.0	135.1	432.1	1424.8	184.9	371.0
Saving	0	0	0	40%	-69%	15%	20%	-4%	20%	42%	-43%	31%

gate in which the back gate voltage of n-type FinFET for pull down network is V_{low}=-0.2V and V_{gbs} of p-type FinFET for pull up network is V_{high}=V_{dd}+0.2. If a voltage higher than 0.2 in V_{high} or lower than -0.2 in V_{low} is used for V_{gbs} the delay will be an issue [25]. Figure 6.c shows the IG NAND gate which benefits from the independent feature to decrease the number of transistors. This configuration has three transistors in contrast to conventional NAND gates; therefore, it has the lowest area [25]. LP/IG NAND gate is shown in Figure 6.d which has less delay and leakage current than LP and IG modes, respectively.

All used FinFET transistors in pull up and pull down networks are the same. Characteristics for IGFinFET HP PTM-MG library of NAND gates are listed in Table 2. Delay has been extracted using HSPICE under three different loads: unloaded and with loads of four (FO4) and twenty (FO20) minimum-sized SG-mode FinFET inverters. T_{Rise} and T_{Fall} are the time when output goes from 0.1*V_{dd} to 0.9*V_{dd} and from 0.9*V_{dd} to 0.1*V_{dd}, respectively. Propagation delay (T_p) is the average of T_{plh} and T_{phl} which is the time difference between input hitting 50% of V_{dd} and output hitting 50% of V_{dd}. T_{Rise} and T_{Fall} of all inputs are 5ps. Leakage current has been reported for standby mode of different input combinations.

For HP library the LP mode has 73% delay more than that of SG mode, while its leakage power is 89% lower. In IG mode only one gate has been used in pull up network and all of transistor sizes are the same. T_{Rise} and T_{Fall} are equal for this configuration of NAND gate for different fan outs. IG leakage power is 26% lower than that of SG mode. When back gate of pull down transistors is biased negatively in IG/LP mode, driving current plunges and T_{Fall} rises. Therefore, it is not a suitable choice for the critical path of circuits. The leakage power for this mode is 67% lower than that of SG mode.

4.3 ISCAS'85 Benchmarking Results

Four different types of standard cell from basic gates in SG, IG, LP, and IG/LP modes have been provided. Gates of each standard cell include NAND, NOR, AND, and OR, that have 2 to 4 inputs, XOR and XNOR that have two inputs, and Inverter and Buffer that only have one input. Table 3 shows power, delay, and Power Delay Product (PDP) of synthesizable circuits of ISCAS'85 benchmark for four standard cells in HP library. Critical paths of ISCAS suite circuits have been extracted using a synthesis tool, then implemented using HSPICE simulator. To attain the propagation delay, inputs are set to have the most possible delay. All transistors used in the libraries have the same size. T_{Rise} and T_{Fall} of all inputs are equal to 5ps and the simulation temperature is 70 °C. Interconnects and fan outs of the circuits are modeled using [26][27]. Total power has been extracted under these conditions: activity factor = 0.1, input clock frequency = 500MHz, supply voltage = 0.9V.

Results in Table 3 show that the best mode of HP library is IG/LP, which improves power by 41%. Although the delay increases by 40%, the PDP is enhanced by 30% in this mode.

Area of the circuits corresponds to the number of transistor fins needed for the implementation which is shown in Table 4. According to subsection 4-2 the number of fins or area could be decreased by 24% for IG and IG/LP mode in comparison to LP or SG mode.

Table 4: Area Estimated

Design		SG - LP	IG - IG/LP
	No. Gates	No. Fins	No. Fins
C432	160	842	683
C499	202	1260	1432
C880	383	2057	1495
C1908	880	4272	3259
C2670	1193	6278	5013
C3540	1669	8412	6606
C5315	2307	12363	9891
C6288	2406	12496	7728
c7552	3512	17042	13036
Average	1412.4	7224.7	5460.3
Saving	-	0	24%

5. CONCLUSION AND FUTURE GUIDELINES

In this work we presented a model for IGFinFET based

on PTM-MG. The implemented model is based on a TCAD developed from silicon data and previously published experimental results of CMGFinFET. Standard cells for HP library have been developed in SG, LP, IG, and IG/LP modes using the implemented model. For LP, IG, and IG/LP NAND gates, the leakage power is reduced by 89%, 26%, and 67%, respectively in comparison to SG. To show the ability of the model to converge, for large circuits, ISCAS'85 suite has been used. The results show that for independent gate in HP PTM-MG library, on average we can save up to 24% in the number of cells and lower the total power by 42%. For future works process and temperature variations can be investigated and the trend of IGFinFET for beyond 20nm node technology will be explored.

6. REFERENCES

[1] D. Frank *et al.*, "Device scaling limits of Si MOSFETs and their application dependencies," *Proceedings of the IEEE*, vol. 89, pp. 259 –288, mar 2001.

[2] G. Neudeck, "An overview of Double-Gate MOSFETs," in *University/Government/Industry Microelectronics Symposium, 2003. Proceedings of the 15th Biennial*, pp. 214 – 217, june-2 july 2003.

[3] S. Shigematsu *et al.*, "A 1-V high-speed MTCMOS circuit scheme for power-down application circuits," *Solid-State Circuits, IEEE Journal of*, vol. 32, pp. 861 –869, jun 1997.

[4] M. Rostami *et al.*, "Dual- Independent-Gate FinFETs for Low Power Logic Circuits," *Computer-Aided Design of Integrated Circuits and Systems, IEEE Transactions on*, vol. 30, pp. 337 –349, march 2011.

[5] S. Tawfik *et al.*, "High speed FinFET domino logic circuits with independent gate-biased double-gate keepers providing dynamically adjusted immunity to noise," in *Microelectronics, 2007. ICM 2007. Internatonal Conference on*, pp. 175 –178, dec. 2007.

[6] D. Frank *et al.*, "Monte Carlo modeling of threshold variation due to dopant fluctuations," in *VLSI Circuits, 1999. Digest of Technical Papers. 1999 Symposium on*, pp. 171 –172, 1999.

[7] D. Fried *et al.*, "Improved independent gate N-type FinFET fabrication and characterization," *Electron Device Letters, IEEE*, vol. 24, pp. 592 –594, sept. 2003.

[8] M. Masahara *et al.*, "Four-Terminal FinFET Device Technology," in *Integrated Circuit Design and Technology, 2007. ICICDT '07. IEEE International Conference on*, pp. 1 –4, 30 2007-june 1 2007.

[9] Y. Cao *et al.*, "New paradigm of predictive MOSFET and interconnect modeling for early circuit simulation," in *Custom Integrated Circuits Conference, 2000. CICC. Proceedings of the IEEE 2000*, pp. 201 –204, 2000.

[10] [Online]. Available: ptm.asu.edu.

[11] W. Zhang *et al.*, "Physical insights regarding design and performance of independent-gate FinFETs," *Electron Devices, IEEE Transactions on*, vol. 52, pp. 2198 – 2206, oct. 2005.

[12] F. J. G., , *et al.*, "A process/physics-based compact model for nonclassical CMOS device and circuit design," *Solid State Electronics*, vol. 48, pp. 919 –926, Jun. 2004.

[13] S. Sinha *et al.*, "Exploring sub-20nm FinFET design with Predictive Technology Models," in *Design Automation Conference (DAC), 2012 49th ACM/EDAC/IEEE*, pp. 283 –288, june 2012.

[14] J. Colinge, *FinFETs and Other Multi-Gate Transistors*. Springer, 2008.

[15] A. Tsormpatzoglou *et al.*, "Threshold Voltage Model for Short-Channel Undoped Symmetrical Double-Gate MOSFETs," *Electron Devices, IEEE Transactions on*, vol. 55, pp. 2512 –2516, sept. 2008.

[16] M. Masahara *et al.*, "Experimental Investigation of Optimum Gate Workfunction for CMOS Four-Terminal Multigate MOSFETs (MUGFETs)," *Electron Devices, IEEE Transactions on*, vol. 54, pp. 1431 –1437, june 2007.

[17] Z. Jaksic *et al.*, "Enhancing 6T SRAM cell stability by back gate biasing techniques for10nm SOI FinFETs under process and environmental variations," in *Mixed Design of Integrated Circuits and Systems (MIXDES), 2012 Proceedings of the 19th International Conference*, pp. 103 –108, may 2012.

[18] Y. Yasuda *et al.*, "Design Methodology of Body-Biasing Scheme for Low Power System LSI With Multi- Vth Transistors," *Electron Devices, IEEE Transactions on*, vol. 54, pp. 2946 –2952, nov. 2007.

[19] A. Bonnoit *et al.*, "Reducing variability in chip-multiprocessors with adaptive body biasing," in *Low-Power Electronics and Design (ISLPED), 2010 ACM/IEEE International Symposium on*, pp. 73 –78, aug. 2010.

[20] C. Chang-Yun *et al.*, "A 25-nm gate-length FinFET transistor module for 32nm node," in *Electron Devices Meeting (IEDM), 2009 IEEE International*, pp. 1 –4, dec. 2009.

[21] J. Kedzierski *et al.*, "High-performance symmetric-gate and CMOS-compatible Vt asymmetric-gate FinFET devices," in *Electron Devices Meeting, 2001. IEDM '01. Technical Digest. International*, pp. 19.5.1 –19.5.4, 2001.

[22] S. Chaudhuri *et al.*, "3D vs. 2D analysis of FinFET logic gates under process variations," in *Computer Design (ICCD), 2011 IEEE 29th International Conference on*, pp. 435 –436, oct. 2011.

[23] B. Raj *et al.*, "Analytical modeling for the estimation of leakage current and subthreshold swing factor of nanoscale double gate FinFET device," *Microelectronics International*, vol. 26, no. 1, pp. 53 –63, 2009.

[24] H. V. Nguyen *et al.*, "Novel methodology for speeding up IC performance in 32nm FinFET," *IEICE Electronics Express*, vol. 9, pp. 227 –233, Feb. 2012.

[25] A. Muttreja *et al.*, "CMOS logic design with independent-gate FinFETs," in *Computer Design, 2007. ICCD 2007. 25th International Conference on*, pp. 560 – 567, 7-10 Oct. 2007.

[26] G. Petley, "VLSI and ASIC technology standard cell library design.," [Online]. Available: vlsitechnology.org.

[27] D. Sylvester *et al.*, "Getting to the bottom of deep submicron," in *Computer-Aided Design, 1998. ICCAD 98. Digest of Technical Papers. 1998 IEEE/ACM International Conference on*, pp. 203 – 211, nov 1998.

A Self-Tuning Multi-Objective Optimization Framework for Geometric Programming with Gate Sizing Applications

Amin Farshidi, Logan Rakai, Laleh Behjat, David Westwick
University of Calgary
2500 University Dr NW
Calgary, Alberta, Canada
{afarshid, lmrakai, laleh, dwestwic}@ucalgary.ca

ABSTRACT

Most engineering problems involve optimizing different and competing objectives. To solve multi-objective problems, normally a weighted sum of the objectives is optimized. However, how the weights are assigned can greatly affect the outcome. Therefore, many designers have to resort to producing the Pareto surface—a time-consuming procedure. In this paper, we propose a framework for solving multi-objective geometric programming problems where weights in the objective are optimally calculated during the optimization problem without having to produce the Pareto surface. It is shown that the proposed self-tuning multi-objective framework can be applied to geometric programming gate sizing problems. Then, the efficacy of the proposed framework is proven using the clock network buffer sizing problem as an application. The problem is first formulated as a geometric programming (GP) problem with the objectives of reducing power, skew, and slew. The problem is solved using ISPD09 circuits. The power, skew and slew of the optimized networks are calculated using ngspice. The results show on average 52% reduction in power and 28% reduction in skew compared to the original networks. The self-tuning multi-objective solution is shown superior to any single objective solution with no impact on runtime.

Categories and Subject Descriptors

G.1.6 [**Optimization**]: [Nonlinear Programming]
; J.6 [**Computer-Aided Engineering**]: [Computer-Aided Design (CAD)]

General Terms

Algorithms, Design

Keywords

Clock network, Multi-objective optimization, Geometric programming

1. INTRODUCTION

Multi-objective optimization is at the heart of decision making in fields spanning engineering, finance, and business. In these fields, the nature of the problems being solved involve many competing forces. When selecting a solution to a problem, everything must be considered simultaneously. A classic example in the field of VLSI is gate sizing, where it is desired to have power, area, and timing metrics (e.g. delay, and slew) simultaneously minimized [15].

To solve such problems, normally two methods are utilized: One method considers a single objective as the target of the optimization while the other objectives are dealt with as problem constraints. For example, sometimes in the gate sizing problem, area or power are considered as the objective while the delay and slew are the problem constraints. This method requires the designer to determine the trade-off between the objectives which requires time-consuming experiments. The second method is to optimize a weighted sum of the objectives. Assigning the weights however is complex and depends on the nature of the problem. Therefore, most designers have to dedicate a significant amount of time to generate the Pareto surface in order to determine a desirable compromise amongst the objectives. This becomes more complicated when more than three objectives are involved and direct visualization is impossible.

In this paper, a self-tuning multi-objective framework for Geometric Programming (GP) is proposed. The weights forming a weighted sum of objectives are considered as optimization variables instead of constants. This way the optimal weights are found automatically. The proposed framework is then applied to the general gate sizing applications where timing constraints and power are optimized simultaneously. Then, as a specific application, clock network buffer sizing is considered. This problem is formulated in GP format in [23]. Since in [26], it is shown that area is well correlated to power, in this paper, a self-tuning multi-objective framework is proposed which optimizes area, skew and slew at the same time. This can result in better quality design by automatically finding a good compromise and saving time by avoiding Pareto surface exploration.

The rest of this paper is organized as follows: In Section 2, background material is presented on multi-objective optimization, geometric programming, and gate sizing. The proposed self-tuning multi-objective framework for geometric programming is presented in Section 3. Results of experiments performed to validate the proposed framework in the context of sizing are given in Section 4. Finally, the paper is concluded and future directions of the work are mentioned in Section 5.

2. BACKGROUND

2.1 Multi-Objective Optimization

Multi-objective optimization problems can be expressed as:

$$\begin{array}{ll} \min. & [f_1(\mathbf{x}) \ \cdots \ f_O(\mathbf{x})]^T \\ \text{s.t.} & \mathbf{x} \in X \end{array} \tag{1}$$

where the goal is to minimize a vector-valued objective $[f_1(\mathbf{x}) \cdots f_O(\mathbf{x})]^T$ over the feasible set X. The objective components are often competing meaning that all components cannot be simultaneously minimized. One vector $\mathbf{x}_a \in X$ is said to *dominate* another $\mathbf{x}_b \in X$ if

$$f_i(\mathbf{x}_a) \leq f_i(\mathbf{x}_b), \ i = 1, \ldots, O,$$

with at least one of the inequalities being strict. A vector that is in X and not dominated by any other vector in X is said to be *Pareto optimal* and exists on the *Pareto front* when $O = 2$ and the *Pareto surface* when $O > 2$. If a vector $\mathbf{x}_a \in X$ neither dominates nor is dominated by a vector $\mathbf{x}_c \in X$, the two vectors are not comparable. These concepts are illustrated in Figure 1(a) where the dotted line is a convex Pareto front, i.e. the feasible set is on and above the line.

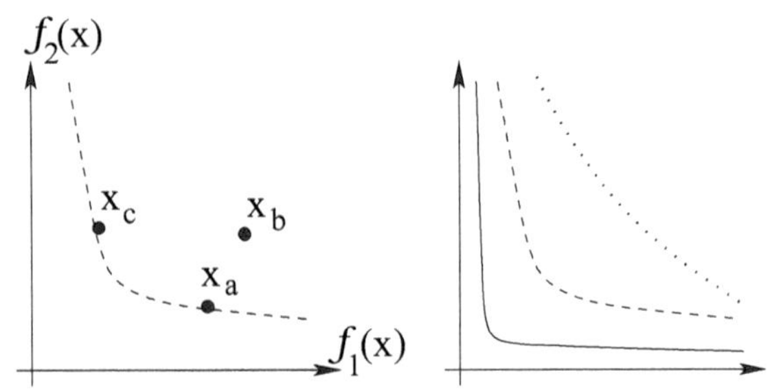

(a) $\mathbf{x}_a$ dominates $\mathbf{x}_b$, but is not comparable to $\mathbf{x}_c$.

(b) Pareto front curves.

Figure 1: Illustration of multi-objective concepts.

A designer must decide on which solution to choose from the many optimal solutions on the Pareto surface. This can be a very difficult task. Three different Pareto fronts are illustrated in Figure 1(b). In the most favorable case of the solid line, although every point on the line is Pareto optimal, designers are naturally drawn to the *knee* of the curve. The knee is the point closest to the often unattainable *utopia point* which is the vector in the multi-objective solution space with each element equal to the minimum value of each objective [22]. When the knee is not very pronounced, as in the case of the dotted line, the designer has a more difficult decision to make. Decreasing one objective comes at the cost of a proportionate increase in the other. The analysis is further complicated when more than two objectives are considered. The shape of the Pareto surface is problem-dependent, but this paper is limited to convex problems.

The most common method of optimizing a multi-objective problem involves constructing a scalar objective by forming a weighted sum of the vector-valued objective's components [18, 22]. The weights are typically non-negative and sum to one. The optimal solution obtained by minimizing this weighted sum is a single point on the Pareto surface. Weights are typically selected based upon the designer's priorities or on heuristics [5,17,32]. In an extreme but common case, designers select a single objective and recast the other objective components as constraints [7, 26]. This technique can be improved by iteratively alternating single objectives and constraining the others, but the order and tightness of constraints are heuristically determined [18]. A solution can also be chosen after a potentially time-consuming search using several different weights to gain insight into the shape of the Pareto surface [18]. Using uniformly distributed weights in the input space may not lead to uniformly distributed points in the solution space. In addition, the units and magnitudes of objective components are often different further compounding the difficulty. The normal boundary intersection method [22] obtains a uniform solution space distribution but cannot guarantee Pareto optimality. Metaheuristics have also been used to carry out the search [19,28]. However, metaheuristics can make slow progress and often involve setting weights themselves [12, 22]. Another weighted method is known as compromise programming [17]. In this method, targets are set a priori for each objective and a weighted error norm of the objective components and their targets is minimized. Ideally the target is set to the utopia point, but this point is not known a priori.

2.2 Geometric Programming

Geometric programs (GPs) are a class of convex optimization problems involving functions referred to as monomials and posynomials [7]. Monomials are of the form

$$cx_1^{a_1} x_2^{a_2} \cdots x_v^{a_v},$$

where $c, x_i > 0$, $i = 1, \ldots, v$, $a_i \in \Re, i = 1, \ldots, v$, and v is the number of variables in the problem. Posynomials are sums of monomials. For a problem to be a GP, the objective and inequality constraints must be posynomials and all equality constraints must be monomials. The non-negativity constraints on the variables are often not written in GP formulations but are implied. The convexity of GPs ensures a globally optimal solution can be attained efficiently.

An example using weighted sums to scalarize the objective of a multi-objective GP is shown in the following:

$$\begin{array}{ll} \min. & \alpha_1 f_1(\mathbf{x}) + \cdots + \alpha_O f_O(\mathbf{x}) \\ \text{s.t.} & g_i(\mathbf{x}) = 1, \ i = 1, ..., M \\ & h_j(\mathbf{x}) \leq 1, \ j = 1, ..., N \end{array} \tag{2}$$

where M and N are number of equality and inequality constraints, respectively. The multi-objective weights are shown with $\alpha_1, \ldots, \alpha_O > 0$ which are often constants and selected by an expert based on the importance of the objectives and nature of the problem. To keep the GP format, $f_1(\cdot), \ldots, f_O(\cdot)$ and each $h_j(\cdot)$ are posynomials while each $g_i(\cdot)$ is a monomial. The scalarized objective remains a posynomial because a sum of posynomials is posynomial.

2.3 Gate Sizing

The gate sizing problem, as well as closely related transistor, device, and wire sizing problems, can be formulated as GPs [7, 16, 26]. In these works, one of area, power, or delay is chosen as the objective and the remaining objectives are made into constraints along with minimum gate size constraints. Although the Elmore delay model is often used, more accurate posynomial delay models are available and have proven to be effective [16]. An important type of sizing problem is the clock network buffer/inverter sizing problem. The clock network buffer sizing problem adds clock skew as an objective compared to standard gate sizing. Clock network sizing has recently been formulated as a GP [23]. Buffer sizing in clock networks is the sizing application highlighted in this work and is explained further in what follows.

Clock networks are typically divided into two categories: clock trees and clock meshes. Mesh networks are usually used in applications that require highly reliable networks [13]. This is because the mesh networks are robust to the process variations as they provide multiple clock paths for a single sink [20, 27]. However, mesh networks are very costly in terms of capacitance and power consumption. Therefore, clock trees remain common in low power circuit designs [15]. The classic deferred-merge embedding [9] algorithm underlies several modern design techniques [10, 21]. Although clock trees consume much less power compared to clock meshes, they are very sensitive to process variations. Therefore, cross links are added to the clock trees to improve their robustness by providing multiple paths from the clock source to each sink [24]. However, adding cross links is a complicated process and several rules have been proposed to determine where the cross links should be added [25]. Online tunable delay buffers have been shown to be effective at reducing skew at the cost of increases in area and power [8].

Skew has been an objective of the buffer sizing problem. In [29], a dynamic programming technique, Van Ginneken's algorithm, is proposed to determine the sizes of buffers among a set of potential buffer locations while minimizing the maximum delay. However, the delay is calculated using the Elmore delay model. In [11], closed form solutions are proposed for sizing a buffered wire segment to minimize Elmore delay. The technique can be extended to tree networks through an expensive iterative scheme which sizes each segment in the tree. In [26], gate sizing for signal nets with the objective of minimizing maximum delay is investigated using robust geometric programming. A sequential programming skew minimization problem with power constraints that considers buffer and wire size variations is proposed in [14]. This method outputs continuous buffer sizes and discretization is not performed. Lack of SPICE simulations makes it hard to judge the effectiveness of this technique.

Power consumption in a clock tree is also an important design factor although direct minimization of clock tree power is $\mathcal{NP}$-Hard [30]. However, buffer sizing problems can focus on reducing power while keeping skew and/or slew under a limit [31]. In the 2010 ISPD clock network synthesis contest [2], capacitance, as a proxy for power, was the objective.

In [26], it is shown that area is well correlated with power consumption. Hence, in [23], the clock network buffer sizing power minimization problem is expressed in GP format as:

$$\begin{array}{ll} \min. & \mathrm{Area}(\mathbf{x}) = \sum_{\mathrm{b}\in\mathbf{B}} \mathrm{x}_{\mathrm{w_b}} \mathrm{x}_{\mathrm{l_b}} \\ \text{s.t.} & \mu \leq (t_{\text{skew}} + d_{min}) \\ & \mathrm{slew_k}(\mathbf{x}) \leq \mathrm{t_{slew}} \;, \forall \mathrm{k} \in \mathrm{B} \cup \mathrm{S} \\ & d_i(\mathbf{x}) \leq \mu, \;\; \forall i \in \mathbf{S} \\ & l_{min} \leq x_{l_b}, \;\; b \in \mathbf{B} \\ & w_{min} \leq x_{w_b}, \;\; b \in \mathbf{B} \end{array} \tag{3}$$

where, $[x_{w_b}, x_{l_b}]$ are the width and length of buffer b belonging to the set of all buffers, $\mathbf{B}$, in the clock tree. It should be mentioned that the solution for x_{l_b} is always l_{min} but it is still included in the formulation for generality. Variable $\mathbf{x}$ is the vector representing the buffer widths and lengths of all the buffers in the clock tree and Area($\mathbf{x}$) is the total area of the buffers. $d_i(\mathbf{x})$ is the delay from source to clock sink i in the set of all sinks $\mathbf{S}$ and target clock skew is represented by t_{skew}. The slew at each node k in the set of all nodes of the clock tree, $\mathbf{N}$, is shown by $\mathrm{slew_k}(\mathbf{x})$. Maximum allowed slew in the clock tree is represented by t_{slew} and l_{min} and w_{min} are the minimum length and width of a buffer. μ is a dummy variable used to prepare the problem in GP format.

3. THE PROPOSED SELF-TUNING MULTI-OBJECTIVE FRAMEWORK FOR GEOMETRIC PROGRAMMING

In this paper, the multi-objective weights in (2) will be treated as variables so that there will be no need for a somewhat arbitrary selection of the weights. This is very effective since the best trade-off between the objectives will be found automatically for convex problems. Adding the weights as variables instead of constants can be complex since it can change the nature of the problem. For example, a linear programming problem will become a nonlinear problem by adding variable weights. This is even worse for a quadratic programming problem where the problem becomes non-convex when variable weights are added. In this paper, the objective is to add the variable weights while retaining the GP problem structure. To account for the dependency between the objectives and achieve the best trade-off, the weights are also considered to be dependent. Forcing the weights to sum to one, e.g. $\alpha_1 + \alpha_2 = 1$, violates the GP structure, as only monomial equality constraints can be included. However, constraining the weights to multiply to one, e.g. $\alpha_1\alpha_2 = 1$, maintains the GP format. Then, the self-tuning scalarized multi-objective problem is formulated as:

$$\begin{array}{ll} \min. & \alpha_1 f_1(\mathbf{x}) + \cdots + \alpha_O f_O(\mathbf{x}) \\ \text{s.t.} & g_i(\mathbf{x}) = 1, \;\; i = 1, ..., M \\ & h_j(\mathbf{x}) \leq 1, \;\; j = 1, ..., N \\ & \alpha_1 \cdots \alpha_O = 1 \end{array} \tag{4}$$

where $\alpha_1, \ldots, \alpha_O > 0$ are the variable multi-objective weights. This problem not only optimizes all of the objectives but also the variable weights. The problem is still in the GP format and can be solved by geometric programming techniques.

In a typical gate sizing problem, power and timing constraints such as delay, skew and slew are the objectives that are optimized. All these objectives have been formulated in GP format and can be solved using convex optimization techniques to obtain a globally optimal solution [26]. Therefore, gate sizing problem can be formulated using the proposed self-tuning multi-objective framework as follows:

$$\begin{array}{ll} \min. & \alpha_1 \mathrm{power}(\mathbf{x}) + \alpha_2 \mathrm{timing}(\mathbf{x}) \\ \text{s.t.} & \alpha_1\alpha_2 = 1 \\ & \mathrm{technicalConstraints}(\mathbf{x}) \leq 1 \end{array} \tag{5}$$

where $\mathbf{x}$ is the vector of all lengths and widths of the gates. power($\mathbf{x}$) is the total power consumption of the gates in the circuit and timing($\mathbf{x}$) includes the timing metrics of the circuit such as delay, skew or slew. technicalConstraints($\mathbf{x}$) consists of other considerations such as minimum sizes of the gates. In Section 3.1, clock network buffer sizing is considered as a specific type of gate sizing and is formulated in detail within the self-tuning multi-objective framework.

3.1 Self-Tuning Multi-Objective GP Formulation for Buffer Sizing

In this section, an application of the proposed self-tuning multi-objective framework for clock network buffer sizing is proposed. In buffer sizing, area/power, skew and slew of the clock network should be minimized simultaneously. In [23], this problem is formulated in GP format. However, only area/power is minimized and the skew and slew are just considered as constraints. This results in over-optimizing area/power and the designer needs to choose the skew and slew limits manually. If the design is not satisfactory, the limits should be changed to achieve a tighter design. This technique has the same shortcomings as choosing constant weights in multi-objective optimization.

The proposed self-tuning clock network buffer sizing problem is formulated as:

$$\begin{array}{ll} \min. & \alpha_1 \mathrm{Area}(\mathbf{x}) + \alpha_2 v_{\mathrm{skew}} + \alpha_3 v_{\mathrm{slew}} \\ \mathrm{s.t.} & d_i(\mathbf{x}) \le d_{min} v_{\mathrm{skew}},\ \forall i \in \mathbf{S} \\ & \mathrm{slew_k}(\mathbf{x}) \le v_{\mathrm{slew}}\ ,\forall \mathrm{k} \in \mathrm{B} \cup \mathrm{S} \\ & l_{min} \le x_{l_b},\ b \in \mathbf{B} \\ & w_{min} \le x_{w_b},\ b \in \mathbf{B} \\ & \alpha_1 \alpha_2 \alpha_3 = 1 \end{array} \quad (6)$$

where $\alpha_1, \alpha_2, \alpha_3 > 0$ are the variable multi-objective weights. v_{skew} and v_{slew} are the upper bounds variables of skew and slew, respectively. The total area of the buffers is represented as Area($\mathbf{x}$). $d_i(\mathbf{x})$ is the delay from the clock source to clock sink i which is calculated using the posynomial delay model used in [26]. In this formulation, skew is determined using a multiplicative factor of the minimum delay ($v_{\mathrm{skew}} > 1$), since the usual difference in delays is not posynomial. $\mathrm{slew_k}(\mathbf{x})$ is the slew at each node k in the clock network which is calculated using the model in [6].

Solutions obtained by solving (6) are nearer to the utopia point than solutions obtained by considering a single objective. Furthermore, costly searches to uncover the Pareto surface are avoided. Experiments to illustrate these benefits are presented in the following section.

4. EXPERIMENTAL RESULTS

In this paper, the clock networks from the 2009 ISPD clock network synthesis contest [1] are used in the experiments. These circuits are chosen since the clock networks produced by all the teams have been released. Experiments are performed on a 2.4GHz Dual-Core AMD Opteron processor with 4GB of memory. All the implementations are written in C++. The process technology is 45nm and the technology files from the 2009 ISPD clock network synthesis contest were used. In all experiments, skews and powers are calculated using ngspice 24 [4]. It should be noted that the powers calculated using ngspice include leakage, short circuit and switching powers. Mosek 6.0 [3] is used to solve the GPs.

To prove the effectiveness of the proposed self-tuning multi-objective framework for clock network buffer sizing, several experiments are performed. In the first experiment, clock networks are sized using the proposed self-tuning Multi-Objective (MO) GP formulation proposed in (6). These networks are also sized when only one single objective, i.e Area Objective (AO), SKew Objective (SKO), or SLew Objective (SLO), is considered when solving. In AO, the objective is area while skew and slew are considered to be constraints. In SKO, skew is targeted as the objective and area and slew are the constraints while in SLO, slew is minimized and skew and area are the constraints. The target skew using posynomial delay in solving AO and SLO is set to 1ps. In SKO and AO, the slew limit is set to 100ps as stated in the 2009 ISPD clock network synthesis contest. Solving GPs results in clock networks with continuous buffer sizes while only certain discrete sizes are available for the buffers. Therefore, the continuous sizes are rounded down to the nearest available discrete sizes.

The resulting networks are compared to the input networks in terms of area and power in Table 1(a) and skew and average slew in Table 1(b). The average slew is calculated by finding the mean of the slew at all the buffers and clock sinks in the circuit. In Table 1(a), in Column 2, the area of the input network is presented. In Columns 3 to 10, area percentage improvements of the resulting networks over the input networks for single objectives AO, SKO, SLO and the proposed self-tuning MO are presented for networks with continuous-size buffers (C) and discrete-size buffers (D). The power consumptions of the input networks are shown in Column 11 while power percentage improvements are presented in Columns 12 to 19 in the same order as Columns 3 to 10. Skew and average slew are compared in Table 1(b) which has the same format as Table 1(a).

All the formulations result in improvements in the area where AO provides the maximum improvement, as expected. The proposed MO framework also results in significant improvements for all the circuits with an average improvement of 53% for continuous and 56% for discrete solutions. As power is correlated with the area, the AO formulation has the maximum power improvement while SKO and SLO formulations result in slight power increases. However, the proposed MO improves the power for all the circuits by about 50%, on average, for both continuous and discrete solutions.

Considering skew, the proposed MO results in improvements for all the circuits but one while it leads to the average improvement of around 26% that is the maximum among all the formulations. Even though SKO is expected to improve the skew significantly, the experiments show degradation. This is due to the differences in the approximations of the delay model used in the GP formulations and the more detailed ngspice delay simulations. The slew results might bring up the same question, as SLO leads to slew degradation compared to the input networks. This is also explained by considering the inaccuracies in the posynomial slew model used in the formulation compared to the SPICE accurate simulations. This issue can be addressed by using more accurate posynomial delay formulations [26]. The average slew results show degradation for all the formulations. This is because the slew limit is set to be 100ps in the 2009 ISPD clock network synthesis contest while the average slews for the input network are very low, i.e. only 31ps, on average, with the additional cost of area and power. Therefore, the proposed self-tuning multi-objective framework gives an optimal solution where all the area, power, skew and slew are optimized and the slew at all nodes is less than the 100ps slew limit.

In the second experiment, the Pareto front of the problem is drawn in order to show how close the solution is to the utopia point. As visualization in two dimensions is more explicit, a bi-objective version of the proposed self-tuning multi-objective GP framework for buffer sizing in (6) is used to generate the Pareto front for Circuit 11 shown in Figure 2. The objective is the weighted sum of area and

(a) Area and Power Comparisons

Cir.	Input Area (μm^2)	Area (% Improvement) AO		SKO		SLO		MO		Input Power (μW)	Power (% Improvement) AO		SKO		SLO		MO	
		C	D	C	D	C	D	C	D		C	D	C	D	C	D	C	D
11	536	67	71	12	15	12	17	46	46	19668	65	65	11	11	-1	9	43	46
12	527	71	75	10	13	13	17	54	57	19685	69	69	1	3	12	14	50	53
21	541	69	72	8	12	16	19	52	55	20631	64	64	7	9	3	7	45	48
22	320	65	69	10	13	8	11	41	44	12125	66	66	10	14	6	9	42	45
31	1177	78	79	8	12	5	8	69	72	42976	66	66	-37	-33	-2	3	59	61
32	895	71	76	7	10	11	14	51	54	33392	67	67	-7	-4	6	10	49	51
nb1	192	76	81	10	13	1	4	59	62	5234	76	76	-11	-5	-54	-51	53	58
Ave.	598	71	75	9	13	9	13	53	56	21959	68	68	-4	-1	-4	0	49	52

(b) Skew and Slew Comparisons

Cir.	Input Ave. Skew (ps)	Ave. Skew (% Improvement) AO		SKO		SLO		MO		Input Ave. Slew (ps)	Ave. Slew (% Improvement) AO		SKO		SLO		MO	
		C	D	C	D	C	D	C	D		C	D	C	D	C	D	C	D
11	48	27	-19	16	21	-339	-17	36	37	33	-31	-36	-19	-20	-26	-23	-32	-38
12	91	-4	-19	26	35	13	10	48	53	32	-41	-50	-20	-21	-20	-21	-39	-46
21	51	18	-10	24	20	-449	-394	16	24	34	-43	-57	-17	-17	-30	-33	-39	-48
22	27	0	-81	-27	-7	-59	-42	-20	-15	32	-26	-30	-19	-20	-20	-20	-29	-34
31	228	29	4	-156	-153	69	64	33	31	32	-123	-125	-18	-18	-18	-18	-90	-125
32	143	15	-29	-75	-69	-13	-7	41	59	33	-44	-60	-18	-18	-22	-21	-38	-46
nb1	25	-16	-40	-35	-26	-782	-770	25	5	24	-84	-96	-20	-20	-32	-34	-76	-91
Ave.	88	10	-28	-32	-26	-223	-165	26	28	31	-56	-65	-19	-19	-24	-24	-49	-61

Table 1: Comparison of area, power, skew and average slew of initial networks and the resultant networks sized by single area objective (AO), single skew objective (SKO), single slew objective (SLO) and the proposed multi-objective (MO) problems. Power, skew and average slew are calculated using ngspice.

skew while slew is considered as a constraint with an upper limit of 100ps. The solution is shown with blue dot in Figure 2. This experiment shows that the proposed self-tuning multi-objective framework finds a solution on the knee of the Pareto front and close to the utopia point.

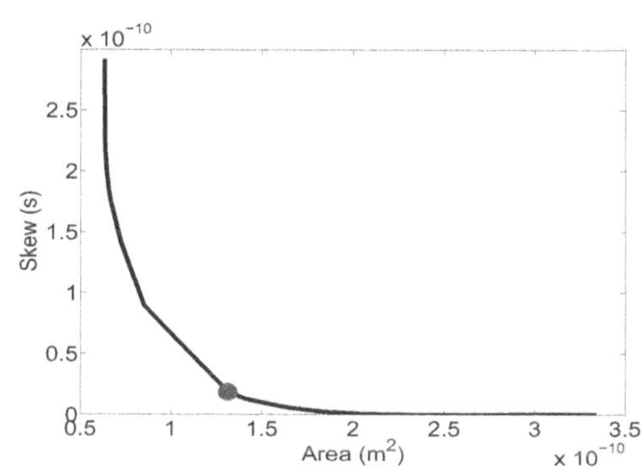

Figure 2: Pareto front curve for Circuit 11.

Finally, Figure 3 is provided to show that the additional cost of solving the self-tuning multi-objective formulation is insignificant. It can be observed that the time to solve the self-tuning multi-objective formulation is not substantially different from that of other formulations. This is not unexpected given that only a few variables and a single constraint have been added to obtain the self-tuning multi-objective solution.

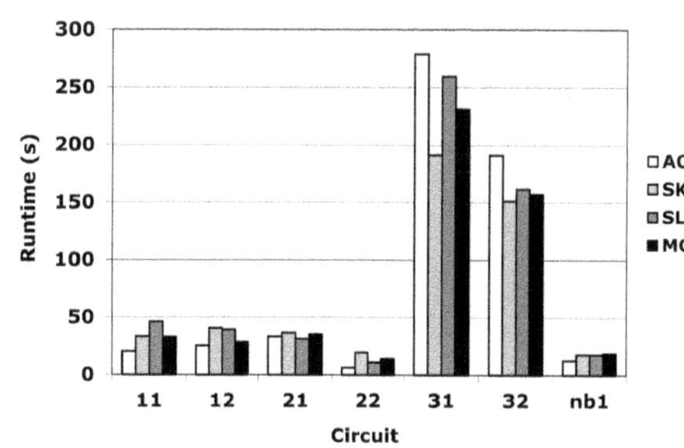

Figure 3: Runtime results to solve single area objective (AO), single skew objective (SKO), single slew objective (SLO) and the proposed multi-objective (MO) problems.

5. CONCLUSIONS AND FUTURE WORK

In this paper, a self-tuning multi-objective framework for geometric programming is proposed. The proposed framework optimally assigns the multi-objective weights and avoids the time-consuming solution space search. The general gate sizing application is presented while a self-tuning multi-objective GP framework for clock network buffer sizing is proposed. The experiments show that improvements are achieved in the area, power and skew of the circuits with no significant cost of runtime. An experimental Pareto front is generated and it is shown that the solution found by the proposed framework is close to the utopia point.

Future work may include using more accurate posynomial delay models. In addition, designing a customized rounding algorithm may provide discrete solutions that are very close to the optimal continuous solutions.

6. ACKNOWLEDGMENTS

This research is supported by the Natural Sciences and Engineering Research Council of Canada and Alberta Innovates Technology Futures. This research has been enabled by the use of computing resources provided by WestGrid and Compute/Calcul Canada.

7. REFERENCES

[1] ISPD 2009 clock network synthesis contest. `http://www.sigda.org/ispd/contests/09/ispd09cts.html`.

[2] ISPD 2010 high performance clock network synthesis contest. `http://www.sigda.org/ispd/contests/10/ispd10cns.html`, date visited: Mar. 4, 2010.

[3] Mosek 6.0. `http://www.mosek.com`.

[4] ngspice 24. `http://ngspice.sourceforge.net/`.

[5] N. Azmard and D. Auvergne. Pops: A tool for delay/power performance optimization. *Journal of Systems Architecture*, 47(34):375–382, 2001.

[6] H. Bakoglu. *Circuits, Interconnections, and Packaging for VLSI*. Addison–Wesley, 1990.

[7] S. Boyd, S.-J. Kim, L. Vandenberghe, and A. Hassibi. A tutorial on geometric programming. *Optimization and Engineering*, 8(1):67–127, 2007.

[8] A. Chakraborty, K. Duraisami, A. Sathanur, P. Sithambaram, L. Benini, A. Macii, E. Macii, and M. Poncino. Dynamic thermal clock skew compensation using tunable delay buffers. *IEEE Trans. on VLSI*, 16(6):639–649, 2008.

[9] T. Chao, Y. Hsu, J. Ho, K. Boese, and A. Kahng. Zero skew clock routing with minimum wirelength. *IEEE Tran. on CAS*, 39(11):799–814, 1992.

[10] S. Chou, C.-S. Han, P.-K. Huang, K.-F. Tien, and T.-Y. Ho. An effective and efficient framework for clock latency range aware clock network synthesis. *Trans. on CAD*, 30(7):1045–1057, 2011.

[11] C. Chu and D. Wong. Closed form solution to simultaneous buffer insertion/sizing and wire sizing. *TODAES*, 6(6):343–371, 2001.

[12] N. Drechsler, R. Drechsler, and B. Becker. Multi-objective optimization in evolutionary algorithms using satisfiability classes. In *Proc. of Intl. Conf. on Computational Intelligence, Theory and Applications: Fuzzy Days*, pages 108–117, 1999.

[13] G. Flach, G. Wilke, M. Johann, and R. Reis. A Mesh-Buffer Displacement Optimization Strategy. In *Proc. of ISVLSI*, pages 282–287, 2010.

[14] M. Guthaus, D. Sylvester, and R. Brown. Clock buffer and wire sizing using sequential programming. In *Proc. of DAC*, pages 1041–1046, 2006.

[15] A. Kahng, J. Lienig, I. Markov, and J. Hu. *VLSI Physical Design: From Graph Partitioning to Timing Closure*. Springer, 2011.

[16] K. Kasamsetty, M. Ketkar, and S. Sapatnekar. A new class of convex functions for delay modeling and its application to the transistor sizing problem [CMOS gates]. *IEEE Trans. on CAD*, 19(7):779–788, 2006.

[17] F. Kashfi, S. Hatami, and M. Pedram. Multi-objective optimization techniques for VLSI circuits. In *Proc. of ISQED*, pages 156–163, 2011.

[18] I. Kim and O. de Weck. Adaptive weighted-sum method for bi-objective optimization: Pareto front generation. *Structural and Multidisciplinary Optimization*, 29(2):149–158, 2005.

[19] H. Kita and S. Kobayashi. Multiobjective optimization by genetic algorithms: a review. In *Proc. of ICEC*, pages 517–522, 1996.

[20] N. Kurd, J. Barkarullah, R. Dizon, T. Fletcher, and P. Madland. A multigigahertz clocking scheme for the Pentium® 4 microprocessor. *IEEE Journal of SSC*, 36(11):1647–1653, 2001.

[21] S.-K. Lim. Low power clock routing for 3D IC. In *Design for High Performance, Low Power, and Reliable 3D Integrated Circuits*, pages 99–128. Springer New York, 2013.

[22] R. Marler and J. Arora. Survey of multi-objective optimization methods for engineering. *Structural and Multidisciplinary Optimization*, 26(6):369–395, 2004.

[23] Omitted for blind review.

[24] F. Qian, H. Tian, and E. Young. Crosslink insertion for variation-driven clock network construction. In *Proc. of GLSVLSI*, pages 321–326, 2012.

[25] A. Rajaram, J. Hu, and R. Mahapatra. Reducing clock skew variability via cross links. In *Proc. of DAC*, pages 18–23, 2004.

[26] J. Singh, V. Nookala, Z. Luo, and S. Sapatnekar. Robust gate sizing by geometric programming. In *Proc. of DAC*, pages 315–320, 2005.

[27] H. Su and S. Sapatnekar. Hybrid structured clock network construction. In *Proc. of ICCAD*, pages 333 –336, 2001.

[28] A. Suppapitnarm, K. Seffen, G. Parks, and P. Clarkson. A simulated annealing algorithm for multiobjective optimization. *Engineering Optimization*, 33(1):59–85, 2000.

[29] L. van Ginneken. Buffer placement in distributed RC-tree networks for minimal elmore delay. In *Proc. of ISCAS*, volume 2, pages 865–868, 1990.

[30] A. Vittal and M. Marek-Sadowska. Power optimal buffered clock tree design. In *Proc. of DAC*, pages 497–502, 1995.

[31] K. Wang and M. Marek-Sadowska. Buffer sizing for clock power minimization subject to general skew constraints. In *Proc. of DAC*, pages 159–164, 2004.

[32] T.-H. Wu, A. Davoodi, and J. Linderoth. GRIP: Global routing via integer programming. *IEEE Trans. on CAD*, 30(1):72–84, 2011.

Low Power 3-D Stacking Multimedia Platform with Reconfigurable Memory Architecture

Po-Han Huang, Huang-Lun Lin, Hsien-Ching Hsieh, Chi-Hung Lin, Shui-An Wen and Yi-Fa Sun
Information and Communications Research Laboratories
Industrial Technology Research Institute, Hsinchu, Taiwan
{pohan, hl_lin, RandyHsieh, chlin, shuianwen, YFSun}@itri.org.tw

ABSTRACT

In this paper, a low power and high performance three-dimensional (3-D) stacking multimedia platform called "3D-PAC" is proposed. This platform is a heterogeneous integration composed of a low power design logic layer (2D-PAC) and a reconfigurable memory tier via 3-D technology. After extensive 3-D architecture exploration with Electronic System Level (ESL) simulation, there is a 54% performance speedup compared with the former 2-D architecture for certain multimedia applications. This chip is fabricated in TSMC 90nm generic CMOS technology. The area of 2D-PAC is about 7880 x 7880 μm^2 and the SRAM layer is about 3880 x 3880 μm^2. Both layers are combined with 1,886 TSVs.

Categories and Subject Descriptors

C.1.2 [**Processor Architectures**]: Multiple Data Stream Architectures (Multiprocessors) – Parallel processors; C.1.3 [**Processor Architectures**]: Other Architecture Styles – Heterogeneous systems

Keywords

Three-dimensional IC (3-D IC), Electronic System Level (ESL), Through-Silicon Via (TSV), heterogeneous

1. INTRODUCTION

In this work, ESL virtual platform is adopted to explore the possibility of implementing 3D-PAC, a 3-D version of 2D-PAC. After the architecture exploration, the stacking memory is directly integrated with the instruction memory unit (IMU) and data memory unit (DMU) of PACDSP core because of the performance and number of TSV to build up the 3-D multimedia platform.

2. HARDWARE ARCHITECTURE

2D-PAC, which is composed of an ARM926EJ-S core and two digital signal processors called PACDSP, is a heterogeneous multi-core SoC. ARM is in charge of the operating system and system control related works. On the other hand, PACDSP is responsible for the heavy computation of multimedia data processing [1].

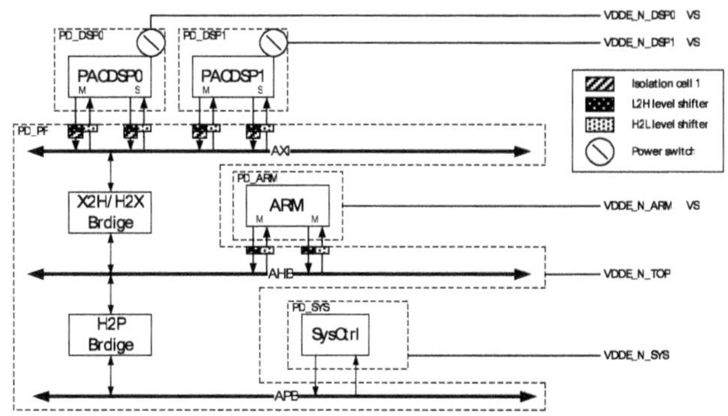

Figure 1. Low power hardware architecture of 2D-PAC

The main hardware architecture of 2D-PAC is shown in Figure 1. It consists of three major blocks: AXI, AHB and APB subsystem. The system controller (SysCtrl) on APB bus handles the important functions of power management, reset and clock generation.

Common Power Format (CPF) of Cadence is adopted as the low power design flow of 2D-PAC SoC. In this CPF description, there are power domain definition, level shifter, isolation cell and other low power cell locations, which are shown in Figure 1. It also describes the operating mode of each power domain.

3. ARCHITECTURE EXPLORATION

For 3-D system architecture exploration, related system-level design approaches have been proposed in many papers [2] [3]. This work presents an ESL simulation framework to explore a 3-D IC system that consists of a heterogeneous multi-core SoC with extended stacking memory.

After the evaluation by ESL, the stacking architecture shown in Figure 2 is adopted. The stacking memory is directly connected with the data and instruction memory unit (DMU & IMU). It equally extends the internal instruction cache and data memory size of both DSP cores.

Two real applications, the multi-channel H.264 decoder and JPEG decoder, are chosen to analyze the impact and efficiency of extending PACDSP local memory through 3-D stacking. The experimental results of 4-channel H.264 decoder shows that after applying some parallelism and optimization, the 3-D system can get about 54% performance improvement compared with 2-D version as shown in Table 1.

GLSVLSI'13, May 2–3, 2013, Paris, France.
ACM 978-978-1-4503-1902-7/13/05.

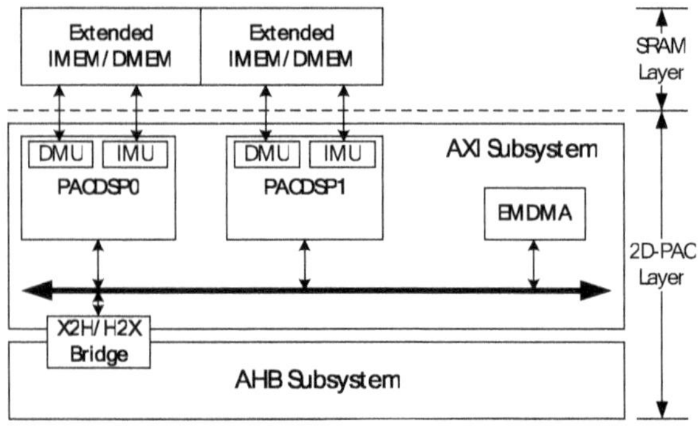

Figure 2. The stacking model of architecture exploration

Table 1. ESL simulation results for 4-channel H.264 decoder

H.264 Version	Total Cycle	FPS	Improve.
Non-Parallel			
2D	194,019,671	11.56	-
3D with 4 binary	151,524,469	14.80	28.02%
Parallel			
2D	132,557,335	16.92	-
3D with 2 binary	86,001,705	26.09	54.19%

The other application, JPEG decoder, shows that there is only little system improvement by extending the instruction cache for JPEG decoder because of the low ratio of cache misses. Otherwise, enlarging the local data memory for JPEG decoder makes huge performance improvement because of the high ratio of external accesses. Based on this analysis, it reveals that suitable memory configuration depends on different applications. This feature is designed in 3D-PAC platform. In Figure 3, Each DSP core owns its private SRAM block (256KB) which is reconfigurable. It allows programmers to configure different architectures for different applications. More detail design of instruction memory and data memory configuration mechanism with tier selection when stacking multiple memory layers are already described in another publication [4].

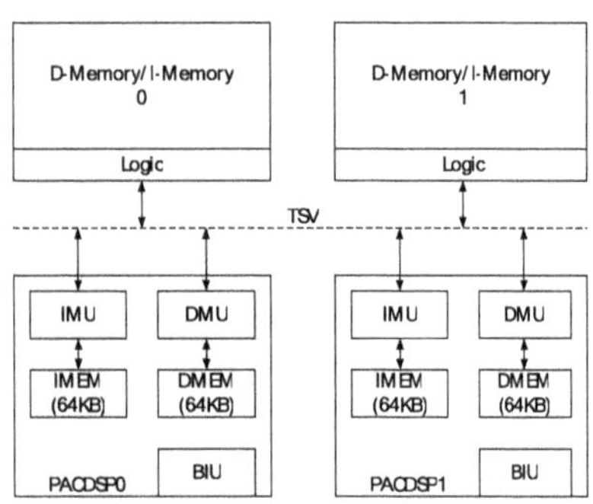

Figure 3. Reconfigurable stacking memory architecture

4. IMPLEMENTATION RESULTS

The 3D-PAC platform is fabricated in TSMC 90nm generic CMOS technology. For 2D-PAC SoC, the maximum frequency can reach 300MHz at both AXI bus and DSP core, and the corresponding chip area is 7880 x 7880 μm^2. The SRAM tier can reach 300MHz for the configuration circuit and the chip area is about 3880 x 3880 μm^2. Figure 4 shows the layout view of 3D-PAC which reveals that logic and memory layers are integrated through 1,886 TSVs located in the middle of chip.

Figure 4. The layout view of 3D-PAC and 4-channel H.264 decoder demo of 2D/3D-PAC

The demo of 4-channel H.264 decoder with both 2D-PAC and 3D-PAC evaluation boards is also shown in Fig. 4. Based on this application, the real chip performance measurement is summarized in Table 2. Under the same resolution, 3D-PAC can achieve about 60% performance improvement.

Table 2. Real chip performance measurement

Target	Resolution	Performance	Improvement
2D-PAC	QCIF 300 frames	18.99 fps	-
3D-PAC		30.30 fps	59.55%

5. CONCLUSION

3D-PAC is a heterogeneous integration of the logic and memory chip through 3-D technology. The stacking architecture is defined by extensive architecture exploration with ESL simulation. For the final proposed architecture, there is about 54.19% performance improvement over the 2-D architecture. Reconfigurable stacking SRAM improves the programming flexibility, where the performance can be optimized for different applications, and the corresponding logic overhead is negligible. The design is fabricated in TSMC 90nm generic CMOS technology and total 1,886 TSVs have been utilized for three-dimensional stacking. Based on the application of 4-channel H.264 decoder, the real chip performance can achieve about 60% improvement close to the result from ESL simulation.

6. REFERENCES

[1] Tay-Jyi Lin, Chun-Nan Liu, Shau-Yin Tseng, Yuan-Hua Chu, and An-Yeu Wu, "Overview of ITRI PAC project – from VLIW DSP processor to multicore computing platform," in *Proc. International Symposium on VLSI Design, Automation and Test (VLSI-DAT)*, pp. 188-191, Apr. 2008

[2] Y. F. Tsai, F. Wang, Y. Xie, N. Vijaykrishnan, and M.J. Irwin, "Design Space Exploration for 3-D Cache," *IEEE Transactions on Very Large Scale Integration Systems*, pp. 444-455, Apr. 2008.

[3] A. Richard, D. Milojevic, F. Robert, A. Bartzas, A. Papanikolaou, K. Siozios, and D. Soudris, "Fast Design Space Exploration Environment Applied on NoC's for 3D-Stacked MPSoC's," in *Proc. Architecture of Computing Systems (ARCS)*, pp. 1-6, Feb. 2010

[4] Hsien-Ching Hsieh, Po-Han Huang, Chi-Hung Lin, Huang-Lun Lin, "Stacking Memory Architecture Exploration for Three-Dimensional Integrated Circuit in 3-D PAC," in *Proc. IEEE International SoC Conference*, pp. 317-321, Sep. 2012

High Frame-Rate Low-Power Compressive Sampling CMOS Image Sensor Architecture

[Extended Abstract]

Nikola Katic
LSM Laboratory, EPFL
CH-1015 Lausanne, Switzerland
nikola.katic@epfl.ch

Mahdad Hosseini Kamal
LTS2 Laboratory, EPFL
CH-1015 Lausanne, Switzerland
mahdad.hosseinikamal@epfl.ch

Mustafa Kilic
LSM Laboratory, EPFL
CH-1015 Lausanne, Switzerland
mustafa.kilic@epfl.ch

Alexandre Schmid
LSM Laboratory, EPFL
CH-1015 Lausanne, Switzerland
alexandre.schmid@epfl.ch

Pierre Vandergheynst
LTS2 Laboratory, EPFL
CH-1015 Lausanne, Switzerland
pierre.vandergheynst@epfl.ch

Yusuf Leblebici
LSM Laboratory, EPFL
CH-1015 Lausanne, Switzerland
yusuf.leblebici@epfl.ch

ABSTRACT

A novel compressive sampling scheme suitable for highly scalable hardware implementation is presented. The prototype design is implemented in a $0.18\mu m$ standard CMOS technology and utilizes compressed acquisition to achieve high frame rates and maintain low power consumption. Specialized pixels, convenient for Comparator-Based Switched Capacitor readout are developed for this purpose. A custom measurement matrix generation algorithm is implemented which reduces in-pixel hardware complexity and performs measurement matrix generation in a single clock cycle. Per column Differential Cyclic-ADCs based on the Zero-Crossing Detection (ZCD) technique are used to convert the analog image measurements. Physical IC design issues such as the device noise, mismatch and non-linearity, are analyzed and their effects on compressed image acquisition are presented and discussed. The final simulation results show that the proposed 256×256 pixels architecture consumes 1.45mW at 250fps and 26.2mW at 8000fps. The proposed architecture can easily be scaled towards newer technology nodes and higher image resolutions.

Categories and Subject Descriptors

B.7.1 [**Hardware**]: Integrated Circuits—*VLSI (Very Large Scale Integration)*

Keywords

CMOS Image Sensor, Image Acquisition, Compressive Sampling, High Frame Rate, Low-Power, Cyclic ADC

1. SYSTEM OVERVIEW

Compressive Sampling (CS) exploits signal sparsity to perform an accurate acquisition from a small number of linear measurements. Due to reduction in the image acquisition cost, CS has recently attracted significant attention [3]. Several image sensor designs based on CS have been presented in literature [5] [7] [6] [8] [9]. In this paper, a compressive sampling measurement technique similar to Bernoulli/Rademacher is introduced,

$$\mathbf{\Phi}_{\mathbf{m}\times\mathbf{n}} = \begin{bmatrix} \Xi & 0 & 0 & \cdots & 0 \\ 0 & \Xi & 0 & \cdots & 0 \\ \vdots & \vdots & \vdots & \ddots & \vdots \\ 0 & 0 & 0 & \cdots & \Xi \end{bmatrix}, \Xi = \begin{bmatrix} \phi_{1,1} & \phi_{1,2} & \cdots & \phi_{1,N} \\ \phi_{2,1} & \phi_{2,2} & \cdots & \phi_{2,N} \\ \vdots & \vdots & \ddots & \vdots \\ \phi_{M,1} & \phi_{M,2} & \cdots & \phi_{M,N} \end{bmatrix} \quad (1)$$

where $\phi_{ij} = \pm 1/N$, and $m = M \times N$, approximately maintaining the same performance. The proposed measurement scheme applies the same random vector to each image column to form the linear random measures. In the conventional case, at each iteration, the reconstruction algorithm requires matrix multiplication with the dense matrices $\mathbf{\Phi}$ and $\mathbf{\Phi}^T$ (order $O(N^4)$). Using the same random vector for

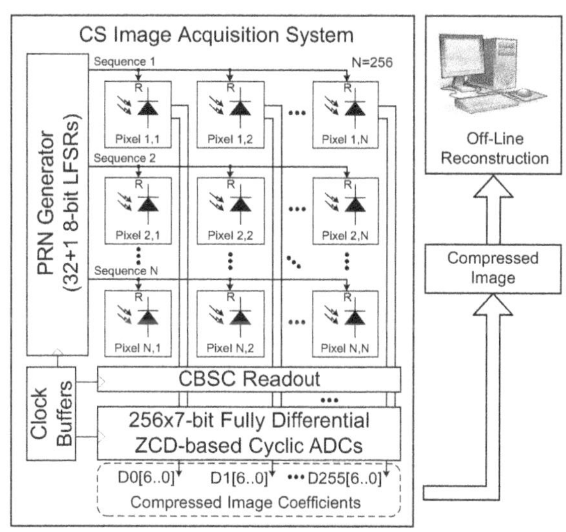

Figure 1: Top Level View

every image column results in the block diagonal structure of the proposed measurement matrix, reducing the multi-

GLSVLSI'13, May 2–3, 2013, Paris, France.
ACM 978-1-4503-1902-7/13/05.

plications with Φ and Φ^T into N multiplications with Ξ and Ξ^T (order $O(N^3)$). Therefore, the required program memory and the reconstruction time of the algorithm are substantially reduced. In addition, the column-by-column measurement technique allows a straightforward hardware implementation, reduces in-pixel hardware complexity, and performs a single clock cycle random matrix generation. On the receiver side, the image x is recovered from the linear measurements, by solving the Basis Pursuit (BP) problem:

$$\underset{x \in \mathbb{R}^N}{\operatorname{argmin}} \|x\|_{TV} + \frac{\lambda}{2}\|y - \Phi x\|_2^2, \tag{2}$$

where $\lambda = 0.1$ is an empirically derived regularization parameter and $\|x\|_{TV}$ is the Total Variation (TV) semi norm [2]. In this work, the BP problem is solved by using the fast-converging FISTA [1] algorithm.

An overview of the system is given in Figure 1. A Pseudo-Random Number Generator (PRNG) generates uncorrelated random binary sequences, one per each pixel row. Thus, at every sampling cycle, all the pixels within the same row receive the same bit R, resulting in each column receiving the same random vector. Using the same vector for each pixel column simplifies the matrix generation hardware, reduces the required connections within the pixel array, and finally results in better system scalability.

For each 8 rows of the pixel array, outputs of two neighboring LFSRs are cross-connected as shown in Figure 2. The output $D[j]$ of one register is XOR-ed with the output

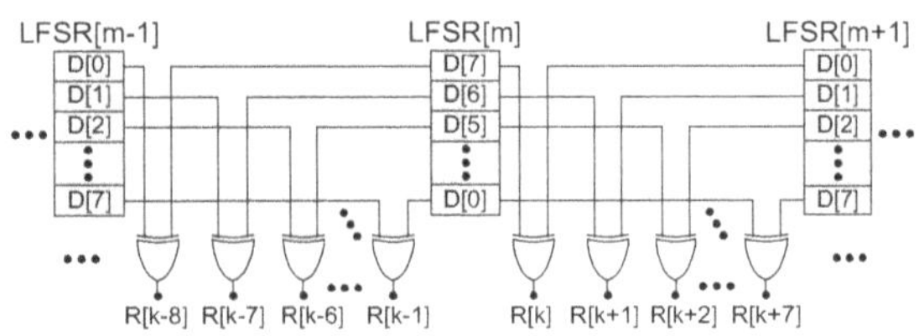

Figure 2: Pseudo-Random Sequence Generator

$D[7-j]$ of the neighboring register. The result of this operation is the generation of 8 mutually uncorrelated pseudo-random sequences. As a consequence, 33x8-bit LFSRs are required for 256 rows, i.e., one register per 8 rows, plus one to be cross-connected with the last LFSR in an array.

2. SIMULATION RESULTS

To obtain the architecture performance, both statistical transistor level and top-level simulations were performed. The final simulation results of the reconstructed sample image (cameraman) for the minimum and the maximum frame rate are shown in Figure 3.

3. CONCLUSION

An image acquisition architecture based on CS is presented that demonstrates the potential of compressed sampling in low-power imaging. It performs single-shot imaging CS by using compact LFSR-based matrix generation circuit and CBSC readout [4]. For each measurement, every pixel column uses the same random vector, which reduces the amount of metal connections within the pixel array (allowing better system scalability), and increases the speed of the

(a) (b)

Figure 3: Reconstructed Image: (a)@250fps (b)@8000fps

reconstruction algorithm. Dynamic CBSC readout provides additional power efficiency by limiting the current consumption to the value required for the complete charge transfers. The proposed design consumes 26.2mW at a 8000fps (1MHz sampling, 8MHz internal clock) and 1.45mW at a 250fps rate (32kHz sampling, 256kHz internal clock).

4. REFERENCES

[1] A. Beck and M. Teboulle. Fast Iterative Shrinkage-Thresholding Algorithm for Linear Inverse Problems. *SIAM Journal on Imaging Sciences*, pages 183–202, 2009.

[2] A. Chambolle. An Algorithm for Total Variation Minimization and Applications. *Journal of Mathematical Imaging and Vision*, 20:89–97, 2004.

[3] E. J. Candes, M. B. Wakin. An Introduction to Compressive Sampling. *IEEE Signal Processing Magazine*, 25:21–30, March 2008.

[4] J. K. Fiorenza, H. S. Lee et al. Comparator-Based Switched-Capacitor Circuits for Scaled CMOS Technologies. *IEEE Journal of Solid-State Circuits (JSSC)*, 41:2658–2668, December 2006.

[5] M. F. Duarte et al. Single-Pixel Imaging via Compressive Sampling. *IEEE Signal Processing Magazine*, 25:83–91, March 2008.

[6] R. Robucci, J. Romberg, P. Hasler et al. Compressive Sensing on a CMOS Separable-Transform Image Sensor. *Proceedings of the IEEE*, 98:1089–1101, Jun 2010.

[7] V. Majidzadeh, P. Vandergheynst, Y. Leblebici et al. A (256x256) Pixel 76.7mW CMOS Imager/Compressor Based on Real-Time In-Pixel Compressive Sensing. *IEEE International Symposium on Circuits and Systems (ISCAS)*, pages 2956–2959, May 2010.

[8] Y. M. Chi, A. Abbas et al. An Active Pixel CMOS Separable Transform Image Sensor. *IEEE International Symposium on Circuits and Systems (ISCAS)*, pages 1281–1284, May 2009.

[9] Y. Oike, A. E. Gamal. A 256x256 CMOS Image Sensor with ΣΔ Based Single-Shot Compressed Sensing. *IEEE International Solid-State Circuits Conference (ISSCC)*, pages 386–388, February 2012.

A 9-bit 50MSps SAR ADC with Pre-charge V_{CM} -based Double Input Range Algorithm

Gong Chen The University of Kitakyushu
1–1 Wakamatsu, Kitakyushu
Fukuoka, Japan
r0dcb401@eng.kitakyu-u.ac.jp

Bo Yang Waseda University
2–7 Wakamatsu, Kitakyushu
Fukuoka, Japan
yang@da-lab.com

Yu Zhang The University of Kitakyushu
1–1 Wakamatsu, Kitakyushu
Fukuoka, Japan
r0dcb401@eng.kitakyu-u.ac.jp

Qing Dong The University of Kitakyushu
1–1 Wakamatsu, Kitakyushu
Fukuoka, Japan
nakatake@kitakyu-u.ac.jp

Shigetoshi Nakatake The University of Kitakyushu
1–1 Wakamatsu, Kitakyushu
Fukuoka, Japan
nakatake@kitakyu-u.ac.jp

ABSTRACT

This paper presents an energy efficient algorithm for 1.2V 9-bit 50MSps Successive Approximation Register (SAR) ADC. This conversion mechanism is based on pre-charge V_{CM} and achieves the twice of the input signal range beyond the supply voltage. Besides, this algorithm requires (i)a dynamic comparator with dual outputs and (ii)an additional comparison stage before each redistribution stage in the pre-conversion. We can reduce the switching power over 10% when compared to the merged capacitor switching (MCS) method with double input range of supply voltage. We design the SAR ADC following our idea on 65nm CMOS technology. The simulation results show 2.5dB better signal-to-noise-and-distortion ratio (SNDR) than the conventional method. Our ADC consumes a total energy of 0.504 mW at a 1.2V supply and 50MSps.

Categories and Subject Descriptors

B.7 [**Hardware**]: Integrated Circuits; B.7.1 [**Integrated Circuits**]: Types and Design Styles—*Algorithms implemented in hardware*

Keywords

pre-charge V_{CM}-based double input range algorithm, saving power switching procedure, 1.2V supply

1. INTRODUCTION

In recent years, as scaling down the feature sizes of CMOS devices, a lower supply voltage causes the limitation of signal swing. Therefore, an advanced SAR ADC must be able to convert input signals that are significantly higher than its supply voltage level.

In this work, we present the analysis and the simulation results for a 9bit 50MSps SAR ADC designed on 65nm/1.2V CMOS technology. The proposed pre-charge V_{CM}-based algorithm is capable of expanding to the twice input of the full supply voltage. The circuit is implemented including a dynamic comparator of a single input and dual outputs. The comparison range can be restricted from the ground voltage to the supply voltage in the redistribution mode. As [V. Hariprasath et al. 2010] and [C. Liu et al. 2010], the total capacitance in the DAC capacitor array is reduced by 50% compared with a conventional 9 bit SAR ADC.

2. SAR ADC ARCHITECTURE

Fig. 1 and Fig. 2 show the proposed SAR ADC and the timing diagram, respectively. During the additional Com mode, V_{in+} and V_{in-} make the first comparison. If V_{in+} is higher than V_{in-}, the output of the most significant bit (MSB, i.e. b8) is 1 and the switch S_{EX} makes V_{DD} and GND keeping in the V_{in+} side and the V_{ref-} side, respectively. If MSB is 0, then the reference voltage of V_{DD} and GND are exchanged sides by S_{EX}.

Fig. 3 demonstrates 3-bit examples of the conventional and proposed switching methods. Fig. 3 (a) shows all possible conversions. The quantitative energy consumption of each switching phase is also shown in the figure. The conventional switching sequence is efficient when all the attempts are successful as in the upper case. However, the switching sequence consumes a lot of energy when attempts are unsuccessful as in the lower case. As seen in Fig. 3 (b), the subsequent switching sequence of the proposed method can be adopted with both top-plate input and bottom-plate input ADC structures, and it is more efficient than that of the conventional one. For a 3 bit case, the conventional switching procedure consumes $8.75CV_{DD}^2$ while the proposed method does only $1.75CV_{DD}^2$. Thus, the proposed method requires 80% less energy than that of the conventional one. Fig. 4 shows a schematic of the dynamic comparator.

3. RESULTS

Fig. 5 shows the dynamic performance versus the input frequency simulated at 50MSps sampling rate and the dynamic performance versus sampling rate with an 1-MHz sine wave input. The restriction of bias voltage range in the MSB comparison mode the proposed ADC achieves 2.5dB better SNDR performance than the conventional one. The performance of the ADC is summarized in Table I.

Table 1: Summary of Performance

Technology	65-nm CMOS
Resolution	9-bit
Sampling Rate	10-MS/s
Supply Voltage	1.2-V
Full Scale Analog Input	2-V_{pp} differential
LSB	4.6875-mV
DNL	+0.42/-0.13LSB
INL	+0.31/-0.1LSB
SNDR (f_{in} 5MHz)	51.2 dB
SNDR ($f_{sampling}$ 50MHz)	50.8dB
ENOB	8.15 bit
$FOM = POWER/2^{ENOB} * f_{Sampling}$	29.84fJ/conversion-step
Power Consumption	0.504-mW

4. CONCLUSION

We present a 9-bit SAR ADC, which is capable of an input range equal to twice the supply voltage. Compared with the conventional method, it is shown that our pre-charge V_{CM}-based method can improve the energy efficiency for every code.

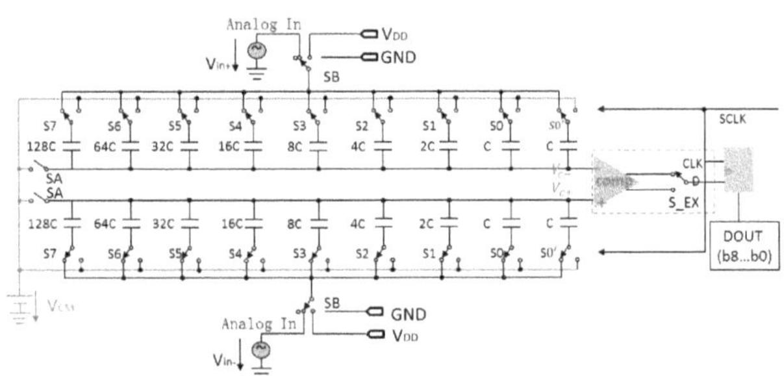

Figure 1: The proposed SAR ADC architecture

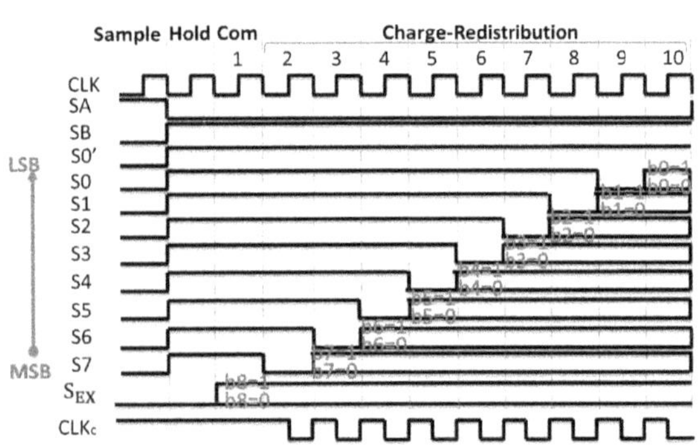

Figure 2: The proposed Timing diagram

5. REFERENCES

[V. Hariprasath et al. 2010] V. Hariprasath, J. Guerber, S.-H. Lee, U. Moon 2010. "Merged capacitor switching based SAR ADC with highest switching energy-efficnecy," *Electron. Lett., vol. 46*, (Apr.), 620-621.

[C. Liu et al. 2010] C. Liu, S. Chang, G. Huang, and Y. Lin 2010. "A 10-bit 50-MS/s SAR ADC with a monotonic capacitor switching procedure," *IEEE J. Solid-State Circuits, vol. 45*, 4 (Apr.), 731-740.

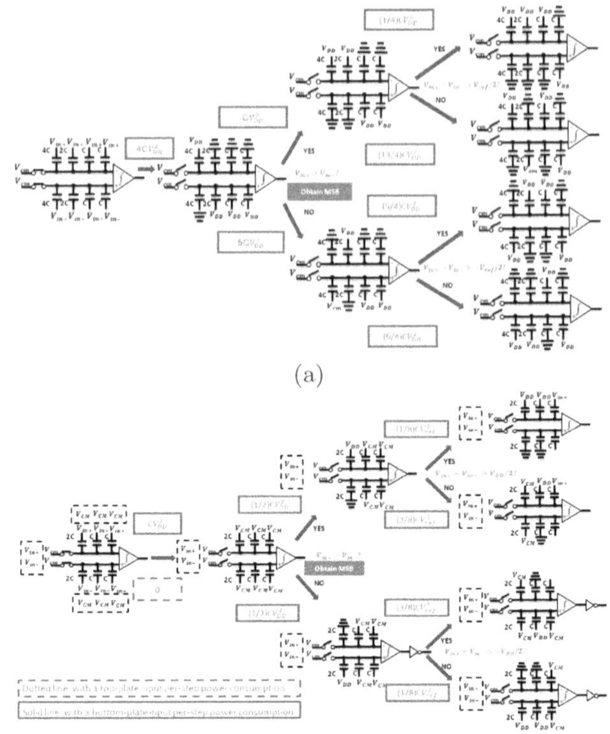

(a)

(b)

Figure 3: (a)Conventional switching procedure. (b)Proposed pre-charge V_{CM}-based switching procedure.

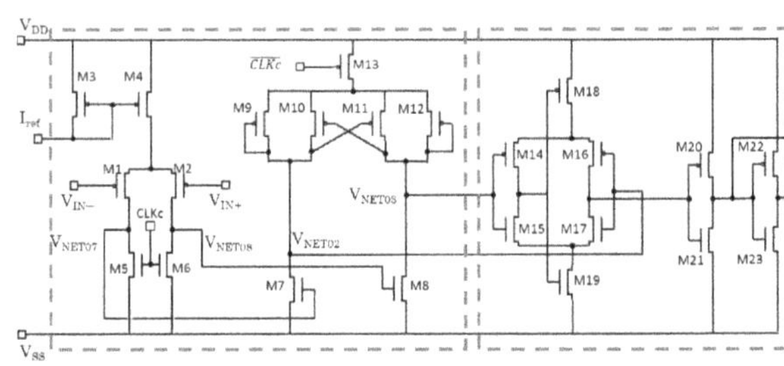

Figure 4: Circuit schematic of the dynamic comparator with dual inverter structure

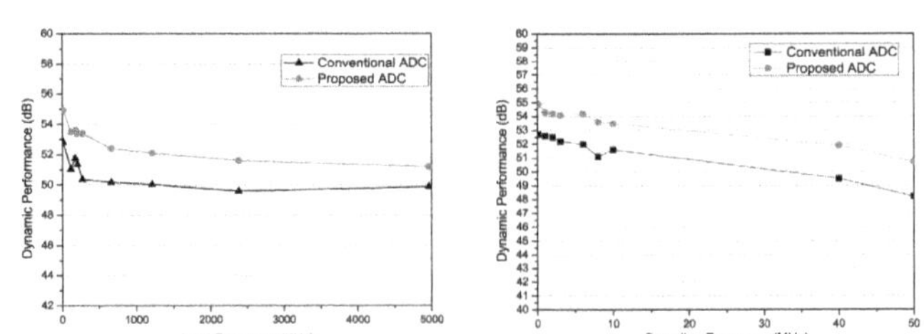

Figure 5: A comparison of SNDR between the proposed method and the conventional method versus to input frequency and sampling frequency

An Evaluation of an AES Implementation Protected against EM Analysis

Paolo Maistri[1], Sebastien Tiran[2], Phillippe Maurine[2], Israel Koren[3], Régis Leveugle[1]

[1] TIMA Laboratory
(Grenoble INP, UJF, CNRS)
Grenoble, France
{Paolo.Maistri,Regis.Leveugle}@imag.fr

[2] LIRMM
(UM2, CNRS)
Montpellier, France
{Sebastien.Tiran,pmaurine}@lirmm.fr

[3] University of Massachusetts
ECE Department
Amherst, MA, USA
koren@ecs.umass.edu

ABSTRACT

EM emissions can be a rich source of leakage for side-channel analysis of cipher implementations. In this paper, we describe a set of novel countermeasures based on dynamic spatial relocation and dynamic mappings, and validate the protection provided by them against EM attacks. The countermeasures improve significantly the security of the circuit.

Categories and Subject Descriptors

B.7.1 ***[Types and Design Styles]****: Algorithms implemented in hardware.*

Keywords

AES, EM, CEMA, CPA, side-channel analysis, countermeasures.

1. INTRODUCTION

Many current secure implementations use the Advanced Encryption Standard (AES) [1]. There has been a large effort in analyzing the robustness of AES. Side-channel attacks can be the most efficient way (considering the required resources) for an attacker to extract the secret information. Using Differential Power Analysis [2], for instance, the attacker can collect a few thousands traces of the power consumption of the device inexpensively, and guess the secret key after performing a simple statistical analysis of the collected data. Several other attacks have been proposed, either based on more advanced data analysis [3], or on different channels leaking information, such as Electro-Magnetic (EM) emissions [4]. EM emissions have recently started to be considered as a serious means of attack: they are contactless, thus monitoring EM emissions is easier and less detectable than power monitoring; they are a very rich source of information, in the spatial, temporal, and frequency domains; and the required knowledge to mount theses attacks is becoming more common.

In this paper we propose several countermeasures against EM attacks for an implementation of AES. The main contribution of this paper is a dynamic data relocation scheme working at two levels, one within and one across the round computations. A similar approach was proposed for a DES implementation in [5]. Our solution differs as it takes advantage of the AES structure and operates on multiple levels. Masking schemes are also employed, in order to improve the robustness of the design and illustrate the compatibility of the proposed solution with the state-of-the-art.

GLSVLSI'13, May 2–3, 2013, Paris, France.
ACM 978-978-1-4503-1902-7/13/05.

2. ARCHITECTURE IMPLEMENTATION

The AES algorithm [1] is round-based with rounds consisting of the following operations:

- *SubBytes*: a nonlinear byte substitution.
- *ShiftRows*: rotation of the rows of the internal state.
- *MixColumns*: each column is processed through polynomial multiplication in binary field $GF((2^8)^4)$.
- *AddRoundKey*: a round-dependent key addition.

The basic building block of the implemented architecture is a round and is derived from [6]. It is a 32-bit-wide design, with only four substitution boxes and four scaling units in a single *MixColumns* element. This iterative single-round implementation takes about 1200 slices on a Xilinx Spartan3.

In order to increase the global throughput of the circuit, four round instances are implemented. However, these blocks are not serially pipelined as it is usually done when implementing unrolled ciphers. All instances are hierarchically equal and are connected to a common bus. The number of lanes depends on the number of round instances in the circuit.

3. SPECIFIC COUNTERMEASURES

3.1 Dynamic Column Relocation (*DCR*)

To implement a dynamic relocation of the columns, the *ShiftRows* has to operate in a slightly different way with respect to the algorithm specifications. Rows are still rotated by an amount proportional to their index; however, the first row is no longer untouched, but it is instead, rotated by a random amount (from 0 to 3). Further rows are rotated accordingly, in order to preserve consistency within the state. Hence, four different configurations are possible after shuffling. The number of possible configurations is limited, due to the constraints imposed by the AES round functions. The columns are correctly relocated at the end in order to produce a result conforming to the standard, so any AES implementation can be used to decrypt the ciphered text.

3.2 Dynamic Block Relocation (*DBR*)

In order to increase the number of spatial configurations of the encrypted text, data is also moved randomly at each round between the different round instances. This means that after the *MixColumns* has been computed, data is not rerouted within the same round circuit instance, but rather sent on the external bus to be transferred to another instance. The destination is chosen randomly, thus at each round the configuration may change. Writing on the bus lanes is managed by a dedicated bus controller. This unit must ensure that all the transfers are performed correctly, i.e., no data is lost or overwritten.

3.3 Dynamic Mapping (*DynMap*)

S-Boxes constitute a large source of information leakage and are often the target of side-channel attacks, hence they are usually protected. In this paper we propose a solution based on composite fields. These fields are often used to implement small substitution boxes, and mapping from one field to the other (and vice versa) is needed. The approach followed in this paper consists of having several mappings from the regular to the composite field instantiated in the design. Then, at computation time, any of them can be randomly chosen to compute the *SubBytes* operation.

3.4 Linear Masking (*LinMsk*)

The linear layer of the AES still operates on unmasked data. For this reason, a simple additive masking is used. A 32-bit mask is added at the output of the *SubBytes* operation, before storing the values into a register. The complexity is kept low by using the same mask for each row of the matrix. This has the advantage that the mask is not affected by *MixColumns*, since the same byte mask is applied to each byte of the same column.

3.5 Preliminary Cost Evaluation

The countermeasures implemented have an associated cost. The first and most apparent cost is related to the number of round instances that are implemented at the same time in the design, which affects also the number of bus lanes. The remaining overhead is mainly due to the additional S-Box mappings, the module generating the bus configuration, a PRNG, and the linear masking. Overall, our design uses 5694 slices on a Spartan3 for a 4-round implementation.

4. ROBUSTNESS AGAINST EM ANALYSIS

Three sets of EM traces were collected. The first set of EM traces included 100k one-shot EM traces corresponding to AES encryptions of messages with all countermeasures disabled. The second set of EM traces corresponds to 200k AES encryptions with only the Linear Masking countermeasure activated. The third set consists of 400k EM traces obtained with all countermeasures activated. To test the improvement in terms of resistance against first order side-channel-attacks, Correlation Power Analyses (CPAs) [3] were applied to the three sets of traces. The attacks were targeting the last round of the AES using a Hamming Weight model. All the 16 S-boxes were attacked and we used the Guessing Entropy [7] to evaluate the efficiency of the different countermeasures presented in this paper.

Figure 1 shows the evolution of the mean Guessing Entropy with the number of processed traces. When no countermeasures are activated, the mean Guessing Entropy of the 16 correct sub-keys decreases monotonically with a significant slope. However, 100k curves are not enough to fully recover the key. When only the linear masking is activated, we observed a decrease in the mean Guessing Entropy, but the rate of decrease is 2.2 times lower. This suggests that at least 2.2 times more traces (roughly 320k traces) are necessary to fully reveal the round key. Finally, with all countermeasures activated, we notice that the mean Guessing Entropy slowly converges towards the value 121 suggesting that a very large number of traces is necessary to fully disclose the round key. Still, the disclosing of one sub-key (one byte discovered out of 16) after the processing of 400k traces demonstrates that even with all countermeasures, the protected AES is not free of first-order leakage. Even if the rank of sub-keys is slowly decreasing and one byte has been found, the entire key is far from being discovered after the processing of 400k curves with a CPA attack. Considering these results, we conclude that the proposed countermeasures do not completely suppress first order leakages but make them considerably more difficult to exploit.

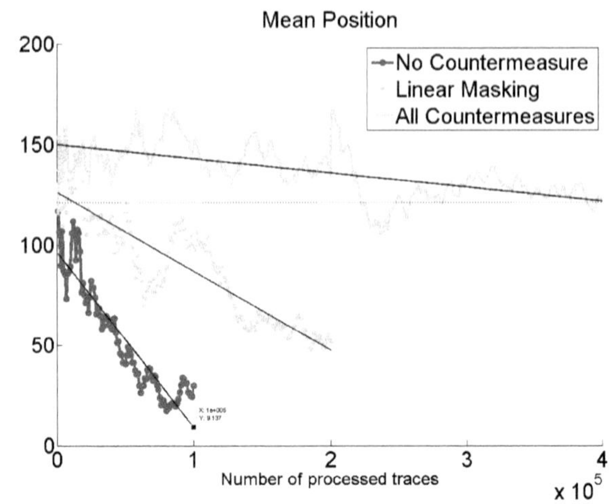

Figure 1. Evolution of the mean Guessing Entropy

5. CONCLUSION

Side-channel analysis is a major threat for secure implementations of cryptographic algorithms. In this paper, we have presented several novel countermeasure schemes. EM traces were collected and subjected to correlation analysis to evaluate the effectiveness of the considered countermeasures. Our results indicate that the complete set of proposed countermeasures provides a very good resistance against EM attacks demonstrating that a very large number of traces is required in order to break the security.

6. ACKNOWLEDGMENTS

This work is partly supported by the French Ministry of Research, through the ANR project EMAISECI (act ANR-2010-SEGI-012-03).

7. REFERENCES

[1] National Institute Standards and Technology (NIST), "FIPS-197: Advanced Encryption Standard," Nov. 2001.

[2] P. C. Kocher, J. Jaffe, B. Jun, "Differential Power Analysis," CRYPTO 1999, pp. 388-397.

[3] E. Brier, C. Clavier, F. Olivier, "Correlation Power Analysis with a Leakage Model," CHES 2004, pp.16-29.

[4] J.-J. Quisquater and D. Samyde: "Electromagnetic analysis: Measures and counter-measures for smart cards," Proc. of E-smart, LNCSvol. 2140, pp. 200-210, Springer-Verlag, 2001.

[5] F. Poucheret, L. Barthe, P. Benoit, L. Torres, P. Maurine, M. Robert: "Spatial EM jamming: A countermeasure against EM Analysis?" VLSI-SoC 2010, pp. 105-110.

[6] N. Pramstaller et al., "Efficient AES Implementations on ASICs and FPGAs," 4h Int'l Conf. Advanced Encryption Standard (AES '04), pp. 98-112, 2004.

[7] F.-X. Standaert et al., "A unified framework for the analysis of side-channel key recovery attacks," EUROCRYPT, vol. 5479 of LNCS, page 443-461. Springer, 2009

PBD: Packet Buffer DVFS

Eyal-Itzhak Nave, Ran Ginosar
Electrical Engineering Department
Technion – Israel Institute of Technology
Technion City, Haifa, 32000, Israel
+972-54-4691461; +972-4-829-4645
eyal.nave1@gmail.com; ran@ee.technion.ac.il

ABSTRACT

The internet is rapidly becoming a major consumer of electricity with measurable economic and environmental impact. DVFS is employed in processors for balancing performance and power consumption. We apply DVFS to network controllers to reduce unnecessary power dissipation in networks. When receive and transmit buffers are relatively empty, energy is saved by lowering the packet processing rate. A simulation environment was developed to assess the energy savings achievable with this approach. Simulations show an average energy savings of 29% across various traffic loads and 19% across various congestion levels.

Categories and Subject Descriptors

C.2.1 **[Computer-Communication Networks]**: Network Architecture and Design - *Network communications, Packet-switching networks;* C.4 **[PERFORMNANCE OF SYSTEMS]**: Modeling techniques.

Keywords

DVFS; Low Power; Network Controller

1. INTRODUCTION

In this research we investigate the potential for power saving in computer networks. As opposed to modern processors, which handle quite well idle periods, network controllers consume the same power whether they are idle or fully utilized. This problem stems from the desire to maintain open network links even when they are not utilized, thus preventing shutdown of relevant components. However, power consumption during low utilization periods can be reduced by lowering the voltage and frequency, which enable maintaining the links open and even handling low traffic loads. Higher voltage and frequency should be used only when needed, i.e. higher traffic loads which require higher processing rate. The frequency and voltage at any specific time point can be adjusted to the work-load. In network controllers, the work-load is indicated by the number of packets in it's transmit and receive buffers. In this research we examine a two-state (high / low) DVFS system: When using PBD, if the packet buffer is filled above a predefined threshold, a high power mode is used, enabling higher performance at the cost of higher power consumption. Once the packet buffer is emptied below a predefined threshold, a low power mode is selected, enabling lower power consumption at the cost of lower performance. In this research we examine the expected energy saving of PBD in a network system and how do network system conditions affect PBD energy saving.

GLSVLSI'13, May 2–3, 2013, Paris, France.
ACM 978-978-1-4503-1902-7/13/05.

2. SIMULATION MODELING

A configurable DVFS simulation environment was developed, simulating different traffic patterns transmitted via a TCP session with different network conditions and comparing energy consumption of different DVFS policies. We assume separate DVFS work-points (High/Low) and transmission rates for TX and RX. The simulated system is described in Figure 1.

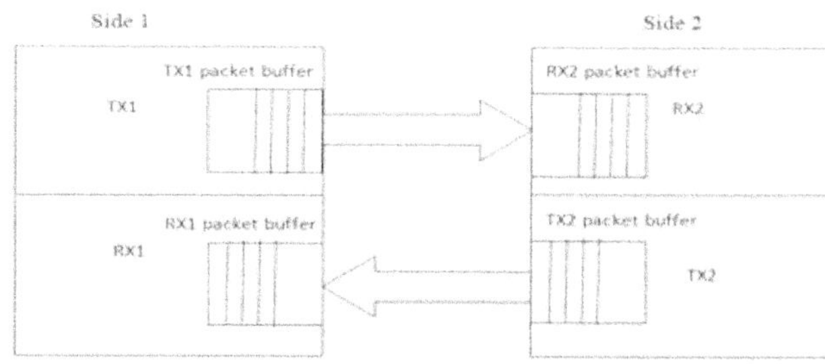

Figure 1. Simulation architecture scheme

Simulated PBD behavior can be observed in Figure 2. At time units 6, 55, 63, 77, 162 and 301 the packet buffer size crosses the high threshold (20), causing a transition to high power mode. At time units 51, 60, 72, 159 and 297 the packet buffer size reaches the low threshold (10), affecting a transition to low power mode. All simulation runs start with empty packet buffers, and employ a real trace [1] that provides packet arrivals during 1600 seconds. The simulation proceeds beyond 1600 seconds until all packets are delivered and acknowledged. The network enters a congestion period 400 seconds after the end of the previous congestion period and stays congested for 200 seconds, in which time a packet is lost every 50 seconds. When summing up the energy dissipated during the packet transmission sessions in both TX and RX of both sides, energy consumption with PBD is 4,695J, compared to 5,217J without DVFS, saving 522J or 10% energy.

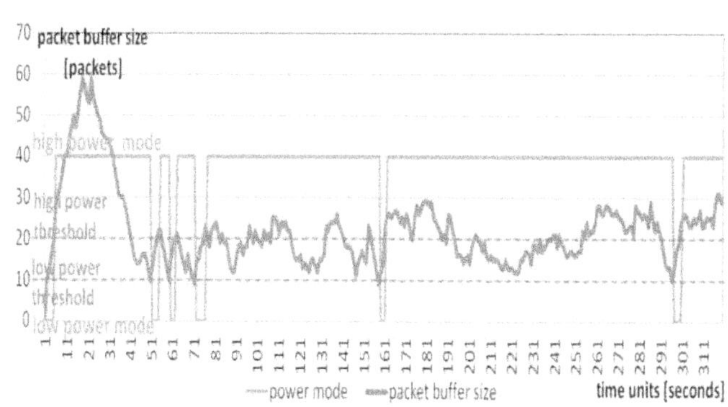

Figure 2. Packet Buffer DVFS simulation behavior

3. ENERGY SAVING ACROSS TRAFFIC LOADS AND CONGESTION LEVELS

The effect of different traffic load levels on DVFS and energy savings is examined by modeling the packet arrival rate using Poisson distribution. We use a benchmark of random Poisson distributed packets arriving during 1600 seconds with a mean average of λ packets per second. Figure 3 compares the energy consumption with and without PBD across various traffic loads.

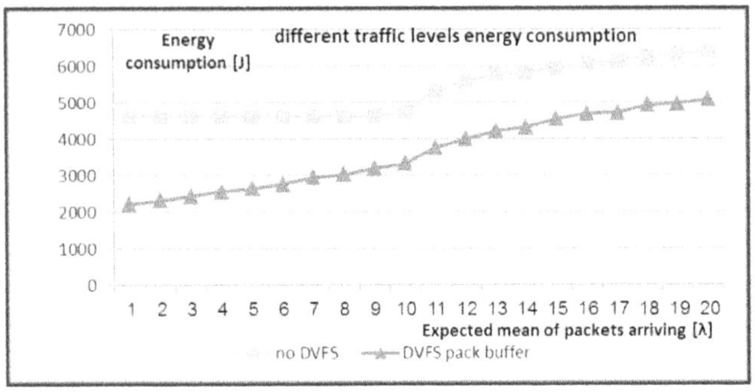

Figure 3. Various traffic levels energy consumption

Higher performance is required for high traffic loads; therefore, high power mode enabling high transmission rate at the cost of higher power/energy consumption is used. Higher transmission rates result in more packets being transmitted. During network congestion, some packets are lost. Each lost packet requires the network controller to retransmit the lost packet as well as all its following packets. The more packets transmitted in parallel the more packets required to be retransmitted in case of a lost packet. Every packet retransmission consumes energy in the transmission circuits and postpones shut-down. When the network is congested, lost packets prevent the actual progress of transmission, as the transmitted packets do not reach the other side, causing the performance advantage of high power mode to be wasted.

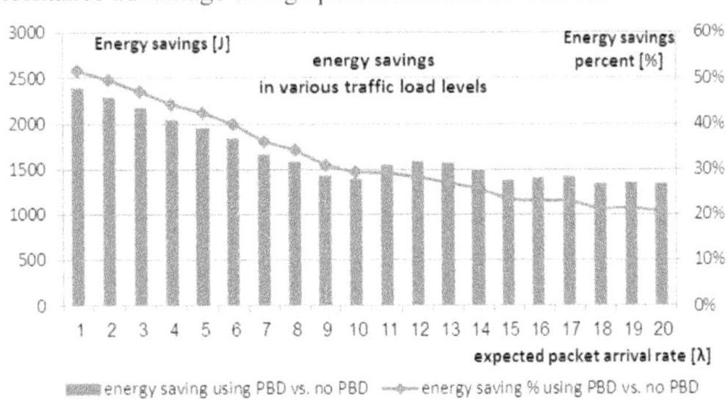

Figure 4. Various traffic levels energy saving

According to Figure 4, the maximum energy saving of ~2.4KJ, 52% of the energy consumed with no PBD, is achieved at λ=1. A traffic load increase causes an energy saving decrease because high traffic loads fill the packet buffer above the threshold causing a transition to high power mode, consuming more energy. The total energy saved by using low power mode is at least 1.3KJ. PBD energy savings percentage across various traffic loads is between 21% and 52%, with a median of 29%.

We now observe the impact of PBD at various levels of network congestion. Ten congestion levels 1-10 were simulated, where 1 is the least congested network level and 10 is the most congested one. The packet arrival distribution is the same for all simulation runs to isolate network congestion effect on PBD energy saving.

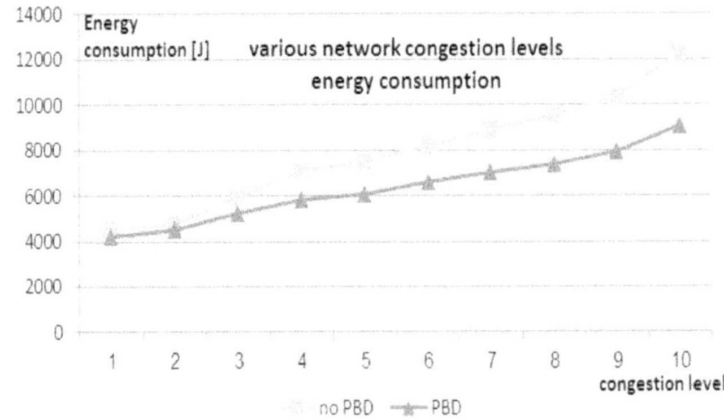

Figure 5. Various congestion levels energy consumption

The energy saving benefit increases with congestion, starting with 200J during low congestion and reaching above 3KJ during high congestion. At the highest congestion level (10), PBD incurs lower energy consumption of 9KJ whereas no PBD dissipates more than 12KJ, resulting in more than 3KJ saving, about 26%.

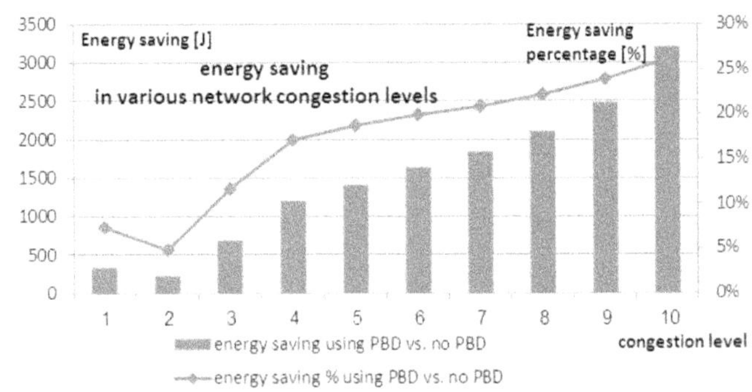

Figure 6. Various congestion levels energy saving

Figure 6 highlights the advantage in energy saving achieved with PBD, which increases with congestion. PBD energy savings percentage across various network congestion levels is between 5% and 26%, with a median of 19%.

4. SUMMARY AND CONCLUSIONS

We apply DVFS for the first time to network controllers and introduce a novel power management circuit, Packet Buffer DVFS (PBD). The DVFS operating-point is determined based on work-load as indicated by the number of packets in the receive and transmit buffers. A simulation-based study of energy savings benefits is reported, comparing a network system with and without PBD. The effects of various traffic loads and network congestion levels were examined as well. We show that PBD can achieve up to 52% energy savings. On average, it achieves 29% savings across traffic loads and 19% across congestion levels. A new simulation environment was developed for comparing DVFS mechanisms at varying conditions providing a base-ground for many future research works on various network DVFS policies.

5. REFERENCE

[1] TIPC-over-TCP_disc-publ-inventory_sim-withd.pcap, "SampleCaptures - The Wireshark Wiki". http://wiki.wireshark.org/SampleCapture

A Memory Mapping Approach for Network and Controller Optimization in Parallel Interleaver Architectures

Aroua Briki Cyrille Chavet Philippe Coussy
Lab-STICC laboratory / Université de Bretagne Sud
Centre de recherche Christiaan Huygens, 56100 Lorient (France)
{aroua.briki@univ-ubs.fr; cyrille.chavet@univ-ubs.fr; philippe.coussy@univ-ubs.fr }

ABSTRACT

Recent communication standards and storage systems uses parallel architectures for error correcting codes (LDPC or Turbo-codes) to reliably transfer data between two equipments. However, parallel architectures suffer from memory access conflicts. In this paper, we present a method that finds a conflict-free memory mapping for any interleaving law and any parallelism. The proposed approach always complies with the interconnection network topology the designer wants to infer. Moreover, the resulting architecture is optimized by reducing the cost of network and controller (network and memory controller) architectures.

Categories and Subject Descriptors

J.6 **Computer-Aided Engineering**: Computer-aided design (CAD); B.7 **Integrated Circuits**: General

Keywords

Parallel architectures, interleavers, memory mapping.

1. INTRODUCTION

Error correcting codes i.e. LDPC (Low Density Parity Check) and Turbo-codes are the foundation of communication. Standards like digital video broadcasting (DVB), high-speed wired links (ADSL...), wireless accesses (WiMAX, Wifi...), and telecommunications systems (HSPA, LTE...) all rely on it. LDPC and Turbo Codes are well-known, near Shannon limit, coding/decoding approaches that are able to achieve very low bit error rates for low Signal-to-Noise Ratio (SNR) applications, that's why they are also used in storage systems. Decoding principle is based on message passing algorithm in which different processing elements iteratively exchange information in order to improve the error correction performance of these codes. In order to design high data rate applications, parallel architecture are needed [1]. To increase memory bandwidth, main memory is divided into different memory banks to provide concurrent parallel access to all the processing elements. This allows to reduce the latency and thus to increase the throughput of decoders. Figure 1 represents an example of typical parallel decoder architecture with three processing elements (i.e. PE_0, PE_1, PE_2) connected through a dedicated Interconnection Network to three memory banks (i.e. RAM_A, RAM_B, RAM_C) and a dedicated Control Unit that drives the architecture. The network interleaves the data exchanged by the PEs according to a rule named *interleaving law* or *permutation law* defined by the standard or the application to design. Unfortunately, depending on both the interleaving law and the memory mapping (i.e. data placement in memory banks), different PEs may try to simultaneously access the same memory bank which results in memory conflicts [2].

GLSVLSI'13, May 2–3, 2013, Paris, France.
ACM 978-978-1-4503-1902-7/13/05.

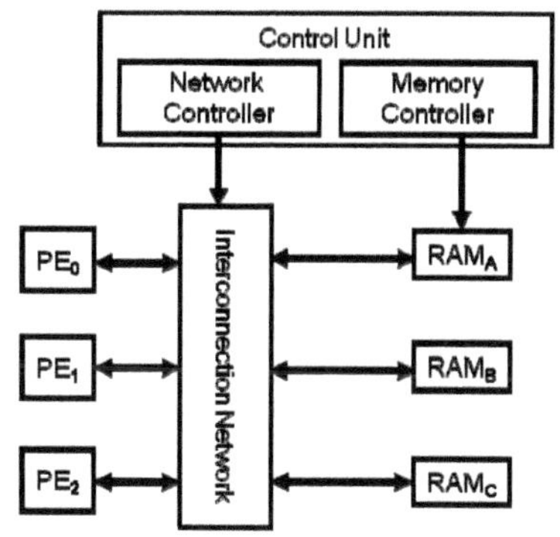

Figure 1. Architecture overview

Three kinds of solution exist to avoid or minimize memory access conflicts: (1) Define an interleaving law that automatically maps data in different memory banks so that all processing elements can access them without any conflict at each time instance [3]. This kind of approach is interesting only when the designer is free to choose the permutation law; (2) Simply store data elements in different memory banks without considering conflicting accesses and then use different complex topologies and additional buffers to mange conflicts on runtime [6]. This increases the cost and latency of the system. (3) Use memory mapping algorithms [4] [5] [7] to map data in different memory banks so that each processing elements can access them without any conflict. This kind of approach either results in a non-optimized network [7], or considers the network as a constraint for in-place turbo-codes decoders only [4] and fails to find a network-oriented memory mapping when the interleaving law is incompatible with the given network. In addition, ROM resources are needed in the controller to store the memory mapping (i.e. control words to address memory banks and to drive interconnection networks). Unfortunately, cost of the controller is not considered for optimization [5].

To our best knowledge, this paper is the first attempt to propose a design approach that enables (1) to generate a conflict-free memory system for any interleaving law, (2) to define the target interconnection network and (3) to reduce the cost of the controller architecture.

2. PROPOSED APPROACH

Our proposed design flow is presented in Figure 2. It is based on two main steps: first, starting from the set of constraints (i.e. the interleaving law, the parallelism and the targeted interconnection network), it generates a conflict free memory mapping and an *Address Conflict Graph ACG*. At this step, the data are assigned to memory banks (i.e. Bank Mapping), but their addresses in banks are still unknown (i.e. Address Mapping is not yet defined). In ACG, vertices represent data accesses in their respective memory banks, and edges represent conflicts between these accesses (e.g. write a new data at a currently used address). The memory mapping is performed thanks to a dedicated heuristic that aims to both generate a conflict free memory mapping and minimize the number of conflict in the generated ACGs. Indeed, when two data are stored in the same memory bank, and if their respective "lifetimes" in this memory bank are overlapping, two different memory locations (i.e. addresses) are required. Hence the bank mapping step strongly impacts the final ACGs, by generating more or less address conflicts.

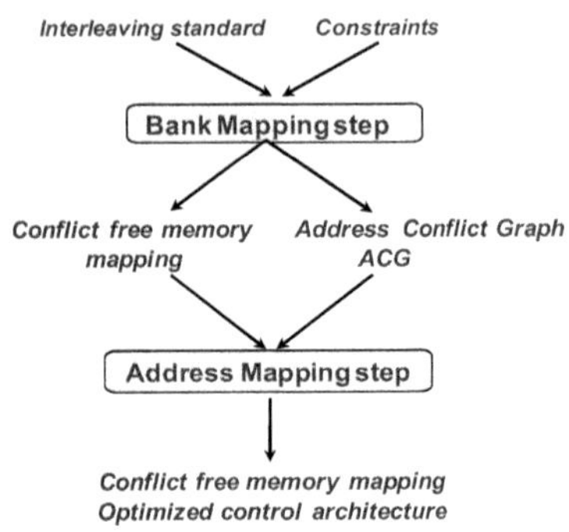

Figure 2. Proposed design flow

Then, memory mapping information and ACG are used during the *Address Mapping* step to explore the memory addressing in order to optimize the final cost of the memory controller. Finally, our design flow generates the final VHDL architecture with a conflict free memory mapping and the associated optimized control units.

3. EXPERIMENTS

Our approach has been applied to explore the design space of several test cases. The generated VHDL by our approach has been synthesized through Xilinx ISE Design Suite 2012, targeting a Virtex6 platform. Different test cases have been used: a High Speed Packet Access (HSPA) [9], an Ultra-WideBand interleaver (UWB) [8], and a Wimax application [10]. These experiments have been performed for several parallelisms and radix. Our results are compared to state of the art approaches [4] and [7] depending on their best target application (Turbo Codes or LDPC). Since the proposed approach is able to target any interconnection network, two families of such network have been explored: Barrel-Shifter network (for UWB test case) and Butterfly network (for HSPA and WiMAX).

Table 1. Synthesis results compared to [7]

Test Case	Frame length	Parallelism	Area		Gain
			[7]	Our solution	
HSPA 1024 Radix 2	*1024*	*4*	967	625	**35,37%**
HSPA 1024 Radix 2	*1024*	*8*	1309	1051	**19,71%**
HSPA 1024 Radix 4	*1024*	*8*	1320	1057	**19,92%**
HSPA 5120 Radix 2	*5120*	*8*	6804	3886	**42,89%**
HSPA 5120 Radix 2	*5120*	*16*	9735	7210	**25,94%**
HSPA 5120 Radix 4	*5120*	*16*	9668	9312	**3,68%**

The synthesis results (cf. Table 1&2) are given in slices and obtained thanks to the ISE design Suite. Since the rest of the design will be the same, only the relevant parts of the results have been taken into account: controller and additional multiplexers and registers. The results show significant controller area reduction (from 3,6% up to 42%), with an average gain of 24,5%. It can be observed that the larger is the frame length compared to the parallelism, the better are the results.

Table 2. Synthesis results compared to [4]

Test Case	Frame length	Parallelism	Area		Gain
			[4]	Our solution	
UWB	*1200*	*8*	2215	1544	**30,29%**
UWB	*1200*	*4*	1519	1038	**31,67%**
UWB	*600*	*8*	1346	981	**27,12%**
UWB	*600*	*4*	994	686	**30,99%**
UWB	*300*	*8*	949	762	**19,70%**
UWB	*300*	*4*	593	394	**33,56%**
WiMAX 2by3	*80*	*10*	553	532	**3,80%**

4. CONCLUSION

In this paper we proposed the first parallel memory design space exploration methodology that is able to optimize both interconnection network, its controller and the memory controller. It relies on a formal modeling describing data communication constraints and address conflicts. This methodology allows the generation of a conflict-free memory mapping. Resulting architectures respect the designer architectural constraints in any case and the controllers are strongly optimized (24,5% in average and up to 42%).

5. REFERENCES

[1] A. Seznec and J.L. enfant, "Interleaved parallel schemes", *IEEE Trans. Parallel Distrib. Syst., vol.5, no.12,* 1994.

[2] A.G iulietti, L.Van Der Perre, M.Strum, "Parallel turbo coding interleavers :avoiding collisions in accesses to storage elements", *Electronics Leters, vol. 38, no. 5,* Feb. 2002.

[3] D.Gnaedig, E.Boutillon, M.Jezequel, V.C.Gaudet, and P.G.Gulak, "On multiple slice turbo codes", in proc.3rd *Int. Symp.Turbo Codes, related Topics,* Brest, 2003.

[4] C. Chavet, P. Coussy, E. Martin and P. Urard, "Static Address Generation Easing: a Design Methodology for Parallel Interleaver Architectures", *proc of the International Conference on Acoustics, Speech, and Signal Processing (ICASSP), pp. 1594-1597,* Dallas, March 2010.

[5] A. Briki, C. Chavet, P. Coussy, E. Martin, " A Design Approach Dedicated to Network-Based and Conflict-Free Parallel Interleavers", *proc of the ACM Great Lakes Symposium on VLSI (GLSVLSI),* Salt Lake City, 2012.

[6] A. La Rosa, C. Passerone, F. Gregoretti, L. Lavagno, "Implementation of a UMTS turbodecoder on dynamically reconfigurable platform", *DATE*, 2004.

[7] A.Tarable, S.Benedetto, and G.Montorsi, "Mapping interleaving laws to parallel turbo and LDPC decoder architectures", *IEEE Trans. Inf. Theory, vol.50, no.9,* Sept. 2004.

[8] IEEE 802.15.3a, WPAN High Rate Alternative

[9] 3GPP, "*Technical specification group radio access network; multiplexing and channel coding (FDD)*" (25.212)

[10] IEEE 802.16, WiMAX - Worldwide Interoperability for Microwave Access

Post-Route Refinement for High-Frequency PCBs Considering Meander Segment Alleviation

Tsun-Ming Tseng
tsun-ming.tseng@tum.de

Bing Li
b.li@tum.de

*Tsung-Yi Ho
tyho@csie.ncku.edu.tw

Ulf Schlichtmann
ulf.schlichtmann@tum.de

Technische Universitaet Muenchen
Arcisstrasse 21, 80333 Munich, Germany

*National Cheng Kung University
No. 1, University Rd., Tainan City, Taiwan

ABSTRACT

In this paper, we propose a post-processing framework which iteratively refines the routing results from an existing PCB router by removing dense meander segments. By swapping and detouring dense meander segments the proposed method can effectively alleviate accumulating crosstalk noise, while respecting pre-defined area constraints. Experimental results show more than 85% reduction of the meander segments and hence the noise cost.

Categories and Subject Descriptors

J.6 [**Computer-aided Engineering**]: Computer-aided design (CAD)

Keywords

PCB routing; crosstalk; post-processing

1 Introduction

In high-frequency printed circuit boards, delay matching between bus signals has become a mainstream problem in modern PCB routers [1, 2, 5, 7]. In these papers, meander lines are commonly used to extend wire lengths to satisfy wire delay constraints, based on the concept that equal wire length should match the delays between different wires. Figure 1 illustrates a wire containing eight meander segments. However, it is shown in [3, 4, 6] that a signal across meander segments can have a speedup effect under high frequency circumstances, caused by the crosstalk noise of the same wire. In this paper, we propose a post-route refinement method to alleviate these dense meander segments in a given PCB design using swap, detour and shift techniques.

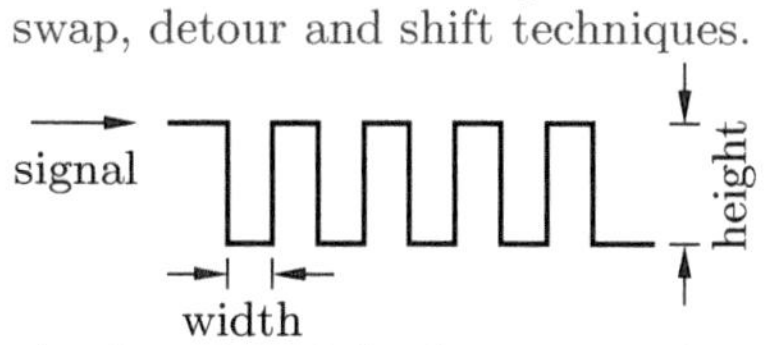

Figure 1: A wire with eight dense meander segments.

2 Proposed Method

In the following, we call the meander segments with small width *dense meander segments (dms)*, or *dense segments* if no ambiguity will ensue. The goal of the the proposed method is to reduce the number of the dense meander segments in a given PCB routing without changing the wire lengths or violating the given fixed outline. The concepts of the three techniques of the proposed method, swap, detour and shift, are explained in Section 2.1–2.3 and the complete flow applying these techniques is presented in Section 2.4.

GLSVLSI'13, May 2–3, 2013, Paris, France.
ACM 978-1-4503-1902-7/13/05.

2.1 Meander Segment Location Swap

In the first step, we swap the positions of the dense meander segments to lower their number on the wires. An example of this operation is shown in Figure 2, where the positions of the dense meander segments *dms*2 and *dms*3 are swapped. After this operation, *dms*1 and *dms*4 in Figure 2(a) are alleviated because the distances of their parallel wire segments are enlarged. During the swap step, no area overhead is required because we only take advantage of the area already used before.

2.2 Detour Generation

The detour technique increases wire lengths using free spaces so that dense meander segments elsewhere on the wires can be removed. An example of this technique is shown in Figure 3, where the dense meander segments are relaxed in (b) or (c) of Figure 3 by growing the required wire length with large distance between segments along the wires, called *detours* in the following. In addition, it pushes multiple wires at the same time to avoid unnecessary area consumption. Consequently, *detour groups* are created on wires to replace the original dense meander segments, as shown in Figure 3.

In generating a detour, the simplest case of a pushable segment is a straight wire, which can be pushed upward or downward freely. However, the appearance of perpendicular wire segments, shown as dashed lines in Figure 4 and 5, may prevent the simple push operation, because the pushed segments may form new dense meander segments with the perpendicular wires. In Figure 4 the perpendicular segments are at the same direction of the straight horizontal wire. If the horizontal wire is short, one or more perpendicular segments are involved in this push operation, as shown in Figure 4(b)-(d). Note that the moving direction in Figure 4(d) can only be upward, because a downward push does not increase the wire length. In Figure 5 the perpendicular segments are at different sides of the straight horizontal wire. Similar to Figure 4(b) and (c), we can push the wires along one of the perpendicular segments. However, the push operation in Figure 4(d) can not be applied in this case, because moving the straight wire upward or downward can not compensate the shortened wire length, if some dense meander segments are removed elsewhere. In implementation, multiple wires are pushed at the same time to generate detour groups. Figure 6(a)-(d) exemplify this process using five wire segments,

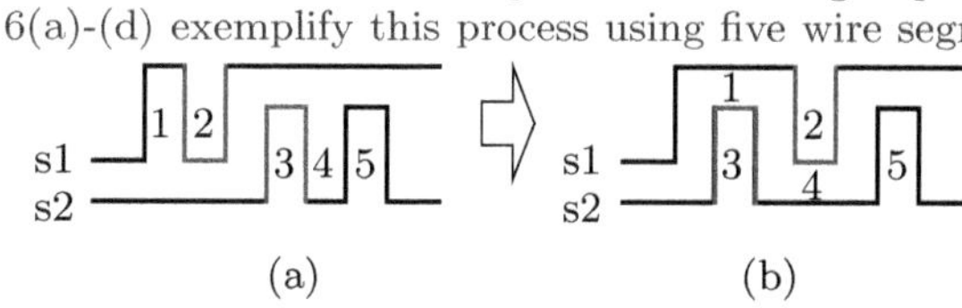

Figure 2: Meander segment swap.

Table 1: Results of routing refinement

	initial routing			100 iterations					500 iterations				
	wires	*dms*	sum_H	*dms*	sum_H	*dms*(%)	sum_H(%)	*runtime*(*s*)	*dms*	sum_H	*dms*(%)	sum_H(%)	*runtime*(*s*)
case1	17	42	581	16	270	38.10	46.47	8.02	9	149	21.43	25.65	31.88
case2	14	30	276	7	41	23.33	14.86	1.57	7	41	23.33	14.86	5.08
case3	92	237	1789	117	869	49.37	48.57	18.56	88	566	37.13	31.64	82.53

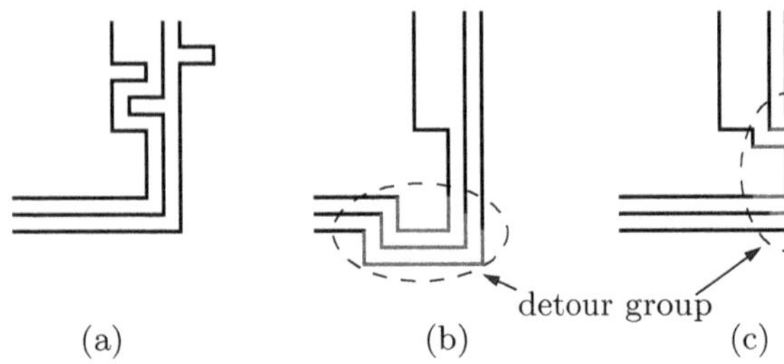

Figure 3: Example of detour generation.

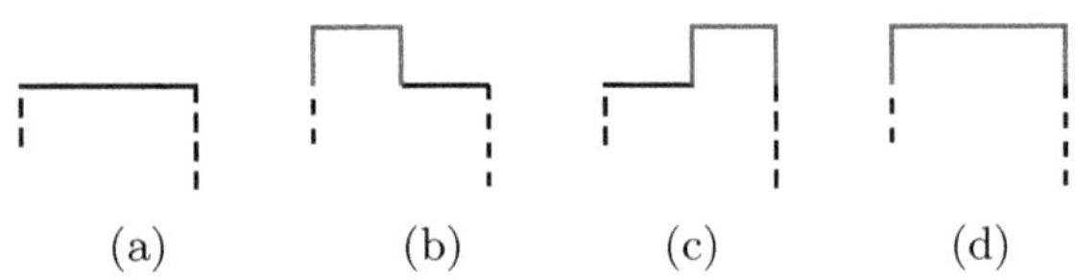

Figure 4: Pushable segment with perpendicular wires at the same side.

where wires are pushed downwards at the same and minimum wire distance between different wires are maintained. In Figure 6(d) the trim operation reduces the pushable region of the wire *s3* from the left side to relax the occupied area. The retract operation in Figure 6(d) works similarly except that all the three wires, *s1*, *s3* and *s5* have their pushable regions reduced from the right side.

2.3 Segment Shifting

After the swap and detour operations, some wire segments may be moved to save routing area. For example, there is some space between the two leftmost wires in Figure 3(b) and (c). Therefore, the upper horizontal segments at the leftmost wire can be moved upward to free the space at the left side of this wire group, while keeping the length of the complete wire unchanged. After scanning the positions of wire segments, the shift operation contracts the movable wire segments in the area which may be needed by further detour groups to release more free space.

2.4 Complete Flow

In the implementation we first scan the given routing to identify existing dense segments using the coordinates of the bending points on the wires. For any pair of dense segments, we exchange their positions using the swap operation illustrated in Figure 2 if the overall speedup cost can be reduced. The detour and shift operations described in Section 2.2 and 2.3 are thereafter applied. The process described above runs many times until the given number of maximum iterations is reached or no improvement can be achieved in the current iteration.

3 Experimental Results

The experiments are performed using a computer with a 2.2 GHz CPU and 8 GB memory. The test cases are from [5]. The results of the three test cases after running the proposed method with 100 and 500 iterations are shown in Table 1. Beyond 500 iterations, no significant improvement was achieved

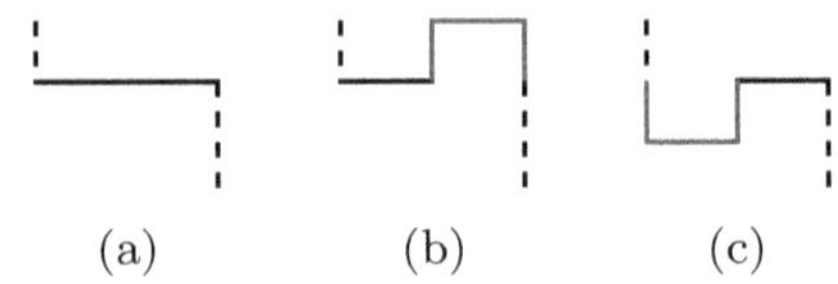

Figure 5: Pushable segment with perpendicular wires at different sides.

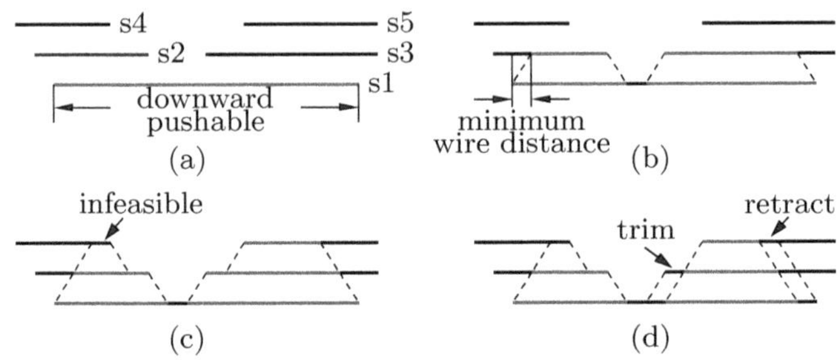

Figure 6: Finding a detour group.

further. In Table 1 the column *wires* shows the number of wires and the columns *dms* represent the numbers of dense meander segments in the routings. The sums of heights of these dense meander segments in the original routing and the routings improved by the proposed method are shown in the columns sum_H. *dms(%)* is the ratio of *dms* after applying the proposed method over *dms* in the original routing, and sum_H(%) is the ratio of the sums of heights. From these results, we can see that both the numbers of dense meander segments and the sums of their heights are reduced effectively.

4 Conclusion

In this paper we have explained a method to alleviate dense segments using techniques including swap, detour and shift. Experimental results confirm that the proposed method can effectively reduce the number of these dense meander segments and hence the speedup effect up to 85%.

5 References

[1] Y. Kubo, H. Miyashita, Y. Kajitani, and K. Takeishi. Equidistance routing in high-speed vlsi layout design. *Integr. VLSI J.*, 38(3):439–449, 2005.

[2] M. M. Ozdal and M. D. F. Wong. A length-matching routing algorithm for high-performance printed circuit boards. *IEEE Trans. Comput.-Aided Design Integr. Circuits Syst.*, 25(12):2784–2794, 2006.

[3] O. Ramahi. Analysis of conventional and novel delay lines: A numerical study. *J. Appl. Comput. Electromag. Soc.*, 18(3):181–190, 2003.

[4] B. Rubin and B. Singh. Study of meander line delay in circuit boards. *IEEE Trans. Microw. Theory Tech.*, 48(9):1452–1460, 2000.

[5] T.-Y. Tsai, R.-J. Lee, C.-Y. Chin, C.-Y. Kuan, H.-M. Chen, and Y. Kajitani. On routing fixed escaped boundary pins for high speed boards. In *Proc. Design, Autom., and Test Europe Conf.*, pages 1–6, 2011.

[6] R.-B. Wu and F.-L. Chao. Laddering wave in serpentine delay line. *IEEE Trans. Compon., Packag., Manuf. Technol. B*, 18(4):644–650, 1995.

[7] T. Yan and M. Wong. Bsg-route: A length-constrained routing scheme for general planar topology. *IEEE Trans. Comput.-Aided Design Integr. Circuits Syst.*, 28(11):1679–1690, 2009.

On-Chip Area-Efficient Binary Sequence Storage

Nan Li
Royal Institute of Technology
Forum 120, 164 40 Kista Sweden
nan3@kth.se

Elena Dubrova
Royal Institute of Technology
Forum 120, 164 40 Kista Sweden
dubrova@kth.se

ABSTRACT

On-chip storage of binary sequences normally require the use of Read-Only Memories (ROMs). However, ROMs do not exploit of the fact that the stored information is accessed sequentially. This paper presents an area-efficient sequence storage technique based on state machines. Experimental results show that the presented method significantly outperforms previous approaches. The resulting state machines are on average 54% smaller than ROMs storing the same sequence.

Categories and Subject Descriptors

B.6.3 [**Logic Design**]: Design Aids

Keywords

sequence generation, ROM, LFSR, binary machine, built-in self-test, cryptography

1. INTRODUCTION

Sequences are one of the most natural structures to present and store information. In electronics industry, binary sequences are used for various purposes. For example, in cryptography, binary sequences are used in stream cipher-based encryption. A stream cipher produces a keystream by combining a pseudo-random sequence with a message, usually by the bit-wise addition [5]. In wireless communications, pseudo-random sequences are used for *scrambling* and *spreading* of the transmitted signal. Today's wireless communication systems typically use Linear Feedback Shift Register (LFSR) sequences, or sequences obtained by linearly combining pairs of LFSR sequences, such as Gold codes [3]. In manufacturing test process of integrated circuits, test patterns are normally generated by an Automatic Test Equipment (ATE) and fed into a chip being tested. These test patterns are also binary sequences.

Binary sequences can be processed by digital circuits. Very often these sequences are stored temporarily in the circuits, using structures such as flip-flops, registers files and on-chip RAMs. However, in certain cases, permanent on-chip storage of these sequences is required.

On-chip permanent storage of sequences are usually done in the same way as randomly accessed data. Read-Only Memories (ROMs) are the most common non-volatile devices to store data. However, for sequence storage, it is possible to further exploit the sequential nature of the stored information and reduce the hardware cost.

GLSVLSI'13, May 2–3, 2013, Paris, France.
ACM 978-1-4503-1902-7/13/05.

In this paper, we present a new method for storing binary sequences on-chip using logic circuits, i.e. state machines, instead of ROMs. Our method has three main advantages compared to ROMs:

1. The area of a state machine is normally much smaller than the area of a ROM generating the same sequence.
2. Our method is capable of storing incompletely specified sequences efficiently without any compression, which is especially important for test applications, since since almost 99% of the test data are "don't-care" bits [4].
3. Our method provides more flexibility than ROMs. A state machine can generate sequences of any size, with any number of bits per clock cycle, while ROMs normally come with specific sizes and data widths.

2. STATE ENCODING FOR BINARY MACHINES

The proposed method is based on binary machine synthesis [2]. A binary machine is a state machine with the output sequence directly encoded into its states. The *degree of parallelization* of a binary machine is the number of output bits generated by each state. If several states have the same output, extra bits are added for differentiation [2].

We made an important observation that the size of the updating function, which dominates the overall circuit size, depends greatly on the number of inputs to the function. In the case of a binary machine, this number is the degree of parallelization plus the number of extra bits. Since degree of parallelization is defined by the sequence requirement and can potentially be a very large number, the resulting binary machine can be unnecessarily big.

We present a new encoding technique aiming at minimizing the number of inputs to the updating function. In the new scheme, the extra bits alone are sufficient to differentiate the states. The resulting binary machine is called *Reduced Binary Machine*. Such encoding allows the updating function to depend on inputs only from the extra bits, but not anymore from the sequence bits.

Since only the extra bits are used to distinguish states, with k extra bits, the state machine can have up to 2^k states. Therefore, a reduced binary machine with n states requires at least $\lceil \log_2 n \rceil$ extra bits. The overall number of bits needed for state encoding is

$$c_{\mathrm{RBM}} = \lceil \log_2 n \rceil + p,$$

where p is the degree of parallelization. The state count n, if not given, can be calculated by

$$n = \frac{s}{p},$$

where s is the overall sequence size in bits.

The extra bit vectors for each state can be selected in any order, as long as they are unique. Some state sequences generated by simple state machines are the best candidates for extra bits, such as LFSR states, binary counter states, and so on. This is because they minimize the part of the updating function that updates extra bit registers.

The number of inputs to the updating function is the same as the number of extra bits, which is $\lceil \log_2 n \rceil$, where n is the number of states in the reduced binary machine. The advantage is that the number of inputs to the updating function does not anymore depend on the degree of parallelization. In fact, this number is minimal, since fewer bits are not sufficient for the updating function to differentiate the states.

The area can be further reduced by simply removing the sequence bit registers and output from the updating function directly. In this case, the size of state register is the same as the number of extra bits, which is

$$c_{\text{RBMx}} = \lceil \log_2 n \rceil.$$

By doing so the reduced binary machine still generates the same sequence, only a clock cycle in advance. However, if the output is directly fed into another combinational block, the critical path delay may increase, limiting the maximum clock frequency of the circuit.

3. EXPERIMENTAL RESULTS

3.1 Comparison to Original Binary Machines

We compare the area of Reduced Binary Machines (RBM), synthesized using the presented method to the area of the original Binary Machines (BM) constructed using the algorithm [2]. We apply both algorithms on the same set of fully specified random sequences. The updating functions are optimized using ABC [1] and mapped into *mcnc.genlib* library. In this library, the unit of area is normalized, where an inverter takes area of 1. All experiments are carried out on a PC with Intel i5 3.20 GHz quad-core CPU and 4GB RAM.

The results are summarized in Table 1. Each result is the average from 20 different random sequences. The area of RBMs is at least 37% smaller than the results of [2]. Especially, for sequences with degree of parallelization one, RBMs takes only one sixth of the area of traditional binary machines.

Table 1: Circuit area compared to binary machine synthesis [2].

DoP*	Seq. size, bits	Area (BM)	Area (RBM)	$\frac{\text{Area(RBM)}}{\text{Area(BM)}}$
1	1K	4327.3	751.9	0.17
	10K	35865.9	5261.4	0.15
	100K	289956.3	38356.7	0.13
	1M	2473573.5	295803.9	0.12
10	1K	1730.2	735.2	0.42
	10K	9971.2	5056.7	0.51
	100K	58996.9	35858.8	0.61
	1M	458130.0	269382.1	0.59
100	1K	1245.2	363.8	0.29
	10K	7865.3	4811.6	0.61
	100K	58918.0	37074.2	0.63
	1M	453999.9	278008.1	0.61

* DoP stands for degree of parallelization.

3.2 Comparison to ROMs

As a means to store data, it is meaningful to compare RBMs to memories. In the following experiment, we make such a comparison to general purpose on-chip ROMs.

For the ease of comparison, we introduce the following parameters:

$$\alpha = \frac{\text{average size of a logic gate}}{\text{average size of a ROM cell}},$$

$$\beta = \frac{\text{number of bits in the sequence}}{\text{number of gates in the RBM generating the sequence}}.$$

Then, the area ratio of a RBM to a ROM can be calculated using these two parameters as:

$$\frac{\text{size of RBM}}{\text{size of ROM}} = \frac{\alpha}{\beta}.$$

Note that α is technology dependent, while β is technology independent.

We extract α parameters for ROMs of different sizes for 90nm CMOS technology. Parameters β are measured by applying our algorithm to random sequences of length from 4K to 512K to construct RBMs with the degree of parallelization 32.

The results are summarized in Table 2. We can see that RBMs are consistently smaller than ROMs since $\alpha/\beta < 1$. The smaller the sequence size, the more we gain from RBMs. When the sequence size is 4 Kbits, RBMs take only one sixth the size of ROMs. This can be explained by the fact that in small memories the control logic (address decoders, etc.) takes a higher fraction of the overall memory size than in large memories.

Table 2: α and β values for different sequence sizes.

Sequence size, bits	α	β	α/β
4K	0.54	3.25	0.17
8K	0.82	3.62	0.23
16K	1.30	4.00	0.32
32K	1.72	4.31	0.40
64K	2.30	4.54	0.51
128K	2.77	4.79	0.58
256K	3.46	5.07	0.68
512K	4.05	5.38	0.75
Average			0.46

4. ACKNOWLEDGEMENT

This work was supported in part by the project No 2011-03336 from Swedish Governmental Agency for Innovation Systems (VINNOVA).

5. REFERENCES

[1] Berkeley Logic Synthesis and Verification Group. ABC: A system for sequential synthesis and verification, release 70930.

[2] E. Dubrova. Synthesis of parallel binary machines. In *Proc. of ICCAD'2011*, San Jose, CA, USA, Nov. 2011.

[3] R. Gold. Optimal binary sequences for spread spectrum multiplexing (corresp.). *Information Theory, IEEE Transactions on*, 13(4):619 –621, October 1967.

[4] S. Hellebrand, B. Reeb, S. Tarnick, and H.-J. Wunderlich. Pattern generation for a deterministic bist scheme. In *Computer-Aided Design, 1995. ICCAD-95. Digest of Technical Papers., 1995 IEEE/ACM International Conference on*, pages 88 –94, Nov. 1995.

[5] M. Robshaw. Stream ciphers. Technical Report TR - 701, July 1994.

Power Gating Topologies in TSV Based 3D Integrated Circuits

Hailang Wang and Emre Salman
Department of Electrical and Computer Engineering
Stony Brook University
Stony Brook, New York 11794
{haiwang,emre}@ece.sunysb.edu

ABSTRACT

Two topologies are proposed at the physical level to achieve reliable power gating in through silicon via (TSV) based three-dimensional (3D) integrated circuits (ICs). The proposed lumped and distributed power gating topologies address the unique differences among distinct TSV fabrication methods such as via-first, via-middle, and via-last, while achieving, on average, 85% reduction in the leakage power. Related tradeoffs among power supply noise, power gating noise, physical area, and turn-on time are also investigated.

Categories and Subject Descriptors

B.7.m [**Integrated Circuits**]: [Miscellaneous]

Keywords

Power gating; TSV; Noise; 3D IC

1. INTRODUCTION

Through silicon via (TSV) based three-dimensional (3D) integration has emerged as a promising technology for both high performance and low power integrated circuits (ICs). Reliable power gating (both at the plane and block levels) is critical for low power 3D systems-on-chip that have long standby operating modes. Idle planes and idle blocks within a plane should be turned off to minimize leakage current while satisfying specific constraints such as power supply noise, power gating noise, area, and turn-on time.

Sleep transistors (with high threshold voltage) are used to achieve power gating. In high performance ICs with multiple (>8) metal layers, the power supply voltage is initially transmitted from the highest metal layer to the lowest metal layer through a stack of vias, reaching the sleep transistor. If the sleep transistor is on, supply voltage propagates back to the metal layers of the semi-global power grid, finally reaching the local power network that supplies current to a specific circuit block. The sleep transistors and the stack of vias introduce additional impedance to a power distribution network.

In 3D ICs, the power distribution networks for each plane are interconnected with the TSVs. The connection scheme depends partly upon the particular TSV fabrication technique [1]. Referring to Fig. 1, in via-first and via-middle technologies, the TSVs typically connect the lowest metal layer of a plane with the highest metal layer of the previous plane. Alternatively, in a via-last technology, the TSVs typically connect the highest metal layers in both planes, thereby causing signal routing blockages. These differences in the connection schemes are important in developing reliable 3D power gating topologies.

In this paper, two power gating topologies are proposed for TSV based 3D ICs while considering the distinct characteristics of via-first, via-middle, and via-last TSVs. The effect of the size of the sleep transistors and number of TSVs on power supply noise, power gating noise, and turn-on time is investigated for both topologies.

GLSVLSI'13, May 2–3, 2013, Paris, France.
ACM 978-1-4503-1902-7/13/05.

2. APPROACH

While fine-grained power gating is similar to traditional 2D power gating methods, plane- and block-level power gating in 3D ICs exhibits different characteristics due to the TSVs. Considering the coarse grain power gating problem, two power gating topologies are proposed based on, respectively, lumped and distributed placement of sleep transistors, as illustrated in Fig. 1 for a three plane 3D IC.

In the lumped power gating topology, as proposed for via-first and via-middle TSVs, all of the sleep transistors are fabricated on the top-most plane. As mentioned earlier, via-first/middle TSVs connect the lowest metal layer of a plane with the highest metal layer of an adjacent plane. Thus, sleep transistors are placed between the lowest metal layer of the top-most plane and TSVs which also land on this layer, as depicted in Fig. 1(a). Two advantages of the lumped topology are (1) maintaining all of the control signals for sleep transistors within a single plane and (2) ensuring higher voltage at the source terminal of the sleep transistors during normal operation.

Alternatively, in the distributed power gating topology, sleep transistors are distributed on each plane. The TSVs are constantly connected to the power supply. Thus, in the distributed scheme, additional stack of vias that transmits the virtual supply voltage (after the sleep transistor) back to the TSV is avoided, thereby enhancing the power supply noise during normal operation.

Another issue related with the power gating topologies in 3D ICs is the sizing of the sleep transistor and the number of the TSVs. The sensitivity of the power supply noise varies depending not only on TSV type, power gating topology, but also on specific design parameters such as the effective width of the sleep transistor and number of TSVs. The

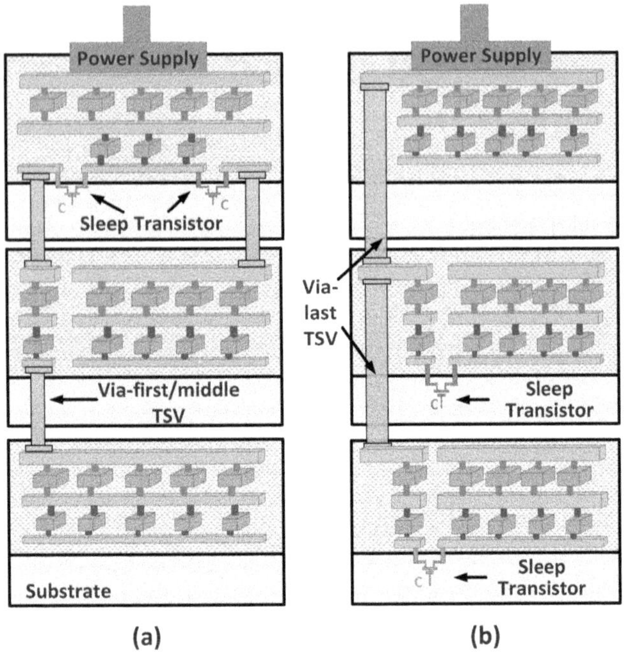

Figure 1: Physical representations of the proposed power gating topologies: (a) lumped topology for via-first and via-middle TSVs, and (b) distributed topology for via-last TSVs.

additional physical area occupied by the sleep transistors and TSVs is an important concern in low power, low cost 3D ICs. A closed-form expression is therefore developed to allocate available area to sleep transistors and TSVs, which minimizes power supply noise during normal operation.

3. PRIMARY RESULTS AND CONCLUSIONS

To evaluate the proposed power gating topologies, simulations are performed in HSPICE using an industrial 65 nm CMOS technology with a nominal supply voltage of 1 V. Electrical models are developed to represent the three plane 3D ICs, as illustrated in Fig. 1.

Distributed and lumped power gating topologies are analyzed for both via-middle and via-last TSVs. A critical step to model the power distribution network is extracting the impedance of the power grid on each plane. The number of interdigitated power/ground lines is obtained from the pitch and width of the lines. The field solver FastHenry is used to extract the effective impedance of the power distribution grid. The package level parasitic impedances are extracted from experimental measurement results [2]. The width of sleep transistor and number of TSVs are varied to evaluate the effect of these parameters on several criteria related to power integrity such as power supply noise, power gating noise, and turn-on time.

To analyze power supply noise, a current switching profile is used during the normal operating of the logic circuit. The worst case power supply noise at the farthest plane is measured. For the analysis of the power gating noise and turn-on time, step voltages are applied to the sleep transistors. The analysis results are qualitatively summarized in Table 1.

Table 1: Qualitative Analysis Results

	Via-middle		Via-last	
	Dist.	Lumped	Dist.	Lumped
Supply noise	Large	Small	Small	Large
Gating noise	Large	Small	Large	Small
Turn-on time	Long	Short	Short	Long

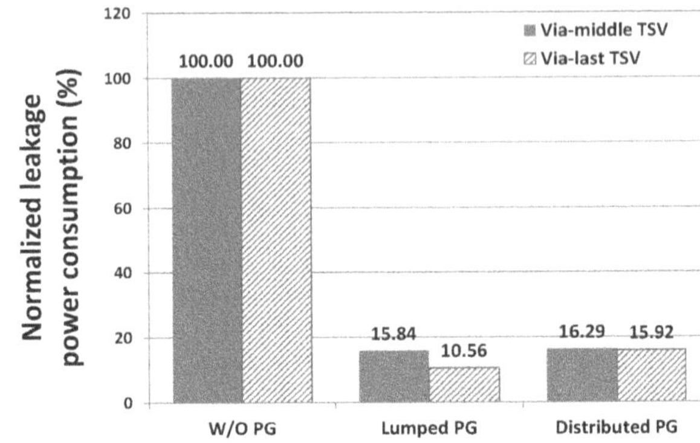

Figure 2: Normalized leakage power consumption achieved by the proposed power gating topologies.

According to Table 1, lumped topology is generally preferable for via-middle TSVs, whereas a distributed topology is preferred for via-last TSVs. The reason is the additional resistance due to stack of vias if lumped topology is used for via-last TSVs and distributed topology is used for via-middle TSVs, as mentioned before. Distributed topology exhibits slightly more power gating noise than the lumped topology for both via-middle and via-last TSVs. This characteristic is because when a plane is turned on, higher current is provided by the neighboring planes in a distributed topology since the impedance between the planes is relatively smaller. Thus, the neighboring planes exhibit slightly more power gating noise in the distributed topology.

The leakage power reduction achieved by the proposed two power gating topologies is also evaluated, as shown in Fig. 2. The leakage power is normalized based on the leakage power consumption of the 3D IC without power gating. For via-middle TSVs, the lumped and distributed power gating topologies can save, respectively, 84.16% and 83.71% leakage power. Alternatively, for via-last TSVs, the leakage power is reduced, respectively, by 89.44% and 84.08%, demonstrating the efficacy of the proposed power gating topologies.

4. REFERENCES

[1] S. M. Satheesh and E. Salman, "Power Distribution in TSV Based 3D Processor-Memory Stacks," *IEEE Journal on Emerging and Selected Topics in Circuits and Systems*, vol. 2, no. 4, pp. 692–703, December 2012.

[2] M. Gupta, J. Oatley, R. Joseph, G.-Y. Wei, and D. Brooks, "Understanding Voltage Variations in Chip Multiprocessors using a Distributed Power-Delivery Network," in *Proceedings of The Design, Automation Test in Europe Conference Exhibition*, April 2007, pp. 1–6.

Architecture and 3D Device Simulation of a PIN Diode-based Gamma Radiation Detector

Amr Elshennawy
Department of Electrical & Computer Engineering
Texas A&M University

Craig M. Marianno
Department of Nuclear Engineering
Texas A&M University

Sunil P. Khatri
Department of Electrical & Computer Engineering
Texas A&M University

ABSTRACT

In this paper, we present a new IC-based gamma radiation detector. We report 3D simulation results for the PIN diode structure which is used in this detector, along with a discussion of the architecture of the readout electronics for this detector. Gamma detection traditionally consists of three parts – i) a scintillator, ii) a detector unit and iii) the readout electronics. Our PIN diode based detector uses a traditional scintillator and a bialkali photocathode to convert the scintillated light into an electron current. This current is detected by the PIN diodes in the detector, which is fused to the photocathode. This structure has many advantages. Sensitivity is enhanced by placing the photocathode and detector IC on all six faces of the NaI scintillating crystal. Our detector is small, has a low-dead-time and it consumes very low power. The sensitive area of the chip is $\sim$ 75%, which is higher than any existing solid-state gamma detector. We have verified the operation of each component of the system by performing circuit-level simulations. The detector is implemented in a low-cost 0.5 μm process. The optimal parameters for the PIN diode detector are explored through 3D device simulations. Our results show that a detector sensitivity better than 20 keV is easily achievable with a PIN diode of dimensions 2.5 μm $\times$ 2.5 μm.

Categories and Subject Descriptors: B.7 [Integrated Circuits]: General

Keywords: radiation detection, solid-state, PIN diode

1. INTRODUCTION

Gamma particle detectors are very important in non-proliferation and homeland security applications. The efficiency of a gamma detector is measured along several axes, such as size, power requirements, energy discrimination ability, the ability to provide radiation counts, and dead-time.

Gamma detection works on the principle of scintillation. There are typically three components to a gamma detector:

- A scintillating crystal to absorb the gamma radiation and convert the gamma energy into visible light.
- A detection mechanism, which consists of either a photocathode and a photo-multiplier tube (PMT), or a solid-state detection scheme. In the first scheme, a photocathode is used to convert the light output from scintillating crystal into a photocurrent. The PMT is used to amplify the photocurrent to detectable levels. The PMTs are typically bulky and have complex power requirements, with bias voltages between 400V and 800V. In solid-state detectors, the scintillation output is directly converted to a detectable electrical signal.
- The readout electronics perform signal processing, including calibration, generation of particle counts, etc.

In this paper, we build towards a vision of a fully-integrated, low footprint, low power, highly sensitive gamma detector IC. This IC is envisioned to have a 2-dimensional array of reverse-biased PIN diodes ($\sim 10^6$ such diodes in an area of 1 cm^2). This detector IC would still be used in conjunction with an NaI scintillating crystal. Instead of directly converting the scintillation output into a detectable electrical signal, we utilize a photocathode, whose electrical signal is amplified by the detector IC. Also, the size of the entire detector would be just slightly more than the size of the crystal, and the bulkiest component of the traditional detector – the PMT – would be eliminated. Also, since there are a million of PIN diode detectors in each IC, the dead-time is practically reduced million-fold, since the probability of a photon striking the same spot on the photocathode is practically zero. The proposed design is better than existing solid-state gamma radiation detectors because it uses regular PIN diodes rather than photodiodes. This makes the fabrication cost for our detector lower, and also requires significantly simplified power delivery.

2. OUR APPROACH

2.1 System Overview

Our proposed IC based gamma detector consists of $\sim 10^6$ radiation sensitive areas (RSAs) which are implemented as PIN diodes. When the photocathode releases electrons as a consequence of scintillation light, the RSA undergoes a change in its voltage. This change (and its magnitude, which is proportional to the energy of the gamma particle) is detected by a radiation hardened area (RHA), which is a watchdog section of the detector IC. The functionality of this section of the chip does not get altered by the photocathode current induced by the gamma particle strike, since it is implemented in a radiation hardened manner [1], with a metal fill on a layer above the RHA circuitry. In this way, the detector combines a photocurrent sensitive RSA and a watchdog RHA, to detect and amplify the effect of a gamma particle strike in the crystal without requiring a PMT.

2.2 PIN Diode based RSA

The PIN diode detector circuit cell is shown in Figure 1(a). Under nominal conditions, I_{gamma} is zero, and the diode is reverse biased. If there is a gamma particle interaction with the photocathode material in the vicinity of the detector cell, a current will be injected into the diode, which is modeled by I_{gamma}. In order to compute I_{gamma}, we assume that we are trying to detect a gamma particle from Cs-137 (which has an energy of 662 keV), which is roughly in the center of the gamma energies of practical interest. An NaI crystal produces 43 photons/keV [2], with a time constant of 230 ns. As a result, a 662 keV gamma would produce 28466 photons. Now the bialkali photocathode has a *quantum efficiency* of

GLSVLSI'13, May 2–3, 2013, Paris, France.
ACM 978-1-4503-1902-7/13/05.

~0.25 [3]. The quantum efficiency is the ratio of the number of emitted electrons to the number of incident photons. As a result, 7116 electrons are produced by the 662 keV gamma particle. Since the time constant of the gamma particle is 230 ns, we can obtain the effective current induced in the photocathode by the 662 keV gamma as 4.95 nA. As a result, I_{gamma} is 4.95 nA for a 662 keV gamma. We assume that this current has a Gaussian distribution, with a peak value of 4.95 nA and $2\sigma = 230$ ns.

Since SbKCs bialkali photocathodes exhibit linear behavior up to 10 nA [2], it would not be a problem for our detector cell to detect gamma energies to 1 MeV, and the current would be about 7.5 nA. In our simulations, we test the energy discrimination of our detector cell by subjecting it to gamma energies of 662 keV and 682 keV. The corresponding photocathode currents are 4.95 nA and 5.10 nA respectively. The difference in the RSA output voltage for these two gamma particles should be detectably large, so that the two energies may be discriminated.

2.3 3D Modeling of PIN Diode based RSA

The detector cell was simulated in Sentaurus. The PIN diode was modeled in 3D, while the resistor and current source of the PIN diode based RSA (Figure 1(a)) were modeled as discrete circuit elements in the same Sentaurus deck. The resistor R is implemented in the highly resistive POLY2 layer. Our goal is to be able to produce these detectors extremely inexpensively so that they may be used ubiquitously. Hence it is essential to utilize a readily available, inexpensive process technology. Also, in our simulations, we utilize bias voltages that do not exceed the rated supply voltages of a 0.5 μm process.

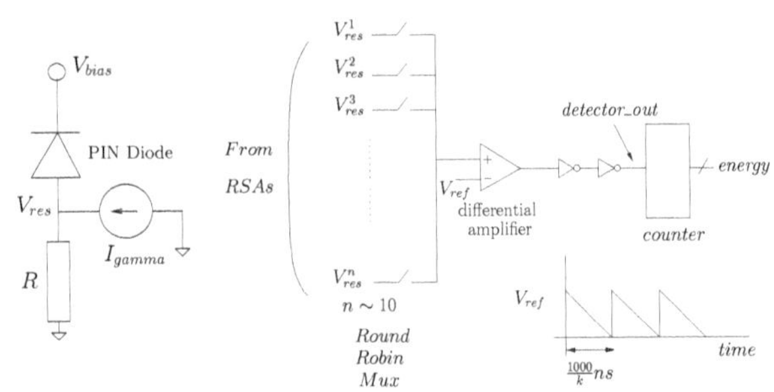

(a) RSA Unit (b) Sensing Unit in the RHA

Figure 1: Circuits for RSA and RHA

2.4 RHAs and Readout Electronics

We partition the array of PIN diodes (RSAs) into clusters. Each cluster has $n \sim 10$ PIN diodes and an RHA to detect a gamma strike on any of the n PIN diodes in the cluster. This clustering makes the sensitive area fraction of the detector upwards of 75%. Figure 1(b) shows the structure of an RHA. The RHA consists of one comparator, which compares the V_{res} signal of the n RSAs with a reference waveform V_{ref}. This comparison is done in a round-robin fashion over the n RSAs. For each RSA, we take k samples of its output waveform in a period of 1 μs to report the maximum energy sampled for the RSA over the duration of a gamma strike. A value of $k \sim 100$ is utilized to ensure a high fidelity sampling, while ensuring the circuit is easily realized.

The comparator in the RHA is implemented with a very low area overhead, in the form of a CMOS differential amplifier circuit. This circuit was simulated in HSPICE and it is not shown for brevity. The reference voltage V_{ref} is a global signal in the detector IC, and has a saw-tooth waveform as shown in Figure 1(b). Figure 2(a), which is obtained from our 3D PIN diode simulations, shows that the relationship between V_{res} and I_{gamma} is almost linear (with a maximum error of 2%) for all I_{gamma} values less than or equal to 8 nA, which is the current corresponding to gamma particles with an energy of 1.07 MeV. Note that the value of V_{res} from Figure 2(a) ranges from 1.76 V to 2.86 V. The saw-tooth waveform generator circuit was also simulated in HSPICE and shown to produce voltages in the range of 1.7 V to 3 V.

The height of the V_{res} signal of the currently sampled RSA results in a pulse at *detector_out*, whose width is proportional to the height of the V_{res} signal. The *detector_out* signal drives the enable signal of a counter, whose count (referred to as *energy* in Figure 1(b)) is proportional to the gamma particle energy. The RHA sends maximum value of *energy* from among all the k waveforms, to a central unit on the detector IC. The central unit on the chip receives the total counts reported by all sensing units, converts them to energies, and continuously reports them over the IO interface of the chip.

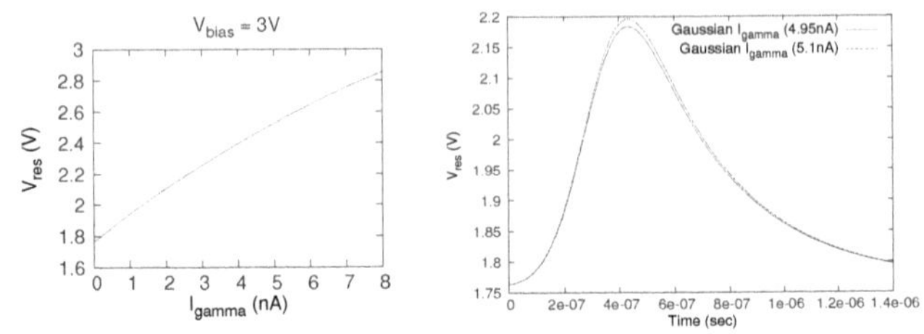

(a) V_{res} and I_{gamma} Characteristics (b) Dynamic Value of V_{res}, with I_{gamma} of 5.10 nA and 4.95 nA

Figure 2: RSA Characteristics

3. EXPERIMENTAL RESULTS

We assume that the detector is implemented in an ON Semiconductor 0.5μm process. Using Sentaurus, we experimented with PIN diodes of various sizes, and present the results for 2.5 μm × 2.5 μm, since it results in highest sensitivity, lowest power and small area compared to other PIN diode sizes. The diode dimensions are comparable to the size of the beam spot induced by the interaction of the gamma particle with the photocathode [4]. We also vary the values of R and V_{bias} in our experiments to find the best configuration for the RSA.

Our PIN diode based RSA has a good discriminating ability. For example, for a 20 keV difference in particle energy, the dynamic differential V_{res} value for the diode is as high as 12 mV (for $V_{bias} = 3$V and $R = 200$ MΩ). Also, we note that the static current for the diode (with $I_{gamma} = 0$) is about 5 nA under these conditions. With a detector array having 10^6 PIN diodes, the static current would be ~5 mA. Using 2 AA batteries (with a rating of about 2000 mAh), this detector would be able to operate for a day. This is a dramatic improvement over current gamma detectors.

Figure 2(b) reports the value of V_{res} for Gaussian I_{gamma} pulses with peak values of 4.95 nA and 5.10 nA respectively for the preferred RSA configuration. Note that the static V_{res} values are near mid-rail, ensuring that the differential amplifier circuit is biased in the sensitive portion of its region of operation. We verified that our differential amplifier was able to detect this difference by yielding a differential pulse width at the *detector_out* signal, which results in different values for the counter output *energy* (see Figure 1(b)).

4. REFERENCES

[1] R. Garg and S. Khatri, "A novel, highly SEU tolerant digital circuit design approach," in *IEEE International Conference on Computer Design (ICCD)*, Lake Tahoe, CA, Oct 2008, pp. 14–20.

[2] H. Photonics, "Photomultiplier tubes: Basics and applications," 2007, third Edition (Edition 3a).

[3] P. Lecoq, A. Annenkov, A. Gektin, M. Korzhik, and C. Pedrini, *Inorganic Scintillators for Detector Systems: Physical Principles and Crystal Engineering*. Springer Publishers, 2005.

[4] M. McClish, P. Dokhale, J. Christian, C. Stapels, E. Johnson, R. Robertson, and K. Shah, "Performance measurements of CMOS position sensitive solid-state photomultipliers," *IEEE Transactions on Nuclear Science*, vol. 57, no. 4, pp. 2280–2286, Aug 2010.

Set-Cover-based Critical Implications Selection to Improve SAT-based Bounded Model Checking

[Extended Abstract]

Mahmoud Elbayoumi
ECE Dept., Virginia Tech
Blacksburg, VA 24061

Michael S. Hsiao
ECE Dept.,Virginia Tech
Blacksburg, VA 24061

Mustafa ElNainay
CSE Dept.,Alexandria Univ
Alexandria, Egypt 21544

ABSTRACT

The effectiveness of Satisfiability (SAT)-based Bounded Model Checking (BMC) critically relies on the deductive power of the BMC instance. Although implication relationships have been used to help SAT solver to make more deductions, too many such implications would result in a large number of clauses and potentially degrade the underlying SAT solver performance. In this paper, we have formulated clause selection as a set-cover problem. Secondly, we have proposed a novel greedy strategy for optimal selection of static logic clauses. This technique maximizes the number of literals that can be deduced by the SAT solver during the Boolean Constraint Propagation (BCP) operation. Our strategy improves BMC by 1.74× against the case where all extended implications were added to the BMC instance. Compared with the original BMC without any implications, up to 55.32× speedup can be achieved.

Categories and Subject Descriptors

D.2.8 [**Hardware**]: Hardware validation—*Functional verification*

Keywords

Bounded Model Checking, Static Implications, Satisfiability

1. INTRODUCTION

In recent years, the scale and complexity of managing larger designs have increased exponentially. Consequently, the validation of the designs has posed as a serious challenge, and verification has now become a bottleneck in Integrated Circuits (IC) design flow. BMC was proposed [1] based on a SAT formulation and gives model checking a great push forward. BMC has enjoyed considerable success with the employment of propositional SAT solvers in recent years. It has been widely applied in bug hunting and property checking within a bounded sequential depth.

One method to improve BMC is to embed clauses in the formula such that the deductive power of the instance is increased [2]. In these methods, a deduced static relations are added as constraint clauses to the original CNF formula for the BMC instance. They aid the SAT solver in deducing a larger set of implied literals at a given decision point. As a result, they can help to prune the search space. However, adding an unnecessarily large number of static implication clauses may degrade the performance.

As far as we know, there is no well-defined criteria for a classification strategy of static logic implication clauses. The only filtering technique used for static logic implication is proposed in [2]. The classification can be regarded as a filter that only includes those constraints that meet the filtering criteria. However, simple filters may inadvertently leave out some key critical relations or fail to remove those less important clauses.

Our contributions are as follows. First, we have formulated the clause filter problem as an elegant set-covering problem. Secondly, we have proposed a novel greedy algorithm based on the Johnson algorithm [3] to find the optimal set of clauses that would accelerate BMC solution. Our clause selection/filtering technique have shown a 1.74× acceleration in BMC solution over the previous technique. Compared with the original BMC without any implications, up to 55.32× speedup can be achieved.

GLSVLSI'13, May 2–3, 2013, Paris, France.
ACM 978-1-4503-1902-7/13/05.

2. BACKGROUND

In BMC, the Circuit Under Verification (CUV) is checked against a certain property p on a k time-frame window. SAT-BMC Checking is performed by constructing the following CNF Boolean Formula (Eq. 1).

$$\varphi = I(s_0) \wedge \bigwedge_{0 \leq i < k} T(s_i, s_{i+1}) \wedge \parallel \phi \parallel_k \qquad (1)$$

where: $T(s_i, s_{i+1})$ is the transitive relation; hence, $\bigwedge_{0\leq i<k} T(s_i, s_{i+1})$ is k-timeframe Iterative Logic Array (ILA) of CUV; $I(s_0)$ is a legal initial state; $\parallel \phi \parallel_k$ is a monitor circuit for some property p (which we want to check). The monitor circuit is constructed such that φ will be satisfiable if and only if the CUV would *violate* p.

Given a BMC instance φ, an implicit exploration of the reachable states to verify whether the property assertion is satisfiable within k frames is performed by a SAT solver. If φ is satisfiable, a trace (counter-example) is generated which exposes the violation of p in the CUV. Otherwise, the unsatisfiability of the instance indicates that no counter-example of length k from the given initial state exists.

Static logic implications [4] describe relationship between different gates in a CUV that hold for all the states of the sequential circuits (whether it is legal or illegal states) . It can be obtained be asserting and propagating logic '0' (or '1'), to every gate in CUV. A direct implication is the result of assigning the value on the target gate in their direct fan-out gates (forward implications) and direct fan-in gates (backward implications). Indirect implication is performed by propagating direct implication transitively throughout the circuit / formula. Extended Backward Learning [4] is performed on unjustified implicants (which are gates whose inputs are not sufficient to justify the output) in the implication list of target gate (which is the gate we need to find its implications). Note that with BCP, extended backward implications cannot be learned. In this sense, clauses derived from extended backward implications can offer additional learning/deductive power to the formula.

3. GREEDY SET-COVER FORMULATION

Our clause selection criteria is based on selecting the clauses that can bring about the maximum number of assignments (aka implications) during BCP. Consider a CUV with eight signal assignments, labeled as eight nodes $S = \{1, 2, ..., 8\}$. Let the set of direct and indirect implications for each of the nodes be, $M_1 = \{2, 7, 8\}$, $M_2 = \{1, 4, 6\}$, $M_3 = \{5\}$, $M_4 = \{7, 8\}$, $M_5 = \{2, 6, 7, 8\}$, $M_6 = \{4, 7\}$, $M_7 = \{1, 3, 6\}$, and $M_8 = \{3, 5\}$. So, $M = \{M_1, M_2, M_3, M_4, M_5, M_6, M_7, M_8\}$ We want to find the min Cover for (S, M). As M_5 has the maximum size (of 4), our greedy algorithm will choose it and remove nodes $2, 6, 7$ and 8 from the other sets, and we will have 4 remaining non-empty sets: $M_2 = \{1, 4\}$, $M_3 = \{5\}$, $M_7 = \{1, 3\}$, $M_8 = \{3, 5\}$. Now, the algorithm has the choice between M_7 and M_8 (as both are of max. size of 2). Let M_8 be selected, and it will remove nodes 3 and 5 from all other set, and we will have 2 remaining non-empty sets. $M_2 = \{1, 4\}$ and $M_7 = \{1\}$. M_2 is selected as it is of max. size of 2. Now, all sets are empty, and implications $7 \to 2$, $7 \to 5$, $7 \to 8$ are chosen to cover all the transitive relations of S. However, instead of using the above greedy approach, we formulate a set-covering solution as follows.

Problem statement: Given an implication graph $G(V, E)$, for each node $i \in V$, we have a set of implications S_i (direct, indirect and extended implications) and M_i (direct and indirect implications only). As BCP can already deduce direct and indirect implication efficiently, we want to select the minimal number of nodes $k \in S_i$ such that their direct and indirect implications (M_k) cover the entire S_i. Formally, let M be the family of direct and indirect implication of nodes in S_i; that is, $M = \bigcup_{w \in S_i} M_w$. The problem is to find a minimum-sized subset $C \subseteq M$ whose implications cover all of S_i, and hence, $S_i = \bigcup_{M_k \in M} M_k$. (S_i, M) is a set-cover instance for clause selection. Fig. 1 describes our greedy algorithm for the selection of clauses.

for all node $i = (N, v)$ in the implication graph **do**
  $FL_i = \emptyset$
  if i is not a constant node **then**
    repeat
      get node $j = (M, v') \in S_i$ such that $|M_j|$ is max.
      $FL_i = FL_i \cup \{j\}$
      $S_i = S_i \setminus M_j$
      for all node $k \neq j$ **do**
        $M_k = M_k \setminus M_j$
      end for
      get node $l = (M, \overline{v'})$ and node $t = (N, \overline{v})$
      $S_l = S_l \setminus M_t$
    until all clauses in S_i is covered
  end if
end for

Figure 1: Greedy-Set-Cover Filtering Algorithm.

4. PRELIMINARY RESULTS

The proposed greedy selection framework has been developed with C++ and their performance were evaluated on HokieOne (SGI UV). We have used 36 dedicated 2.66 GHz Intel Xeon cores, with total of 190.8 GB of RAM, running SUSE Linux 11 to compute the implications in parallel. We have compiled our program with g++ under -O3 option. In our approach, we need the original circuit to deduce the implications. In addition, many of the available BMC are dated and trivially easy. Thus, we have developed a set of BMC benchmark based on ISCAS89 circuits that require a large computational cost to solve.

In Table 1, we report the time used to solve the CNF formula. We have used Minisat 2.0 to solve different CNF formulas on a single core machine. k is the number of unrolled time frames. we list the time when set-covering selection is used (SC), and when extended implication clauses is used (EI). In addition, the time to solve the original CNF formula without any added static implication clauses is reported (Orig.). We can get an average 1.74× speed up due to our selection algorithm (compared with extended implication approach). Also, we gain an average 63.52× speed-up compared to the original CNF formula. Moreover, we have an average 55.32× overall speed-up in the original problem.

Table 1: BMC Cost in seconds

ckt	k	SC	EI	Orig.
s298	4000	**17.29**	57.84	2107.00
s344	4000	**242.86**	527.96	1910.00
s382	2000	**33.88**	46.60	1365.00
s400	2000	**24.37**	44.37	1303.00
s444	2000	**26.27**	36.06	2145.00

5. CONCLUSIONS

In this paper, we have formulated the clause selection problem as a set-cover problem, and we have proposed a novel low-cost greedy set-covering based selection algorithm for choosing static implications. In comparison with extended implication approach, we got an average 1.74× speed up due to our selection algorithm. Moreover, we have an average 55.32× overall speed-up compared to the original problem.

6. REFERENCES

[1] A. Biere, A. Cimatti, E. Clarke, M. Fujita, Y. Zhu. Symbolic Model Checking using SAT procedures instead of BDDs, Proc. ACM Design Automation Conf., 1999.

[2] V. Vimjam and M. S. Hsiao,Increasing the deducibility in CNF instances for efficient SAT-based bounded model checking, Proc. IEEE High Level Design Validation and Test Workshop, 2005, pp. 184-191.

[3] D. S. Johnson, Approximation Algorithms for Combinatorial Problems, Journal of Computer Systems Science, vol. 9, pp 256 âĂŞ278, 1974.

[4] J.-K. Zhao, E. M. Rudnick, and J. H. Patel, "Static logic implication with application to redundancy identification," Proc. VLSI Test Symp., pp. 288-293, 1997.

Circuit Design of a Novel Adaptable and Reliable L1 Data Cache

Azam Seyedi[†‡], Gulay Yalcin[†‡], Osman S. Unsal[†], Adrian Cristal[†°]
[†]BSC-Microsoft Research Centre [‡]Universitat Politècnica de Catalunya
[°]IIIA - Artificial Intelligence Research Institute CSIC - Spanish National Research Council
{azam.seyedi, gulay.yalcin, osman.unsal, adrian.cristal }@bsc.es

ABSTRACT
This paper proposes a novel adaptable and reliable L1 data cache design (Adapcache) with the unique capability of automatically adapting itself for different supply voltage levels and providing the highest capacity. Depending on the supply voltage level, Adapcache defines three operating modes: In high supply voltages, Adapcache provides reliability through single-bit parity. In middle range of supply voltages, Adapcache writes data to two separate cache-lines simultaneously in order to use one line for error recovery when the other line is faulty. In near threshold supply voltages, Adapcache writes data to three separate cache-lines simultaneously in order to provide the correct data based on bitwise majority voter.

Categories and Subject Descriptors
B.3.2 [**Memory Structure**]: Design Styles

General Terms
Performance, Design, Verification, Reliability

Keywords
Cache Design, Low Power Design, Fault Tolerance

1. INTRODUCTION
A very effective approach in reducing the energy consumption is to reduce the supply voltage close to the transistor's threshold. However, the energy reduction in the low-power mode comes with a drastic increase in the number of memory cell failures especially in large memory structures such as on-chip SRAM memories. This motivates us to design a cache which is resilient to large number of cell failures and operates at lower supply voltages.

In this paper, we propose a novel highly reliable L1 data cache for three distinct supply voltage ranges. Our proposed cache, Adapcache, automatically adapts itself for different supply voltages in order to provide the most possible capacity with minimum energy consumption and access time. Adapcache operates in three states: (1) Single Copy State (SCS), (2) Double Copy State (DCS) and (3) Triple Copy State (TCS). When the supply voltage is relatively high, Adapcache operates in SCS to satisfy high performance execution. In this state, Adapcache provides reliability only based on single-bit interleaved parity. DCS is utilized in medium low supply voltage ranges in which error rate starts to increase. Adapcache writes data to two selected cache-lines simultaneously. At every read instance, Adapcache compares both those values through XOR circuits to check their equality. This approach is similar to Dual Modular Redundancy technique (DMR) but unlike DMR, the utilization of parity bits enables Adapcache to decide the correct value upon an error. When the supply voltage is at near-threshold and the error rate increases drastically, Adapcache operates in TCS. Adapcache writes data to three selected cache-lines simultaneously to provide very high reliability. In read, the correct value is decided through bit-wise majority voters similar to Triple Modular Redundancy technique (TMR) but the error correction capability is improved by parity bits participation in majority voting.

2. ADAPCACHE CICUIT DESIGN
Our sub-bank design is based on the structure that authors proposed in [1]. Figure 1 shows the block diagram of each sub-array structure. Whenever the supply voltage is high, Adapcache operates in SCS and only one cache-line is activated at each access time. For writing the selected cache-line, signal IEU1 is high and activates input buffers and data can transfer to the selected cache-line via Bus4 and Bus1; and similarly for reading the selected line, signal OEU1 is high and output buffers are active and data is transferred from Bus1 to Bus3. The error protection is based on bit-parity calculation in order to achieve higher speed but less accuracy.

In medium supply voltage range Adapcache operates in DCS where at each cache-access two cache-lines are activated simultaneously. At writing time data is written to two selected lines at the same time. Parity calculator circuits generate parity bits and write them in parity bit cells as well. For reading the two selected lines, Bus1 is divided into two parts and output buffers transfer two selected data each from separate sub-array slice groups (sub-array slices 0 to 3 and sub-array slices 4 to 7) to the XOR circuit to check their identically. If there is at least one bit flip, signal EN10 activates two parity calculator circuits to calculate parity bits of selected lines and compare with the original parity bits of each selected-lines. Whenever one of comparator shows equality, the related output buffer transfers its data to Bus3. In TCS, for writing the selected cache-lines signal IEU1 is high and data is transferred and written in three selected lines simultaneously. At the reading time, Bus1 is divided into three parts and three selected lines each from separate sub-array slice group (sub-array slices 0, 1, 2, and sub-array slices 3, 4, 5, and sub-array slices 6, 7 and extra slice) are transferred to a majority voter. The majority voter output for cache-lines is DataM and for their parity-bits is ParityM. The parity bits of DataM are calculated and compared with ParityM. If there are any differences, the parity bits of selected lines should be calculated and compared with their original parity bits. Whenever one of comparator shows equality, the related output buffer transfers its data to the output. We designed a new decoder circuit to activate one, two or three word-line addresses at each access time and described its detail in [2].

GLSVLSI'13, May 2–3, 2013, Paris, France.
ACM 978-978-1-4503-1902-7/13/05.

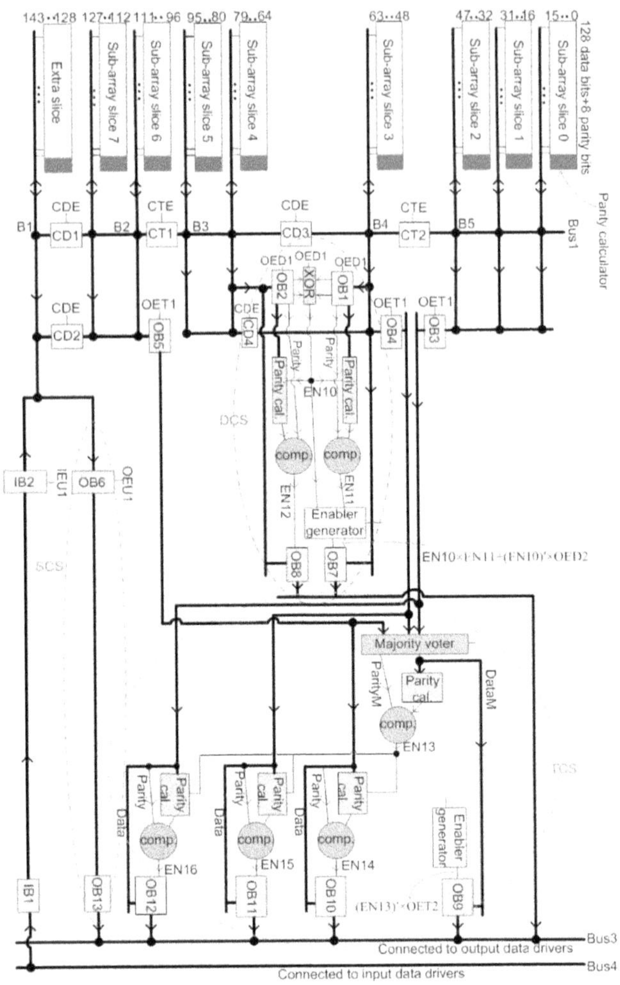

Figure 1: Sub-array block diagram.

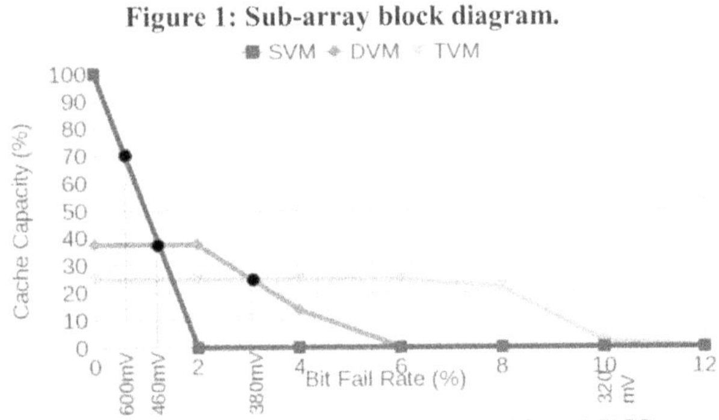

Figure 2: Comparison between SCS, DCS and TCS.

3. EVALUATION

We evaluate useful cache capacity, access time and energy consumption of Adapcache and compare to DMR and TMR caches and also to Parichute [3]. We inject random persistent faults according to bit failure rate [0%-12%] by repeating each experiment 100 times. We calculate the useful cache capacity as the undisabled portion of the cache as can be seen in Figure 2. Bit failure rate for persistent failures in the given Vcc is examined by Miller et al [3] which we reference in our work. Figure 2 presents the Vcc intervals that SCS, DCS and TCS can be used efficiently according to persistent bit failure rate. The intersections of the three lines indicate the optimum useful points of SCS, DCS and TCS according to Vcc. Thus, we decide using SCS when Vcc=[1V-600mV], DCS when Vcc=]600mV-400mV] and TCS when Vcc=]400mV-320mV].

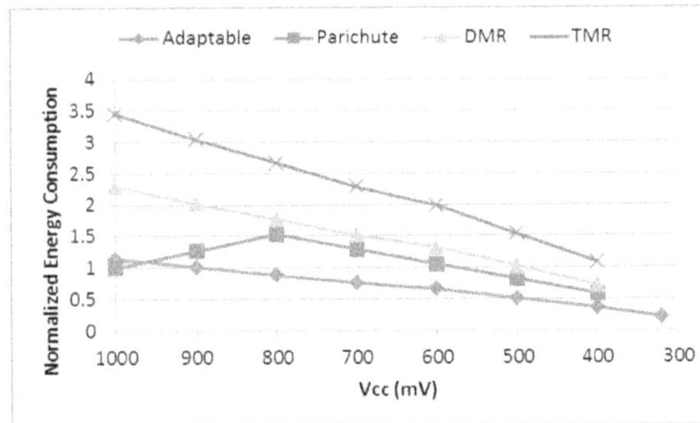

Figure 3: Normalized energy consumption of caches.

We simulate a 64-KB L1 data Adapcache with Hspice 2003.03 using HP 45-nm Predictive Technology Model at 2GHz processor frequency for (1V-0.6V) supply voltages, at 900MHz for (0.6V-0.4V) supply voltages, and at 400MHz for (0.4V-0.32V) supply voltages. We calculate the energy consumption and latency of the Adapcache versus the Vcc levels and compare it with three 64-KB caches which deploy TMR, DMR and Parichute techniques. Parichute, can reduce baseline energy consumption by 64%. In comparison, Adapcache can reduce baseline energy by 74% with a 7X less access latency than Parichute. Figure 3 depicts the normalize energy consumption of each cache structure (We normalized based on a single parity protected cache in Vcc=1V). We calculate the latency of all structures for a single cache-line. In Adapcache, if a line is faulty, it takes 3 clock cycles to read and correct it. For DMR case it takes 5 clock cycles to read selected lines and detect the errors. Similarly, it may take 7 clock cycles for TMR structure to read selected cache-lines and check and correct the errors. Finally for Parichute, it may take up to 20 cycles if the bit failure rate is very high. This latency is almost two times the L2 cache latency which may not be acceptable.

4. ACKNOWLEDGMENTS

This work was partially supported by the cooperation agreement between the Barcelona Supercomputing Center and Microsoft Research, by the Ministry of Science and Technology of Spain and the European Union (FEDER funds) under contracts TIN2007-60625 and TIN2008- 02055-E, and by the European Network of Excellence on High-Performance Embedded Architecture and Compilation (HiPEAC), and by the European Community's Seventh Framework Programme [FP7/2007-2013] under the ParaDIME Project (www.paradime-project.eu), grant agreement no. 318693.

5. REFERENCES

[1] A. Seyedi, et al. Circuit design of a dual-versioning L1 data cache for optimistic concurrency. In GLSVLSI 2011

[2] A. Seyedi, et al. Circuit Design of a Novel Adaptable and Reliable L1 Data Cache, Technical Report UPC-DAC-RR-CAP-2013-3 , UPC, 2013

[3] T. N. Miller. Parichute: Generalized Turbocode-Based Error Correction for Near-Threshold Caches. In MICRO 2010.

Physical Design Exploration of 3D Tree-based FPGA Architecture

Vinod Pangracious, Emna Amouri and Habib Mehrez
UPMC, Sorbonne University
Laboratoire d'informatique
Paris VI, France
Vinod.Pangracious@etu.upmc.fr

Zied Marrakchi
FlexRas Technologies
Paris France
Zied.Marrakchi@Flexras.com

ABSTRACT

An innovative 3D physical design exploration methodology for Tree-based FPGA architecture is presented in this paper. In a Tree-based FPGA architecture, the interconnects are arranged in a mult-dimensional network with the logic unites and switch blocks placed at different levels, using a Butterfly-Fat Tree network topology. A 3D physical design exploration methodology leverage on Through Silicon Via (TSVs) using a horizontal break-point to re-distribute the Tree interconnects into multiple stacked active silicon layers proposed in this paper.

Categories and Subject Descriptors

B.7.1 [**INTEGRATED CIRCUITS**]: Types and Design Styles–*Advanced technologies*

Keywords

3D Integration Circuits, Tree-based FPGA, TSV

1. INTRODUCTION

Mesh is the most studied and used industrial topology. Considerable amount of research work [2] and industrial applications has already been implemented. One of the major disadvantage of Mesh-based FPGA architecture is, that the re-configurable interconnect consume almost 90% of the total die area [7]. The 3D integration technology has the potential to reduce the programmable interconnects length by bringing the logic components close together, which leads to significant improvement in performance and silicon die area. 3D integration is particularly useful to improve density and performance of FPGA systems. It can address problems pertaining to routing congestion, low resource utilization, long wire delays, thermal issues etc. This paper focus primarily on the development of physical design and exploration methodologies to develop 3D Tree-based FPGA, where the scope covers all activities of design and layout exploration.

2. RELATED WORK

A new FPGA creation is a challenging task because of the significant amount of time required for the design and validation. An automated layout design and verification tools for 2D Mesh-based FPGA reported in [4] shows that, a Mesh-based FPGA design process starting from design to tapeout is done with limited amount of time, however the tools reported to have a huge area penalty in cost of process automation. The electrical optimization of interconnect of an SRAM based 2D Mesh FPGA reported in [3] concludes that the best way to improve speed is to perform partial optimization on routing interconnect and remainder sections can be optimized for density. However this type of layout optimization leads to increase irregularity in interconnect design.

The interconnect and logic wiring requirements for 3D Mesh-based FPGA was reported in [7] and deals with 3D integration of Mesh-based FPGA with 3D switch blocks. However the vertical interconnections of switch box are assumed to be same as the wire length between two adjacent LUTs on the same layer may lead to unrealistic timing results. A design framework for 3D Mesh-based FPGA architecture level exploration methodology is presented in [8], to explore the vertical interconnect distribution, however this leads to usage of 2D and 3D switch blocks intermittently, which may lead to a number of design and manufacturing issues. In this paper, we present the issues associated to 2D physical design and propose a new 3D physical design exploration methodology for Tree-based FPGA in order to improve performance and density.

3. TREE-BASED FPGA ARCHITECTURE

A re-programmable Tree-based FPGA architecture is proposed in [5]. The main motivation for the Tree-based FPGA architecture is to achieve the best performance by balancing interconnect and logic block utilization, where logic blocks and routing resources are sparsely partitioned into a multilevel clustered structure [6]. In a Tree-based FPGA architecture, the LBs (Logic Blocks) are grouped into clusters located at different levels of the Tree. Each cluster contains a switch block to connect local LBs. A switch block is divided into MSBs (Mini switch Blocks). The Tree-based FPGA architecture unifies two unidirectional upward and downward interconnection networks using a *Butterfly-Fat-Tree* topology in order to connect DMSBs (Downward MSBs) and UMSBs (Upward MSBs) to LB's inputs and outputs.

4. PHYSICAL DESIGN EXPLORATION

The proposed methodology for physical design exploration of 3D Tree-based FPGA architecture is illustrated in Figure 1. The HDL generator is designed to generate VHDL

GLSVLSI'13, May 2–3, 2013, Paris, France.
ACM 978-1-4503-1902-7/13/05.

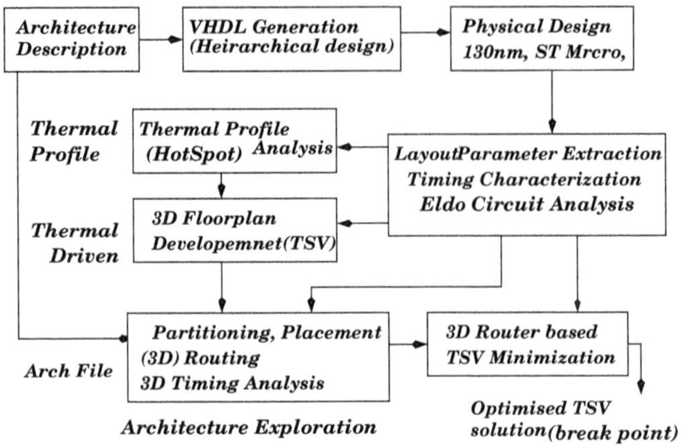

Figure 1: Physical Design exploration methodology

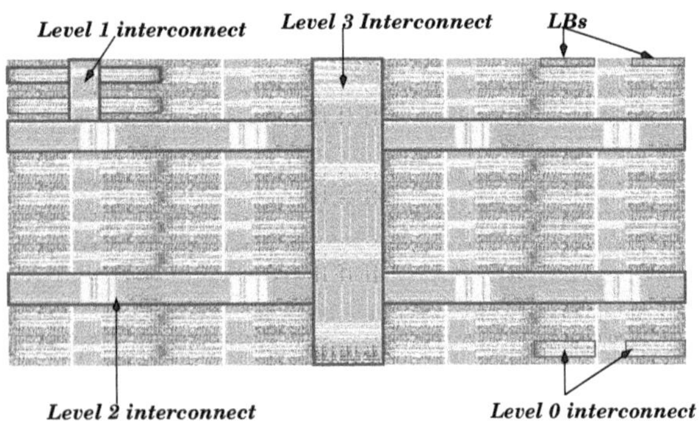

Figure 2: Tree-based FPGA with Arity 4: VLSI Layout of *3 level Tree*

code based on a hierarchical design approach that partitions the design into smaller sections, implement them separately and assemble them together at the final design phase. The physical design experiments are performed based on the layout generated using ST Micro's 130nm technology node. Figure 2 illustrates the layout of Tree-based FPGA with 3 tree levels. Mentor's circuit simulator *Eldo* is used to estimate the wire delay and power consumption of switches and interconnection networks at different tree levels. The thermal model presented in [1] is used to extract the thermal profile of the multi-layer chip based on layout geometrical features and power consumption of the functional unites.

To evaluate the performance of the proposed 3D Tree-based FPGA architecture, we place and route the largest set of MCNC[1] benchmark circuits, and compare with the 3D Mesh-based FPGA [8]. The performance analysis of 2D Tree-based FPGA layout versus 3D Tree-based FPGA with TSV design is presented in figure 3. The vertical axis shows the critical path delay measured for each MCNC benchmark design. It is evident from the graph presented in Figure 3, that for all designs the 3D implementation of Tree-based FPGA outperformed compared to the 2D counterpart. An average speed improvement of 68.7% is recorded in this experiment compared to our 2D design. The gain obtained in performance is due to the optimized wire delay at higher levels of the Tree interconnect by re-arranging them in the 2

[1] http://er.cs.ucla.edu/benchmarks/ibm-place.

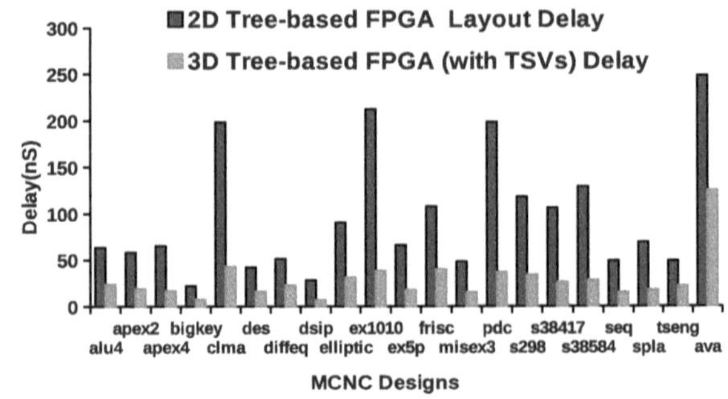

Figure 3: Performance analysis of 2D vs 3D Tree-based FPGA

layer 3D chip and using TSVs to connect *level 3 and 4*. The multidimensional nature of Tree interconnect offers more flexibility in 3D design compared to rigid nature of Mesh-based interconnect. This is the main reason for Tree-based FPGA to outperform in 3D implementation. Based on Figure 3, it is evident that the Tree-based FPGA is a consistent architecture for building 3D re-configurable systems.

5. CONCLUSION

A systematic physical design implementation methodology for 3D Tree-based FPGA is presented. The performance analysis reveals the 3D Tree-based FPGA has 68.7% advantage in performance compared to 2D Tree-based FPGA. This places 3D Tree-based FPGA as a viable alternative to build 3D re-configurable systems compared to 3D Meash-based FPGA.

6. REFERENCES

[1] J. L. Ayala, A. Sridhar, V. Pangracious, D. Atienza, and Y. Leblebici. Through Silicon Via-Based Grid for Thermal Control in 3D Chips. *NanoNet Springer*, pages 90–98, June 2009.

[2] V. Betrz, A. Marquardt, and J. Rose. A New Packing Placement and Routing Tool for FPGA Research. *Inter' Workshop on FPGA.*, 15(5):213–222, Nov 1997.

[3] V. Betz and J. Rose. Circuit Design, Transistor Sizing and Wire Layout of FPGA Interconnect. *IEEE Custom Integrated Circuit Conference*, 1999.

[4] I. Kuon, A. Egier, and J. Rose. Design Layout and Verification of FPGA using Automated Tool. *FPGA 2005*, pages 215–226.

[5] Z. Marrakchi, H. Mrabet, E. Amouri, and H. Mehrez. Efficient Tree Topology for FPGA Interconnect Network. *ACM GLSVLSI.*, pages 321–326, 2006.

[6] Z. Marrakchi, H. Mrabet, U. Farooq, and H. Mehrez. FPGA Interconnect Topologies Exploration. *Inter' J' of Reconfigurable Comput.*, 15(5):795–825, Nov 2009.

[7] A. Rahman, S. Das, A. Chandrakasan, and R. Reif. Wiring Requirments and Three-Dimensional Integration of Field Programmable Gate Arrays. *SLIP ACM*, March 2001.

[8] K. Siozios, A. Bartzas, and D. Soudris. Architecture Level Exploration of Alternative schmes Targeting 3D FPGAs: A Software Supported Methodology. *Intern' Journal of Reconfigurable Computing*, 2011.

Co-optimization of TSV Assignment and Micro-channel Placement for 3D-ICs

Bing Shi, Caleb Serafy and Ankur Srivastava
University of Maryland, College Park, MD, USA
{bingshi, cserafy1, ankurs}@umd.edu

ABSTRACT

The three dimensional circuit (3D-IC) brings forth new challenges to physical design such as allocation and management of through-silicon-vias (TSVs). Meanwhile, the thermal issues in 3D-IC becomes significant necessitating the use of active cooling schemes such as micro-channel liquid coolings. Both TSVs and micro-channels go through the interlayer regions of 3D-IC resulting in potential resource conflict. This paper investigates the co-optimization of TSV assignment to interlayer nets and micro-channel allocation such that both wirelength and micro-channel cooling energy are co-optimized. We propose a multi-commodity flow based formulation to solve the co-optimization. The experimental results show that, our approach achieves 51% cooling power savings or 6.08% wire length reduction compared with the approaches that assign TSVs and allocate micro-channels separately.

Categories and Subject Descriptors

B.7.2 [**Integrated Circuits**]: Design Aids

Keywords

3D-IC, micro-channel, liquid cooling, TSV assignment

1. INTRODUCTION

3D-IC contains two or more layers of active silicon stacked vertically. It provides several advantages including faster on-chip communication and heterogeneous integration, etc. Several research directions are being pursued that attempt to develop effective tools for 3D-IC physical design. One noteworthy problem in this regard is allocation and management of TSVs that enable communication between layers.

Despite significant potential performance improvement, stacked 3D structures bring forth new challenges pertaining to the removal of high heat density resulting from several stacks of chips. Micro-channel based liquid cooling, where micro-scale channels are embedded in the interlayer regions of 3D-ICs, is capable of removing significantly higher power levels by pumping coolant through these channels. Micro-channel uses extra energy for performing chip cooling. Its effectiveness strongly depends on the structure and locations of micro-channels. Recently some research has been done to address the problem of micro-channel placement to satisfy thermal constraints at minimum pumping energy while accounting for constraints imposed by existing TSVs [1][2]. These works show that the cooling effectiveness of micro-channels critically depend on the locations of TSVs, and hence properly allocating TSVs can help reduce the required pumping power.

TSVs impose significant constraints on how and where the micro-channels can be located, and form obstacles to the micro-channel placement since the micro-channels cannot be placed at the locations of TSVs. Conventional approaches for interlayer net to TSV assignment do not consider the possibility of adding micro-channels in the interlayer regions. Such approaches can have the following detrimental effects:

1. If TSVs are placed surrounding thermally critical areas, micro-channels would fail to reach these areas thereby resulting in hotspots.
2. In order to fix the thermal hotspots in areas where micro-channels cannot reach, we need to increase the fluid flow rate or place more micro-channels in the surrounding regions, resulting in a significant increase in cooling energy.

The existing works on micro-channel allocation assume that the TSV locations have already been decided before micro-channel allocation and then place micro-channels in the remaining areas [2]. Existing works on TSV allocation ignore this possible resource conflict with micro-channels [3][4]. Two trivial approaches for allocating TSVs to nets and micro-channels to interlayer regions *together* can be conceived as follows: TSV first approach and Micro-channel first approach. If micro-channels are allocated before TSVs, there is a possibility of increase in wirelength since the available whitespace for TSVs shrinks due to the existence of micro-channels which deter allocation of TSVs in those areas. A TSV first approach also has disadvantages. If TSVs are placed only to minimize the wirelength (as is the case in previous approaches in [3][4]), they might be placed in some thermally critical regions, and therefore micro-channels will fail to reach these regions thereby resulting in thermal violation or hotspots. In order to cool the hotspot area, we may need to increase the flow rate through micro-channels, which results in increase in cooling energy consumption. The location of TSVs is essentially decided by the allocation of interlayer nets to TSVs. The exiting works for Post Placement TSV allocation (which ignore the possibility of allocating channels) and micro-channel placement (which assume the TSV locations to be fixed) do not consider the possibility of combining these steps for obtaining better results.

In this paper, we investigate co-optimization of TSV assignment and micro-channel allocation simultaneously such that the total wirelength is minimized, and maximizing the micro-channel cooling effectiveness. TSV assignment corresponds to allocating interlayer nets to TSVs and deciding their locations. To the best of our knowledge, this is the first work that explores the electrical and cooling aspects of 3D-IC together. To solve this complex optimization problem, we propose a multi-commodity min-cost flow based algorithm.

2. CO-OPTIMIZATION ALGORITHM

We use multi-commodity min-cost flow (MCMCF) to solve the TSV assignment and micro-channel placement co-optimization. Given: a) the 3D-IC structure, b) potential locations of TSVs and micro-channels, c) the interlayer netlist and d) thermal criticality factor $g(i,j)$ of each micro-channel (i,j), the MCMCF problem is illustrated in Fig. 1 and 2. Fig. 1 illustrates a 3D-IC grid structure with three active layers and two interlayer nets along with four potential TSVs and sixteen potential micro-channels. We instantiate a MCMCF formulation as follows.

For each net, we allocate one unit of unique commodity flow. The flow network has one node for each terminal of the nets and potential TSV locations. We assume that the net terminal in the higher layer is the source of this one unit flow and net terminal in the lower layer is the sink. For the example in Fig. 1, the flow network is illustrated in Fig. 2 which indicates that *net*1 and *net*2 terminals in the top layer are sources. They are connected by directional edges to the TSV nodes in that active layer. If the nets span multiple layers, the TSVs in this layer would connect to the TSVs in the layer just below to transfer the signal. Finally sink terminals of nets are also connected to TSVs in that layer. Let us ignore the presence of micro-channels for now. The problem of allocating nets to TSVs can be modeled as a MCMCF (illustrated in Fig. 2(a)). Let the cost of each net-TSV or TSV-TSV edge be the half perimeter bounding box between the two. Let each

GLSVLSI'13, May 2–3, 2013, Paris, France.
ACM 978-1-4503-1902-7/13/05.

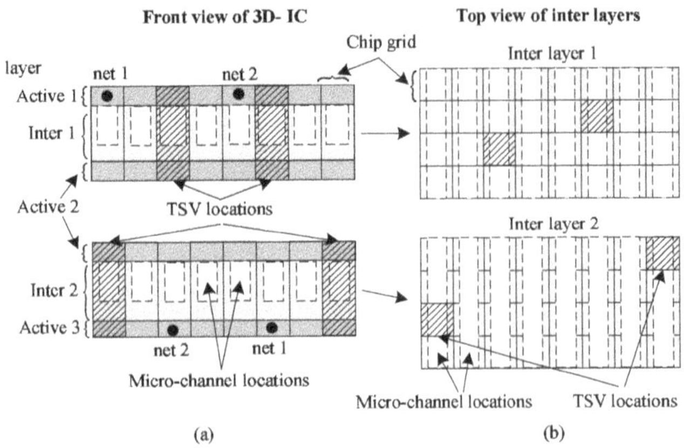

Figure 1: Potential TSV and micro-channel locations

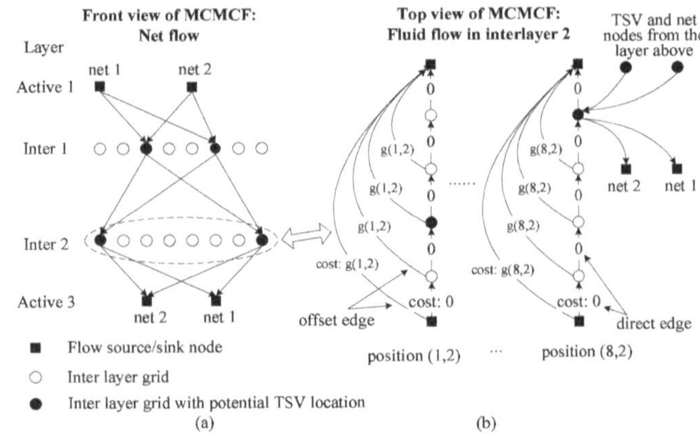

Figure 2: Corresponding MCMCF formulation

Table 1: Comparison between our, TSV first and channel first approach(Q_{pump} : W, WL : $\#grids$, temperature: ^oC)

Circuit	#layer	#TSV	# inter layer nets	Air cooling T_{peak}	TSV first			Micro-channel first			Co-optimization						
					WL	Q_{pump}	Below T_{max}	WL	Q_{pump}	Below T_{max}	WL	Q_{pump}	Below T_{max}	WL change w.r.t TSV first	WL change w.r.t MC first	Q_{pump} change w.r.t. TSV first	Q_{pump} change w.r.t. MC first
1	2	190	80	121.97	3010	n/a	N	3402	3.90	Y	3165	4.96	Y	+5.15%	-6.97%	n/a	+27%
2	2	348	100	110.25	8324	25.52	Y	8818	6.38	Y	8419	9.57	Y	+1.14%	-4.52%	-63%	+50%
3	2	652	125	128.83	10194	19.14	Y	10838	6.38	Y	10180	9.57	Y	-0.14%	-6.07%	-50%	+50%
4	3	511	80	154.06	6084	76.56	Y	6492	20.73	Y	6162	22.33	Y	+1.28%	-5.08%	-71%	+8%
5	3	714	100	152.39	10311	35.09	Y	10852	22.68	Y	10059	25.52	Y	-2.44%	-7.29%	-27%	+12%
6	3	1111	200	161.05	19932	38.28	Y	21423	19.86	Y	20016	22.33	Y	+0.42%	-6.57%	-42%	+12%
Avg														+0.90%	-6.08%	-50.6%	+27%

TSV node have a capacity of 1. Also all edges have an individual commodity capacity as 1 and a total capacity as 1. The MCMCF solution that sends the unique commodity flow from each net source to the corresponding net sink on the network in Fig. 2(a) at the minimum total cost corresponds to the net to TSV assignment with minimum total wirelength. A total unity capacity for each TSV node ensures that only one net is allocated to it.

The micro-channel is allocated an additional flow commodity. Fig. 2(b) indicates the process of accounting for micro-channels in the flow network of Fig. 2(a). The figure shows the top view of an interlayer region where the micro-channels are located and potentially conflict with TSVs. The micro-channels span the entire length of the chip in the fluid flow direction. If there is even one TSV allocated in this path, a channel cannot be allocated. We instantiate a source at the beginning of each potential channel location and a sink at the end. This source contains a unique commodity corresponding to the fluid flow. Several paths exist between the source and sink for each channel. The simplest path is the one that goes through all the grids that span the entire length of the chip. Some of these grids are potential TSV locations and have other nets/TSVs connected to them by way of directional edges (as indicated in Fig. 2(a)). The fluid flow edges are directed longitudinally while the net interconnection edges are directed vertically. As indicated earlier, the TSV nodes have a total capacity of one. The longitudinal edges that represent fluid flow have unit capacity for the fluid flow commodity while 0 capacity for all the net commodities. The fluid flow edges that connect the adjacent grids (*direct edges* in Fig. 2(b)) have a cost of 0. Now each intermediate grid in this direct path is also connected directly to the fluid sink for that channel through *offset edges*. This edge also has a fluid commodity capacity of 1 and net commodity capacity of 0. The cost of this edge $g(i,j)$ represents the cooling demand for that channel. All the edges that represent net-TSV or TSV-TSV connection have a fluid flow commodity capacity as 0. Now the cheapest way of sending a fluid flow commodity from source to sink is to follow the simple path that spans all the adjacent nodes. The cost of this path is 0 but it forces us to use all the potential TSV nodes on the path since they have a total capacity as 1. Hence none of these potential TSV nodes can be used for net interconnection. On the other hand, if any one of these nodes has been allocated to a net, a fluid commodity cannot go through this simple path and has to take *offset edges* to the sink. Any such alternative path has a cost of $g(i,j)$ which represents the price that we pay by *not* having a channel at that location since we would rather use some of the TSV on the channel path for routing nets.

Sending min-cost multi-commodity flow on this network results in an allocation of nets to TSVs and micro-channels to channel locations such that the total cost is minimized. This cost is a combination of $g(i,j)$ and bounding box wirelength, and represents a balance between cooling and wirelength.

3. EXPERIMENTAL RESULTS

In our experiment, we tested both two-layer and three-layer 3D-ICs. We use IBM-PLACE 2.0 circuits with placement information as the benchmark [5]. For each test, we choose two or three circuits from $ibm01 - ibm10$ circuits, each circuit corresponds to one 3D-IC layer. Based on the placement information, we find the whitespace between layout, which are basically the potential TSV locations. The number of potential TSV locations ranges from around 50-1000. We also randomly generate 30-200 interlayer nets. To obtain the power profiles for each layer, we randomly assign a value for each cell as the power density for the cell. The chip dimension is $9 \times 9mm^2$. The micro-channel width × height is $100 \times 100\mu m^2$, and the diameter of TSV is $10\mu m$. The maximum temperature constraint T_{max} is 85°C.

For each benchmark, we test three approaches to compare the wirelength and pumping power:our *co-optimization* approach, *TSV first* approach and *Micro-channel (MC) first* approach. For each approach, once we obtained the TSV assignment result, we route the interlayer nets to the TSVs in each layer separately using Labyrinth 2D router [6] to obtain the total wirelength (WL). We also estimate the pumping power Q_{pump} based on the number of channels used and the given pressure drop.

Table 1 shows the comparison of wirelength and micro-channel cooling power for the three approaches. Compared with *TSV first* approach, our approach achieves 50.6% pumping power savings only at a cost of 0.9% wirelength increase. Compared with *MC first* approach, our approach achieves 6.08% wirelength reduction at a cost of 27% pumping power increase. Moreover, for benchmarks where thermal violations occur using *TSV first* approach, using our approach could reduce the temperature below thermal constraints without consuming excessive pumping power.

4. REFERENCES

[1] D. B. Tuckerman and R. F. W. Pease, "High-performance heat sinking for VLSI," *Electron Device Letters*, pp. 126–129, 1981.

[2] B. Shi, A. Srivastava, and P. Wang, "Non-uniform micro-channel design for stacked 3D-ICs," in *DAC*, pp. 658–663, 2011.

[3] X. Liu, Y. Zhang, G. Yeap, and X. Zeng, "An integrated algorithm for 3D-IC TSV assignment," in *DAC*, pp. 652–657, 2011.

[4] T. Zhang, Y. Zhan, and S. S. Sapatnekar, "Temperature-aware routing in 3D ICs," in *ASP-DAC*, pp. 309–314, 2006.

[5] "Ibm-place 2.0 benchmark," in *http://er.cs.ucla.edu/benchmarks/ibm-place2/.*

[6] "Labyrinth global router," in *http://cseweb.ucsd.edu/ kastner/research/labyrinth/.*

Lifetime Reliability Assessment with Aging Information from Low-Level Sensors

Yao Wang
Delft University of Technology
Mekelweg 4, 2628CD
Delft, The Netherlands
Yao.Wang@tudelft.nl

Sorin D. Cotofana
Delft University of Technology
Mekelweg 4, 2628CD
Delft, The Netherlands
S.D.Cotofana@tudelft.nl

Liang Fang
National University of Defense Technology
Changsha 410073, China
lfang@nudt.edu.cn

ABSTRACT

Aggressive technology scaling has led Integrated Circuits (ICs) suffer from ever-increasing wearout effects. As a consequence, Dynamic Reliability Management (DRM) becomes an essential approach to assure IC's lifetime reliability. Accurate and efficient reliability modeling from low-level aging sensor measurements is critical to DRM systems. This work presents a Time-Sharing Sensing (TSS) method for V_{th}-sensor based DRM to assess the dynamic NBTI-induced degradation experienced by the circuit under monitoring. SPICE simulation results suggest that the proposed TSS method can accurately capture the circuit reliability status under random stress conditions.

Categories and Subject Descriptors

B.8 [**Performance and Reliability**]: Reliability, Testing, and Fault-Tolerance.

Keywords

Dynamic Reliability Management, NBTI, Reliability.

1. INTRODUCTION

Due to the aggressive MOSFET technology scaling that took place in the past decades, reliability has been becoming a rising concern for processor designers. The increasing power densities and operating temperatures made integrated circuits (ICs) suffer from multiple intrinsic failure mechanisms during their servicing time. As a result, device performance degrades gradually and consequently might lead to fails during IC's expected lifetime [3]. To combat the gloomy outlook of reliability situation, a variety of techniques have been proposed to ensure ICs' lifetime reliability. Dynamic Reliability Management (DRM) techniques [5, 6] attempt to hide the inherent pessimistic reliability landscape while maintaining the system performance and lifetime expectation within the desired range. This work presents a Time-Sharing Sensing (TSS) method, for threshold voltage (V_{th}) sensor based DRM systems to assess NBTI-induced degradation in ICs. Our simulations demonstrate that it can accurately capture the dynamic reliability status from the circuits under monitoring with a limited number of aging sensors, which is crucial for DRM practical implementations.

GLSVLSI'13, May 2–3, 2013, Paris, France.
ACM 978-1-4503-1902-7/13/05.

2. DELAY SHIFT DUE TO AGING

NBTI has been considered as the most dominant failure mechanism in current and future CMOS technology nodes [3], and it is well understood that $|V_{th}|$ increases due to NBTI stress. A long-term V_{th} shift under dynamic NBTI stress is given by [2]:

$$\Delta V_{th} = \left(\frac{\sqrt{K_v^2 \alpha t_{clk}}}{1 - \beta_t^{1/2n}} \right)^{2n} = A \cdot (\alpha t)^n, \tag{1}$$

where $n = 1/6$ is the power-law time constant, K_v is the accelerating coefficient (including temperature and electrical field), T_{clk} is clock period, α ($0.1 \leq \alpha \leq 0.9$) is the stress probability of NBTI, i.e. NBTI duty-cycle, β_t is a coefficient reflecting the recovery effect of NBTI, and A is the NBTI time-evolution dependence.

According to the alpha-law model [4], the first order of delay shift due to V_{th} degradation can be expressed as:

$$\Delta \tau_{pd} = \frac{\alpha \tau_0}{V_{dd} - V_{th0}} \cdot \Delta V_{th} = k_\tau \Delta V_{th} \cdot \tau_0, \tag{2}$$

where k_τ is the ΔV_{th} correlation efficient. In a V_{th}-sensor based DRM system, aging sensors are designed to replicate the NBTI stress of a device in the critical paths (CPs) or potential critical paths (PCPs), such that the aging status (i.e., delay shift) of CPs and PCPs can be estimated from the aging sensors output by Eq. (2).

According to Eq. (2), the gate-level delay shift for the NAND and NOR gate can be expressed as:

$$\Delta \tau = \begin{cases} \Delta \tau_{pd,i}, & \text{for NAND,} \\ \sum_{i=1}^{n} \Delta \tau_{pd,i}, & \text{for NOR,} \end{cases} \tag{3}$$

where $\tau_{pd,i}$ is the delay shift of the ith PMOS device in the gate. In other words, the delay shift of the NAND gate is determined by the delay shift of the PMOS device in the CP or PCP, while the delay shift of the NOR gate is the sum of delay shifts of all the PMOS devices in the gate.

Assuming that the NBTI effect on NMOS is negligible, the delay shift of a CP (or PCP) $\Delta \tau_{cp}$ is then the sum of delay shifts of the Pull-Up Gates (PUGs). From Eq. (2), $\Delta \tau_{cp}$ can be derived as:

$$\Delta \tau_{cp} = \sum_{i \in PUGs}^{l} \Delta \tau_i = k_\tau \sum_{i=1}^{h} \Delta V_{th,i}, \tag{4}$$

where l is the CP depth, $\Delta \tau_i$ is the delay shift of a single logic gate, and h equals the number of PMOS devices in

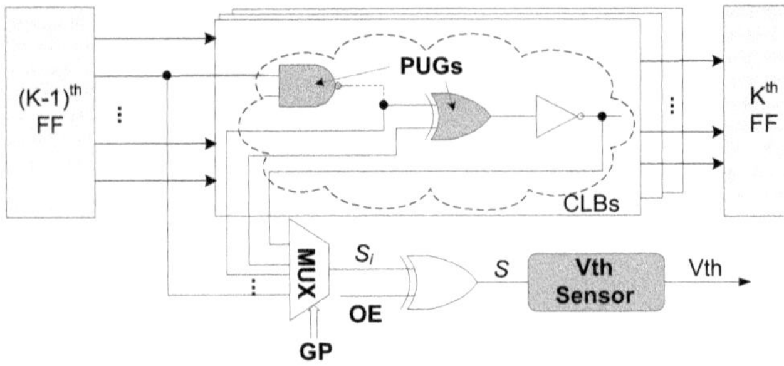

Figure 1: Schematic of Time-Sharing Sensing.

the PUGs. Notice that only half of the logic gates (i.e., odd-/even-stage gates) in the CP is in the PUGs during an input signal transition.

3. TIME-SHARING SENSING SCHEME FOR AGING ASSESSMENT

In order to obtain an accurate $\Delta\tau_{cp}$ estimation, we propose a TSS scheme for V_{th} aging sensors, as illustrated in Figure 1. TSS scheme allows the V_{th} sensor to probe the stress on different nodes at different time moments. This is achieved by routing all the necessary signals (including all inputs of NORs in the PUGs) from the CP to a MUX, and the MUX selects one signal from the inputs and sends it to the V_{th} sensor. The the input signal selection is controlled by the *GP* signal based on a round-robin algorithm, and the selection bit shifts at every time interval Δt.

Based on the proposed scheme, the ΔV_{th} of aging sensor after N intervals can be expressed by:

$$\Delta V_{th} = \sum_{i=1}^{N} \Delta V_{th,i} = \sum_{i=1}^{N} A_i(\alpha_i \Delta t_i)^n, \tag{5}$$

where $\Delta V_{th,i}$ is the parameter degradation during the ith time interval. Further assume $N = k \cdot h$, then the time intervals can be divided into k sub-sequences $\Delta T_j = h \cdot \Delta t$, where $j = 1, 2, ...k$. For the jth sub-sequence, assume the sampled activity ratio of the mth logic gate is the activity ratio for the entire time span, i.e, $\bar{\alpha}_{t\in T_j} \approx \alpha_{t\in(T_{j-1}+m\Delta t)}$, then according to Eq. (1), the total ΔV_{th} of the mth logic gate at the end of sub-sequence ΔT_j can be written as:

$$\Delta V_{th,t\in\Delta T_j} = \left(\frac{\Delta T_j}{\Delta t}\right) \cdot \Delta V_{th,t\in\Delta t_m} = h\Delta V_{th,t\in\Delta t_m}, \tag{6}$$

where t_m is the mth interval Δt in ΔT_j. Insert Eq. (6) into Eq. (4), the delay shift of the entire CP during a large time interval ΔT_j can estimated as:

$$\Delta\tau_{cp,t\in\Delta T_j} = hk_\tau V_{th,t\in\Delta T_j}. \tag{7}$$

Then, the accumulated delay shift can be estimated as:

$$\Delta\tau_{cp} = \sum_{j=1}^{k} \Delta\tau_{cp,t\in\Delta T_j} = hk_\tau V_{th}. \tag{8}$$

Eq. (8) suggests that by introducing the TSS scheme into a DRM system, the long-term delay shift of a critical path is proportional to the V_{th} degradation measured from the associated aging sensor. In other words, the V_{th} sensor measurement indicates the NBTI-induced degradation of the entire CP in TSS scheme, instead of the degradation of one single device in the conventional configuration, which is a significant improvement in terms of area and power efficiency.

Table 1: TSS Based Delay Shift Estimation for ISCAS85&89 Circuits

Name	Estimation Error by TSS Scheme relative to time 0 delay (%)							
	Δt = 1000 clk		Δt = 2500 clk		Δt = 5000 clk		Δt = 10000 clk	
	Cst.	Dvt.	Cst.	Dvt.	Cst.	Dvt.	Cst.	Dvt.
C432	2.34	3.28	2.31	2.73	2.26	2.77	2.19	2.32
C499	4.73	6.17	4.45	5.09	4.27	3.46	4.11	3.89
C880	2.89	2.21	2.64	3.17	2.37	2.55	2.17	2.25
C1908	4.18	3.96	4.73	8.07	4.65	4.12	4.53	4.79
C3540	4.46	2.95	4.37	3.51	4.28	3.29	4.25	5.03

4. EXPERIMENTAL RESULTS

In order to validate the proposed reliability modeling framework, simulations are conducted for a set of ISCAS85 and ISCAS89 benchmark circuits. A PTM [1] 32nm technology library is used for SPICE simulation. The technology library consists of 5 different cells: INVERTER, 2-input NAND, 3-input NAND, 2-input NOR and 3-input NOR. The simulations for SPICE are performed in HSPICE at an accelerated condition, i.e., $T = 120°C$, with an equivalent lifetime of 10-year operation at normal condition.

Table 1 presents the absolute errors of the proposed TSS scheme for delay shift estimation. Four different sampling intervals are investigated, i.e., Δt = 1000, 2500, 5000, and 10000 cycles, respectively. Two kinds of input signal are evaluated: *Cst.* corresponds to a constant activity ratio ($\alpha = 50\%$) signal, and *Dvt.* corresponds to variable activity ratio signals ($\alpha = 50\% \pm 10\%$). The results indicate TSS scheme can accurately capture the delay shift due to NBTI at circuit level.

5. CONCLUSIONS

In this work we presented a time-sharing sensing method for existing V_{th}-sensor based DRM systems to assess NBTI-induced degradation in ICs. SPICE simulation results indicate that the proposed TSS method can achieve accurate reliability assessment under random stress conditions.

6. REFERENCES

[1] Predictive Technology Model (PTM). `http://ptm.asu.edu/`.

[2] S. Bhardwaj, W. Wang, R. Vattikonda, Y. Cao, and S. Vrudhula. Predictive Modeling of the NBTI Effect for Reliable Design. In *Proceedings of Custom Integrated Circuits Conference*, pages 189–192, 2006.

[3] J. W. McPherson. Reliability Challenges for 45nm and Beyond. In *Design Automation Conference*, pages 176–181, 2006.

[4] T. Sakurai and A. Newton. Alpha-Power Law MOSFET Model and Its Applications to CMOS Inverter Delay and Other Formulas. *IEEE Journal of Solid-State Circuits*, 25(2):584–594, 1990.

[5] J. Srinivasan, S. Adve, P. Bose, and J. Rivers. The Case for Lifetime Reliability-Aware Microprocessors. In *Proceedings of International Symposium on Computer Architecture*, pages 276–287, 2004.

[6] R. Teodorescu and J. Torrellas. Variation-Aware Application Scheduling and Power Management for Chip Multiprocessors. In *Proceedings of International Symposium on Computer Architecture*, pages 363–374, 2008.

A Hardware-Efficient Architecture for Embedded Real-Time Cascaded Support Vector Machines Classification

Christos Kyrkou
KIOS Research Center,
University of Cyprus
kyrkou.christos@ucy.ac.cy

Theocharis Theocharides
KIOS Research Center,
University of Cyprus
ttheocharides@ucy.ac.cy

Christos-Savas Bouganis
Imperial College London
christos-savvas.bouganis@imperial.ac.uk

ABSTRACT
This work presents an optimized architecture for cascaded SVM processing, along with a hardware reduction method for the implementation of the additional stages in the cascade, leading to significant improvements. The architecture was implemented on a Virtex 5 FPGA platform and evaluated using face detection as the target application on 640×480 resolution images. Additionally, it was compared against implementations of the same cascade processing architecture but without using the reduction method, and a single parallel SVM classifier. The proposed architecture achieves an average performance of 70 frames-per-second, demonstrating a speed-up of 5× over the single parallel SVM classifier. Furthermore, the hardware reduction method results in the utilization of 43% less hardware resources, with only 0.7% reduction in classification accuracy.

Categories and Subject Descriptors
B.2.1 [**Arithmetic and Logic Structures**]: Design Styles—*parallel*; C.3 [**Special-Purpose and Application-Based Systems**]: Real-time and embedded systems

General Terms
Performance, Design

Keywords
FPGA, Parallel Architecture, Support Vector Machines.

1. INTRODUCTION
Support vector machines (SVMs) [1] are a powerful supervised pattern recognition algorithm which has been used in object detection, amongst other applications, demonstrating high classification accuracies [1-3]. However, for large scale problems the high classification accuracy rates of SVMs come with the cost of longer classification times. The reason for this is that the run-time complexity of the algorithm is proportional to the number of support vectors; which are samples from the training set that specify the separating hyperplane that the SVM algorithm selects in order to construct its classification decision function. As such, embedded SVM-based classification systems with hundreds of support vectors, find it difficult to meet real-time processing demands, without sacrificing accuracy.

It is possible to speed-up SVM-based classification systems for a variety of applications, including object detection, by exploiting the facts that: (a) the majority of the samples presented to the classifier do not belong to the object class and (b) the majority of those samples can be easily distinguished from samples belonging to the object class. Cascade SVM classifiers, take advantage of these two observations by utilizing stages of classifiers, which are sequentially applied to the input data. The early stages, usually linear SVMs, have low complexity, meaning that they require less training data to be processed, and as such take less time to process. Conversely, the latter stages, typically RBF and polynomial kernels, have higher complexity as they require more training data to be processed, and hence have a longer classification-time. If at any stage the input data is classified as a non-object the classification process stops and the next sample is processed, otherwise it must go all the stages to be classified as an object. Under this scheme the stages at the beginning of the cascade discard a large amount of input samples very fast, resulting in significant speedups over single SVM classification [2], and is highly suitable for embedded applications such as object detection where a lot of data are generated from a single image and need to be classified.

This paper proposes a dedicated hardware architecture for cascaded SVM processing and a method for reducing the hardware implementation requirements of cascaded SVMs. The hardware architecture was implemented on a Virtex-5 FPGA platform, and evaluated using face detection on 640×480 resolution images. The implementation of the proposed hardware architecture for processing the adapted SVM cascade demonstrates an average performance of 70 frames-per-second, which is a 5× speedup over a parallel single SVM classifier implementation. Furthermore, by applying the hardware reduction method, the adapted hardware architecture for the implementation of the cascade consumes 43% less custom logic resources, with only a 0.7% reduction in classification accuracy.

2. PROPOSED ARCHITECTURE AND HARDWARE REDUCTION FRAMEWORK

2.1 Cascade SVM Hardware Reduction
The proposed hardware reduction method, outlined in Figure 1, is to round-off the support vector and alpha values of the low complexity kernels with the nearest power of two values. This will result in all the multiplication operations in the SVM classification phase (the kernel dot-product calculations and computations related to the alpha coefficients) becoming shift operations. Additionally, since the support vectors and alpha coefficients are now power of two values there is no need to store the binary representations of decimal numbers but only shift data (shift amount, shift direction, and number sign). Hence, this results in an *adapted cascade SVM* with reduced storage and computational demands.

2.2 Cascade SVM Hardware Architecture
The nature of the cascaded SVM classifiers means that each stage will have less input data to process and more support vector data to process than the previous. Accordingly then, the proposed hardware architecture for the cascaded SVM classifier consists of

GLSVLSI'13, May 2–3, 2013, Paris, France.
ACM 978-978-1-4503-1902-7/13/05.

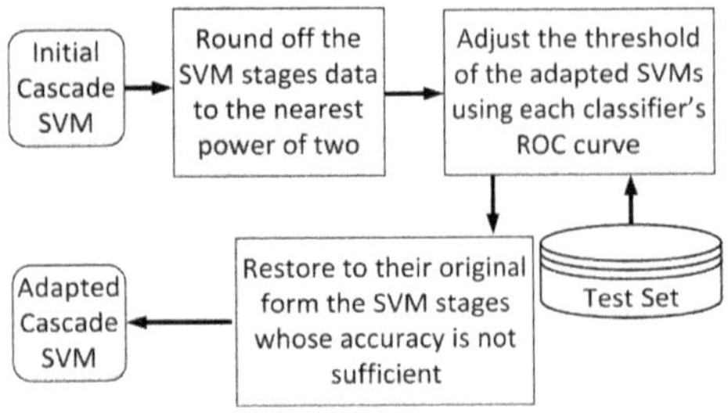

Figure 1. Cascade Hardware Reduction Method

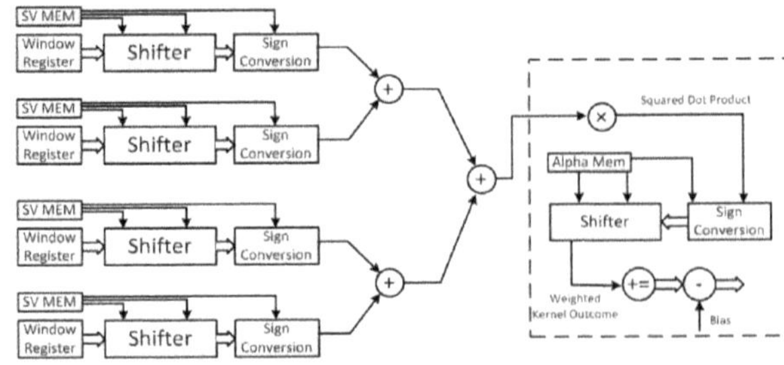

Figure 2. Architecture for the Parallel Processing Module

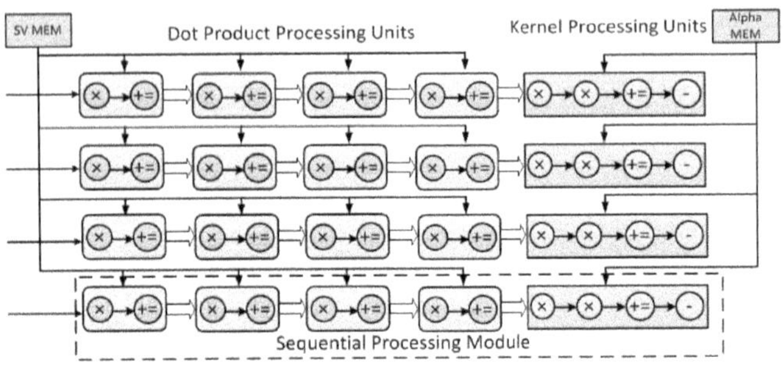

Figure 3. Architecture of single parallel SVM and Sequential Processing Module

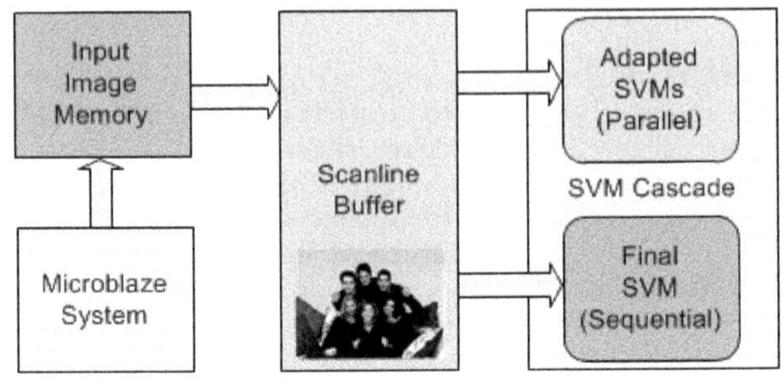

Figure 4. Cascaded Hardware Architecture for Object Detection

two main processing modules, which provide different parallelism with respect to the input data, in order to meet the different demands of the cascade stages. The first is a fully parallel module (Figure 2) which performs the processing necessary for all the adapted SVM stages and the parallelism focuses on processing all the input vector elements in parallel. The second is a sequential processing module (Figure 3), optimized for the high complexity SVM stages which demand processing a large number of support vectors and thus parallelism focuses on processing more support vectors in parallel. In addition, a shift register structure is used to provide sequential and parallel data access to the two processing modules, and also to take advantage of potential data overlap and reduce memory I/O.

3. EXPERIMENTAL PLATFORM AND RESULTS

The proposed hardware architecture (Figure 4) is evaluated using the embedded application of face detection on 640×480 resolution images [4,5] for a cascade comprised of four stages (2 linear SVMs followed by two non-linear SVMs). It is evaluated in terms of the detection accuracy, frame-rate, as well as requirements in terms of computing resources. Additionally, the proposed architecture for the *adapted cascade* is compared against two other systems in terms of the above metrics. The first system, which will be referred to as the *initial cascade* and is an implementation of the architecture that can process the same cascade SVMs, but without applying the hardware reduction method, and thus the parallel processing module is implemented using multipliers. The second system, referred to as *single parallel classifier*, consists of multiple sequential processing modules and implements only the most accurate SVM in the cascade, typically the final cascade stage, but in parallel (Figure 4). All systems were implemented on a Xilinx ML505 Virtex 5-LX110T FPGA and all use a Microblaze-based I/O system for I/O and verification. The following sections detail the evaluation process and results.

Through this optimization the total memory demands to store the SVM training data for the first three stages of the SVM cascade are reduced by 40%. Considering both the reduction in processing and storage resources together the adapted cascade implementation requires 43% less FPGA Look-Up Table (LUT) resources compared to the initial cascade implementation. Overall, both the adapted cascade and initial cascade implementations achieve a performance of 70 fps, resulting in a 5× speedup over the single parallel SVM classifier implementation which achieved 14 fps. Finally, the reduction method resulted in only a 0.7 % drop in overall classification accuracy resulting in a 84% of correct detections rate. Overall, the detection accuracy of the adapted cascaded SVM is similar to software works which range from 75% to 88% [1-3].

4. ACKNOWLEDGMENTS

This work was co-funded by the European Regional Development Fund and the Republic of Cyprus through the Research Promotion Foundation (Project NEA ΥΠΟΔΟΜΗ/ΣΤΡΑΤΗ/0308/26)

5. REFERENCES

[1] Osuna E., Freund R., and Girosi F., 1997, Training Support Vector Machines: An Application to Face Detection, Proc. IEEE Conf. Computer Vision and Pattern Recognition, pp. 130-136.

[2] B. Heisele, T. Serre, S. Prentice, and T. Poggio. 2003, Hierarchical classification and feature reduction for fast face detection with support vector machines, Pattern Recognition, 2007–2017.

[3] I. Kukenys, B. McCane, 2008, Classifier cascades for support vector machines, Intl. Conf. on Image and Vision Computing, 1-6.

[4] "CBCL Face Database #1" [Online]. Available: http://cbcl.mit.edu/software-datasets/FaceData2.html

[5] Bao Face Database, [Online]. Available: http://www.facedetection.com/facedetection/datasets.htm

Pipeline Template and Scheduling Algorithm for Mapping Multiple Loop Nests on FPGA with Limited Resources

Yazhuo Dong
Naval Academy of Armament,
No. 3 , Wanshou Street, Haidian area, Beijing city, China,100036
0086-010-66974181
xinya0624@gmail.com

Wu Zhan
Naval Academy of Armament,
No. 3 , Wanshou Street, Haidian area, Beijing city, China,100036
0086-010-66974118
WuZhan@gmail.com

Xiqing Ye
Naval Academy of Armament,
No. 3 , Wanshou Street, Haidian area, Beijing city, China,100036
0086-010-66974111
XiqingYe@gmail.com

ABSTRACT
There are large numbers of high-level algorithms consisting of multiple loop nests in image compression, pattern recognition and digital signal processing. FPGA provides a convenient and flexible solution to speed up these loop-intensive algorithms. However, FPGA reconfiguration which needs a long time is inevitable when switching between the loop nests. This paper presents a parameterized pipeline template to execute all the loop nests in sequence without FPGA reconfiguration. Five steps are designed to decide the parameters. Experiments show that the pipeline template can achieve a comparative execution cycles for a loop comparing with the special hardware structure.

Categories and Subject Descriptors
B.1.2 Control Structure Performance Analysis and Design Aids

Keywords
FPGA, Loop, Pipeline Temple

1. INTRODUCTION
There are large numbers of high-level algorithms consisting of multiple loop nests in image compression, pattern recognition and digital signal processing. If the space utilization exceeds the device capacity, these loop nests have to be mapped onto the target hardware respectively. Reconfiguration which needs a long time is inevitable when switching between the loop nests. Our paper will focus on this question to generate a pipeline template and scheduling algorithm for mapping multiple loop nests on FPGA with limited resources. All loop nests in an application can be mapped onto one hardware template in sequence without the process of reconfiguration.

In this paper, we restrict input programs to multiple loop nests computation on array and scalar variables (no pointers), where all subscript expressions are affine with a fixed stride. The loop bounds must be constant and loops with control flow are supported. The target architecture is a single FPGA with multiple external memories.

GLSVLSI'13, May 2–3, 2013, Paris, France.
ACM 978-978-1-4503-1902-7/13/05.

2. PIPELINE TEMPLATE
Given an application and a FPGA with limited resources, this paper presents a hardware pipeline template which can map all the loop nests on it one by one without FPGA reconfiguration. Figure 1 illustrates the pipeline template. It is designed to exploit high degree of parallelism available with a large number of function units (FUs). Each FU implements a type operation such as add or

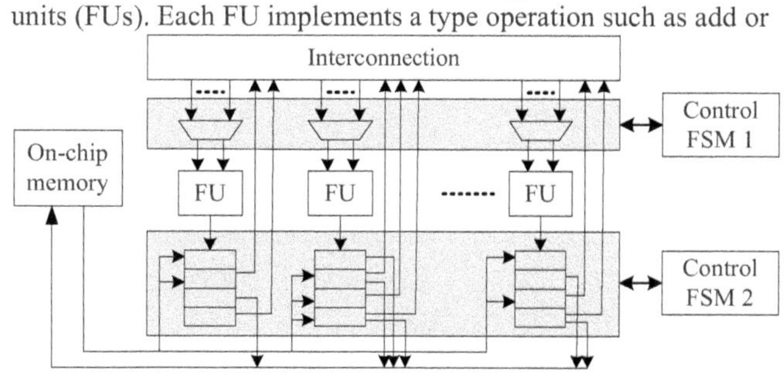

Figure 1. The parameterized pipeline template

multiply etc. Multiple registers may be connected to each FU input: a multiplexer (MUX) which is controlled by control FSM1 is used to select the appropriate input. Since the operations executing in a loop are periodic, the selector for this MUX is simply a modulo counter. The original data is read into registers directly by load operation, and results are sent into on-chip memory by store operation. Control FSM2 is used to send the appropriate intermediate data in the expanded virtual registers back to the interconnection. The loop accelerator begins execution when a start signal is asserted by the host processor. When the loop execution is complete, a done signal is fed back to the host processor.

The pipeline template is a parameterized hardware architecture, where the parameters are decided by the loops' characteristics. For different loop nests, the parameters are different. The target of the work in section 3 is to decide a set of uniform parameters for all loop nests in an application. The parameters are as follows:

- The type of a FU and the amount of it: FU_i, Num(FU[i]);
- The number of virtual registers connected to each FU: Num reg_1, Num_reg_2, ... Registers of the ith FU are denoted as reg_i^1, reg_i^2, . . .;
- The number of input ports of each FU: Num_$inport_1$, Num $inport_2$, ... The input ports of ith FU are denoted as: $Inport_i^1, Inport_i^2, \ldots$;

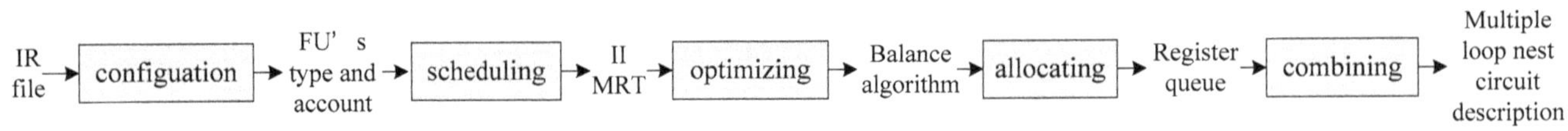

Figure 2. Parameters confirming steps

begin
if($Res_{max} \leq S_{FPGA}$) //Resources are enough generate the hardware pipelining architecture, get Res= Res_{max};
else begin
if($Res_{min} > S_{FPGA}$) //The resources on FPGA are so absent that the loops can not implement on them.
else begin
Res= Res_{min};
until(Res>S_{FPGA})
repeat begin
Get the execution time of each loop under the current FUs: $Time_1', Time_2', \cdots, Time_n'$

Define $delay_i = \frac{Time_i'}{Time_i}$; assme there are p loops who content

$$delay_{i1} = delay_{i2} = \cdots = delay_{ip} = \max\{\frac{Time_1'}{Time_1}, \frac{Time_2'}{Time_2}, \cdots, \frac{Time_n'}{Time_n}\}$$

The execute time of these loops are delayed mostly by the limited resources; adjust the type and number of FUs for these loops.
For(i=1;i<p;i++)
begin
for each loop i, compute $II_i = \frac{NUM_{op}^i}{NUM_{FU}^i}$;

$$\max II = \max\left\{\frac{NUM_{op}^1}{NUM_{FU}^1}, \frac{NUM_{op}^2}{NUM_{FU}^2}, \cdots, \frac{NUM_{op}^n}{NUM_{FU}^n}\right\}$$

Assume there are q number of FUs whose , increase the number of these FUs from one the the II_i is debased, and define:

$$\cos t(i) = A_1 FU_1 + A_2 FU_2 + \cdots + A_n FU_n$$

end
Get the loop i whose cost(i) is least. Add the FUs of loop i in the Res:
Res_{old}=Res; Res=Res+{n*FU;
end
Res= Res_{old};
end
end
end

Figure 3. The FU configuration algorithm

• The connection aggregates between virtual registers and FUs' input ports are denoted as: $connect_1$, $connect_2$, ...For example, $\{(reg_2^1, In\_port_1^3)\} \in connect_2$ means the connection relationship between the first register of the second FU with the third input port of the first FU.

3. CONFIRMING PARAMETERS

There are five steps designed to confirm the parameters which are shown in figure 2. The main part is the FU configuration,which is shown in figure 3. The FU configuration algorithm is used to decide the number of FUs and their capabilities in the pipeline template. This is decided by the application's size and chip capacity. Given the operations in the loops and a library of hardware cell capabilities, the FU configuration algorithm is to come up with a mix of FUs that minimizes execution cycles while meeting the capacity constraint.

4. EXPERIMENTS

We chose five benchmarks. Livermore kernel is structured as three loop nests. HLSynth92 is a test-bench for high-level synthesis system which is composed by two loop nests. Then, three loops in mediabench are picked up. Kalman filter is a mathematical power tool including three loop nests. Finally, Three loops are involved in image processing: Sobel edge detection, DT forward and DT backward. Experiments show that the pipeline template may increase the execution cycles of some loop nests in a certain extent.

5. ACKNOWLEDGMENTS

This work is supported by NSFC of China under the NO. 60903057.

REFERENCES

[1] Shail Aditya, Michael S. Schlansker, "ShiftQ: A buffered interconnect for custom loop accelerators." CASES 2001, pp. 158-167.

[2] Vinod Kathail, Shail Aditya, Robert Schreiber, B. Ramakrishna Rau, Darren C. Cronquist, Mukund Sivaraman, "PICO: Automatically Designing Custom Computers." IEEE Computer 35(9), pp. 39-47, 2002.

Study of ASIC Technology Impact Factors on Performance Evaluation of SHA-3 Candidates

Meeta Srivastav
Virginia Tech
Blacksburg, VA, USA
meeta@vt.edu

Yongbo Zuo
AMD Pvt. Ltd.
Fort Collins, CO, USA
zyongbo@vt.edu

Xu Guo
Qualcomm Pvt. Ltd.
San Diego, CA, USA
xuguo@vt.edu

Leyla Nazhandali
Virginia Tech
Blacksburg, VA, USA
leyla@vt.edu

Patrick Schaumont
Virginia Tech
Blacksburg, VA, USA
schaum@vt.edu

ABSTRACT

The main aim of NIST's SHA-3 competition is to choose the new secure hash standard which will become successor of SHA-2. Analysing the strength of candidates in terms of hardware performance is complicated due to presence of different platforms such as FPGA, ASIC. Within them, implementation in an ASIC platform can be challenging due to different choices available during implementation. In this study, we will present these choices and also present an approach to evaluate the impact of these choices on the performance characteristics of SHA-3 finalists. This article will help NIST as well as cryptographers in better understanding the ASIC benchmarking results presented by different research groups.

Categories and Subject Descriptors

B.7.0 [**Hardware**]: Integrated circuitsGeneral

General Terms

Design, Performance, Security

Keywords

SHA-3, ASIC, hash, cryptography, security

1. INTRODUCTION

The objective of the SHA-3 NIST competition is to select, among multiple competing candidates, a standard algorithm for cryptographic hashing. An ideal winner should posses adequate cryptographic properties and good implementation characteristics over a wide range of target platforms including both software and hardware.

We implemented and successfully measured results from likely the first SHA-3 ASIC [4]. Due to limited budget constraint at the time of implementation, we used flat approach [4] where all candidates does not have physical boundaries. Later, we implemented the same netlist with the floorplan which yields highest *Throughput-to-Floorplan Area* for all candidates. We observed 12%-34% change in the performance characteristics. In summary, while doing ASIC benchmarking there are many factors or choices that the designer has, that can change the overall implementation characteristics of SHA-3 candidates. It is difficult to recommend a standard methodology as in the case of FPGA platforms. However, we can study the impact of different choices while implementing SHA-3 candidates in ASIC. As far as we know, there has been no significant effort to investigate the impact on performance characteristics. In this study, we first recommend a comparison methodology and later present our analysis on major choices.

GLSVLSI'13, May 2–3, 2013, Paris, France.
ACM 978-1-4503-1902-7/13/05.

2. RELATED WORK AND BACKGROUND

Several comprehensive studies in SHA-3 ASIC implementations have been reported [1, 2, 3, 4, 5]. Generally, researchers rank the candidates based on performance characteristics such *Throughput*, *Throughput-Area ratio* and energy. Table 1 lists some of the choices made in the process of implementation by all above mentioned groups. From the table, we observe that all implementations differ with each other at so many levels. With the wide-variety of choices now available for ASIC implementation, sometimes the rankings between the candidates that are promoted may not truly reflect the strength of an algorithm, but instead their sensitivity to the choices. Unless we have a study which does an analysis on the impact of these choices on implementation results, it will be very difficult to comprehend them. In this study, we will try to analyze the impact of four major choices such as library, constraints, technology nodes and physical floorplan shape & sizes on performance characteristics of SHA-3 candidates.

3. METHODOLOGY

This section describes the overall methodology used to analyze and compare the effect of choices on the characteristics of each SHA-3 finalist candidates.

3.1 Comparison methodology

To compare the impact of choices, we present two-dimensional comparison methodology namely intra-candidate and inter-candidate comparison.

3.1.1 Intra-candidate comparison

When comparison of an evaluation metrics is done within a candidate, we call this as intra-candidate comparison. Intra comparison will help in evaluating the effect of choices on each candidate's algorithm. This will also help in analyzing the robustness of the algorithm with the change in choices.

Table 1: The related SHA-3 hardware benchmarking work in ASICs.

Group	Henzen[2]	Knezevic[3]	Tillich[5]	Gurkaynak[1]	Gaj[1]	Guo[4]
Technology	90nm	90nm	180nm	65nm	65nm	130nm
Optimization criteria	T	T/A	T	Fixed T	T/A	T/A
Library	UMC	STM	UMC	UMC	UMC	IBM
Constraints	Max-Frequency	Latency/Area	Max-Frequency	Targeted T	Max T/A	Max T/A
Floorplan	Square	NA	Different	Flat	Flat	Flat
(Shape and Size)	(85% util)	(Synthesis results)	shapes	(No shape)	(No shape)	(No shape)

Table 2: Summary of variations.

Choices	Throughput						Throughput/Area					
	Intra-candidate			Inter-candidate			Intra-candidate			Inter-candidate		
Library	2%	26%	30%	0.6%	14%	41%	0%	1%	7%	1%	6%	9%
Constraints	0%	2.3%	24.6%	0.5%	5%	24%	1%	3.2%	9.7%	1%	4%	9%
Technology	-	-	-	2%	17%	56%	-	-	-	1%	9%	17%
Floorplan shape and size	0.8%	6.5%	16.3%	0.3%	9%	17%	1.5%	13.4%	27.4%	1%	19%	29%

For the same algorithm and hardware architecture, lower percentage change due to different choices, implies lower sensitivity to the choice and hence higher robustness of an algorithm.

3.1.2 Inter-candidate comparison

The main aim of the benchmarking is to analyze the relative hardware efficiency of the candidates. Inorder to achieve that, we propose another comparison strategy called inter-candidate comparison. Inter-candidate comparison helps in evaluating relative effect of the same choice on all candidates. Higher the relative percentage change (positive or negative) implies higher sensitivity to the choices and hence lower robustness of an algorithm. We this two-dimensional comparison approach, we can effectively compare the sensitivity of algorithms of all candidates.

3.2 Experimental setup

In this section we will briefly review the experimental setup that is being used for all experiments. To study the impact of four choices, we have considered following, 1)*Library*: We consider two widely used library namely UMC and IBM standard cell.2)*Constraints*: We consider four different implementations obtained using four different synthesis constraints. Mainly Max-Frequency (F), Max-frequency & Min-Area (FA), Max-frequency & Min-Power (FP) and Max-Frequency, Min-Area & Min-Power (FAP) are considered. 3)*Technology Nodes*: We will study the impact while implementing the design on 180nm, 130nm, 90nm and 65nm nodes.4)*Floorplan shape and size*: We consider five different shapes and sizes. They are square, rectangle with an aspect ratio of 1:2 (flat), rectangle 1:3, rectangle 2:1 (standing) and rectangle 3:1.

4. RESULTS

Table 2 shows the summary of variations due to different choices while implementing SHA-3 finalists. From the table, at an average 2.3%-26% variation is seen in the performance characteristics. Change in the library can bring variations of 1%-26%. Candidates are sensitive to the library. Candidates are not sensitive to the synthesis constraints 2.3%-4% and the effect is very uniform on all. Technology scaling can introduce variations of 9%-17%, implying candidates are very sensitive to the technology nodes they are implemented in and that the sensitivity is not uniform. Last, candidates are highly sensitive (6.5%-19%) to the shape they are implemented in. This shows implementation dependency of the algorithms on implementation shapes and the effect due to shapes is not uniform among all.

The summary presented in this study will help cryptographers as well as NIST in analyzing hardware ASIC benchmarking results from various different research groups. Considering the hash algorithms, this study will not only help in SHA-3 competition but various secure hash competitions in future. This study summarises 20% improvement or deterioration in performance characteristics of any candidate may not be because of the algorithm characteristics, but due to the choices made during the implementation instead.

5. CONCLUSIONS

In this paper we summarize our effort in doing *effective* hardware benchmarking in ASIC of five SHA-3 finalists.

6. REFERENCES

[1] F. Gürkaynak and et. al. Lessons learned from designing a 65nm asic for evaluating third round sha-3 candidates. In *3rd SHA-3 Candidate Conference, Washington DC*, 2012.

[2] L. Henzen and et al. Developing a Hardware Evaluation Method for SHA-3 Candidates. In *CHES 2010*.

[3] M. Knezevic and et al. Fair and consistent hardware evaluation of fourteen round two sha-3 candidates. *VLSI Systems, IEEE Transactions on*, 2011.

[4] M. Srivastav and et al. Design and benchmarking of an asic with five sha-3 finalist candidates. *Microprocessors and Microsystems*, 2012.

[5] S. Tillich and et al. Uniform Evaluation of Hardware Implementations of the Round-Two SHA-3 Candidates. In *The Second SHA-3 Candidate Conference*, August 2010.

Timing-Constrained Replacement Using Spare Cells for Design Changes

Zhi-Wei Chen
College of Engineering,
Chung-Hua University,
Hsinchu, Taiwan, R. O. C.

Jin-Tai Yan
Department of Computer Science and Information Engineering, Chung-Hua University,
Hsinchu, Taiwan, R. O. C.

ABSTRACT

Spare cells have been widely used to realize design changes at post-placement stage for functional changes or timing violations. However, many prior works relative to spare cell utilization neglect the timing effect due to spare cell rewiring. In this paper, the problem of timing-constrained cell replacement using spare cells is firstly formulated to consider the timing effect of the rewiring result. Furthermore, a three-phase approach is proposed to replace the changed cells in a combinational circuit with available spare cells while the timing constraints on the changed cells are satisfied. Experimental results show that our approach can efficiently realize the functional changes under the timing constraints for 5 tested cases.

Categories and Subject Descriptors

B.7.2 [**Integrated Circuits**]: Design Aids – *Placement and routing*

General Terms: Algorithms, Design

Keywords: ECO, Functional change, Spare cell, Timing constraint

1. INTRODUCTION

Due to the increase of circuit complexity and the shrink of design cycles, the requirements of design changes are significantly increasing through overall design stage. To avoid involving enormous effort to re-perform processed steps, a useful methodology named as ECO(engineering change order)[1] is introduced. Recently, more and more defects are found after the design has been taped out[2]. It leads to a considerable amount of design changes required to be realized at post-mask stage. Since the changes occur at post-mask stage, the manufactured mask set must be modified. Because the re-spin cost is very expensive and transistor masks take 68% cost overall in a mask set, an ECO technique named as metal fix is proposed by using some extra logic cells, named as spare cells. In general, spare cells are placed in the transistor layer at placement stage and used to realize the changes by spare cell rewiring, which only the metal layers are modified. Because the cost of re-spin can be effectively saved, spare cell utilization has been widely applied to perform ECO at post-mask stage.

According to the purpose of performing ECO, ECO can be classified into functional and timing ECOs. Researches relative to spare cell utilization for functional or timing ECO have been extensively studied. However, prior works only consider the functional or timing ECO problem. Recently, the functional and timing ECO problems in a complex design are concurrently considered in [3][4]. However, both of their works neglect the timing effect due to spare cell rewiring such that timing violation may be generated after performing functional ECO. An example of the spare cell rewiring is illustrated in Figure 1. Assume that two spare cells, s_1 and s_2, are available to replace the cell, g, for a functional change. Once the result of total wirelength minimization is evaluated, the spare cell, s_2, is selected to replace the cell, g. However, such a result may generate the timing violation on the path from the input signal, i_1, to the spare cell, s_2, due to insufficient timing slack on the signal, i_1. In contrast, replacing by the spare cell, s_1, incurs more total wirelength but the timing violation can be avoided.

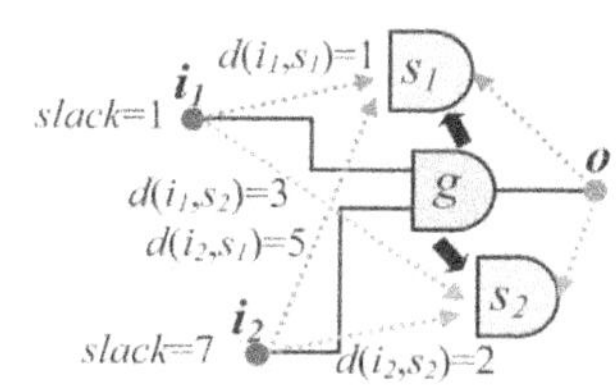

Figure 1 Spare cell rewiring

In this paper, the problem of timing-constrained replacement using spare cells for functional ECO is firstly formulated. Firstly, the timing-constrained feasible region of any changed cell is introduced. Next, the candidates of the available spare cells for any changed cell can be found based on its feasible region. Furthermore, a matching-based approach is proposed to replace all changed cells with a set of available spare cells while the timing constraints on the changed cells are satisfied. Experimental results show that our approach can efficiently realize the functional changes under the timing constraints for 5 tested cases.

2. PROBLEM FORMULTATION

General speaking, the rewiring process of any spare cell is to accordingly rewire all the signal connections from a changed cell to a spare cell, that is, reconstructing the signal connections on the changed cell. Therefore, except for the functionality match, the changed cell can be replaced by a spare cell if all signal connections are reconstructed without exceeding to their respective timing slacks. Basically, the available slack of reconstructing a signal connection on any changed cell can be divided into *local slack* and *global slack*. For a reconstructed connection, the local slack is the connection delay from its connected input signal to output signal via the changed cell and the global slack is the allowable delay tolerance on its timing path. Therefore, the available slack can be defined as the *timing constraint* of reconstructing a signal connection.

Given a combinational circuit with a set of p inputs, $I = \{i_1, i_2,...,i_p\}$ and a set of b outputs, $O = \{o_1, o_2,...,o_b\}$, a set of n changed cells, $G = \{g_1, g_2,...,g_n\}$, $1 \le i \le n$, in a functional ECO list, a set of r connections, $N = \{n_1, n_2,\ldots,n_r\}$, and a set of m spare cells, $S = \{s_1,$

GLSVLSI'13, May 2–3, 2013, Paris, France.
ACM 978-978-1-4503-1902-7/13/05.

$s_2, \ldots, s_m\}$, which is evenly distributed in the placement plane, it is assumed that the number of specific spare cells required by functional changes is sufficient. Furthermore, the timing constraints on any changed cell can be evaluated by analyzing all timing paths on the circuit. The problem of realizing the given functional changes by using spare cells is to replace all changed cells with available spare cells while the timing constrains of reconstructing their connections on all changed cells are satisfied.

3. TIMING-CONSTRAINED CELL REPLACEMENT FOR FUNCTIONAL CHANGE

For timing-constrained cell replacement in a combinational circuit, a three-phase approach, including *Extraction of timing-constrained feasible regions of changed cells*, *Construction of an edge-weighted bipartite graph with replacement relations*, and *Matching-based cell replacement*, is proposed and the design flow is shown in Figure 2. Firstly, based on the timing constraints of the input and output pins on any changed cell, the timing-constrained routing lengths of the input and output pins can be extracted. Furthermore, the timing-constrained feasible region of any changed cell using spare cells can be constructed according to the timing-constrained feasible regions of the input and output pins. In the second phase, based on the extracted feasible regions of all the changed cells, the candidates of the spare cells for any changed cell can be found by using a sweeping-line algorithm. Next, the replacement relations between the changed cells and the candidates of the spare cells can be constructed and further modeled as an edge-weighted bipartite graph. Finally, a matching-based approach is proposed to achieve a minimum-weighted matching result in the edge-weighted bipartite graph.

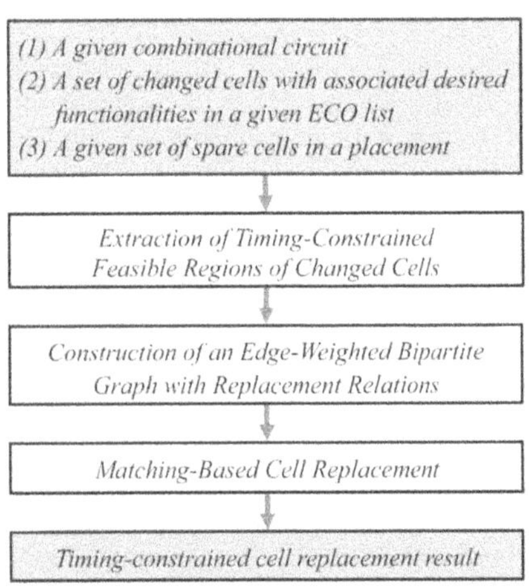

Figure 2 Design flow of timing-constrained cell replacement using sparse cell for design changes

4. EXPERIMENTAL RESULTS

For timing-constrained cell replacement by using spare cells, our proposed approach has been implemented by standard C++ language and run on an Intel Core2 Quad 3.4GHz machine with 2G memory. Three modified circuits, testcase1A, testcase2A, and testcase3A, and two original circuits, testcase4 and testcase5, from [5] are used as tested cases shown in Table I. Due to no timing constraint in the tested cases, we randomly set the timing constraints of signal connections on each changed cell. The parameters in 0.18μm process from NTRS'97 are used to compute the connection delay in Elmore delay model.

TABLE I NUMBER OF PINS, CELLS, NETS, SPARE CELLS, AND FUNCTIONAL CHANGES IN 5 TESTED CASES

Circuits	#Pin	#Cell	#Net	#Spare	#ECO
testcase1A	483	28,591	28,705	350	4
testcase2A	483	28,591	28,705	2,300	9
testcase3A	483	28,591	28,705	2,300	52
testcase4	33	198	181	40	3
testcase5	30	938	850	100	4

Considering the number of the functional changes(#FC) and the incremental ratio of wirelength(INC #WL) in 5 tested cases under various timing constraints in 110%, 120%, 130%, and 140% of the original wiring delay, the experimental results are shown in Table II. In Table II, the incremental ratio of the wirelength in each case is not increased or decreased significantly because the distribution of the spare cells are evenly and the number of the used spare cells is not used overly. Besides that, the results show that all functional changes in all the cases can be completely realized in reasonable CPU time by performing our proposed approach.

5. CONCLUSIONS

The problem of timing-constrained cell replacement by using spare cells for design changes is firstly formulated in this paper. Given a combinational circuit with a set of changed cells, and a set of spare cells in a placement plane, our proposed approach can realize all functional changes under the given timing constraints on all changed cells. Experimental results show that our approach can efficiently realize the functional changes for 5 tested cases.

6. REFERENCES

[1] S. Golson, "The human eco compiler," *Synopsys User's Group Conference*, pp.1-57, 2004.

[2] K. H. Chang and I. L. Markov, "Reap what you sow: spare cells for post-silicon metal fix," *ACM International Symposium on Physical Design*, pp. 103-110, 2008.

[3] H. Y. Chang, H. R. Jiang and Y. W. Chang, "Simultaneous functional and timing ECO," *ACM/IEEE Design Automation Conference*, pp. 140-145, 2011.

[4] J. H. Hung, Y. K. Yeh, Y. S. Tseng and T. M. Hsieh, "A new ECO technology for functional changes and removing timing violations," *IEEE International Symposium on Quality Electronic Design*, pp. 634-638, 2011.

[5] Faraday Corp., "Spare Cell Selection for Functional Change," IC/CAD Contest, Taiwan, 2010. http://cad_contest.cs.nctu.edu.tw/cad11/Problems/97/

TABLE II EXPERIMENTAL RESULTS FOR CELL REPLACEMENT USING SPARSE CELLS UNDER FOUR DIFFERENT TIMING CONSTRAINTS

Circuits	Original WL(μm)	110%			120%			130%			140%		
		#FC	INC #WL	Time(s)	#FC	INC #WL	Time(s)	#FC	INC #WL	Time(s)	#FC	INC #WL	Time(s)
testcase1A	27,207,777,450	1	-0.00005%	2.74	3	0.00104%	2.82	4	0.00223%	2.87	4	0.00223%	2.88
testcase2A	28,686,988,970	1	0.00013%	7.12	7	0.03372%	7.33	8	0.03824%	7.41	9	0.04153%	7.44
testcase3A	28,686,988,970	11	0.05225%	14.74	49	0.26743%	15.17	52	0.28529%	15.28	52	0.28529%	15.34
testcase4	4,302,710	2	0.06508%	0.73	3	0.40347%	0.78	3	0.40347%	0.81	3	0.40347%	0.81
testcase5	41,914,750	2	-0.03073%	0.76	4	-0.64397%	0.87	4	-0.64397%	0.87	4	-0.65733%	0.87

Rotary Traveling Wave Oscillator Frequency Division at Nanoscale Technologies

Ying Teng
Department of Electrical and Computer Engineering
Drexel University
Philadelphia, PA 19104
yt74@drexel.edu

Baris Taskin
Department of Electrical and Computer Engineering
Drexel University
Philadelphia, PA 19104
taskin@coe.drexel.edu

Categories and Subject Descriptors

B.7.1 [**Hardware**]: INTEGRATED CIRCUITS—*Types and Design Styles*

General Terms

Design

Keywords

VLSI; clocking; resonant; frequency divider; oscillator

Resonant clocking is an alternative to the traditional clock generation and network design which provides a low power, low skew solution for clock generation and distribution [1, 2, 4, 6, 7]. Rotary traveling wave oscillator (RTWO) [7] based resonant clocking technology is a particularly promising approach, which provides high frequency oscillation signals with rail-to-rail signal amplitude and multiple clock phases. However, the cost of a resonant oscillator, such as the power consumption and area overhead, grow rapidly along with the decrease of the target resonant frequency. Due to this inability to down-scale the naturally high frequency of resonant clocks, their application at lower frequencies remains limited. For instance, the resonant clocking implementation on AMD Piledriver chip [5] is only used at high frequencies, and the clock network is multiplexed off to be driven by a non-resonant clock source for lower frequency modes.

The only prior work on resonant clock frequency division is in a USPTO-filed patent [3]. The patent [3] describes a circuitry for frequency division of the RTWO-based resonant clocking that uses the "spot-advancing" blocks (SABs) driven by multi-phases of the rotary clock. The circuitry of the spot-advancing block (SAB) is shown in Fig. 1, which is the core cell in the RTWO frequency divider.

The SAB is composed of 6 transistors. When operation starts, the SABs advance a 1 value pulse from left to right through the Spot_in port to the Spot_out port. Suppose Spot_out1 of SAB1 becomes 1, a connection is built between Clk2 and Spot_mid2 in SAB2. When Spot_mid2 discharged to Clk2 and become 0, a path is formed between Clk2 and Spot_out2 in SAB2 meanwhile the path between Clk1 and Spot_out1 in SAB1 is closed. When the Spot_out2 of SAB2 becomes 1, it drives the Spot_out1 of SAB1 (which is also

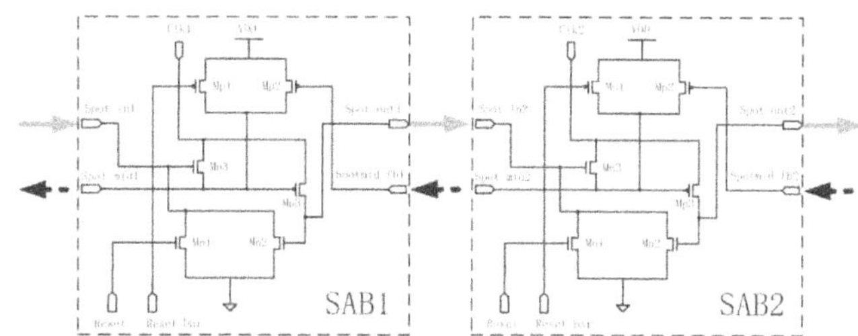

Figure 1: Spot-advancing block in [3].

Spot_in2 of SAB2) to 0 in order to realize a 1 value pulse forwarding from SAB1 to SAB2. The width of the 1 value pulse of Spot_out1 in SAB1 is determined by the the phase difference of the input clock signals Clk1 and Clk2.

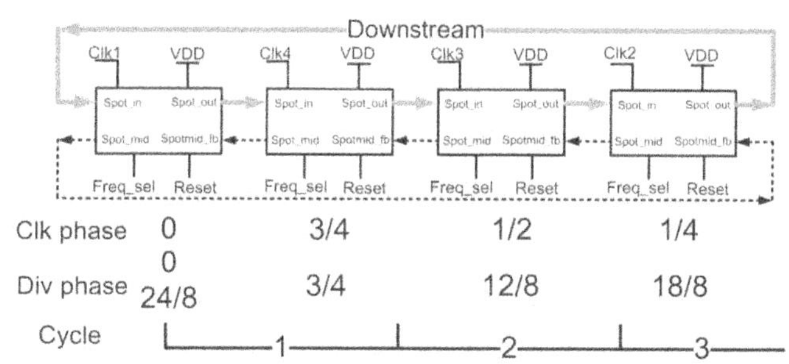

Figure 2: RTWO frequency divider with a division of 3 using 4 SAB blocks.

When the SABs are used to construct a frequency divider, all the SABs are connected to form a loop and the Spot_out port of each SAB is an output of the frequency divider. The topology of an RTWO frequency divider with a division of 3 in [3] is shown in Fig. 2, which is composed of 4 SABs. The 4 multi-phase clock signals Clk1 to Clk4 maintain a constant phase delay of $\frac{1}{4} * 2\pi$ between adjacent signals. For a frequency division of 3, the multi-phase clock signals for each SAB in the downstream direction are Clk{1, 4, 3, 2}. Thus, two adjacent Clk signals maintain a phase delay of $\frac{3}{4} * 2\pi$, which causes the phase delay between adjacent Spot_in/Spot_out signals also as $\frac{3}{4} * 2\pi$. When the Spot_in/Spot_out traverses the loop and comes back to the original SAB, the total phase delay is $4 * \frac{3}{4} * 2\pi = 3 * 2\pi$ so as to divide its input Clk signal by 3.

In this paper, HSPICE simulation-based analyses of 90nm RTWO frequency divider for different division values are performed. These analyses are presented to observe the electrical characteristics of the frequency divided resonant signal. An arbitrarily selected 5GHz RTWO core is selected as the source of the multi-phase Clk signals and 8 Clk signals with $\frac{1}{8} * 2\pi$ between adjacent Clk signals are tapped off, the rising edge of which is restricted to be 5% of the clock cycle. The analysis of the RTWO frequency dividers with different divisions are performed to observe two characteristics: (1) The power consumption; (2) Maximum capacitance loading imbalance among all divider outputs.

Fig. 3 shows the power consumption of 9 RTWO frequency dividers with divisions of 2 to 10. The line plot in Fig. 3 shows the power consumption of the frequency dividers. The

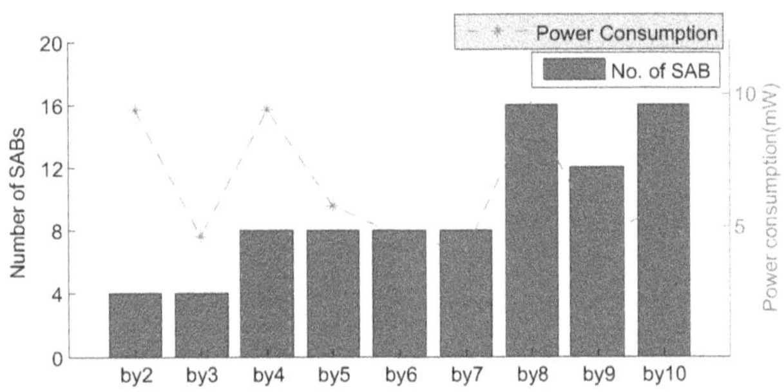

Figure 3: Power consumption of the RTWO frequency dividers with divisions of 2-10.

power consumption of the frequency dividers is influenced by both the number of SABs and the value of division, which is varied between $4.1mW$ to $9.5mW$. As shown in the plot, the power consumption non-monotonically decreases (by as much as 50%) with an increasing value of division. The non-monotonicity is observed as the power reduction depends on the number of SABs in the divider for the selected dividers. The "division of 2, 4, 8" dividers have the highest power consumptions. The number of SABs in these three dividers are 4, 8 and 16, respectively. Since the number of SABs increases at the same pace as the value of division, the power consumption is similar in these three frequency dividers.

The same experimental setup for the nine dividers from 2 to 10 in Fig. 3 is used in order to test the influence of the capacitance loading imbalance on the performance of the RTWO frequency divider. Under this experiment, only one output of a frequency divider is loaded with external capacitance and the other outputs are left with self-loading. The amplitude of the output loaded with (external) capacitance degrades with increasing capacitance loading. For functionality, it is required that the amplitude of the loaded output of the frequency divider should be no less than 90% of the full amplitude and the rising time of the output signals should be less than 10% of the period. As shown in the Fig. 4, the imbalance capacitance driving capability of the frequency dividers are mainly determined by the phase delay between adjacent SABs. Since the process of charging the Spot_out of the current SAB to 1 should finalize before the rising edge of its downstream neighbor Clk occurs, along with increasing loading capacitance, the process of charg-

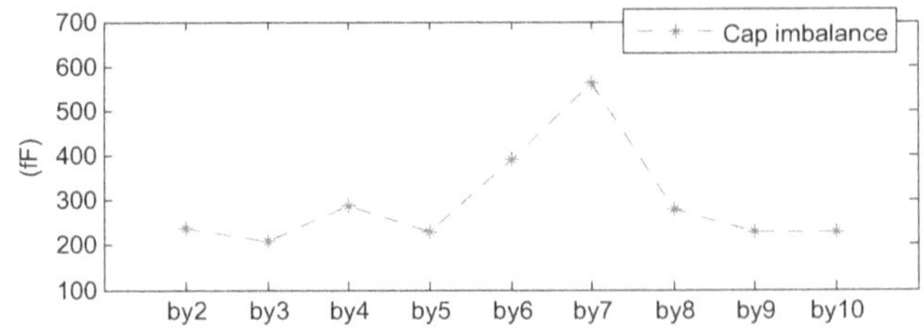

Figure 4: The maximum imbalance capacitance driving capability of the RTWO frequency dividers with division of 2-10.

ing the Spot_out port becomes slower until the next SAB cannot catch the rising edge of its Clk signal. Thus, a frequency divider with a larger phase delay between adjacent SABs possesses a higher driving capability. The frequency divider with a division of 7 possesses the highest imbalance capacitance driving capability of $560fF$. Correspondingly, frequency dividers with same phase delay between adjacent SAB blocks maintain similar capacitance driving capability. As shown in the plot, the frequency dividers with division of 2, 4 and 8 maintain similar imbalance capacitance driving capability.

In summary, simulation-based analyses of the RTWO frequency divider at a nanoscale technology demonstrate the distinct power consumption trends and the capacitance imbalance tolerance of the RTWO frequency divider.

1. REFERENCES

[1] W. Andress and D. Ham. Standing wave oscillators utilizing wave-adaptive tapered transmission lines. *IEEE Journal of Solid-State Circuits*, 40(3):638–651, March 2005.

[2] V. Cordero and S. Khatri. Clock distribution scheme using coplanar transmission lines. In *Design, Automation and Test in Europe (DATE) Proceedings*, pages 985–990, March 2008.

[3] J.Wood. *Frequency Divider*. Patent: US 2008/0258780 A1, Oct. 2008.

[4] J. Lu, V. Honkote, X. Chen, and B. Taskin. Steiner tree based rotary clock routing with bounded skew and capacitive load balancing. In *Design, Automation and Test in Europe (DATE) Proceedings*, pages 455–460, March 2011.

[5] V. Sathe, S. Arekapudi, C. Ouyang, et al. Resonant clock design for a power-efficient high-volume x86-64 microprocessor. In *IEEE International Solid-State Circuits Conference (ISSCC) Proceedings*, pages 68–70, feb. 2012.

[6] Y. Teng and B. Taskin. Synchronization scheme for brick-based rotary oscillator arrays. In *ACM Great Lakes Symposium on VLSI (GLSVLSI) Proceedings*, pages 117–122, May 2012.

[7] J. Wood, T. Edwards, and S. Lipa. Rotary traveling-wave oscillator arrays: A new clock technology. *IEEE Journal of Solid-State Circuits*, 36(11):1654–1665, November 2001.

Reversible Synthesis of Symmetric Boolean Functions based on Unate Decomposition

Arighna Deb,
Debesh K. Das
Comp.Sci.& Engg.,
Jadavpur University, India
arighna87@rediffmail.com,
debeshd@hotmail.com

Hafizur Rahaman
Information Technology
Bengal Engg.& Sci. Univ.,India
rahaman_h@hotmail.com

Bhargab B. Bhattacharya
Nanotechnology Research
Triangle
Indian Statistical Institute,India
bhargab@isical.ac.in

ABSTRACT

In this paper, we introduce a new method to realize symmetric Boolean functions with reversible logic based on unate decomposition. In contrast to earlier synthesis methods, our solution uses a simpler circuit structure of reversible gates, which enables a significant reduction with respect to quantum cost. The resulting design offers an improved solution to reversible synthesis of symmetric Boolean functions.

Categories and Subject Descriptors

B.m [**Hardware**]: Miscellaneous

General Terms

Design

Keywords

Quantum logic, Reversible logic, Symmetric function

1. INTRODUCTION

Symmetric functions have found applications in the field of cryptography, VLSI, and nanoscale designs. Besides, many computing, control and communication circuit functionalities are described by symmetric functions. Reversible synthesis of unate symmetric functions was reported earlier [1]. However, this method requires reversible gates of high quantum cost.

We propose an alternative design that is composed of Peres-CNOT pairs and Toffoli gates in order to realize unate symmetric Boolean functions. Our solution uses simpler reversible gates and is built upon an earlier work [2]. This design offers a significant reduction in quantum cost. The suitability of the proposed design is demonstrated by comparing it with an existing technique [1].

2. PRELIMINARIES

2.1 Reversible Logic

A Boolean function $f : \mathbb{B}^n \to \mathbb{B}^n$ is reversible if and only if it is bijective i.e. it maps each input to a unique output. A reversible circuit is a fan-out free cascade of reversible gates. The common reversible gates include CNOT gate, Toffoli gate and Peres gate.

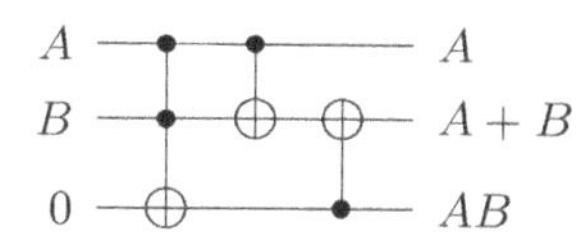

Figure 1: Peres-CNOT pair

In reversible circuit, the quantum cost indicates the total number of elementary quantum operations (CNOT, CV, CV+) required to realize a function. Some of the reversible gates such as CNOT gate, Peres gate and Toffoli gate have quantum cost of $1, 4$ and 5, respectively.

2.2 Symmetric Function

A switching function $f(x_1, x_2, \cdots, x_n)$ is called totally symmetric with respect to the variables $(x_1, x_2, \cdots, x_n)$, if it is invariant under any permutation of the variables. An n-variable symmetric function is denoted as S^n_A, where $A \subseteq 0, 1, 2, \cdots, n$; all the vertices with weight $w \in A$ will appear as true minterms in the function. If each variable of a symmetric function appears in uncomplemented (complemented) form (but not both) in its minimum SOP expression, then the function is called unate symmetric. Such function is expressed as S^n_{k-n} as $u_k(n)$ for $1 \leq k \leq n$.

3. SYNTHESIS OF UNATE SYMMETRIC FUNCTIONS

Our design uses a structure called *Module(n)*, which is a network of Peres-CNOT pair (a Peres gate followed by a CNOT gate) and a number of Toffoli gates. Such a pair (fig. 1) has quantum cost of 4. The structure produces n outputs for any n inputs, where, each output u_i is a positive unate symmetric function given as follows:

$$u_1(n) = S^n_{(1,2,3,\cdots,n)} = \vee x_i \text{ for } i = 1 \text{ to } n;$$
$$u_2(n) = S^n_{(2,3,4,\cdots,n)} = \vee x_i x_j \text{ for } i, j = 1 \text{ to } n;$$
$$\cdots$$
$$u_n(n) = S^n_{(n)} = x_1 x_2 \cdots x_n$$

The design of *Module(n)* is as follows. When n is even (let $n = 2m$), the *Module(n)* consists of two stages, namely stage-A and stage-B. Stage-A has two blocks- Block A-1 and A-2, where, each block is identical to *Module(n/2)*. This

GLSVLSI'13, May 2–3, 2013, Paris, France.
ACM 978-1-4503-1902-7/13/05.

stage realizes the following outputs:

$$a_i = \vee x_{j1}x_{j2}\cdots x_{ji} \text{ for } j1, j2, j3, \cdots, ji = 1 \text{ to } n/2$$
$$\text{and } b_i = \vee x_{j1}x_{j2}\cdots x_{ji} \text{ for } j1, j2, j3, \cdots, ji = (n/2+1) \text{ to } n$$

where, a_i and b_i are the outputs of Block A-1 and A-2, respectively.

The stage-B consists of three sub-stages- B-1,B-2 and B-3. The sub-stage B-1 consists of $n/2$ Peres-CNOT pairs with n-inputs ($= 2m$) (a_i, b_i). The corresponding outputs are given by $c_i = a_i + b_i$ and $d_i = a_i b_i$ for $1 \leq i \leq m$. The first output c_1 represents $u_1(n)$ and the nth output d_m represents $u_n(n)$. The outputs from this stage appear as an inputs to sub-stage B-2 consisting of only Toffoli gates (TOF(a,b;c)). Toffoli gate having input values c_i and d_j with target input equal to zero produces the product $c_i d_j$ and all such Toffoli gates with $i > j$ and $1 < i \leq m$ exist in B-2. All the outputs from sub-stage B-2 appear as inputs to sub-stage B-3 which consists of $(n-3)$ Peres-CNOT pairs. The corresponding outputs are $f(z_1, f(z_2, \cdots, f(z_n)))$. Three outputs $f(z_1, f(z_{n-1}, f(z_n)))$ are obtained by directly passing the signals from the earlier stage, such as $f(z_1) = c_1, f(z_{n-1}) = e_{m,m-1} = c_m d_{m-1}$ and $f(z_n) = d_{n/2}$. Except these three, each $f(z_i)$ for $1 < i < n-1$, is realized by a single Peres-CNOT pair, such that the desired functions are produced. The Peres-CNOT pair realizes the function $f(z_i)$ as follows:

$$f(z_i) = c_i + e_{i-1,1} + e_{i-2,2} + \cdots$$
$$+ e_{\lfloor (i+1)/2 \rfloor, \lfloor (i-1)/2 \rfloor} + \beta d_i \text{ for } i \leq m (= n/2)$$

$$f(z_i) = e_{n/2, i-n/2} + e_{n/2-1, i-n/2+1} + \cdots$$
$$+ e_{\lfloor (i+1)/2 \rfloor, \lfloor (i-1)/2 \rfloor} + \beta d_i \text{ for } i > m (= n/2)$$

where, $\beta = 0(1)$ for $i =$ odd (even).

For $n =$ odd, let us assume $n = 2m + 1$. *Module(n)* will have two stages as shown in the case of $n =$ even. Stage-A contains two parts- one is *Module(m+1)* and other one is *Module(m)*. The outputs $(a_i, b_j)(1 \leq i \leq m+1, 1 \leq j \leq m)$ of stage-A appear as inputs to stage-B consisting of three sub-stages B-1, B-2 and B-3. The sub-stage B-1 consists of $n(= 2m+1)$ inputs, a_i and b_j. There are $(3m+1)$ outputs, where first $2m$ outputs are given by $c_i = a_i + b_i$ and $d_i = a_i b_i$ for $1 \leq i \leq m$. The $(m+1)^{th}$ output is obtained by directly passing a_{m+1} through sub-stage B-1. The last m outputs are obtained by directly passing the inputs b_j through sub-stage B-1. The sub-stage B-2 consists of only Toffoli gates with control input equal to zero. It has $(3m+1)$ inputs from sub-stage B-1, as mentioned above. Its output lines can be classified into two groups-(i) $n(= 2m+1)$ outputs-all the inputs from the outputs of sub-stage B-1 (c_i, d_j, a_{m+1}) are passed through B-2, and (ii) $m(m-1)/2$ outputs realizing the products $c_i d_j (i > j)$ and $a_{m+1} b_j$ and are denoted by e_{ij} for $i > j$. The outputs from Sub-stage B-2 are now given to B-3, which consists of $(n-2)$ Peres-CNOT pairs. At this stage, the outputs are $f(z_1, f(z_2), \cdots, f(z_n))$. Except the two outputs $f(z_1) = c_1$ and $f(z_n) = a_{m+1} b_m$ which are obtained by directly passing the signals from the earlier stage, each $f(z_i)$ for $1 < i < n$, is realized by a single Peres-CNOT pair, such that the desired functions are produced. The Peres-CNOT pair realizes the function $f(z_i)$ as follows:

$$f(z_i) = c_i + e_{i-1,1} + e_{i-2,2} + \cdots$$
$$+ e_{\lfloor (i+1)/2 \rfloor, \lfloor (i-1)/2 \rfloor} + \beta d_i \text{ for } i \leq m$$

$$f(z_i) = a_i + e_{i-1,1} + e_{i-2,2}$$
$$+ \cdots + e_{\lfloor (i+1)/2 \rfloor, \lfloor (i-1)/2 \rfloor} + \beta d_i \text{ for } i = m+1$$

$$f(z_i) = e_{m+1, i-m-1} + e_{m, i-m} + \cdots$$
$$+ e_{\lfloor (i+1)/2 \rfloor, \lfloor (i-1)/2 \rfloor} + \beta d_i \text{ for } i > m (= n/2)$$

where $\beta = 0(1)$ for $i =$ odd (even).

4. RESULTS

The proposed design has been evaluated on unate symmetric functions of different input sizes and also, on some benchmark functions. The results obtained by the proposed approach are compared to the previous work [1] and presented in table 1 and 2.

Table 1: Quantum cost of *Module(n)*

number of inputs	Quantum Cost	
n	As in [1]	Our Design
1	0	0
2	14	4
3	42	12
4	84	20
5	140	46
6	210	48
7	294	93
8	392	105

Table 2: Comparison of quantum cost for some benchmark functions

function			Quantum Cost	
Name	In	Out	As in [1]	Our Design
$rd53$	5	3	145	51
$rd73$	7	3	302	101
$rd84$	8	4	403	116
$sym9$	9	1	505	168
$sym10$	10	1	631	190

5. CONCLUSION

We have reported an alternative method for reversible synthesis of symmetric Boolean functions. As compared to previous approaches to synthesis of symmetric Boolean functions, our method uses simpler gates. We have also studied reversible implementations of some well known benchmark symmetric functions. Experimental results demonstrate that the proposed technique reduces the quantum cost significantly.

Acknowledgement

The work was partly supported by CSIR grant (ref.-22(0590)/ 12/$EMR-II$) and UGC MRP grant (ref.-41−620/2012(SR)).

6. REFERENCES

[1] M. Perkowski, P. Kerntopf, A. Buller, M. Chrzanowska-Jeske, A. Mishchenko, X. Song, A. Al-Rabadi, L. Jozwiak, A. Coppola, and B. Massey. Regularity and symmetry as a base for efficient realization of reversible logic circuits. In *IWLS*, pages 245–252, 2001.

[2] H. Rahaman, D. K. Das, and B. B. Bhattacharya. Implementing symmetric functions with hierarchical modules for stuck-at and path-delay fault testability. *Journal of Electronic Testing: Theory and Applications*, 22(2):125–142, 2006.

Author Index

www.ingramcontent.com/pod-product-compliance
Ingram Content Group UK Ltd.
Pitfield, Milton Keynes, MK11 3LW, UK
UKHW051241300926
15471UKWH00015B/1538

9 781450 322799